# THE INCREDIBLE BONES OF THE NARRENTURM

# THE INCREDIBLE BONES OF THE NARRENTURM

## A Photographic Comparative Atlas of the Pathological and Anatomical Collection of the Fool's Tower

*By*

**ROBERT W. MANN, Ph.D.**

*and*

**EDUARD WINTER**

CHARLES C THOMAS • PUBLISHER • LTD.
*Springfield • Illinois • U.S.A.*

*Published and Distributed Throughout the World by*

CHARLES C THOMAS • PUBLISHER, LTD.
2600 South First Street
Springfield, Illinois 62704



ISBN 978-0-398-09410-2 (hard)
ISBN 978-0-398-09411-9 (ebook)

Library of Congress Catalog Card Number: 2023011009 (print)
2023011010 (ebook)

*With* THOMAS BOOKS *careful attention is given to all details of manufacturing and design. It is the Publisher's desire to present books that are satisfactory as to their physical qualities and artistic possibilities and appropriate for their particular use.* THOMAS BOOKS *will be true to those laws of quality that assure a good name and good will.*

*Printed in India*
*MX-M-5*

**Library of Congress Cataloging-in-Publication Data**

Names: Mann, Robert W., 1949- author. | Winter, Eduard, author.
Title: The incredible bones of the Narrenturm : a photographic comparative atlas of the pathological and anatomical collection of the Fool's Tower / by Robert W. Mann and Eduard Winter.
Description: Springfield, Illinois : Charles C Thomas, Publisher, Ltd., [2023] | Includes bibliographical references and index.
Identifiers: LCCN 2023011009 (print) | LCCN 2023011010 (ebook) | ISBN 9780398094102 (hardback) | ISBN 9780398094119 (ebook)
Subjects: LCSH: Bones--Diseases--Atlases. | Human skeleton--Atlases. | Paleopathology. | Physical anthropology. | Naturhistorisches Museum (Austria). Narrrenturm--Catalogs
Classification: LCC RC930.4 .M34 2023 (print) | LCC RC930.4 (ebook) | DDC 616.7/100222--dc23/eng/20230616
LC record available at https://lccn.loc.gov/2023011009
LC ebook record available at https://lccn.loc.gov/2023011010

*Dedicated to the memory of Nirut Jamklai (1970-2019), beloved father, husband, son, brother and friend.*

# PREFACE

This book is the culmination of decades of personal and scientific curiosity, inquiry, and study of the human body, specifically, its internal architecture: the skeleton. The human body's ability to adjust to insults resulting from disease, genetics, development, the environment, and trauma is beyond 'incredible.' It is more closely akin to the miraculous. Many of these insults are life-changing; some are life-ending. The term "incredible" here applies to the rarity, severity, or unusual location of insult and/or resorption. Equally important is the ability of these bones to educate scientists about the body's capabilities to mitigate the effects of disease and trauma, and of genetic and environmental factors.

*Incredible Bones* is based on the perspective of two experienced scientists. Their shared goal is to better understand the human skeleton and to compile an accurate photographic and historical document of a small portion of the vast osteological collection at the Narrenturm Museum. Commonly referred to as the "Fool's Tower," the Narrenturm is located in Vienna, Austria (Figures 1-6). This comparative atlas records and shares some of the most incredible skeletal malformations and the wide range of variability and severity that can afflict the human skeleton, before and after the advent of antibiotics. It captures examples of disease, malformations, and trauma with little or no surgical or medical intervention and reveals their natural progression, often, without treatment. It is not intended to be a comprehensive atlas of medical or anatomical curiosities.

The format of a photographic atlas was selected to provide examples of disease, trauma, and genetic and congenital malformations that can be used by a diverse audience as exemplars. Exemplars by definition are examples or illustrations of "...something set before one for guidance or imitation" (Merriam-Webster 2011). This book specifically targets the medical, anatomical, archaeological, and anthropological communities who have already, or may encounter similar skeletal conditions. Hopefully, the exemplars in this atlas will assist practitioners in making more accurate interpretations and in making better differential diagnosis of human skeletal remains, whether ancient, modern, or contemporary.

This atlas will also serve as a resource for scientists, oftentimes paleopathologists, in the process of excavating, analyzing, interpreting, and accurately reconstructing skeletal remains in historical contexts. While the Narrenturm contains many "eye-catching and sensationalistic specimens," the subjects included in this atlas were selected because their relevant boney elements were visible for examination, or they conveyed the soft-tissue features of a particular disease or other skeletal condition.

The exemplars, or as we refer to them "the silent teachers," in this book are intended to teach, not so much through words, but through their presentation and contribution of a particular condition in its full form and detail. While they may be stripped of their flesh, these teachers retain their dignity and humanity as valued individuals throughout their afterlife teaching career. Encountering a skeletal condition, whether through archaeological excavation or in consultation with medico-legal authorities, is but the first step in the complicated and often lengthy process of identifying and interpreting a skeletal condition. One of the best ways to accurately identify a disease, trauma or injury, or developmental condition is by comparison with exemplars that are accompanied by historical documentation such as patient records and museum accession cards in conjunction with historical reports, books and textbooks. Reading about a disease, however, is typically not as useful as actually seeing the features and effects of the disease in documented cases.

The skeletal exemplars selected for *Incredible Bones* were chosen to illustrate some of the more severe and unusual manifestations of diseases and malformations, as well as their severity in expression or form. Information written on each individual or museum accession cards is provided in German and, when possible, translated into English using current terminology and interpretations. Many, if not most, of the bones are from individuals of known biological age, sex, cause and date of death, and existing disease or developmental malformation. The information on these individuals is neither consistent nor complete. The vast majority of individuals depicted in this book were not patients in the Narrenturm. Most were transferred to the Narrenturm from hospitals, museums, universities, and physicians who were no longer able to care for them. Today's modern medical systems suffer from inadequate, inconsistent, incomplete, and incompatible record systems. The patient information provided from the Narrenturm is presented in the photo captions. The individuals presented in this atlas were undoubtably as frustrated then as we are today.

Most books that rely on contemporary examples of disease, development, and trauma using small color or black and white photographs provide readers with limited detail or information. Our level of interpretation is based on the "evidence" presented to us visually. *Incredible Bones* presents large, full color photographs augmented with diagnoses based on historical records and sometimes patient histories. Each description and interpretation (diagnosis as it were) in this book are based on medical diagnosis prior to or at the time they were received at the Narrenturm. Additionally, these interpretations are augmented by contemporary diagnoses, perspectives, and descriptions by the first author. The first author, Dr. Bob Mann, has worked in physical and forensic anthropology, anatomy, and skeletal pathology for more than 40 years. He has examined human remains at more than 30 universities, medical schools, hospitals, and osteological collections from around the world, encompassing collections in the United States, Southeast Asia, Europe, and South Africa. The second author, Eduard Winter, is curator of the anatomical and historical documents, including autopsy records, and collections in the Narrenturm spanning more than 15 years. Both authors have extensive experience in the scientific examination of human remains.

Each photograph in *Incredible Bones* is intended to "show" the reader a level of detail and clarity usually only gleaned through physical examination. This detail is difficult, if not impossible, to achieve in written presentation or published form. In effect, the authors have sought to bring the museum to the reader, not the reader to the museum. Bones of individuals and images were chosen and photographed to show the smallest detail and present as many views and perspectives as necessary to properly document and examine/ interpret the specimen for the reader. The authors use high quality and sometimes larger than life photographs so that readers can interpret what they are seeing and perhaps even identify subtle details not provided by the authors. The emphasis is on quality, detailed photographs. Captions provide one perspective of the subject – there are potentially many others and readers are encouraged to focus on those that pique their interest or suit their research. Readers can also compare disease conditions in this book with those they have seen in other anthropological, anatomical, or medical setting. *Incredible Bones*, therefore, serves as a comparative and exploratory photographic atlas of skeletal conditions not usually seen in daily practice or even in most skeletal assemblages, collections, or museums. This photographic atlas is a window to the past and is unique in that many of the bones are not on exhibition at the Narrenturm. As a result, this book provides readers with a behind-the-scenes perspective of some of the most extreme and thought-provoking examples in the Narrenturm pathological-anatomical collection. This book is not intended to present every disease, malformation, anomaly, trauma, or birth defect, but a carefully selected sample from the vast museum collection. Unless stated otherwise, all photographs were taken by the first author.

The format of *Incredible Bones* utilizes a regional approach to the human skeleton with each chapter beginning with the cranium and ending with the feet, a "head to toe" approach. Using a regional approach to the skeleton based on broad categories of conditions that include, for example, tumors and neoplasia, enables the reader to search for a skeletal condition without first having to know the name of the disease, anomaly, or other condition they are seeking to identify. Unlike textbooks divided into chapters based on specific diseases such as tuberculosis, syphilis, or leprosy, regional categories relying on the skeletal elements in question minimizes the necessity of identifying a disease *a priori*. Most so-called minor anatomical variants including foramina, metopic suture, and sutural ossicles that are commonly encountered in clinical, radiological, and anthropological examinations were not included in this atlas. There already exists a vast literature covering these topics.

R.W.M.
E.W.

# INTRODUCTION AND HISTORY OF THE NARRENTURM ("FOOL'S TOWER")

History, like a river, flows through time and space. To appreciate the Narrenturm today it helps to understand a little of its past. The Narrenturm belongs to Vienna, and Vienna belongs to the Danube.

The Danube begins its journey from the confluence of two tributaries in the Black Forest of Germany. It winds its way through ten countries and, 1700 miles later, empties into the Black Sea via the Danube Delta in Romania and Ukraine. It defines the borders of several countries and flows through four capital cities of Europe - one of which is Vienna.

Like many other cities of Continental Europe, Vienna traces its origin to ancient Roman times. In the first century A.D. Imperial Rome appreciated the military significance of this river on its frontier and established a military outpost, now occupied by the Vienna city center. This cantonment was one of many like it that comprised the Danube Limes, that is, the Roman military frontier. The camp grew into a Roman settlement known as Vindobona. The first mention of its current name appears in the ninth century. By the 11th century it had become a trading center. Today the streets of the Inner Stadt (the First Municipal District) reveal the locations of the first Roman walls and moats. Vienna has a long history as a military town and fortification.

Trade produces growth, and immigration exposes inhabitants to more than new cultures, languages, religions, and skills. Vienna has been no stranger to sickness and disease. The city has been devastated repeatedly by severe plague epidemics between the 14th and early 18th centuries. By 1500 the city had some 20,000 inhabitants; by 1754 their number had increased to 175,000. Vienna has had to cope with a veritable population explosion, and the attendant malaises that accompany rapid societal growth.

During this period, anomalous people in need of medical help or those perceived as a public nuisance were locked away in confinements in order to "protect society" and to remove them from public view. Many individuals were incarcerated not because they had committed a crime but because society desired that they be invisible. The public didn't want to see, hear, or deal with the impaired or the impoverished. Detention served to silence both criminals and patients, and assuage the public sensibilities. Incarceration afforded neither treatment nor rehabilitation, but served only to remove the undesirables from society – simply said, they were located "out of sight, and out of mind."

By the mid-eighteenth-century Vienna was witness to revolutionary changes. It was the Age of Enlightenment. Among "the Enlightened Monarchs" (known to others as "the enlightened despots") was the Holy Roman Emperor Joseph II of Austria, son of Maria Theresa and regent from 1780 until his death in 1790. Joseph's contemporary enlightened monarchs included Fredrich the Great of Prussia, Catherine the Great of Russia, and Leopold II of Tuscany.

During his journeys of enlightenment, Joseph II became acquainted with innovations in the field of general nursing, especially in France. He was particularly impressed by the Hôtel Dieu in Paris. Located on the banks of the Seine near Notre-Dame and founded in 651 A.D. by Saint Landry, it is reputedly the oldest, continuously-operated hospital in the world. By the mid-1780s the Hôtel Dieu was already renowned as an educational and training hospital.

Inspired by what he observed in Paris, Emperor Joseph II sought to reform the Austrian health system. Specifically, he set two broad objectives to improve Austria's treatment of the sick and medical science:

- The establishment of a central hospital.
- The reformation of medical studies. To this end, the Josephinum was opened in Währingerstraße in 1785 as the "medical-surgical academy." The Josephinum still exists today, and it houses the museum of the Institute for the History of Medicine.

Under Joseph's reign Vienna's General Hospital, the Allgemeines Krankenhaus (AKH), opened in 1784, as did Vienna's first insane asylum, the Narrenturm. The General Hospital, site of today's university campus, was built on the grounds of Vienna's former Großarmen-und Invalidenhaus, which was constructed in 1693-97 on the orders of Emperor Leopold I (1640-1705). By 1683, after the second Turkish siege of Vienna, the city of Vienna and its immediate surroundings, were occupied by many homeless, refugees, and invalids. These individuals sought shelter in the newly built poorhouse, the Invalidenhof.

The existing Großarmenhaus closed in 1782, and was selected as the site for a central hospital. In 1783 Joseph II announced a competition by court decree, asking Viennese physicians to submit proposals for the sensible conversion of the Großarmenhaus into a large hospital. The winner of this competition was offered the post of director of the General Hospital. Not surprisingly, the winner was Josef Freiherr von Quarin, the emperor's personal physician. He was assuredly the person most familiar with the emperor's concepts of reform.

The large Viennese hospital was based on the Parisian model and was the first of its kind in the German-speaking world. All of the buildings of the Großarmenhaus were adapted to suit contemporary needs. The only new building was the Narrenturm, or as it was called at the time, the "Tollhaus" (i.e., lunatic asylum or madhouse). Its official name was k. k. Irrenanstalt zu Wien (k. k. stands for kaiserlich-königlich i.e., imperial-royal). To the locals it quickly became known as the "Fool's Tower." Joseph II financed the building from his private fortune. Josef Gerl was appointed as the architect for the Narrenturm. Herr Gerl designed a five-story structure in the shape of a tower with an open center. It had a familiar look to the residents of Vienna. Due to its appearance, and probably because of the Viennese fondness for culinary delights, the Viennese mockingly called the building the "Emperor's Gugelhupf" (i.e., the emperor's "Bundt cake.") Even today, the term "Guglhupf" is still used in Viennese for a psychiatric institution.

The Narrenturm stood somewhat apart from the rest of the hospital complex. This physical separation meant that the local population had unchecked access the grounds. Consequently the "Fool's Tower", or rather the patients within, became a kind of public amusement. To protect the patients, the director of AKH, Vienna's General Hospital, Dr. Johann Peter Frank, had a wall erected around the tower in 1796. The wall kept onlookers at bay and afforded patients limited outdoors space, in addition to the existing small inner courtyards. Two courtyards were established: the men's courtyard facing the AKH, and the women's courtyard located on the other side of the tower.

The Narrenturm was informally opened on 19 April 1784. The first patients were transferred from two facilities: 66 people were relocated from the Versorgungshaus zu St. Marx, built in 1394 and considered the oldest hospital in Vienna; and 43 people were moved from the Spanisches Spital (the Spanish Hospital) founded in 1717. The ceremonial opening of the AKH did not take place until four months later, on 16 August 1784.

The Narrenturm was the first building erected solely for the purpose of psychiatry. In the course of the 18th century, the field of psychiatry gradually matured into an independent science. The Age of Enlightenment kindled a generally more humane treatment of the mentally ill, at least by the standards of the time. Of course, the term "mentally ill" is fairly ambiguous. In the 18th century "being insane" was not recognised as an illness. Rather, the patients' disorders were frequently ascribed to their own culpability i.e., their conditions resulted from their dissolute living conditions and their refusal to submit to social norms. Alternative diagnoses were worse – demonic possession. They were generally consigned to the larger faction of the "poor," the "outcasts," and those who languished outside society. Accordingly, they were treated just like other outcasts, like criminals. These unfortunates ended up in prisons, were locked away in cellars, or even exhibited for amusement at fairs. Furthermore, until the end of the 18th century it was alleged that mental disorders were incurable and that the only remedy was restraint and confinement. The idea that the mentally ill were in fact sick individuals who needed "treatment" gradually gained acceptance by the end of the century. The establishment of a separate institution for this group of castaways on the grounds of the hospital can certainly be seen as progress, certainly in a humane sense. Of course, today's standards of compassionate medicine should not be applied here. For example, decrees of the time speak of "forging" the unruly lunatics (that is, reforming the individual and making them compliant), and remnants of chains have been found in some cells. The patients were treated according to the humoral pathology that was common at the time. Humoral pathology was based on the idea that the human body has four important "humors," or fluids: blood, phlegm, black bile and yellow bile. Consequently, fashionable treatments included: the administering of enemas, bloodletting, vomiting therapies, and cold-water cures. No less a figure than King George III of Great Britain and Ireland was subjected to these primitive treatments. Not half way through his 60-year reign (1760-1820), George experienced episodic mental illness, characterized by acute mania. The King was physically restrained and caustic poultices were applied to draw out the "evil humors." By 1811, George has become permanently insane and he lived out his remining years "in seclusion." One must bear in mind that this new Narrenturm preceded by one hundred years the revolutionary psychodynamic theory proposed by another Austrian neurologist. It was 1881 when Sigmund Freud, the founder of psychoanalysis, practiced in Vienna and then penned his "The Interpretation of Dreams" in 1900.

The demand for space at the Narrenturm soon exceeded capacity and additional accommodations were required. Thus the "Lazareth" at the end of Währingerstraße, originally built as a plague hospital and used as a military hospital since 1766, was assigned to the Narrenturm as its own department in 1803. Here, "curable" and "calm" lunatics were accommodated. Another department was the so-called "Dreiguldenstock," which was established in the northernmost part of the AKH. Here, as the name suggests, the wealthier lunatics, those who could pay 3 guilders per day, were admitted. They received special care and meals.

In 1848-53 the new Lower Austrian State Lunatic Asylum was built at Bründlfeld (Vienna). In 1870, Bründlfeld established the first psychiatric university hospital. Additionally, Bründlfeld instituted a completely new concept of hospital accommodation. While the cell corridor system (i.e., individual or pair detention) prevailed in the Narrenturm, Bründlfeld arranged group units or guard rooms. This facility was ultimately demolished in 1974 to make way for the new AKH's construction. The new AKH was officially opened on 7 June 1994.

Beginning in 1866, patients were moved from the Narrenturm to other institutions (Landesirrenanstalt am Bründlfeld, and Ybbs) and in 1869 the last patients had left the Narrenturm tower. After 1870, the Narrenturm was used to house the AKH's service rooms. After the turn of the 20th century service flats were set up here for the AKH's nursing staff and doctors. In 1971, the Pathological Museum moved into the Narrenturm and was initially housed in 25 cell rooms on the B floor. Gradually, additional individual rooms and floors were allocated for use by the museum. In November 1993, the last workshops and service flats were moved out. Since 1993, the entire tower has been used exclusively as a museum. Today the Narrenturm is a national registered building and is now owned by the University of Vienna. In 1993 the old AKH was donated by the City of Vienna for use as a university campus.

Born of military necessity on the banks of the Danube in Roman times, Vienna continues to breathe new life into the fields of anthropology, culture, forensics, history, and medicine. Today the Narrenturm is officially known as the Pathologisches-Anatomisches Bundesmuseum, or the Vienna Pathological-Anatomical Museum. This exceptional institution thrives under the auspices of the National Historical Museum of Vienna. The Narrenturm houses pathological and developmental anomalies previously curated or housed in universities, hospitals, and museums throughout Austria. Its unique collection has grown to more than 50,000 elements. The Narrenturm unarguably houses the largest collection of pathological human remains in the world. It contains a permanent record, inscribed in bone, of Vienna's history. And yet to this day, despite its irreplaceable and inimitable value, it continues to be known as the "Fool's Tower."

Our goal in creating this atlas is to safeguard Vienna's rich past and to reassure researchers that it has a future.

**Figure 1.** The five-story Narrenturm (Fool's Tower) in Vienna, Austria as it appeared in 2018.

**Figure 2a-b.** Some of the wet specimens in glass containers and locked medical cabinets housing skeletal elements and one of the many articulated skeletons in 2018.

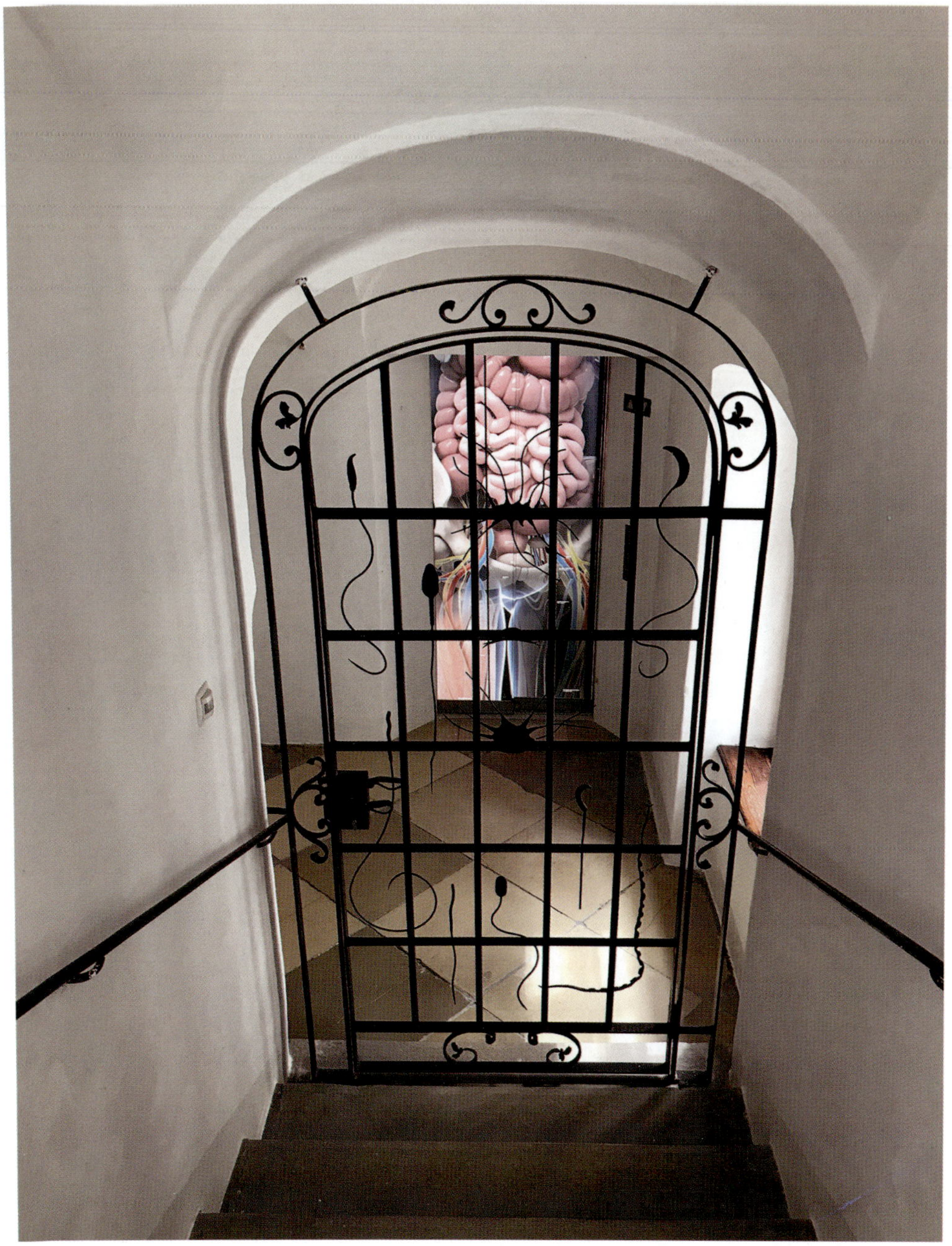

**Figure 3.** One of the decorative doors in the Narrenturm and the newly designed elevator.

**Figure 4a-b.** Some of the many hallway exhibitions and locked medical cabinets in the Narrenturm with its original vaulted ceiling and flooring in 2018.

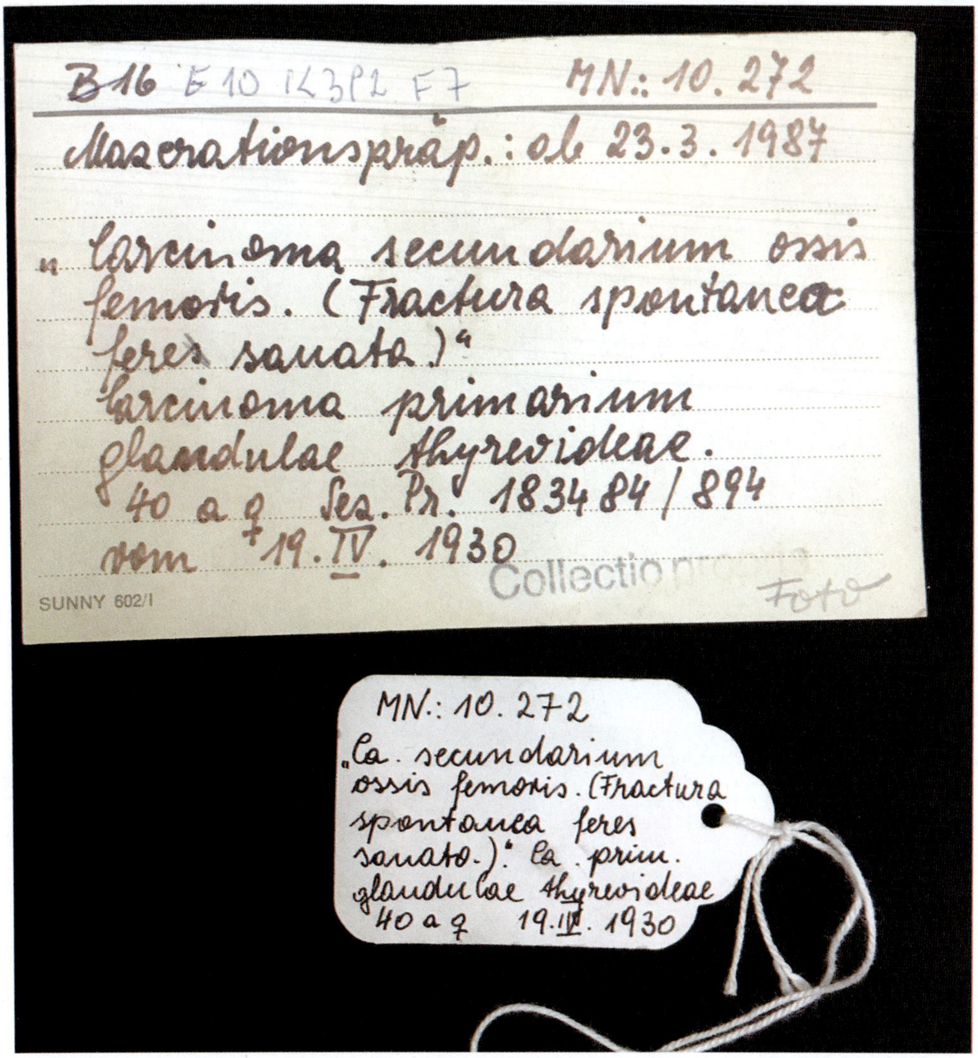


**Figure 5.** An example of one of the museum cards and paper tags that accompanies and provides an overview of each skeleton or skeletal element housed in the Narrenturm. This card and tag are for a 40-year-old female who sustained a fracture of the femur, secondary to thyroid cancer and died in 1930. Most museum specimens are accompanied with more detailed historical, medical, or patient histories.

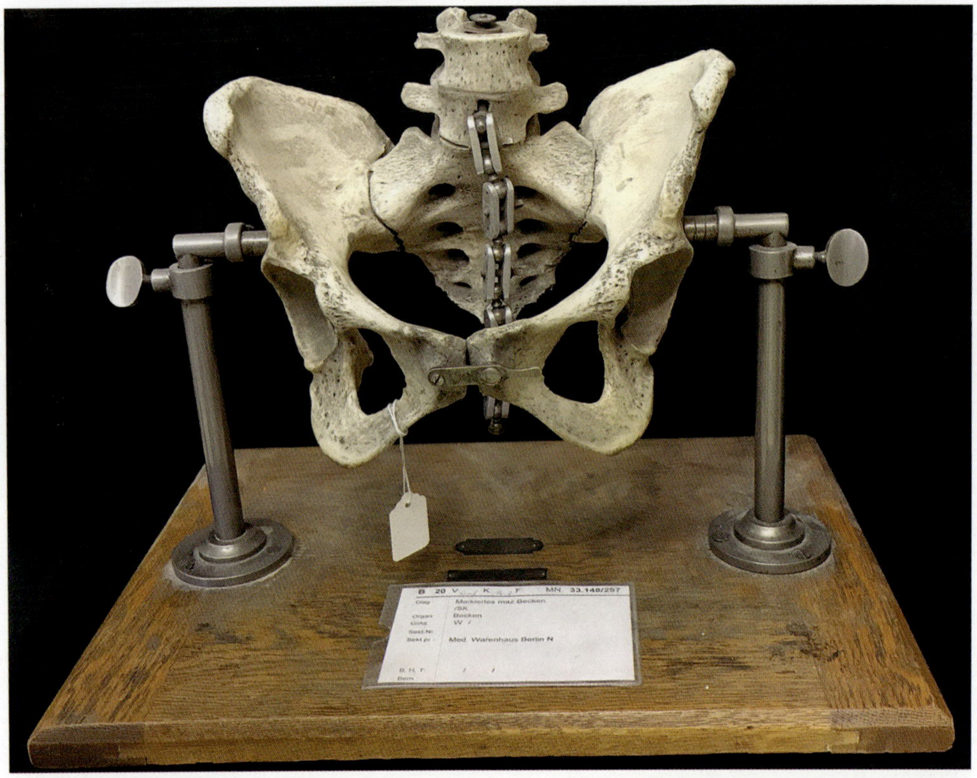


**Figure 6.** Display of a normal female pelvis from the Medical Warehouse in Berlin, Germany. The pelvis is elaborately mounted so that it can be rotated for closer examination.

## ACKNOWLEDGMENTS

The authors thank Dr. Beatrix Patzak, Dr. Verena Hofecker, Mag. Anatole Patzak and Dragica Simecek and from the Natural History Museum, Dr. Katrin Vohland and Mag. Markus Roboch for their support, as well as Prof. Renate Kain from the Medical University of Vienna. We would also like to thank retired pathologist Prof. Feigl.

Additional consultation was provided by other specialists to include orthopedic surgeon Paul Moroz MD; clinical pathologist Helen Thompson MD of the John A. Burns school of Medicine University of Hawaii; forensic odontologist J DeMeo, DDS; dermatologist Douglas W. Johnson MD; and forensic osteologist Sittiporn Ruengdit, PhD of the Department of Forensic Medicine, Chiang Mai University School of Medicine, Thailand. Hanna Polasky assisted with a search of the literature for some exemplars. We are thankful to our editors Professor John R. DeFreytas and Sherrye L. Landrum who took our original manuscript cut it to the bone, eliminated bones of contention, and then fleshed it out to become the book you now hold.

## CONTENTS

# THE INCREDIBLE BONES OF THE NARRENTURM

# Chapter 1

# CONGENITAL AND DEVELOPMENTAL MALFORMATION

## Skull

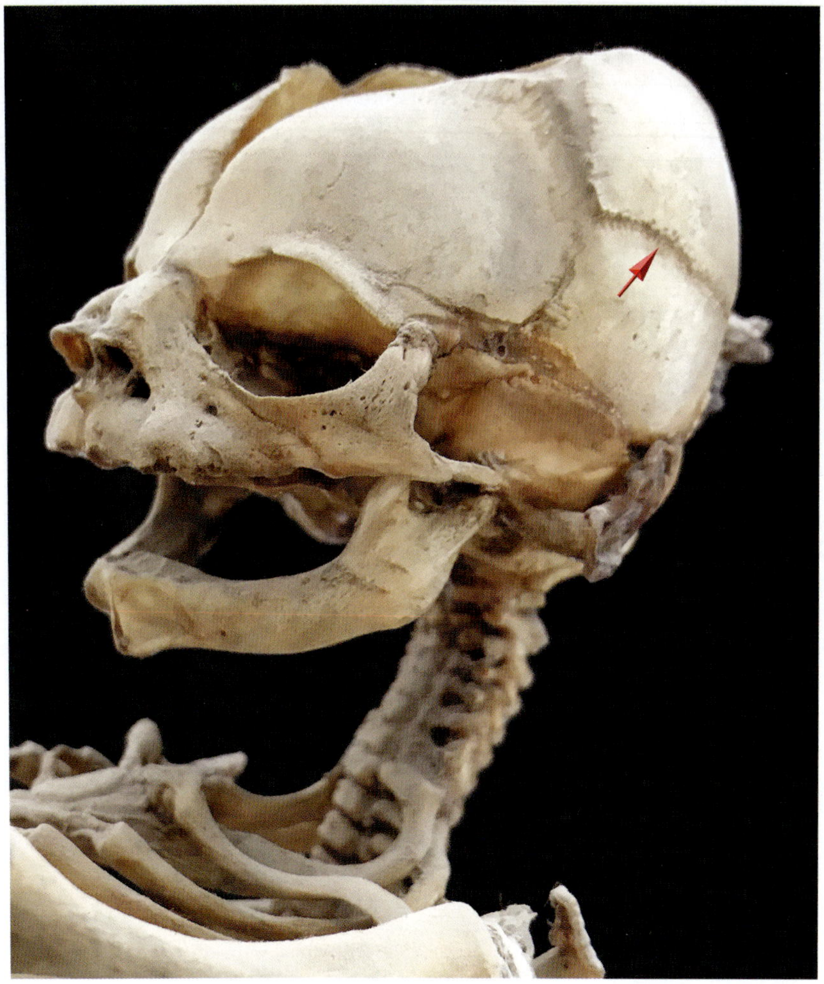

**Figure 7.** Bipartite/divided left parietal bone (arrow) in a fetus with an encephalocele in the occipital region, a rare congenital birth defect occurring in approximately 1 in 3,000 to 1 in 10,000 live births (Markovic et al. 2020; Sheikh et al. 2014) resulting in leakage of cerebrospinal fluid and herniation of the brain. Bipartition, the formation of an accessory suture in the skull, is rare (Weidjik et al. 2016), may be misidentified as a fracture (Mann and Hunt 2019), and occurs as a result of a secondary center of ossification (Mann et al. 2016). The simultaneous occurrence of these two conditions as in this individual is exceptionally rare.

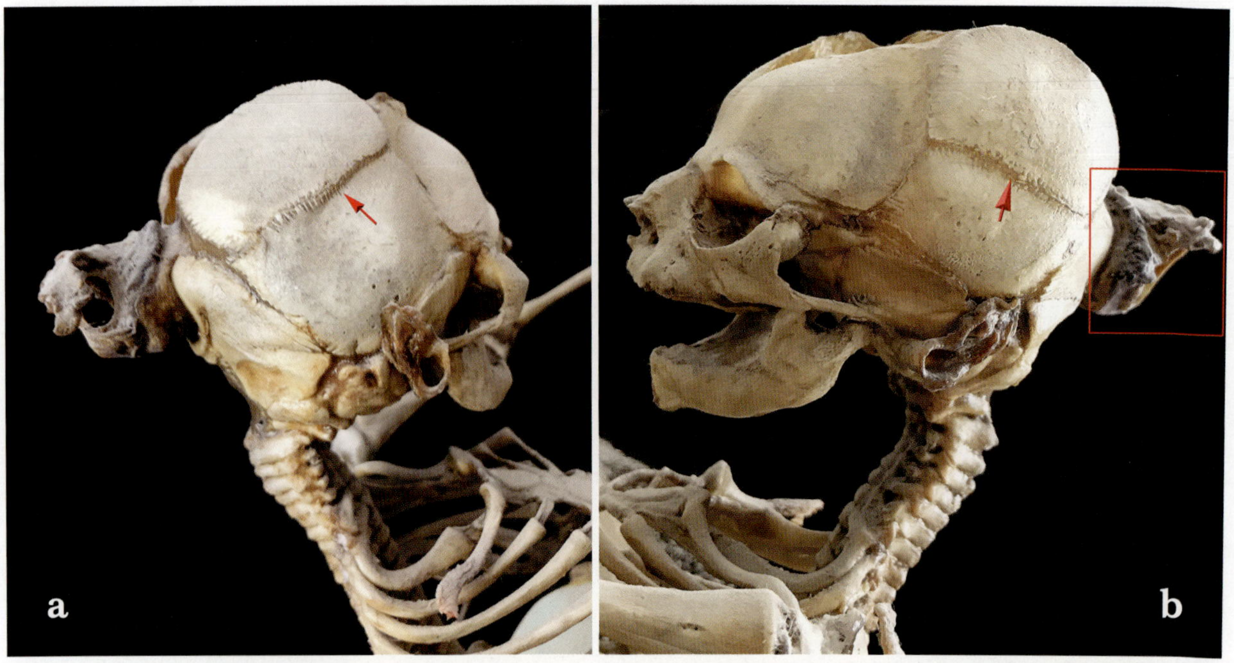


**Figure 8a-b.** (a) Right and (b) left lateral views of bipartite bones, sometimes referred to as the sub-sagittal suture (arrows) (Sanchez et al. 2010) dividing the parietal bones into superior and inferior halves, possibly resulting in plagiocephaly in a fetus. The parietal bone typically ossifies from a single center and rarely from two centers that fail to unite as in this fetus (Mann et al. 2016). Unilateral bipartite, sometimes referred to as divided parietal bone, is one of the rarest bipartitions in the human skull occurring at a frequency of 3 in 25,000 radiographs (Shapiro 1972), or about 1 in 8,300 crania. The frequency of bilateral bipartite bones is unknown but is exceptionally rare. Note that the dura mater containing cerebrospinal fluid and brain matter has herniated (encephalocele) through the occipital region (rectangle).

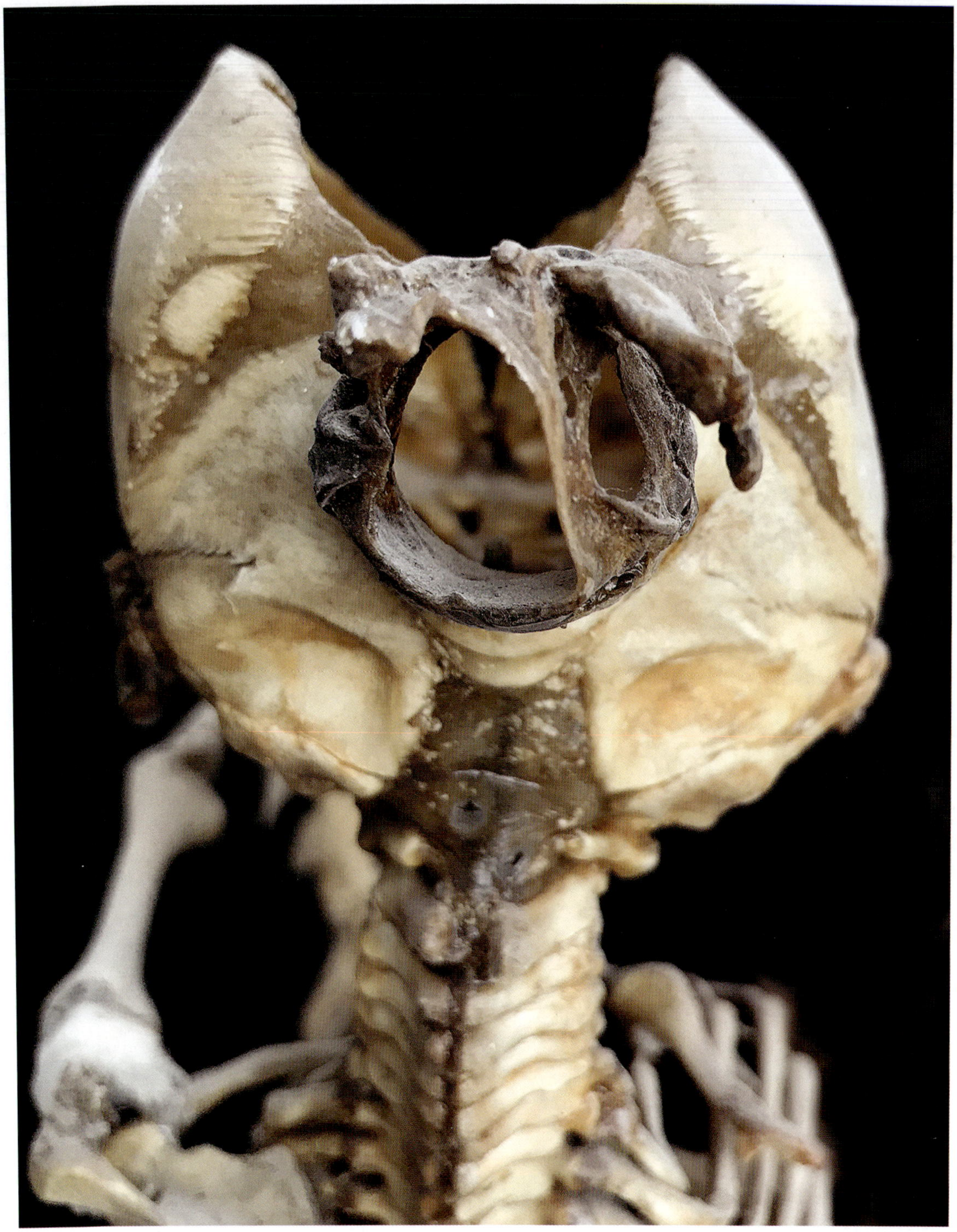

**Figure 9.** Posterior view showing an encephalocele, with herniation of dura mater through the occipital region in a fetus. Encephalocele is a condition within the spectrum of neural tube defects and a rare birth defect consisting of cerebral tissue, sometimes with a partial or complete meningeal covering, through a defect in the skull (see Sheikh et al. 2014).

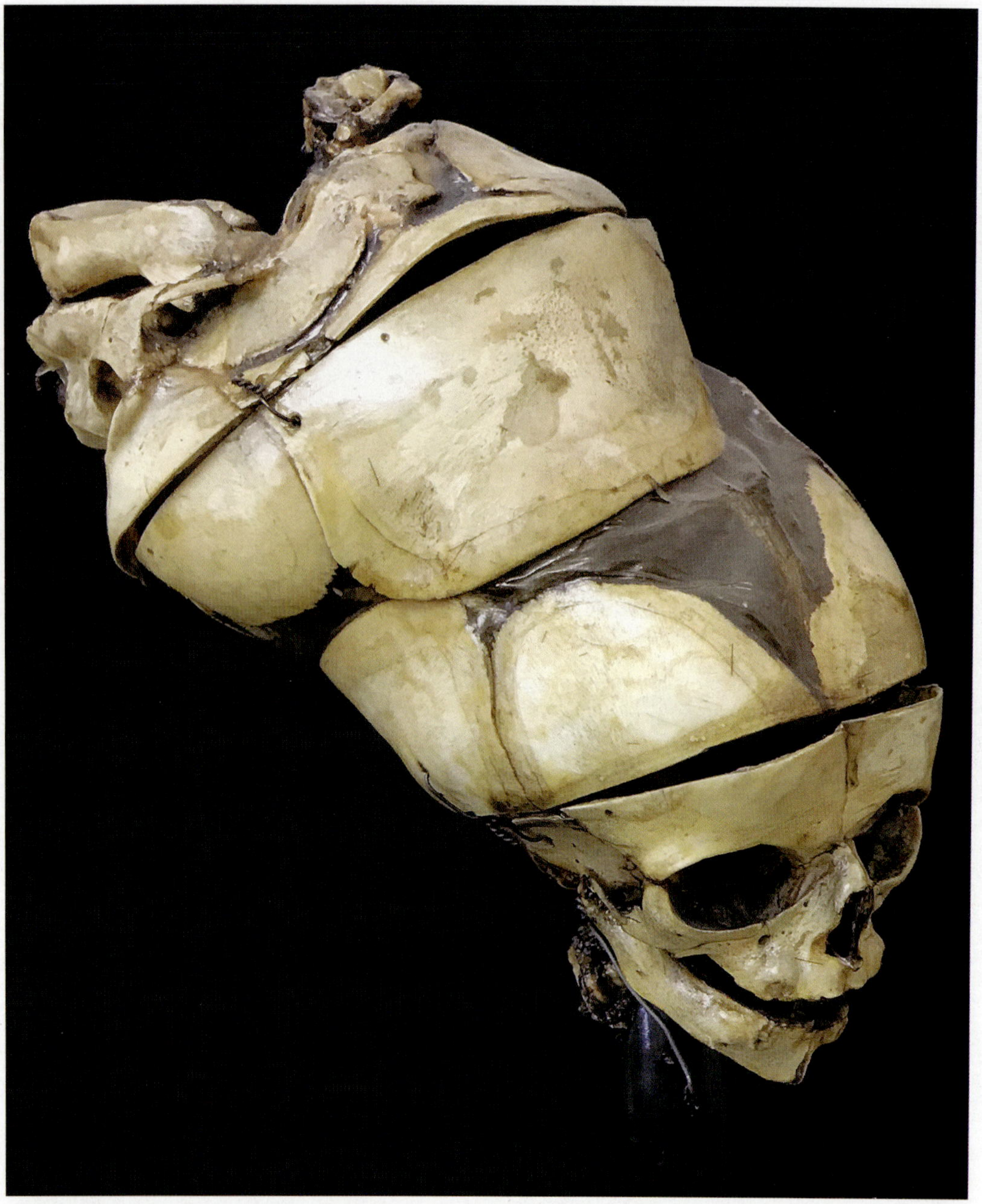

**Figure 10.** Craniopagus conjoined twins (classified as partial Vertical II) fused at the skull (Stone and Goodrich 2006) (MN81.359/4426) who died in 1884. This rare congenital malformation due to incomplete embryonic separation is more common in females and has been reported as occurring once in about every 2.5 million live births (O'Connell 1976; Edmonds and Layde 1982). Conjoined twins are identical twins who are connected physically and, in these two individuals, shared part of the skull and brain tissue. Approximately 25% of craniopagus twins survive and may now be considered for surgical separation (Kaufman 2004). See Stone and Goodrich (2006) for classification of craniopagus twins.

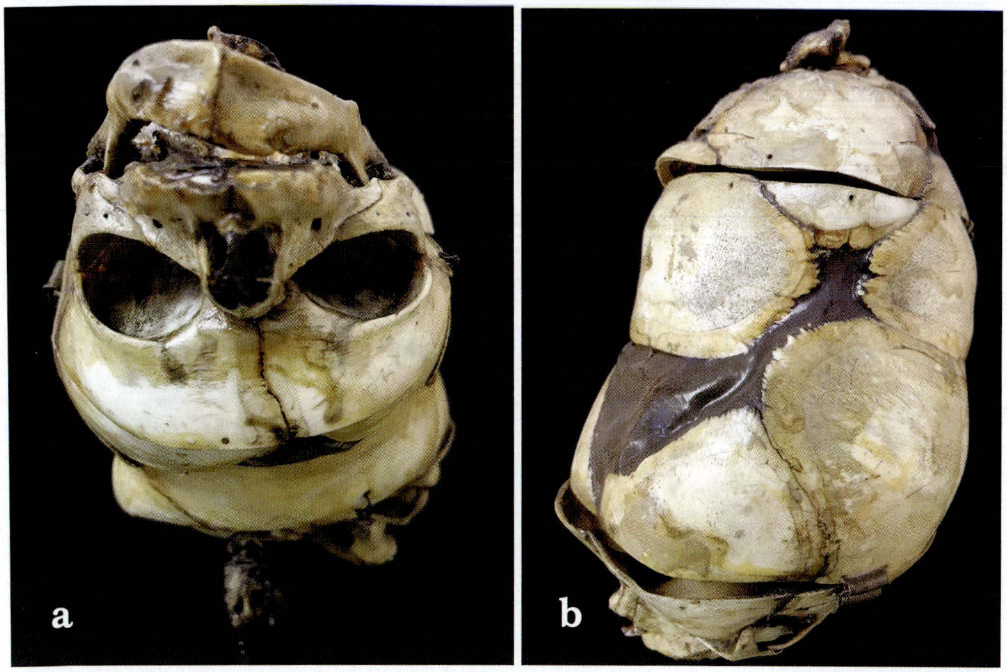


**Figure 11a-b.** Craniopagus conjoined twins, classified as partial Vertical II, fused at the skull (Stone and Goodrich 2006) who died in 1884 (MN81.560/4426). This rare congenital malformation of unknown etiology is more common in females and has been reported as occurring once in about every 2.5 million live births (O'Connell 1976; Edmonds and Layde 1982). Conjoined twins are identical twins who are connected physically and, in these two individuals, shared part of the skull and brain tissue. Approximately 25% of craniopagus twins survive and may now be considered for surgical separation (Kaufman 2004).

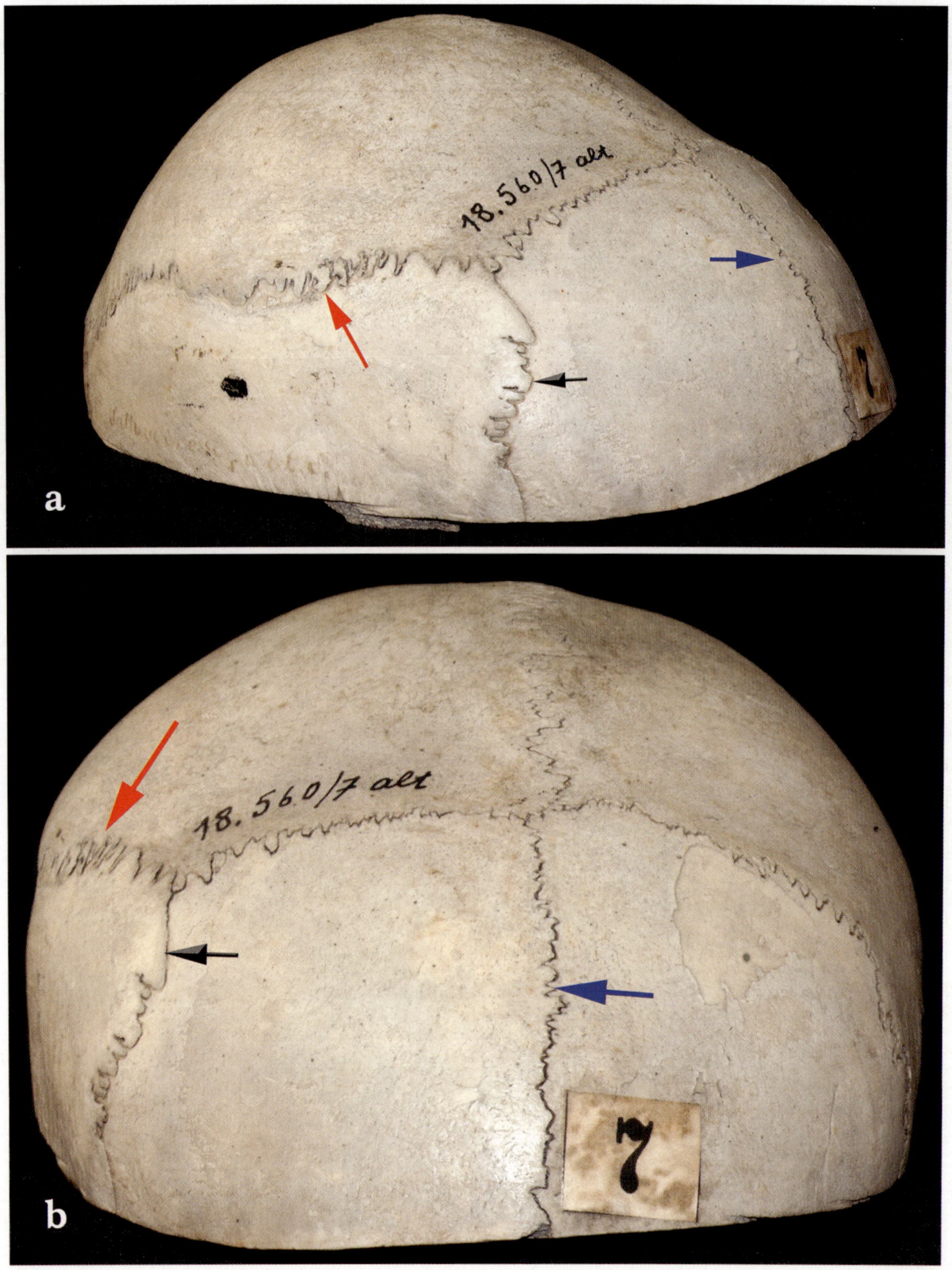


**Figure 12a-b.** (a) Right oblique and (b) anterior views of a bipartite parietal bone (accessory sub-sagittal suture; red arrow) in an unknown adult who died in 1868 (MN18.560/7a). The metopic suture (blue arrow) is open and the coronal suture (black arrow) is positioned more anteriorly in the right half of the frontal bone. Bipartition resulting in partial or complete division of a bone is a very rare developmental defect resulting from a secondary center of ossification (Mann and Hunt 2005; Mann et al. 2016; Shapiro 1972) that may be mistaken for a fracture (Mann and Hunt 2019; Sanchez et al. 2010; Weir et al. 2006). Although the metopic suture may remain open throughout one's lifetime, typical closure has been shown to occur between 3-9 months of age (Vu et al. 2001). An Australian study of children, in comparison, revealed closure of the metopic suture by 19 months (Teager et al. 2019).

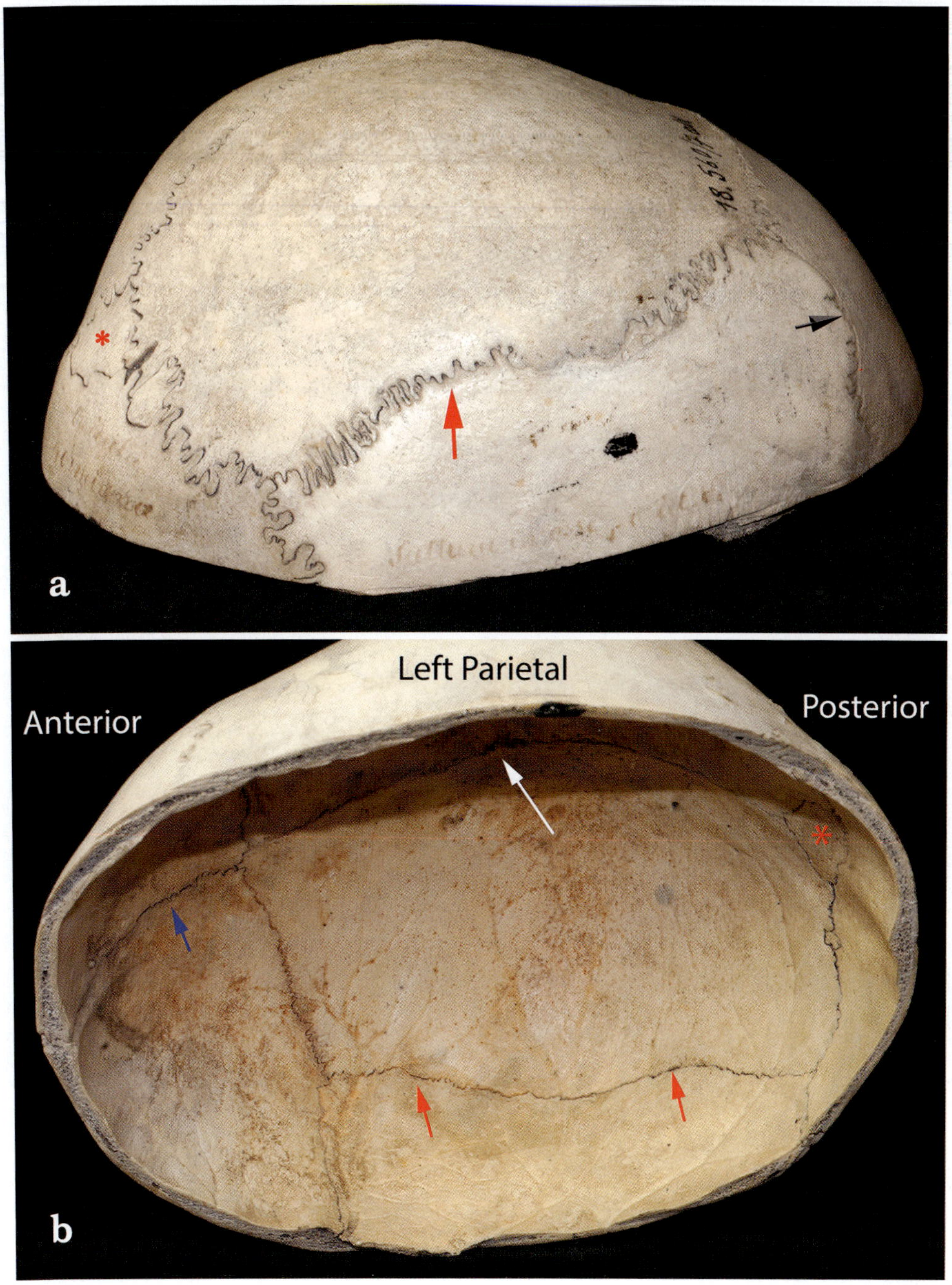


**Figure 13a-b.** (a) Right lateral and (b) left inferior views of the calotte of an adult who died in 1868, showing the sub-sagittal suture (red arrow) and the coronal suture (black arrow) anteriorly positioned in the frontal bone. (b) Inferior endocranial view showing the metopic suture (blue arrow), sub-sagittal suture (red arrows) extending from the coronal suture to the right lambdoidal suture and the sagittal suture (white arrow) (MN18.560/7a). Bipartition is a rare phenomenon due to a secondary center of ossification (Mann et al. 2016; Shapiro 1972). Although the metopic suture may remain open throughout one's lifetime, it typically closes between 3-9 months of age (Vu et al. 2001). Note (a and b) the lambdoidal ossicle (*), a common anatomical variant.

**Figure 14.** Superior view of an open metopic suture (blue arrow), anteriorly positioned coronal suture, sub-sagittal suture (red arrow), also known as interparietal suture and a lambdoidal ossicle (*) in an adult who died in 1868 (MN18.560/7a). Although the metopic suture may remain open throughout one's lifetime, it typically closes between 3-9 months of age (Vu et al. 2001). The right half of the coronal suture (black arrow) is more anteriorly positioned than usual in response to the presence of a sub-sagittal suture dividing the right parietal bone in two pieces (Shapiro 1972).

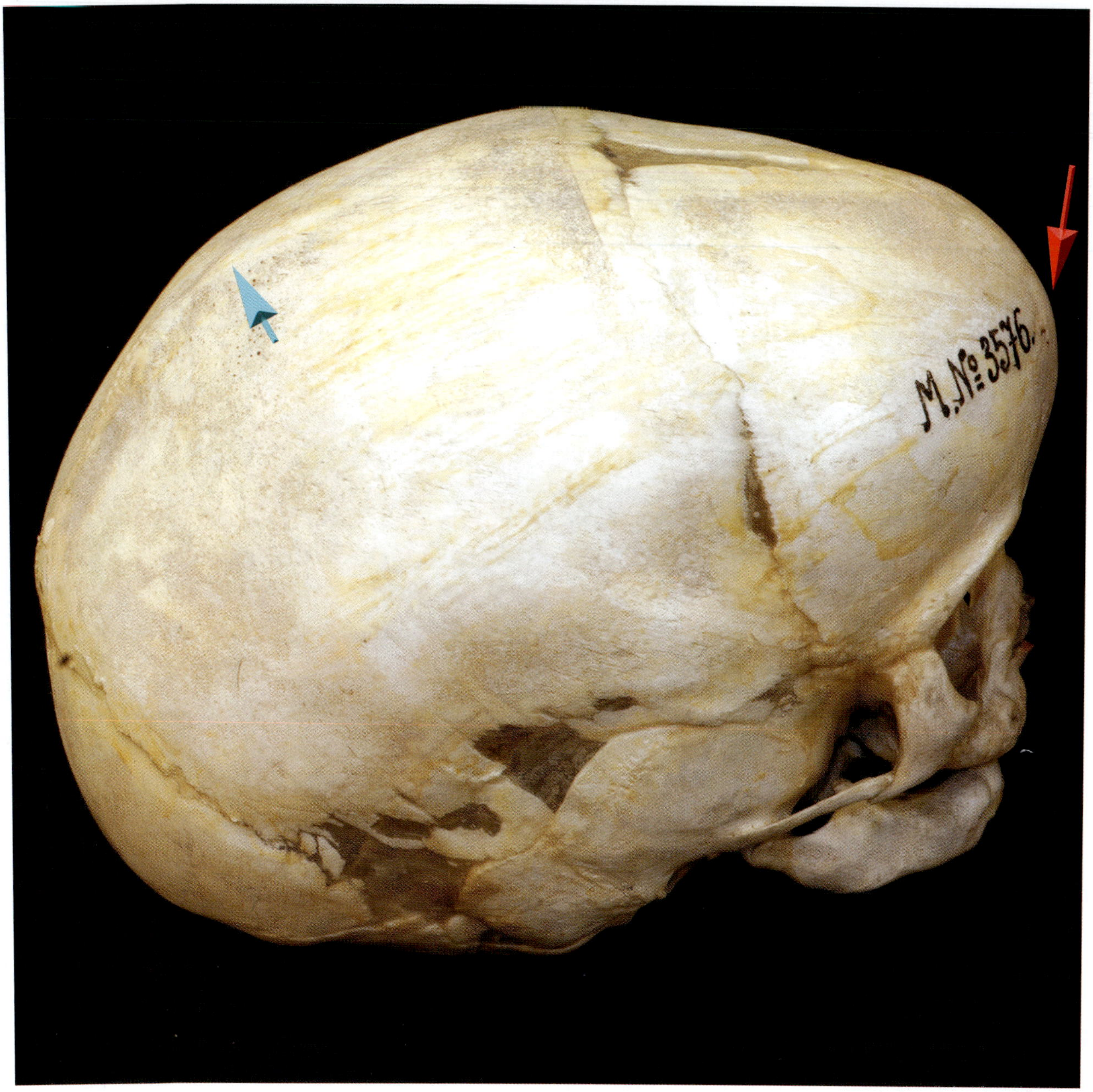


**Figure 15.** Right view of scaphocephaly due to premature closure of the sagittal (blue arrow) and part of the metopic suture (red arrow) in a 9-week-old male who died in 1877 (MN3576). Premature synostosis resulted in anterior-posterior elongation of the skull, a triangular-shaped frontal bone and in this individual, two accessory ossicles (Figure 16) along the posterior foramen magnum and anterior to the Mendosal suture have yet to fuse. Scaphocephaly is one of the more common synostoses present in about 1 in 2,500 births (Edmonds and Layde 1982) and occurring more often in males. Although the metopic suture may remain open throughout one's lifetime, a 3D-CT study of 159 infants and children (without premature craniosynostosis) revealed that it typically closes between 3-9 months of age (Vu et al. 2001). An Australian study of children, in comparison, revealed closure by 19 months (Teager et al. 2019).

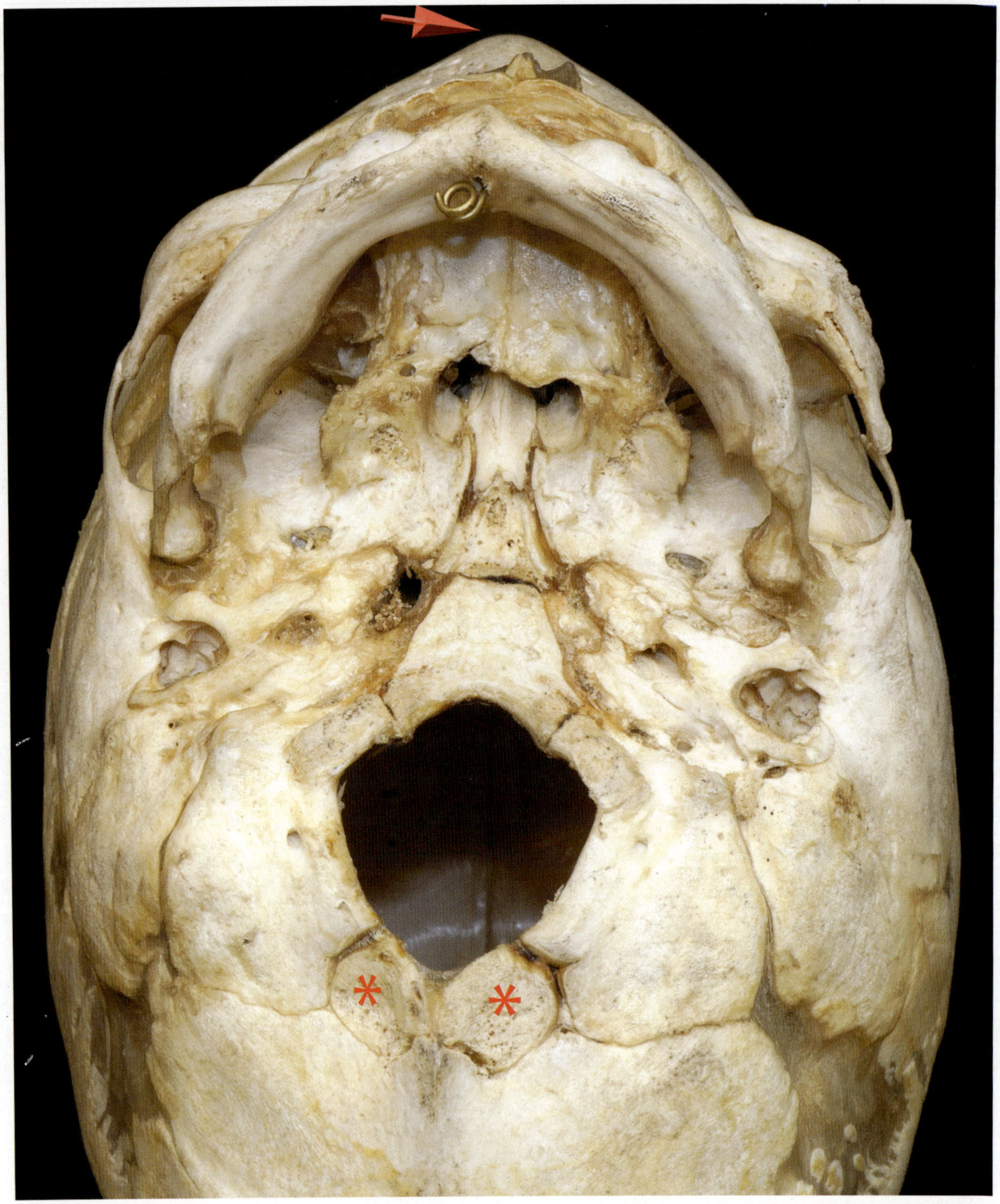

**Figure 16.** Inferior view of scaphocephaly, premature closure of the sagittal and part of the metopic suture resulting in anterior-posterior elongation of the skull, a triangular-shaped frontal bone (arrow), and two accessory ossicles (*) along the posterior foramen magnum in a 9-week-old male who died in 1877 (MN3576). The two accessory ossicles formed anterior-medial to the Mendosal suture and represent interparietal/Inca bones (Wu et al. 2010).

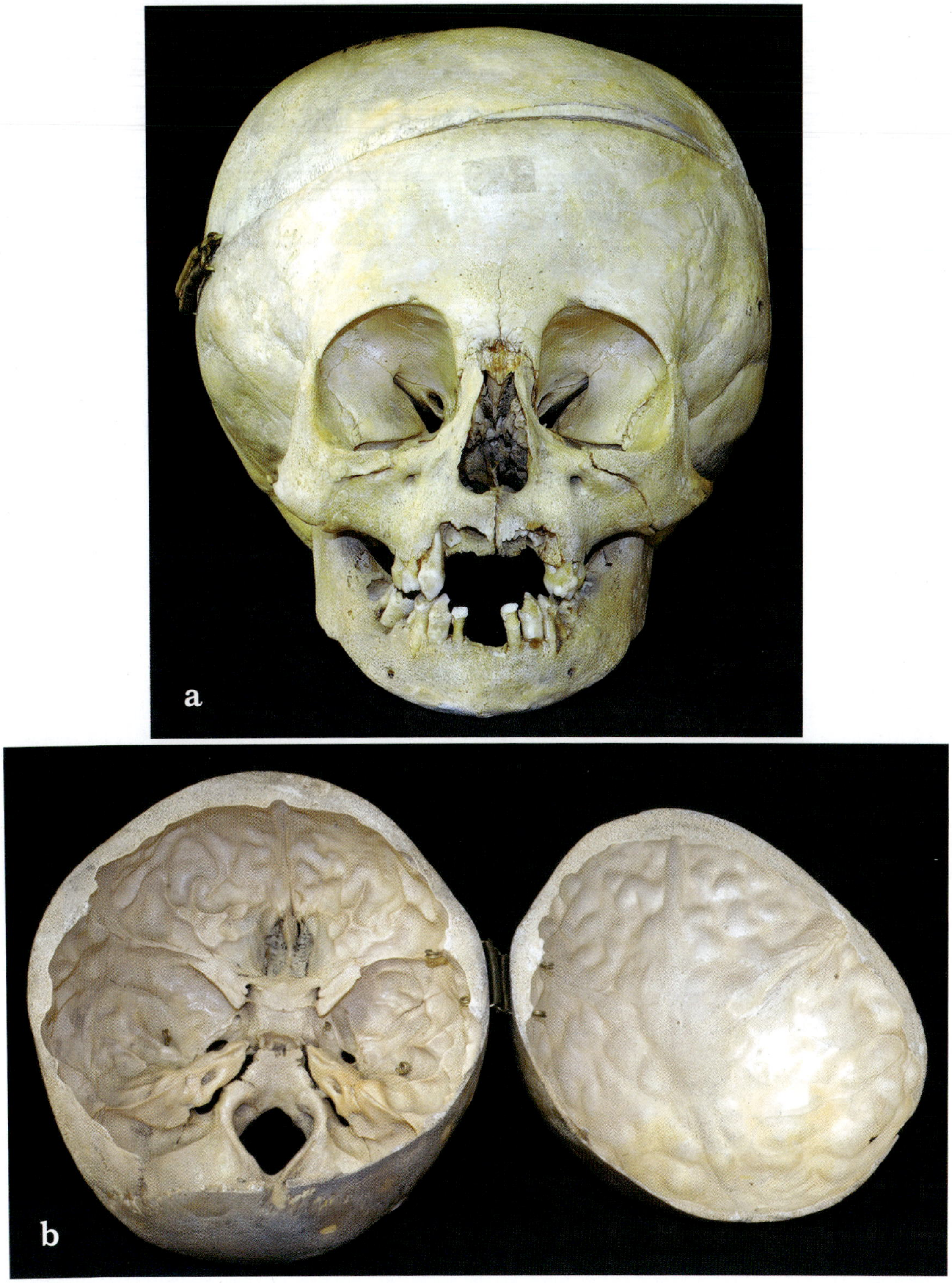


**Figure 17a-b.** (a) Anterior and (b) endocranial views of premature craniosynostosis of the vault sutures resulting in microcephaly in a subadult who died in 1930. (b) The endocranium exhibits an undulating "copper-beaten" appearance (Bogdanovic et al. 2019; Ruhli et al. 2007; Tuite et al. 1996) due to chronic increased intracranial pressure of the brain gyri and cerebrospinal fluid against the inner table in children.

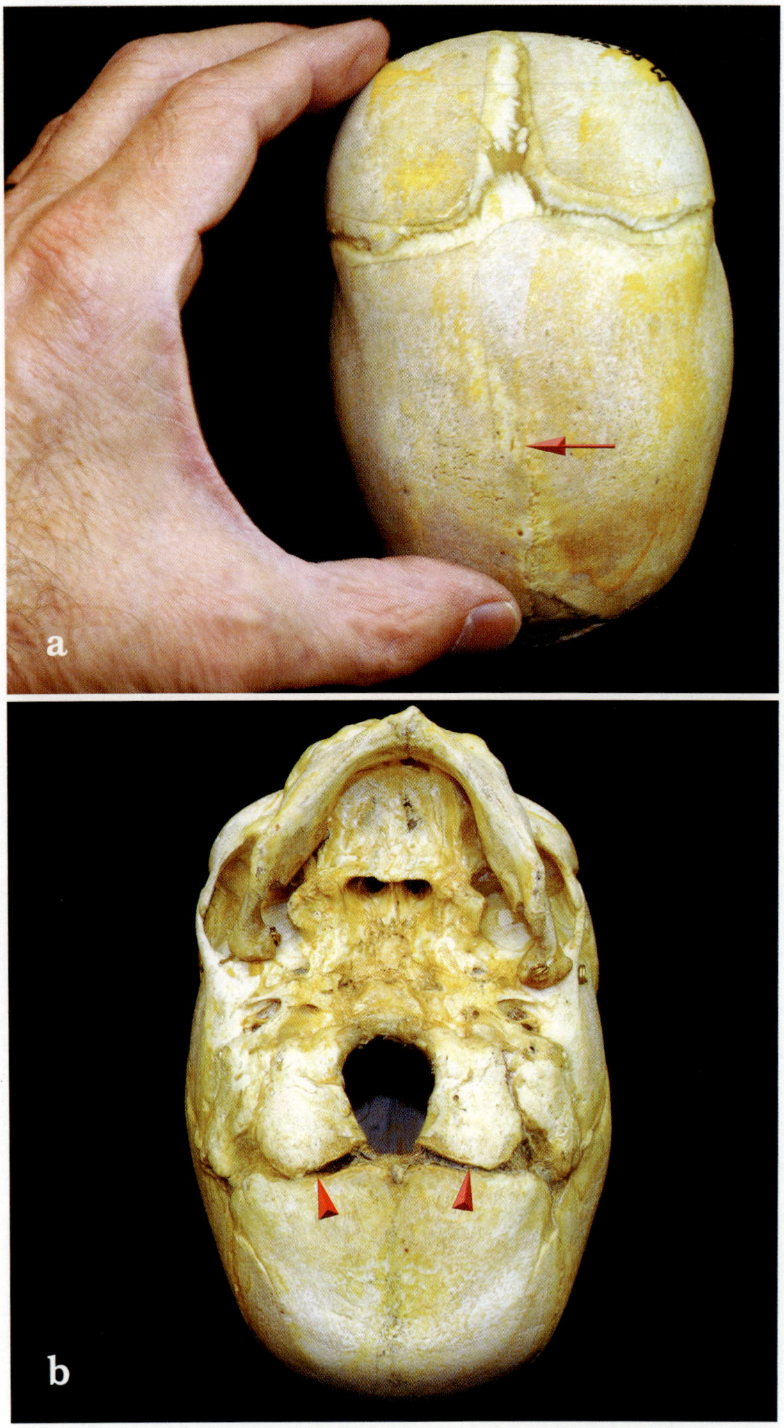


**Figure 18a-b.** (a) Superior and (b) inferior views of premature craniosynostosis of the sagittal suture (a; arrow) resulting in scaphocephaly in a 3-week-old female who died in 1879 (MN3747). (b) Note the Mendosal suture (arrowheads) is still open at this young age. While craniosynostosis refers to closure of a suture, "premature craniosynostosis" refers to early closure/ obliteration of a suture that can result in cranial deformity.

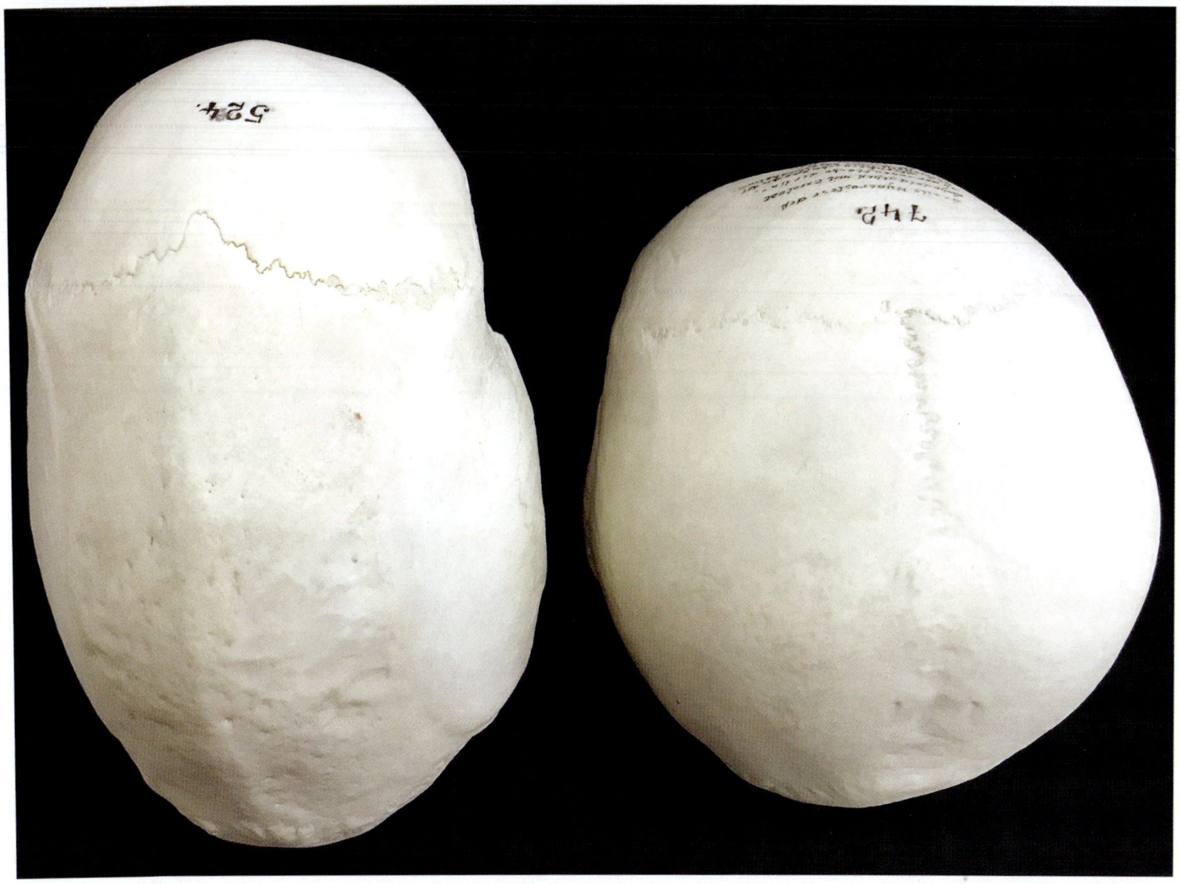


**Figure 19.** Premature craniosynostosis (scaphocephaly) in an unknown individual (MN524) compared to an average sized/shaped adult skull (MN742). Premature synostosis of one or more sutures alters the size and shape of the skull resulting in compensatory growth along the other sutures that are open (non-synostosed). Premature closure of the sagittal suture in this individual resulted in a long and narrow skull known as scaphocephaly, meaning "boat-shaped" in Greek, for its resemblance to an inverted boat and keel (Mantillas-Rivas et al. 2020; Persing et al. 1989).

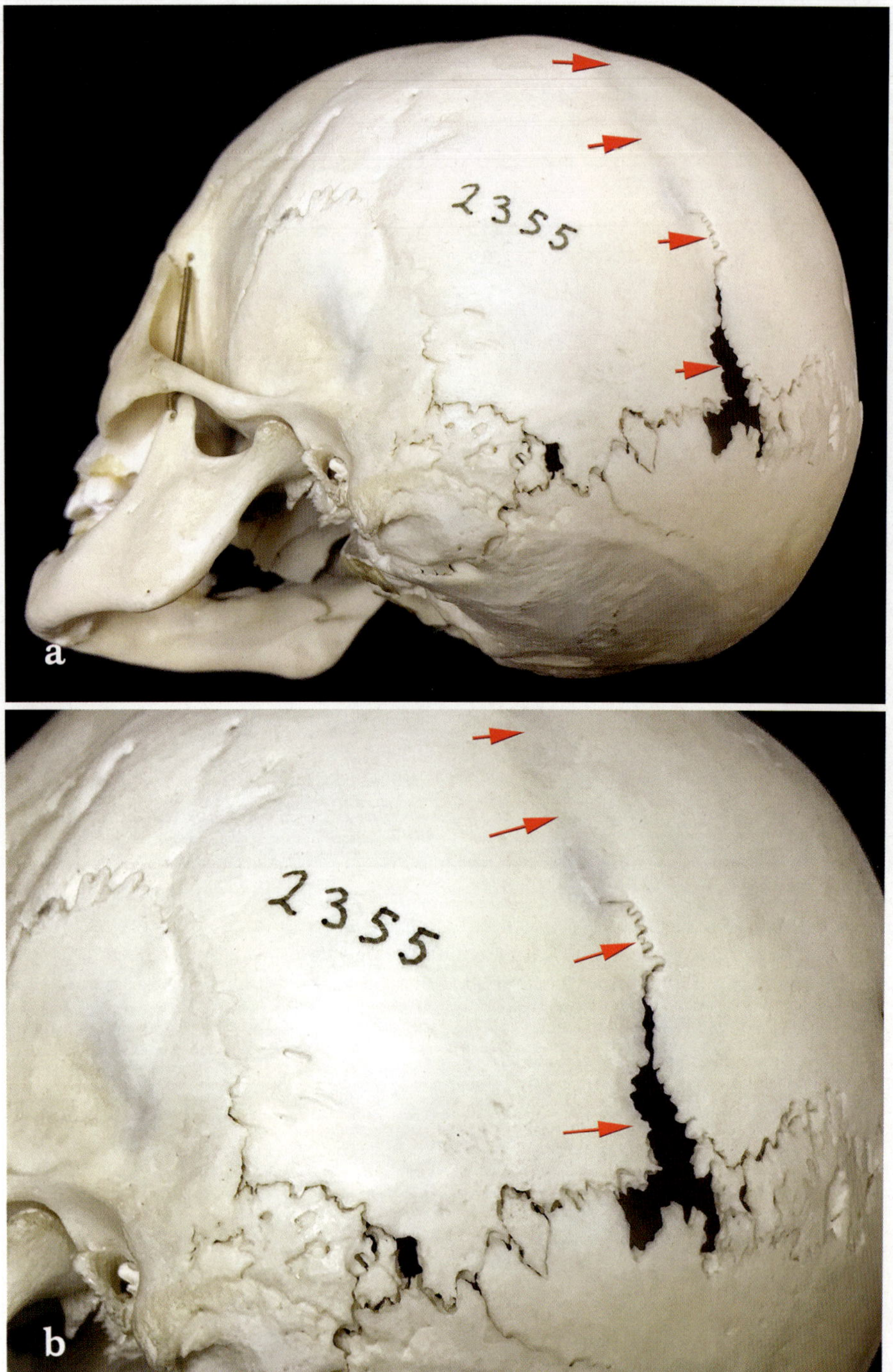


**Figure 20a-b.** (a) Left oblique and (b) magnified views of accessory suture (arrows) extending from the lambdoidal suture to the coronal suture forming a rare bipartite/divided parietal bone in a 6-year-old female who died in 1874 (MN21.055/2355). This serrated suture is still unfused near the lambdoidal suture and closes/obliterates as it approaches the coronal suture. This accessory sub-sagittal suture divides the left parietal bone into upper and lower halves of uneven size and shape (see Berry 1910 for an example of bipartition). Note the numerous lambdoidal ossicles often associated with premature craniosynostosis (Mann et al. 2016).

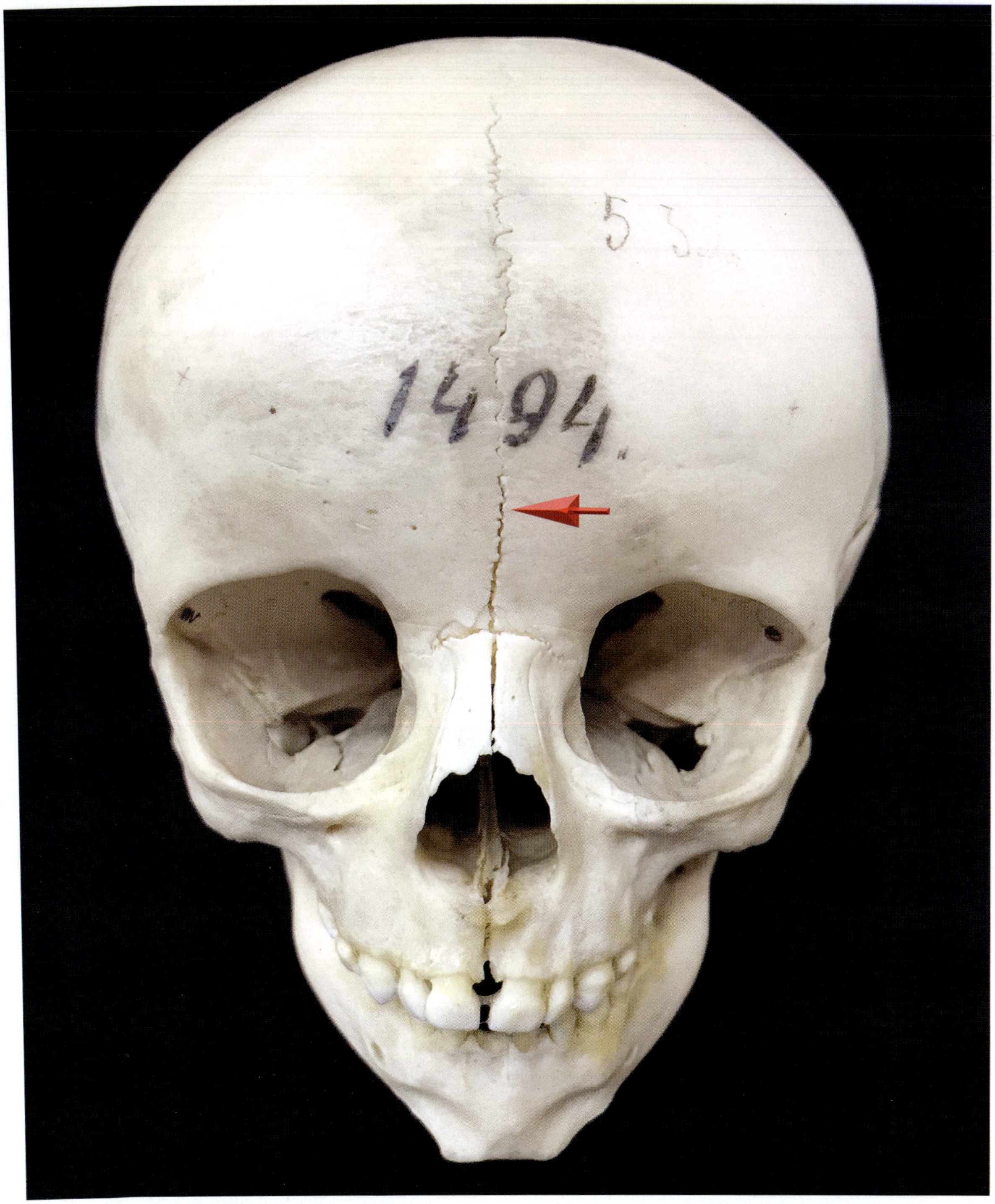


**Figure 21.** An open (patent) metopic suture and partial bipartition (see Figure 22) in the parietal bones of a 5-year-old male who died in 1869 (MN1494). The square shape and indentions along either side of the mental protuberance of the mandible are typical male features, even at this early age. Radiographic (3D-CT) examination of 159 infants and children revealed the earliest age of closure of the metopic suture was 3 months with complete closure at 9 months (Vu et al. 2001). An Australian study of children, in comparison, revealed closure of the metopic suture by 19 months (Teager et al. 2019).

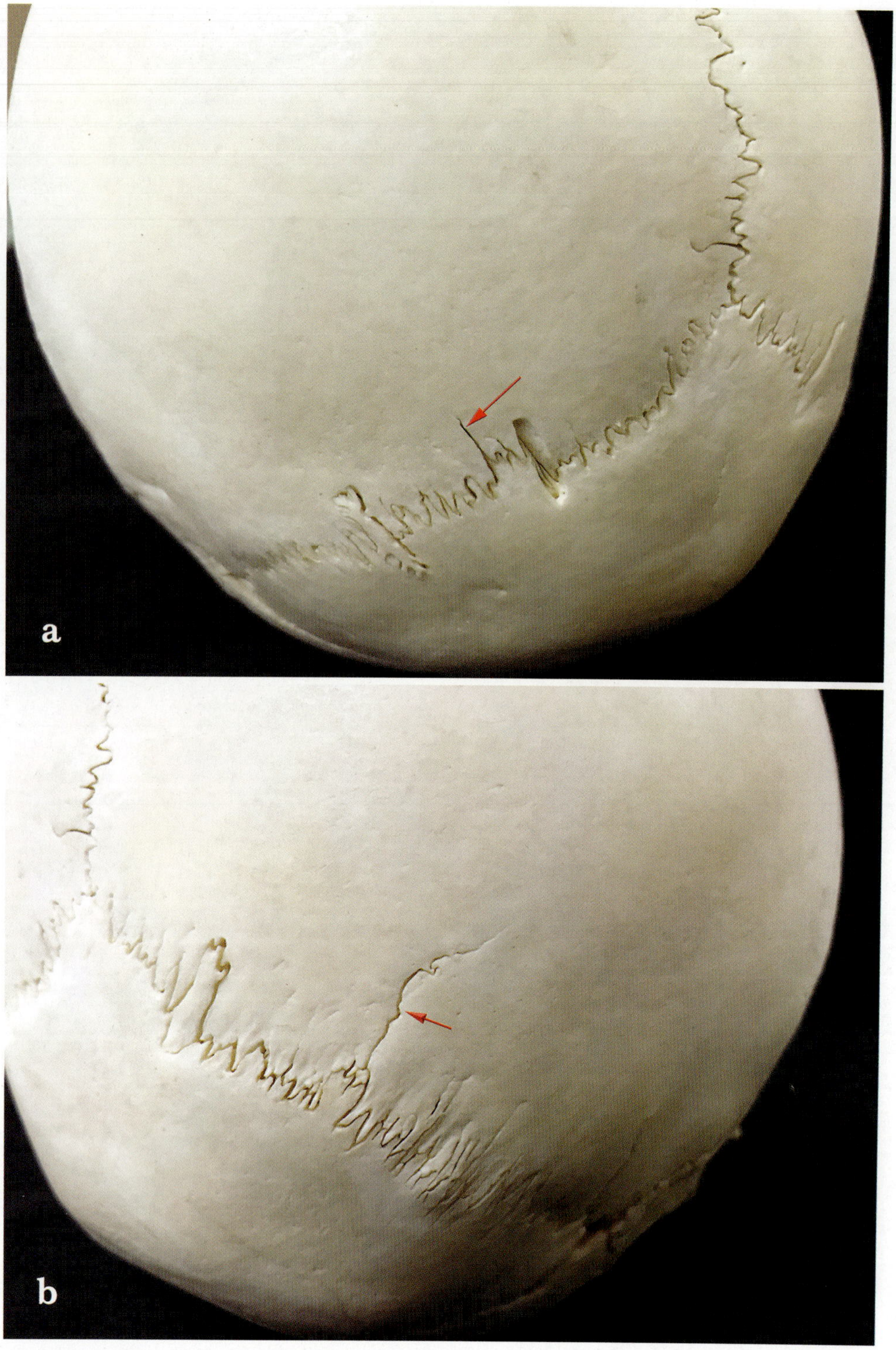


**Figure 22a-b.** (a) Left and (b) right posterior oblique views of rare partial bipartition of the (a) left and (b) right parietal bones in a 5-year-old male who died in 1869 (MN1494). Close examination revealed that these are sutures (arrows) and not lambdoidal ossicles, fractures or postmortem breakage. See Mann and Hunt (2019); Sanchez et al. (2010); Weir et al. (2006).

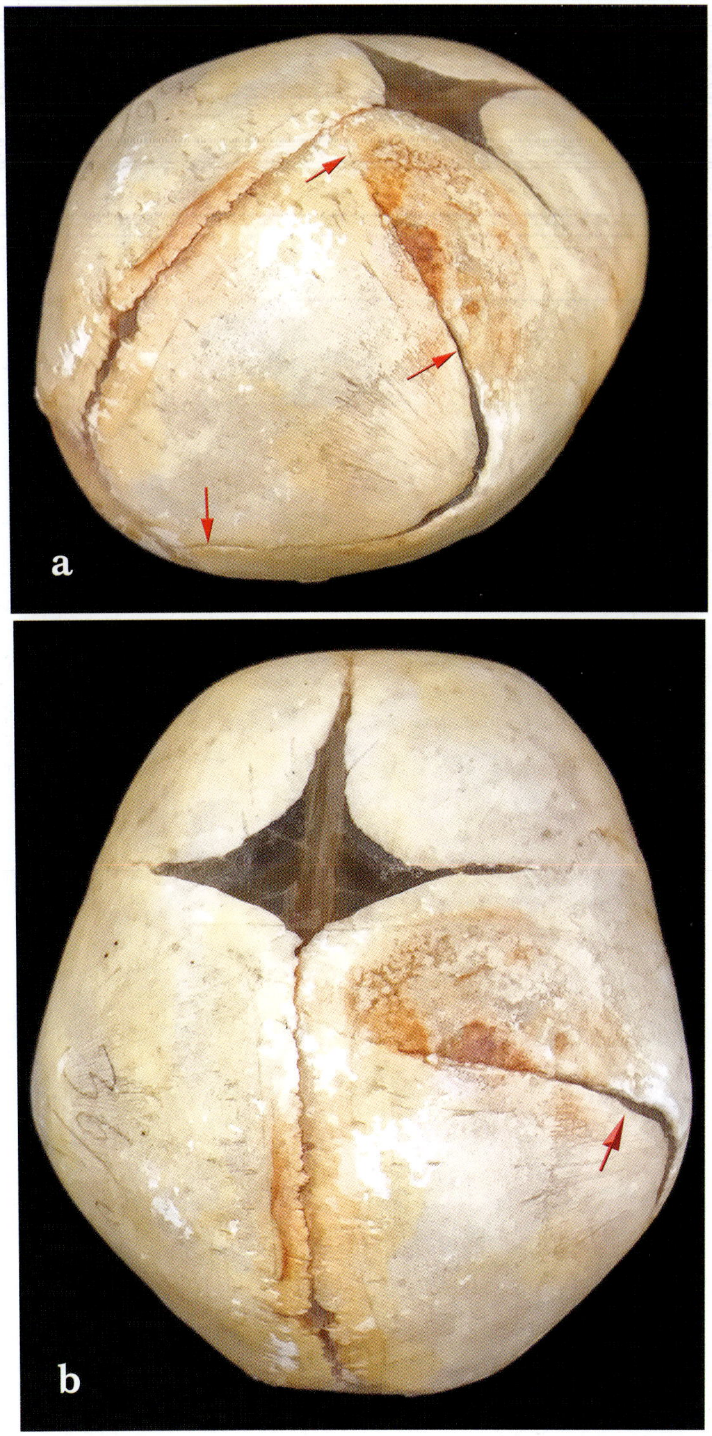


**Figure 23a-b.** (a-b) Right superior-inferior and (b) superior views of divided/bipartite (cleft) right parietal bone (arrows) in a 12-day-old neonate who died in 1862 (MN21.055/636). Bipartite parietal bones usually consist of mildly curved accessory sutures, unlike that in this individual that arcs from the lambdoidal to the sagittal suture. While perimortem trauma is a possibility, the absence of radiating fractures, periostitis or remodeling is more consistent with a suture than a fracture. See Mann and Hunt (2019); Sanchez et al. (2010); Weir et al. (2006).

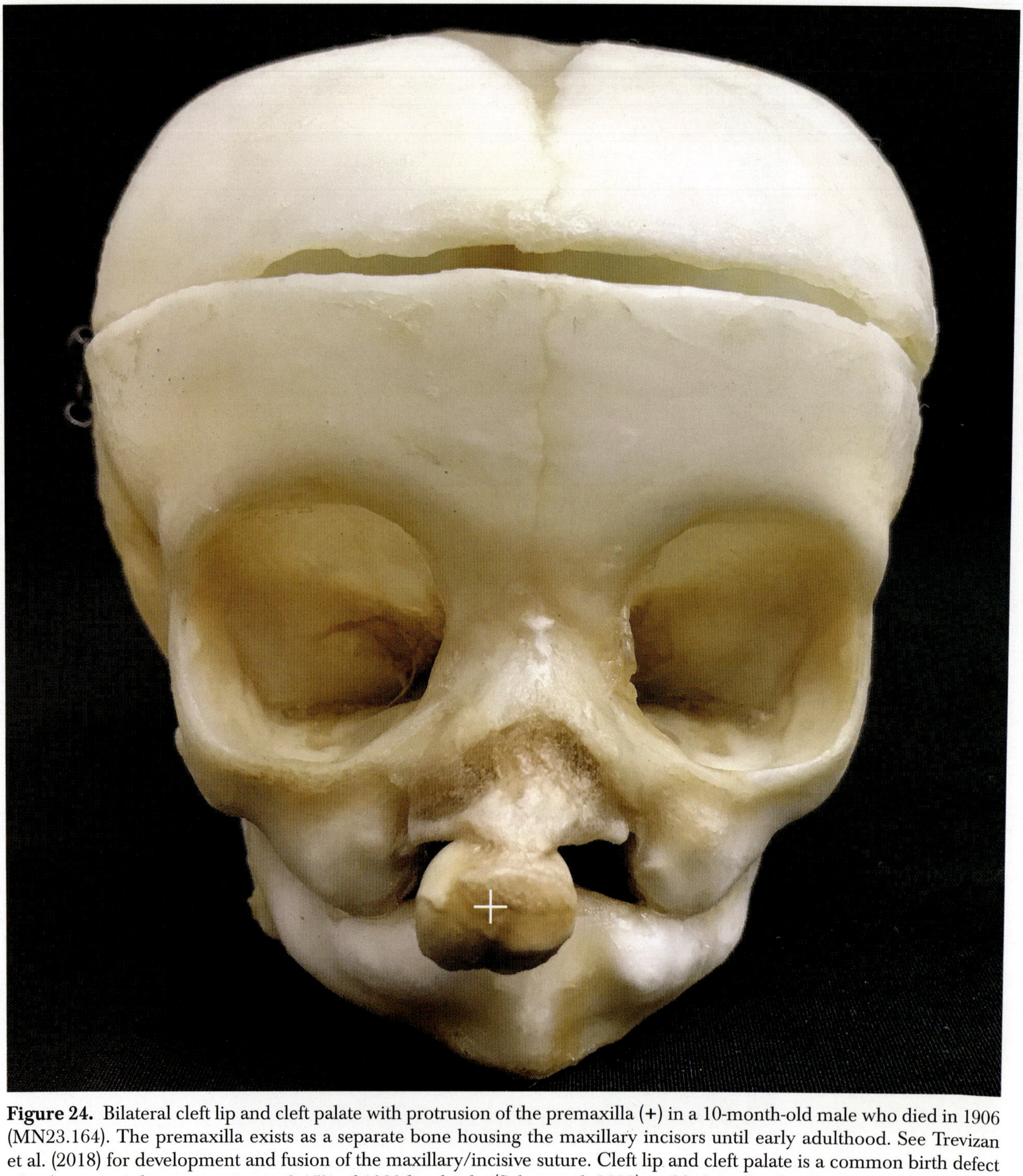

**Figure 24.** Bilateral cleft lip and cleft palate with protrusion of the premaxilla (+) in a 10-month-old male who died in 1906 (MN23.164). The premaxilla exists as a separate bone housing the maxillary incisors until early adulthood. See Trevizan et al. (2018) for development and fusion of the maxillary/incisive suture. Cleft lip and cleft palate is a common birth defect of unknown etiology occurring in 0.45% of 1000 live births (Salari et al. 2022) and believed to be the result of genetic and environmental variables (see Barnes 1994, 2012; Vyas et al. 2020). See Manchester (1983) for archaeological example of cleft palate and Wyszynski (2002) for more on cleft lip and cleft palate clinically and developmentally.

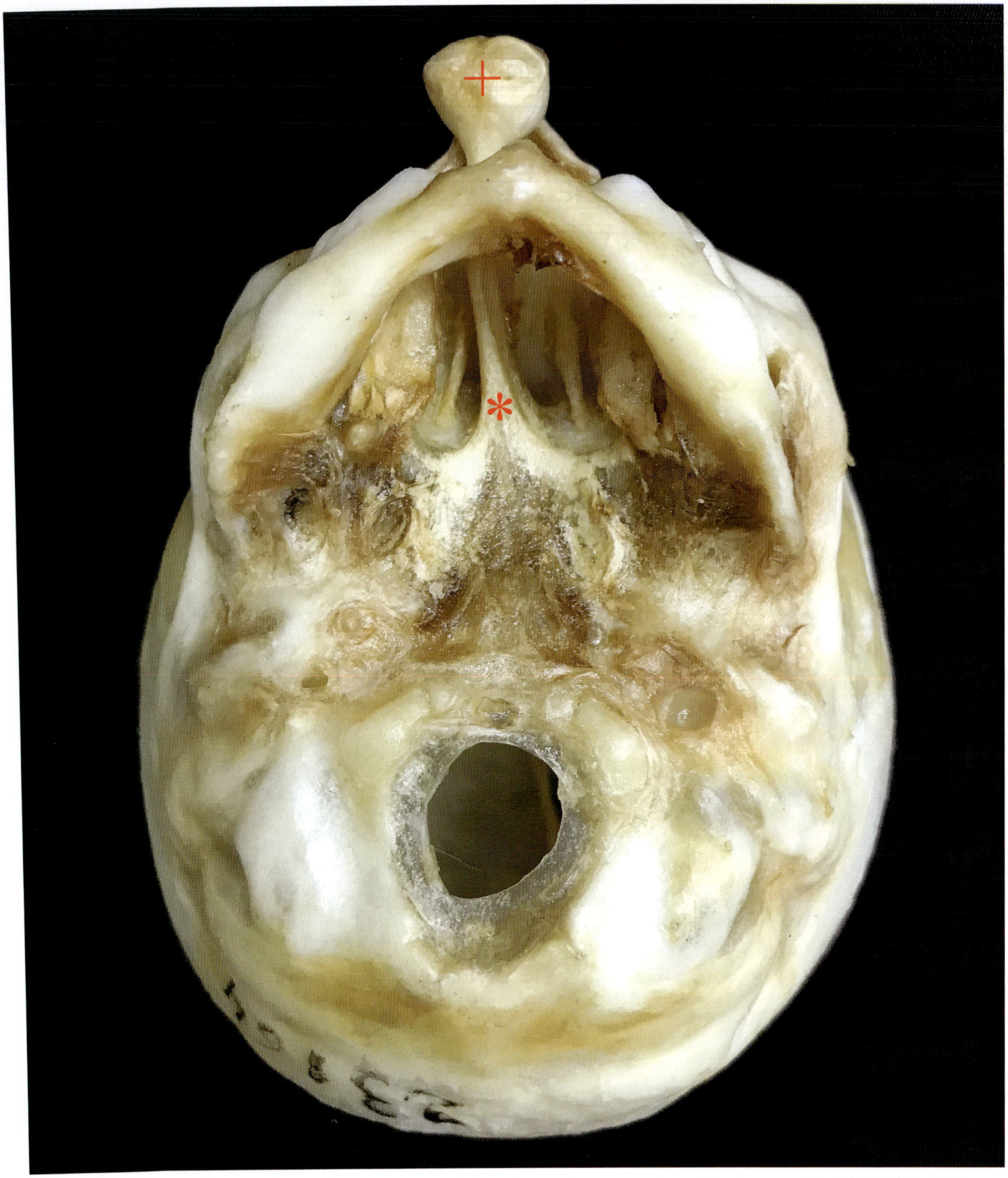

**Figure 25.** Inferior view of bilateral cleft lip and cleft palate with protrusion of the premaxilla (+) (MN23.164) in a 10-month-old male who died in 1906 (see Barnes 1994, 2012; Salari et al. 2022; Vyas et al. 2020; Wyszynski 2002). The vomer (*) joins the premaxilla at the premaxillary-vomerine suture (not visible in photo). Note the presence of the translucent periosteum around the foramen magnum and covering much of the skull.

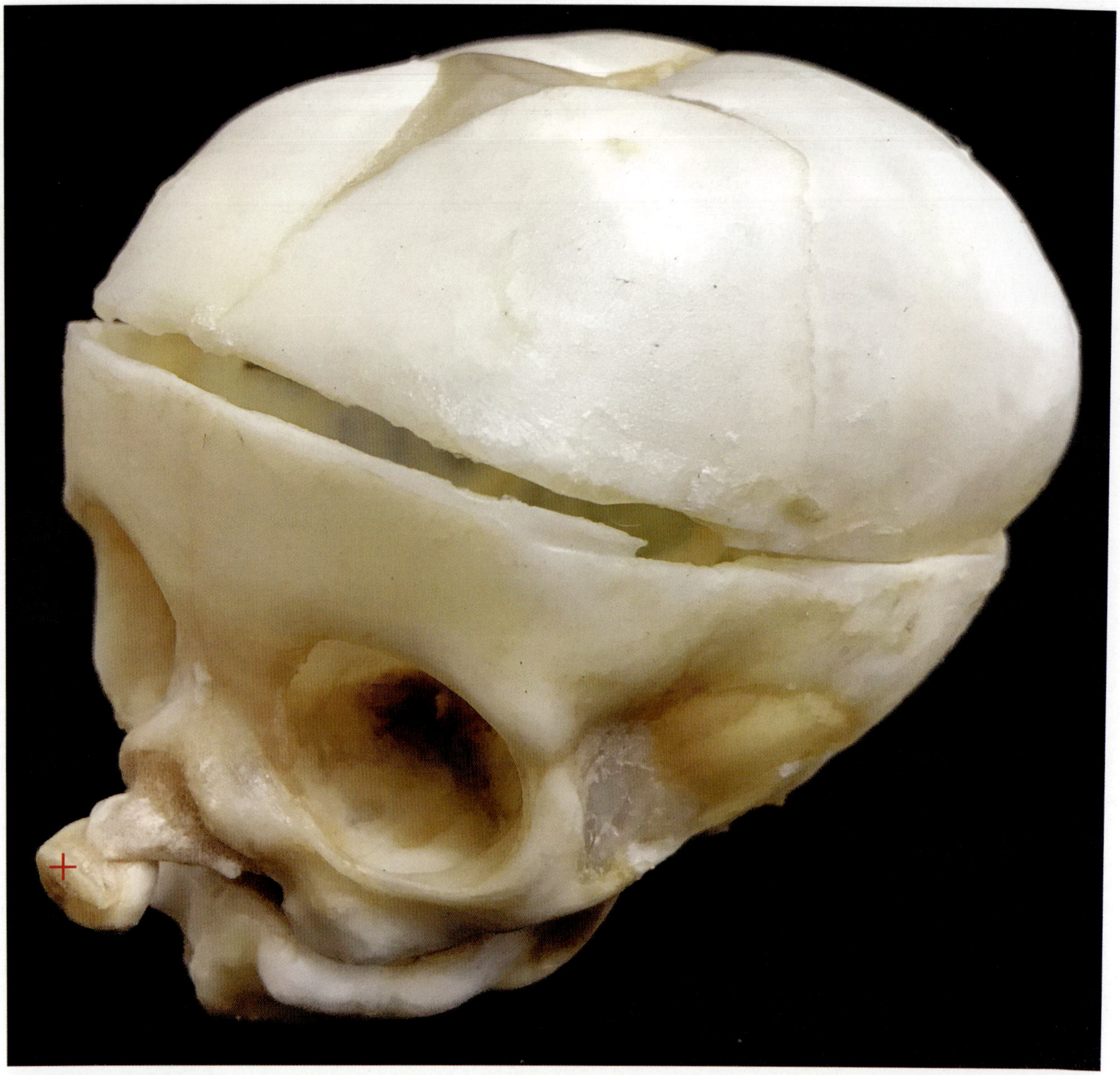

**Figure 26.** Left oblique view of bilateral cleft lip and cleft palate with protrusion of the premaxilla (+) in a 10-month-old male who died in 1906 (MN23.164). See Barnes (1994, 2012); Salari et al. 2022; Vyas et al. 2020; Wyszynski (2002).

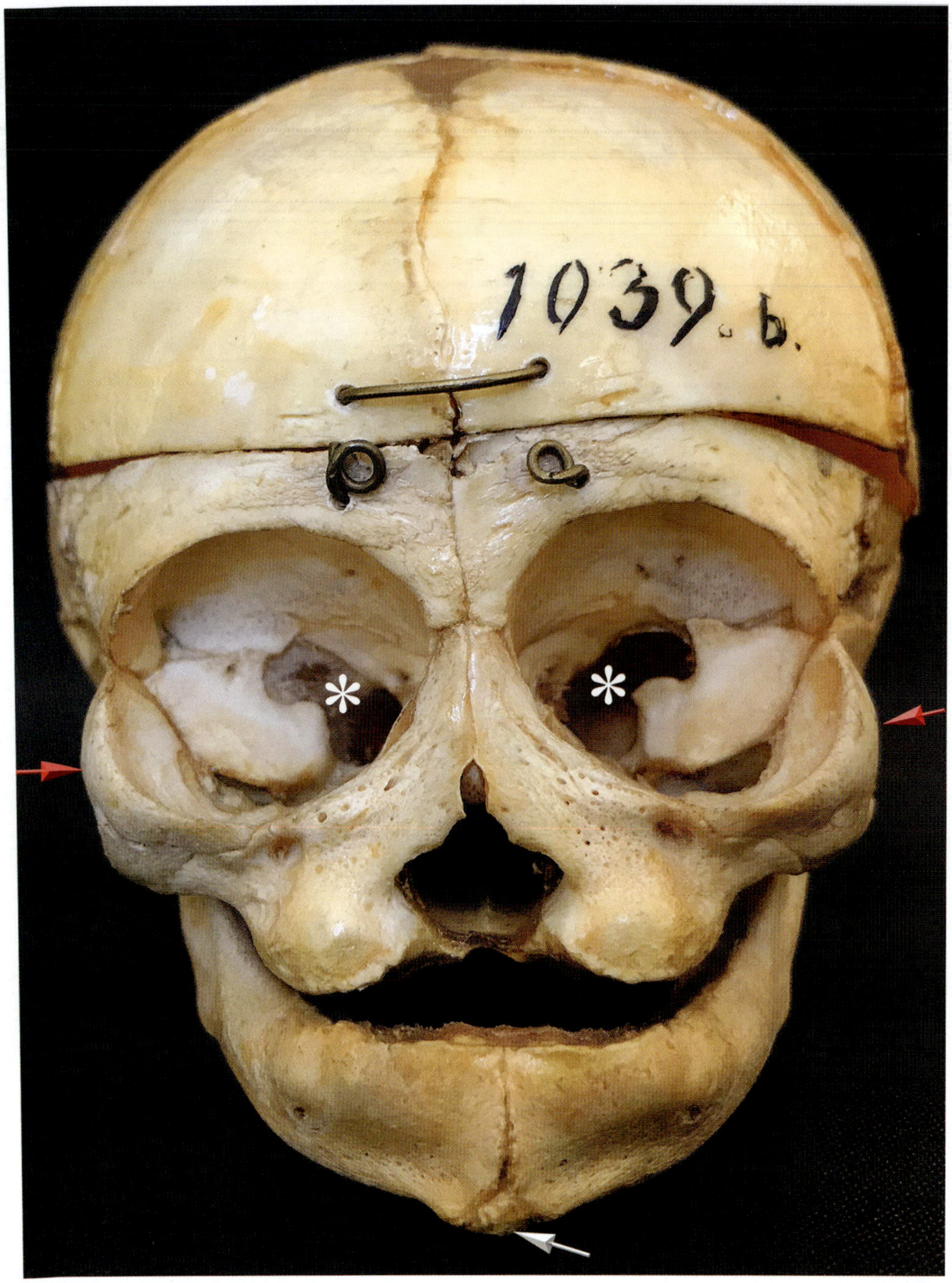


**Figure 27.** Malformation of the maxilla with absence of nasal bones, and cleft palate, and hypotelorism, and midface hypoplasia in a neonate male who died in 1866 (MN21055/1039b). These features suggest midline defects which are associated with holoprosencephaly and trisomy 13 (K. Thompson, personal communication, September 27, 2022). Note the unusually large orbital fissures (*), ossicle (white arrow) at the inferior portion of the mental symphysis and the laterally projecting malar bones (red arrows). See Allan et al. (2014); Barnes (1994, 2012); Vyas et al. (2020). Cf. Wahby (1903) for an illustration of a similar fetus with absence of nasal bones and premaxillae.

**Figure 28.** Left oblique view of a neonate male who died in 1866 with midline defects associated with holoprosencephaly and trisomy 13 (MN2.055/1039b) (K. Thompson, personal communication, September 27, 2022). See Allan et al. (2014); Barnes (1994, 2012); Vyas et al. (2020). Cf. Wahby (1903) for an illustration of a similar fetus without nasal bones or premaxillae.

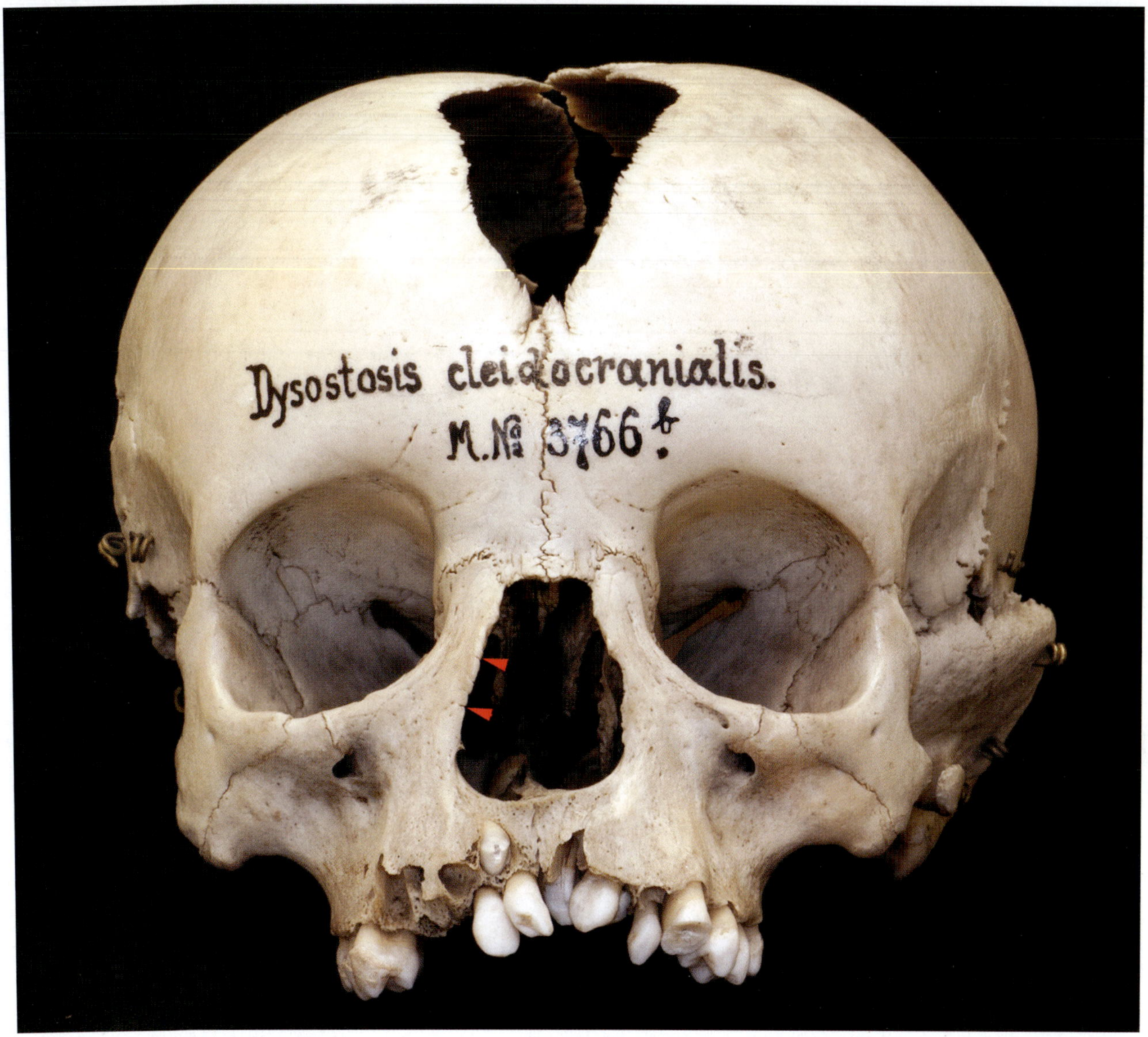


**Figure 29.** Anterior view of cleidocranial dysostosis ("Dysostosis cleidocranialis" on the skull), a rare genetic disorder of bones caused by a defect in intramembranous ossification, often involving the clavicles which are hypoplastic or absent and teeth in an subadult (MN3766b). This individual exhibits widely separated orbits and vault sutures, large anterior fontanelle, supernumerary teeth, and based on the blunt margins and tiny notches (arrowheads) along the nasal aperture, absence of the nasal bones. See Mundlos (1999); Roberts et al. (2013); Schmorl and Junghanns (1959) for more on cleidocranial dysostosis. Cf. Ortner and Putschar (1981) for more on this individual.

**Figure 30.** Posterior view of cleidocranial dysostosis (cleidocranial dysplasia), a rare genetic disorder of bones caused by a defect in intramembranous ossification, often involving the clavicles (hypoplastic or absent) and teeth in a subadult (MN3766b). This individual exhibits a large anterior fontanelle, an unusually complex sagittal suture and widely separated sutures. (Only the skull is present.) See Mundlos (1999); Roberts et al. (2013); Schmorl and Junghanns (1959) for more on cleidocranial dysostosis. Cf. Ortner and Putschar (1981) for more on this individual.

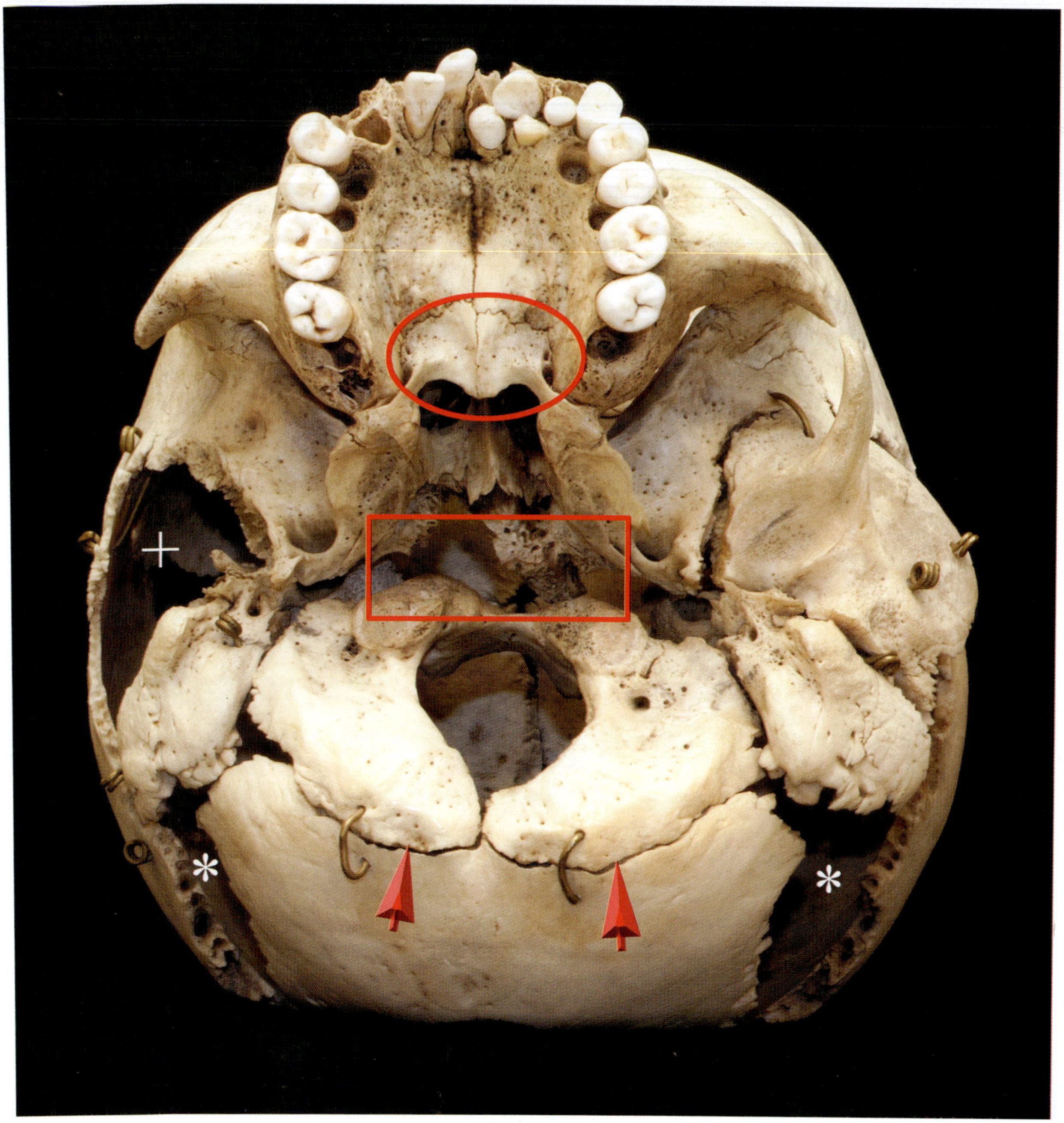


**Figure 31.** Inferior view of cleidocranial dysostosis in a subadult (MN3766b). The skull exhibits a large anterior fontanelle, widely separated orbits and vault sutures, abnormal development and eruption of the teeth to include several supernumerary teeth and absent nasal bones. Note separation of the lateral parts of the basilar bones (exoccipitals) and Mendosal suture (red arrows), widely separated lambdoidal sutures (*) and spheno-occipital synchondrosis (rectangle), parabolic palate with small, narrow palatine bones (oval), delayed development and eruption of several teeth, and supernumerary teeth, several of which are absent postmortem. The right temporal bone (+) is missing postmortem. See Mundlos (1999); Roberts et al. (2013); Schmorl and Junghanns (1959) for more on cleidocranial dysostosis. Cf. Ortner and Putschar (1981) for more on this individual.

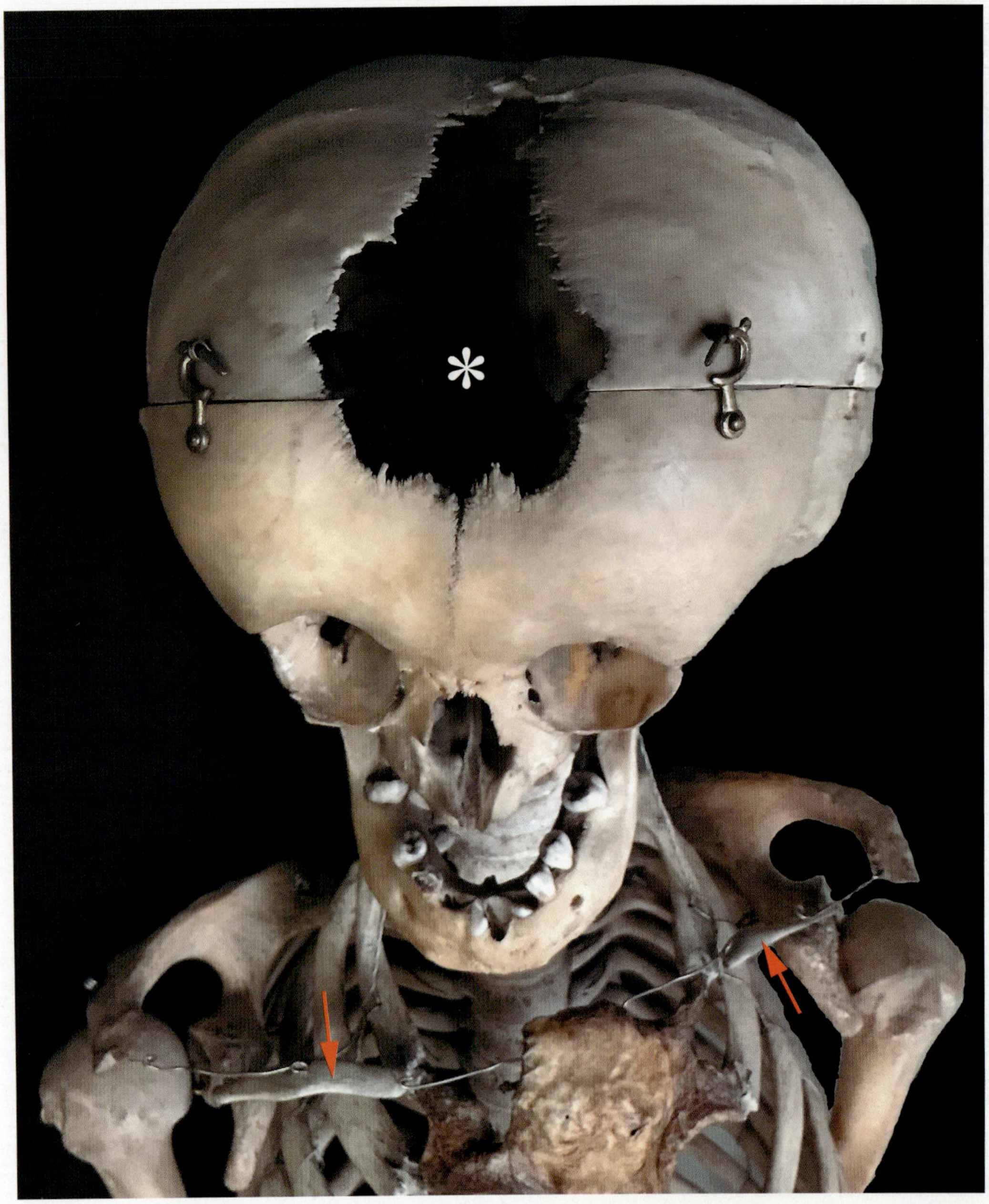

**Figure 32.** Anterior-superior view of cleidocranial dysostosis, macrocephaly, and an unusually large anterior fontanelle (*), thin orbits, and aplasia/absence of the nasal bones in a 25-year-old male who died in 1909 (MN5701). See Mundlos (1999); Roberts et al. (2013) for more on cleidocranial dysostosis. Cf. Ortner and Putschar (1981) for more on this individual.

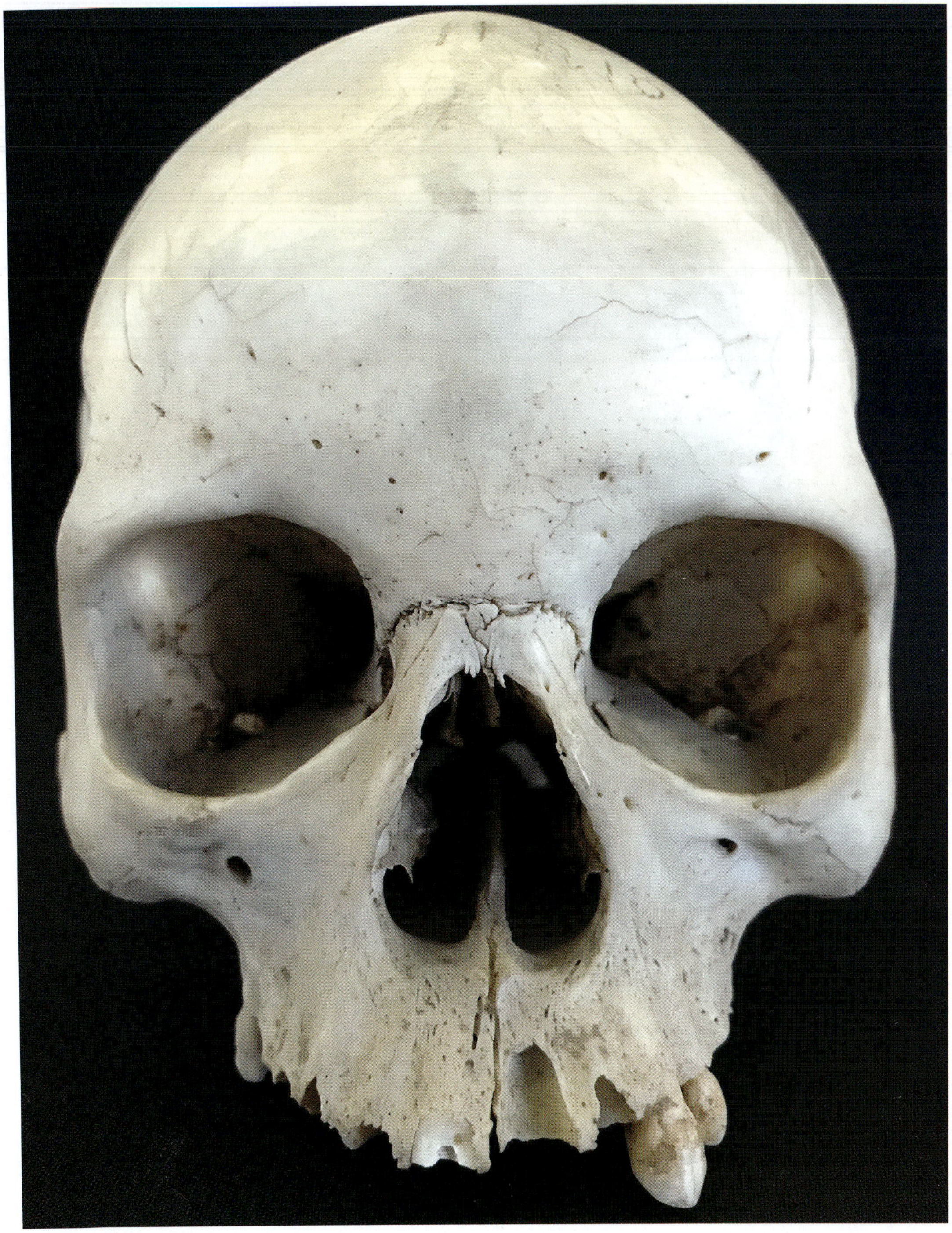

**Figure 33.** Rare absence or antemortem loss of the nasal bones in an 18-year-old male who died in 1909 of actinomycosis, a rare type of chronic bacterial infection (MN5686). The margin of the nasal aperture that holds the nasal bones is smooth, rounded, and exhibits no evidence of articulation with the nasal bones. Actinomycosis often affects the neck and face, but rarely the nasal cavity (see Barnes 2012; Sakuma et al. 2016).

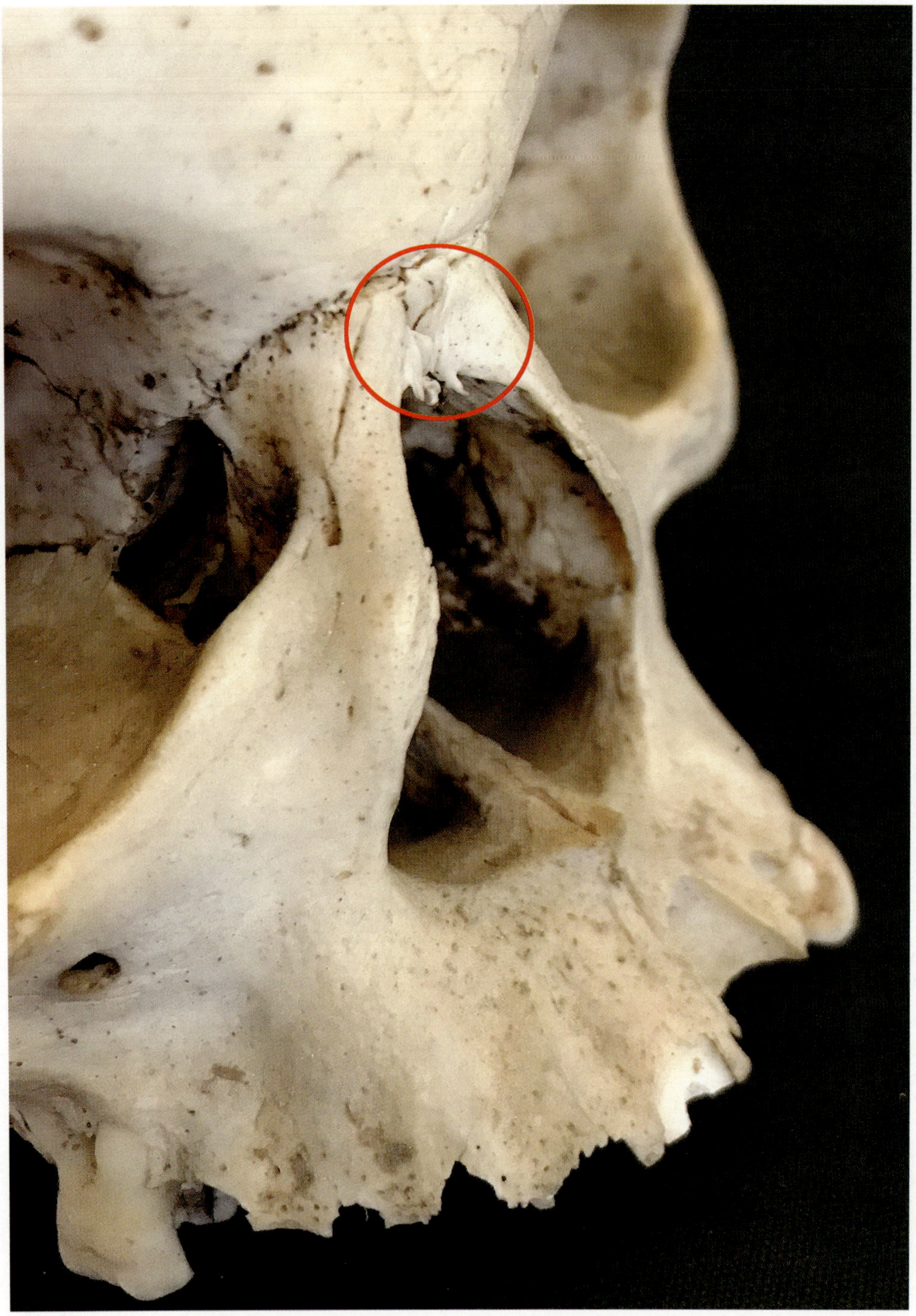

**Figure 34.** Rare absence or antemortem loss of the nasal bones (circle) in an 18-year-old male who died in 1909 with actinomycosis, a rare type of bacterial infection (MN5686).

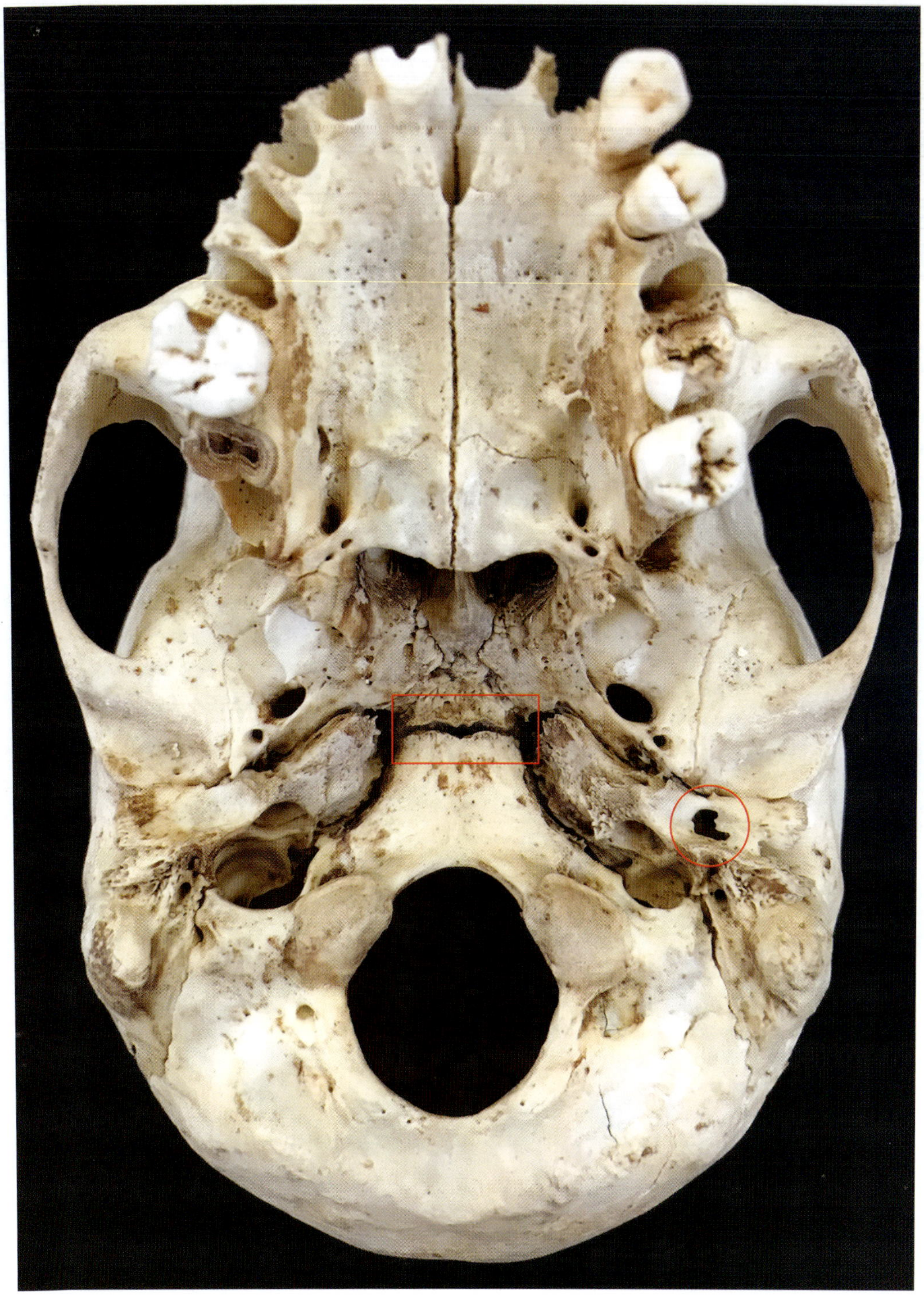

**Figure 35.** Unusually large hard palate/maxilla in an 18-year-old male who died in 1909 with actinomycosis, a bacterial infection. (MN5686). Note the unusually large hard palate/maxilla, wide zygomatic arches, short occipital bone and left tympanic dehiscence (circle), also known as "foramen of Huschke" and foramen tympanicum (Tozoglu et al. 2012). The third molars have not erupted and the spheno-occipital synchondrosis, also known as the basiocciput synchondrosis, has not begun to close (rectangle) (Alhazmi et al. 2017). See Mohamad et al. (2021) for a case report of cranial actinomycosis and review of the literature.

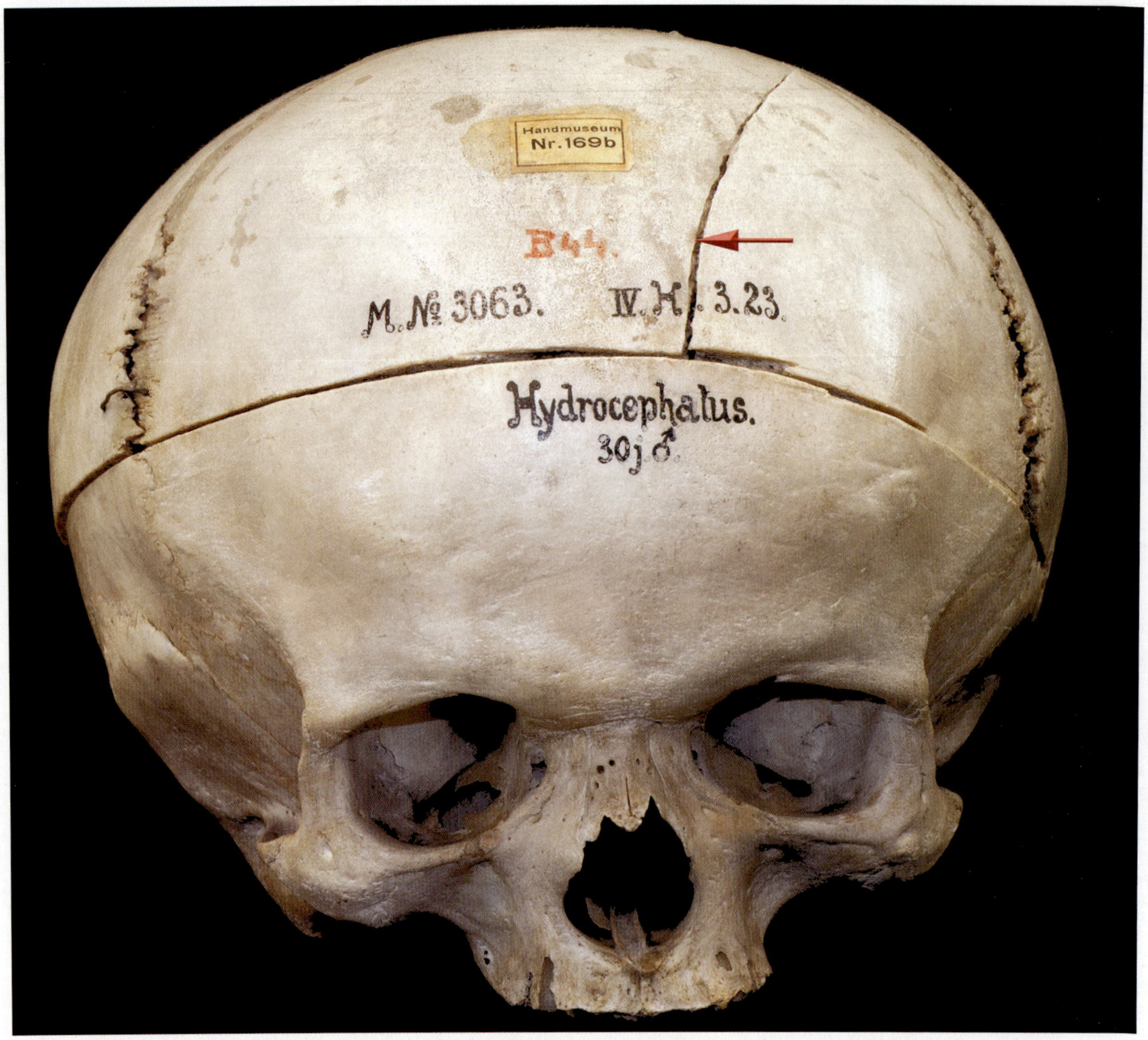


**Figure 36.** Hydrocephalus ("Hydrocephalus" on the skull) in a 30-year-old male who died in 1867 (MN3063) (see Barnes 2012). The vault and internasal breadth are wide, the nasal bones and cavity are small, the zygomas are hypoplastic, the orbits are small in proportion to the vault and are downward facing and the coronal suture is complex and widely separated. Hydrocephalus is a condition that may be present at birth (congenital) or that develops (acquired) after birth and consists of an excess of cerebrospinal fluid surrounding the brain. The underlying cause of hydrocephalus may be a result of genetic and environmental factors that results in an excess of cerebrospinal fluid surrounding the brain. The vertical fracture in the frontal bone (red arrow) was likely caused when the vault/calotte was pried away with a bone chisel.

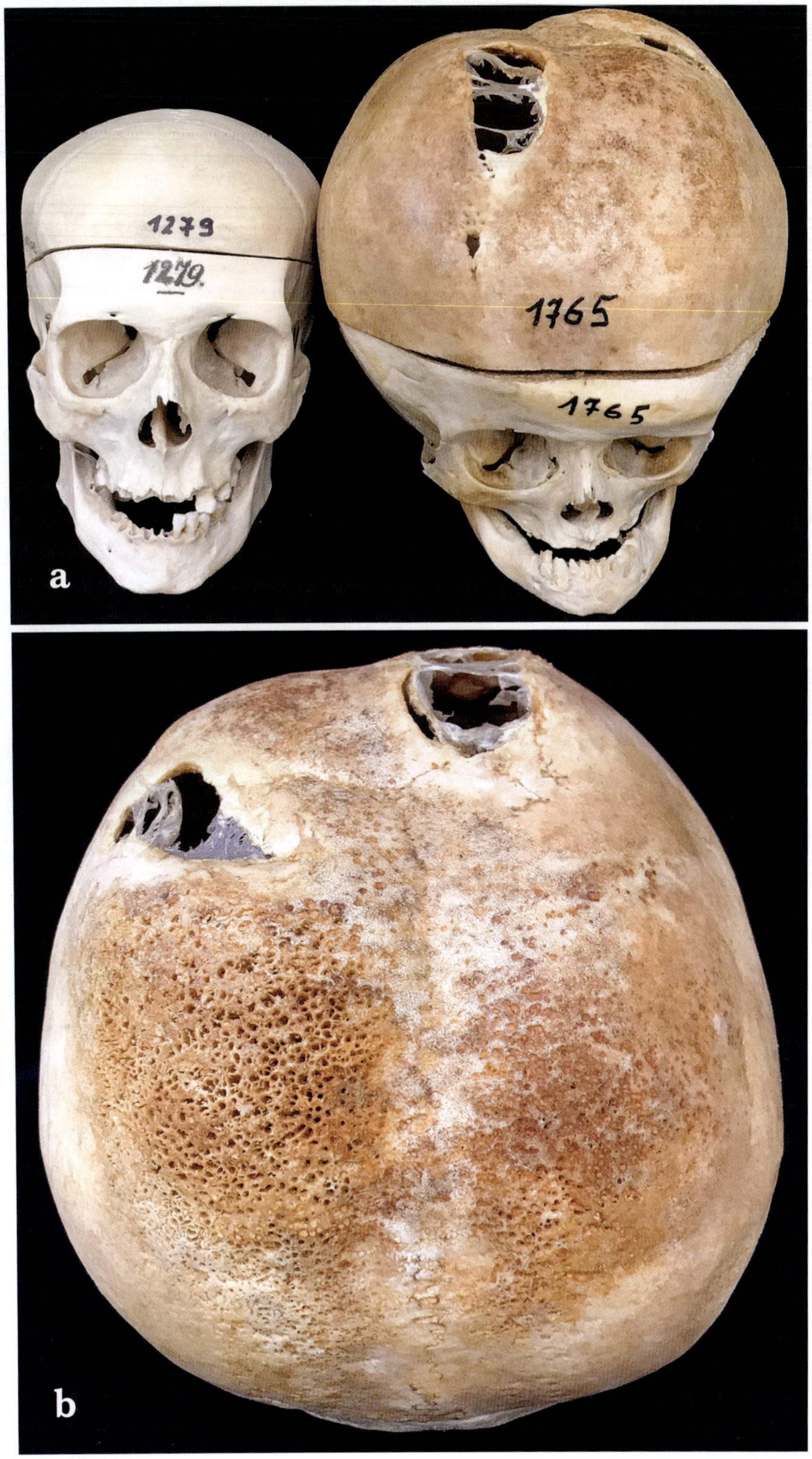


**Figure 37a-b.** (a-b) Hydrocephaly and rachitis (rickets) in the skull of a 34-year-old female who died in 1871 (MN21.055/1765) compared to the skull of a typical sized adult male (MN1279). (b) The parietal bones exhibit raised areas with porosities consistent with porotic hyperostosis possibly due to iron-deficiency anemia (Mann and Hunt 2005) or hemolytic and megaloblastic anemias (Walker et al. 2009) (See Barnes 1994, 2012; Snapper 1957).

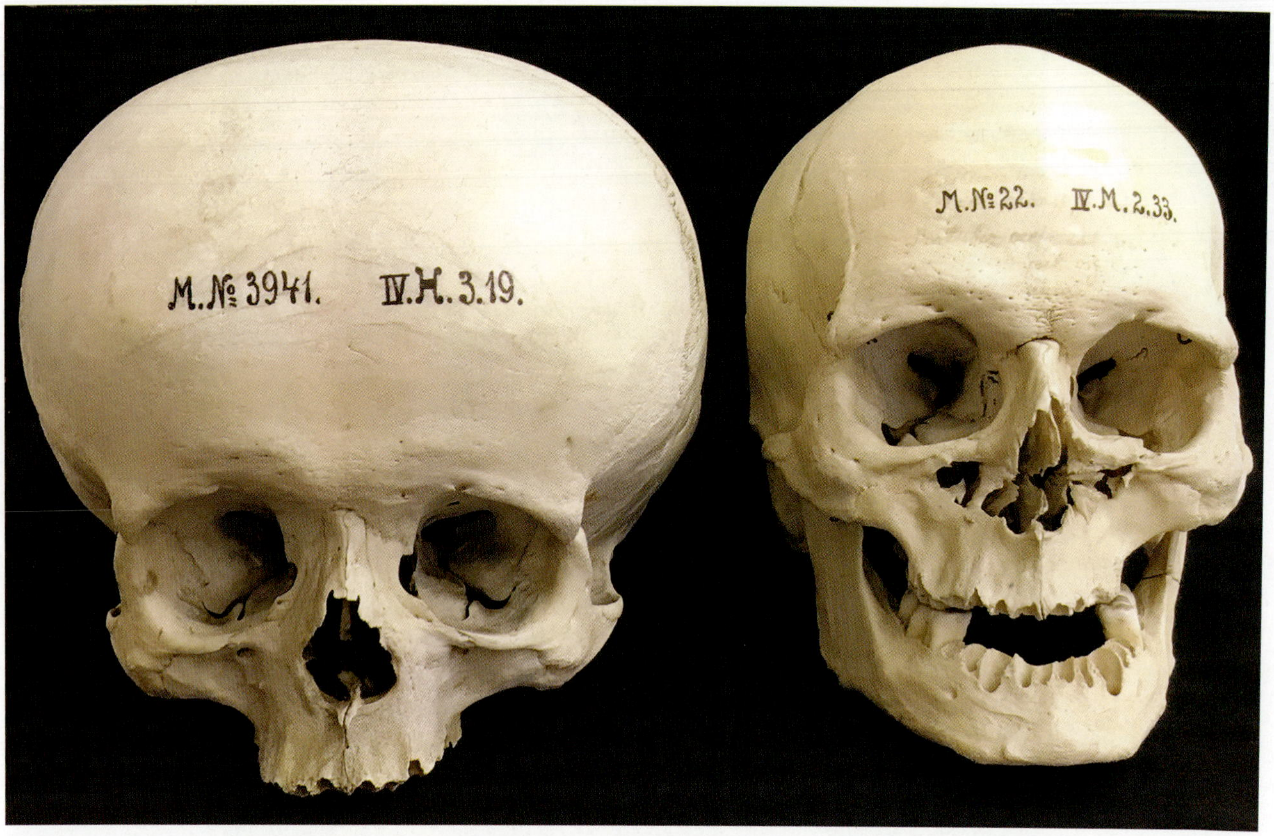


**Figure 38.** Hydrocephalic male skull (left in photo) compared with a typical adult male skull (right in photo) with healed facial trauma (MN3941 and MN22). The hydrocephalic skull exhibits a greatly enlarged vault, wide facial region and interorbital width and a comparatively small maxilla.

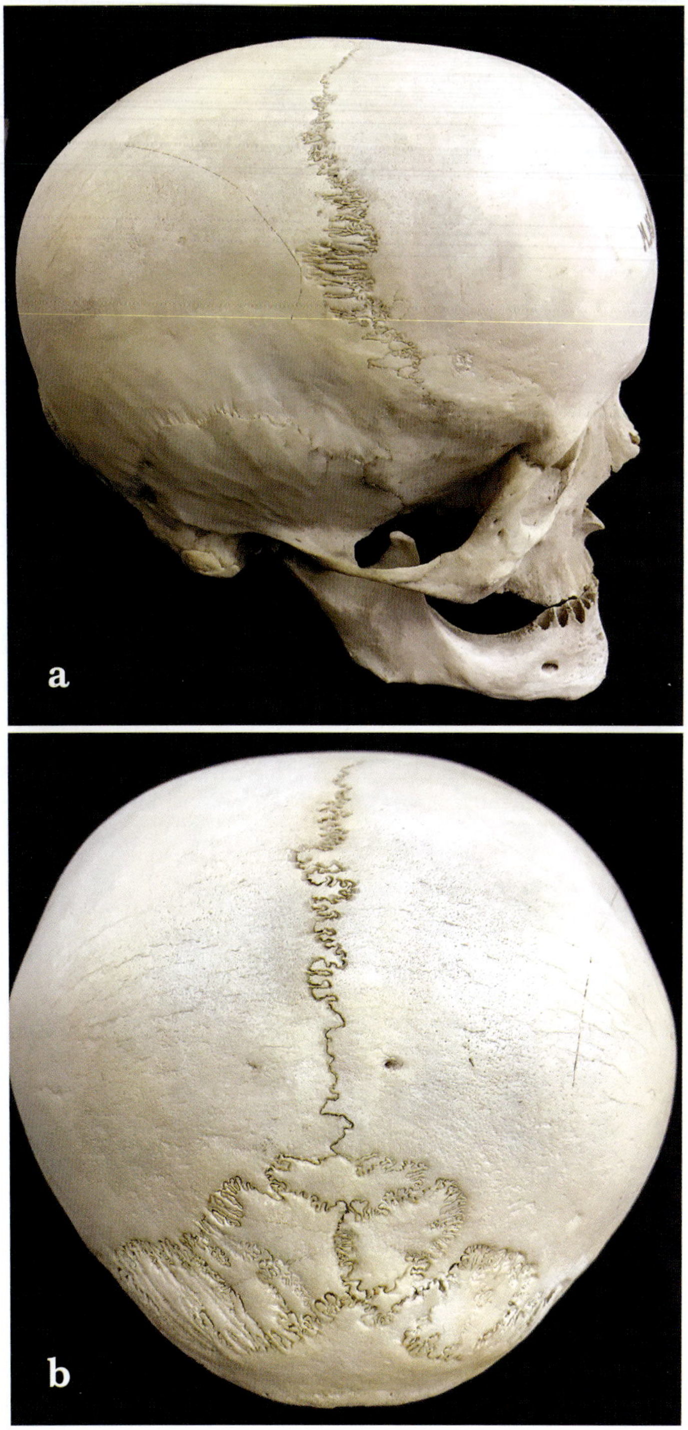


**Figure 39a-b.** (a) Right lateral and (b) posterior views of hydrocephaly in a 34-year-old male who died in 1869 (MN3941). Note the enlarged vault, protruding frontal bone, unfused sutures and numerous large lambdoidal ossicles. Hydrocephaly due to increased cerebral spinal fluid from the ventricles and other conditions associated with rapid growth of the skull often result in highly interdigitated and complex sutures such as these. The presence of numerous ossicles (Wormian/wormian bones) of varying size along the lambdoidal suture is, however, a common anatomical variant and not necessarily indicative of hydrocephaly.

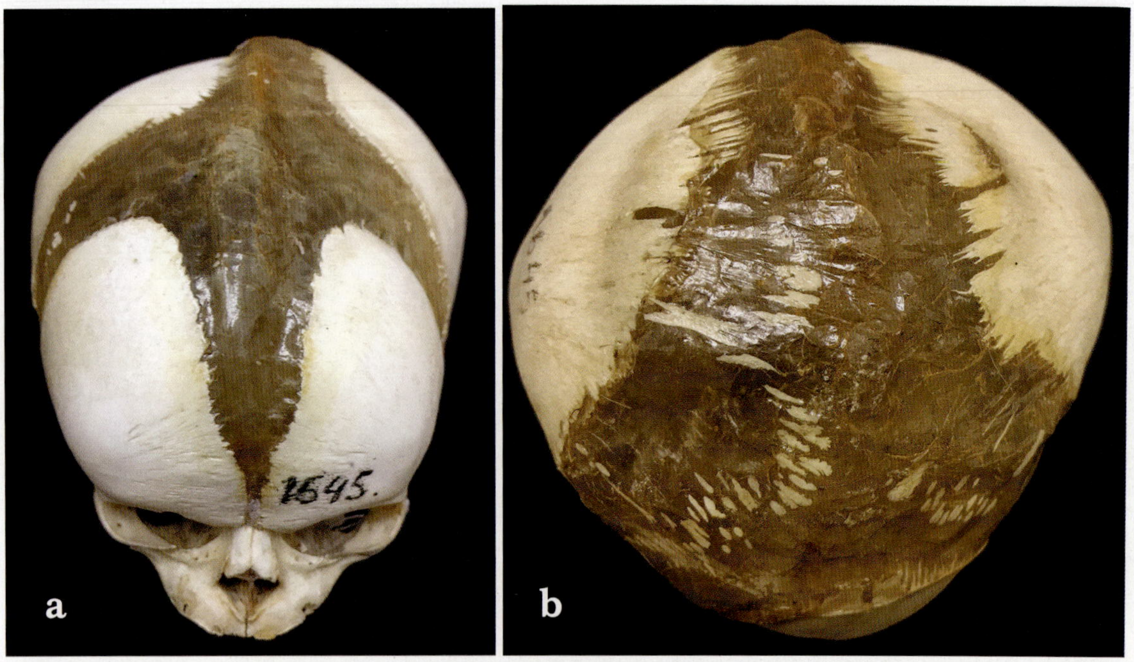


**Figure 40a-b.** (a) Superior and (b) posterior views of hydrocephalus and dolicocephaly in a fetal female (MN21.055/1545). Hydrocephalus or "water on the brain" as it's sometimes referred to, is due to an accumulation of cerebrospinal fluid. Dolicocephaly refers to an unusually long (anterior-posterior) skull in comparison to its width and is often due to premature closure of the sagittal suture or, as in this individual, an excess of cerebrospinal fluid. Note the many ossicles or "bone islands" in the dura mater, the transparent membrane that protects the brain, along the lambdoidal and sagittal sutures that form as a result of rapid enlargement of the skull.

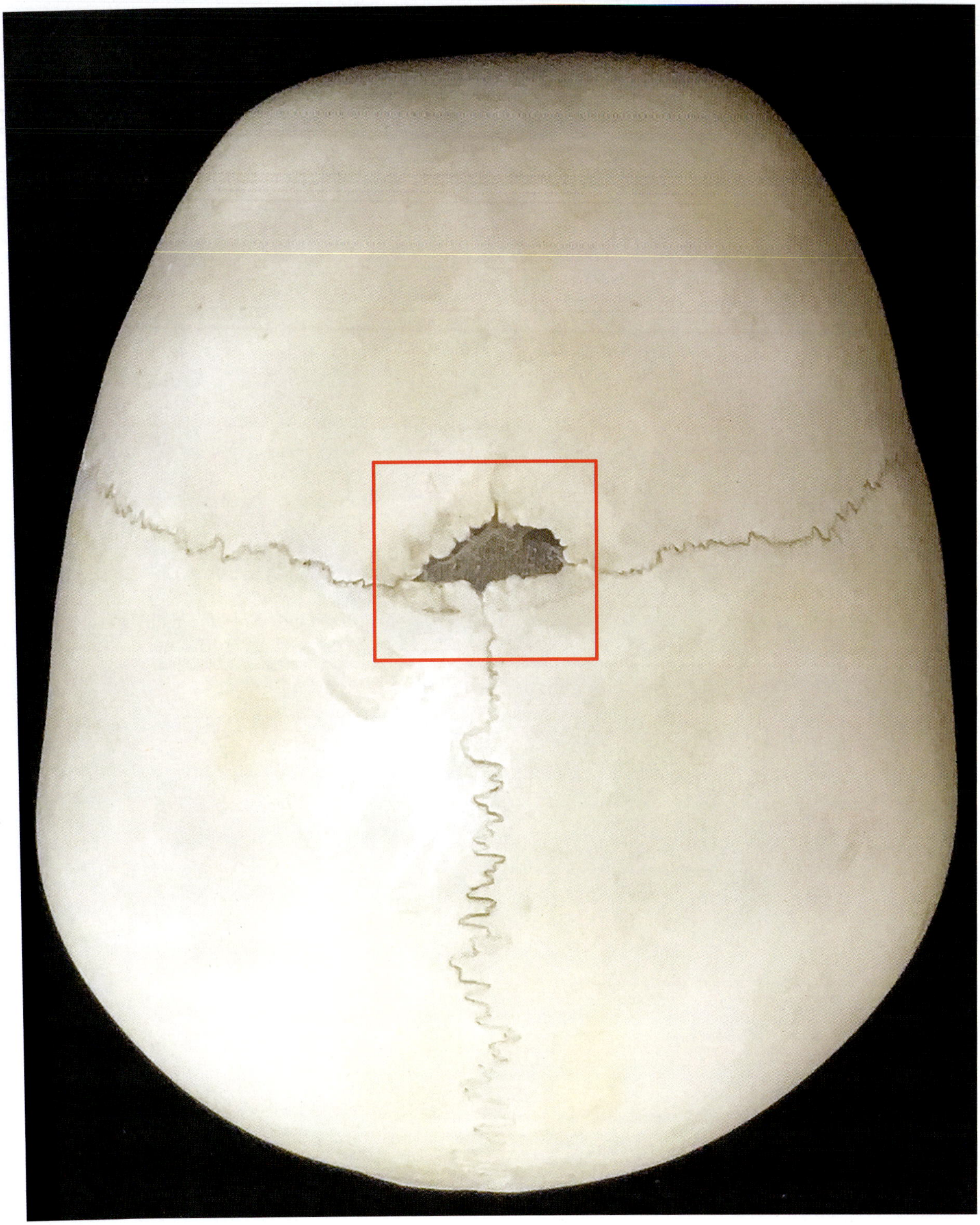

**Figure 41.** Incomplete closure of the anterior fontanelle (square), or "soft spot," in a 3-year-old male with hydrocephaly, a condition caused by excess production of cerebrospinal fluid in or around the brain (MN21.055/2384). Hydrocephaly, a neurological disorder, typically results in an unusually large head, protruding frontal bone ("bossing"), accessory sutural ossicles, and downward pointed ("setting sun sign" or "sunset") eyes clinically (Croft et al. 2016).

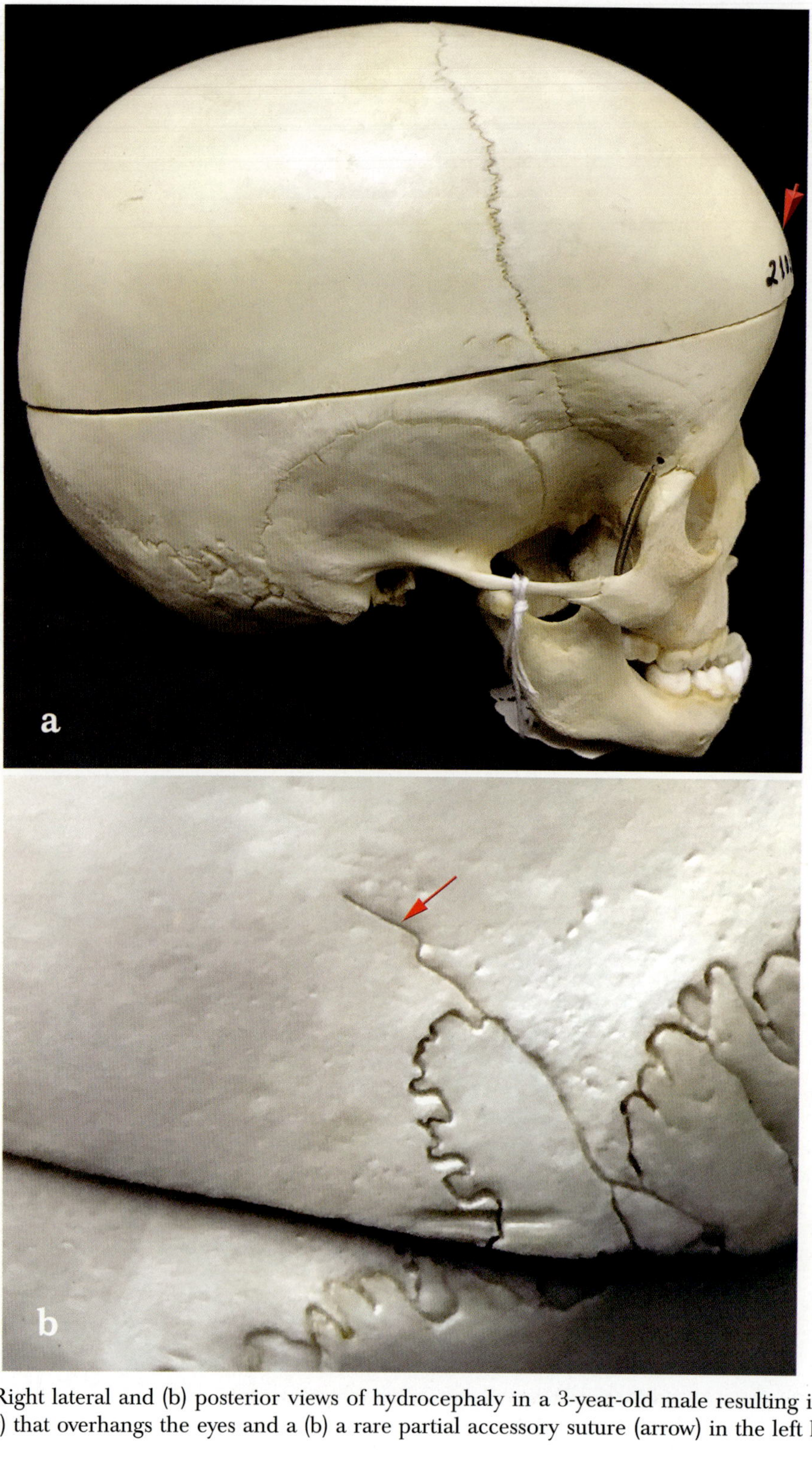


**Figure 42a-b.** (a) Right lateral and (b) posterior views of hydrocephaly in a 3-year-old male resulting in (a) a protruding frontal bone (arrow) that overhangs the eyes and a (b) a rare partial accessory suture (arrow) in the left lambdoidal suture (MN21.055/2384).

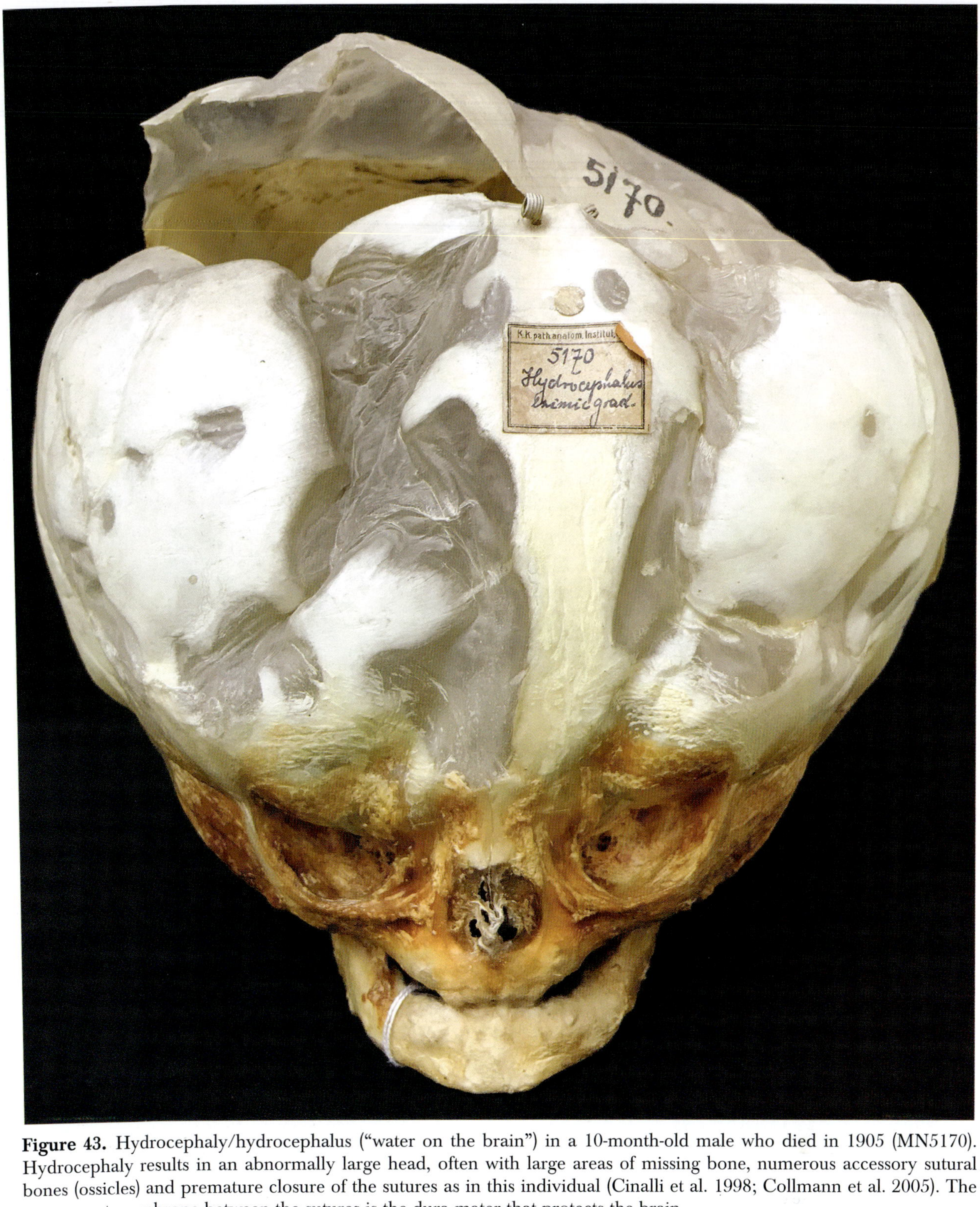


**Figure 43.** Hydrocephaly/hydrocephalus ("water on the brain") in a 10-month-old male who died in 1905 (MN5170). Hydrocephaly results in an abnormally large head, often with large areas of missing bone, numerous accessory sutural bones (ossicles) and premature closure of the sutures as in this individual (Cinalli et al. 1998; Collmann et al. 2005). The transparent membrane between the sutures is the dura mater that protects the brain.

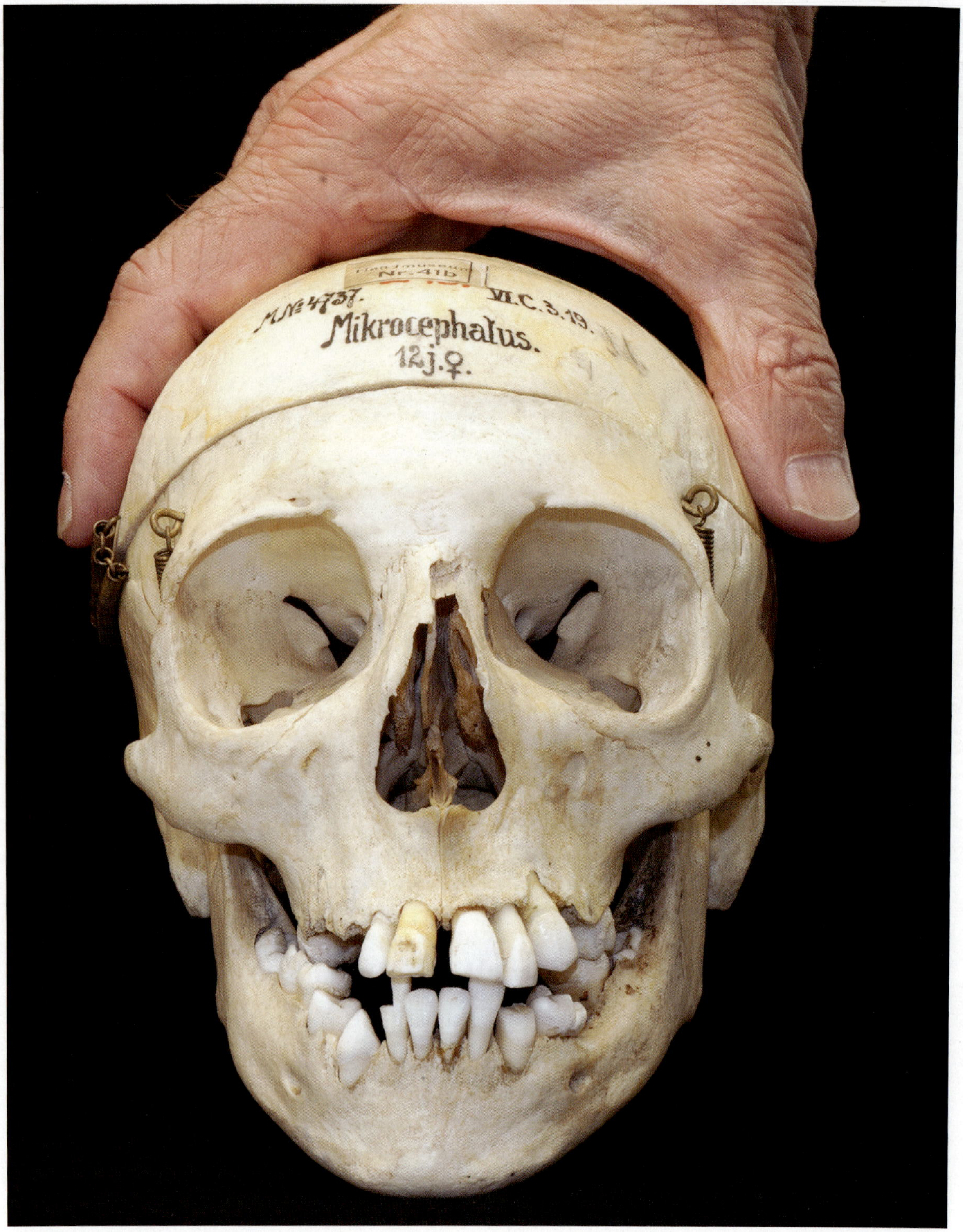


**Figure 44.** Microcephaly ("Microcephalus" on the skull) in a 12-year-old female who died in 1886 (MN4737). Microcephaly is a condition where the circumference of the head is much too small for the body size, has an incidence of 1.3–150/100,000 (Hashmi et al. 2016) and can be genetic, caused by drugs, a virus, maternal malnutrition or alcoholism (Hanzlik and Gigante 2017).

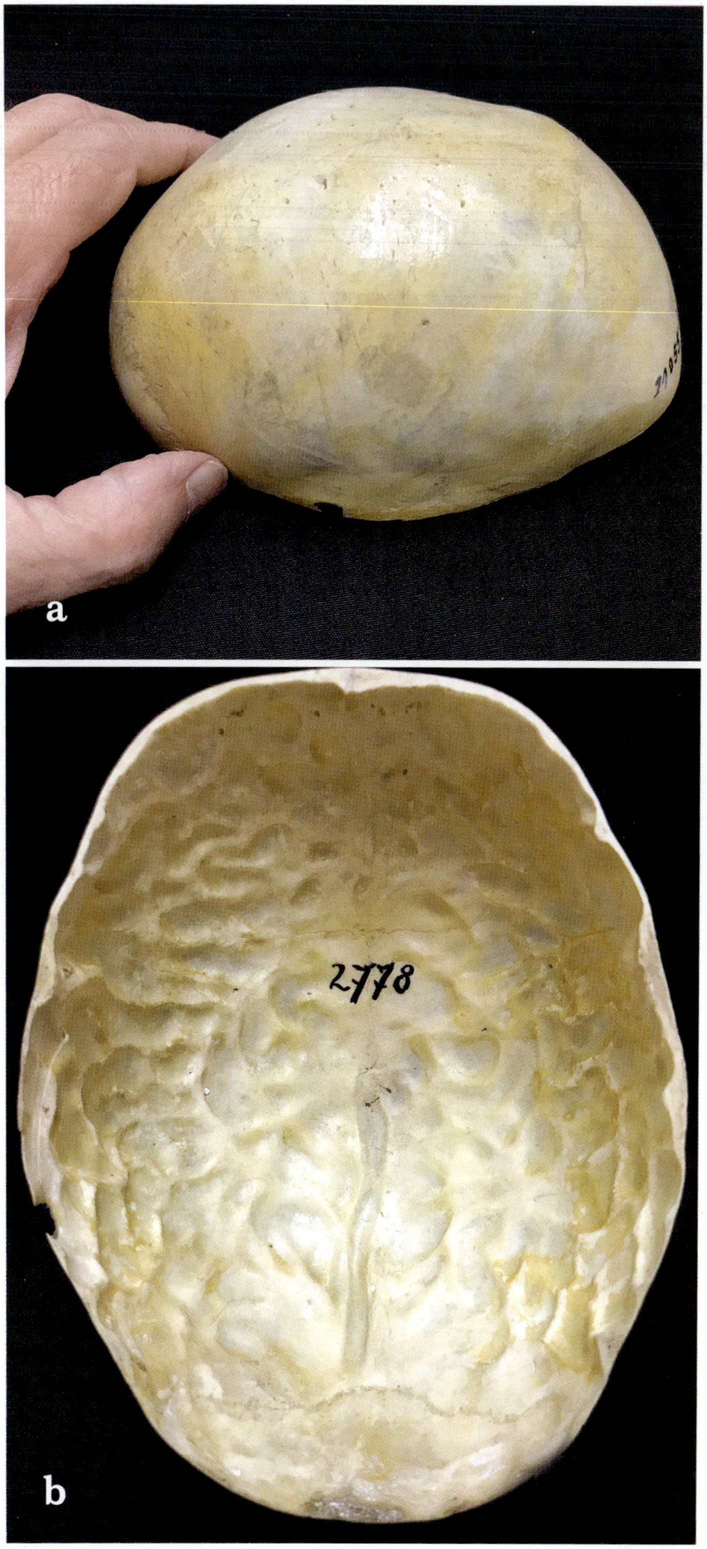


**Figure 45a-b.** (a) Ectocranial and (b) endocranial views of microcephaly and dolicocephaly, an abnormally small and long skull, resulting in gyri (ridges) or convolutions with a "copper beaten" appearance (Desai et al. 2014) along the inner surface of the skull due to increased pressure against the brain in a 2-year-old child who died in 1876 (MN2778).

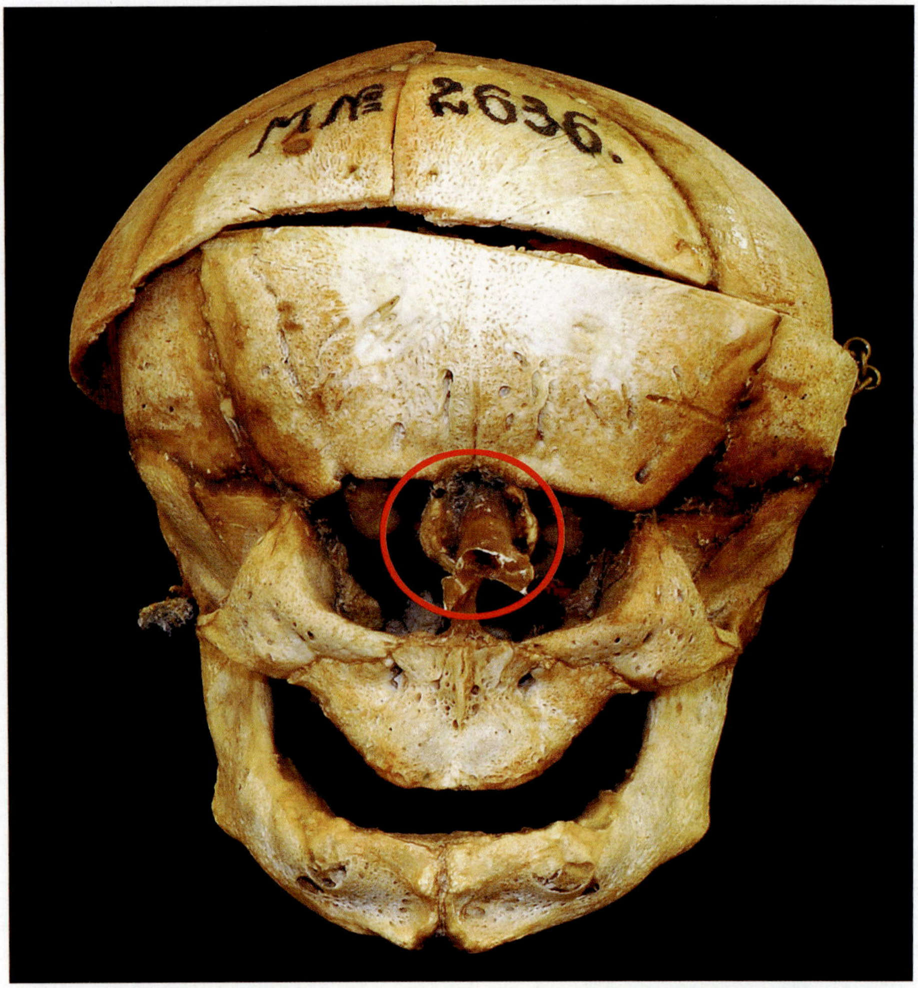


**Figure 46.** Cyclopia ("cyclops"), a congenital birth defect resulting in failure of the two orbits to divide, in a neonate male who died in 1857. Note absence of the nose and rudimentary and malformed nose (proboscis, red circle) at nasion situated superior to the orbit, not inferior to it as sometimes illustrated (MN2636). Cyclopia, also known as cyclocephaly and synophthalmia, a rare form of holoprosencephaly, is a birth defect transmitted by a recessive trait (Kunwar et al. 2021) occurring in about 1 in 100,000 births. Cyclopia results when the forebrain doesn't properly divide into right and left hemispheres during embryonic development. See Smith and Boulgakow (1926) for a detailed description of cyclopia in an Egyptian fetus.

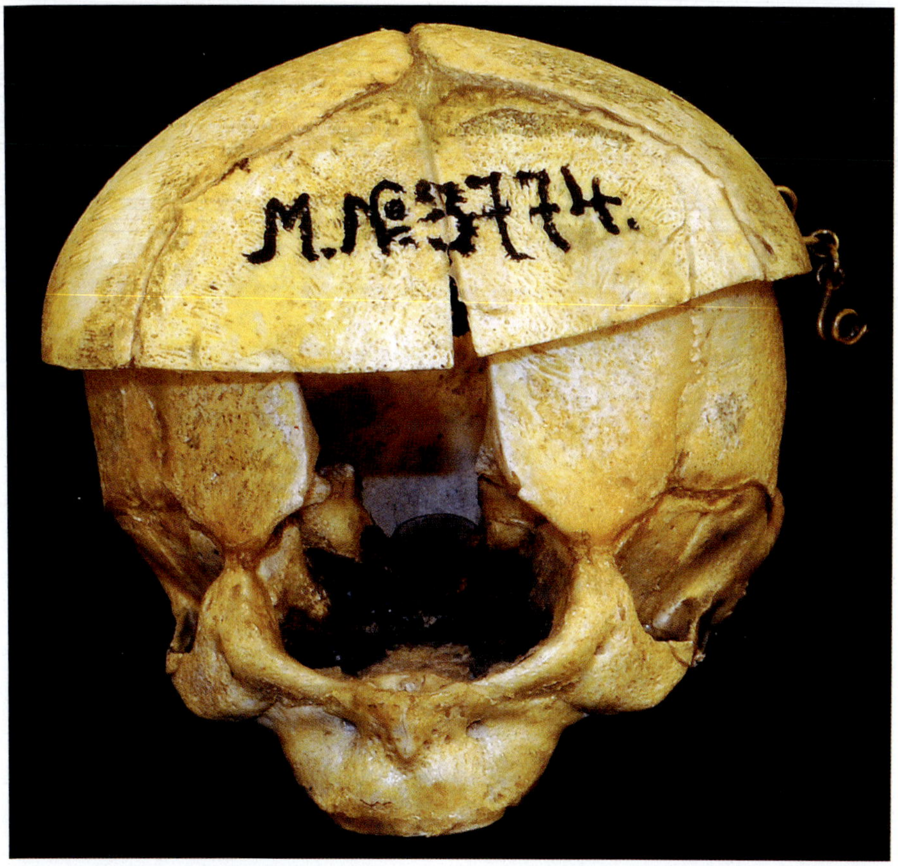


**Figure 47.** Cyclopia ("cyclops"), a congenital birth defect, in a neonate who is missing the central portion of the frontal bone removed postmortem and who died in 1880 (MN3774). Note the structural similarities of the facial region to those in the neonate in Figure 48 that includes a single bony orbit and absence of a nose. See Smith and Boulgakow (1926) for a detailed description of cyclopia in an Egyptian fetus.

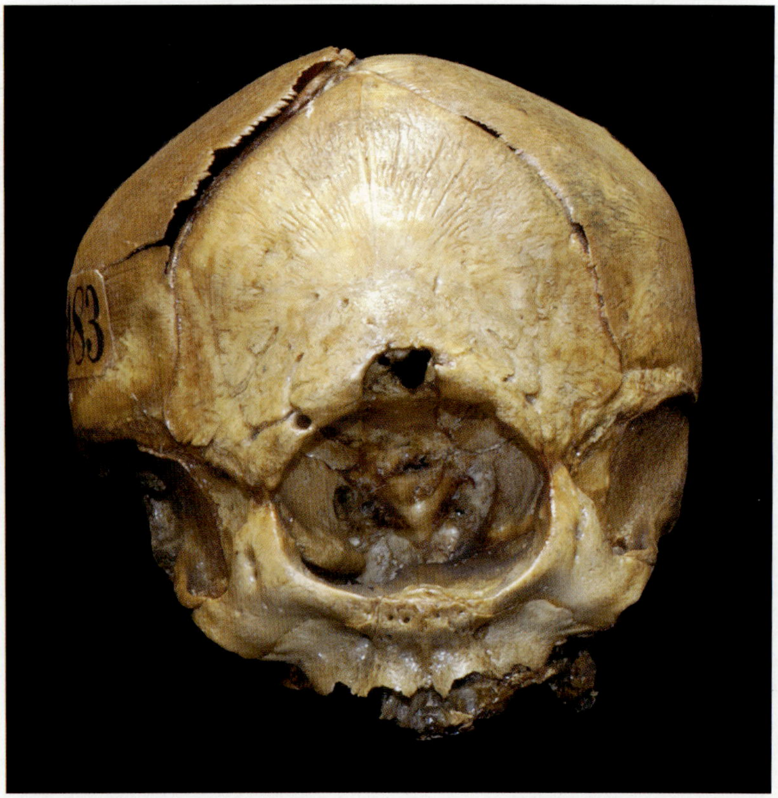

**Figure 48.** Cyclopia ("cyclops"), a rare and lethal form of holoprosencephaly (alobar, meaning that the brain failed to separate into right and left hemispheres), in a neonate who died in 1882 (MN3983). Note the single orbit and bony defect along the superior rim of the orbit, possibly associated with the proboscis (remnant of the nose). See McGahan et al. (1990) for more on alobar and semilobar.

**Figure 49.** Cyclopia (holoprosenchephaly), a rare form of synophthalmia in a female, with fusion of both orbits in one socket (MN28.183) (Kunwar et al. 2021). Note that unlike some inaccurate illustrations of cyclopia, the malformed nose (proboscis) is cephalad/superior to the orbit. Cyclopia is usually associated with holoprosencephaly, failure of the prosencephalon (the embryonic forebrain) to properly divide into the two lobes of the cerebral hemispheres (National Institutes of Health; https://www.ninds.nih.gov/health-information/disorders/holoprosencephaly). See McGahan et al. (1990) for more on alobar and semilobar variants of holoprosenchephaly.

**Figure 50.** Cyclopia, complete and total fusion of both eyes into one orbit in a 10-month-old (gestation) male with arrhinia (absence of the nose) and a large proboscis protruding in the midline cephalad/superior to the eye who died shortly after birth in 1946 (MN18.560/1835) (Kunwar et al. 2021). See McGahan et al. (1990) for more on alobar and semilobar variants of cyclopia.

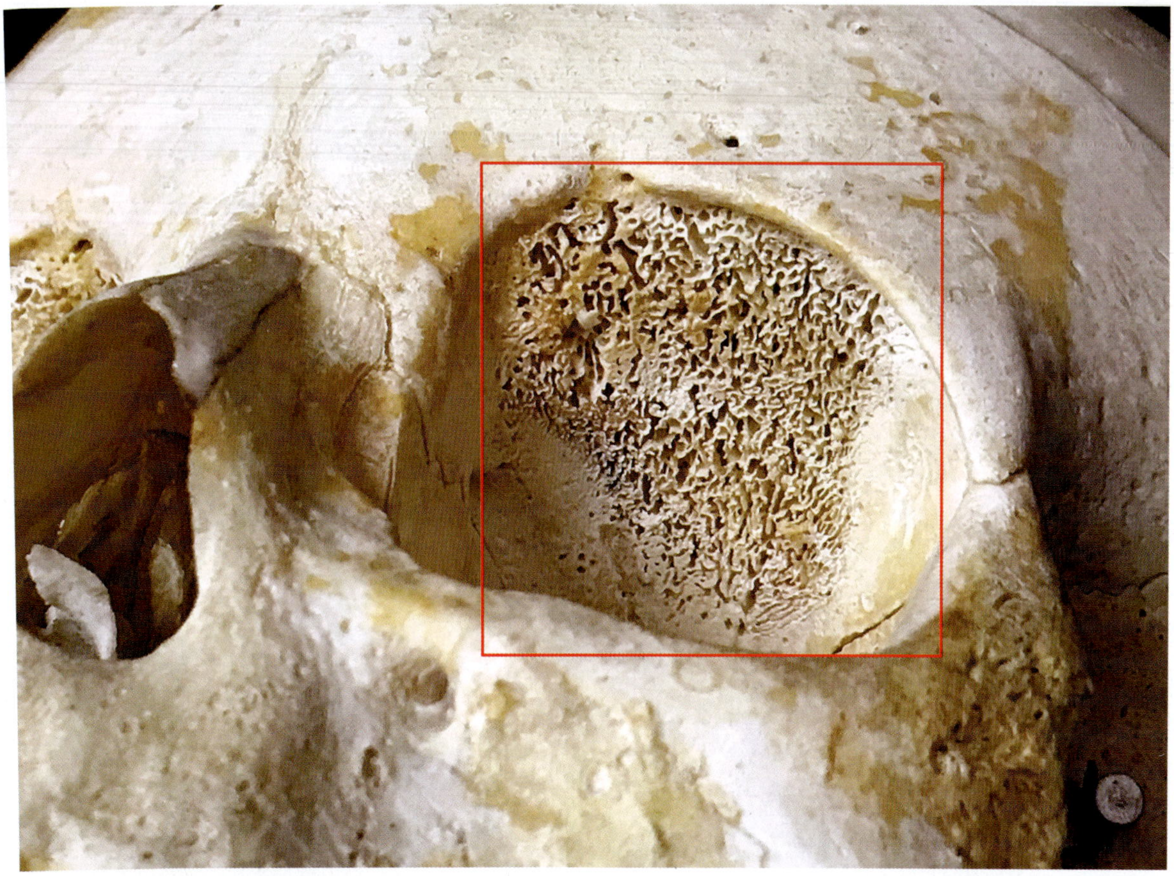

**Figure 51.** Severe bilateral cribra orbitalia in a child with coral-like bony growths involving the superior orbital plates of the frontal bone (square). Although subject to debate (Mann and Hunt 2005; Rivera and Lahr 2017; Rothschild et al. 2021; Walker et al. 2009) this condition is believed to be the result of iron-deficiency anemia (Ortner and Putschar 1981; Steinbock 1976; Stuart-Macadam 1987). Cribra orbitalia ranges in severity and expression from tiny, scattered or coalesced pits to thickened and coral-like superior orbital plates and is often found in association with porotic hyperostosis consisting of thickened bone and diffuse pitting (porosity) in the outer table of the vault. There is no additional information accompanying this individual.

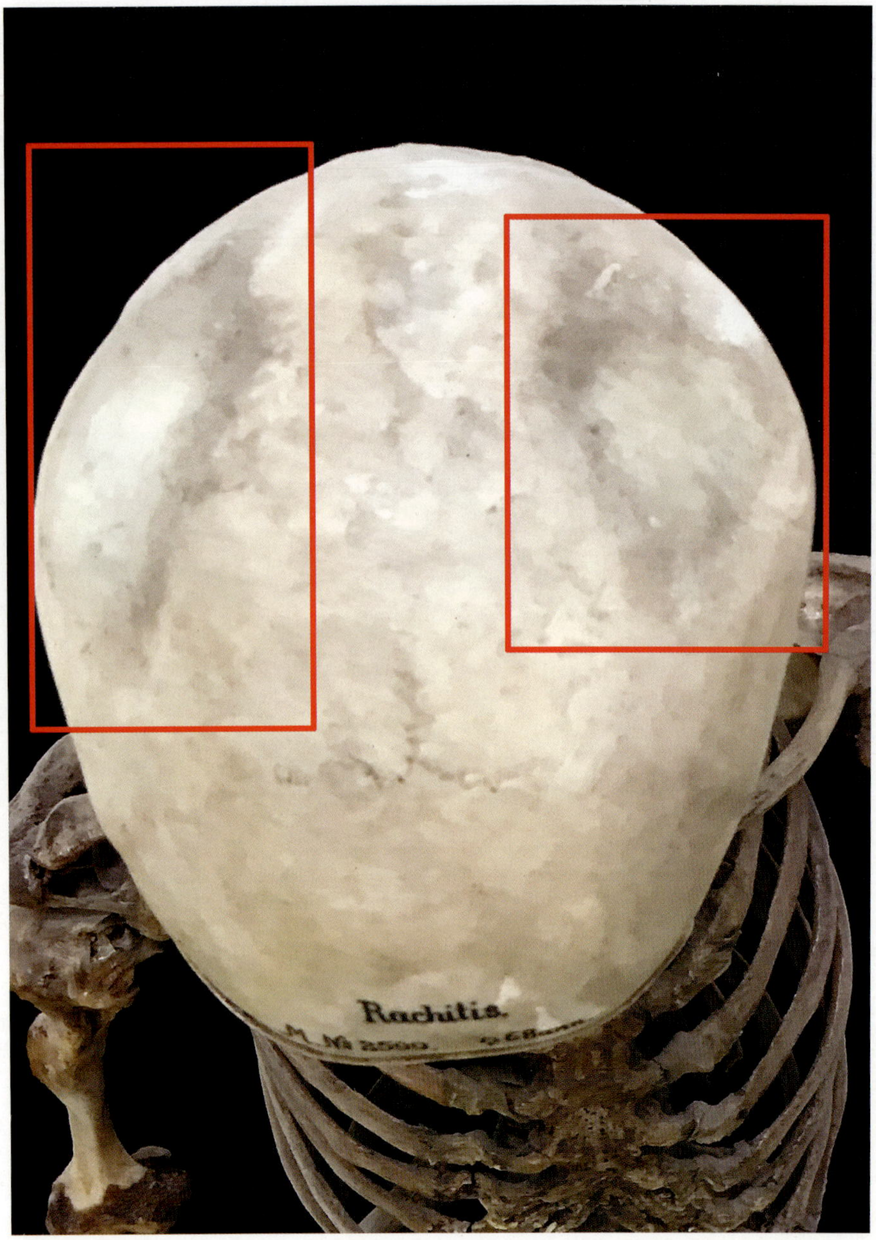

**Figure 52.** Rachitis (rickets) and early-stage bilateral parietal thinning (rectangles) in a 68-year-old female (MN2599). A shortened right humerus with widening of the epiphysis and rachitic rosary are evident. A leading nutritional cause of short stature, also known as dwarfism, is vitamin-D-resistant rickets during childhood. See Durward (1929); Mann and Hunt (2005); Smith (1907); Snapper (1957); Thomas and Kirsten (1985); Watson-Jones (1955); Wheeler et al. (2010).

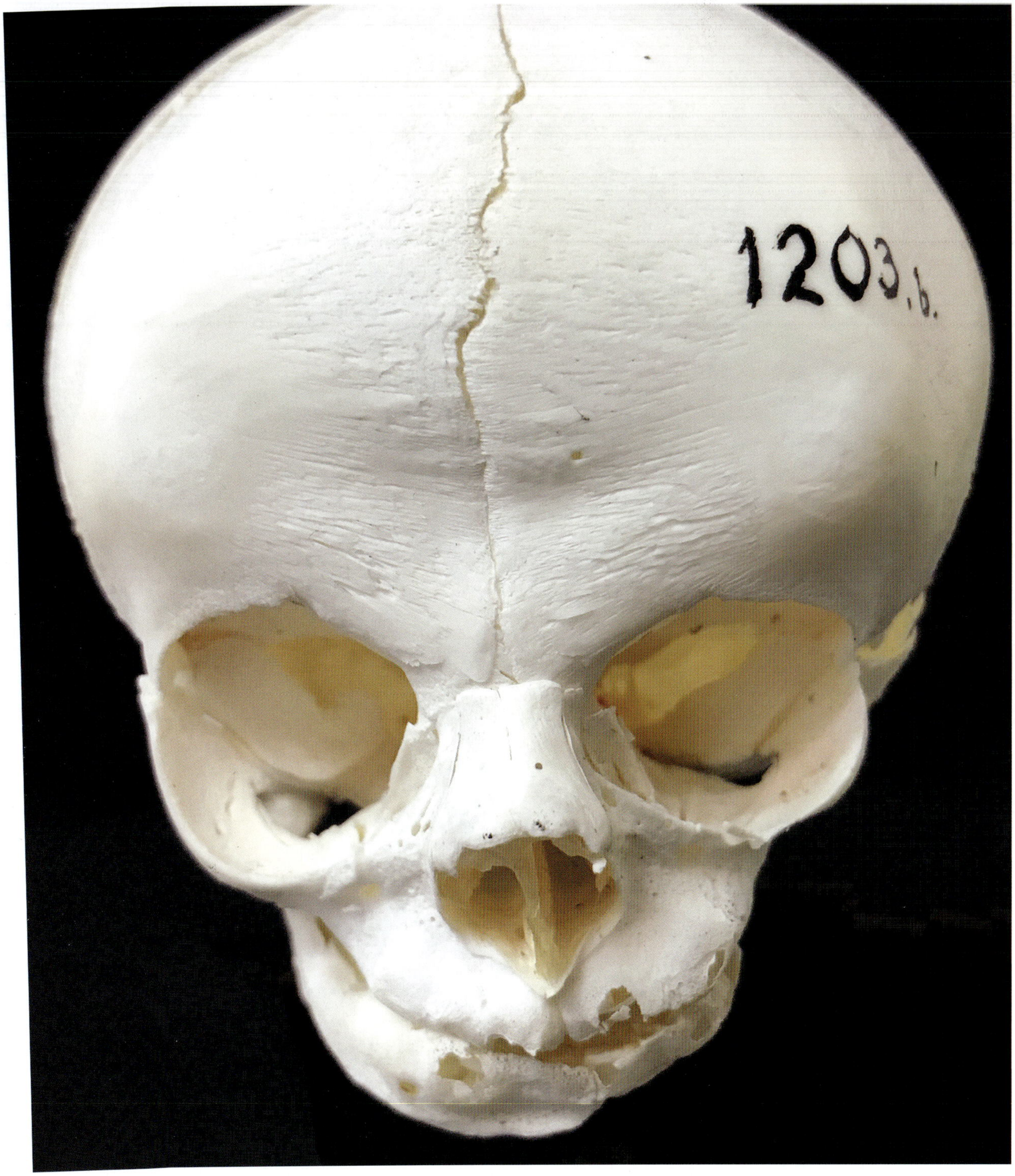


**Figure 53.** A 4-day-old male with agenesis (aural atresia) of the left external auditory/acoustic meatus who died in 1867 (MN1203b). Cf. De Lima (1912) for an example of atresia in a 78-year-old male.

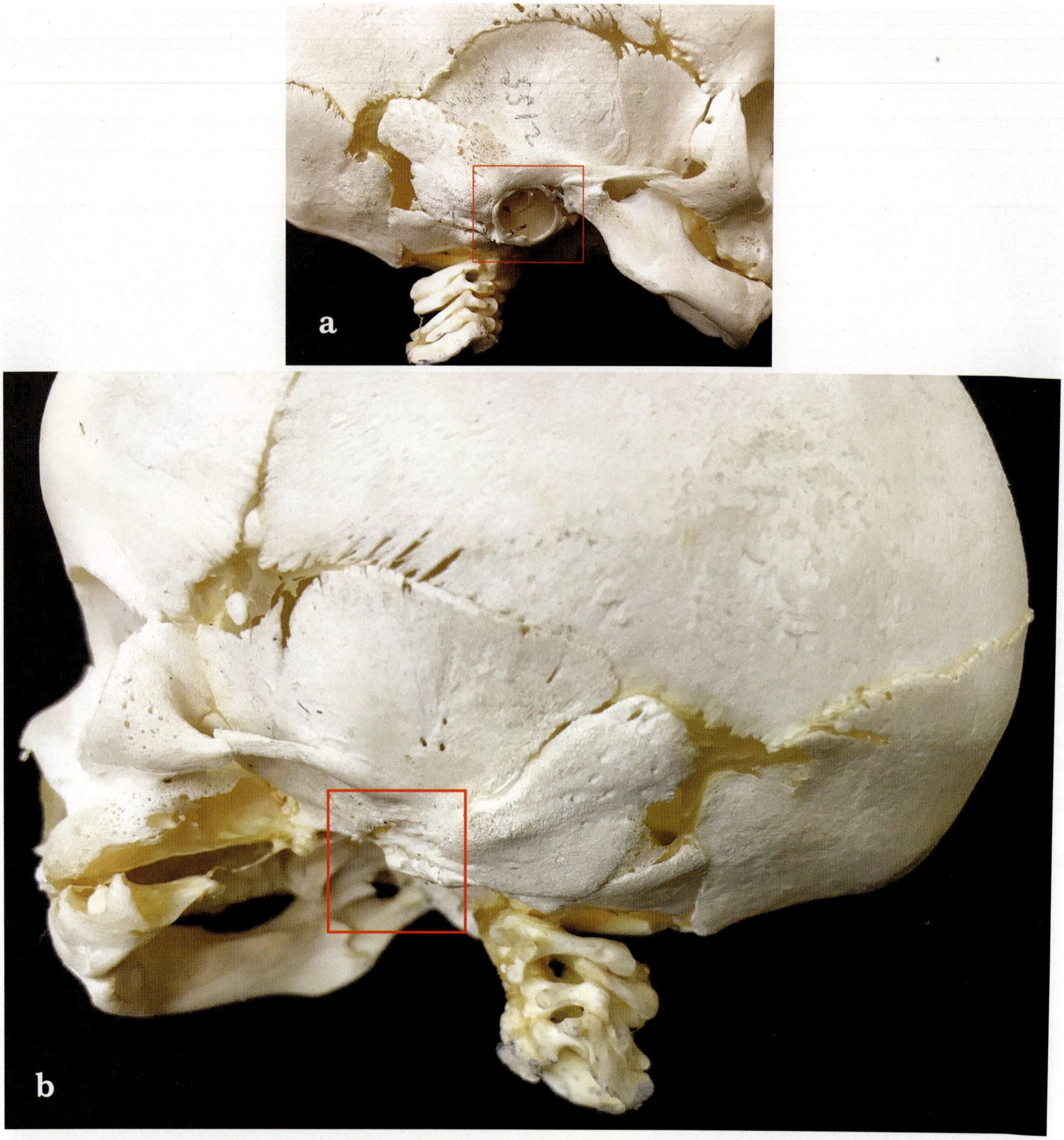


**Figure 54a-b.** (a) Right lateral and (b) left lateral views of 4-day-old male with agenesis (aural atresia) of the left external auditory meatus who died in 1867 (MN1203b). (a) The morphology of the right auditory canal (square) is normal while (b) there is complete absence of the left auditory canal (square) (see Abdel-Aziz 2013). Malformation of the left temporal bone is accompanied by hypoplasia of the left maxilla and mandible. Agenesis of the external auditory meatus in dry bone specimens is an extremely rare condition, this being the first examined by the authors. Cf. De Lima (1912) for an example of atresia in a 78-year-old male.

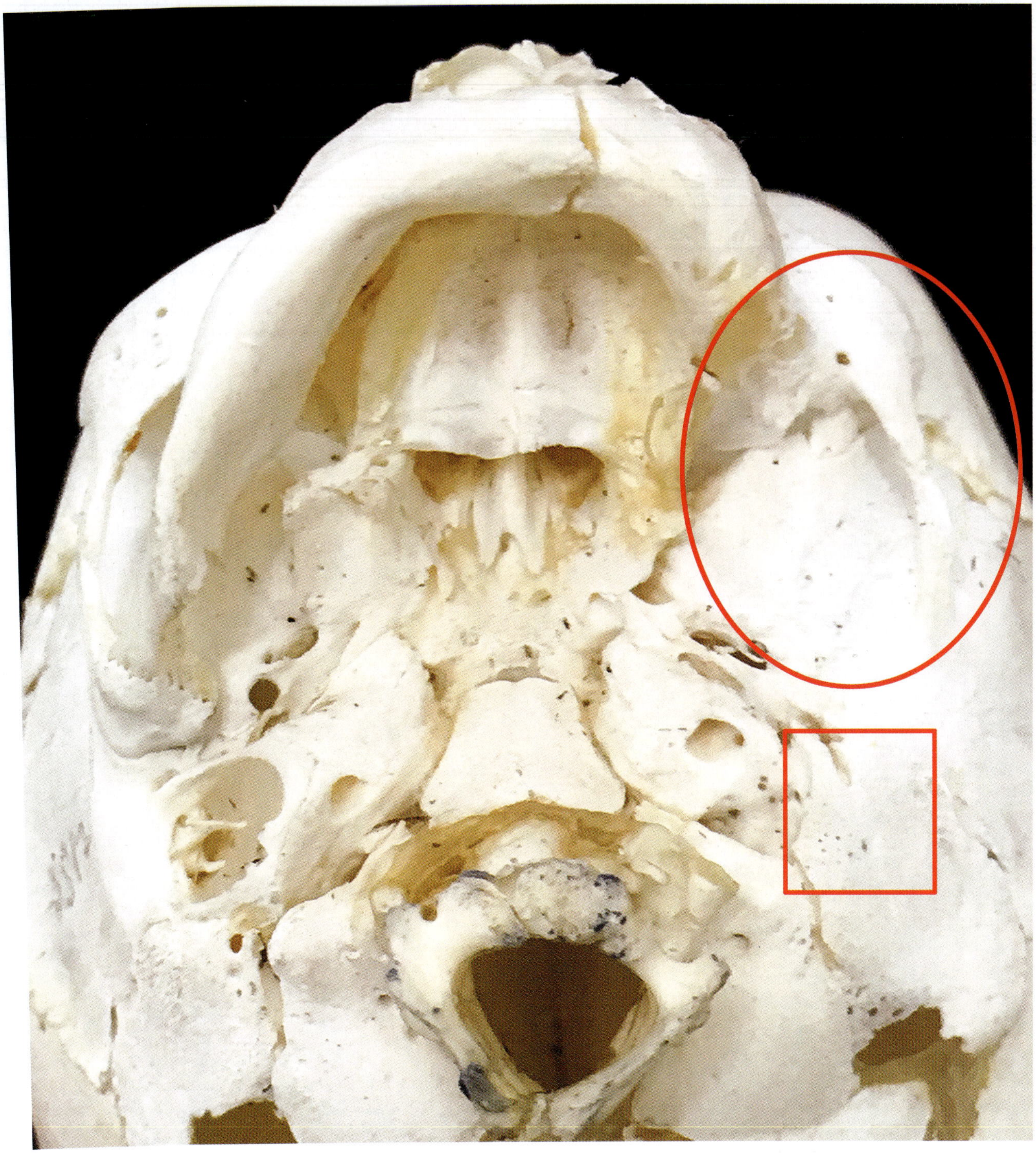

**Figure 55.** Inferior view of skull of a 4-day-old male with agenesis (aural atresia) of the left external auditory meatus who died in 1867 (MN1203b) (see Abdel-Aziz 2013). Note the normal morphology of the right auditory canal, ear ossicles and tympanic ring attached compared to the left temporal bone exhibiting absence of the auditory meatus (square). Malformation (circle) of the left temporal bone in this child is accompanied by hypoplasia of the left maxilla, zygoma, zygomatic arch and mandible. Cf. De Lima (1912) for an example of atresia in a 78-year-old male.

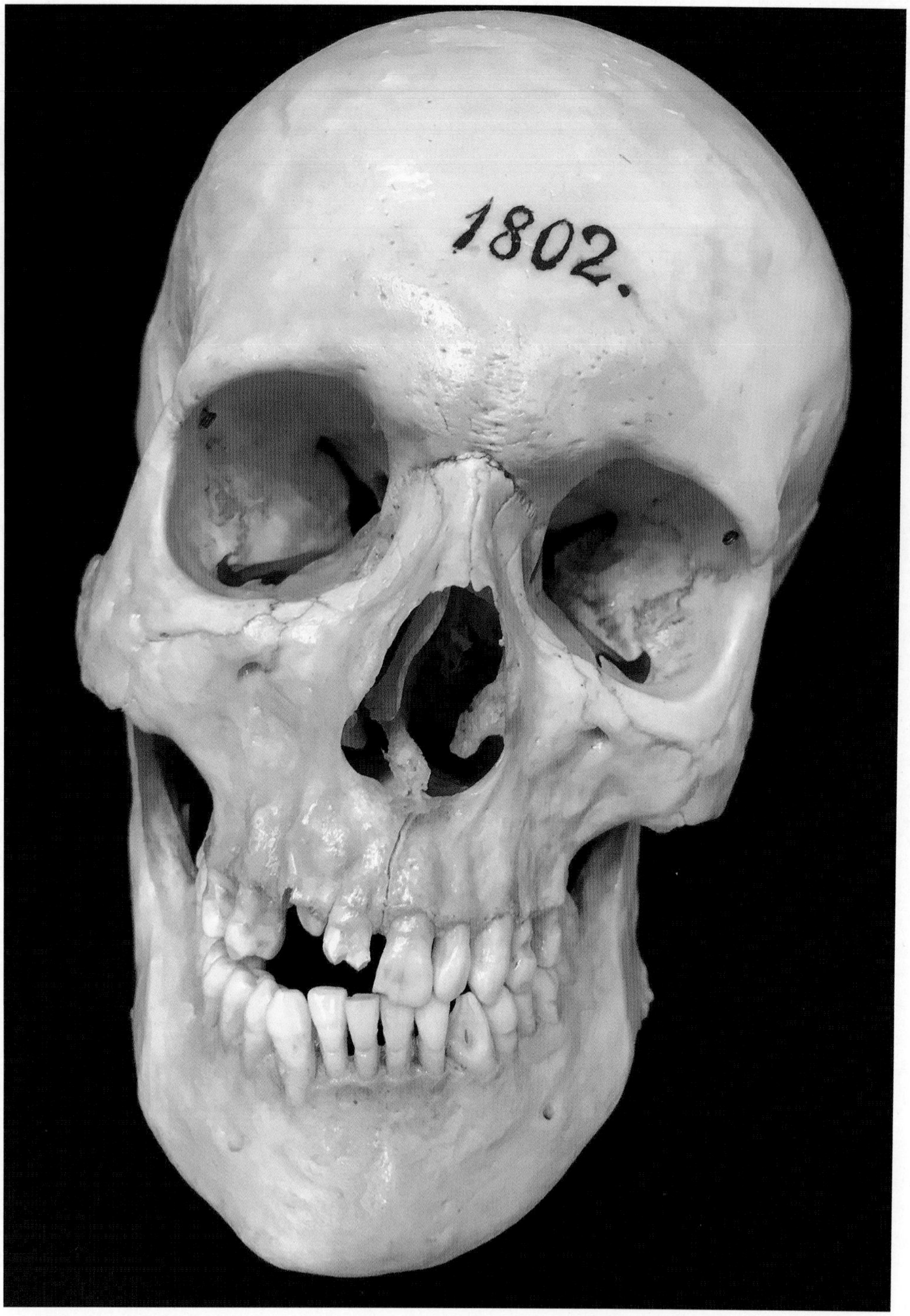


**Figure 56.** Anterior view of bathrocephaly (on the museum card) and apertognathia secondary to asymmetric mandibular hyperplasia (J. DeMeo, personal communication, July 15, 2022) in a 54-year-old male with a deviated nasal septum and asymmetry of the mandibular rami and maxillae who died in 1870 (MN21.055/1802). Apertognathia, or open bite, is a congenital developmental deformity that usually appears in childhood (Ramon et al. 1987).

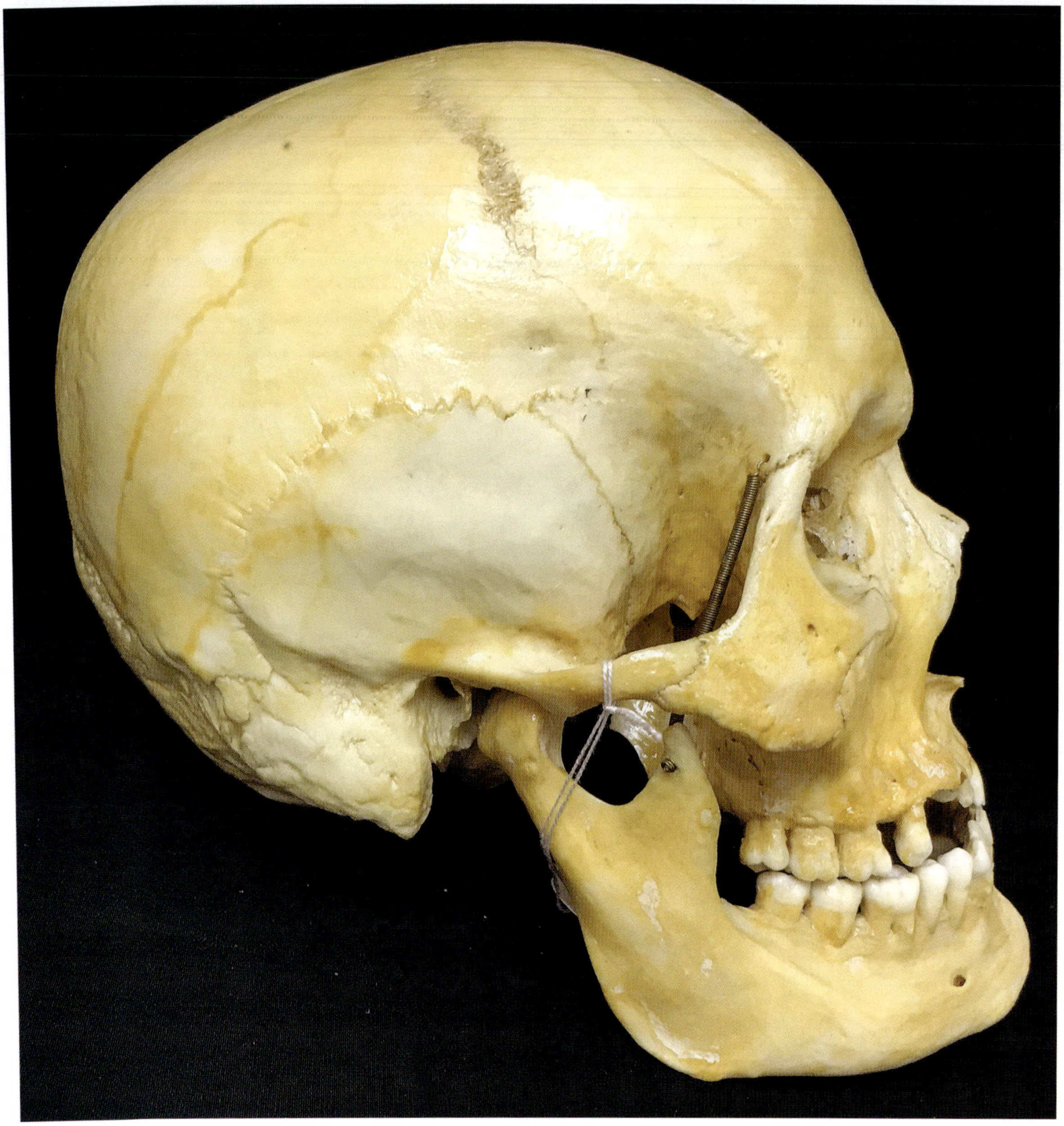

**Figure 57.** Right lateral view of bathrocephaly (on the museum card) and apertognathia secondary to asymmetric mandibular hyperplasia (J. DeMeo, personal communication, July 15, 2022) in a 54-year-old male with deviated nasal septum and asymmetry of the mandibular rami and maxillae who died in 1870 (MN21.055/1802).

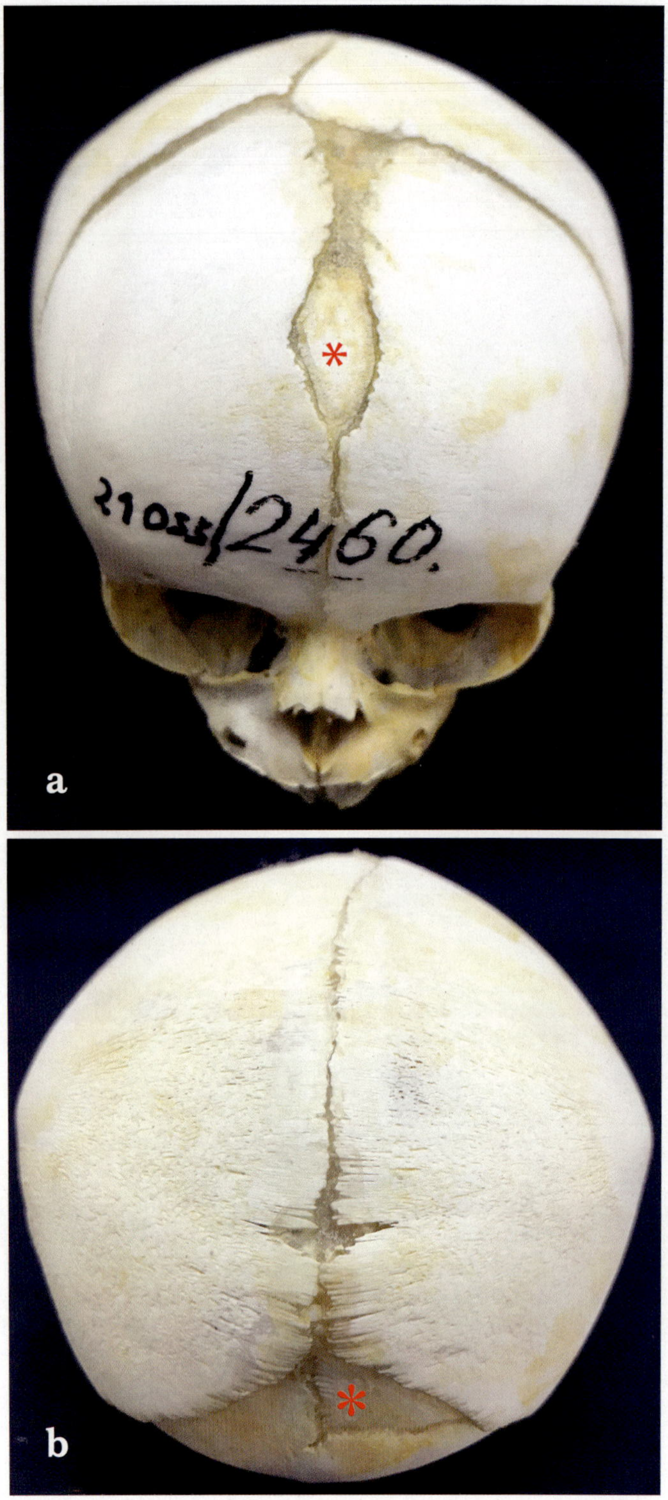


**Figure 58a-b.** (a) Anterior-superior and (b) posterior views of anterior fontanelle with ossicle (*) in the metopic suture and asymmetry of bregma where the sagittal suture joins the coronal suture in a neonate female (MN21.055/2460). (b) Divided/ bipartite (Inca or interparietal bone) occipital bone (*) and asymmetry of sutures are common anatomical variants/features. See Hepburn (1908) for more on the interparietal bone.

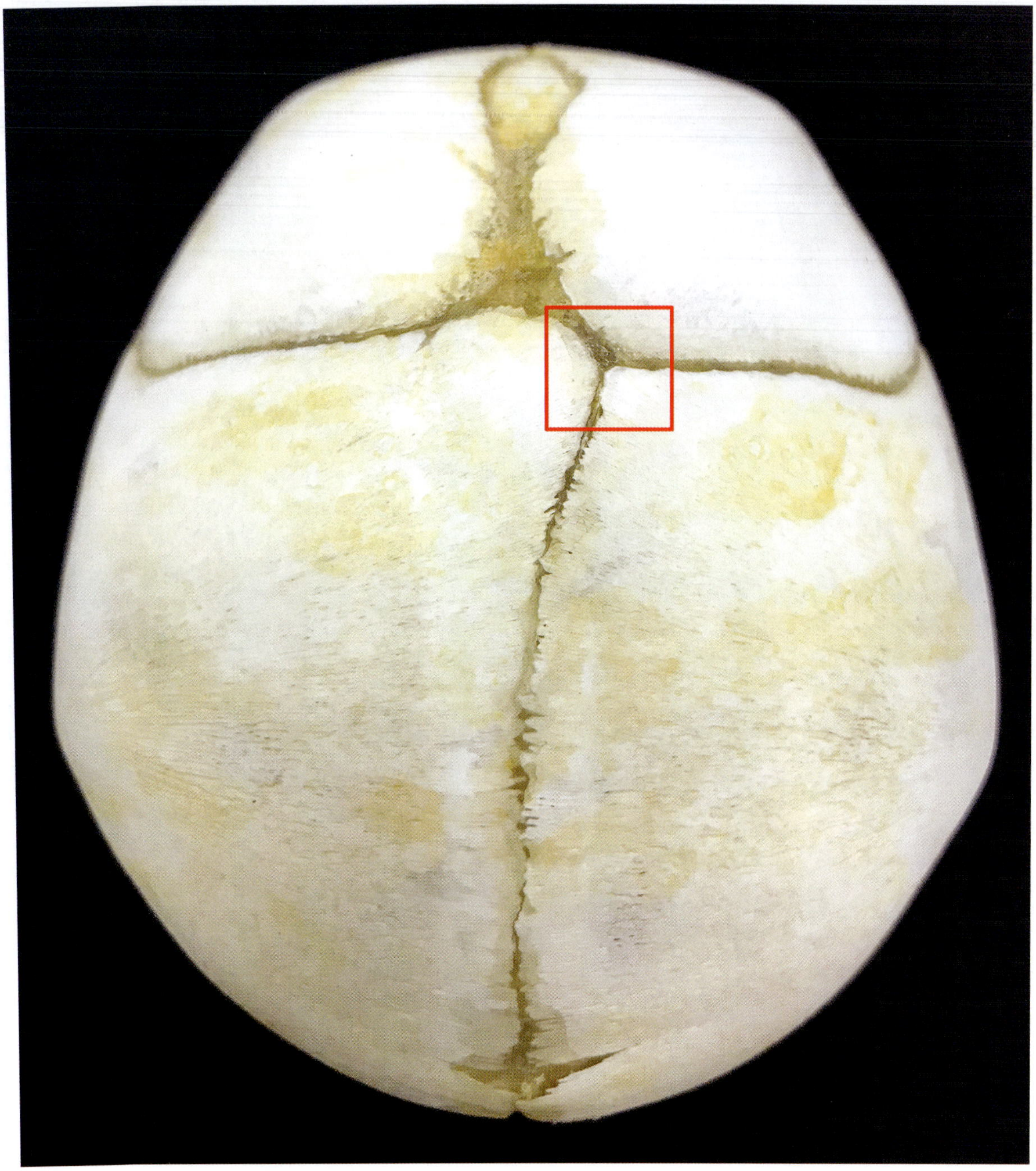

**Figure 59.** Asymmetry of bregma (square) where the sagittal suture joins the coronal suture, likely the result of the presence of the anterior fontanelle and ossicle in a neonate female (MN21.055/2460). The sagittal and metopic sutures often do not align anterior-posteriorly and is a common anatomical feature/variant.

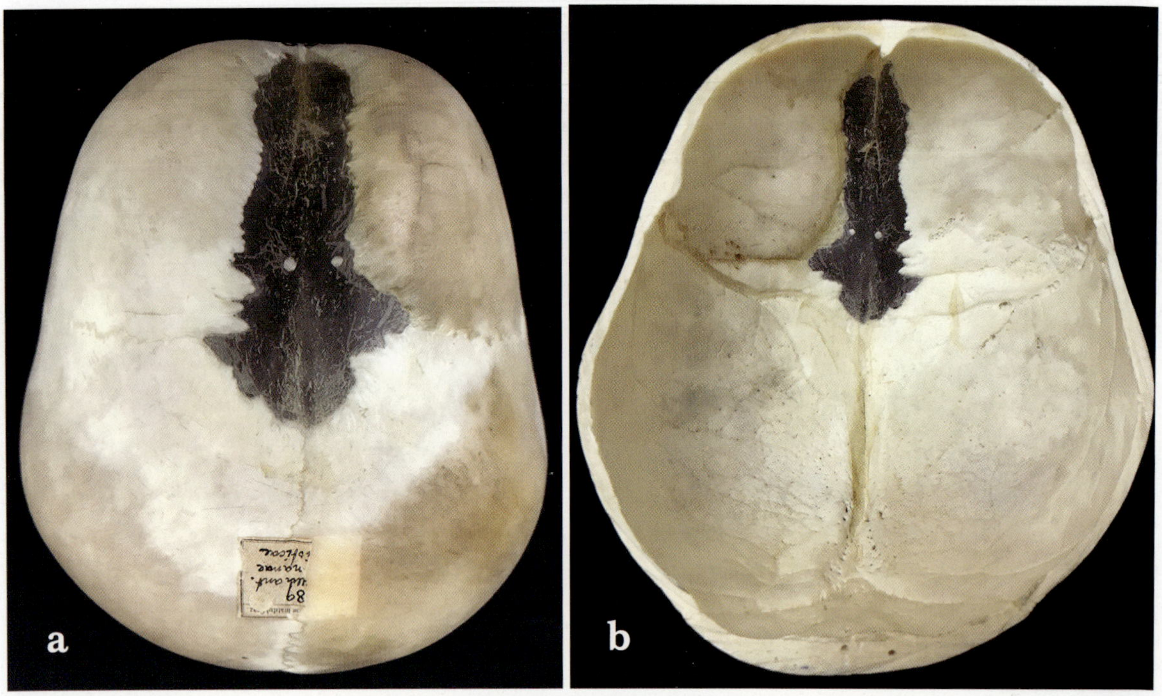


**Figure 60a-b.** (a) Superior ectocranial and (b) inferior endocranial views of persistent anterior fontanelle at the junction of the coronal, sagittal, and metopic sutures in a 27-year-old female who died in 1906 (MN21.055/5289). The museum card reads "Fontanella magna aperta cranii manae cretinisticae" and was due to hypothyroidism at birth that resulted in delayed biological and mental growth. The normal range of closure of the anterior fontanelle is 4 to 6 months (Gaetano and Gaetano 2002; Tunnessen 1990), is considered a common variant in some children and may be asymptomatic in some adults.

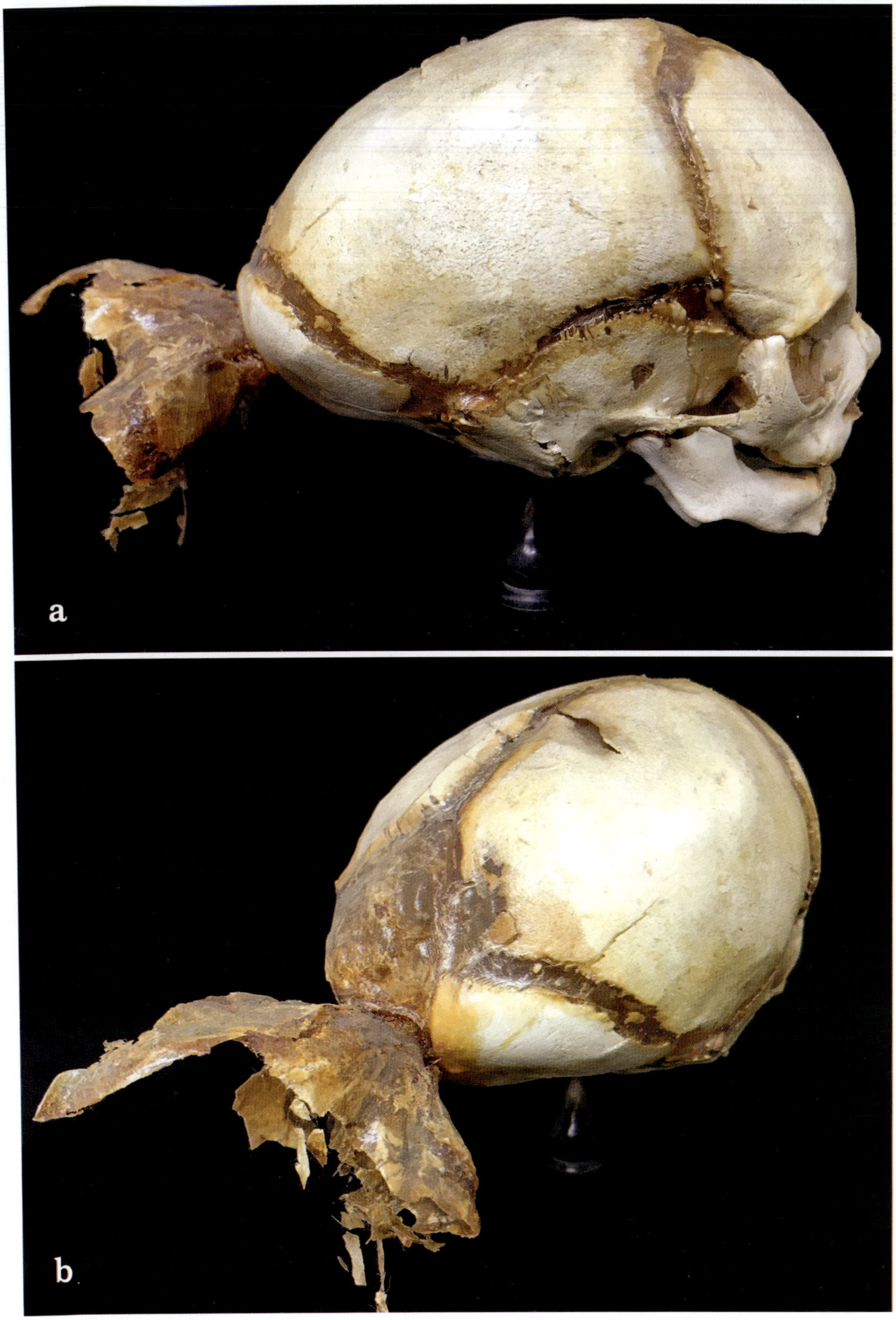

**Figure 61a-b.** (a) Right lateral and (b) right oblique views of posterior occipital encephalocele, the most common type of encephalocele, in a neonate (MN21.055/1771). Note that the dura mater has prolapsed between the occipital bones. Encephalocele is a rare birth defect where the neural tube does not properly close (Ortner and Putschar 1981; Sheikh et al. 2014).

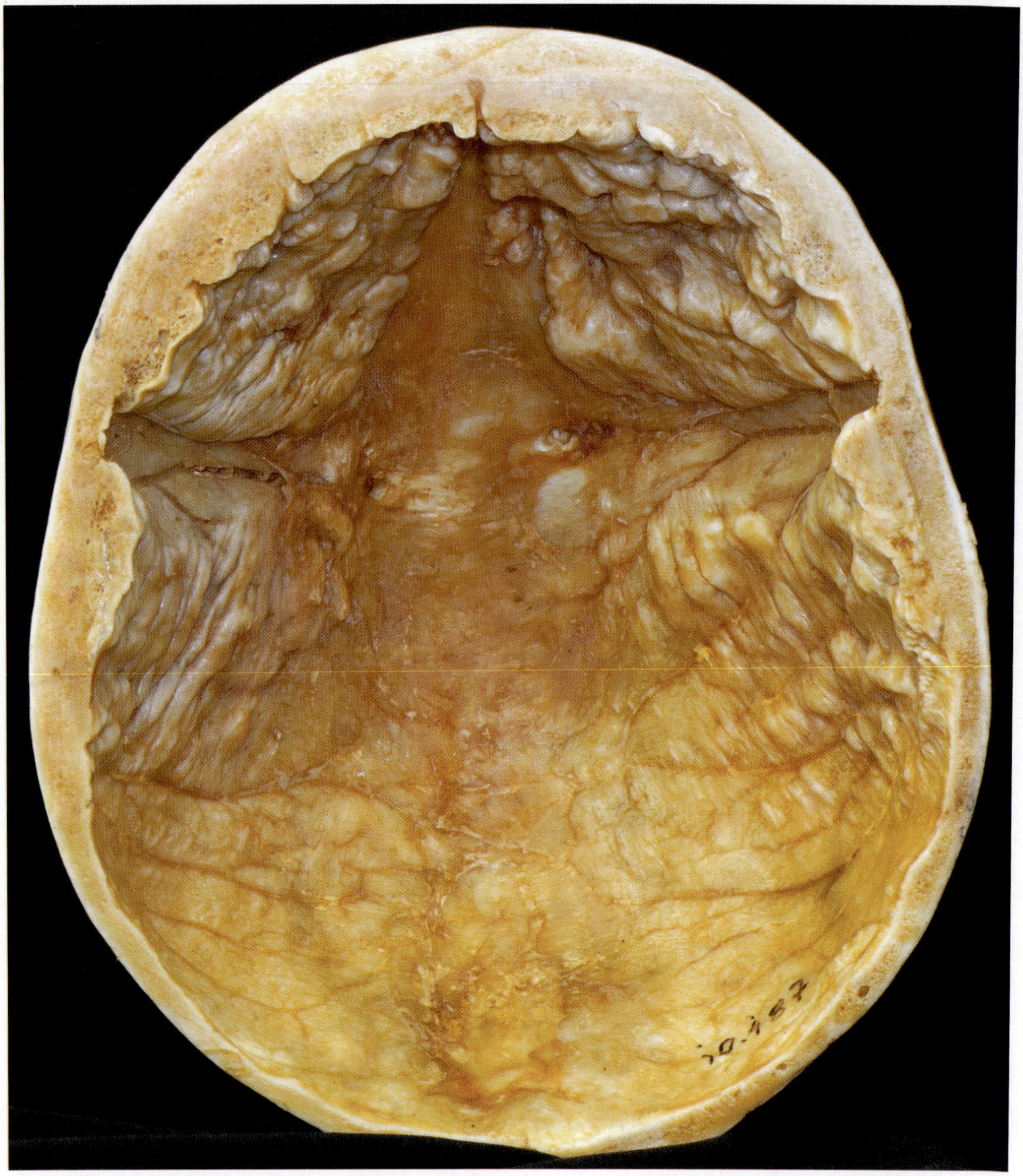

**Figure 62.** Hyperostosis frontalis interna (HFI) in a 62-year-old female (MN10.187). The etiology of HFI is unknown, but is expressed in this individual by bilateral and symmetrical benign bony growths covering the endocranial surface of the frontal and much of the parietal bones, but does not involve the superior sagittal sinus, the shallow groove or sulcus that extends along the midline of the vault. HFI, despite its name, not only involves the endocranial surface of the frontal bone, but can be present in other cranial bones resulting in thick and noticeably heavy bones. The yellow color of this vault is due to the resin used to preserve it. HFI is a common finding in contemporary crania. See Hershkovitz (1999, 2002); She and Szakacs (2004).

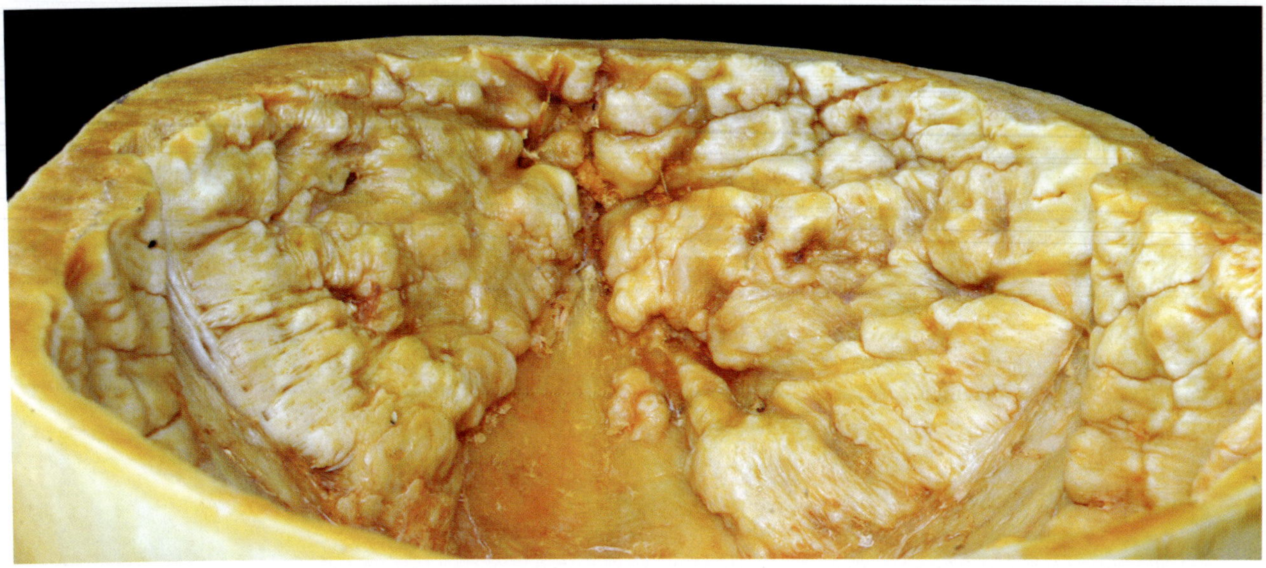

**Figure 63.** Endocranial view of hyperostosis frontalis interna (HFI) represented by dense bilateral and symmetrical growths covering the inner table of the skull in a 62-year-old female (MN10.187). Bilateral and symmetrical benign bony growths (osteophytes) associated with HFI, which is a common finding in contemporary crania, may cover the frontal, parietal, sphenoidal and temporal bones causing them to be unusually thick and heavy. See Herskovitz et al. (1999, 2002) for more on the etiology of HFI and its frequency in dry bones.

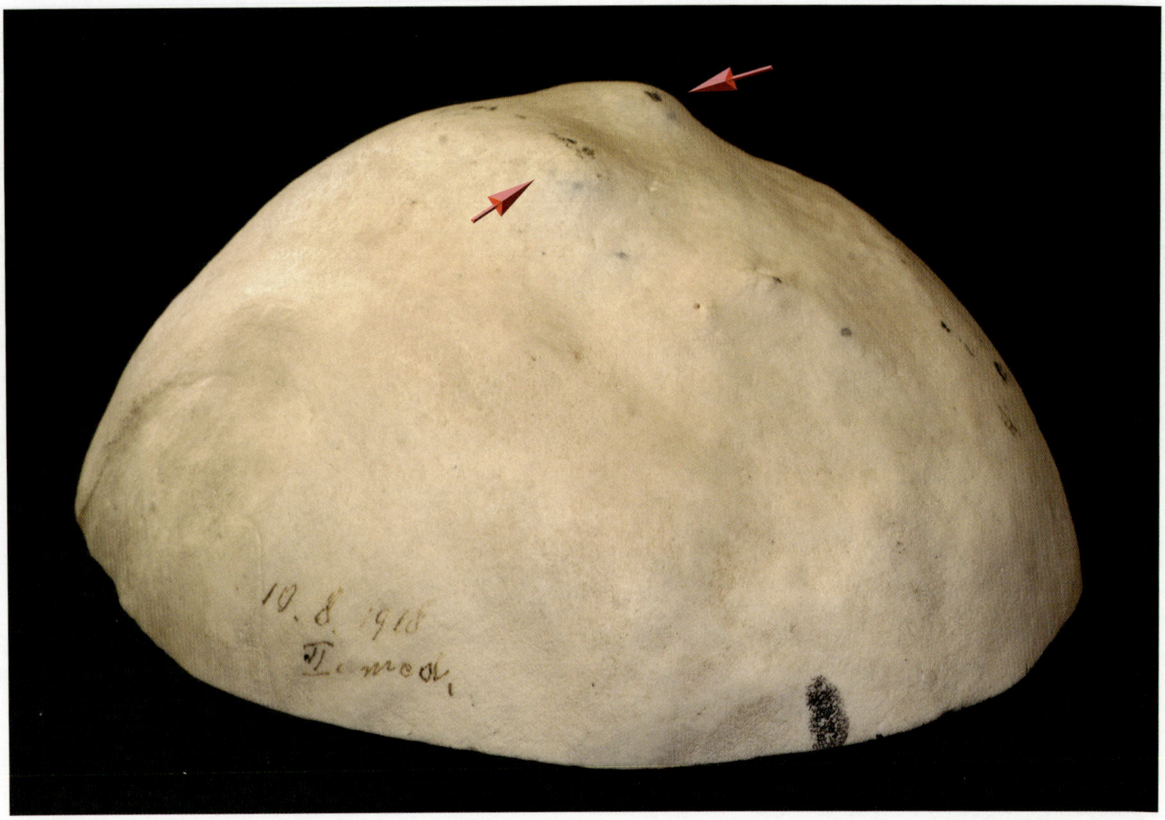


**Figure 64.** Large and deep granular foveolae ("Usura calvariae granulationibus arachnoidae"), also known as Pacchionian pits (Mann et al. 2016) housing arachnoid granulations that resemble clusters of rice situated along either side of the superior sagittal sinus in the parietal bones. Pressure against the inner table of the skull thinned the vault and resulted in the formation of two large protrusions in the outer vault of this unknown adult who died of lung cancer in 1918 (MN18.560/361). See Le Gros Clark (1920) for a description of Pacchionian bodies in the soft tissue and skull.

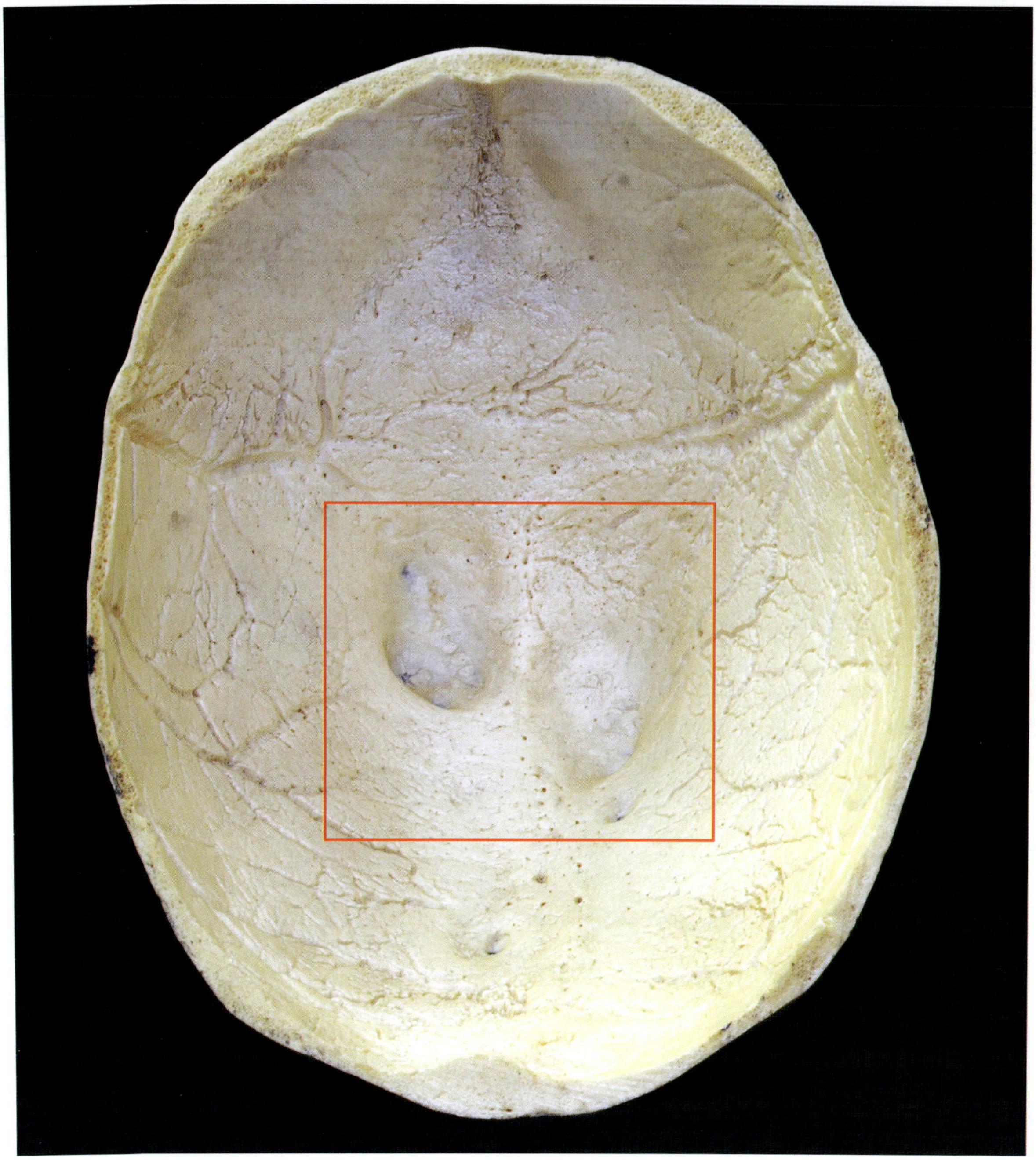

**Figure 65.** Deep granular foveolae (square) along either side of the sagittal suture in an unknown adult who died of lung cancer in 1918 (MN18.560/361). These depressions with thin walls were likely caused by pressure by enlarged, calcified, or unusually large clusters of arachnoid granulations and vessels exerting chronic pressure to the endocranium (Mann et al. 2016). Chronic pressure against the endocranium by arachnoid granulations could result in contact pressure erosion and thinning of the inner and inner table and diploe and remodeling that, in rare cases, perforate the outer table.

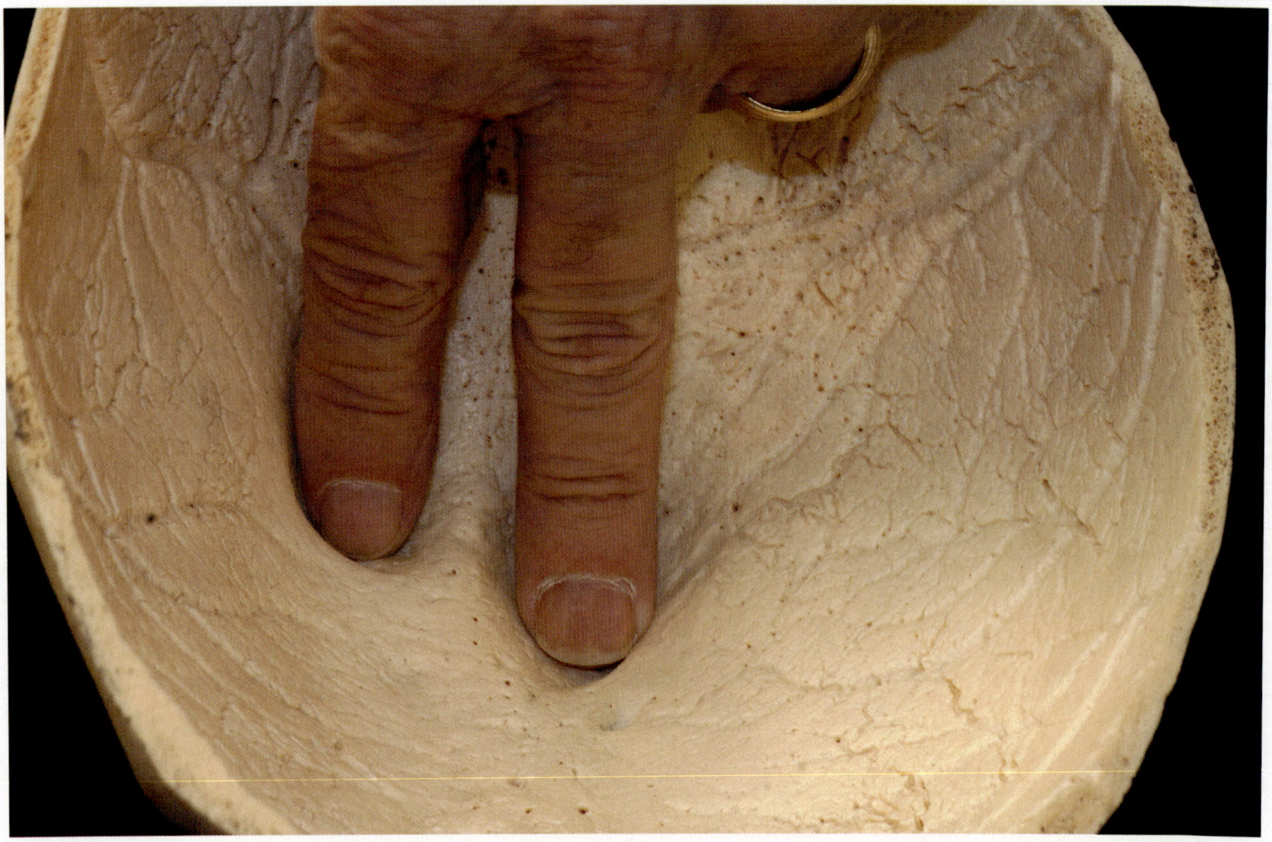

**Figure 66.** The first author demonstrating the large size and depth of the granular foveolae endocranially that resulted in large ectocranial protrusions in this unknown adult who died of lung cancer in 1918 (MN18.560/361) (see Mann et al. 2016).

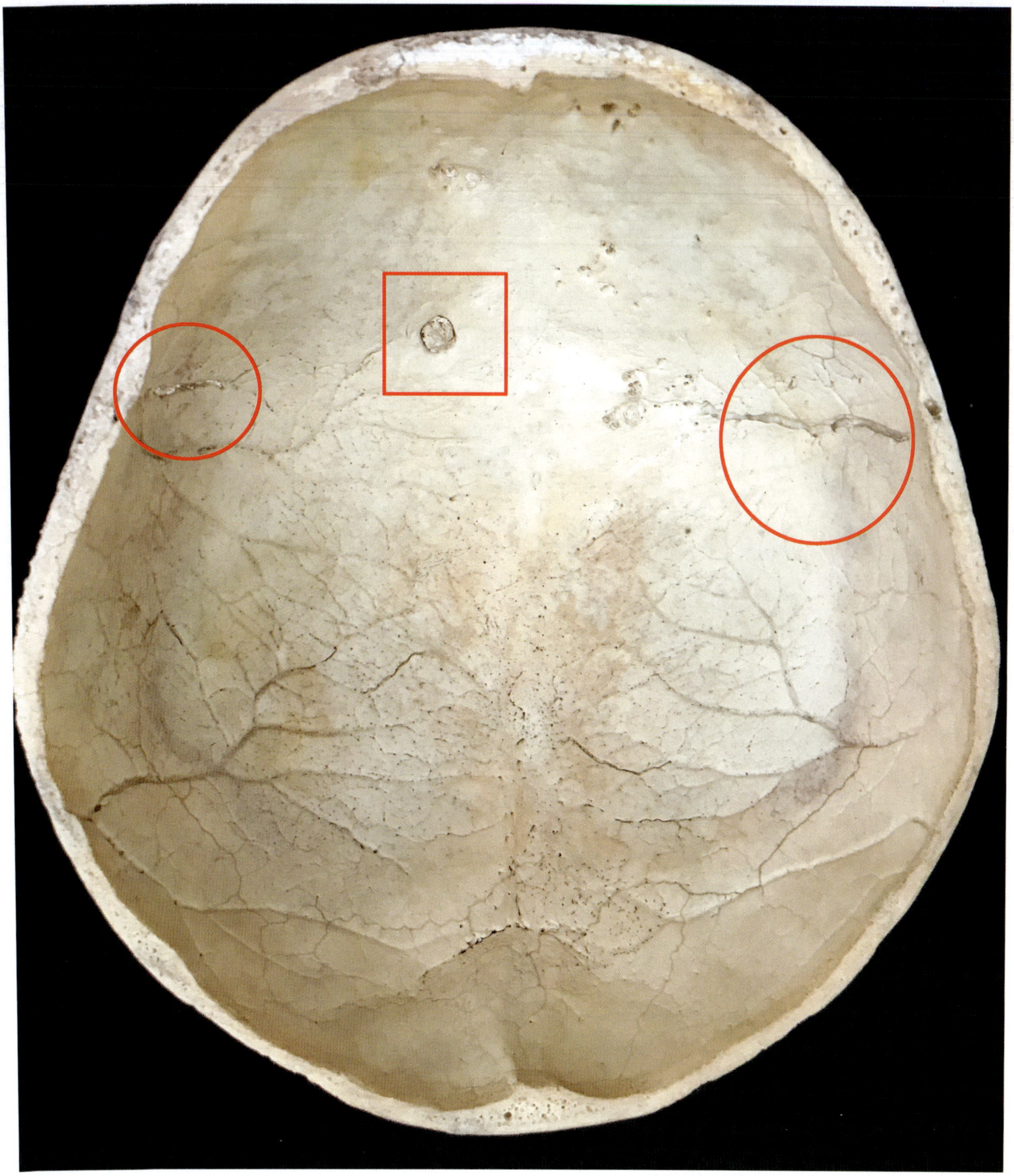

**Figure 67.** Endocranial view of calcified soft tissue filling a Pacchionian pit/depression (square) in the right parietal bone of an unknown individual who died in 1885 (MN21.055/525b). This rare osseous lesion is likely a cluster of arachnoid granulations situated against the endocranium and that can calcify in the elderly. There is a crater-like depression in the outer table of the skull opposite this calcification. Note calcification of the meningeal arteries (circles) that ongoing research by the first author suggests accompanies advanced age and cardiovascular disease.

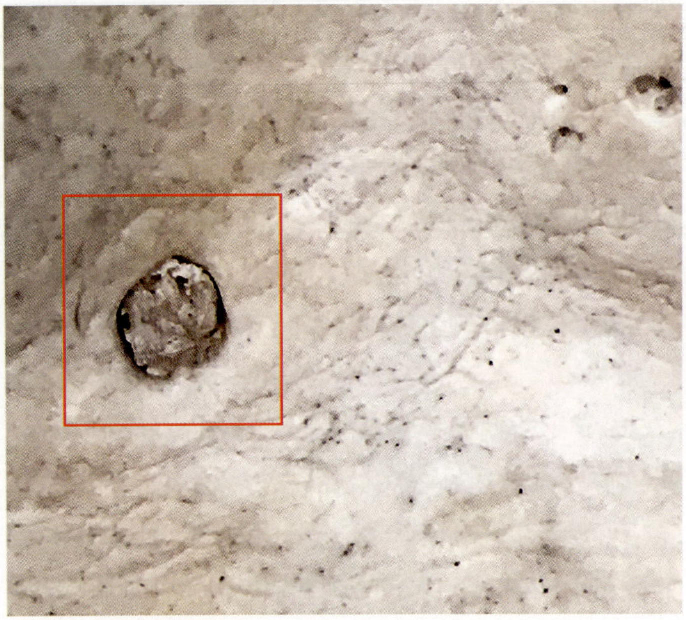

**Figure 68.** Endocranial view of a calcification filling a Pacchionian pit/depression (square) in the right parietal bone in an unknown individual who died in 1885 (MN21.055/525b). This rare lesion is likely a cluster of arachnoid granulations situated against the endocranium that filters cerebrospinal fluid and that may calcify in the elderly. There is a cratered depression of unknown etiology in the outer table of the skull opposite this calcification (Figure 69). It is difficult to be certain of the etiology of this lesion.

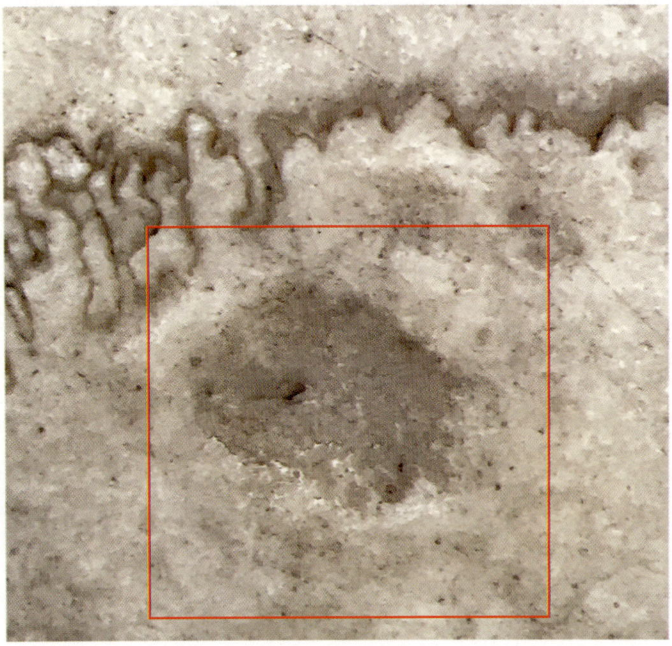

**Figure 69.** Magnified view of the outer table of the right parietal bone showing a cratered depression opposite the inner-table calcification filling a Pacchionian pit in an unknown individual who died in 1885 (MN21.055/525b). It is difficult to be certain of the etiology of this lesion as age related or due to trauma.

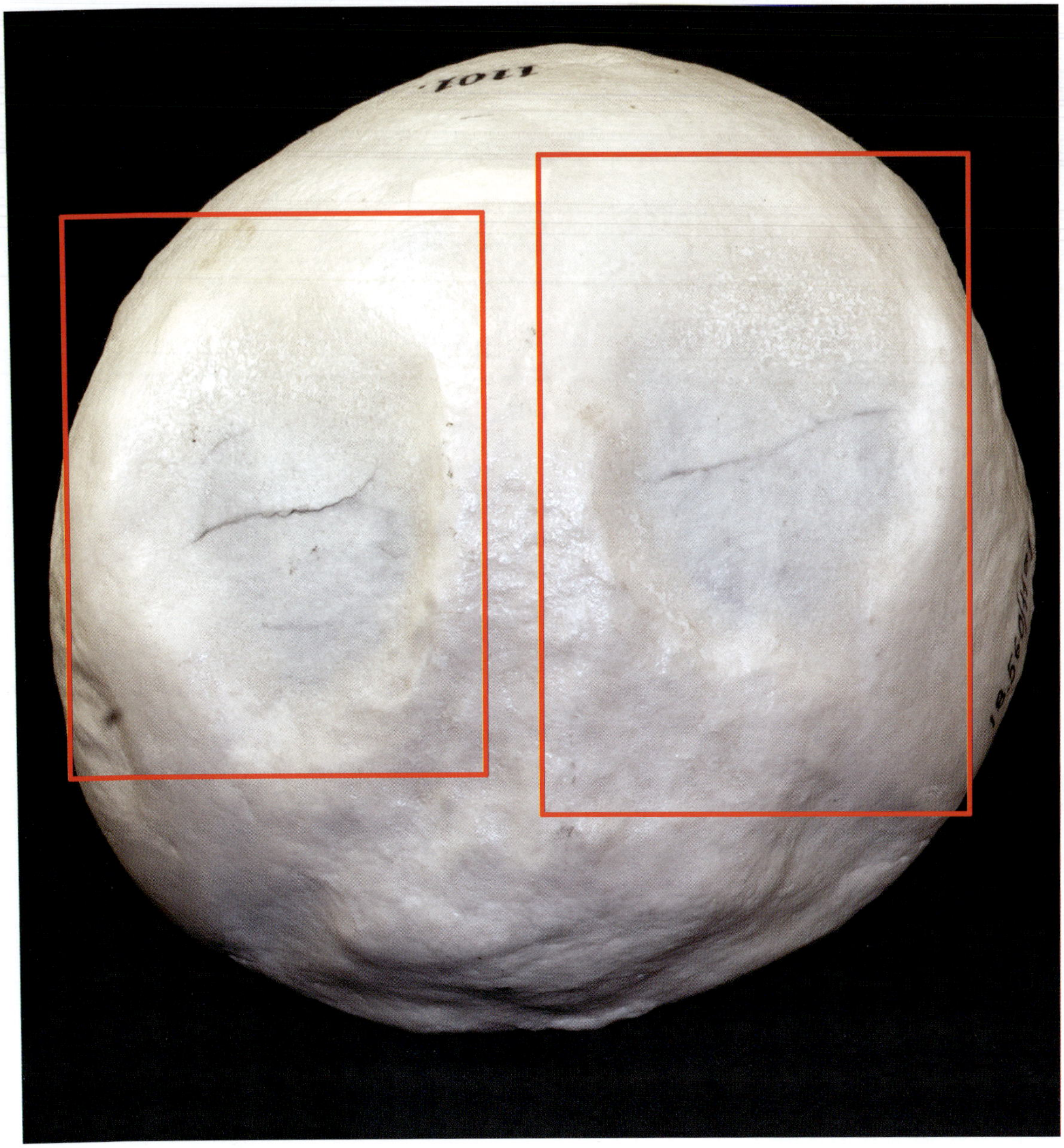

**Figure 70.** Ectocranial view of biparietal thinning (rectangles) in a 64-year-old female (MN18.560/1101). Biparietal thinning begins to develop with resorption of the outer table followed by the diploe and usually spares the inner table and maintains the contour of the inner cranial vault. As is often the case in elderly individuals and exacerbated by biparietal thinning, the meningeal artery grooves in this individual have widened and deepened, making this part of the parietal bones extremely thin – thinner than an egg shell – and susceptible to fracture, bleeding, and possibly death from a fall even from a standing height (Mann et al. 2017). Parietal thinning, also known as senile osteoporosis, biparietal atrophy, osteoporosis circumscripta and biparietal osteodystrophy, among others, is of unknown etiology and can affect one or both parietal bones in middle aged patients with a female predominance Western and Bekvalac (2017). See Mann and Hunt (2005); Ortner and Putschar (1981) for more on biparietal thinning in dry bone.

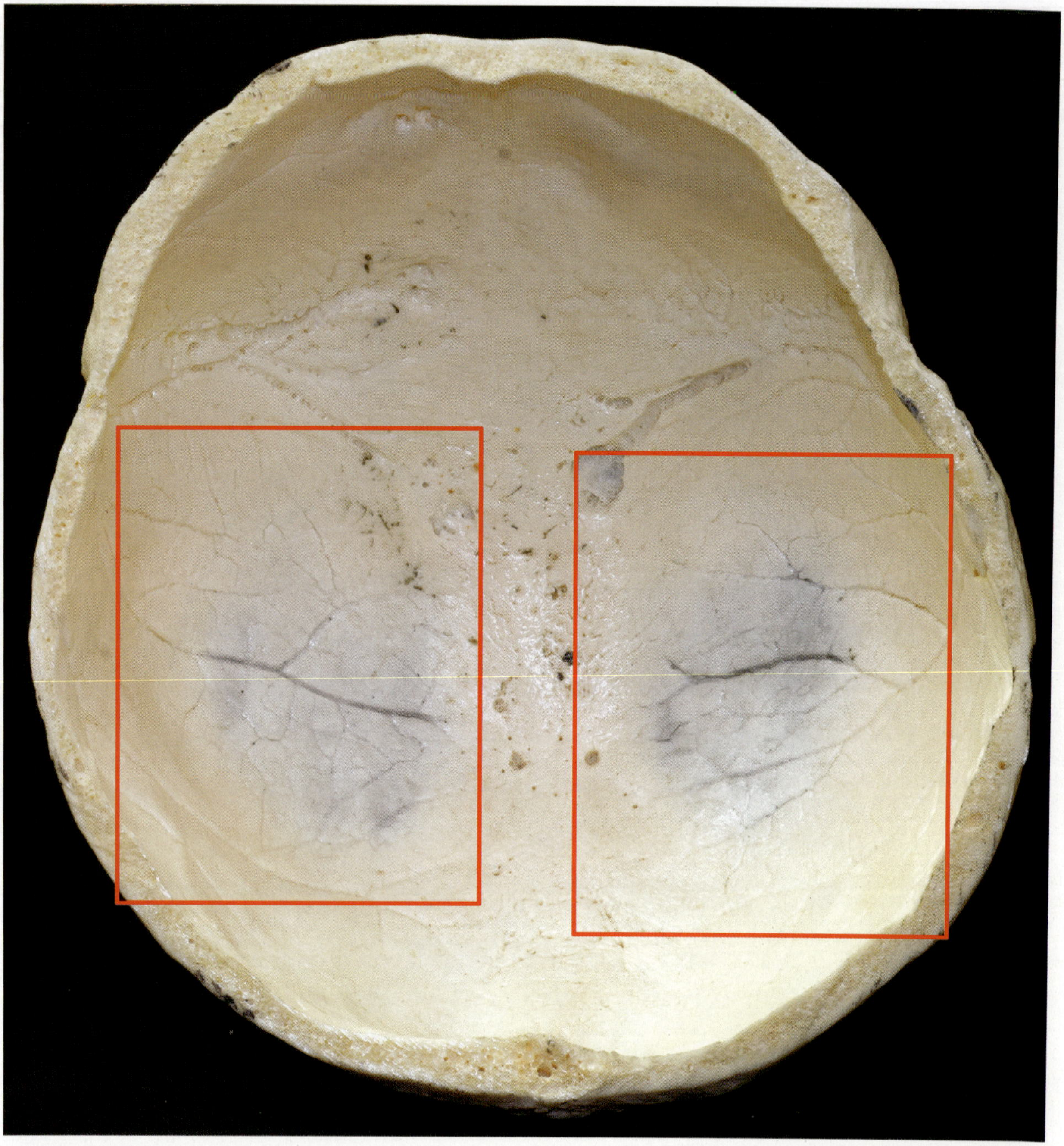

**Figure 71.** Endocranial view of biparietal thinning (rectangles) in a 64-year-old female (MN18.560/1101). Biparietal thinning develops beginning with resorption of the outer table followed by the diploe and usually spares the inner table and maintains the contour of the inner cranial vault. As is often the case in elderly individuals and exacerbated by biparietal thinning, the meningeal artery grooves in this individual have widened and deepened, making this region of the parietal bones extremely thin (thinner than an egg shell) and susceptible to fracture, bleeding, and possibly death even from a fall from a standing height (Mann et al. 2017). See Durward (1929); Mann and Hunt (2005); Ortner and Putschar (1981); Smith (1907) for more on biparietal thinning.

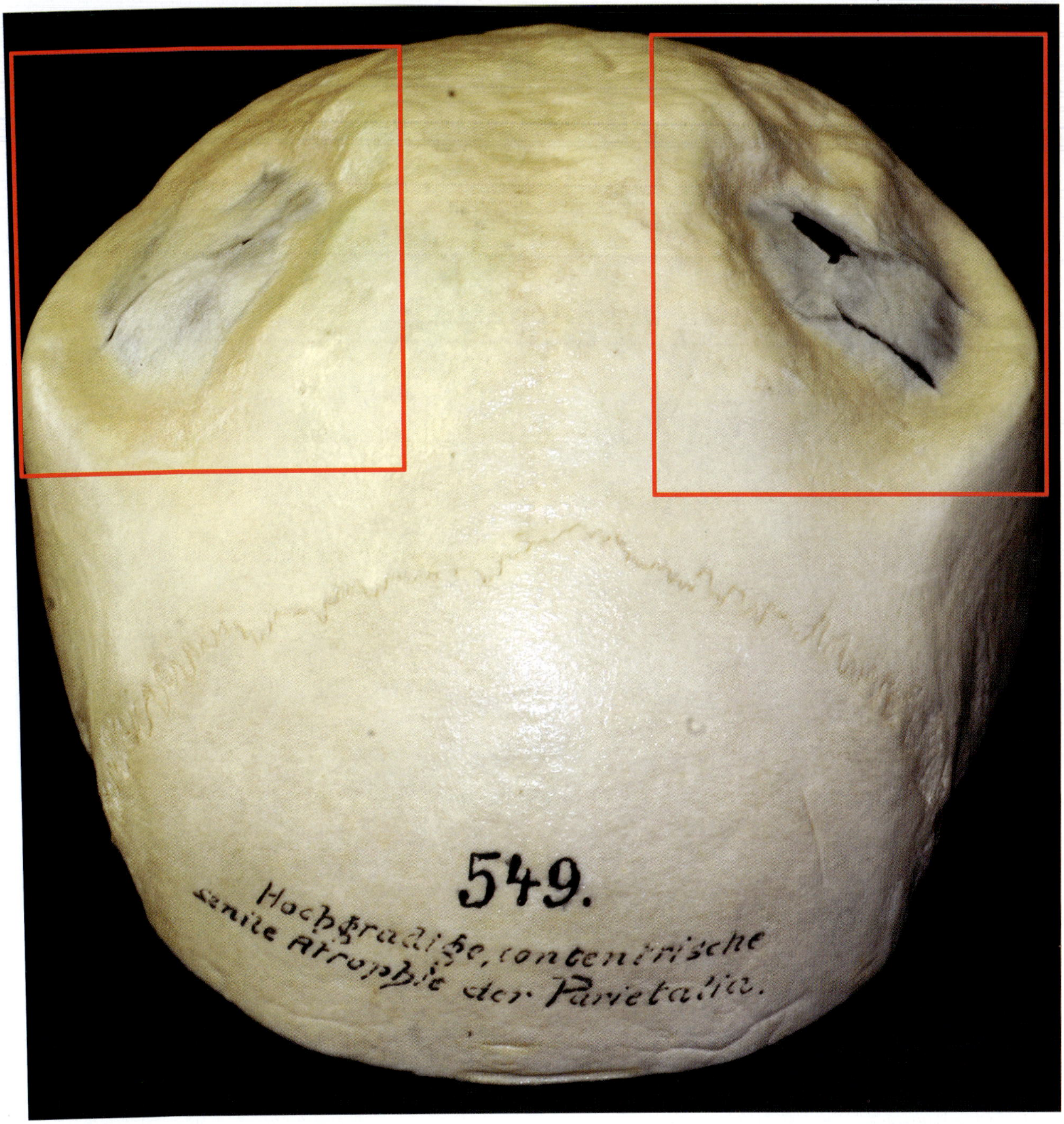


**Figure 72.** Biparietal thinning (translated "Senile Atrophy of the Parietals") in an unknown individual who died in 1916 (MN16.981/549). Note that the egg-shell thin oval-shaped lesions are bounded by normally thick bone and they avoid the sutures and superior temporal line. The fragile areas of broken and missing bone in these lesions are almost certainly broken post-mortem from handling or drying. See Durward (1929); Mann and Hunt (2005); Mann et al. (2017); Ortner and Putschar (1981); Smith (1907) for more on biparietal thinning.

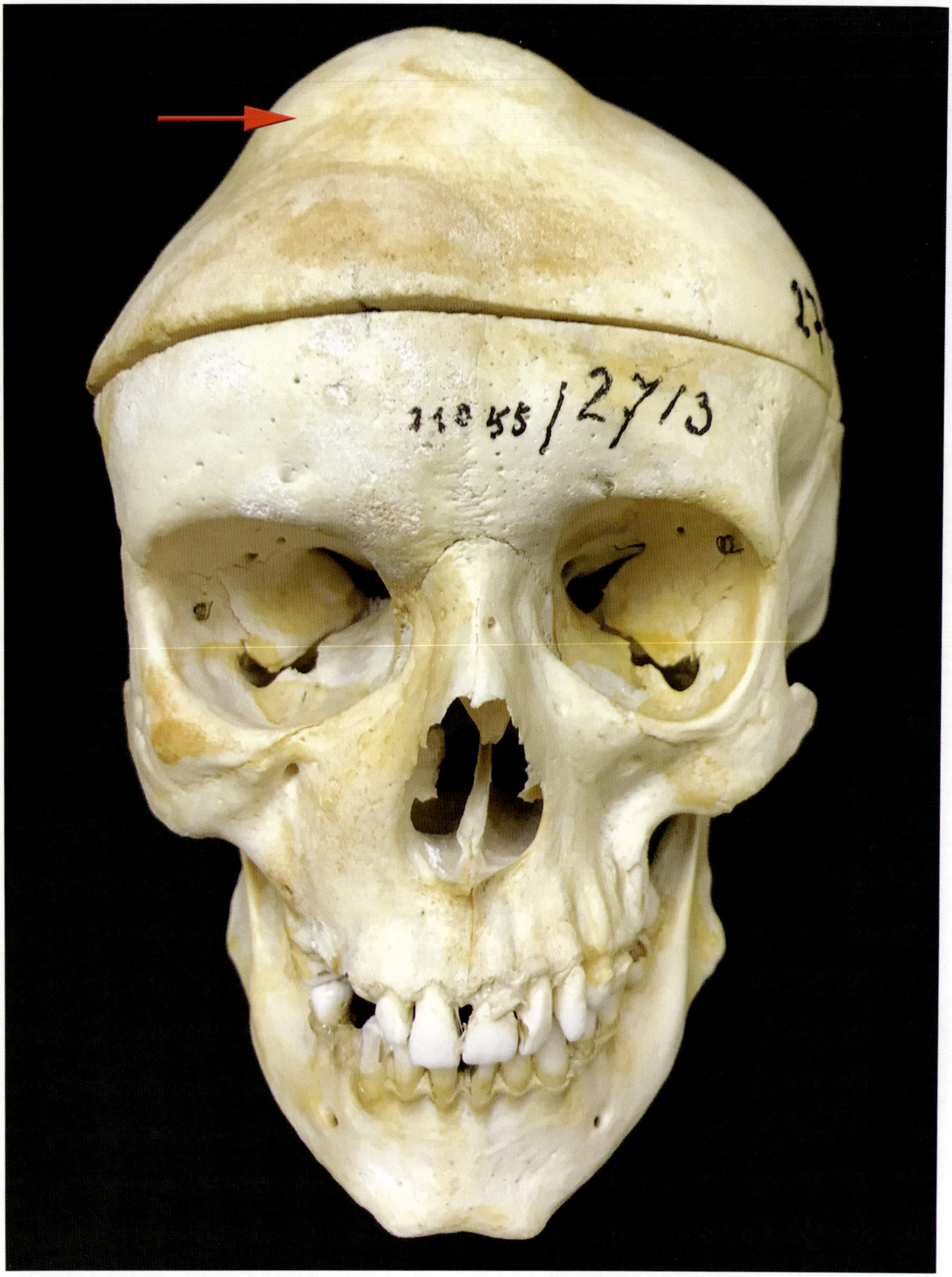

**Figure 73.** Anterior view of a 20-year-old male with premature craniosynostosis of all of the sutures giving the skull nowhere to grow except upward (arrow), resulting in turricephaly (oxycephaly; Bodian 1950) or so-called "tower skull" (MN21.055/2713). See Dias et al. (2020) for the classification of premature craniosynostosis and related disorders.

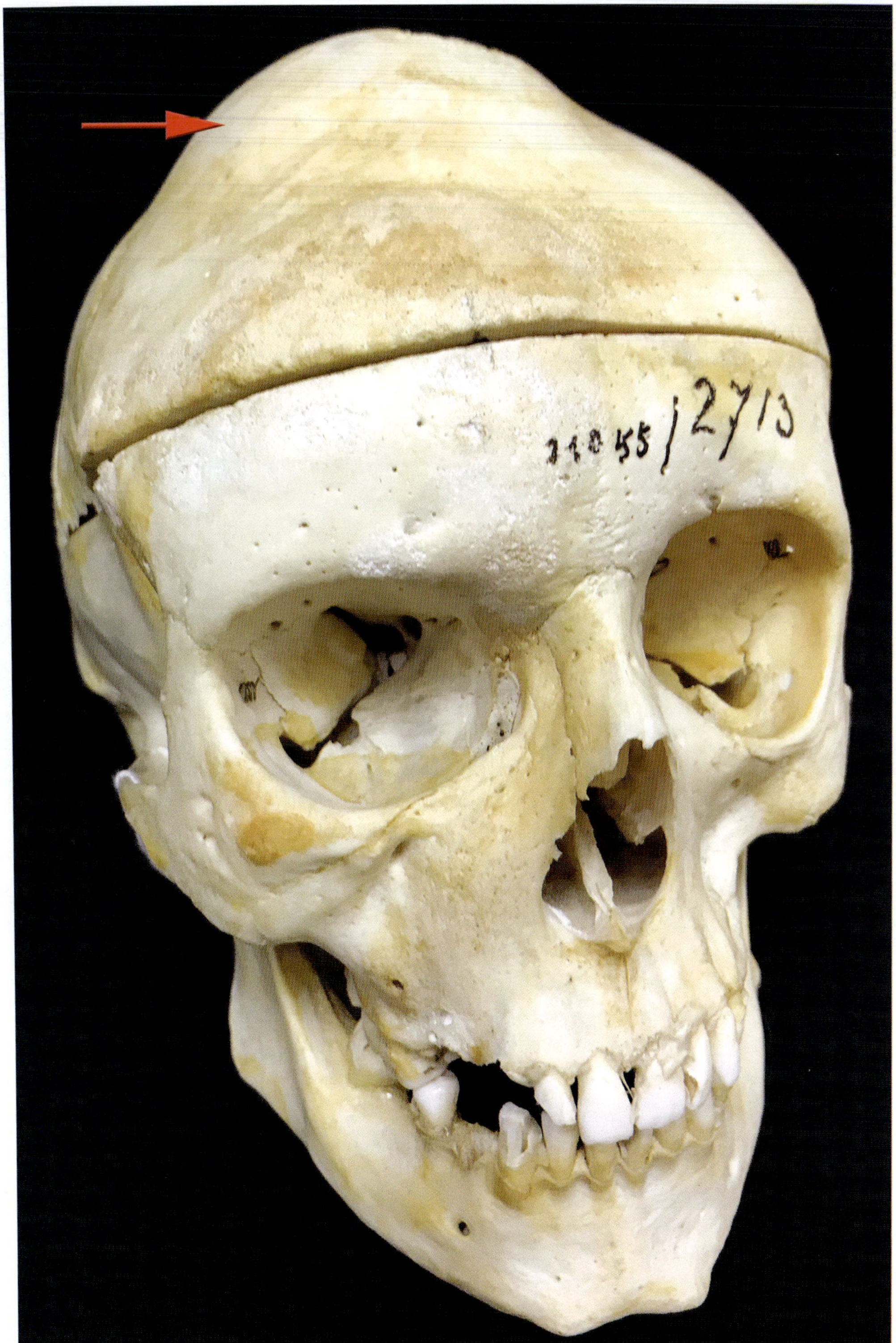

**Figure 74.** Right oblique view of a 20-year-old male with premature craniosynostosis of all of the sutures giving the skull nowhere to grow except upward (arrow), resulting in turricephaly (oxycephaly; Bodian 1950) or so-called "tower skull" (MN21.055/2713). See Dias et al. (2020) for the classification of premature craniosynostosis and related disorders.

## Whole Skeleton

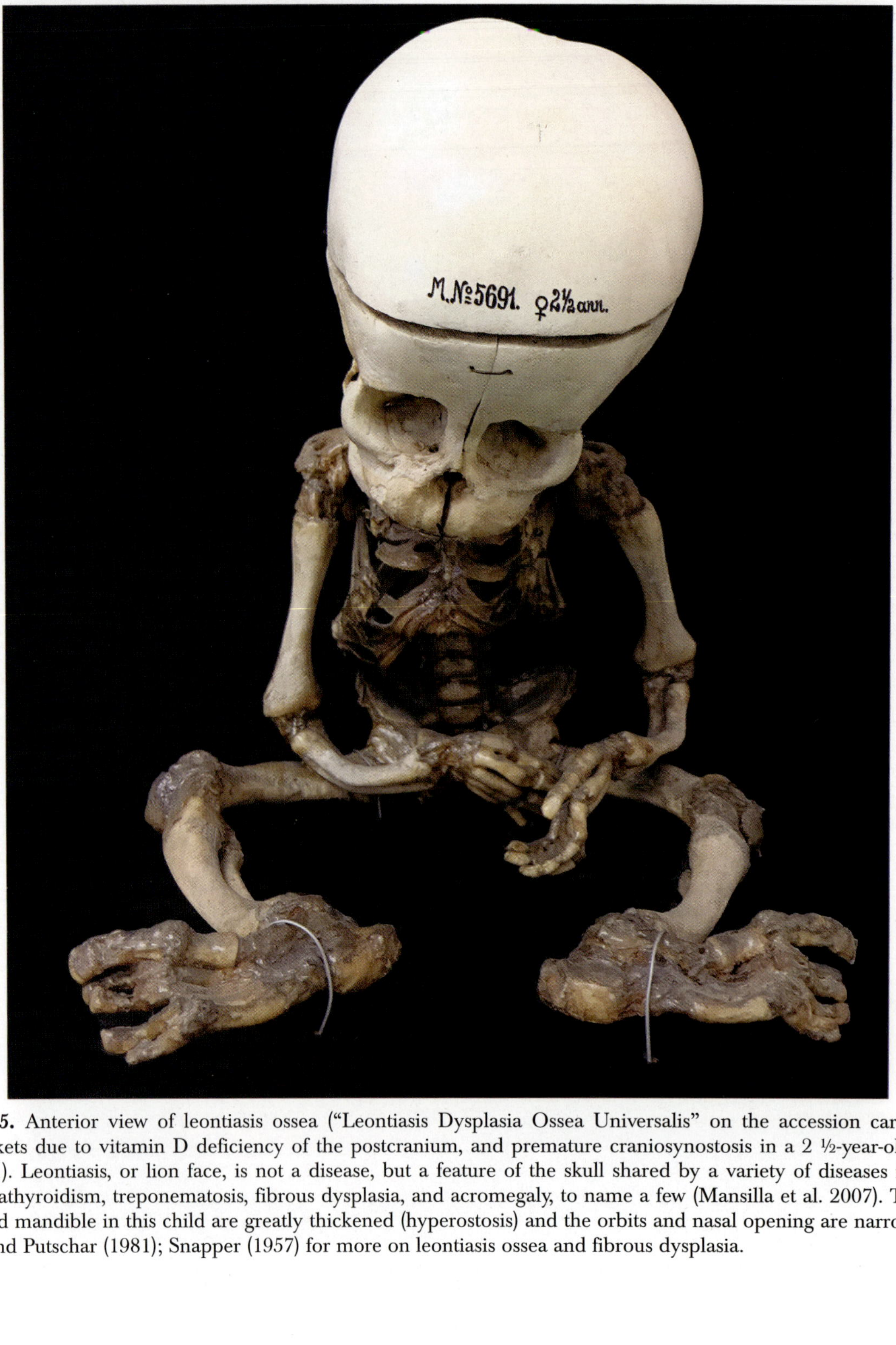


**Figure 75.** Anterior view of leontiasis ossea ("Leontiasis Dysplasia Ossea Universalis" on the accession card) of the skull, rickets due to vitamin D deficiency of the postcranium, and premature craniosynostosis in a 2 ½-year-old female (MN5691). Leontiasis, or lion face, is not a disease, but a feature of the skull shared by a variety of diseases including hyperparathyroidism, treponematosis, fibrous dysplasia, and acromegaly, to name a few (Mansilla et al. 2007). The facial bones and mandible in this child are greatly thickened (hyperostosis) and the orbits and nasal opening are narrowed. See Ortner and Putschar (1981); Snapper (1957) for more on leontiasis ossea and fibrous dysplasia.

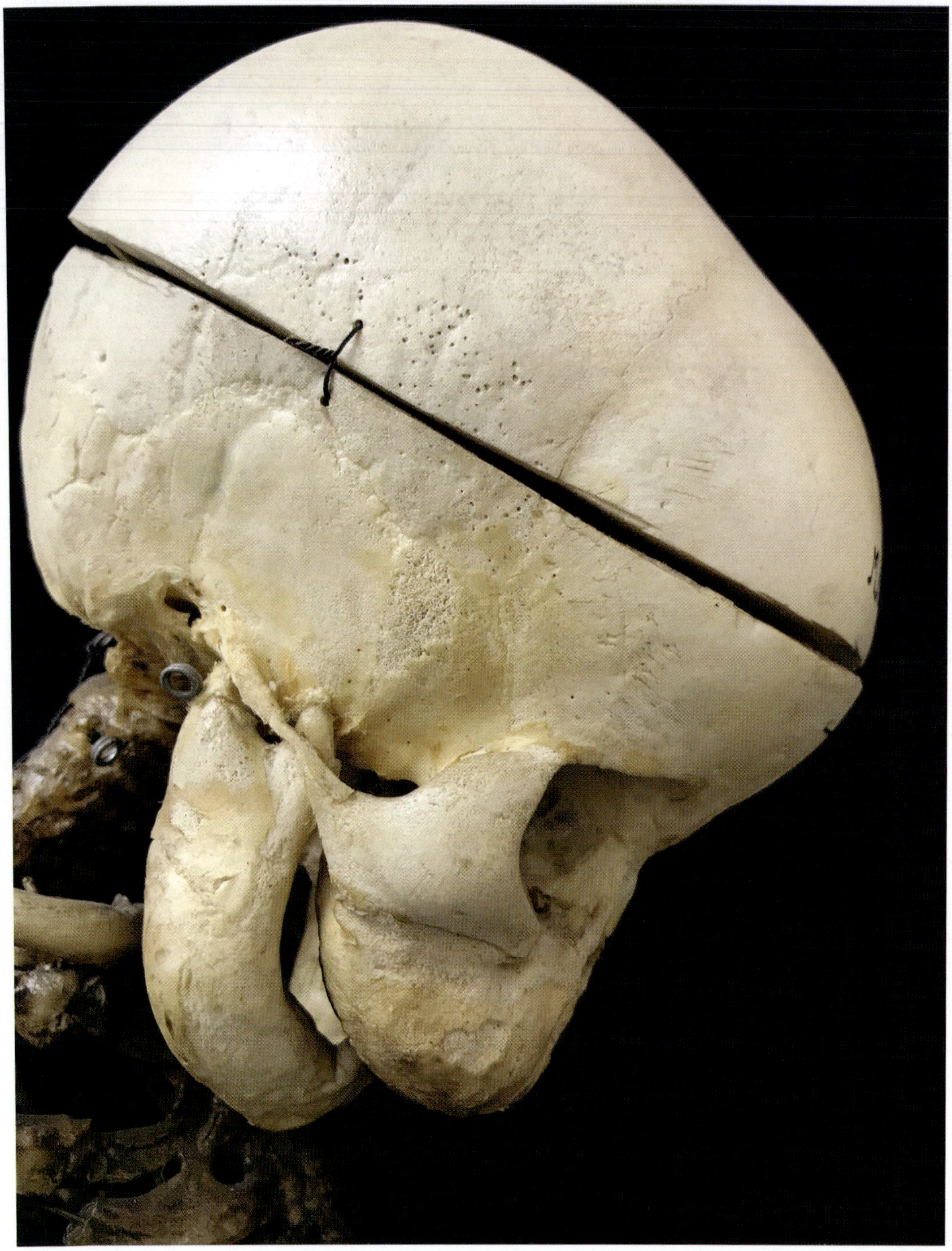

**Figure 76.** Right lateral view of leontiasis ossea ("Leontiasis Dysplasia Ossea Universalis" on the museum card), also known as Virchow's Disease, rickets in the postcranial skeleton, and craniosynostosis of all sutures of a 2 ½-year-old female (MN5691). Note enlarged and thickened (hyperostosis) crania, most notably in the facial region and mandible, rounded edges of the facial bones and pitting in the right parietal and temporal bones. Hyperostosis in leontiasis ossea is a rare condition that is typically so severe that the inner and outer tables can no longer be differentiated (Freedman 1933). See Ortner and Putschar (1981); Snapper (1957) for more on leontiasis ossea and fibrous dysplasia.

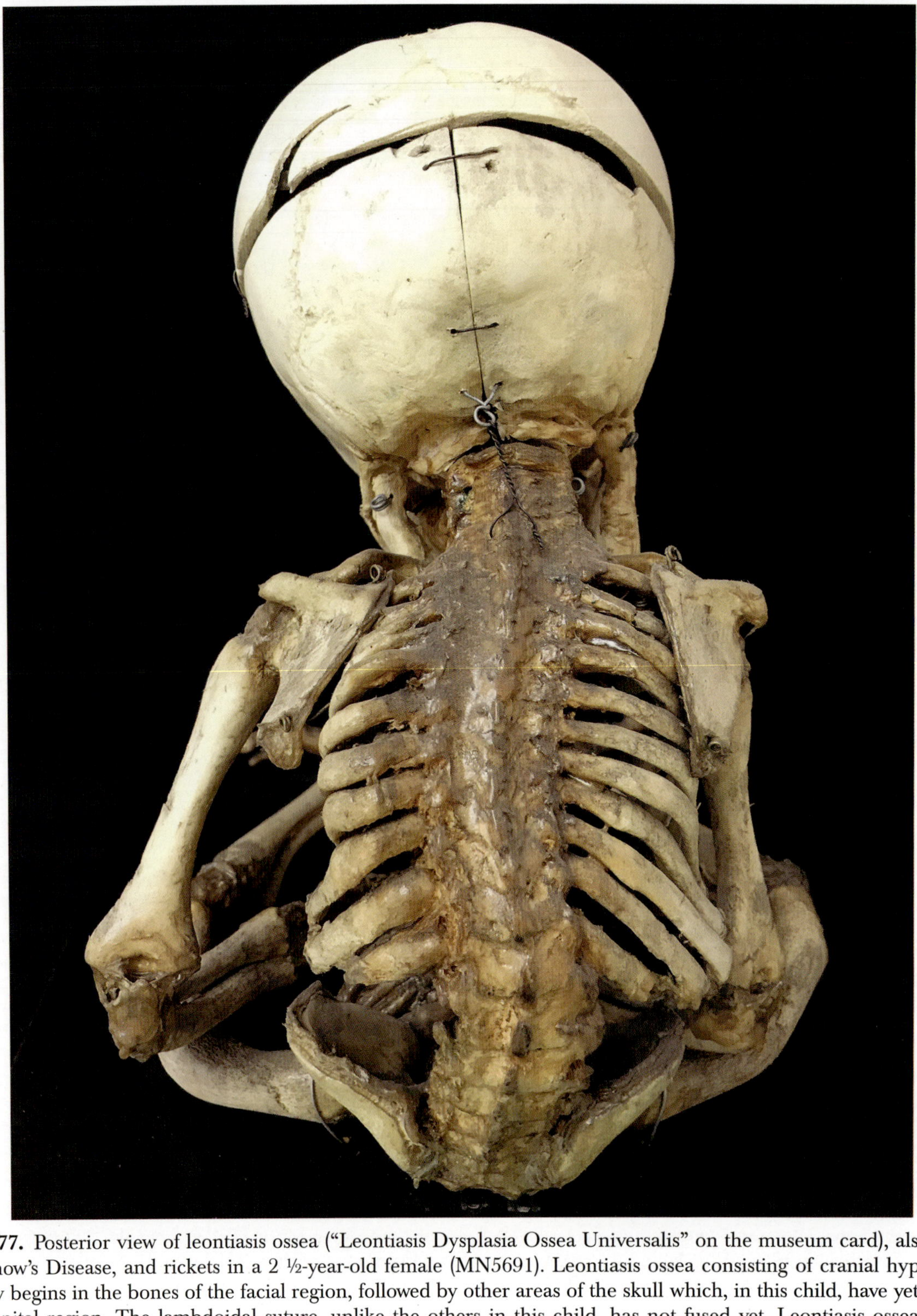

**Figure 77.** Posterior view of leontiasis ossea ("Leontiasis Dysplasia Ossea Universalis" on the museum card), also known as Virchow's Disease, and rickets in a 2 ½-year-old female (MN5691). Leontiasis ossea consisting of cranial hyperostosis typically begins in the bones of the facial region, followed by other areas of the skull which, in this child, have yet to affect the occipital region. The lambdoidal suture, unlike the others in this child, has not fused yet. Leontiasis ossea is not a disease, but a generalized feature shared with many diseases, including Paget's disease and fibrous dysplasia.

**Figure 78.** Cephalothoracopagus, or so-called "Janus" newborn twins, who died in 1868 (MN3086). Janus twins are named after the Roman god Janus, with one face looking forward (anterior) and one facing backward/behind (posterior), an extremely rare congenital anomaly resulting in two faces (craniofacial duplication) in one skull. In this projection there are two orbits, nasal bones and opening, shared parietal bones, conjoined thorax, two spines, and two arms and legs. The reverse view of this skull has cyclopia consisting of one orbit. The incidence of diprosopus is 1 in 180,000 to 1 in 15 million births (D'Armiento et al. 2010).

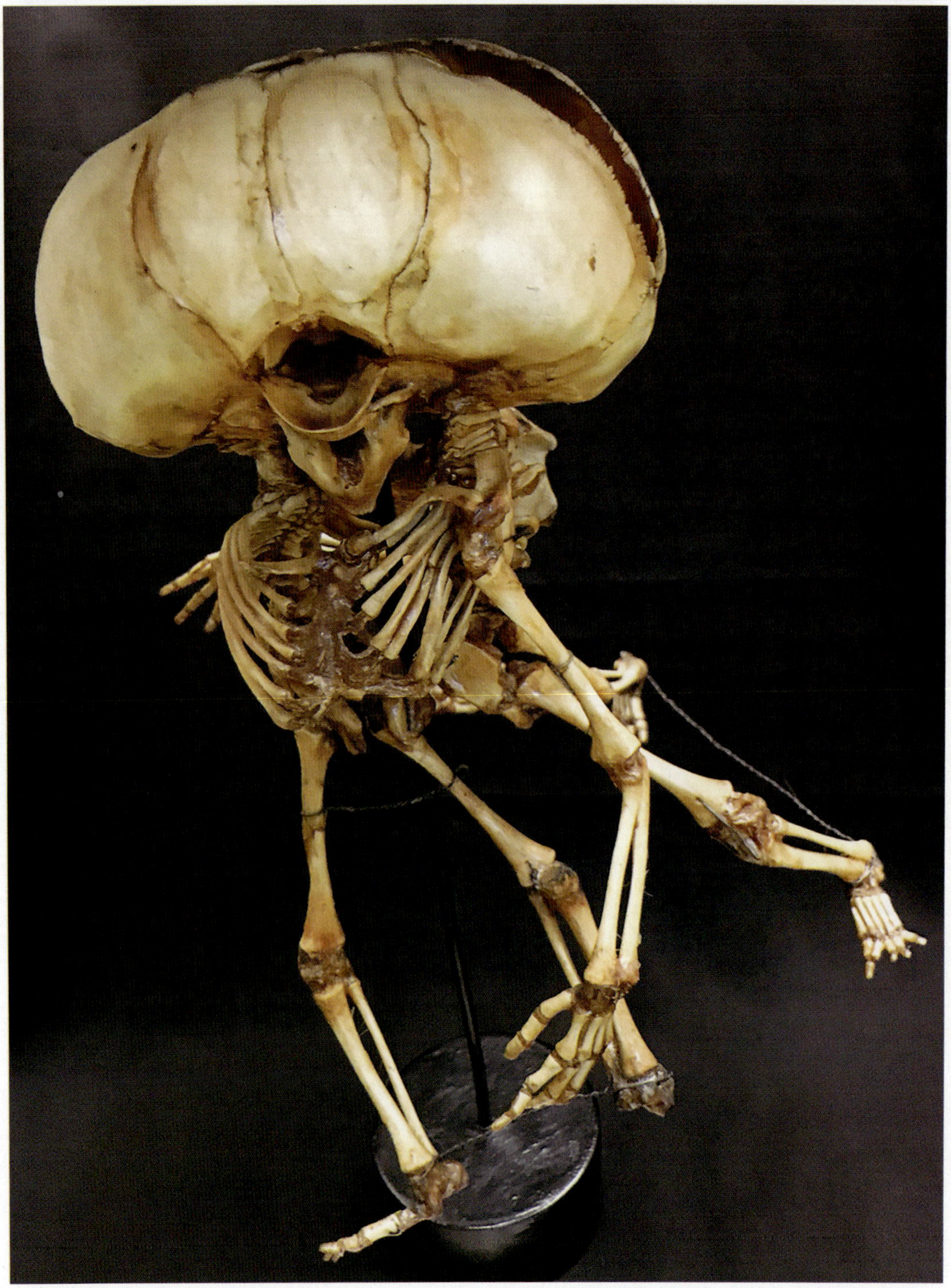

**Figure 79.** Cephalothoracopagus (Janus) newborn twins who died in 1868 (MN3086). In this projection there is one orbit (cyclopia), absence of nasal bones and opening, malformed and narrow mandible, shared parietal bones, conjoined thorax, two spines, and two arms and legs. The incidence of diprosopus is 1 in 180,000 to 1 in 15 million births making it the rarest form of the conjoined twins. (D'Armiento et al. 2010).

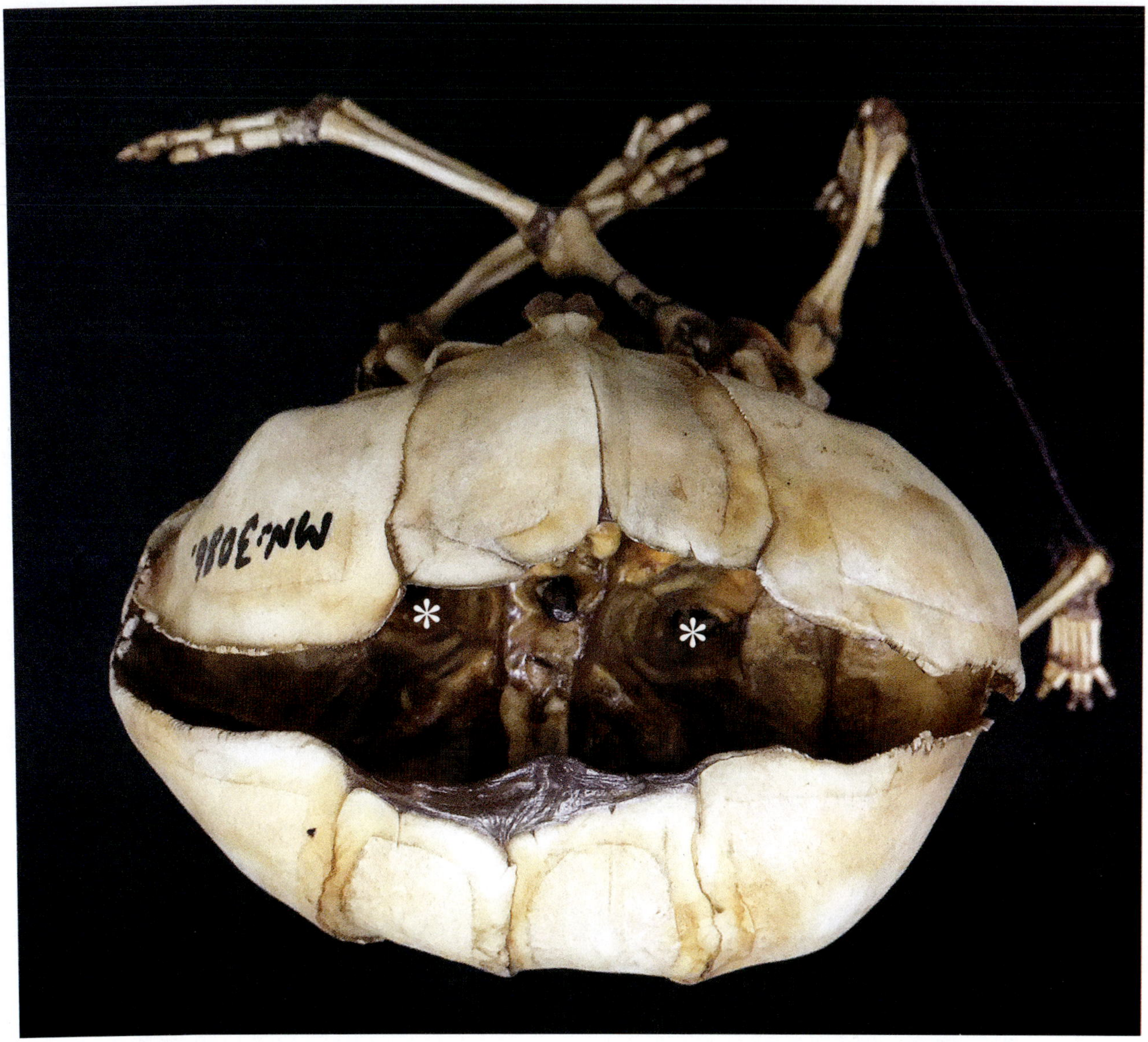

**Figure 80.** Superior view of cephalothoracopagus (Janus) newborn twins who died in 1868 (MN3086). Note there are paired anterior and posterior frontal bones, paired left and right parietal bones, and two foramen magnums (*) for passage of the spinal cord.

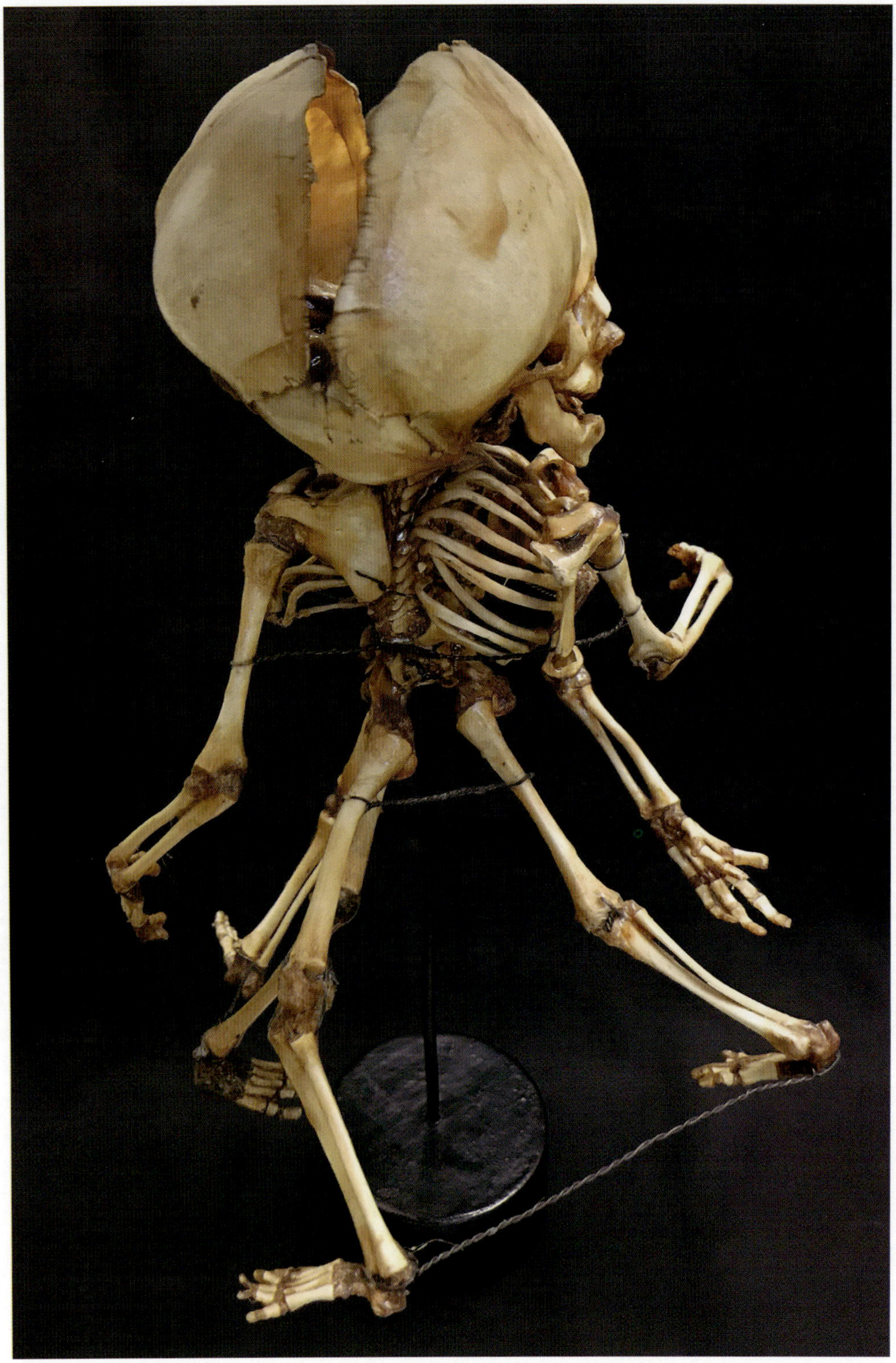

**Figure 81.** Lateral view of cephalothoracopagus (Janus) newborn twins who died in 1868 (MN3086). Note there are paired anterior and posterior frontal bones, paired left and right parietal bones.

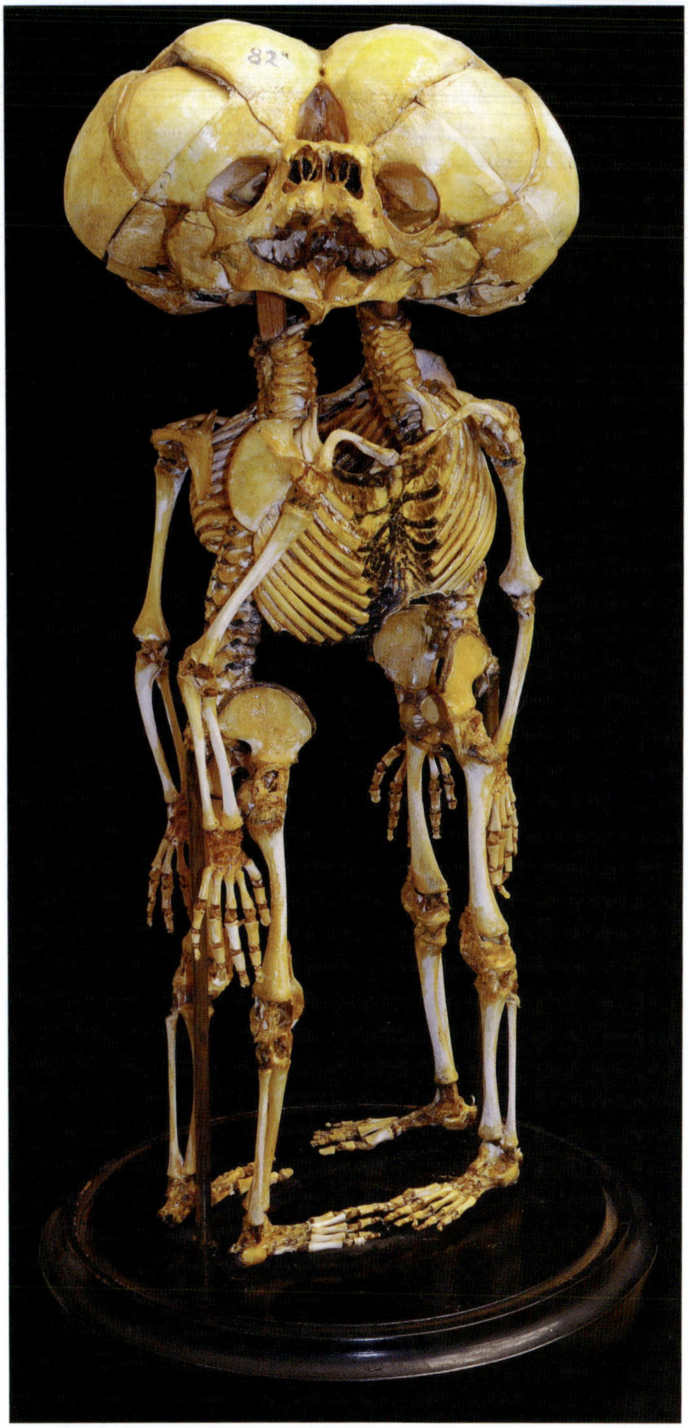

**Figure 82.** Cephalothoracic janiceps monosymmetros joined at the head and thorax (MN82a). See Stone and Goodrich (2006) on the classification of craniopagus twins and Turgut et al. (1998) for more on cephalothoracopagus twins.

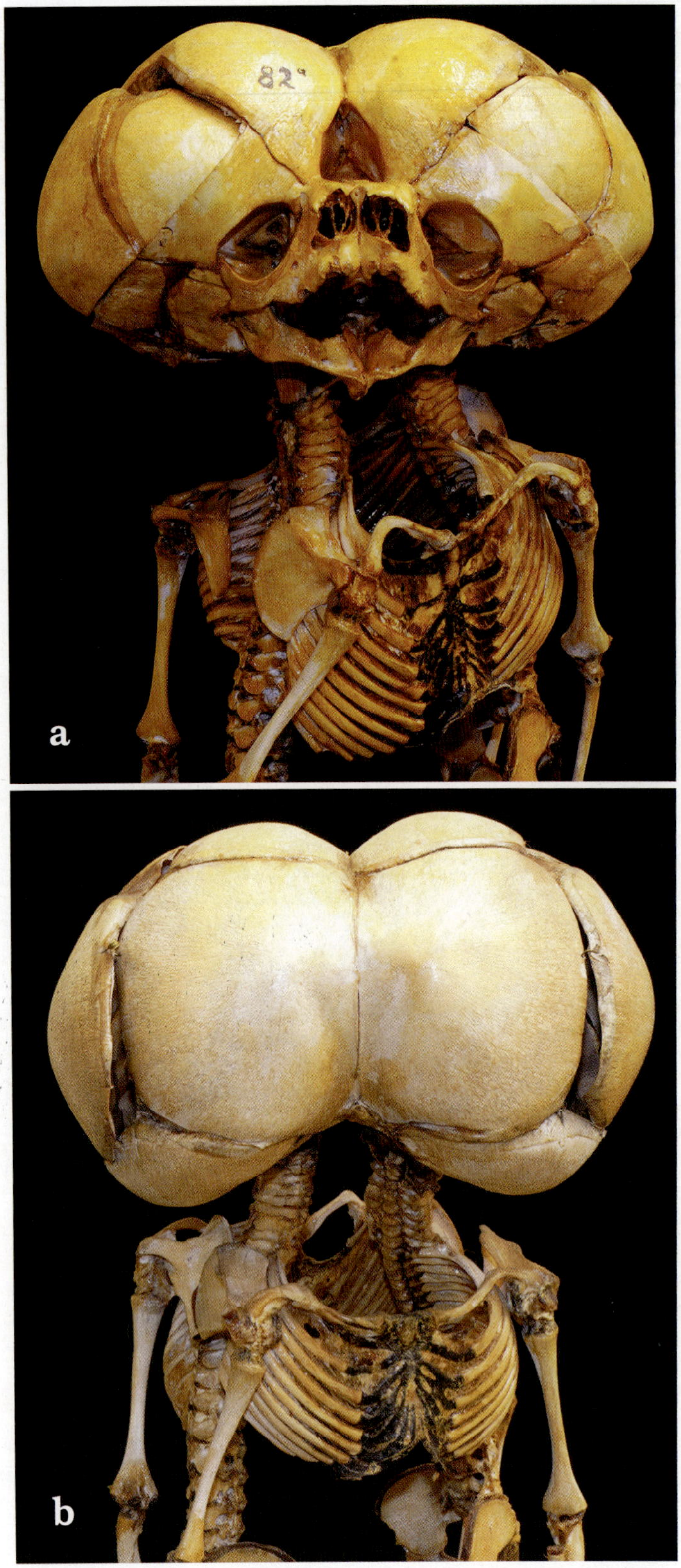


**Figure 83a-b.** (a) Anterior and (b) posterior views of cephalothoracic janiceps monosymmetros (MN82a). See Stone and Goodrich (2006) on the classification of craniopagus twins and Turgut et al. (1998) for more on cephalothoracopagus twins.

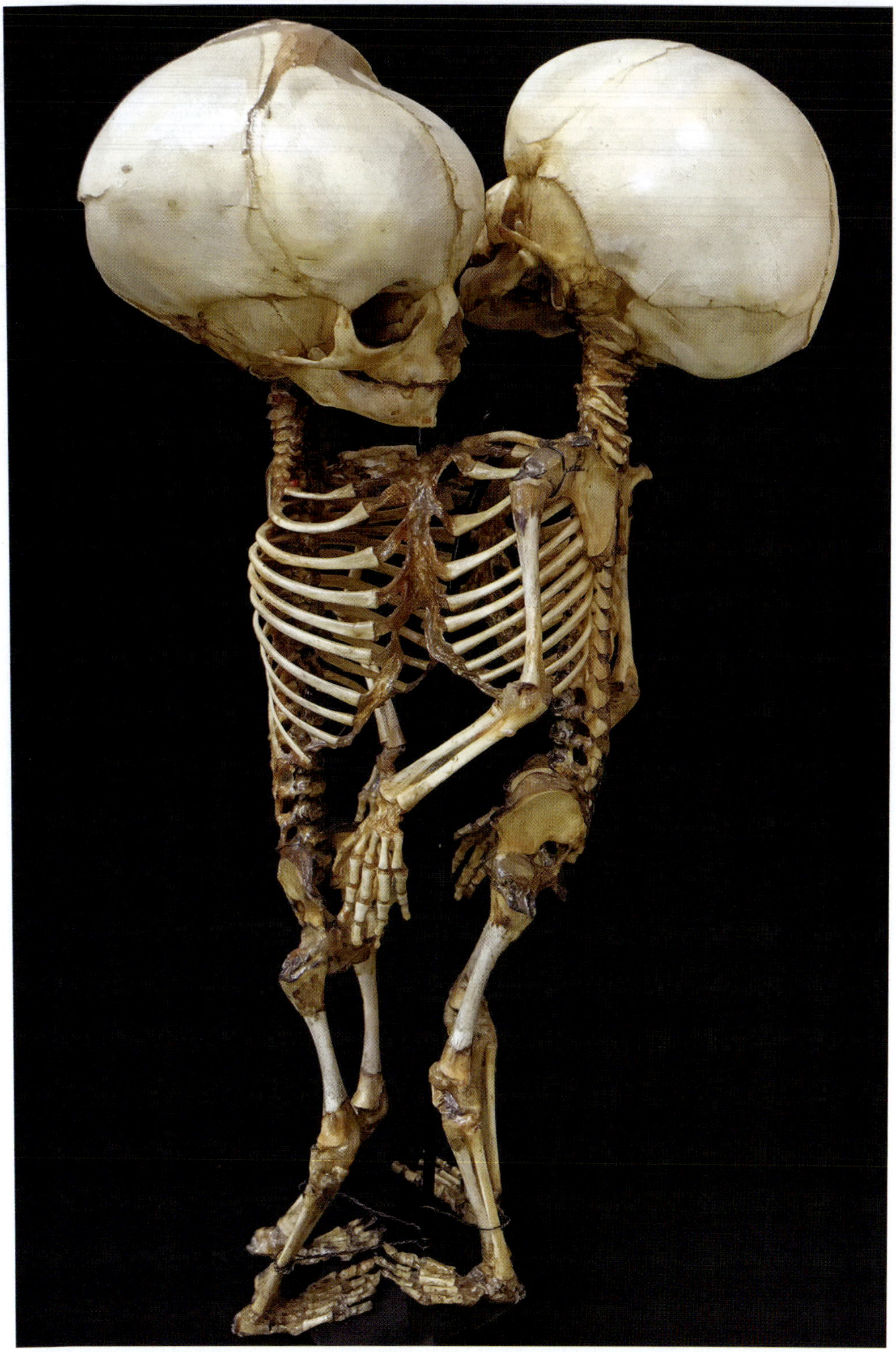

**Figure 84.** Thoracopagus conjoined twins who died before 1920 (MN7767). Thoracopagus refers to conjoined twins fused at the thorax. In this case the twins share two pairs of ribs, sterni, and clavicles, each positioned laterally. See Osmanagaoglu et al. (2011) for more on the classification of thoracopagus conjoined twins.

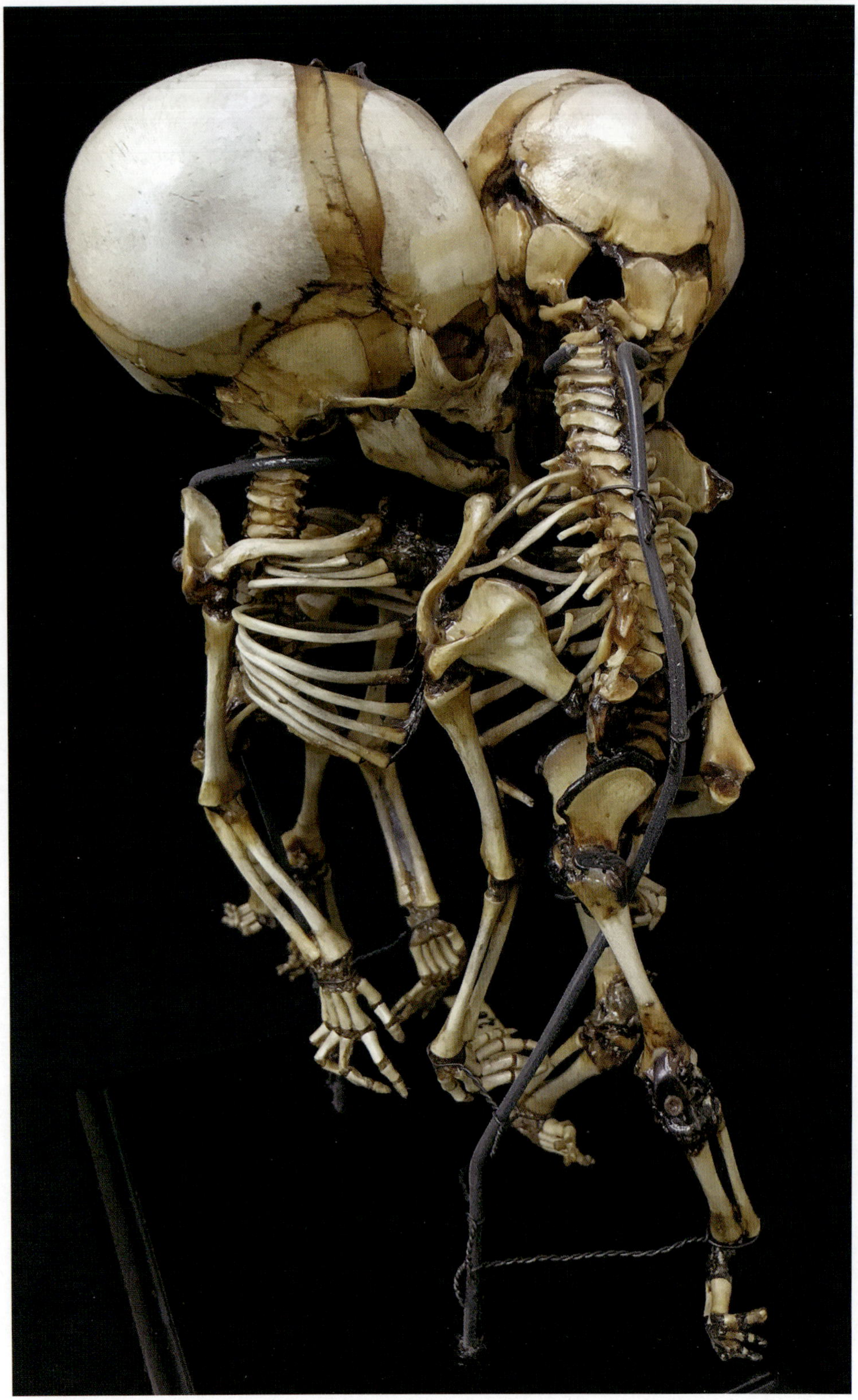

**Figure 85.** Magnified view of thoracopagus conjoined twins who died before 1920 (MN7767). Thoracopagus refers to conjoined twins fused at the thorax. In this case the twins share two pairs of ribs, sterni, and clavicles, each positioned laterally. See Osmanagaoglu et al. (2011) for more on the classification of thoracopagus conjoined twins.

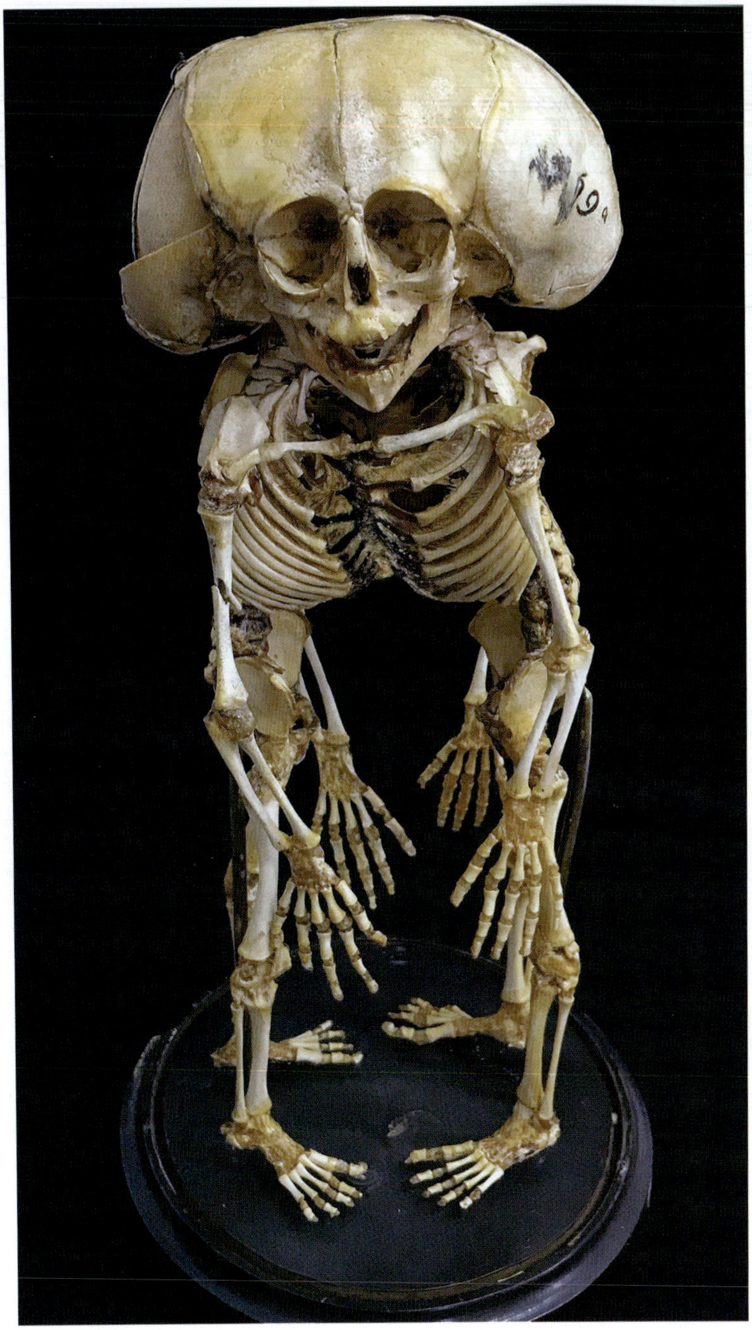

**Figure 86.** Anterior view of cephalothoracopagus or Janiceps ("Janus") conjoined stillborn-twins who died in 1869 (MN21.055/1569a) where the upper bodies are joined and the lower bodies are separate. This condition refers to the Roman god Janus who possessed two faces – one looking forward and the other looking backward or behind ("January" the end of a year and beginning of another). In this view there are two orbits and the reverse view (Figure 88) has a single orbit (cyclopia). Janus twins typically have two separate bodies and two crania that are conjoined, sharing some portions thereof. This individual exhibits an extremely rare form of Janus twins in that they are not only conjoined at the skull, but the chest. Conjoined twins are extremely rare occurring in 1 in 50,000 pregnancies (Amorosa et al. 2017).

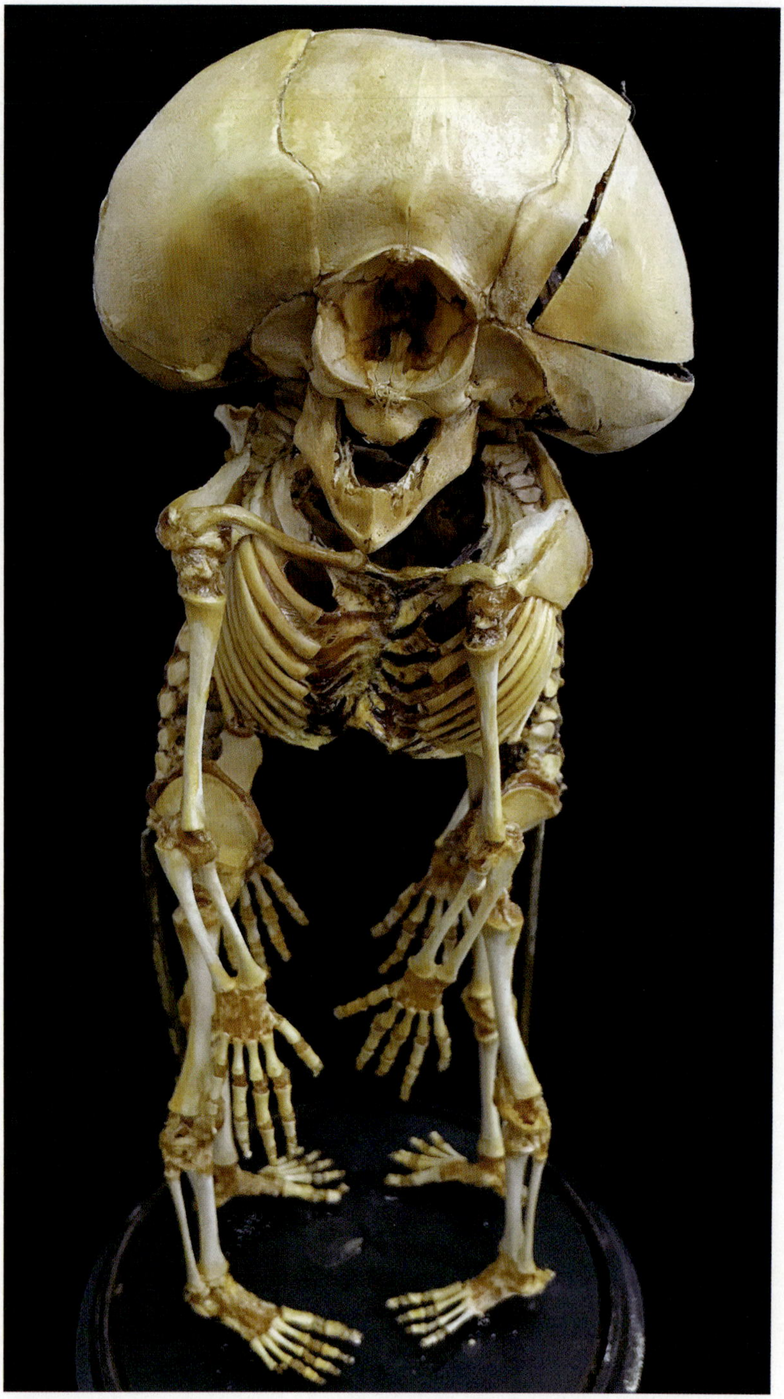

**Figure 87.** Posterior/reverse view of extremely rare cephalothoracopagus or Janiceps ("Janus") conjoined twins with a single orbit (cyclopia) who died in 1869 (MN21.055/1569a). See Stone and Goodrich (2006) on the classification of craniopagus twins and Turgut et al. (1998) for more on cephalothoracopagus twins.

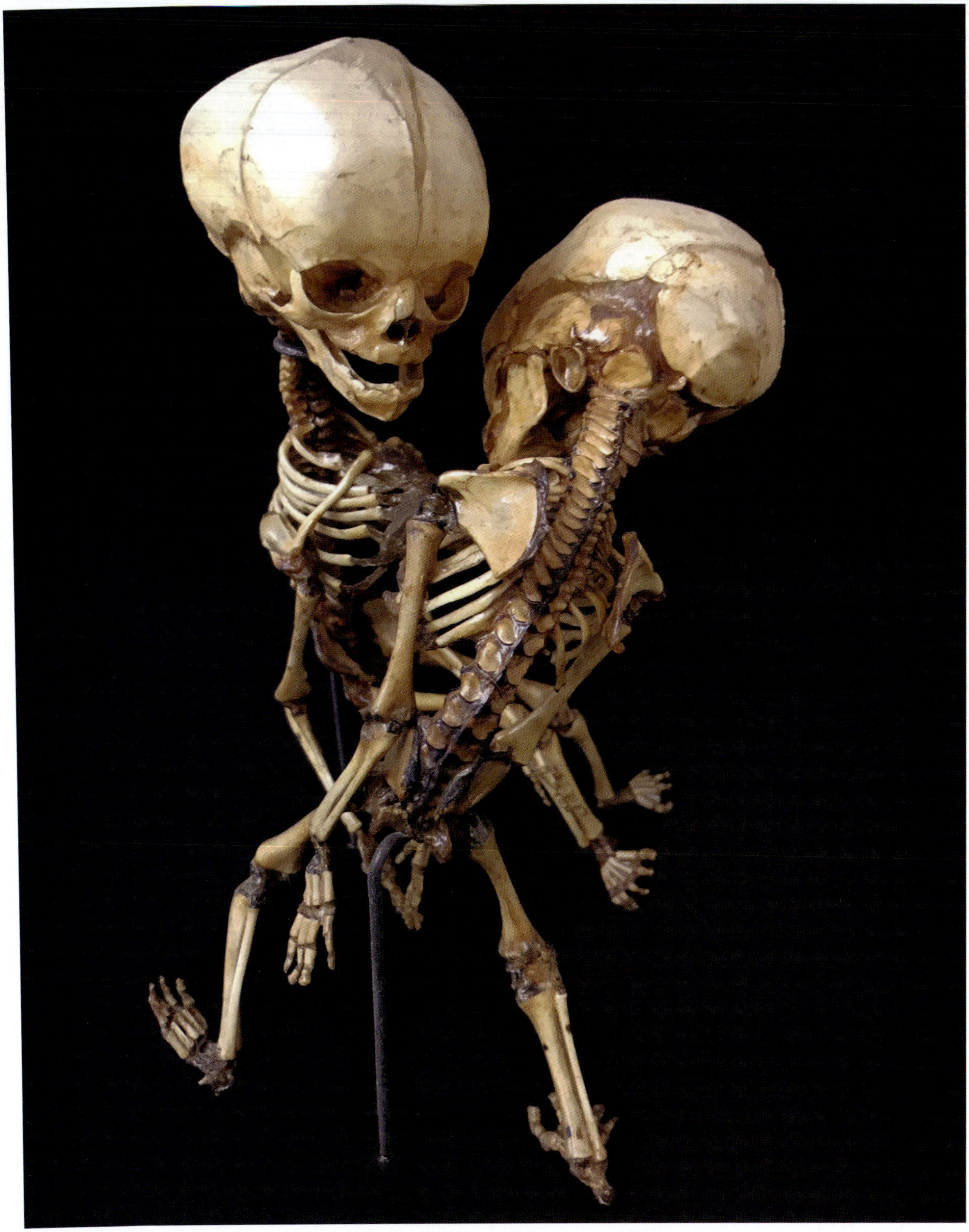

**Figure 88.** Thoracopagus conjoined twins united at the thorax (MN2602). Conjoined twins is a rare congenital malformation present in 1 in 50,000 to 1 in 100,00 live births (Rees et al. 1993). Thoracopagus is the most common form of conjoined twins with the individuals having separate arms and legs. See Ortner and Putschar (1981) for more on conjoined twins.

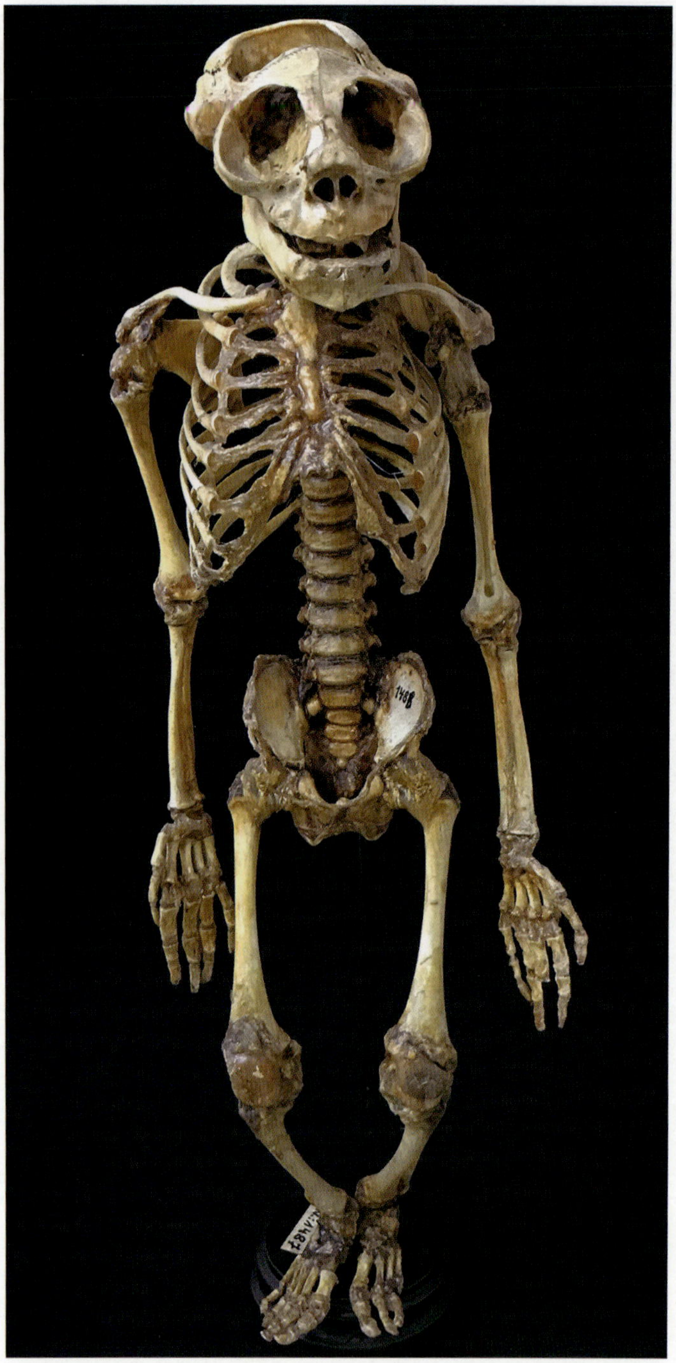

**Figure 89.** Anencephaly in a stillborn male who died in 1833 (MN1487). Anencephaly ("open skull"), a rare neural tube defect of multifactorial etiology, occurs in 1:1000-1:20000 births (Thomas et al. 1976; Shilpa et al. 2018) and results in absence of major portions of the skull, scalp, and brain while in the womb. Note the orbits and nose are unusually large and the vault is open. Children with anencephaly are usually stillborn or die within a few hours or days. See Ortner and Putschar (1981) for more on anencephaly in dry bone and Munteanu et al. (2020) for clinical aspects and classification of anencephaly.

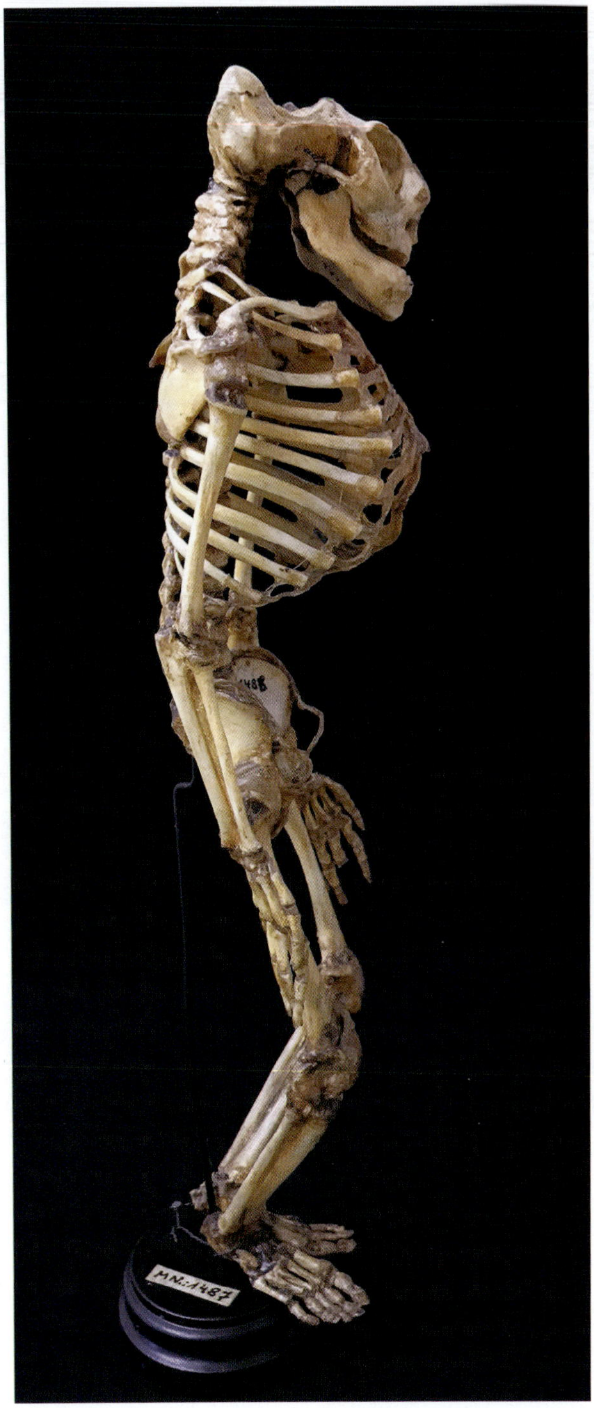

**Figure 90.** Right lateral view of anencephaly, a neural tube defect, showing absence of the cranial vault (and brain) in a stillborn male who died in 1833 (MN1487). See Ortner and Putschar (1981) for more on anencephaly in dry bone and Munteanu et al. (2020) for clinical aspects and classification of anencephaly.

**Figure 91.** Anencephaly (craniorachischisis), a neural tube defect resulting in absence of the cranial vault and brain in a stillborn male who died in 1853 (MN2470). See Ortner and Putschar (1981) for more on anencephaly in dry bone and Munteanu et al. (2020) for clinical aspects and classification of anencephaly.

**Figure 92.** Left oblique view of anencephaly (craniorachischisis), a neural tube defect resulting in absence of the cranial vault and brain in a stillborn male who died in 1853 (MN2470). Note the low position of the clavicles, scapulae and arms. See Ortner and Putschar (1981) for more on anencephaly in dry bone and Munteanu et al. (2020) for clinical aspects and classification of anencephaly.

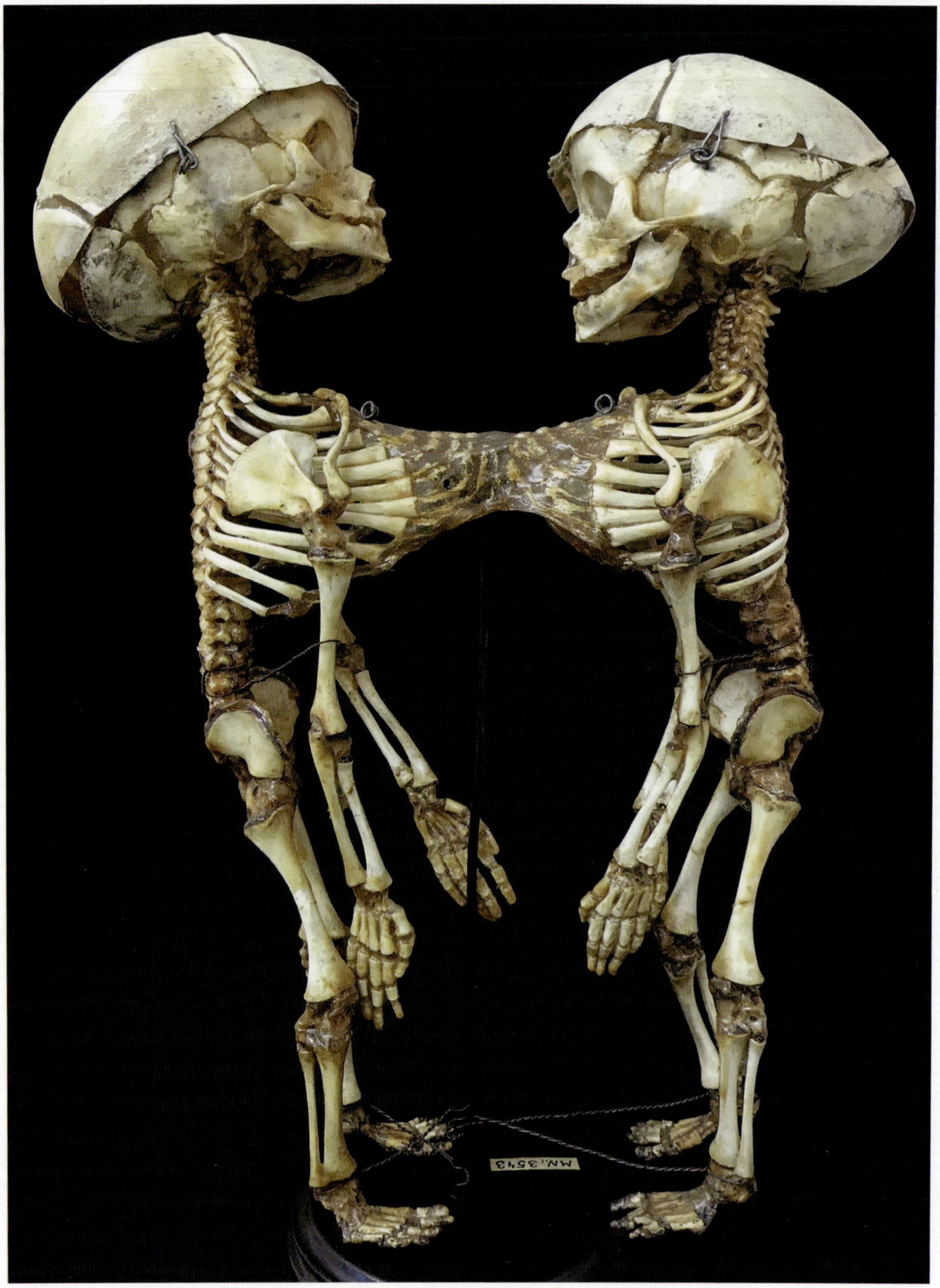

**Figure 93.** Xiphopagus females joined at the chest (MN3543). About 3% of conjoined twins are xiphopagus with Chang and Eng Bunker ("Siamese twins" born in Thailand in 1811 and died in 1874) the first known and documented pair of xiphopagus twins conjoined at the liver. Conjoined twins form when an early-stage embryo only partially divides. See Ortner and Putschar (1981) for more on conjoined twins. Conjoined twins is a rare complication of monozygotic twinning or so-called "identical twins" (De Paepe 2019). See Britannica (2022) for more on the history of Chang and Eng Bunker.

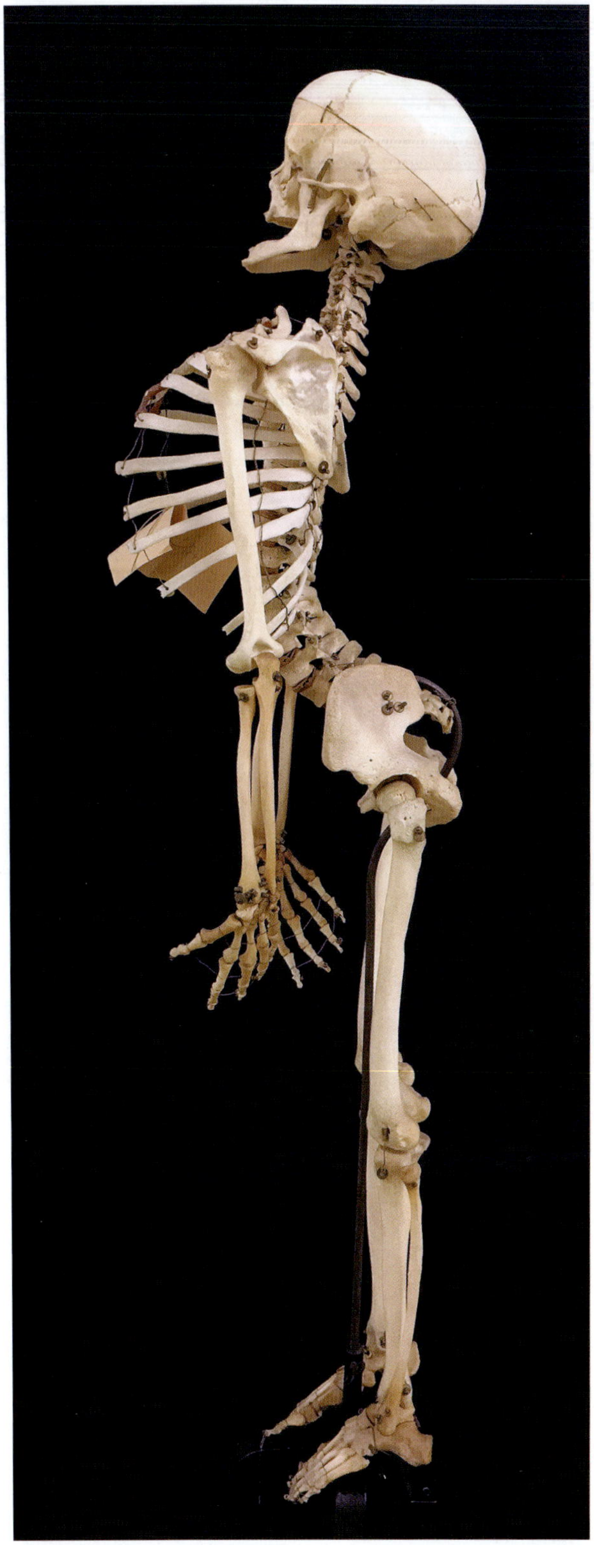

**Figure 94.** Left lateral view of an adult male with osteoporosis and lordosis resulting in anterior curvature of the thoraco-lumbar spine who died before 1921 (MN18.648). See Sorensen et al. (2015) for more on lumbar lordosis.

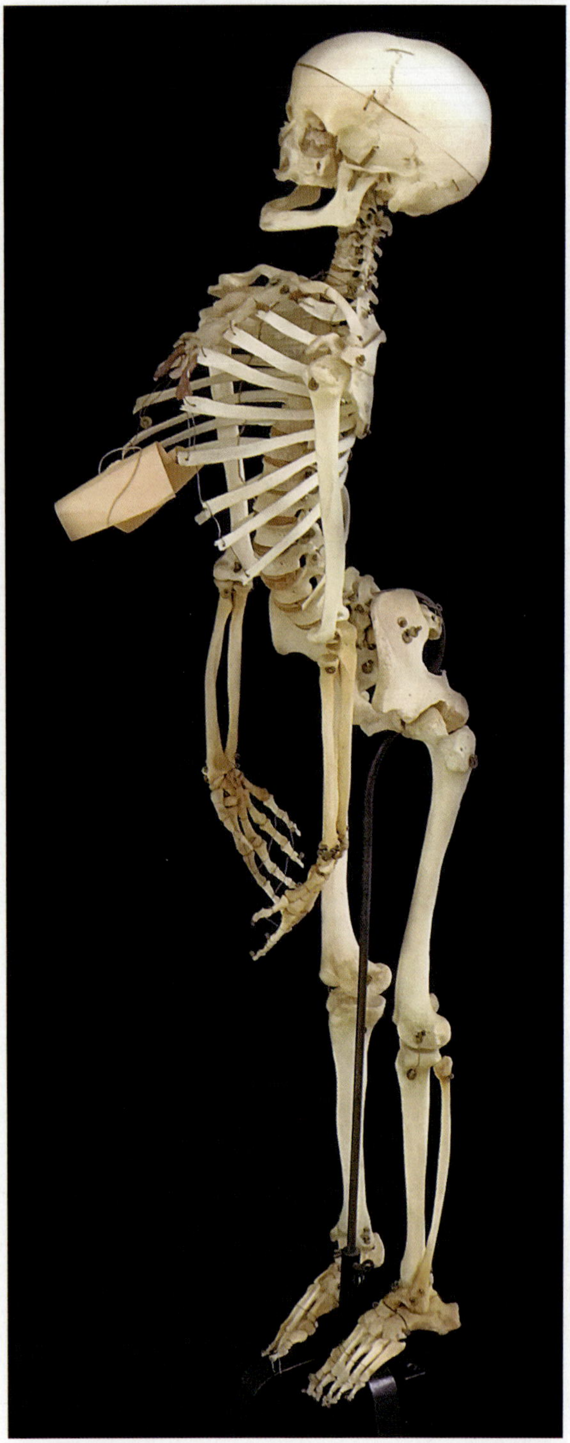

**Figure 95.** Left oblique view of a male with lumbar lordosis, also known as swayback, an anterior curvature of the lower spine who died before 1921 (MN18.648). Lordosis may be congenital, but is often the result of spondylolisthesis, when a lower vertebra "slips" forward from its proper position with the vertebra below it. Other causes of lordosis include posture and rickets. See Sorensen et al. (2015) for more on lumbar lordosis.

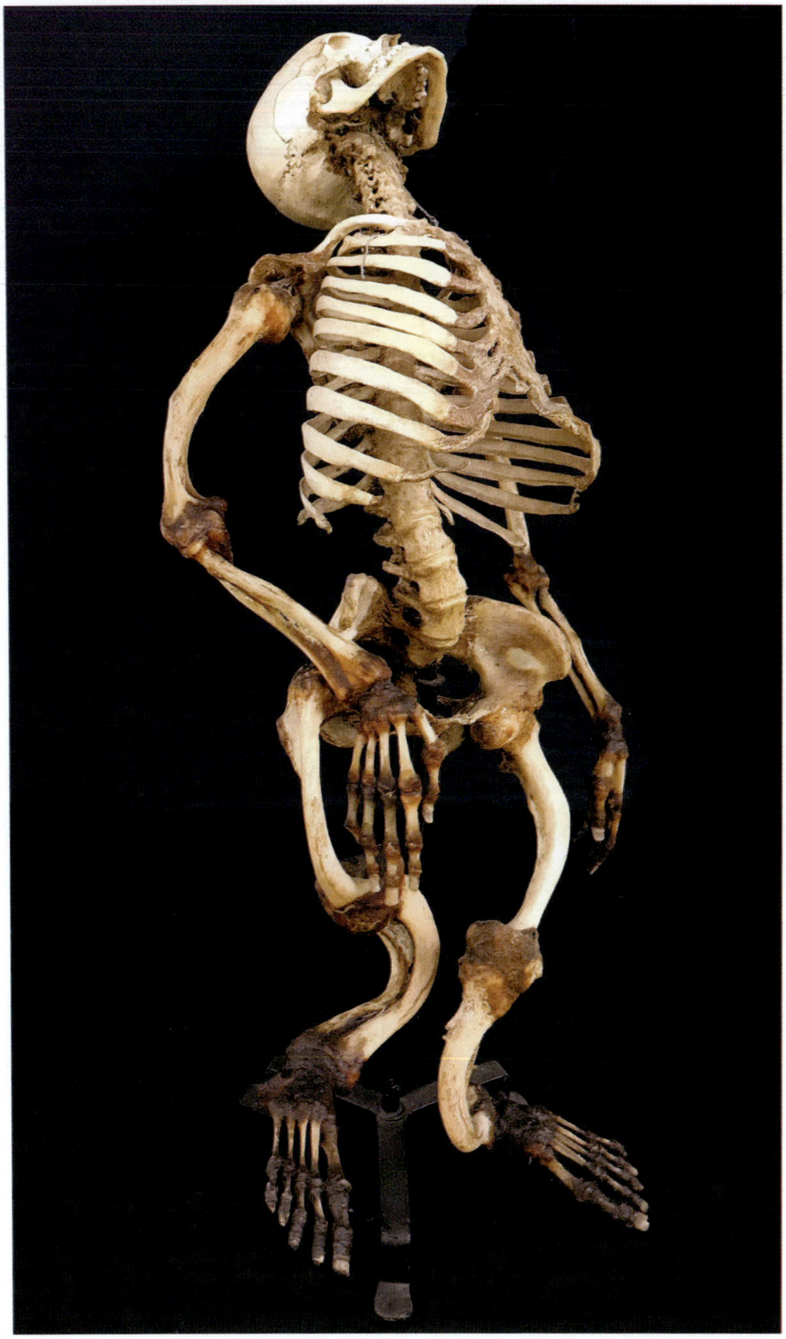

**Figure 96.** Anterior view of a 36-year-old female with childhood rickets due to a deficiency of dietary or environmental (sunlight) vitamin D that results in bones that are weak, short, and curved, and that cannot adequately support the individual's body (MN115) (Snapper 1957). This female died after a Cesarean section in 1813. Childhood rickets mainly affects the growth plates in bones due to improper or impaired mineralization of osteoid resulting in soft bones that do not harden properly. Rickets during infancy may also result in enamel hypoplasia and delayed eruption of teeth (Galhotra et al. 2015). Note the bowed bones in the arms, legs, spine and clavicles in this individual.

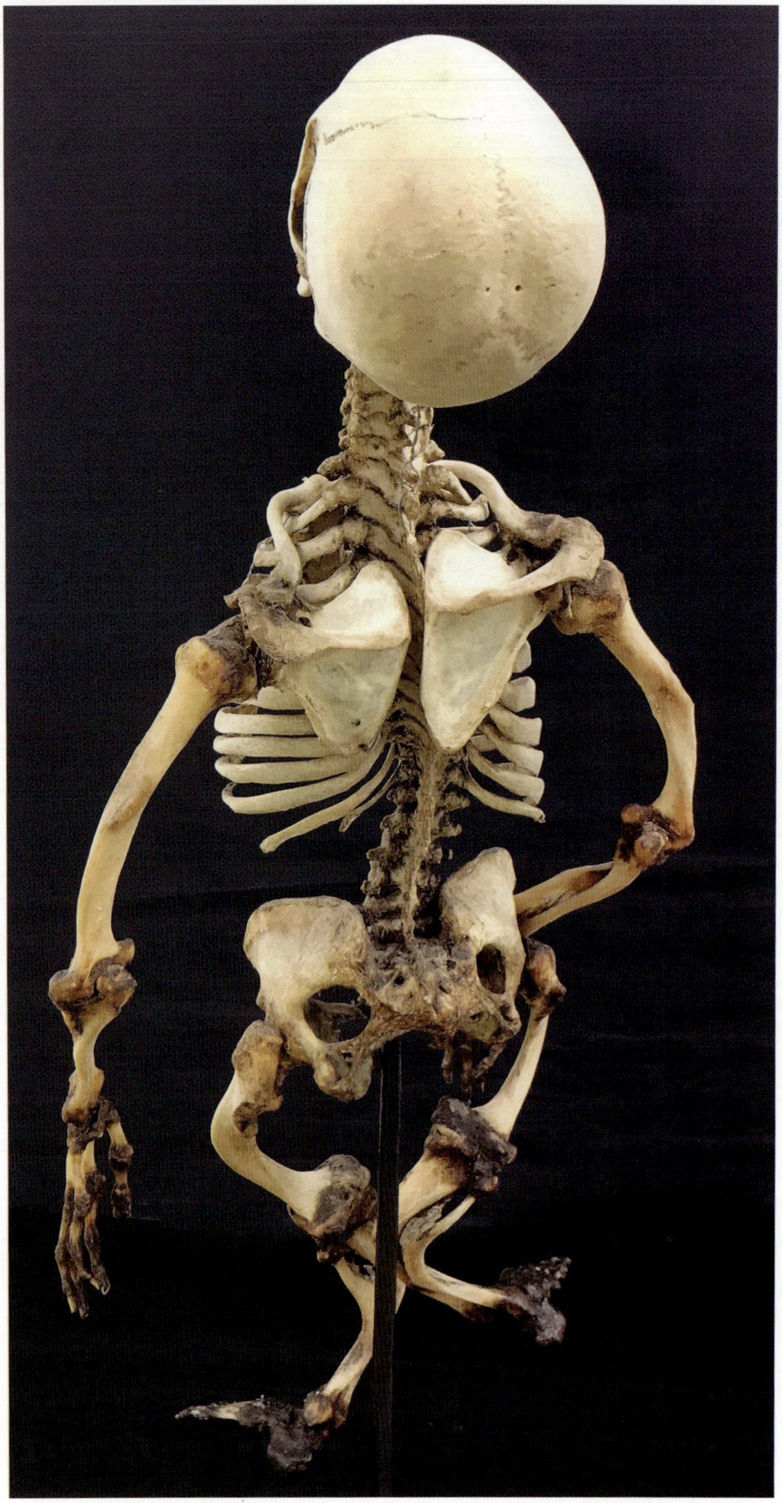

**Figure 97.** Posterior view of a 36-year-old female with childhood rickets who died after Cesarean section in 1813 (MN115). Note scoliosis of the spine and the short, bowed bones in the arms, legs, and clavicles. See Barnes (1994, 2012); Hawes and O'Brien (2006); Kim et al. (2020); Ortner and Putschar (1981); Schmorl and Junghanns (1959) for more on spinal deformities and their development.

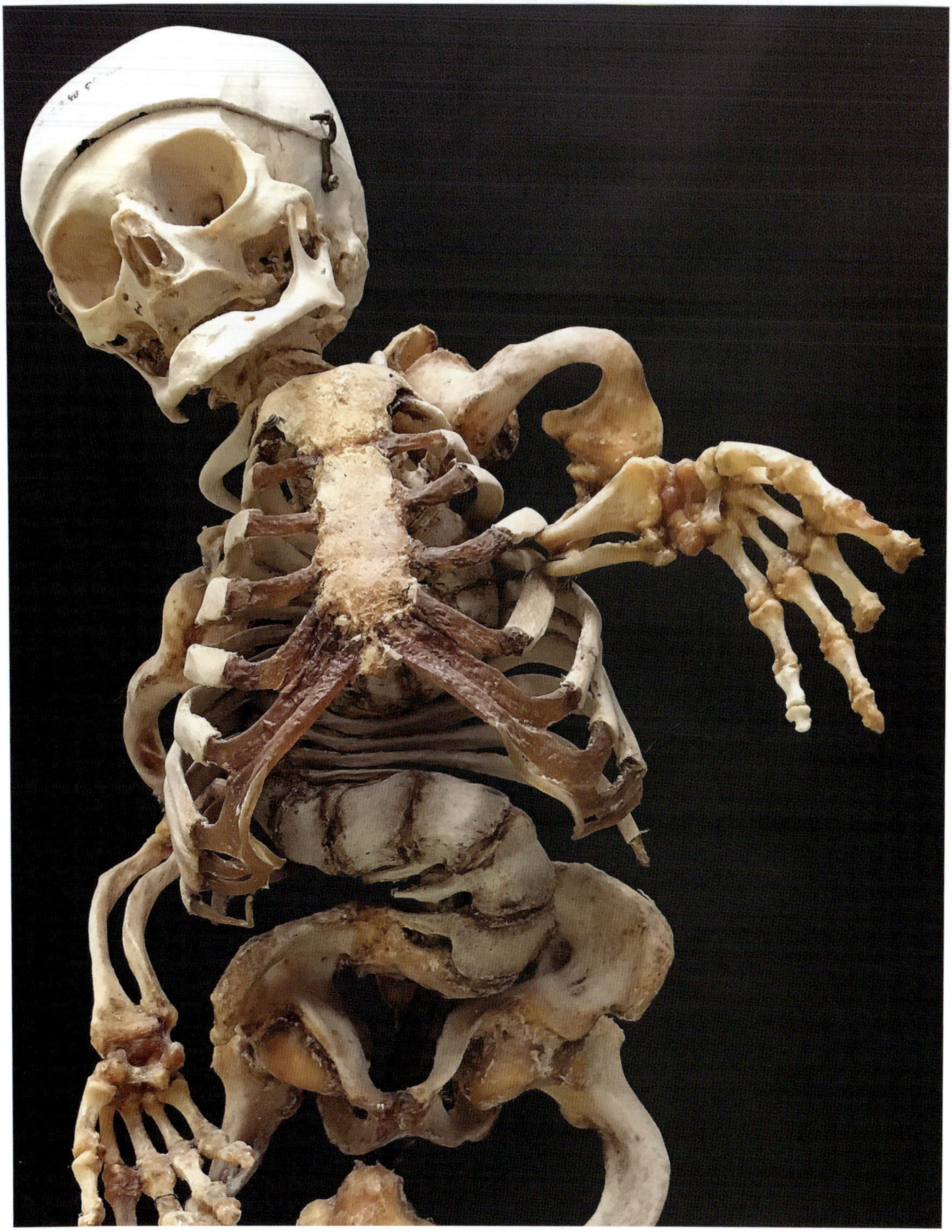

**Figure 98.** Rickets in a 61-year-old female (MN3140). Note the triangular-shaped rib cage and extreme bowing of the long bones of the arms, legs, pelvis, and lower spine (kyphoscoliosis). Some of the soft tissues are still present, especially at the joints. Interestingly, the costal cartilage in this female doesn't exhibit calcification (converting to bone), a feature typically present before age 60. See Barnes (1994, 2012); Kim et al. (2020); Ortner and Putschar (1981); Schmorl and Junghanns (1959) for more on spinal deformities and their development and classification.

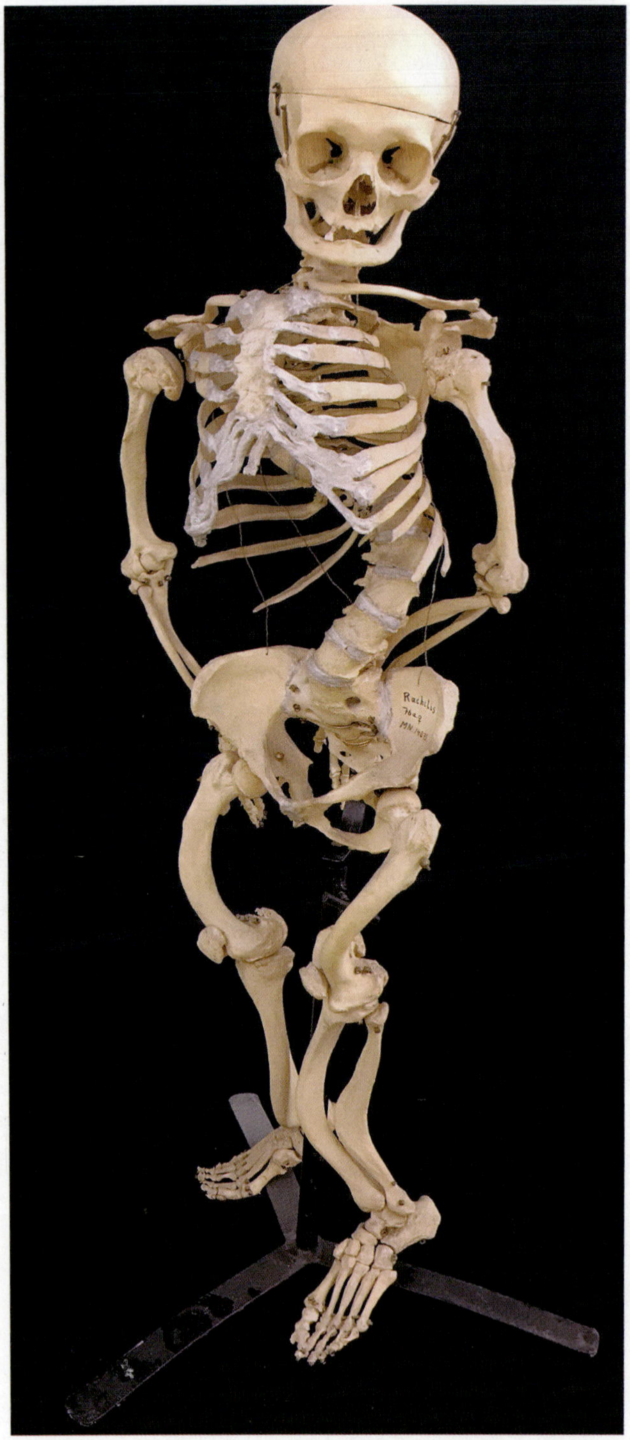

**Figure 99.** Rickets in a 76-year-old female (MN14.073). Most of the skeleton, even non-weight bearing bones are short and deformed due to prolonged vitamin D deficiency resulting in softening of the growing ends (epiphyses) of bones during childhood – long bone diaphyses /shafts become short and their ends are often enlarged. Note flattening of the left tibia and fibula. See Geshchickter and Copeland (1936); Ortner and Putschar (1981); Snapper (1957) for more on rickets and osteomalacia in adults.

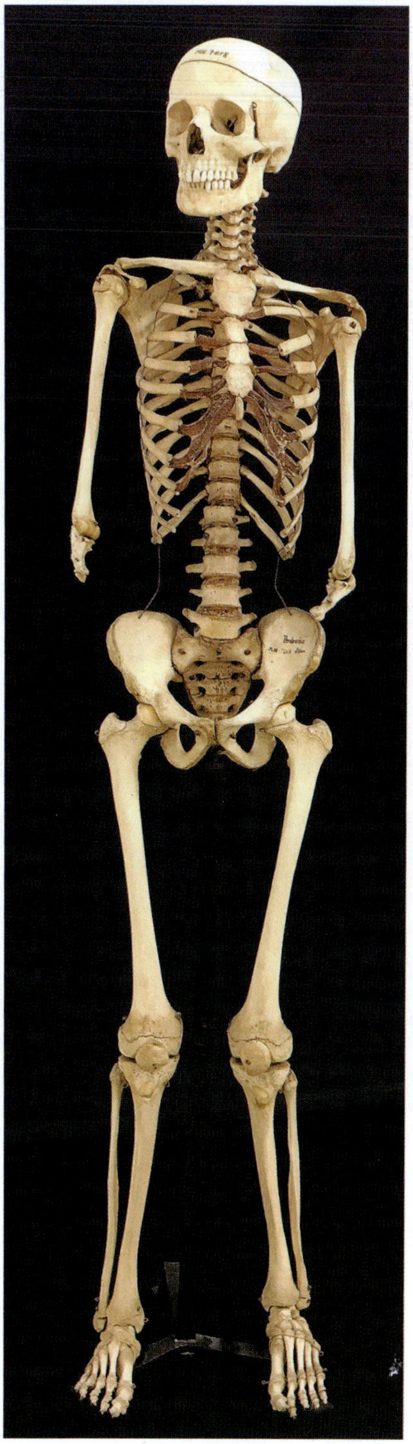

**Figure 100.** Bilateral perobrachius, an outdated term for congenital malformation of the arm or hand, in a 16-year-old male who died in 1879 (MN7078). Note the rudimentary radius and ulna, and absence of the hand, a rare condition known as transverse terminal deficiency of the upper extremity that occurs in 3 of 10,000 births (Gaebler-Spira and Lipschutz 2010). This individual exhibits incomplete hemimelia (Frantz and O'Rahilly 1961) meaning "half a limb."

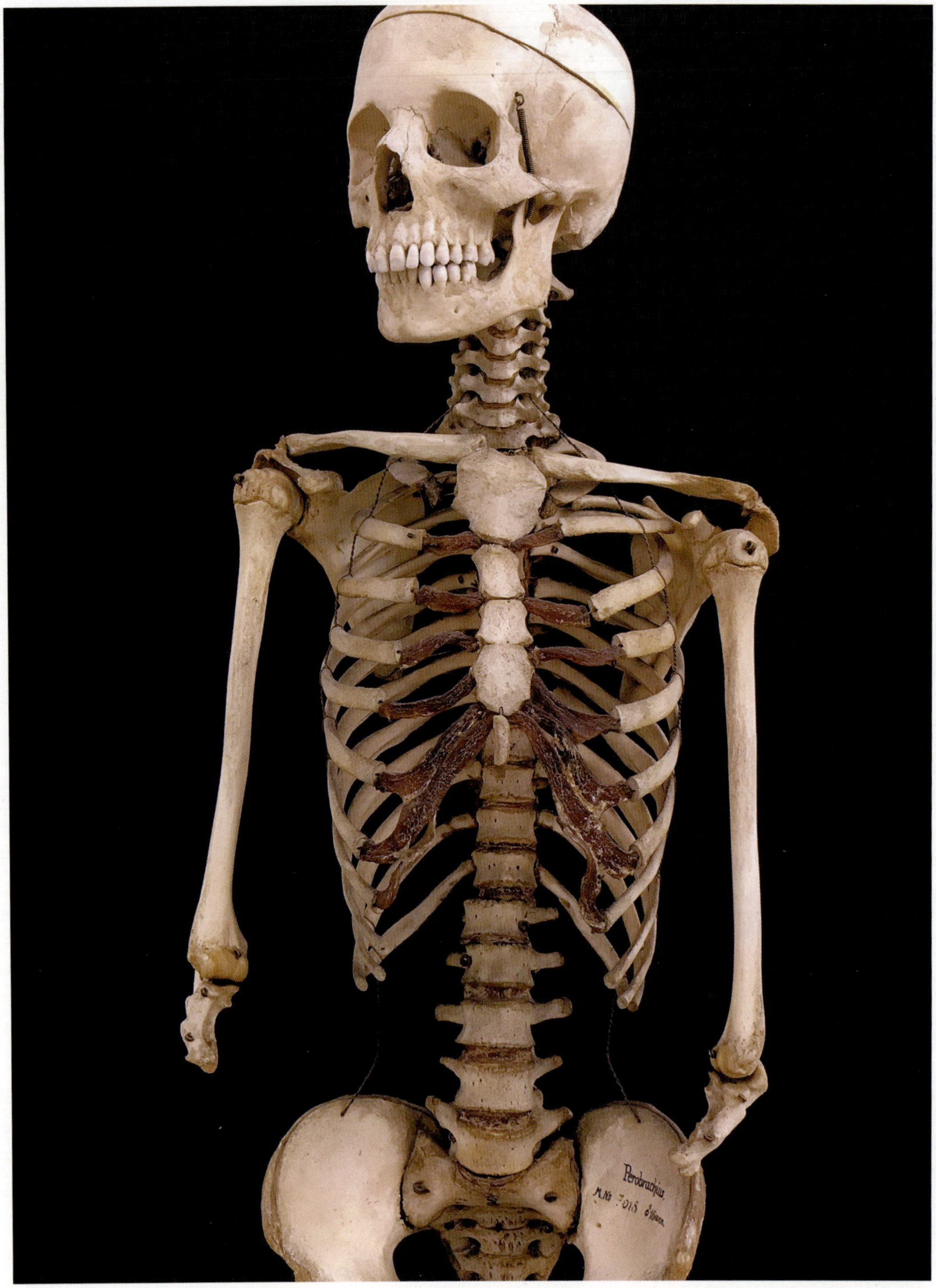


**Figure 101.** Bilateral perobrachius (an old term for congenital malformation of the arm or hand) in a 16-year-old male who died in 1879 (MN7078). This individual exhibits bilateral hemimelia of the upper limbs, a rare congenital birth defect occurring in 3 of 10,000 births (Gaebler-Spira and Lipschutz 2010). Note the many open epiphyses (growth caps) along the iliac crest, proximal humerus, and sternum indicating that he has not reached skeletal maturity (adulthood).

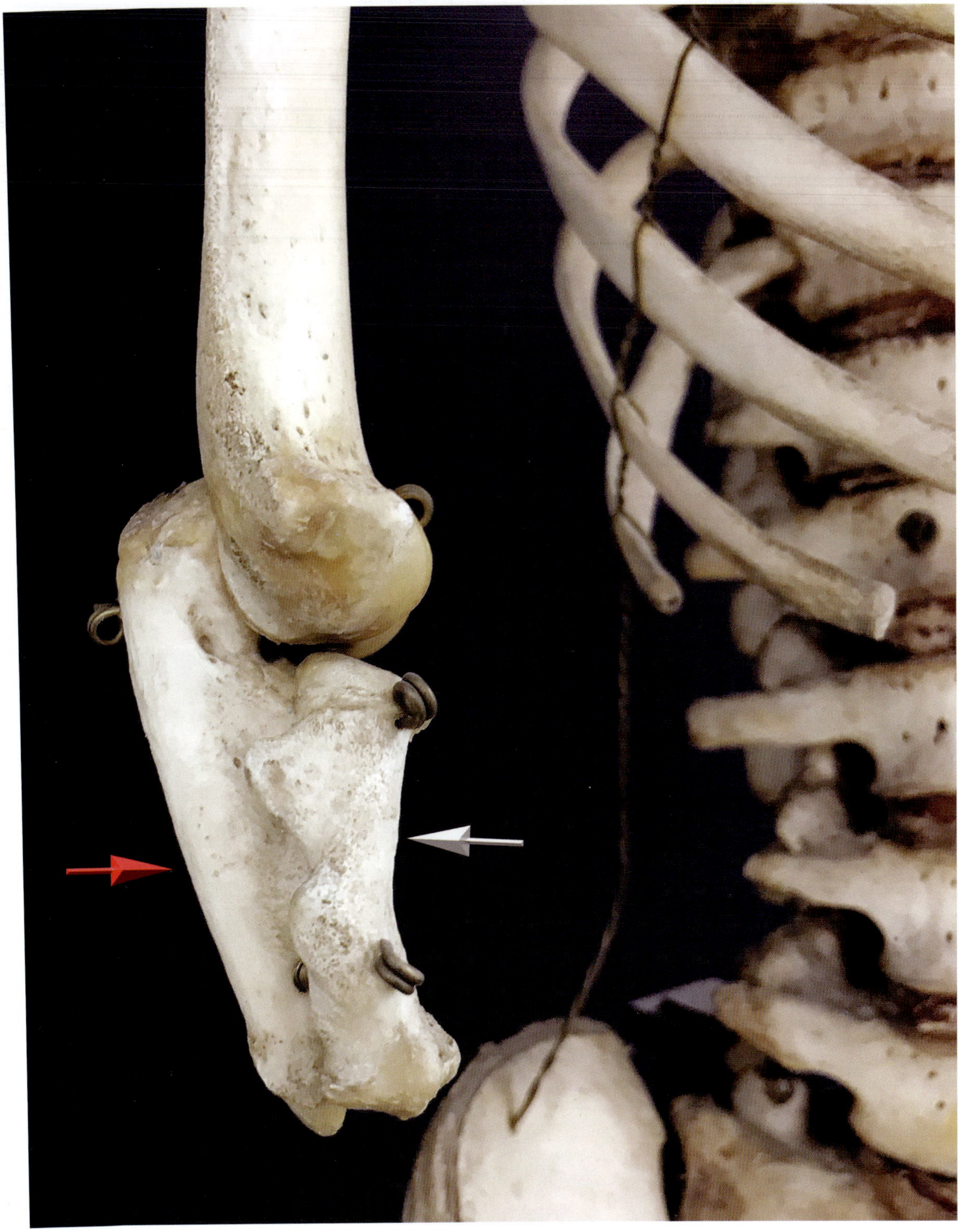

**Figure 102.** Bilateral hemimelia, "Bilateral perobrachius" on the museum card, in a 16-year-old male who died in 1879 (MN7078). Note that the rudimentary radius (white arrow) and ulna (red arrow) are represented only by their most proximal portions, of which the rest of the arm and hand did not develop. This individual exhibits bilateral hemimelia of the upper limbs, a rare congenital birth defect (Gaebler-Spira and Lipschutz 2010).

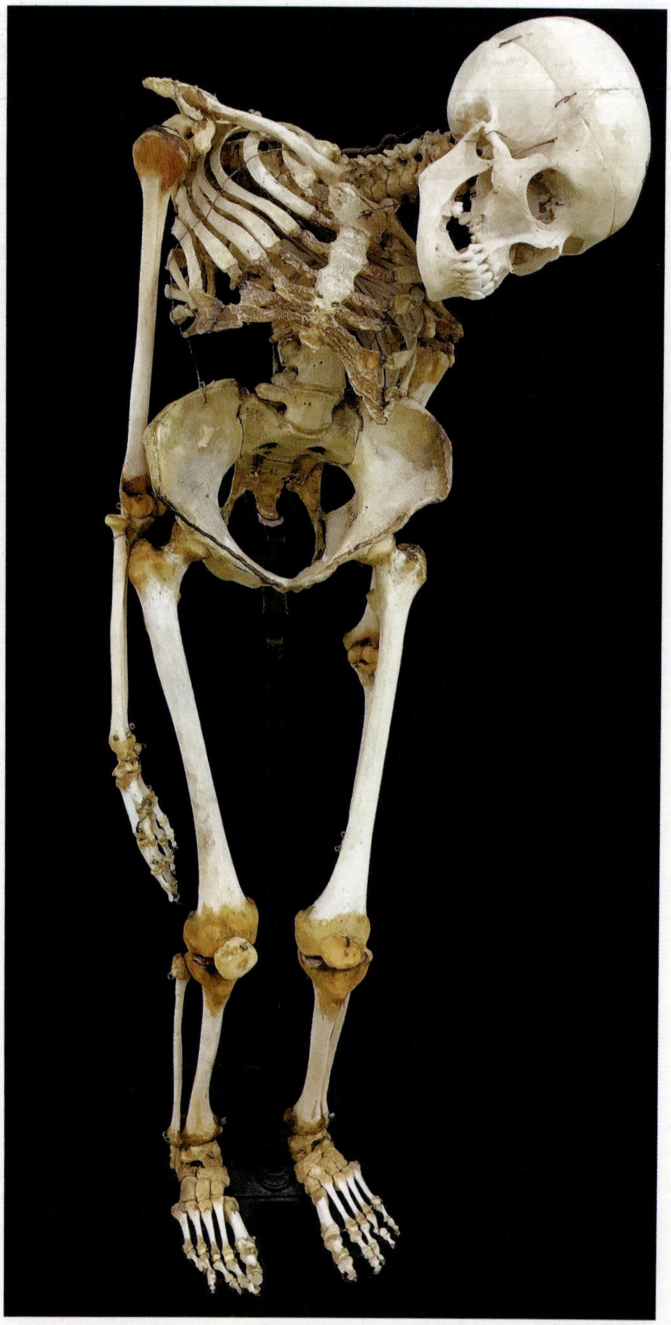

**Figure 103.** Infantile paralysis (polio, poliomyelitis) ("paralysis infantinum" on the museum card) in a 24-year-old male (MN5013). This individual exhibits kyphoscoliosis in its most severe form and atrophied arms and legs due to muscle wasting. In its most severe form, the highly contagious polio virus/infection can invade the brain and spine resulting in nerve injury, paralysis, and death. Paralysis of the legs by the poliovirus often results in atrophied femurs with very slender shafts and normally sized proximal and distal ends. The "iron lung," a mechanical respirator developed in 1928, was used to help patients breathe. See Ortner and Putschar (1981) for more on poliomyelitis in dry bone and Hassin et al. (1915) for a radiographic case report of poliomyelitis.

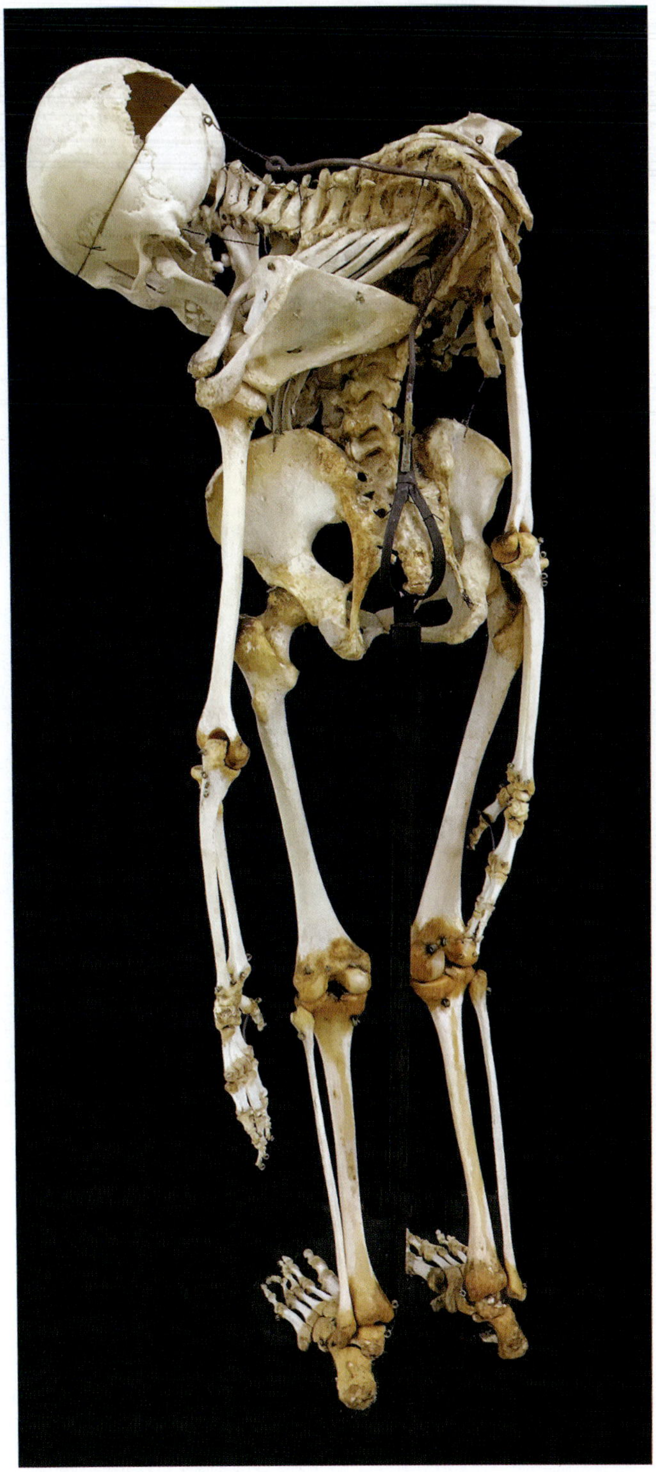

**Figure 104.** Infantile paralysis (polio, poliomyelitis) ("paralysis infantinum" on the museum card) in a 24-year-old male that resulted in severe kyphoscoliosis (MN5013). See Ortner and Putschar (1981) for more on poliomyelitis in dry bone and Hassin et al. (1915) for a radiographic case report on bone loss due to poliomyelitis.

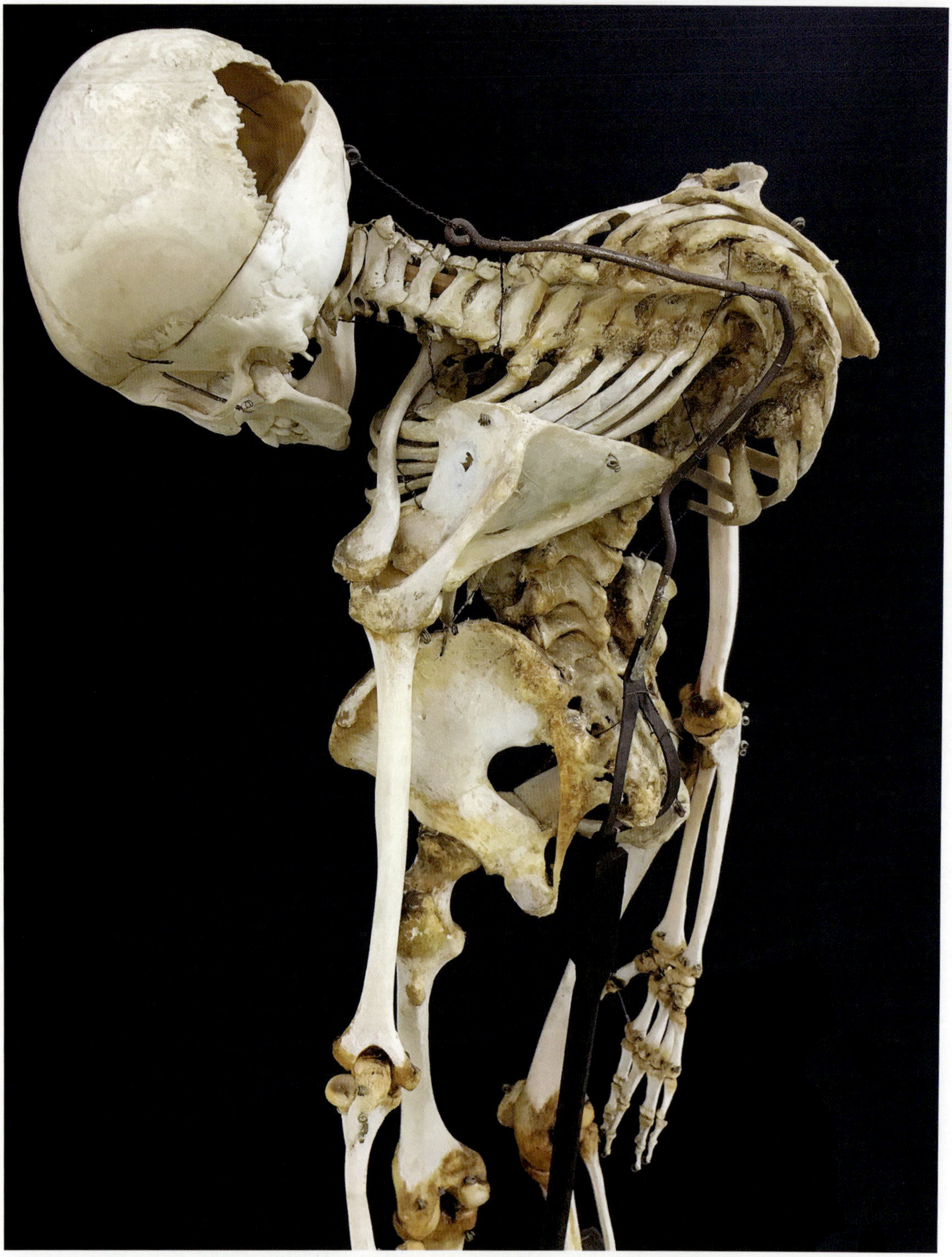

**Figure 105.** Magnified view of infantile paralysis (polio, poliomyelits) in a 24-year-old male showing slender arm bones and severe kyphoscoliosis (anterior and lateral curvature of the spine) (MN5013) (see Hassin et al. 1915; Ortner and Putschar 1981). Polio, a contagious virus, could result in progressive muscle weakness, contraction flexures of the hip and knee, paralysis, difficulty in breathing, and death. A polio vaccine developed by Jonas Salk in 1953 led to the near eradication of the disease worldwide (Rachlin et al. 2022).

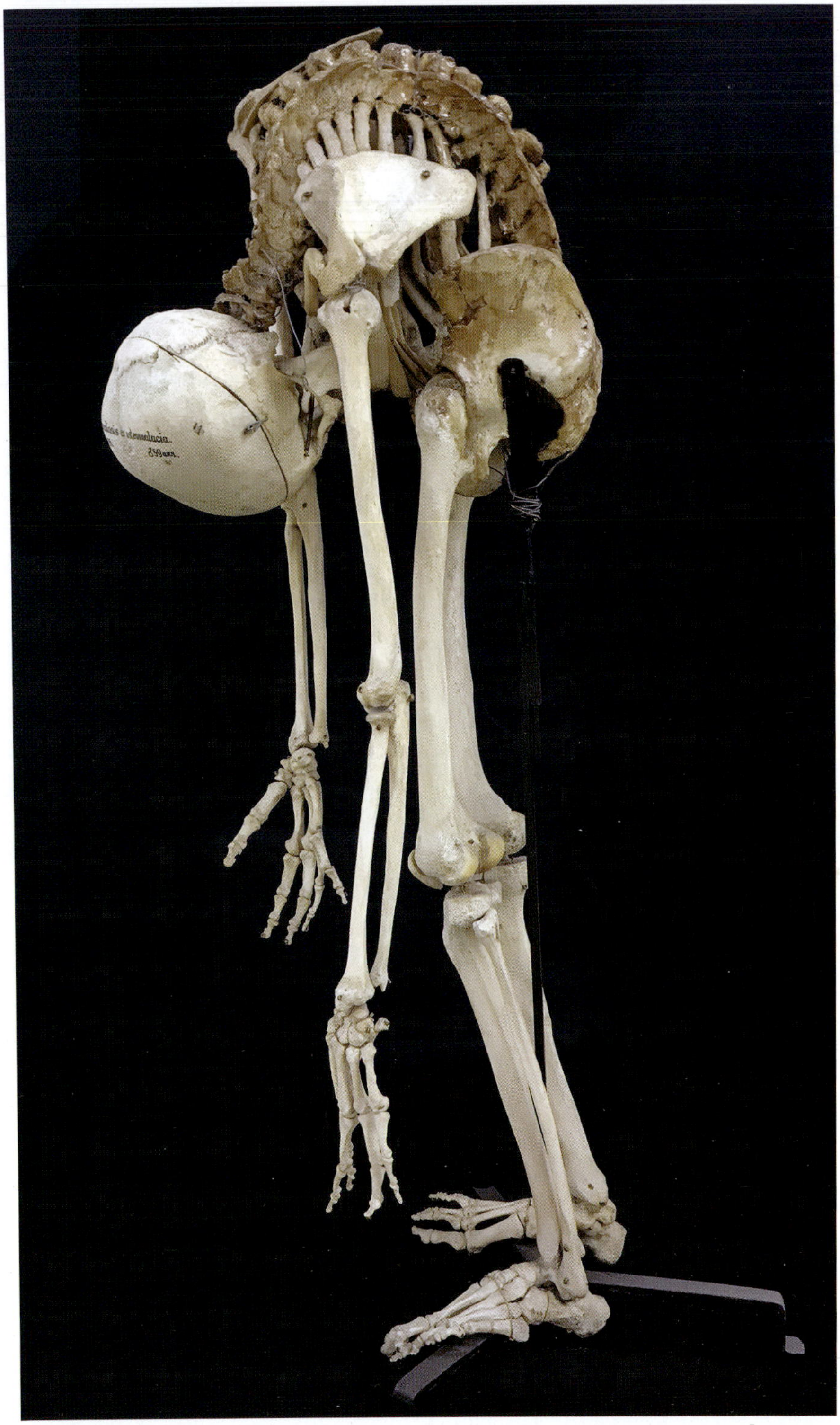

**Figure 106.** Severe kyphosis with fusion of the vertebrae and osteomalacia (adult rickets) in a 59-year-old male who died in 1887 (MN4796). Spinal deformity resulted in extreme flexion and left lateral rotation causing the individual's head to be positioned near the anterior of his pelvis. See Geshchickter and Copeland (1936) for more on osteomalacia and Kim et al. (2020) for more on the development and classification of spinal deformities.

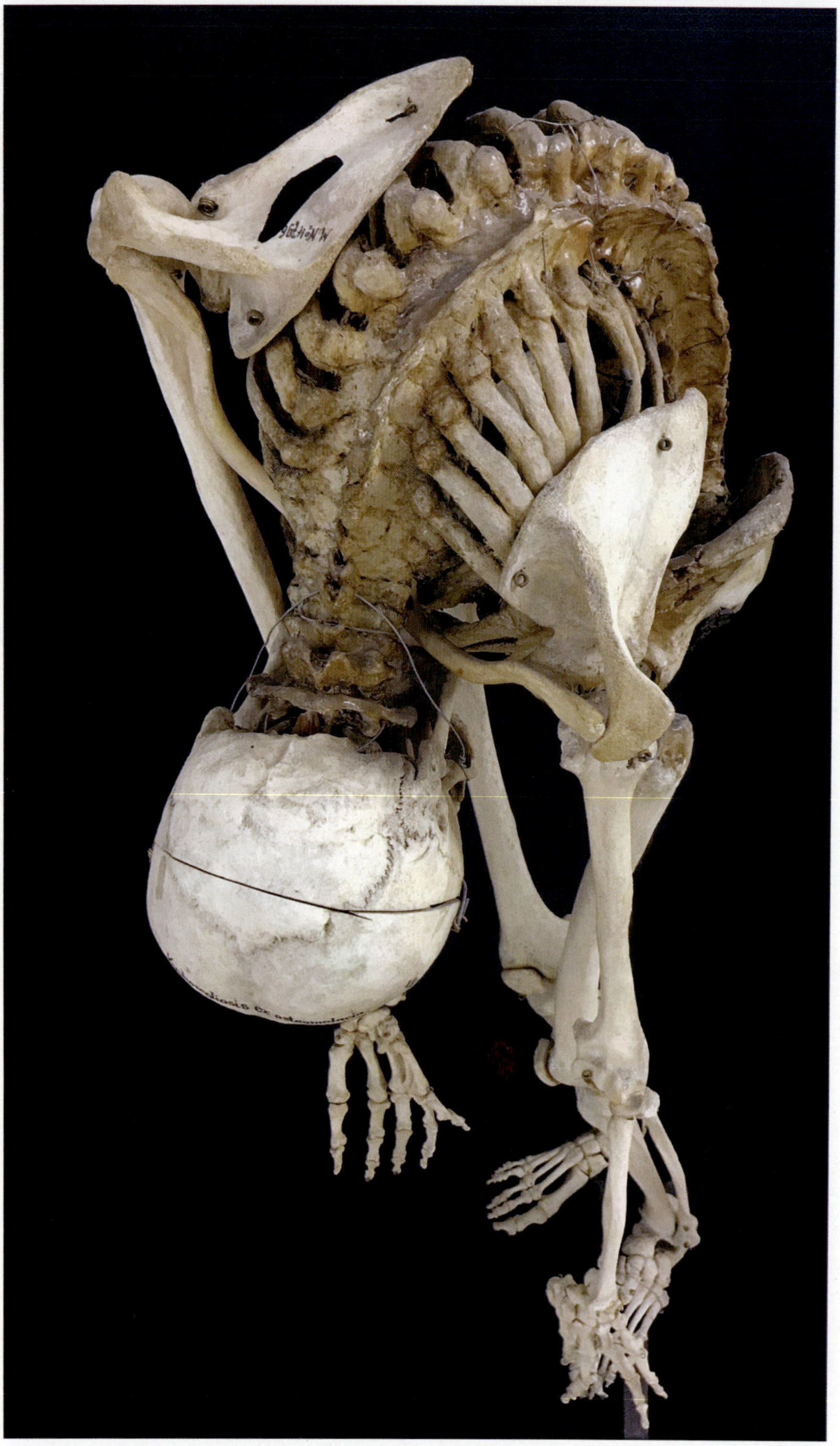

**Figure 107.** Kyphoscoliosis and fused vertebrae greatly reduced movement in the spine and osteomalacia (adult rickets) in a 59-year-old male who died in 1887 (MN4796). See Geshchickter and Copeland (1936) for more on osteomalacia and Kim et al. (2020) for more on the development and classification of spinal deformities.

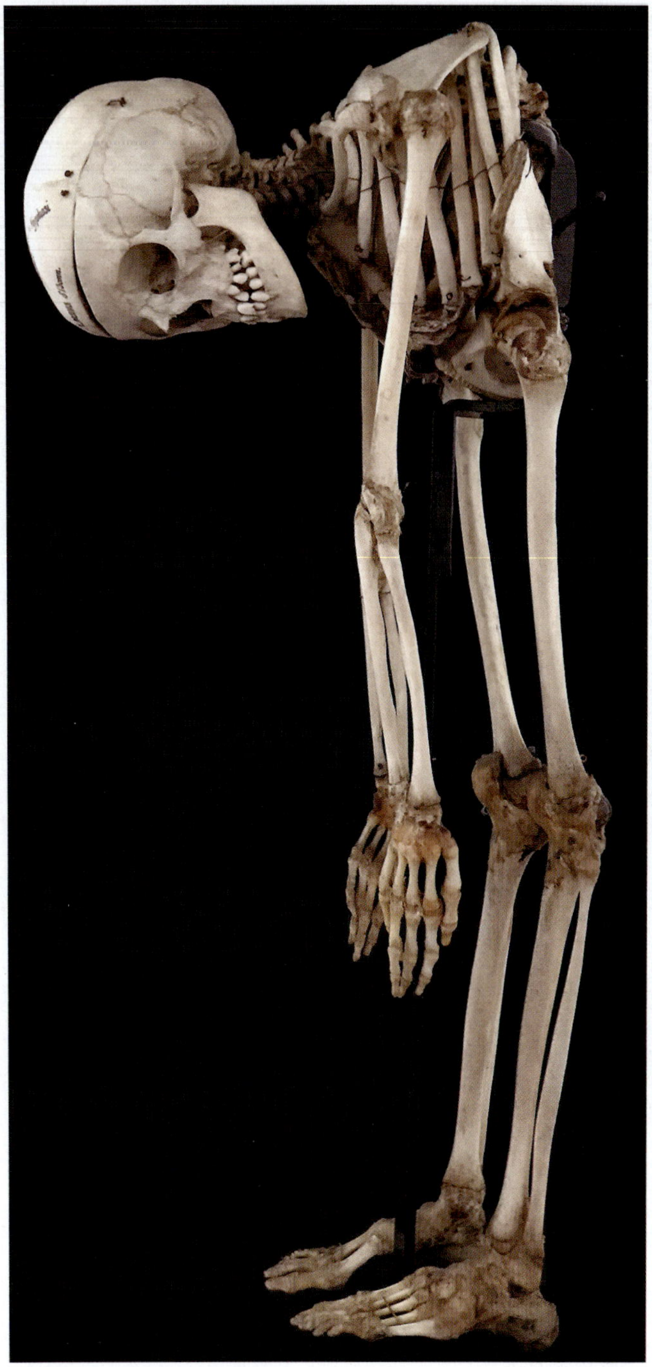

**Figure 108.** Mycobacterium tuberculosis (TB) of the spine resulting in dorsal kyphosis (gibbus formation) and "hunchback" in a 13-year-old male who died in 1895 (MN5193). Invasion of the mycobacterium tuberculosis bacillus often resulted in lytic destruction of vertebral bodies/centra (tuberculosis spondylitis) causing structural weakening and collapse of one or more vertebrae. Tuberculosis shows a predilection for the spine (Pott's disease) and large joints in the body due to the rich blood supply in the centra and growth plates (eipiphyses) of the long bones (Pigrau-Serrallach and Rodríguez-Pardo 2013). See Garg and Somvanshi (2011); Palfi et al. (2015) for more on Pott's disease of the spine and Kim et al. (2020) for classification of spinal deformities.

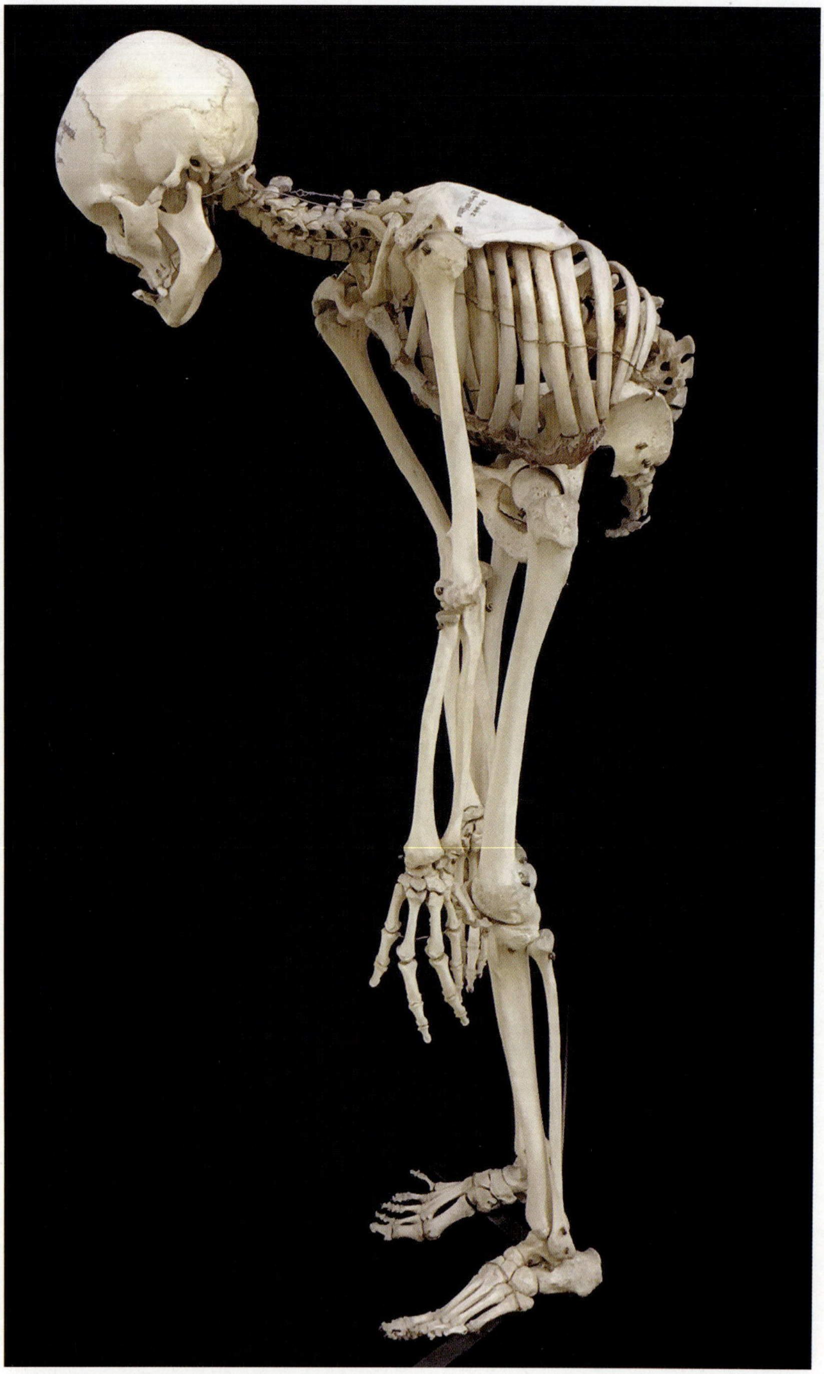

**Figure 109.** Left lateral view of the spine resulting in dorsal kyphosis in a 35-year-old female with tuberculosis (Pott's disease), also known as tuberculosis spondylitis, who died in 1895 (MN10.1278/1135). Sir Percivall Pott first described tuberculosis of the spine in 1779 and again in 1782, noting that several vertebrae could be affected resulting in carious lesions and cartilaginous destruction of centra, paralysis of the legs and spinal curvature leading to pathological deformity, some of which could be misdiagnosed and neglected (Ross 1986; Sternbock 1996). Tuberculosis often affects the lower thoracic and upper lumbar vertebrae, but may affect other parts of the skeleton including weight bearing and non-weight bearing joints.

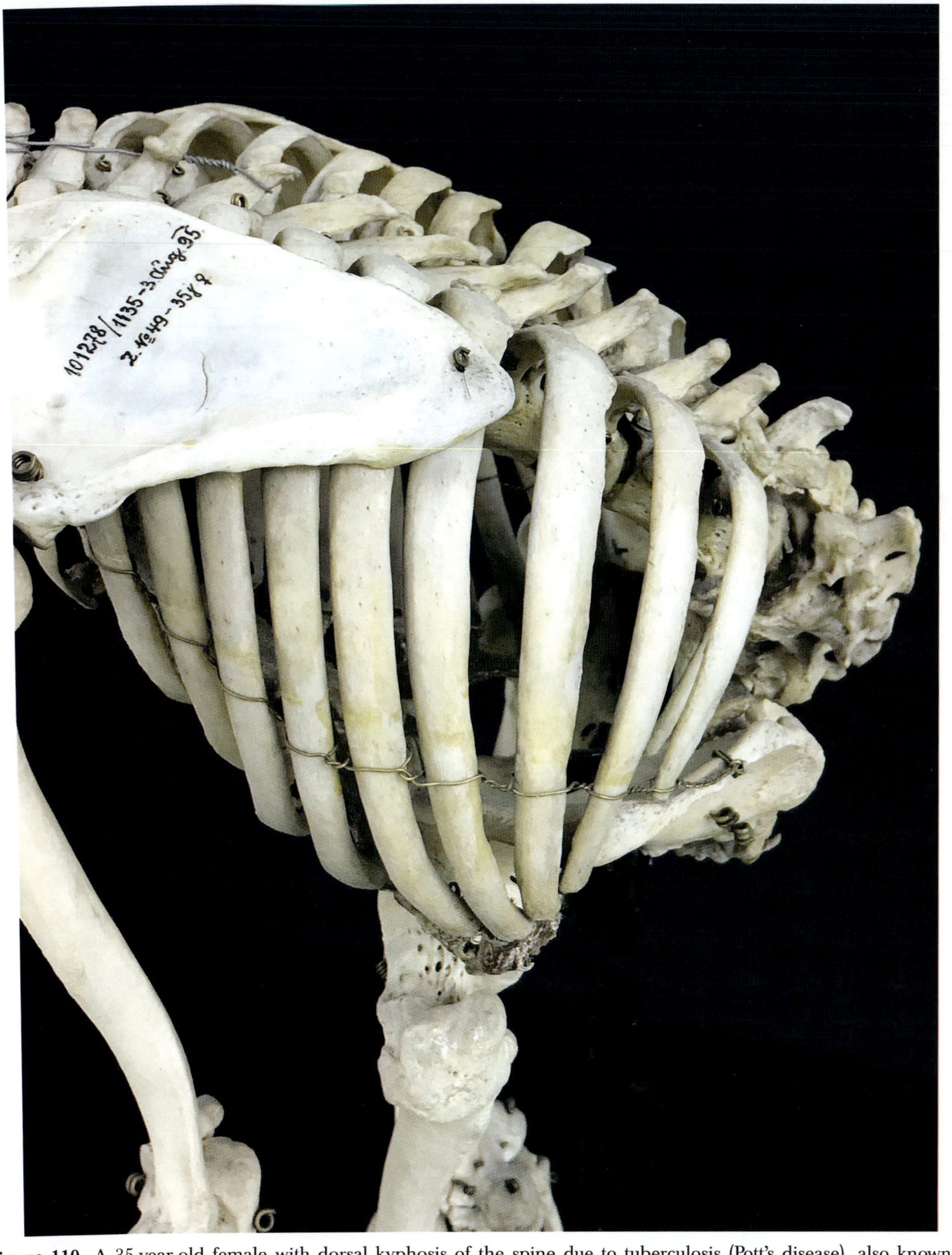


**Figure 110.** A 35-year-old female with dorsal kyphosis of the spine due to tuberculosis (Pott's disease), also known as tuberculosis spondylitis, who died in 1895 (MN10.1278/1135). See Dass et al. (2002); Palfi et al. (2015) for more on Pott's disease, its expression and presence of compression fractures in the spine.

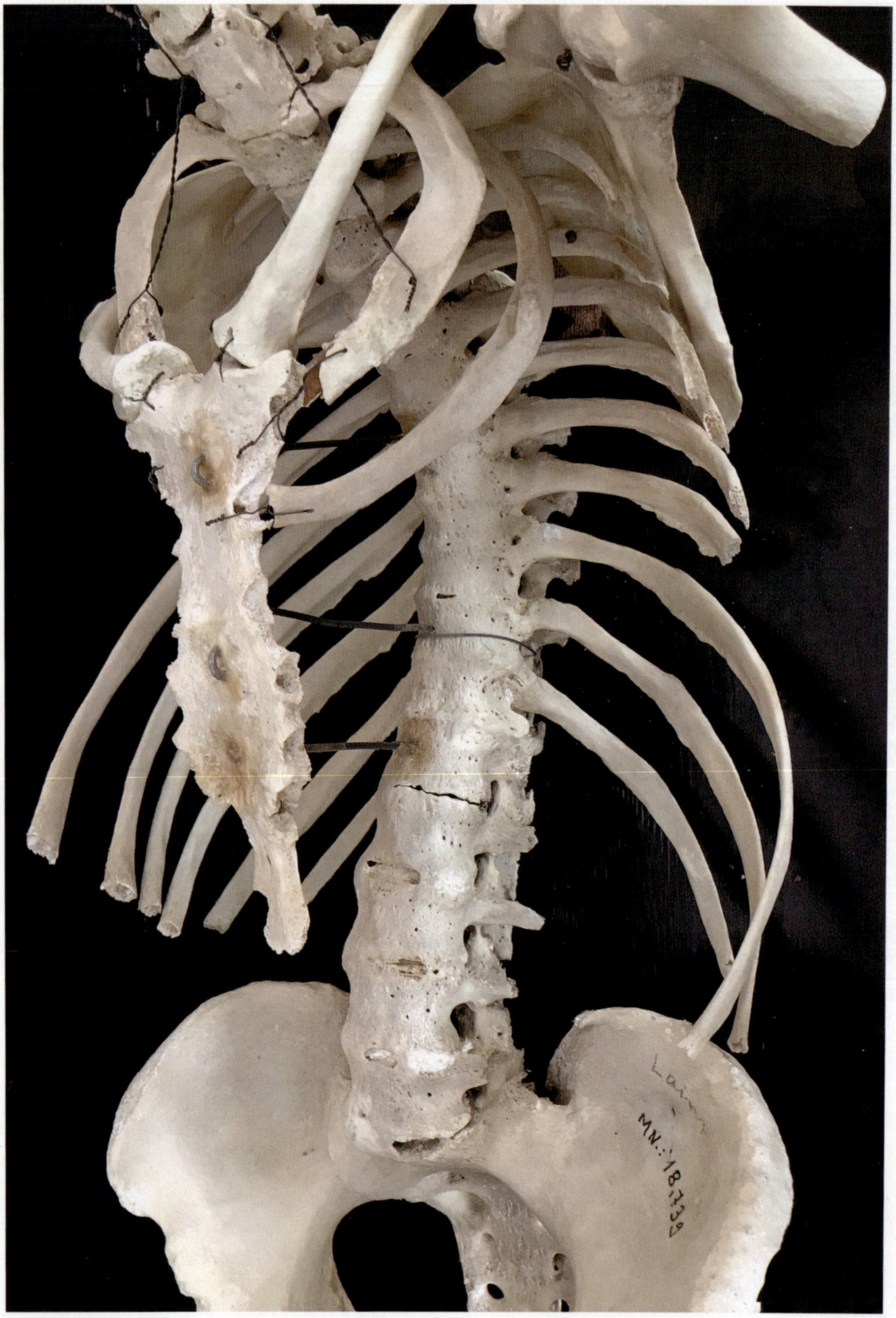

**Figure 111.** Left oblique view of ankylosing spondylitis (AS), also known as bamboo spine, Marie-Strümpell's spondylitis and Bechterew's disease in an adult female (MN18.738). Ankylosing spondylitis is a progressive inflammatory disease characterized by ossification of the anterior longitudinal ligament and fusion of the vertebral bodies, sacroiliac joints, and ribs to the spine. Fusion of the spine and hips results in anterior curvature and limited motion of the spine (Aufterheide and Rodriguez-Martin 1998; Ortner 1981).

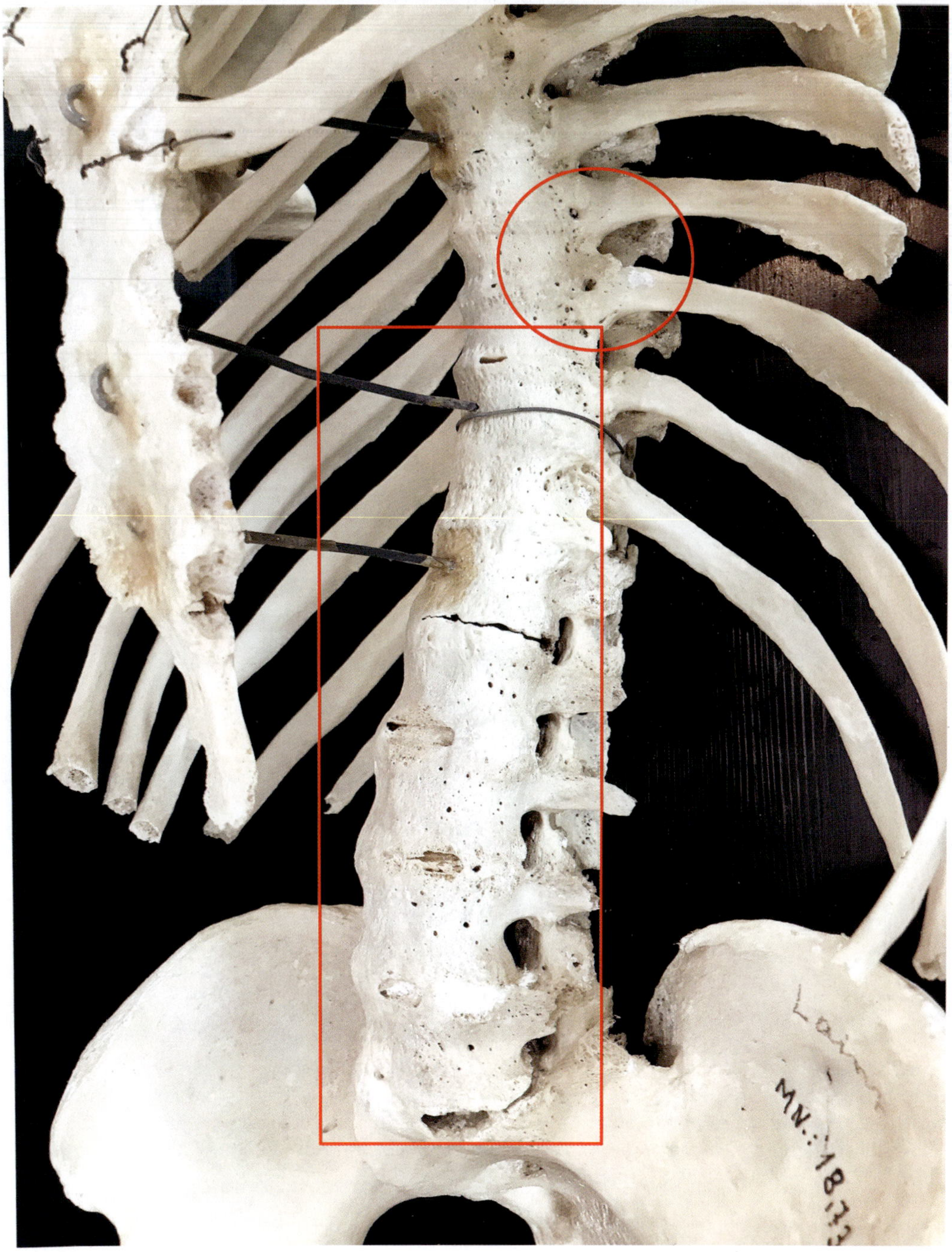


**Figure 112.** Magnified left oblique view of ankylosing spondylitis (AS), also known as bamboo spine, Marie-Strümpell's spondylitis and Bechterew's disease in an adult female (MN18.738). Ankylosing spondylitis is a progressive inflammatory disease characterized by ossification of the anterior longitudinal ligament and fusion of the vertebral bodies (rectangle), sacroiliac joints, and ribs (circle) to the spine. Fusion of the spine and hips results in anterior curvature and limited motion of the spine (Aufterheide and Rodriguez-Martin 1998; Ortner 1981).

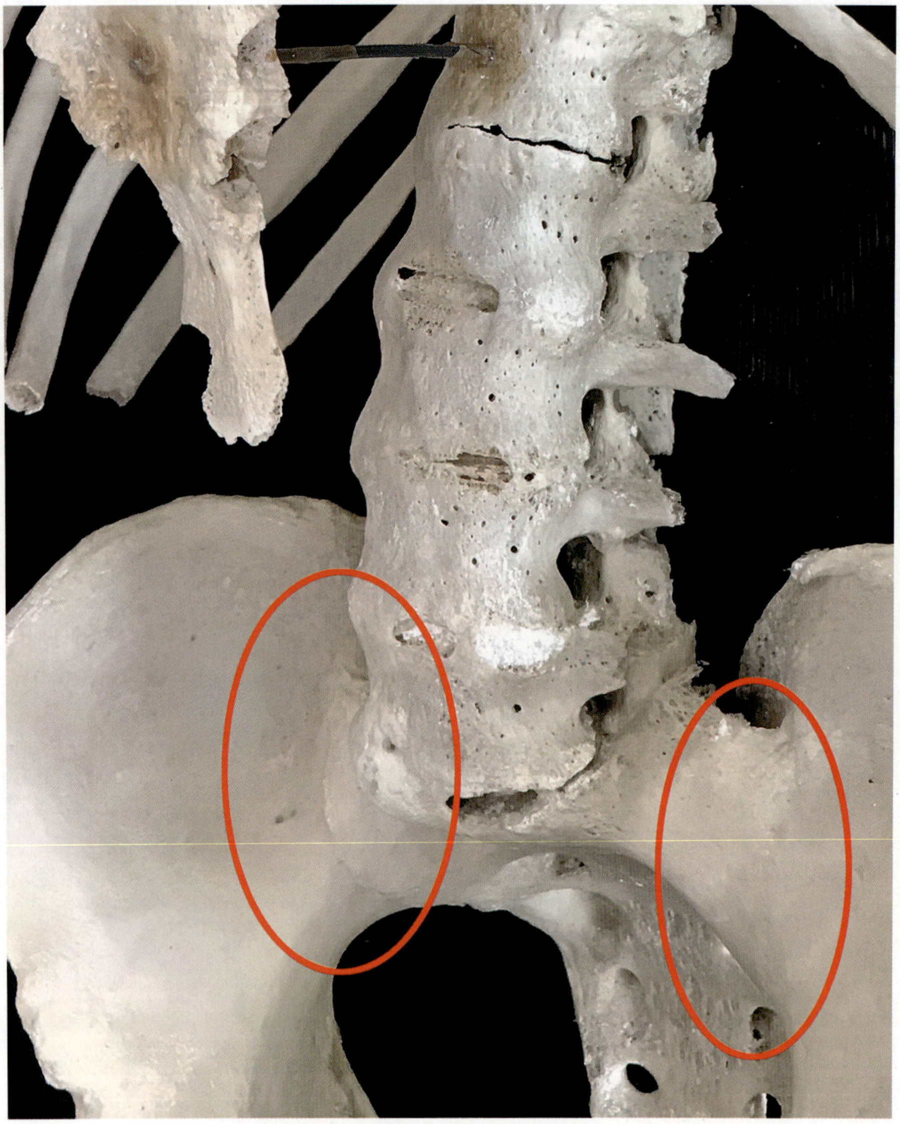

**Figure 113.** Magnified left oblique view of ankylosing spondylitis (AS), also known as bamboo spine, Marie-Strümpell's spondylitis and Bechterew's disease in an adult female (MN18.738). Ankylosing spondylitis is a progressive inflammatory disease characterized by ossification of the anterior longitudinal ligament and fusion of the vertebral bodies, sacroiliac joints (ovals), and ribs to the spine. Fusion of the spine and hips results in anterior curvature and limited motion of the spine (Aufterheide and Rodriguez-Martin 1998; Ortner 1981).

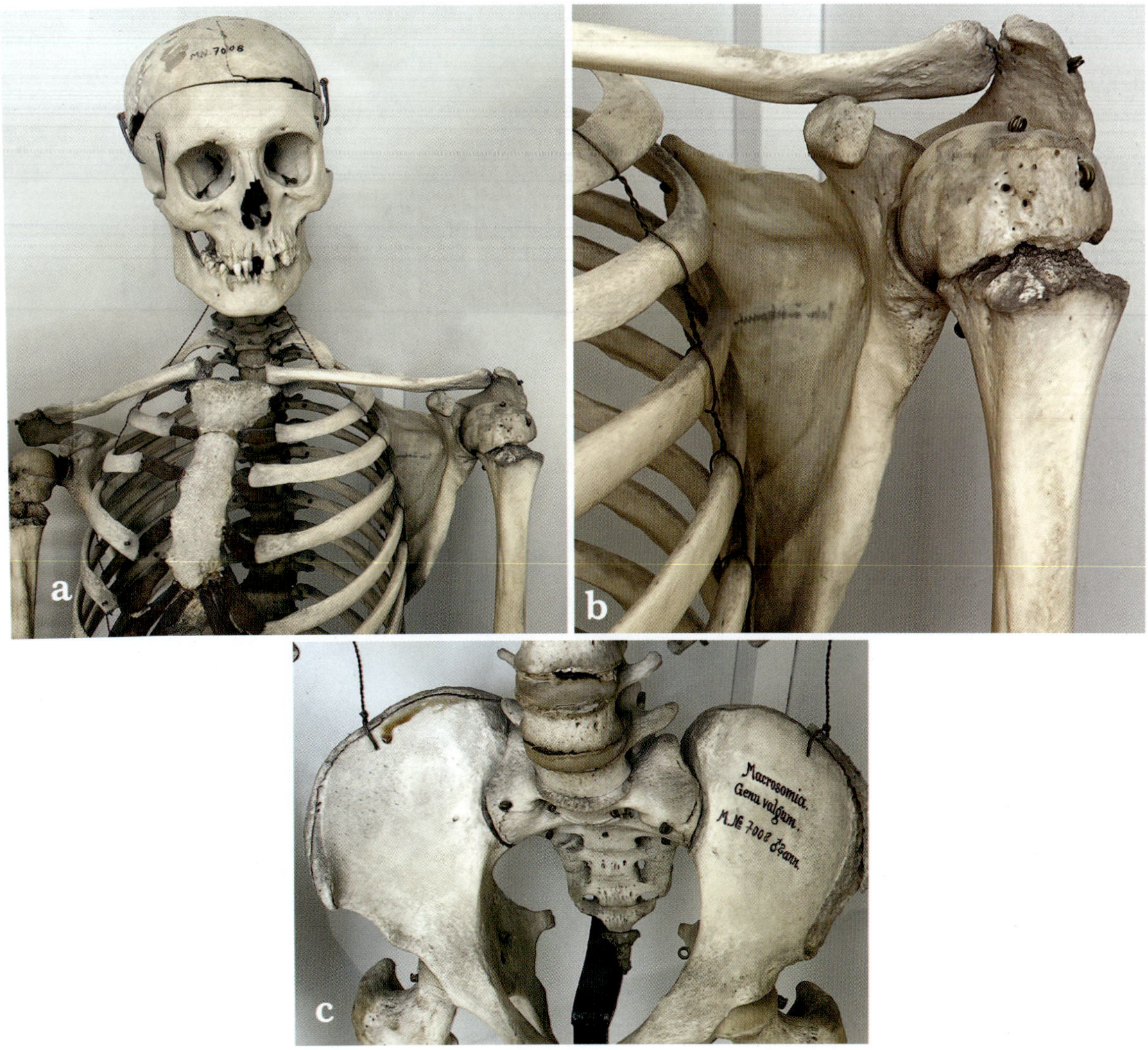


**Figure 114a-c.** Macrosomia with Genu valgum or "knock knees" in a subadult male (MN7008). (a) Although the skull in this individual is well developed, the (b) proximal humerus and (c) iliac crests have not fully fused indicating age as a subadult. Fetal macrosomia ("big body") results in an unusually large baby due to a variety of environmental and developmental variables including maternal diabetes, overdue pregnancy and maternal weight (Akanmode and Mahdy 2022; Jolly et al. 2003; Said and Manji 2016).

## Spine, Ribs and Pelvis

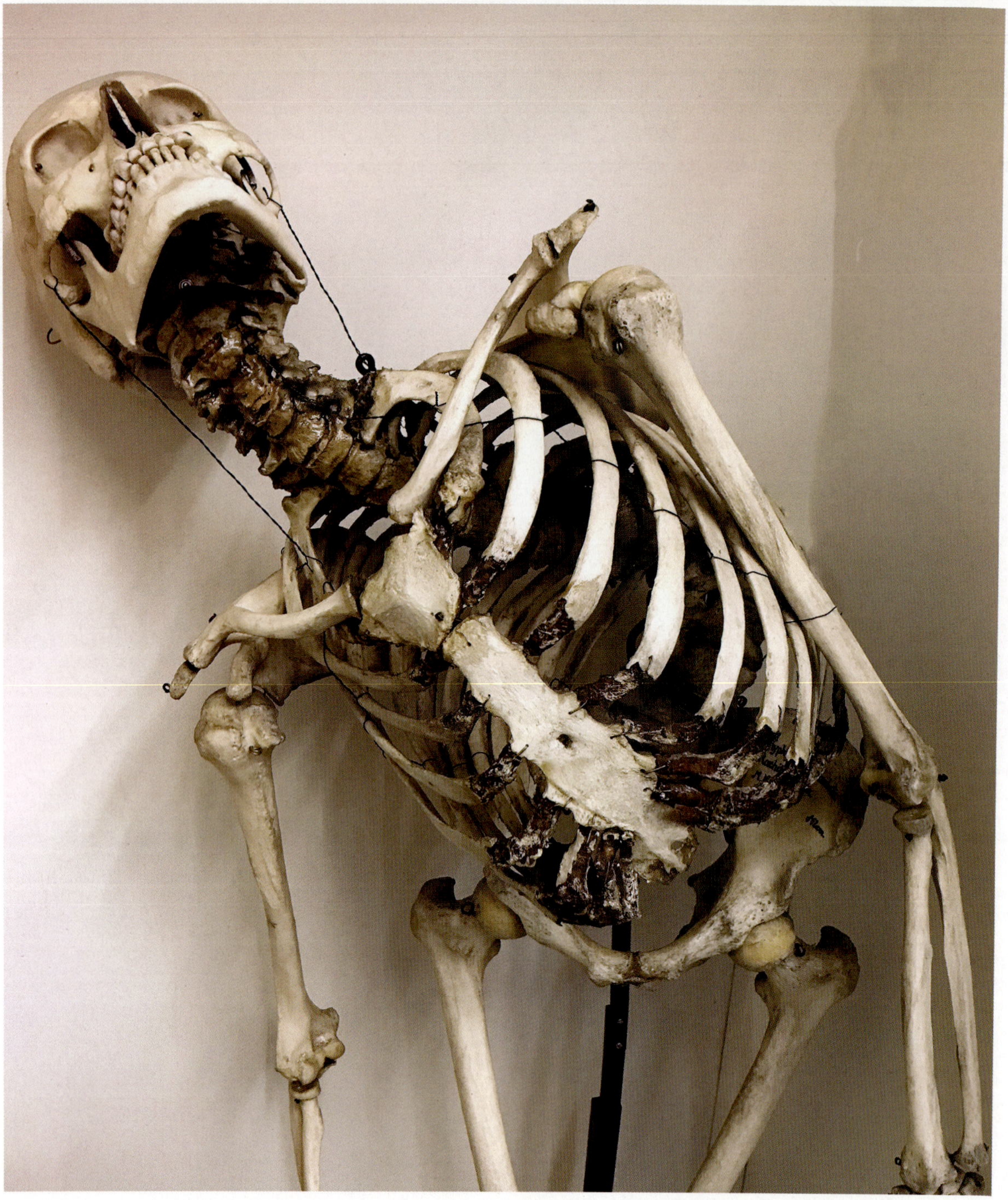

**Figure 115.** Rickets ("kyphoscoliosis rachitis peracta" on the museum card) resulting in angulation of the rib cage to the right side in a 33-year-old female (MN4388). Rickets is caused by vitamin D deficiency that results in softening and malformation of the skeleton, especially the thorax, limbs, and hips. See Ortner and Putschar (1981); Snapper (1957); Thomas and Kirsten (1985); Watson-Jones (1955). (K. Thompson, personal communication, September 27, 2022.)

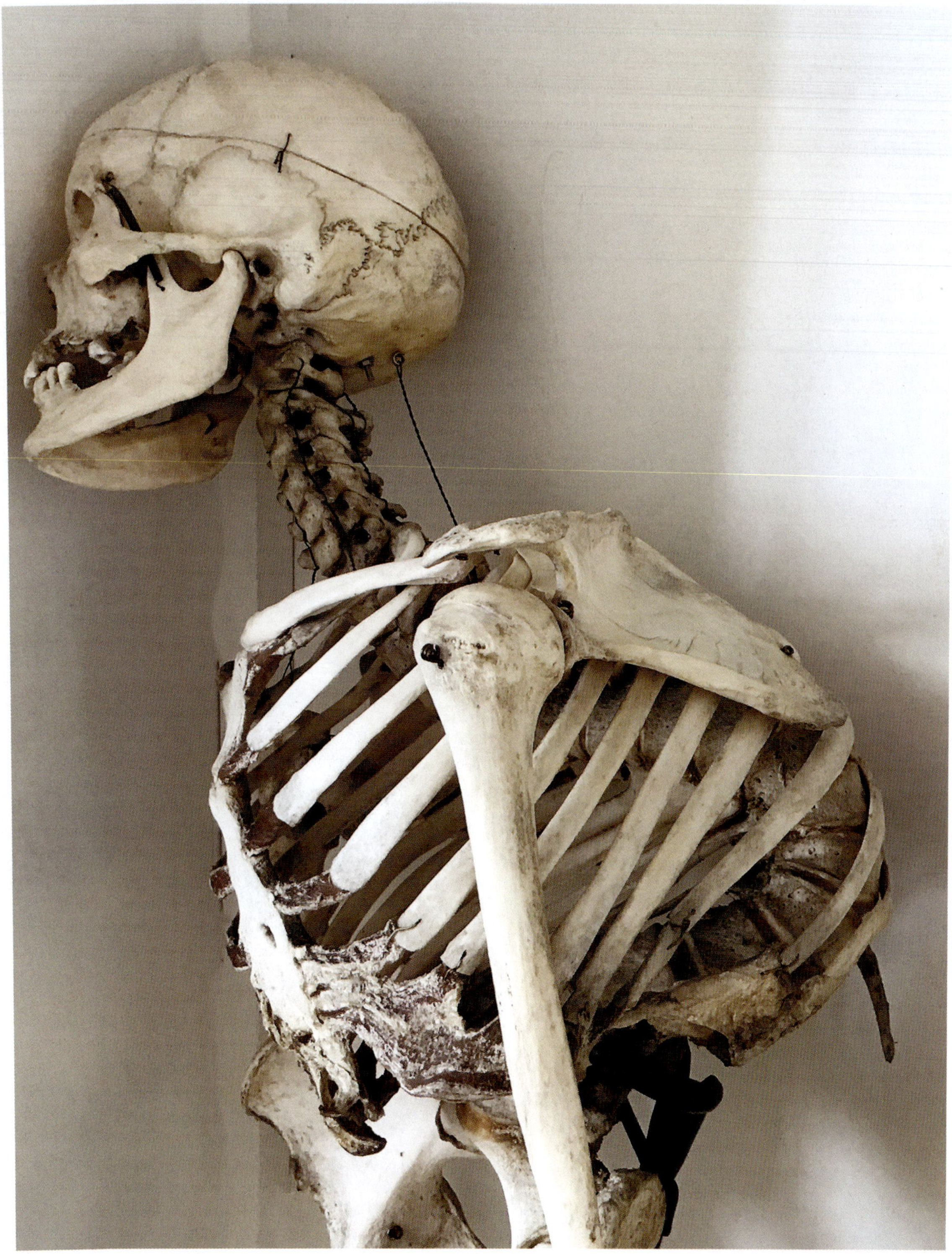

**Figure 116.** Rickets ("kyphoscoliosis rachitis peracta" on the museum card) in a 33-year-old female (MN4388). Rickets is the result of a deficiency of vitamin D resulting in softening and malformation (bowing and twisting) of the skeleton, especially the thorax, limbs, and hips. See Snapper (1957); Thomas and Kirsten (1985); Watson-Jones (1955). Note the extreme lordosis or curvature of the lower spine that alters the shape of the rib cage.

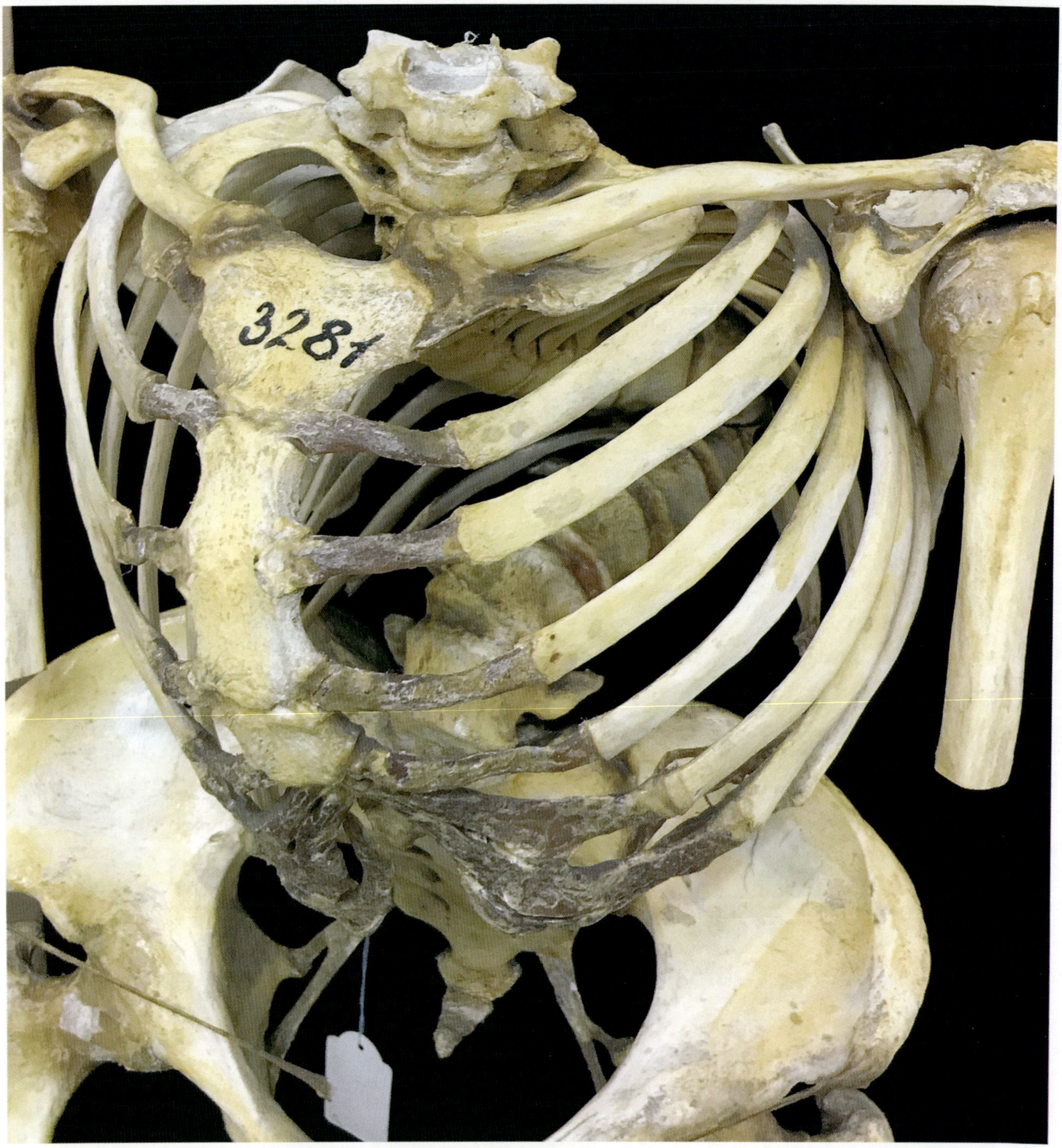


**Figure 117.** Kyphoscoliosis and spina bifida of the sacrum in a 40-year-old female (MN3281). The sternum in this individual is anteriorly curved and the lower thoracic spine exhibits severe kyphosis. See Barnes (1994, 2012); Ortner and Putschar (1981) for more on spinal deformities and their etiology.

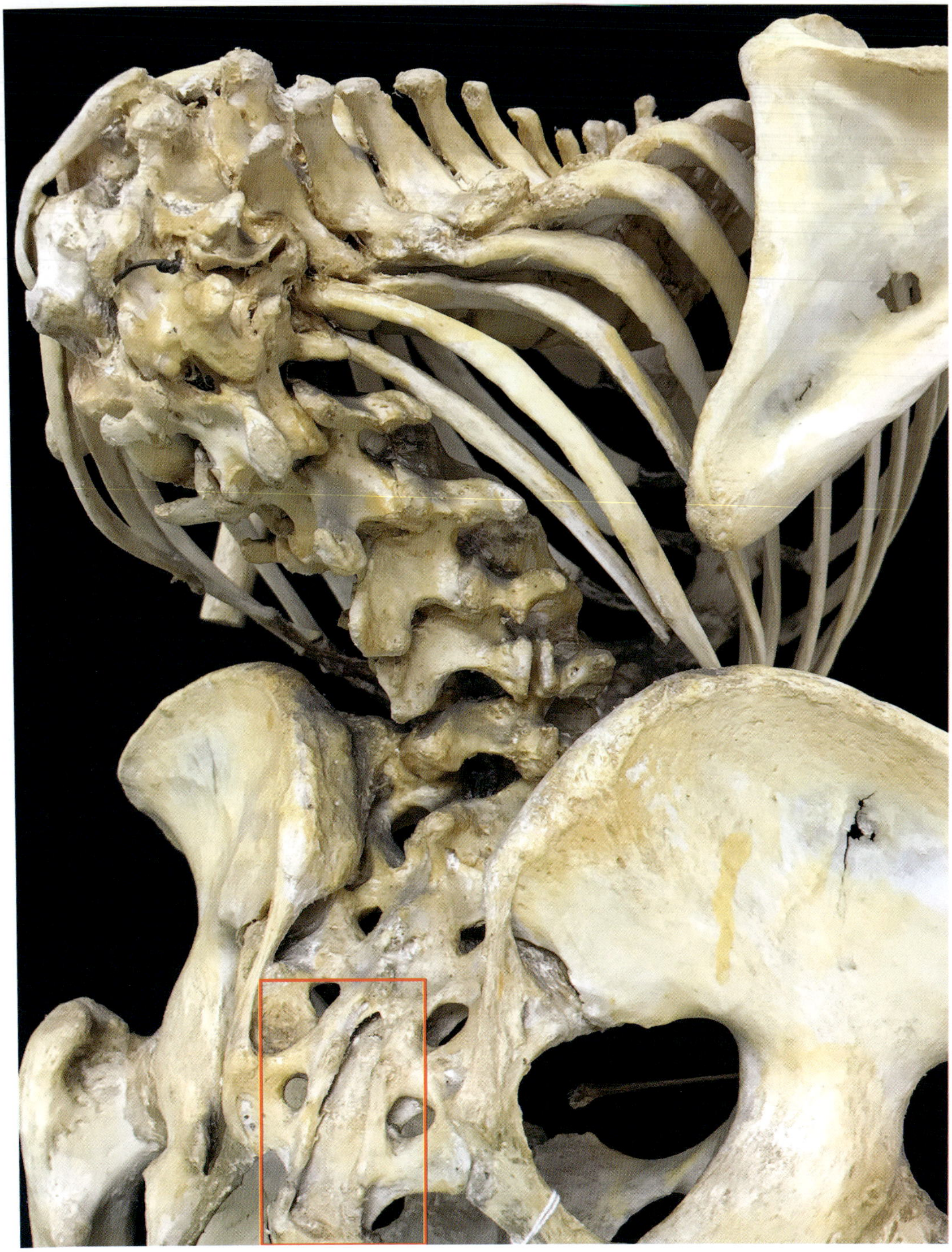

**Figure 118.** Right lateral view of kyphoscoliosis and spina bifida (rectangle) of the sacrum in a 40-year-old female (MN3281). Spina bifida is a congenital defect of the spine or sacrum that may (cystica) or may not (occulta) result in symptoms. Note the extreme kyphosis (curvature) of the lower thoracic vertebrae. Several of the ligaments joining the os coxae and sacrum in this individual have calcified. It is usually very difficult to determine if spina bifida in dry bone resulted in symptoms or not. See Barnes (1994, 2012); Ortner and Putschar (1981) for more on spinal deformities and their etiology.

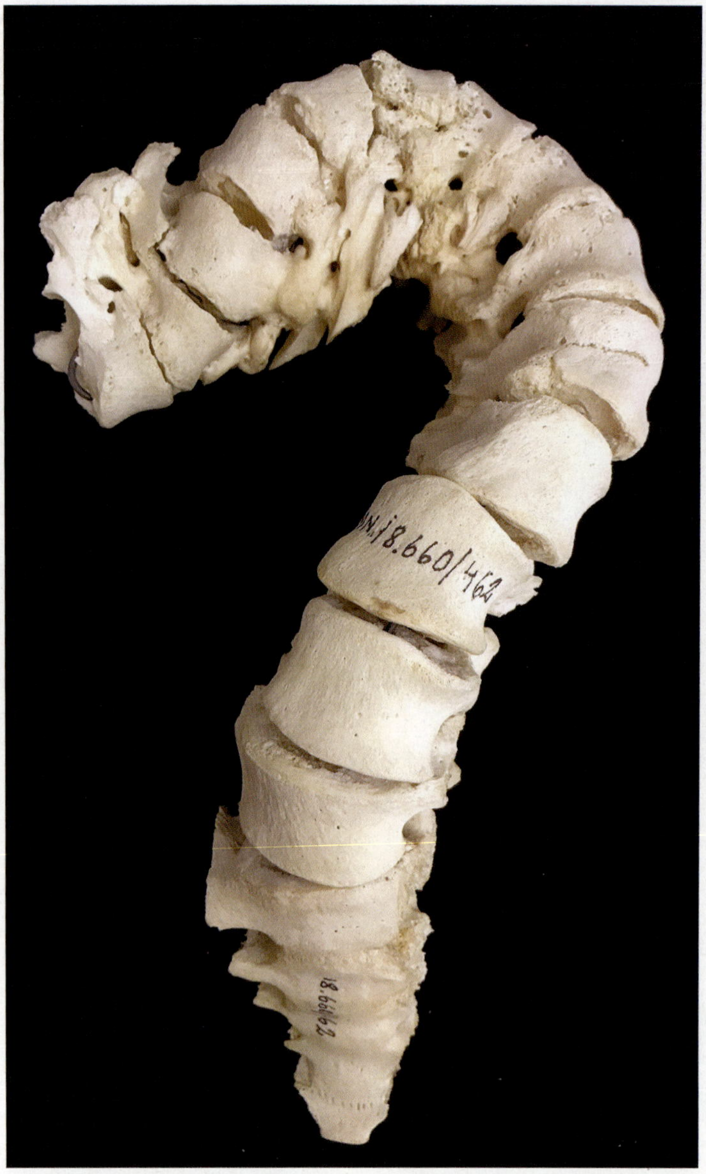

**Figure 119.** Congenital block vertebrae (T9-L2) and severe scoliosis due to improper segmentation of fetal somites in a 40-year-old female who died in 1923 (MN18.660/462). "This is definitely congenital 'scoliosis,' with the lower lumbar deformity being 'hyperlordosic' and the upper thoracic curve in the coronal plane is a severe 3-D kypho-scoliosis. She would have died young due to restrictive lung disease, or "thoracic insufficiency syndrome", because her lung volumes were not big enough to sustain adult sized life. Small lung volumes (since the thorax can't truly expand properly) means small numbers of alveoli and so insufficient lung capacity makes the heart work too hard when you grow to adult size, which then leads to development of right sided heart failure, pneumonia and an early demise" (P. Moroz, personal communication, July 30, 2022). See Barnes (2012) and Chaturvedi et al. (2018) for more on development and segmentation of the spine and *Campbell's Operative Orthopaedics* (1956) on scoliosis.

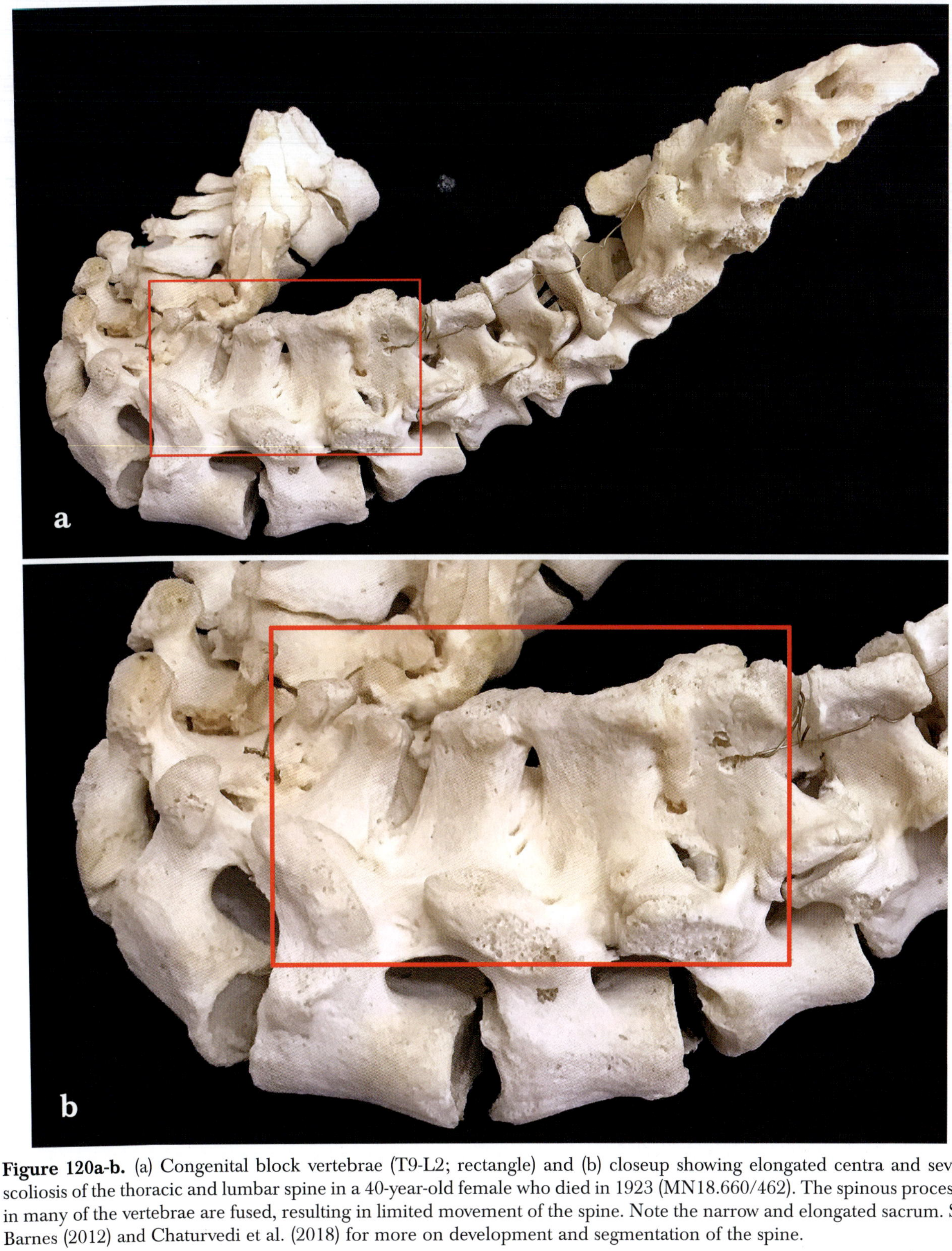

**Figure 120a-b.** (a) Congenital block vertebrae (T9-L2; rectangle) and (b) closeup showing elongated centra and severe scoliosis of the thoracic and lumbar spine in a 40-year-old female who died in 1923 (MN18.660/462). The spinous processes in many of the vertebrae are fused, resulting in limited movement of the spine. Note the narrow and elongated sacrum. See Barnes (2012) and Chaturvedi et al. (2018) for more on development and segmentation of the spine.

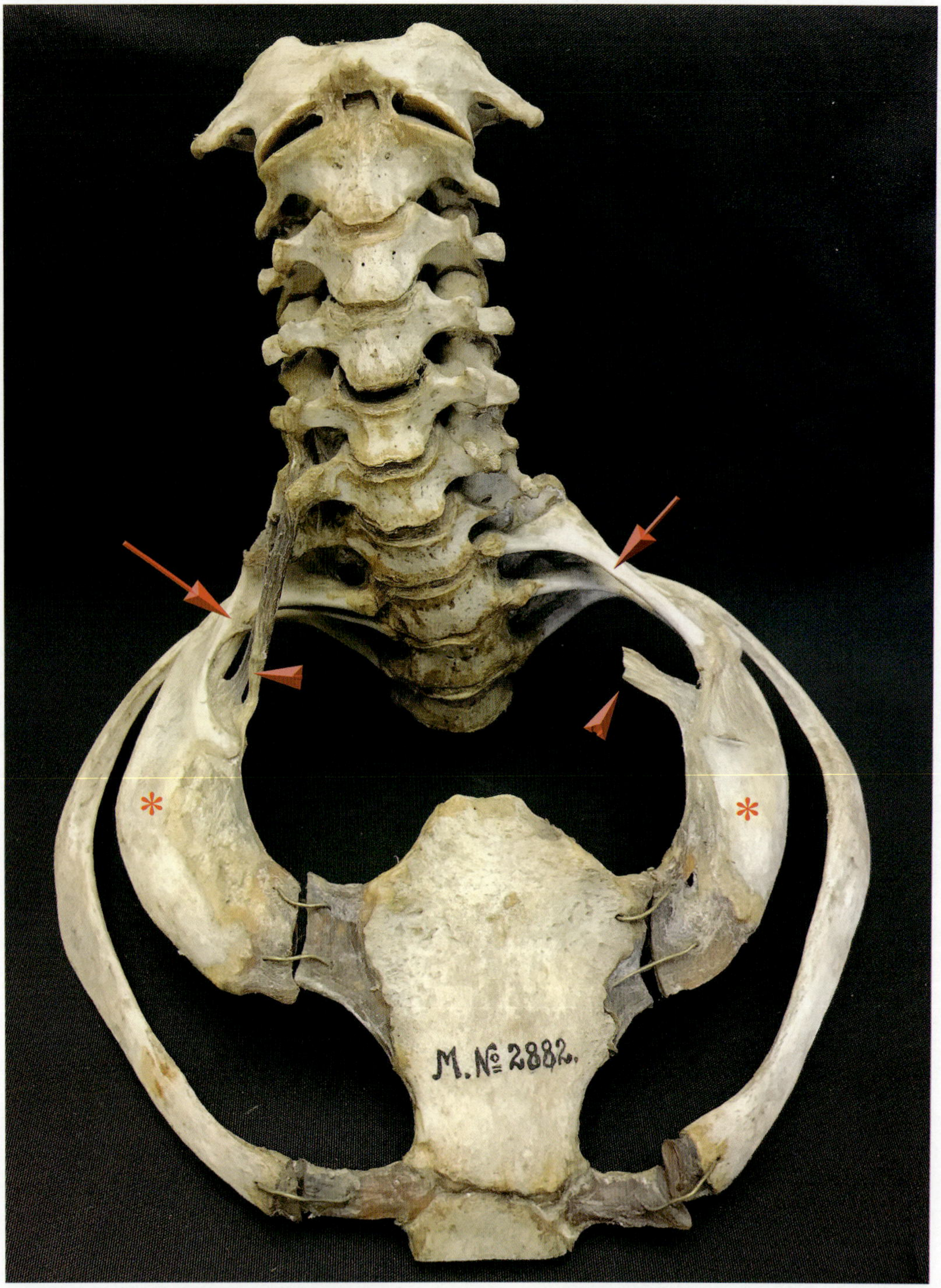


**Figure 121.** Cervical ribs (red arrows) originating from C6 and attached to the first thoracic rib (*); also present are calcified portions of the anterior scalene muscle (arrowheads) that attach to the transverse processes of C4 to C6 vertebra in an unknown adult who died in 1862 (MN2882). Cervical ribs vary greatly in size and shape, are sometimes overlooked or misidentified and appear in about 2% of the population and are more commonly present in females (Barnes 1994, 2012; Viertel et al. 2012). Care must be exercised to distinguish cervical from lumbar ribs in dry remains based primarily on the presence of an articular facet in the latter.

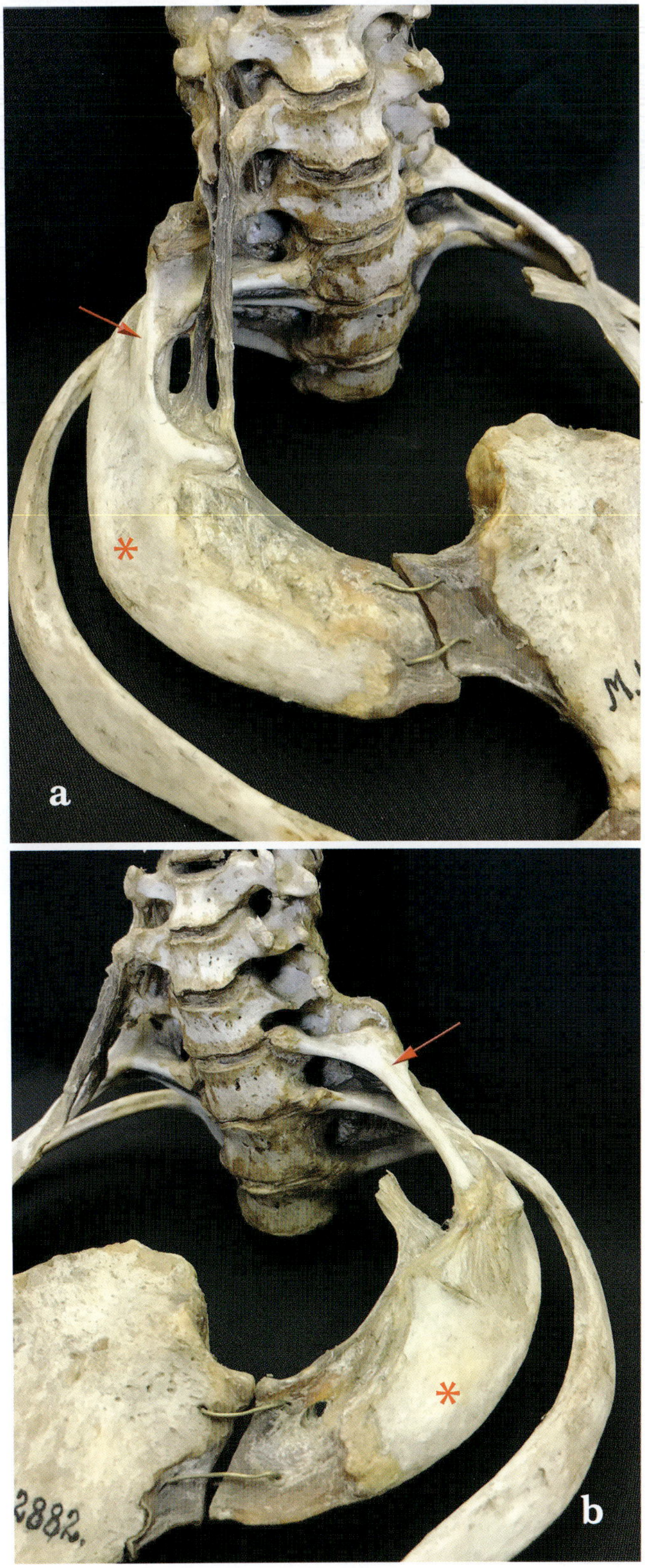


**Figure 122a-b.** (a) Right and (b) left oblique views of cervical ribs (arrows) originating from C7 and attached to the first thoracic ribs (*) in an unknown adult who died in 1862 (MN2882) (Barnes 1994, 2012; Viertal et al. 2012).

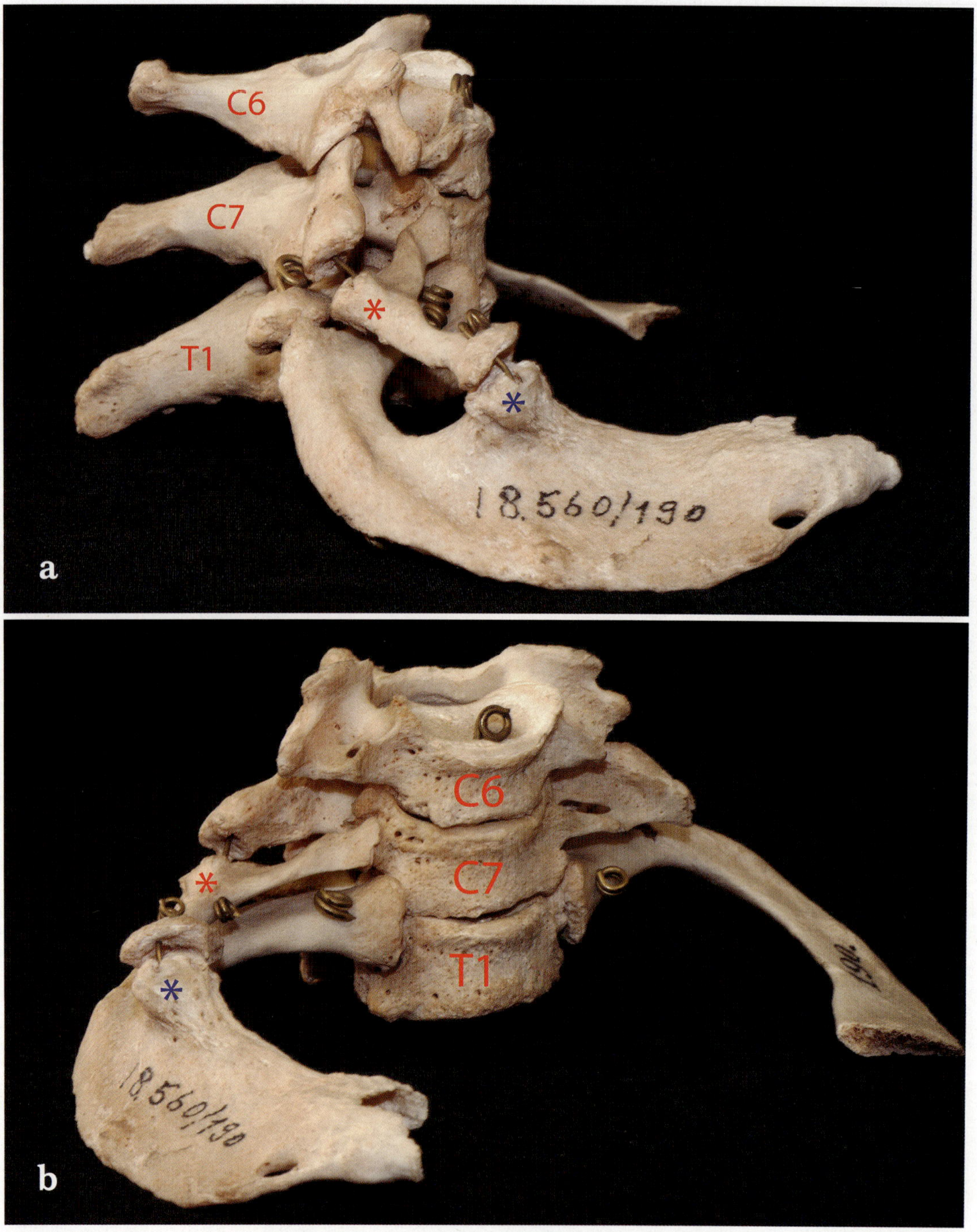


**Figure 123a-b.** (a) Right lateral and (b) oblique views of cervical (C7) and upper thoracic vertebrae (T1) with bilateral cervical ribs in an adult who died in 1918. The right cervical rib (red *) and pseudarthrosis (blue *) articulate with an unusually large right 1st rib in an adult (MN18.560/190). Cervical ribs vary in size and shape, result from persistent ossification of the lateral process(es) of C7, may be unilateral or bilateral, and may have their costal end free, or as in this individual, fused to the first rib. Radiologic (CT) examination revealed 67 of 3,404 (2.0%) patients, predominantly females, in the U.S. with cervical ribs (Viertal et al. 2012). See Barnes (1994, 2012); Mann et al. (2016).

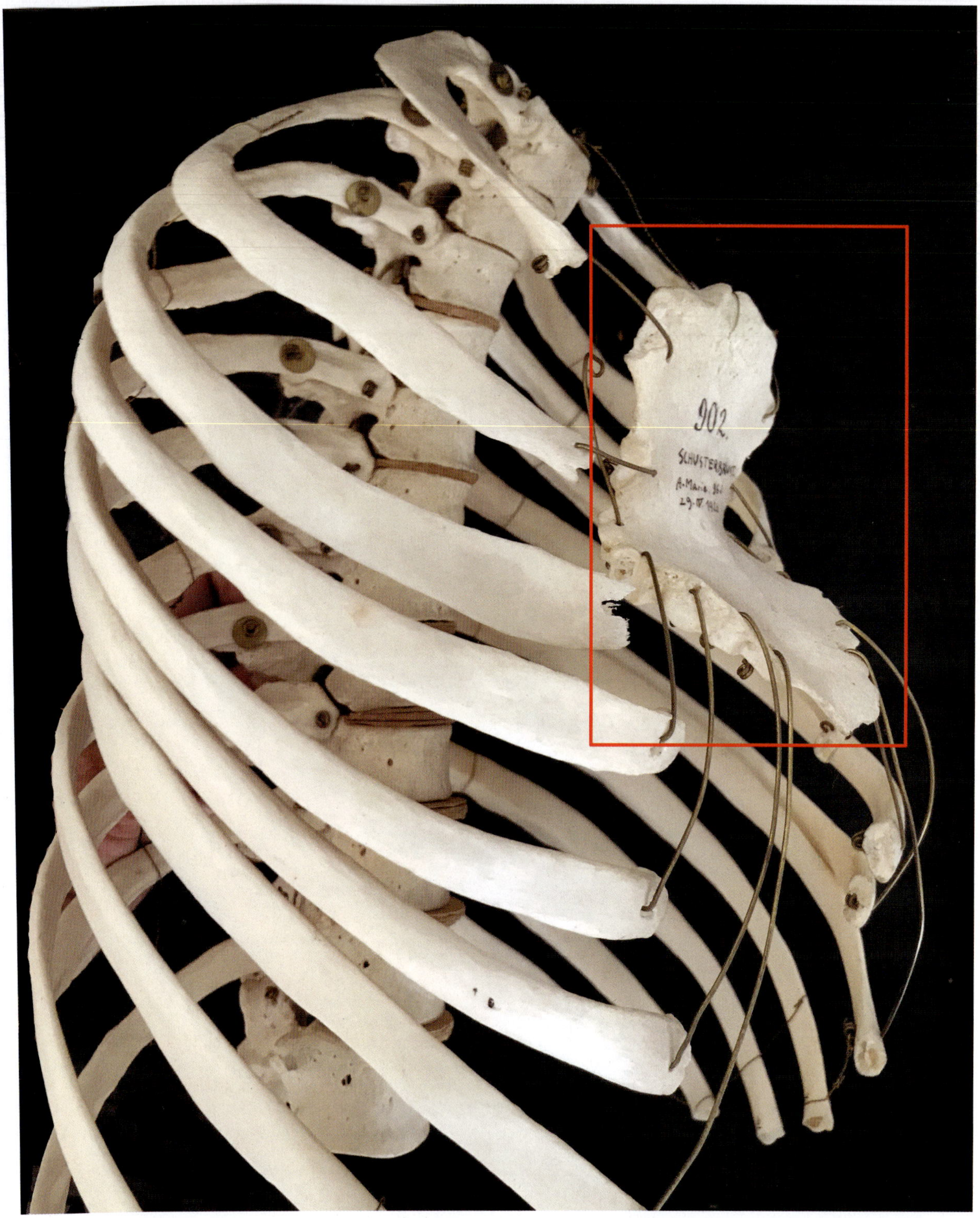


**Figure 124.** Pectus excavatum, sometimes referred to as "sunken chest" (rectangle) when the sternum grows inward, in a 36-year-old female (MN16.981/902). Pectus excavatum is a congenital defect of the costal cartilage that joins the ribs to the sternum, occurs in about 1 in 1,000 people, and is more common in males (Winkens et al. 2009). This individual exhibits a sharp angular deformity between ribs 2 and 4.

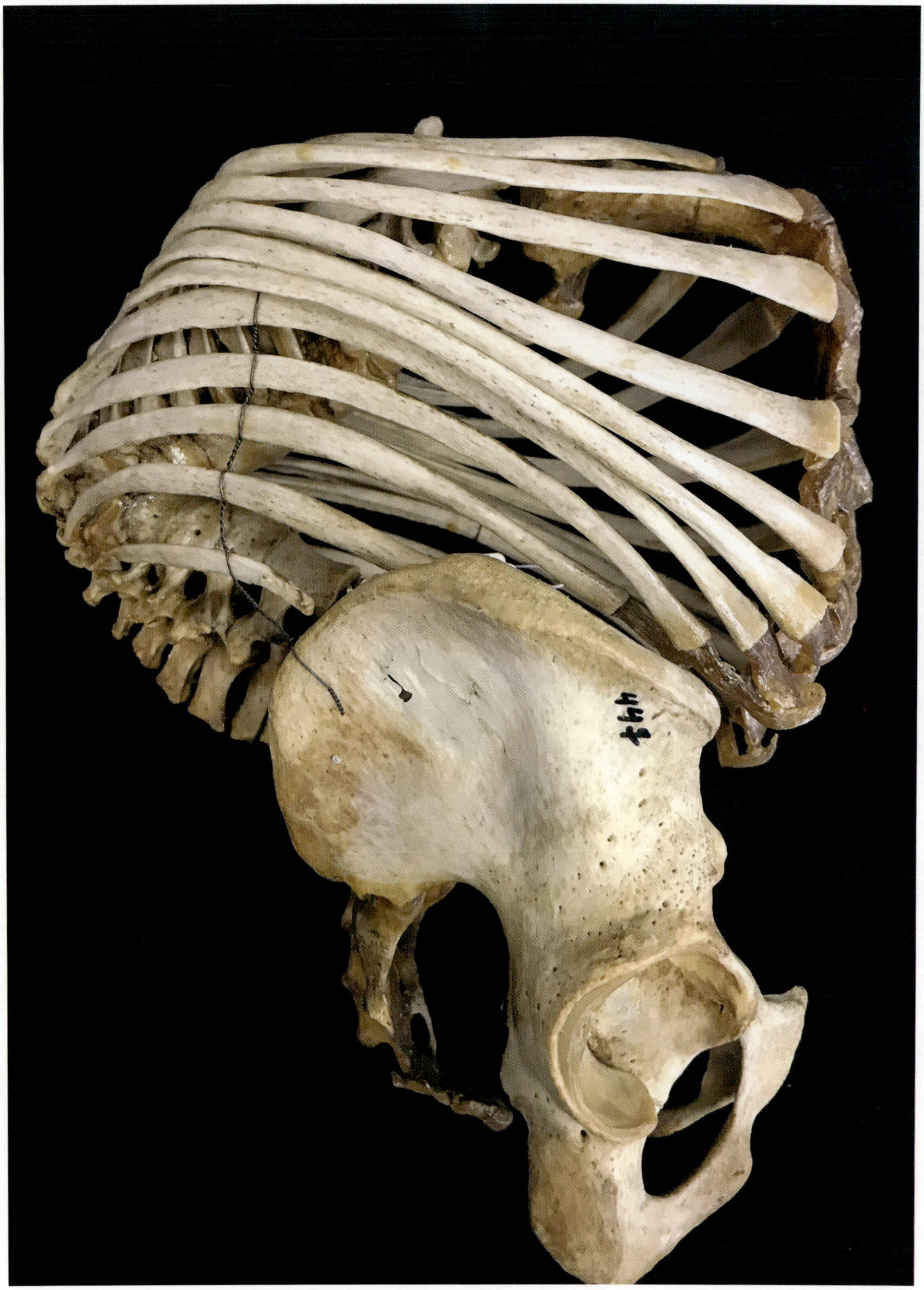

**Figure 125.** Right lateral view of an adult female with extreme kyphosis ("scoliosis" and "buckling of the spine pelvic ring, whole spine, all ribs, sternum" on the museum card) and pectus carinatum of the spine who died in 1884 (MN21.055/447). Note the extreme anterior (convex) protrusion/curvature of the sternum resulting in what some refer to as "pigeon chest."

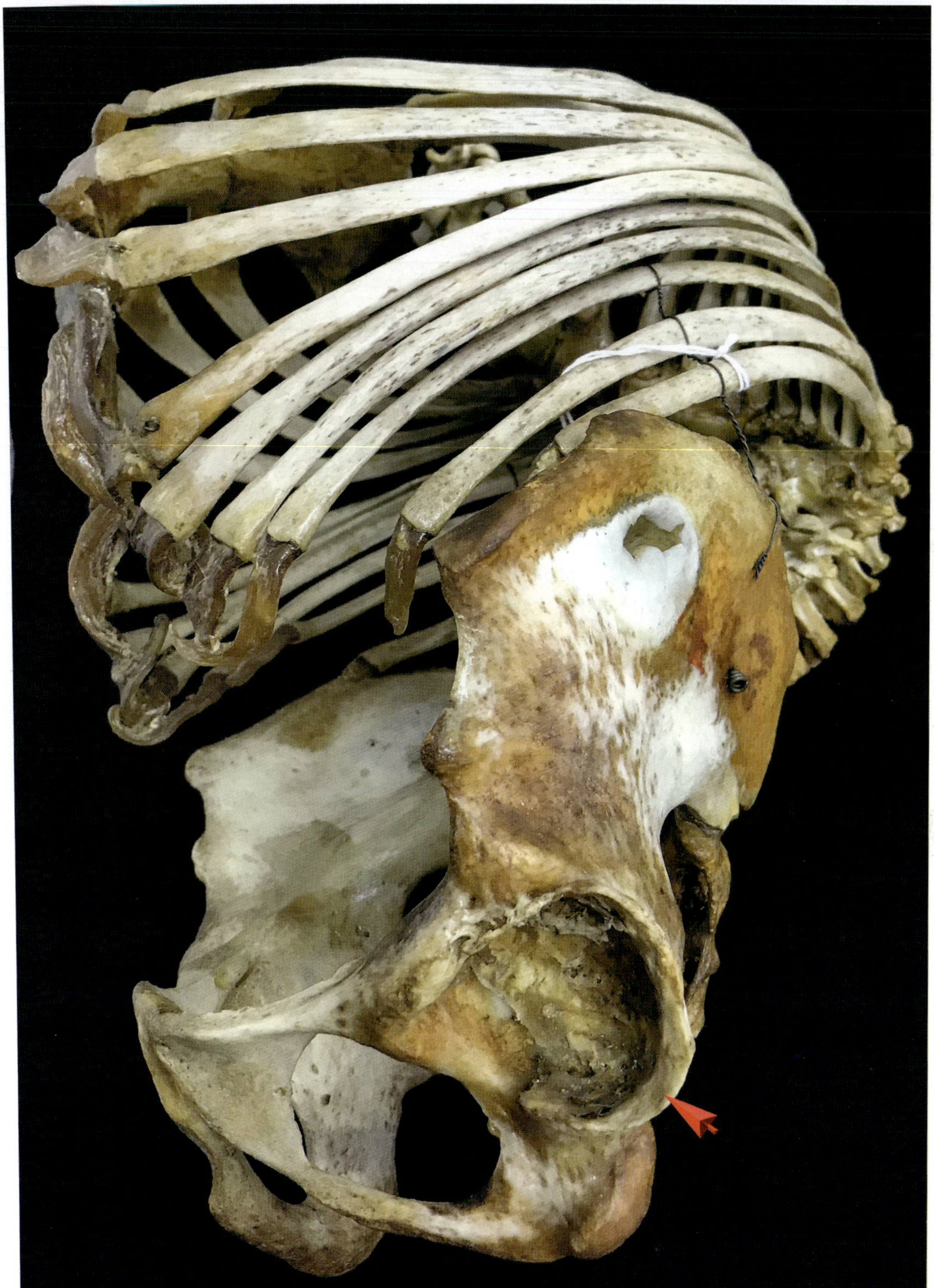

**Figure 126.** Left lateral view of an adult female with scoliosis, kyphosis and pectus carinatum of unknown etiology who died in 1884 (MN21.055/447). Note the extremely thin and elongated/flared left acetabulum (arrow) exhibiting osteoarthritis.

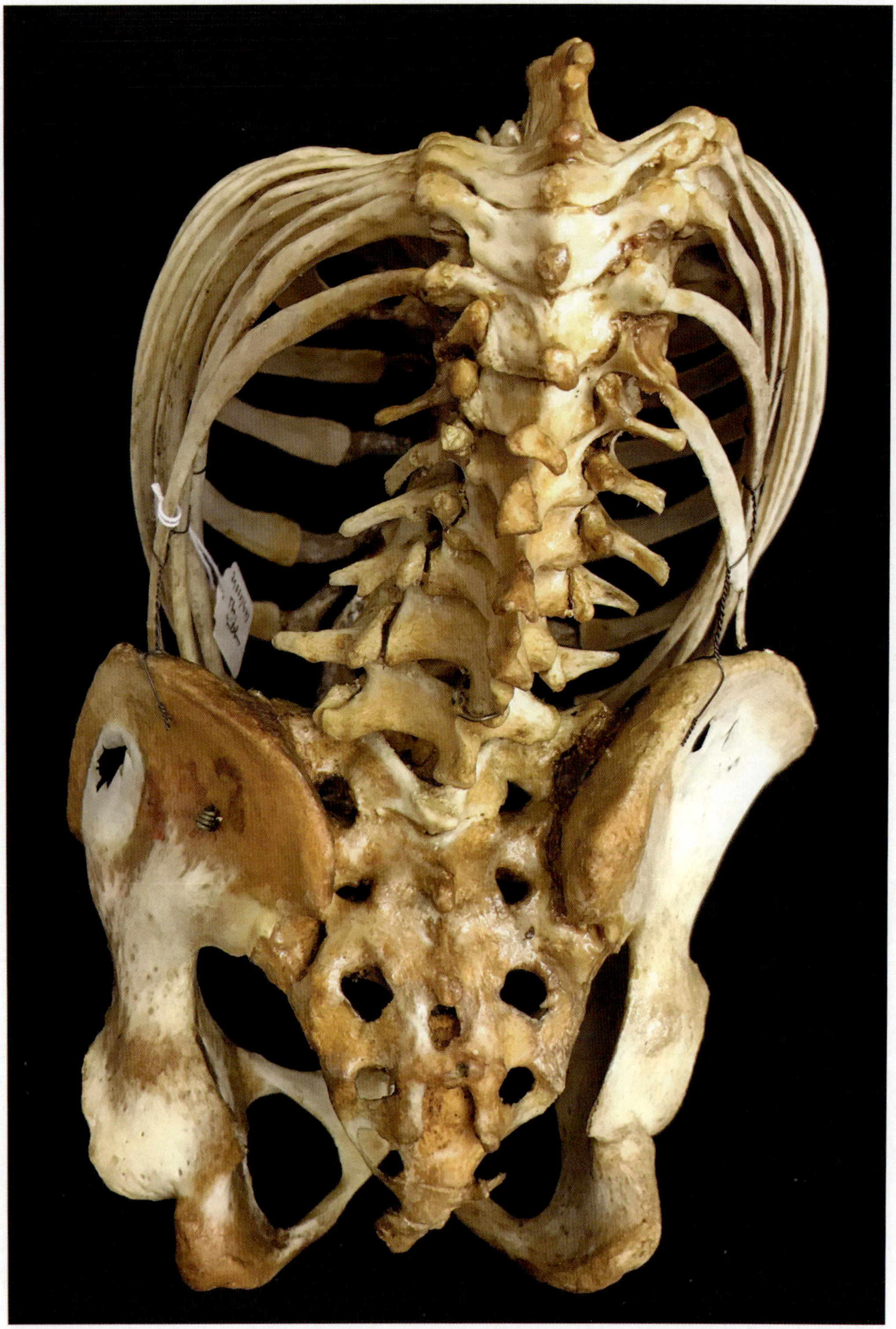

**Figure 127.** Posterior view of an adult female with severe kyphoscoliosis of the spine of unknown etiology who died in 1884 (MN21.055/447). The opening along the posterior portion of the sacrum, possibly reflecting spina bifida, may or may not have been symptomatic. Congenital scoliosis is nearly always due to one or more anomalies of the spine (*Campbell's Operative Orthopaedics* 1956).

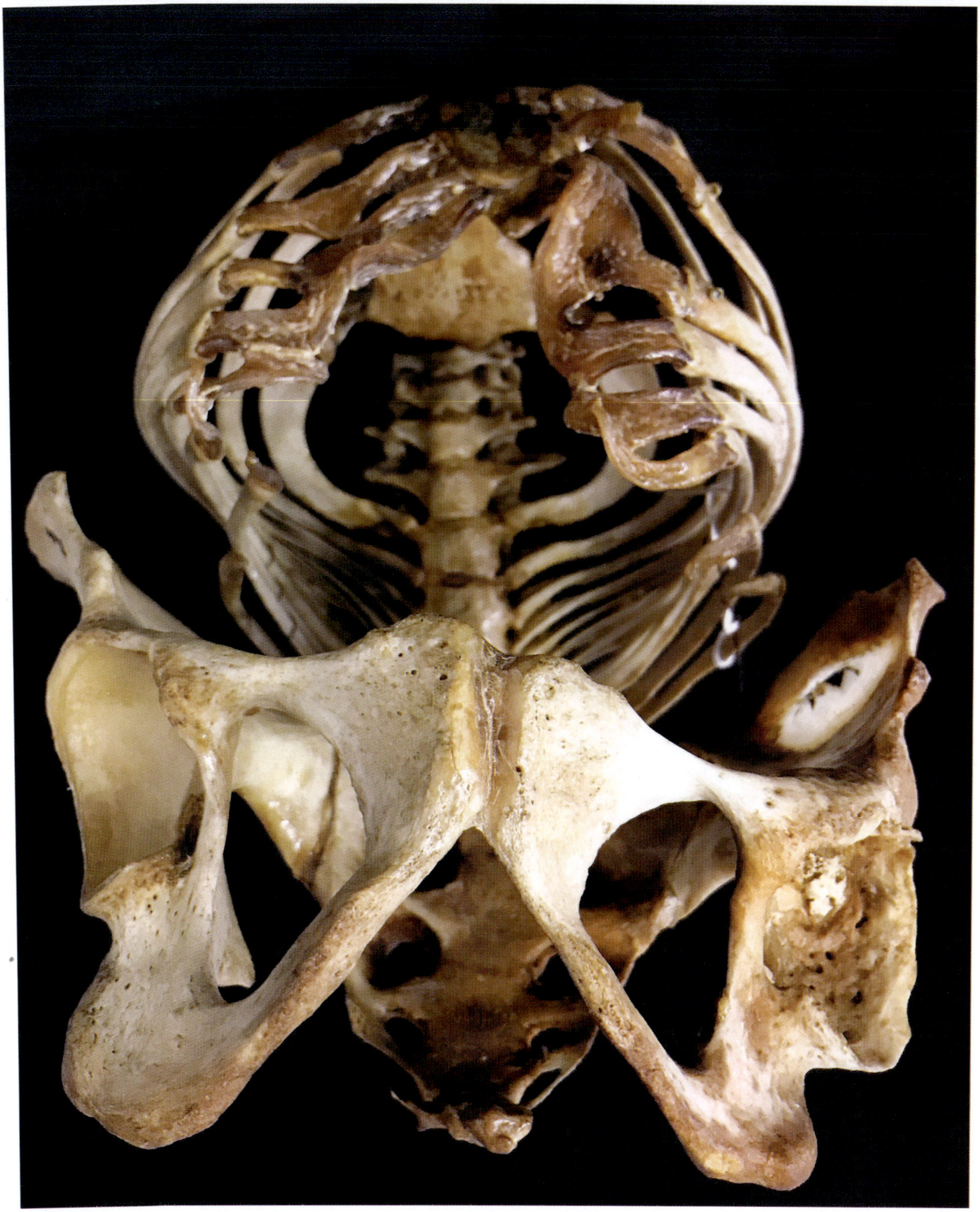

**Figure 128.** Probable female with kyphoscoliosis and "gibbus" formation of the spine T10-T12 vertebrae of unknown etiology who died in 1884 (MN21.055/447). The left acetabulum exhibits degenerative arthritis.

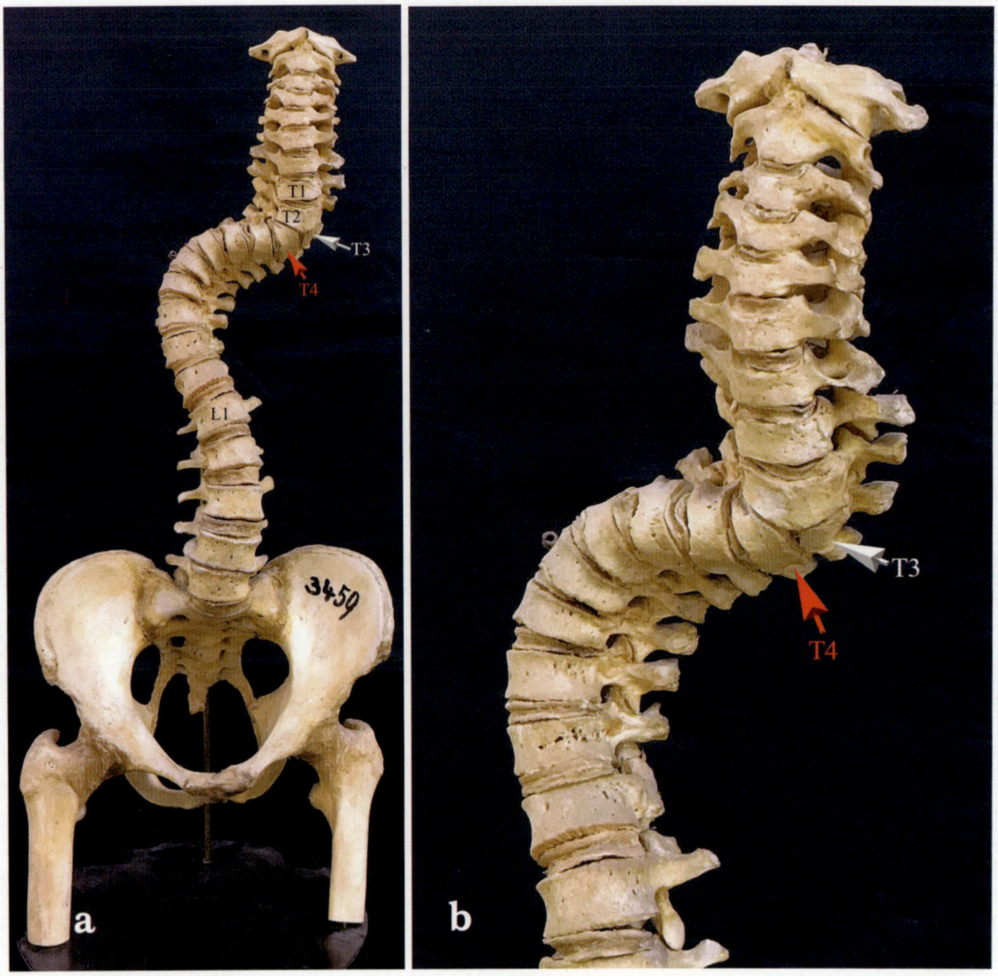


**Figure 129a-b.** (a) Anterior and (b) left oblique views of scoliosis and hemivertebra T3 and T4 in a 72-year-old male (MN21.055/3459). Scoliosis refers to lateral twisting or deviation from the midline and can be congenital or developmental, or due to disease, or trauma. Hemivertebra refers to the incomplete formation (structural development), often in the shape of a wedge or triangle, of one or more vertebrae that cause a "bend" or sharp angular turn in the spine, and that are not the result of disease such as an infection, osteoporosis, or acute trauma. See Barnes (1994, 2012) for more on developmental defects of the spine and Ortner and Putschar (1981) for osteoporosis.

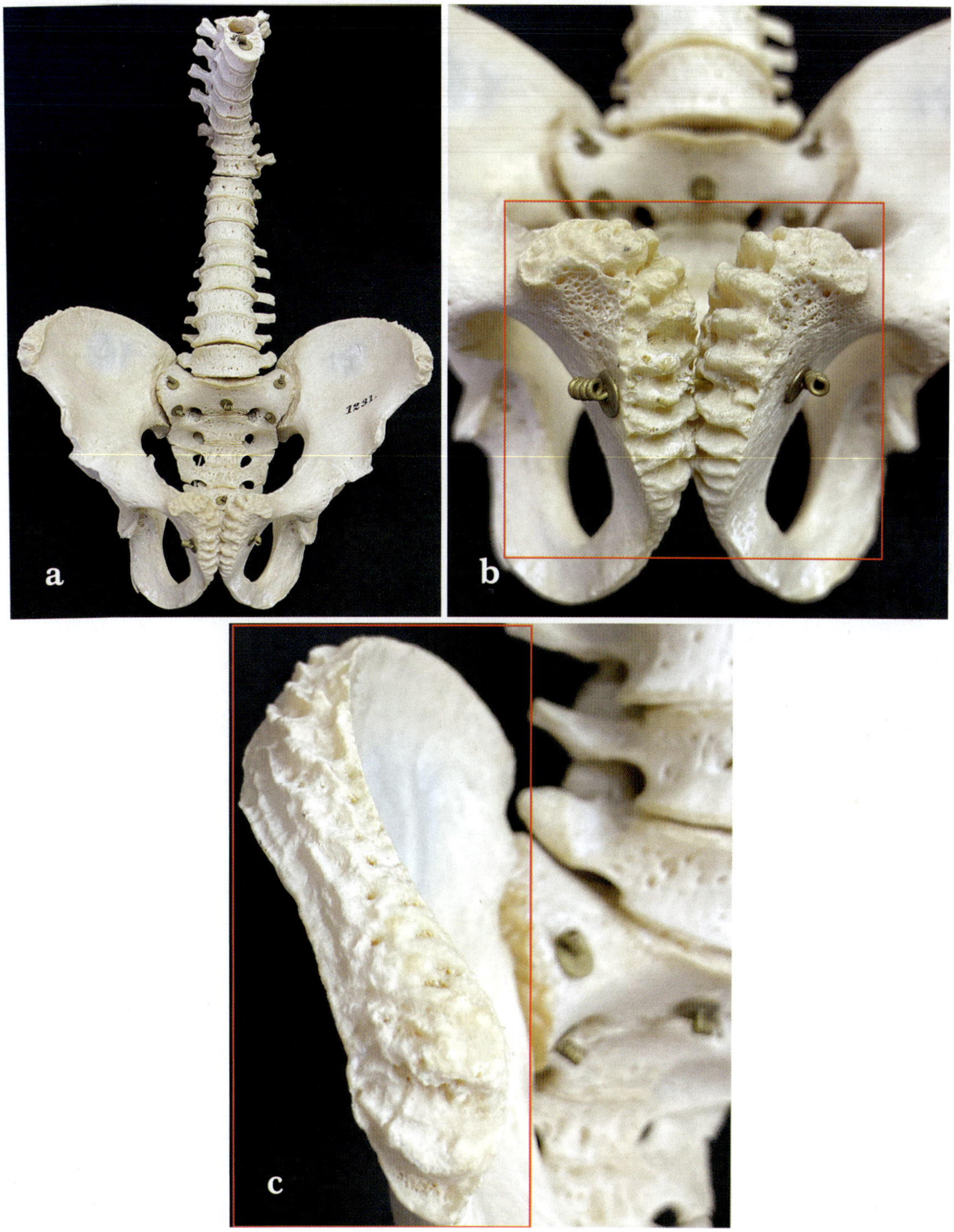


**Figure 130a-c.** (a) Delayed development and maturation in the spine and pelvis of a 32-year-old male who died in 1925 was due to hypopituitarism showing (b) unusually prominent pubic symphysis and (c) right iliac crest (MN18.560/1231). This male also exhibits incomplete fusion/closure of the vertebral endplates. See Kobylinska et al. (2019) for more on the effect of hypopituitarism on the spine and pelvis.

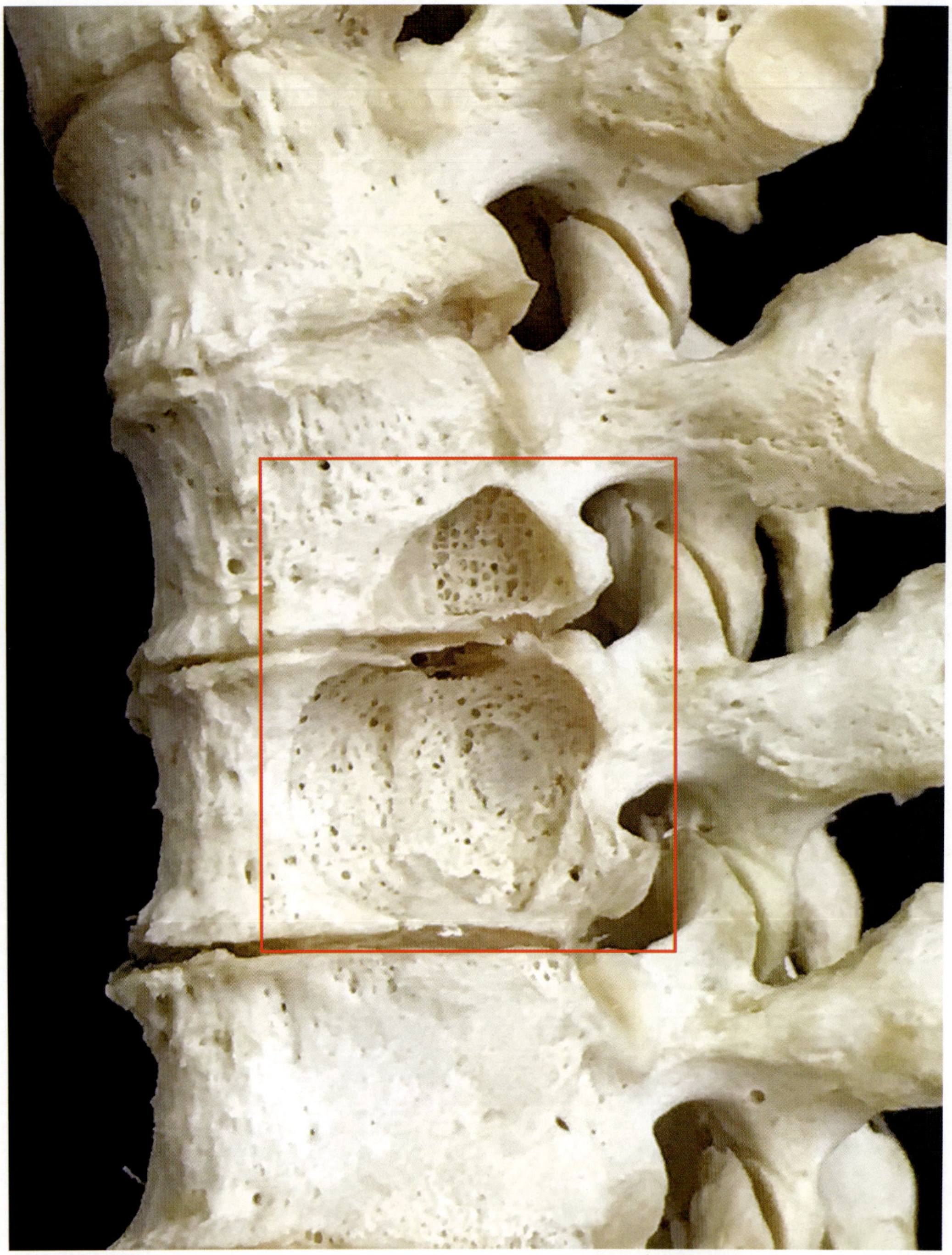

**Figure 131.** Thoracic aortic aneurysm (rectangle) resulting in erosion and cavitations in two adjacent thoracic centra in an unknown female who died in 1920 (MN18.660/455). Note the smooth margins of the cavitations and absence of reactive bone surrounding them. There are a few areas in the human skeleton where contact against a bone by soft tissue (e.g., submandibular salivary gland resulting in Stafne's defect) results in what the first author calls contact pressure erosion that can result in crater-like defects. See Ekin et al. (2018) for more on aortic aneurysm and involvement of the spine. Spinal tuberculosis can result in similar lesions (see Palfi et al. 2015).

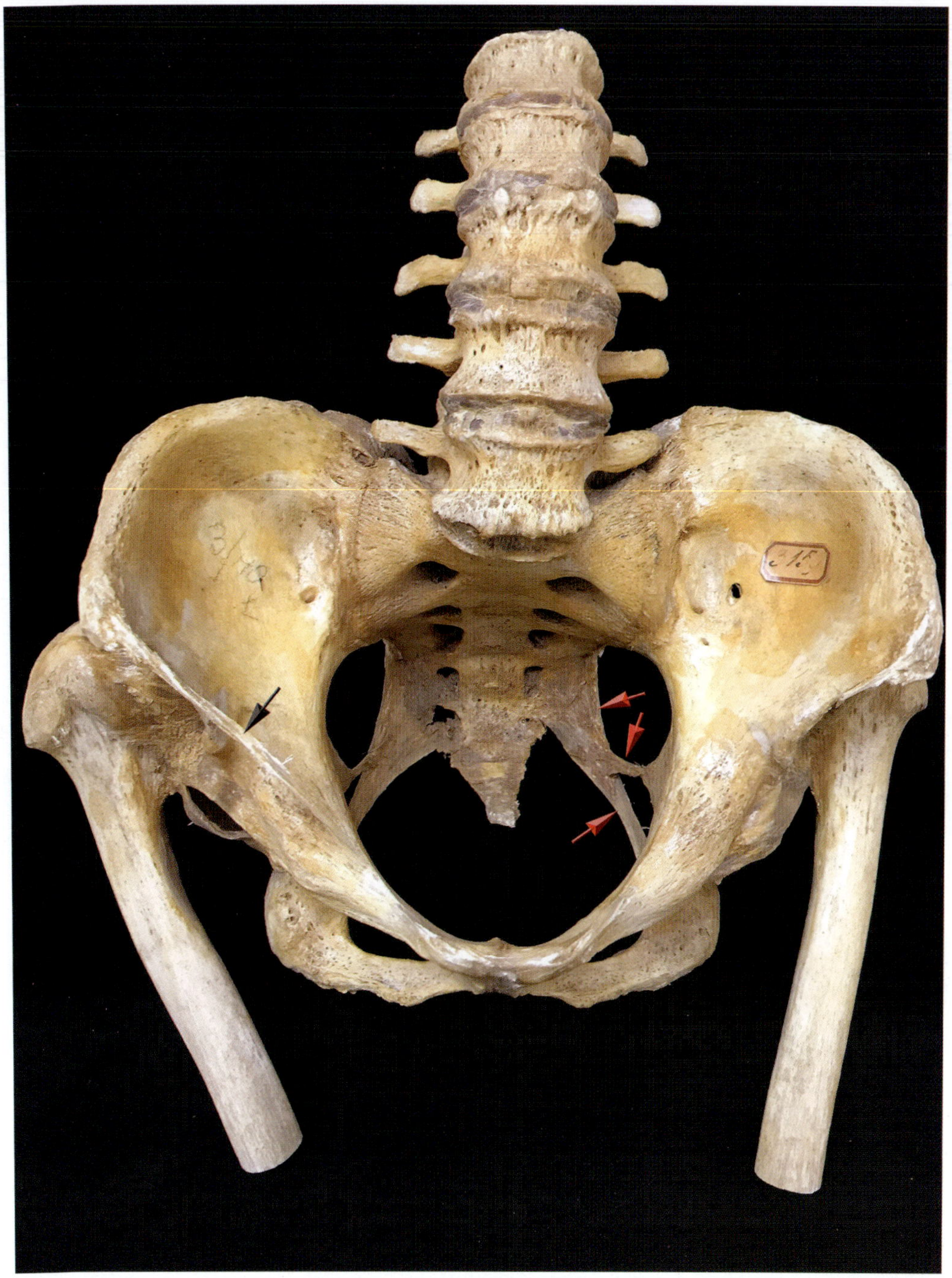

**Figure 132.** Anterior view of probable bilateral developmentally dislocated (dysplasia) hips (DDH) in what otherwise was a normal 54-year-old male who died in 1877 (MN21.055/3159) (P. Moroz, personal communication, November 17, 2022). If this were traumatic, one or both of the femoral heads and acetabuli would be fragmented and healed. Individuals with this condition can walk well into adulthood. Note that many of the ligaments including the inguinal (black arrow), sacrotuberous and sacrospinous ligaments (red arrows) were preserved in this individual. See Aronsson et al. (1994); Storer and Skaggs (2006) for more on DDH and Watson-Jones (1946 and 1955) and *Campbell's Operative Orthopaedics* (1956) for more on fractures, dislocations and other types of injuries of the hip.

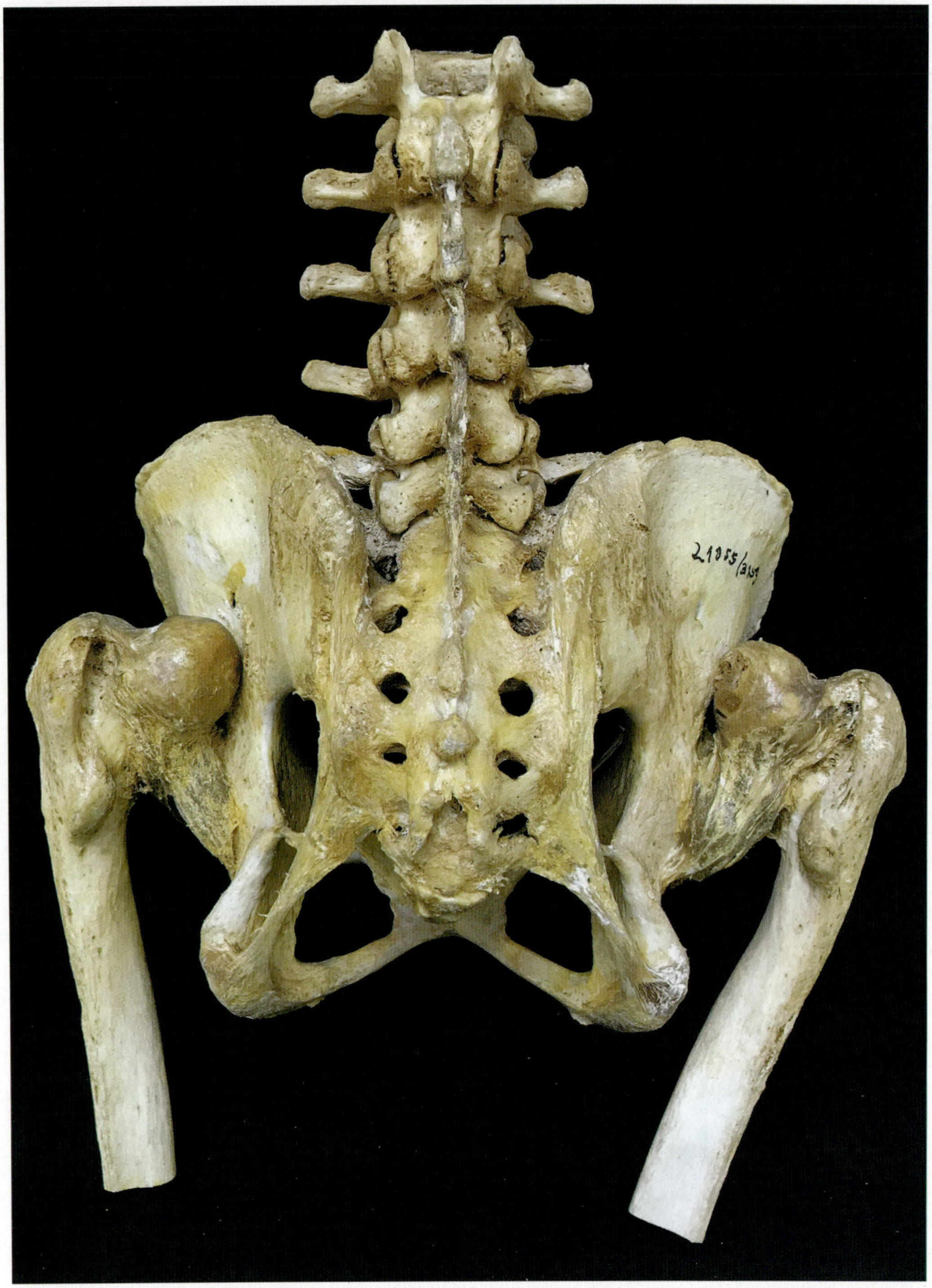

**Figure 133.** Posterior view of probable bilateral developmentally dislocated (dysplasia) hips (DDH) in what otherwise was a normal 54-year-old male who died in 1877 (MN21.055/3159) (P. Moroz, personal communication, November 17, 2022). Dislocations, with the exception of congenital dislocation of the hip (Griffiths 1981) are usually the result of trauma. See Storer and Skaggs (2006) for more on DDH and Watson-Jones (1946 and 1955) and *Campbell's Operative Orthopaedics* (1956) for more on fractures, dislocations and other types of injuries of the hip.

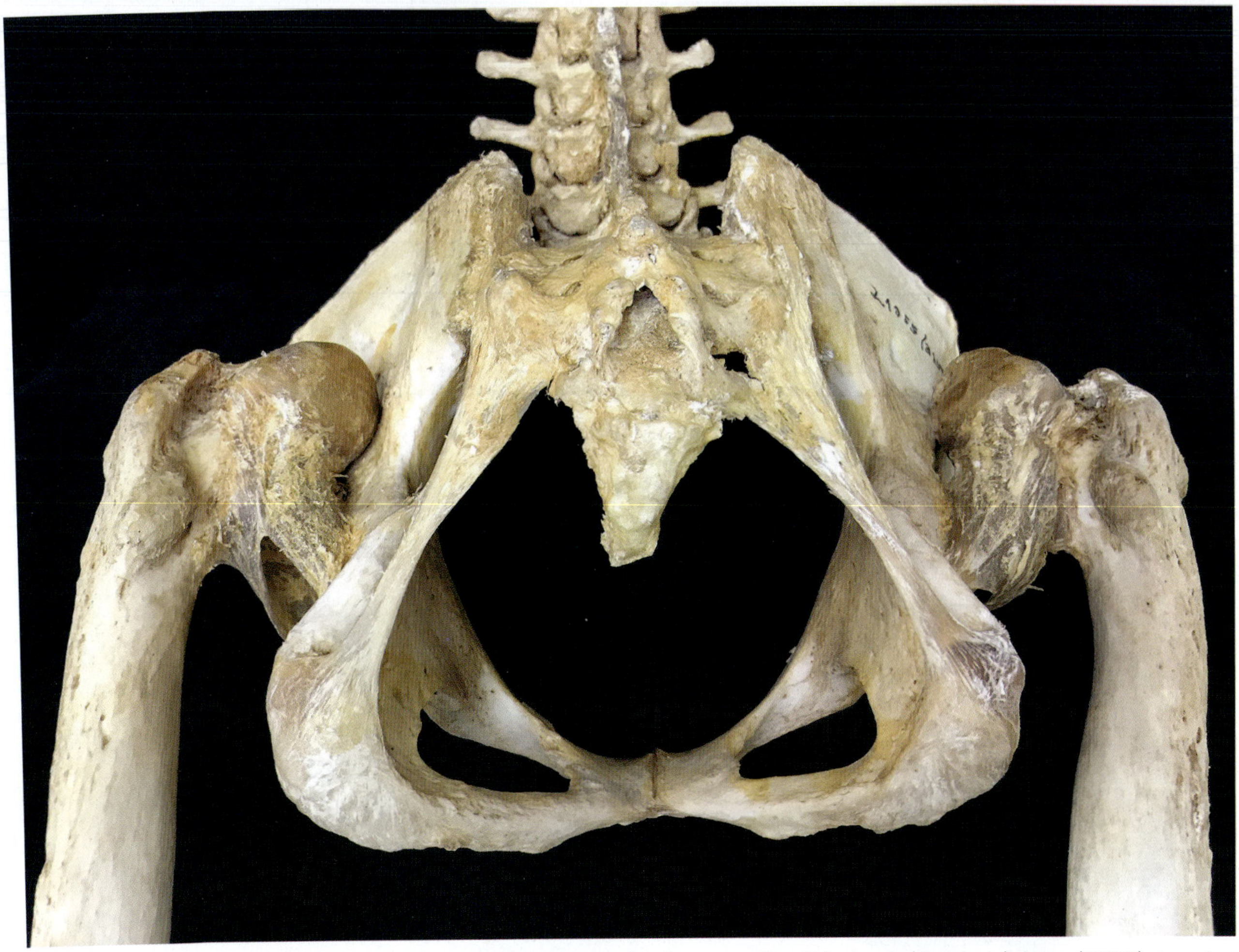

**Figure 134.** Inferior-superior view of probable bilateral developmentally dislocated (dysplasia) hips (DDH) in what otherwise was a normal 54-year-old male who died in 1877 (MN21.055/3159). The femoral heads are not in the acetabuli, but against the external surface of the ilia and was probably a common condition in the late 1800s (P. Moroz, personal communication, November 17, 2022). See Aronsson et al. (1994); Storer and Skaggs (2006) for more on DDH and *Campbell's Operative Orthopaedics* (1956); Griffiths (1981); Watson-Jones (1946 and 1955) for more on fractures, dislocations and other types of injuries of the hip.

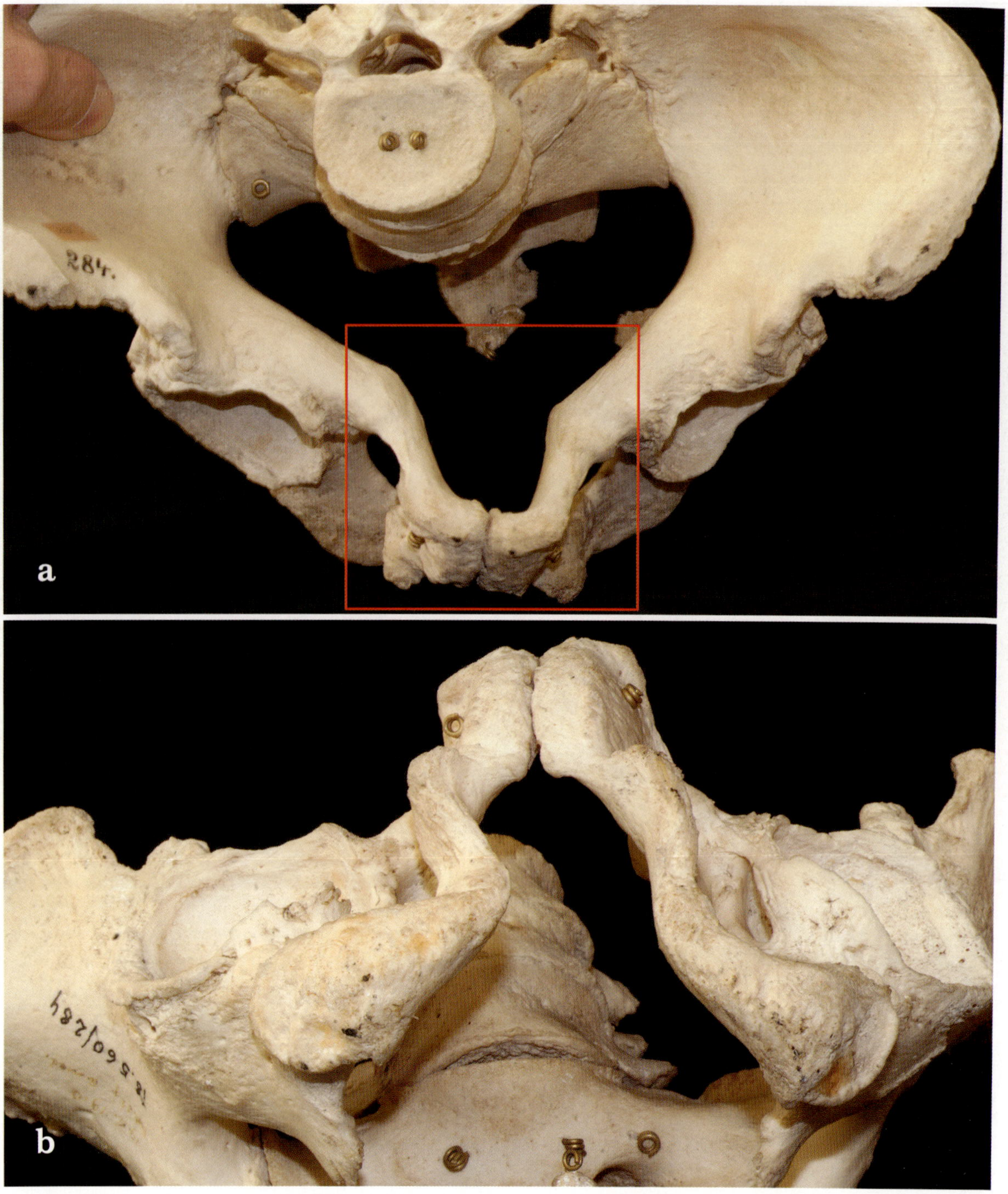


**Figure 135a-b.** (a) Superior-anterior and (b) inferior-anterior views of osteomalacia, a disease primarily in women due to prolonged dietary deficiency of vitamin D or limited exposure to sunlight and malabsorption, rarely appears before age 20 years (Chalmers et al. 1967; Minisola et al. 2021). In this 73-year-old female who died in 1910 (MN18.560/284), softening of the pelvis resulted in buckling of the pubic rami, a condition leading to a "waddling gait." See Geshchickter and Copeland (1936).

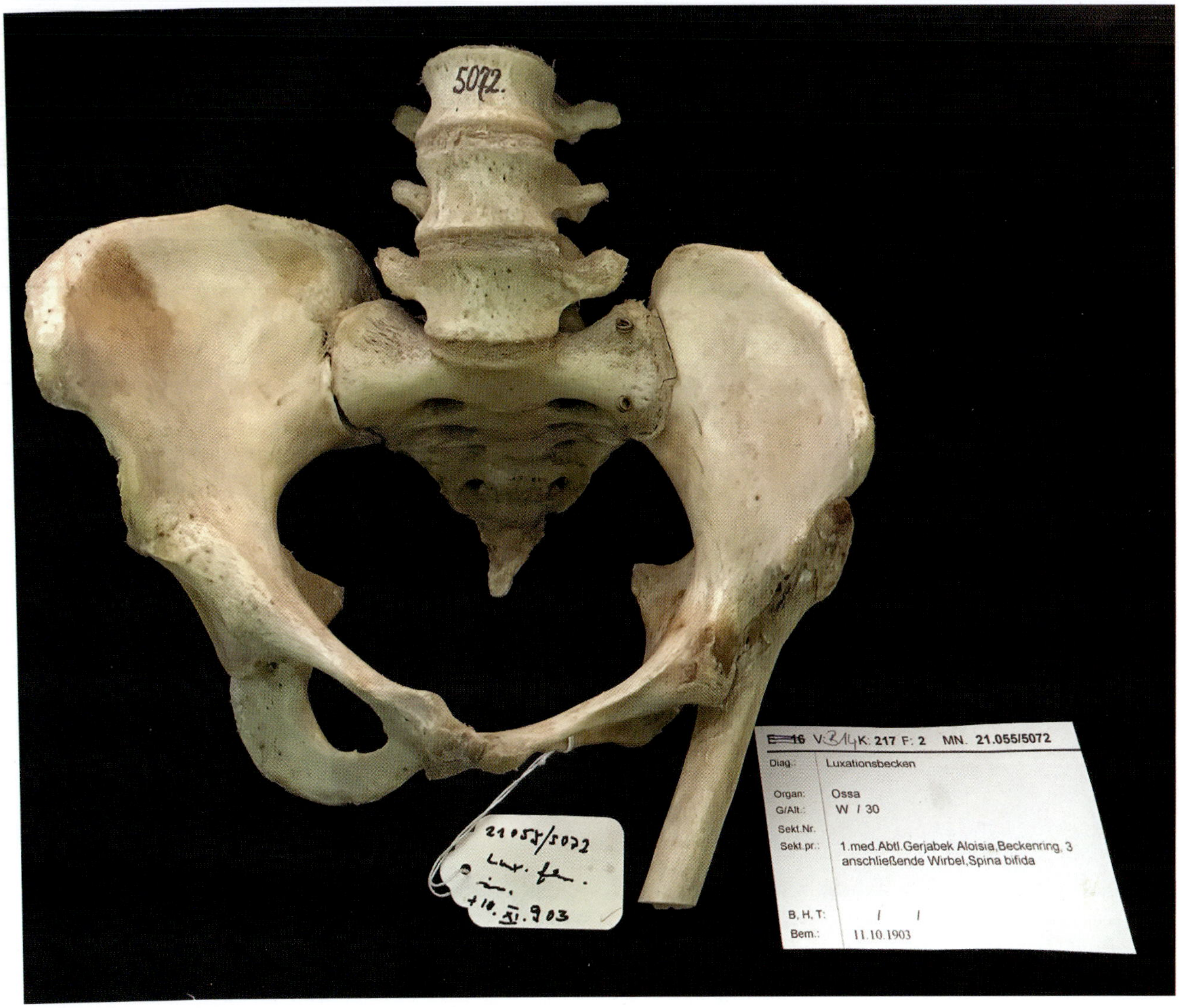


**Figure 136.** Spina bifida, asymmetry of the pubic bones and left ilium, and subluxation of the left hip socket in a 30-year-old female who died in 1903 (MN21.055/5072). The sacrum in this female is unusually wide and triangular in shape. Spina bifida, although commonly used to describe any cleft or opening along the posterior spine, is a neural tube defect that may be cystica (symptomatic as a meningocele) or occulta (asymptomatic) and very difficult to accurately identify in dry bones without supporting or confirmatory evidence in the soft tissues (Copp et al. 2015; Mann et al. 2016).

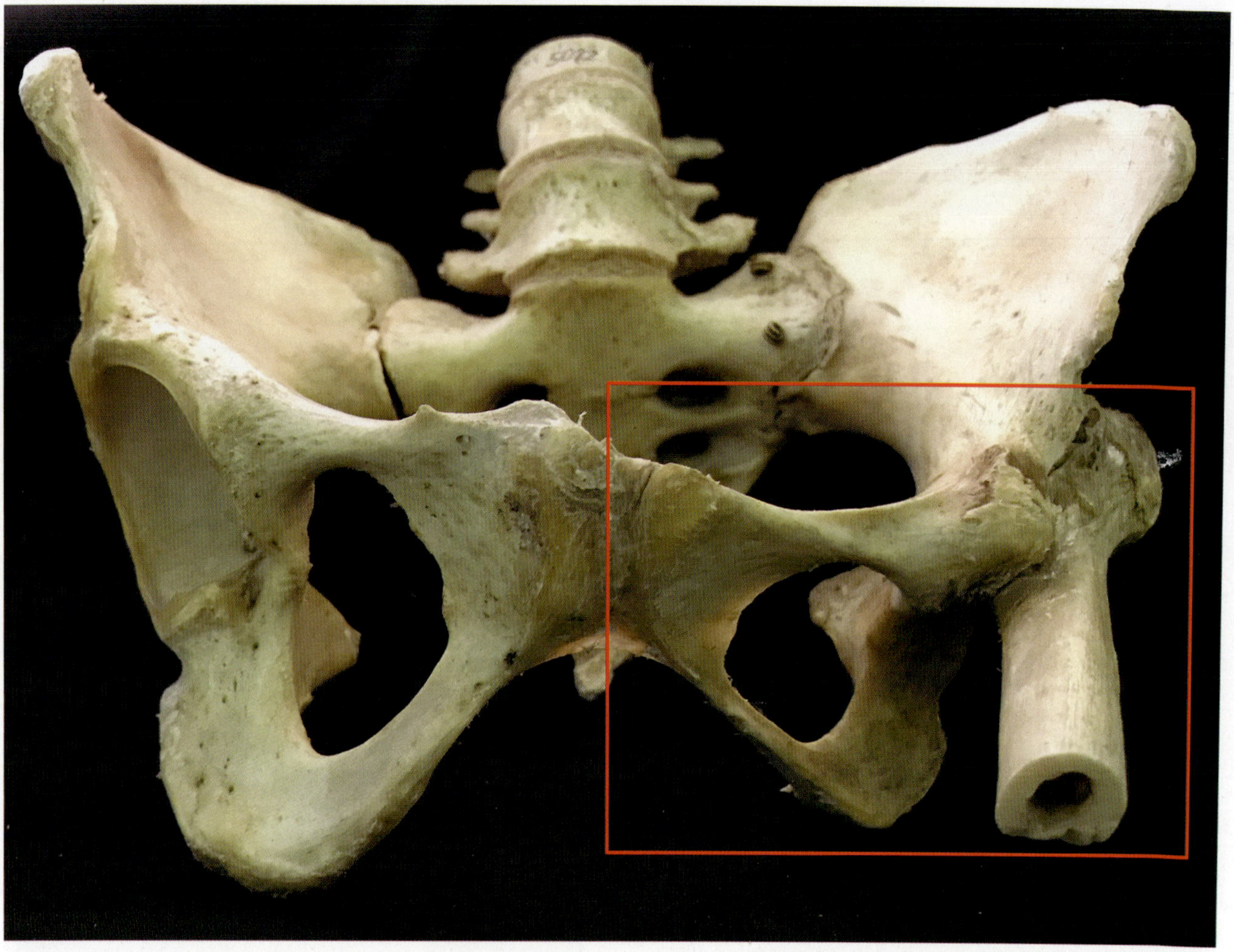

**Figure 137.** Spina bifida, asymmetry of the pubic bones, and subluxation of the left hip socket in a 30-year-old female who died in 1903 (MN21.055/5072). Note anterior extension and articulation of the left femoral diaphysis with the anterior rim of the acetabulum, as well as malformation and atrophy of the left pubic and ischiopubic ramus.

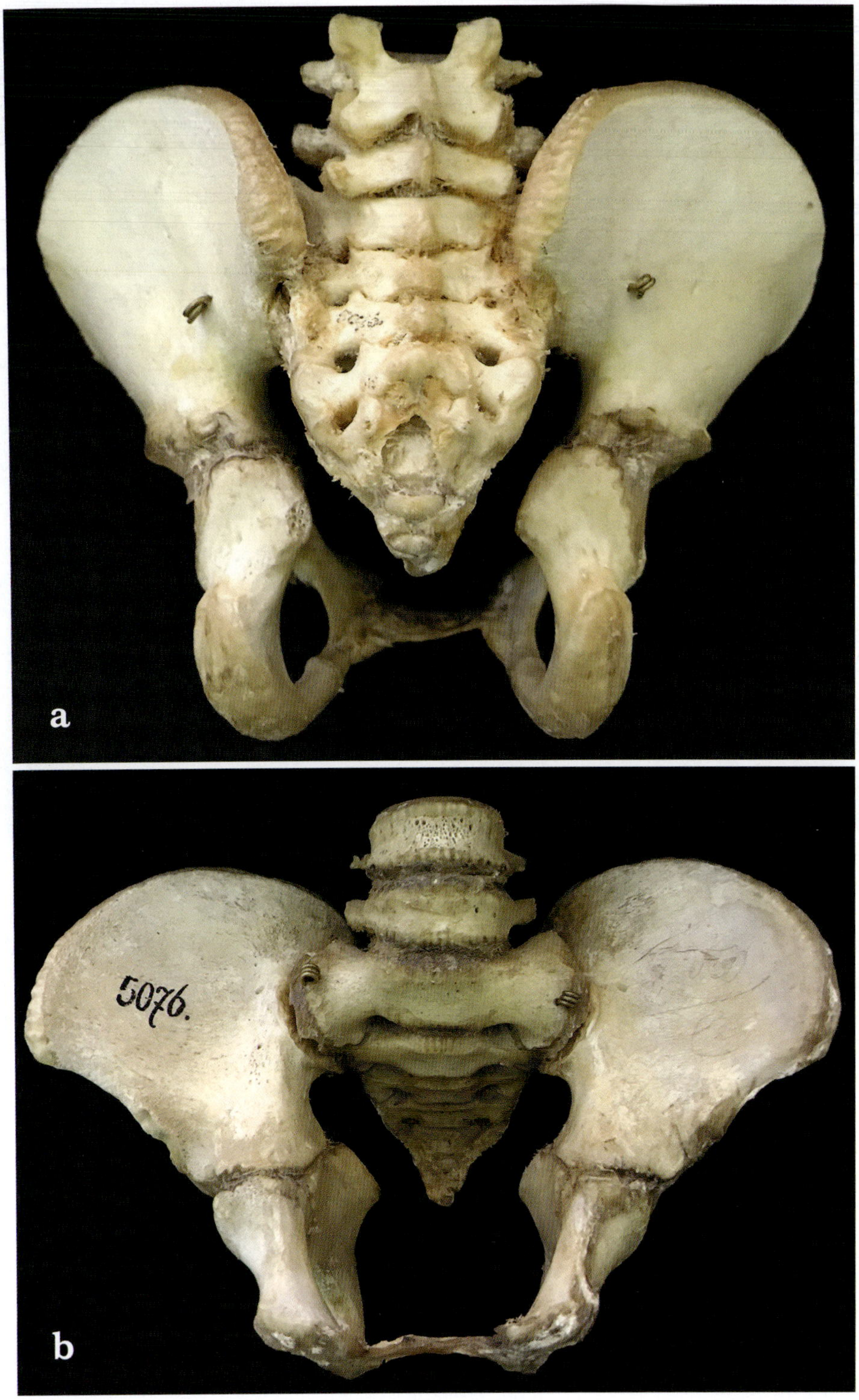


**Figure 138a-b.** (a) Posterior and (b) anterior views of diastasis (wide), separation of the pubic bones, due to spina bifida in a 5-year-old male who died in 1903 (MN21.055/5076). The bones and epiphyses of the pelvis at this young age have not fully fused.

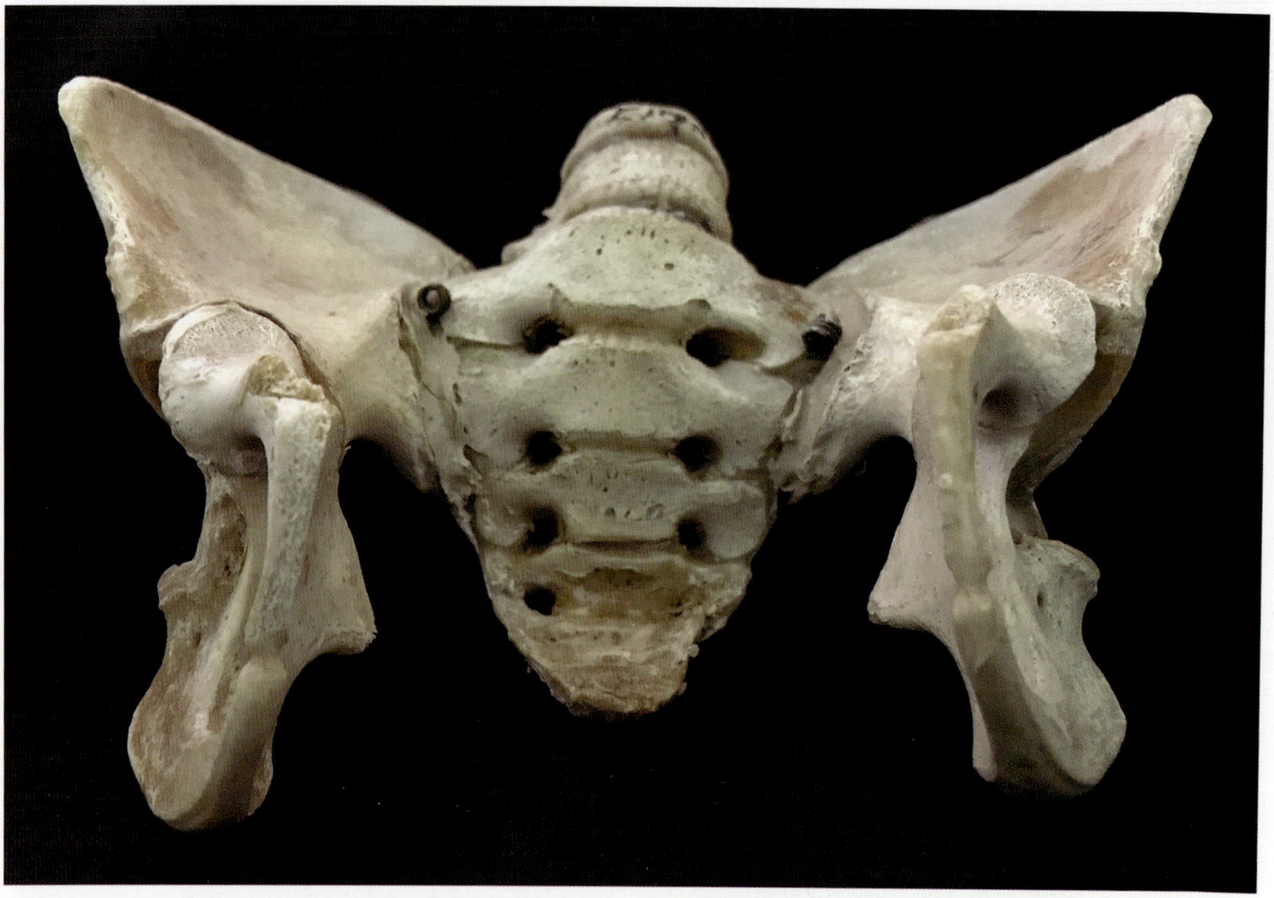

**Figure 139.** Diastasis, unusually wide separation of the pubic bones possibly due to congenital dysplasia, tuberculosis and rickets, among others, in an 8-year-old female who died in 1904 (MN21.055/5175). The normal pubic symphysis has a 4-5 mm gap. See Ortner and Putschar (1981) for an example of cleft pelvis with extrophy of the bladder and sacrum in a 20-year-old male.

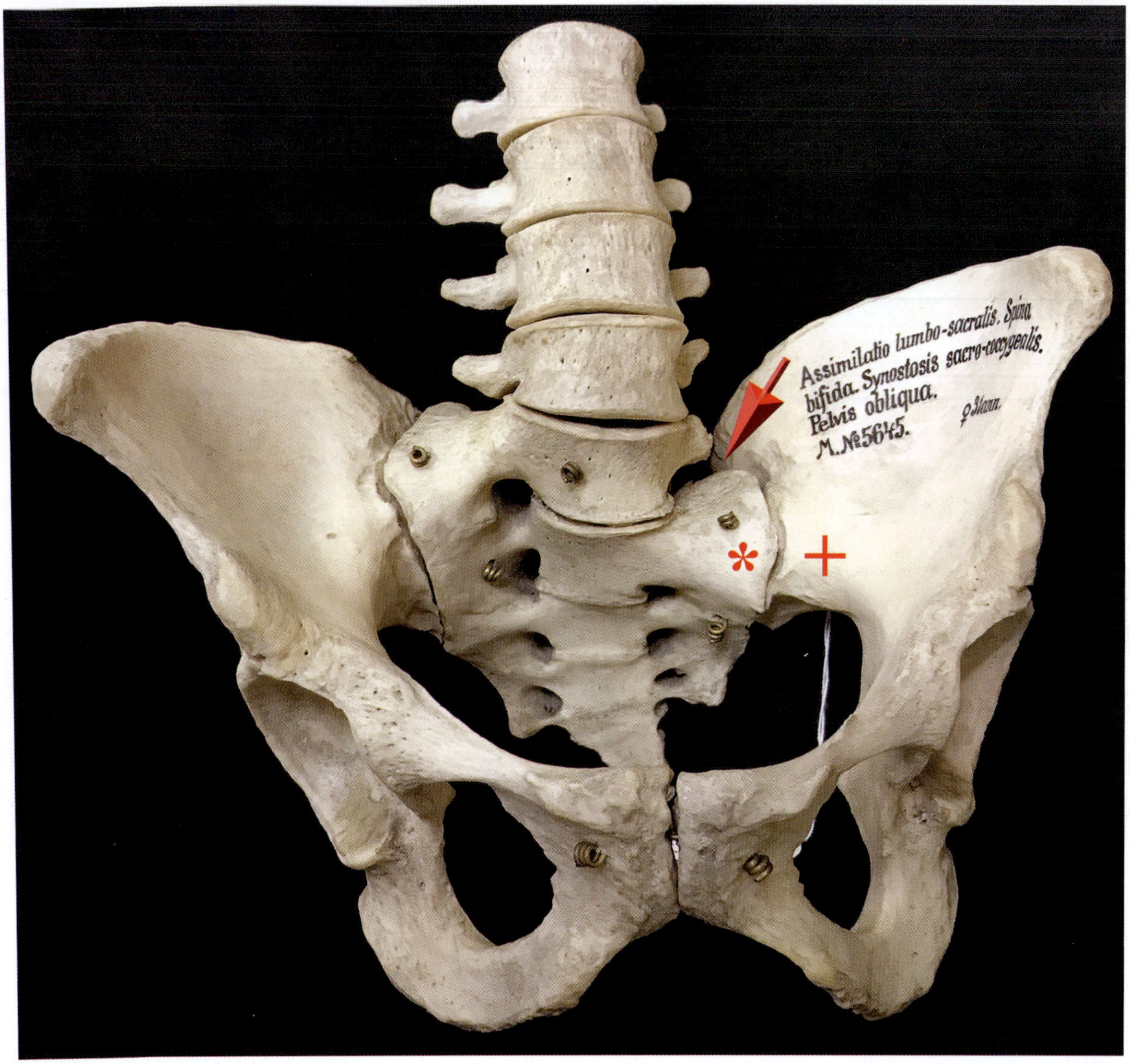


**Figure 140.** Anterior view of pelvis with incomplete development (spina bifida) of the left ala of S1 (arrow) resulting in oblique assimilation of the S2 ala (*) to the ilium (+) in a 31-year-old female who died in 1902 (MN5645). The sacrum has accommodated this unusual articulation by moving anteriorly.

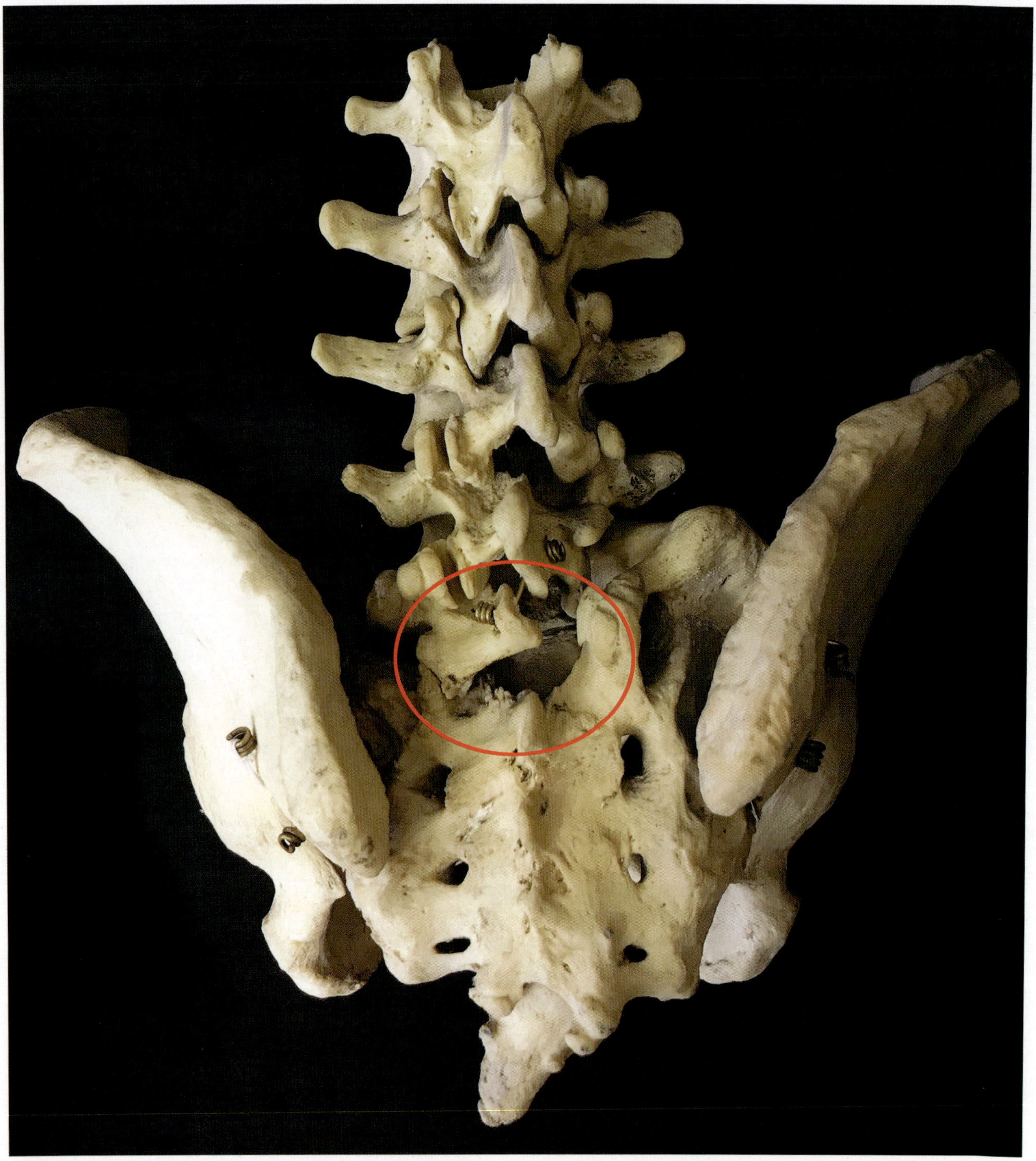

**Figure 141.** Posterior view of pelvis and lower spine showing incomplete development (spina bifida) of the neural arch of S1 (oval) in a 31-year-old female who died in 1902 (MN5645). Note that the sacrum has accommodated the asymmetrical articulation of the left sacral ala to S2 resulting in an inferior shifting of the entire left pelvis, which is lower than the right pelvis – the posterior iliac crest is lower than the right iliac crest.

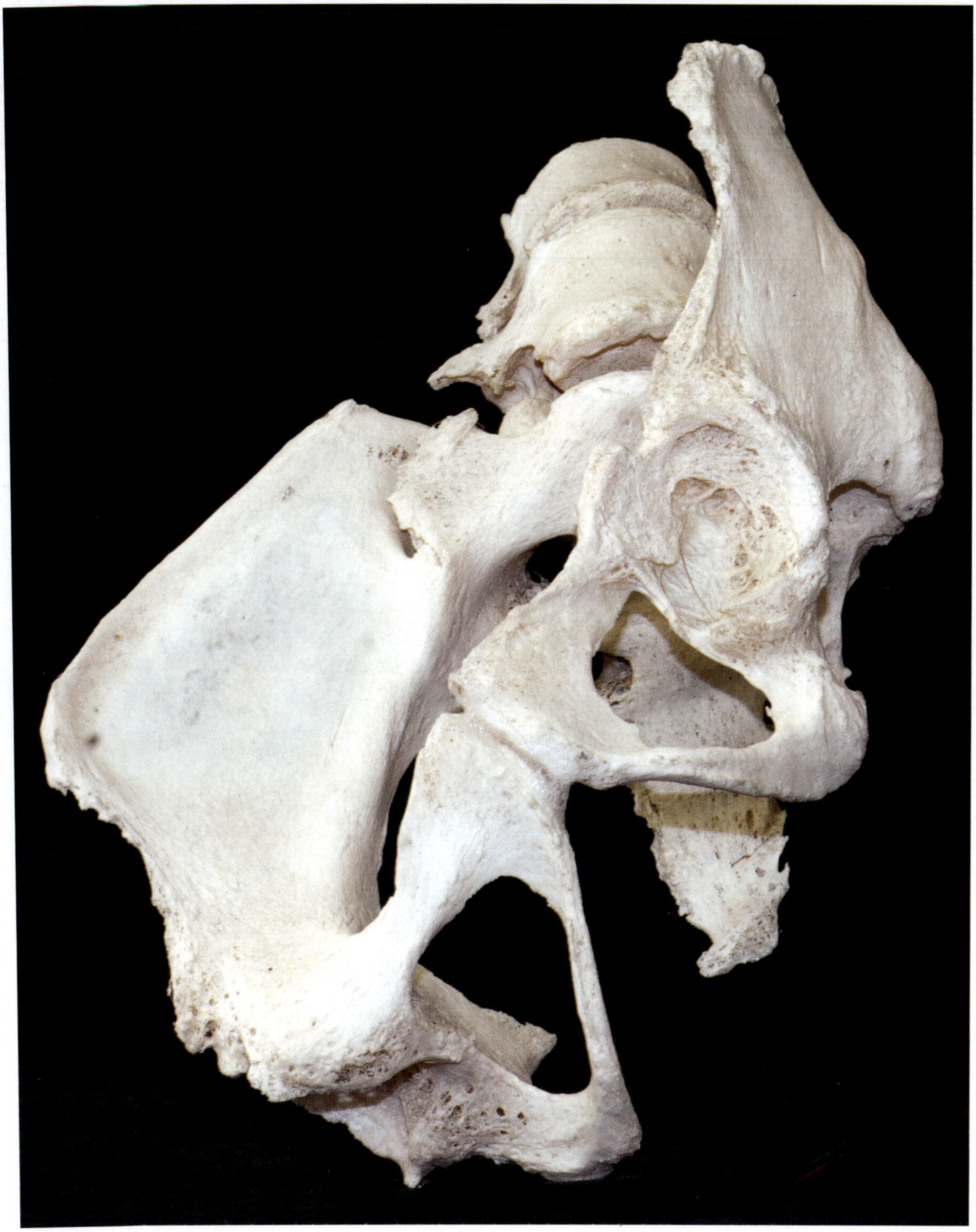

**Figure 142.** Left anterior-inferior view of an adult female with "Obliqua pelvis que eximii gradus e causa ignota" (on the museum card) (MN18.803). Oblique pelvi resulting in flattening can be the result of differences in leg length, pelvic contractures, structural scoliosis, or any combination of these causes (Chan et al. 2019; Winter and Pinto 1976), as well as tumors or rickets (El-Mowafi 2017).

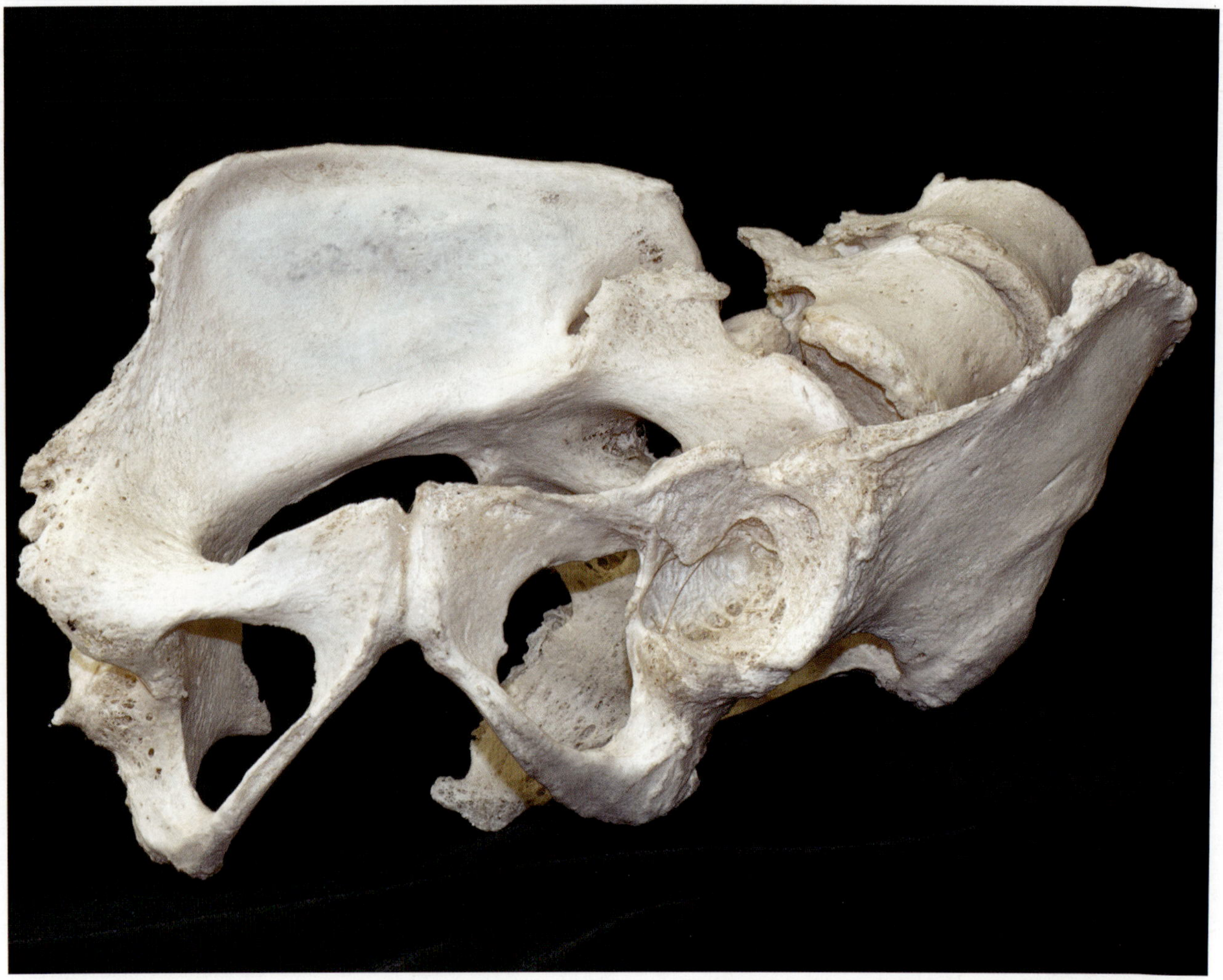

**Figure 143.** Left oblique view of "Obliqua pelvis que eximii gradus e causa ignota" on the museum card (MN18.803). Oblique pelvis can be the result of differences in leg length, pelvic contractures, structural scoliosis, or any combination of these causes (Chan et al. 2019; Winter and Pinto 1976), as well as tumors or rickets (El-Mowafi 2017). Pelvic obliquity is commonly associated with individuals with cerebral palsy (Hagglund et al. 2018) as a result of spastic or neuromuscular scoliosis and presents as asymmetry with one anterior superior iliac spine higher than the other and the individual's pelvis as "falling" (pelvic tilt) to one side or the other (Chan et al. 2019).

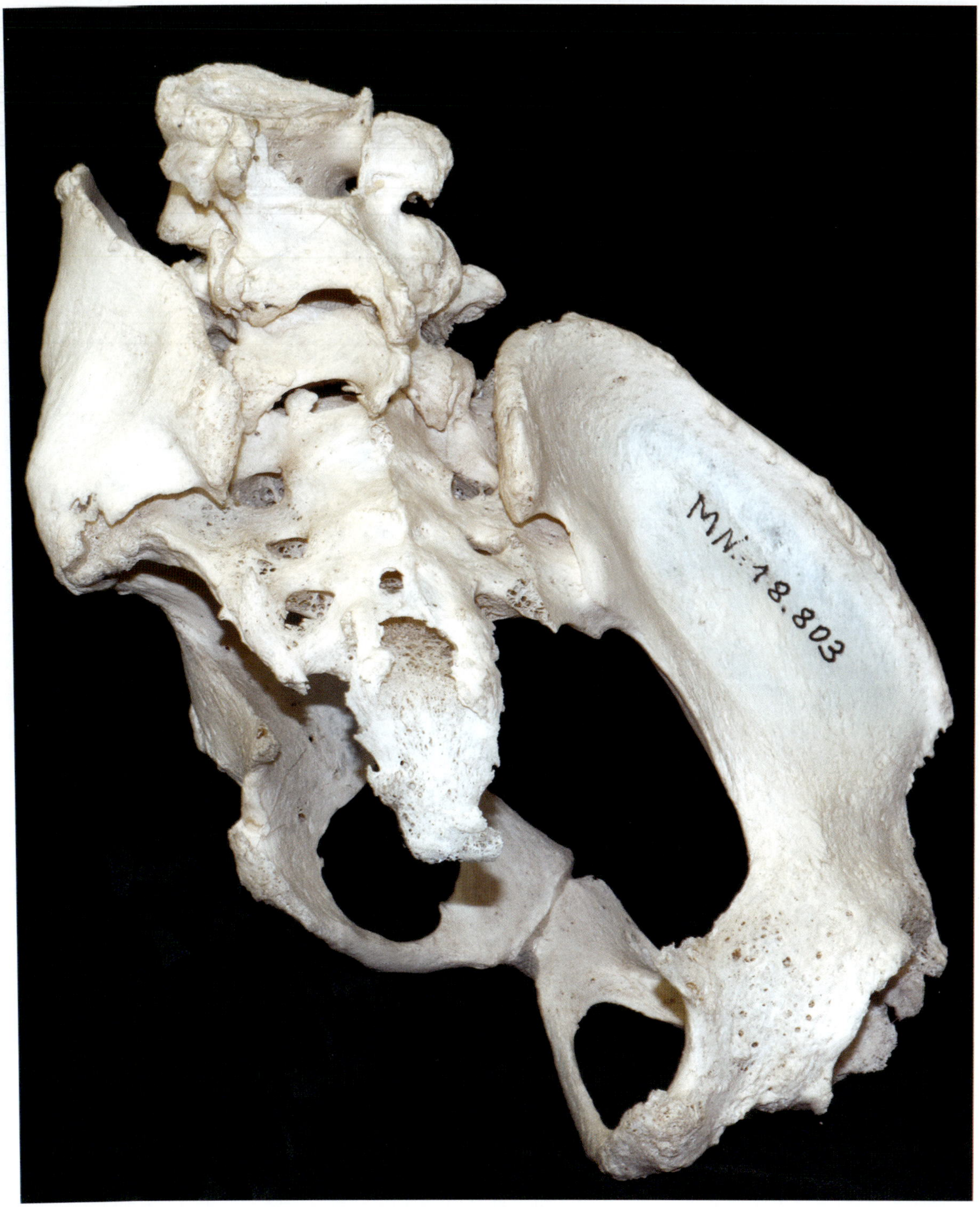


**Figure 144.** Posterior view of "Obliqua pelvis que eximii gradus e causa ignota" on the museum card (MN18.803). Oblique pelvis can be the result of differences in leg length, pelvic contractures, structural scoliosis, or any combination of these causes (Chan et al. 2019; Winter and Pinto 1976), as well as tumors or rickets (El-Mowafi 2017).

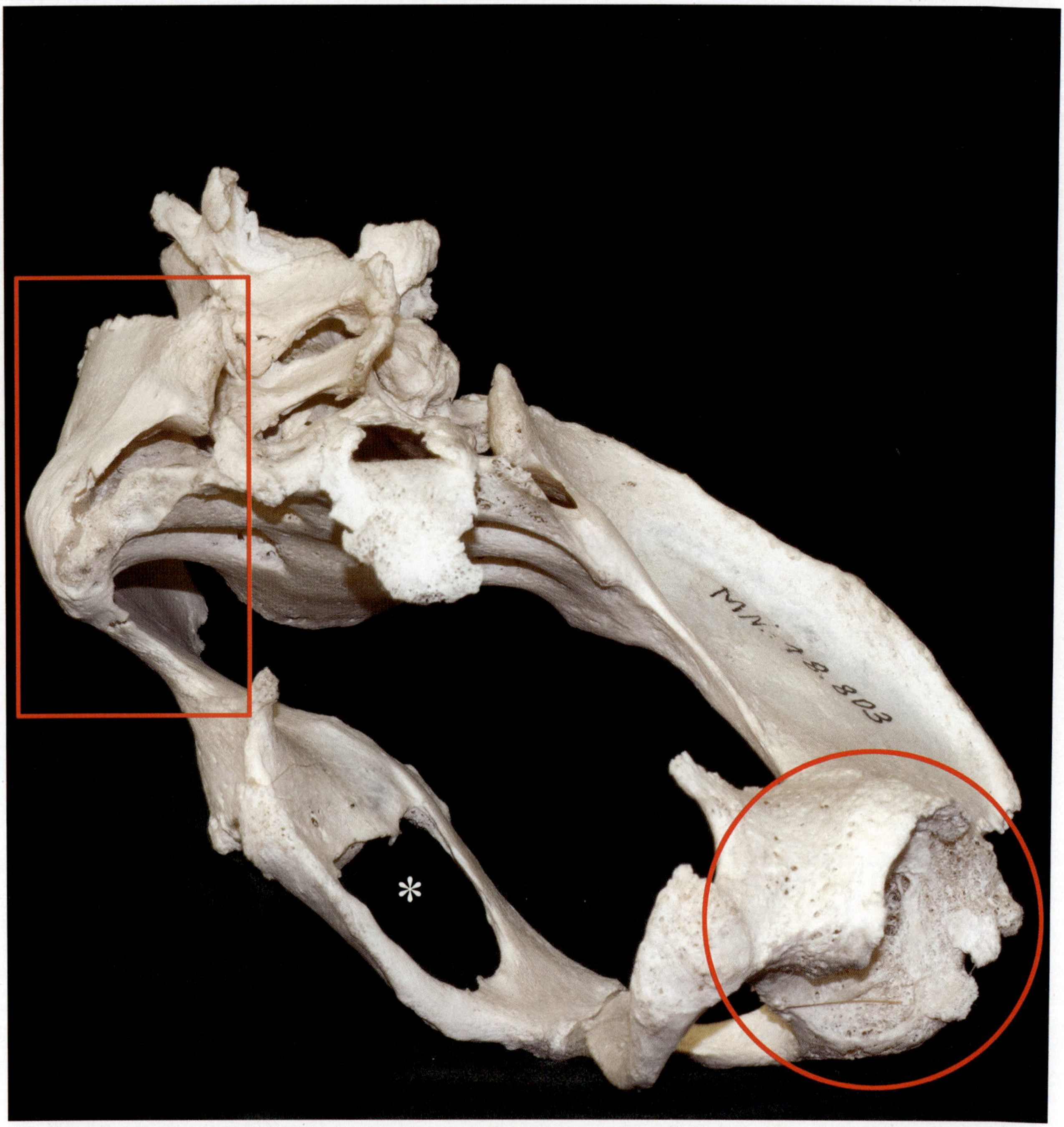


**Figure 145.** Inferior view of "Obliqua pelvis que eximii gradus e causa ignota" on the museum card (MN18.803). Oblique pelvis can be the result of differences in leg length, pelvic contractures, structural scoliosis, or any combination of these causes (Chan et al. 2019; Winter and Pinto 1976), as well as tumors or rickets (El-Mowafi 2017). Oblique pelvis as seen from below (inferior-superior) in this photograph, reveals that the left ilium has shifted right laterally (rectangle), the right acetabulum (circle) is altered by degenerative joint disease, and the left obturator foramen (*) is greatly enlarged and elongated. The lesion in the rectangle looks like it could be a fracture that hasn't completely healed (K. Thompson, personal communication, September 27, 2022).

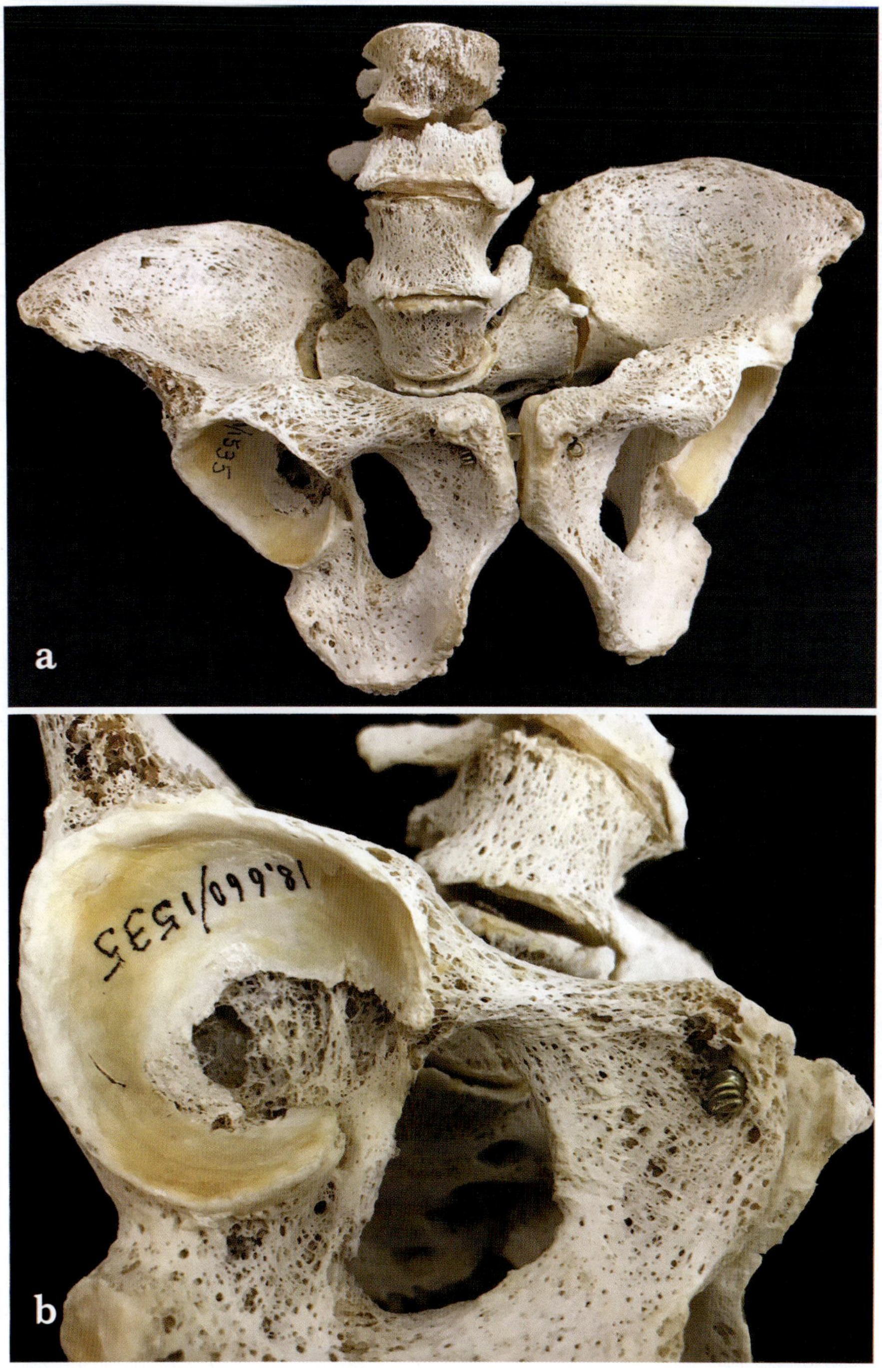


**Figure 146a-b.** (a) Anterior and (b) right oblique views of rickets and osteoporosis due to prostate cancer in a 60-year-old male who died in 1936 (MN18.660/1535). The first author has examined three Thai adult skeletons with patient histories documenting that one of them died with thalassemia and severe anemia, one of unknown causes, and the third of sepsis, all of whom exhibited a similar pattern of bone disease. See Snapper (1957); Steinbock (1976) for more on rickets, thalassemia, and osteoporosis. See Mann and Tuamsuk (2005) for a case report on an adult Thai male exhibiting this pattern of bone disease.

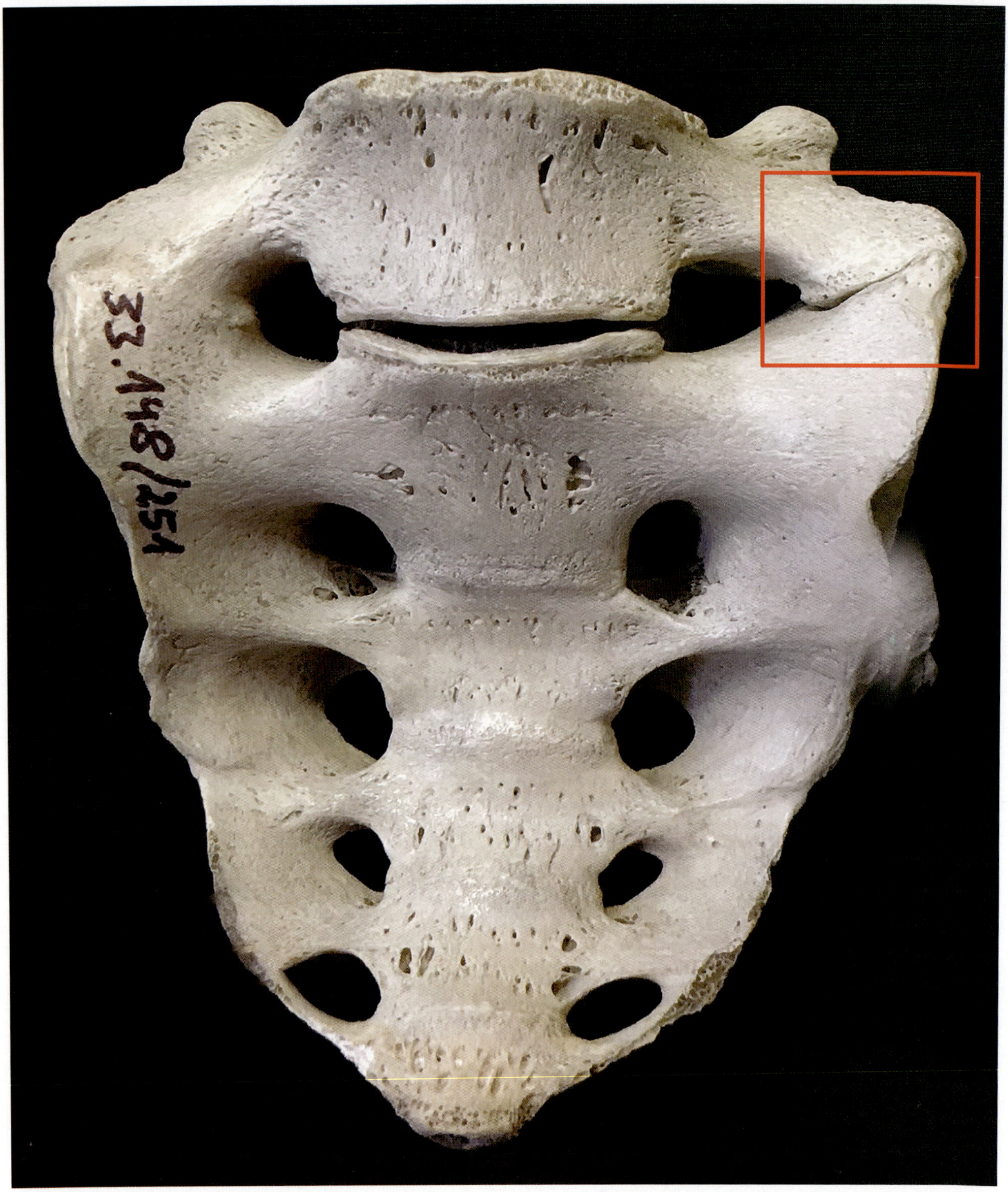


**Figure 147.** Anterior view of a sacralized 5th lumbar vertebra with articulation (square) and partial fusion of the left ala in a 19-year-old male who died in 1901 (MN18.560/300). This condition represents a cranial border shift resulting in the 5th lumbar vertebra assuming the morphology of the first sacral/alar segment (Barnes 1994, 2012).

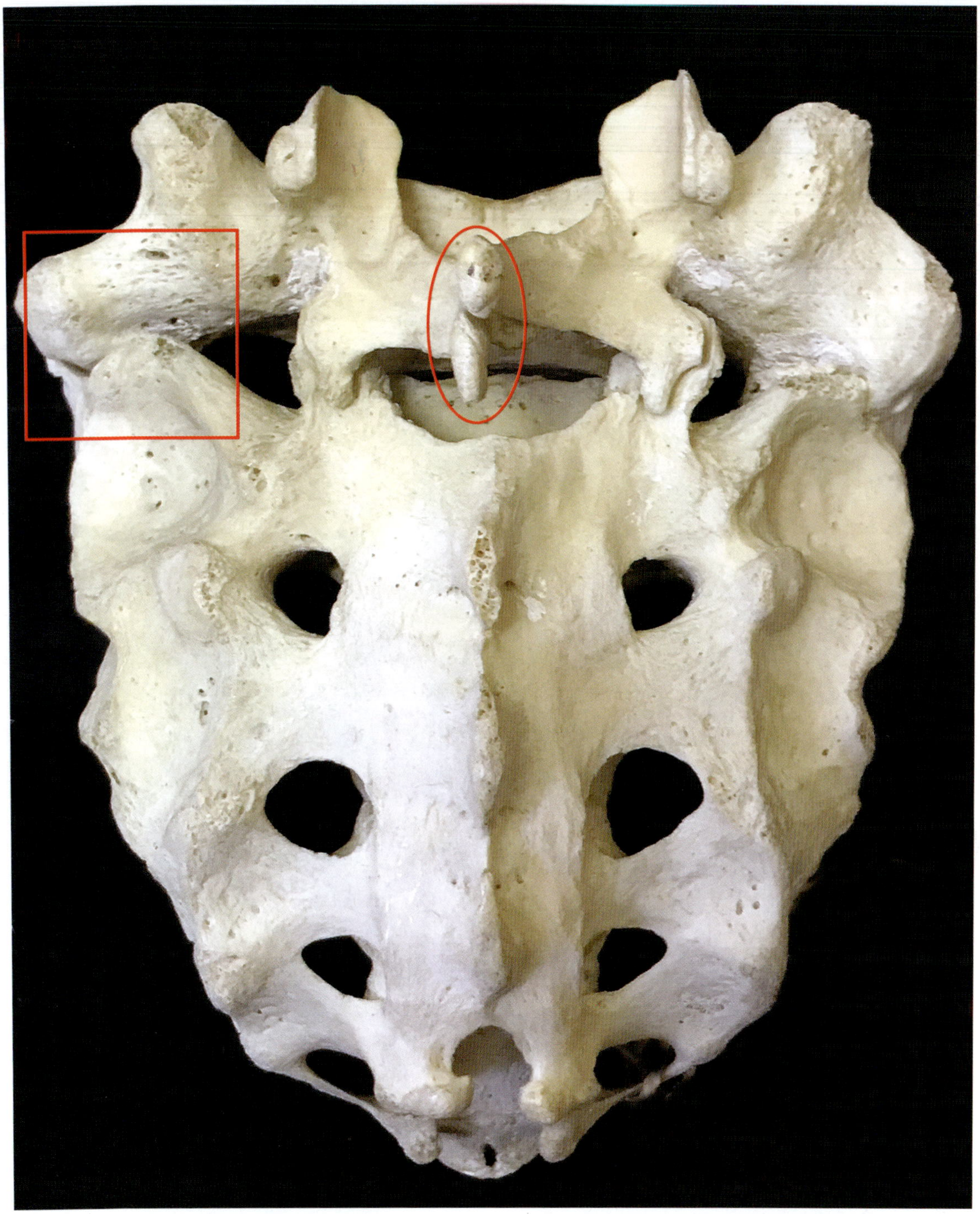

**Figure 148.** Posterior view of sacralized 5th lumbar vertebra, a cranial border shift (Barnes 2012), with articulation and partial fusion (square) of the left ala, unusual morphology and misalignment of the right articular facet and partial cleft leading into an aperture/cleft along the midline of the sacrum in a 19-year-old male who died in 1901 (MN18.560/300). Note malformation and right-left asymmetry ("slanted") of 5th lumbar vertebra (see Barnes 1994, 2012 for more on cranial/caudal border shifting).

## Arm and Hand

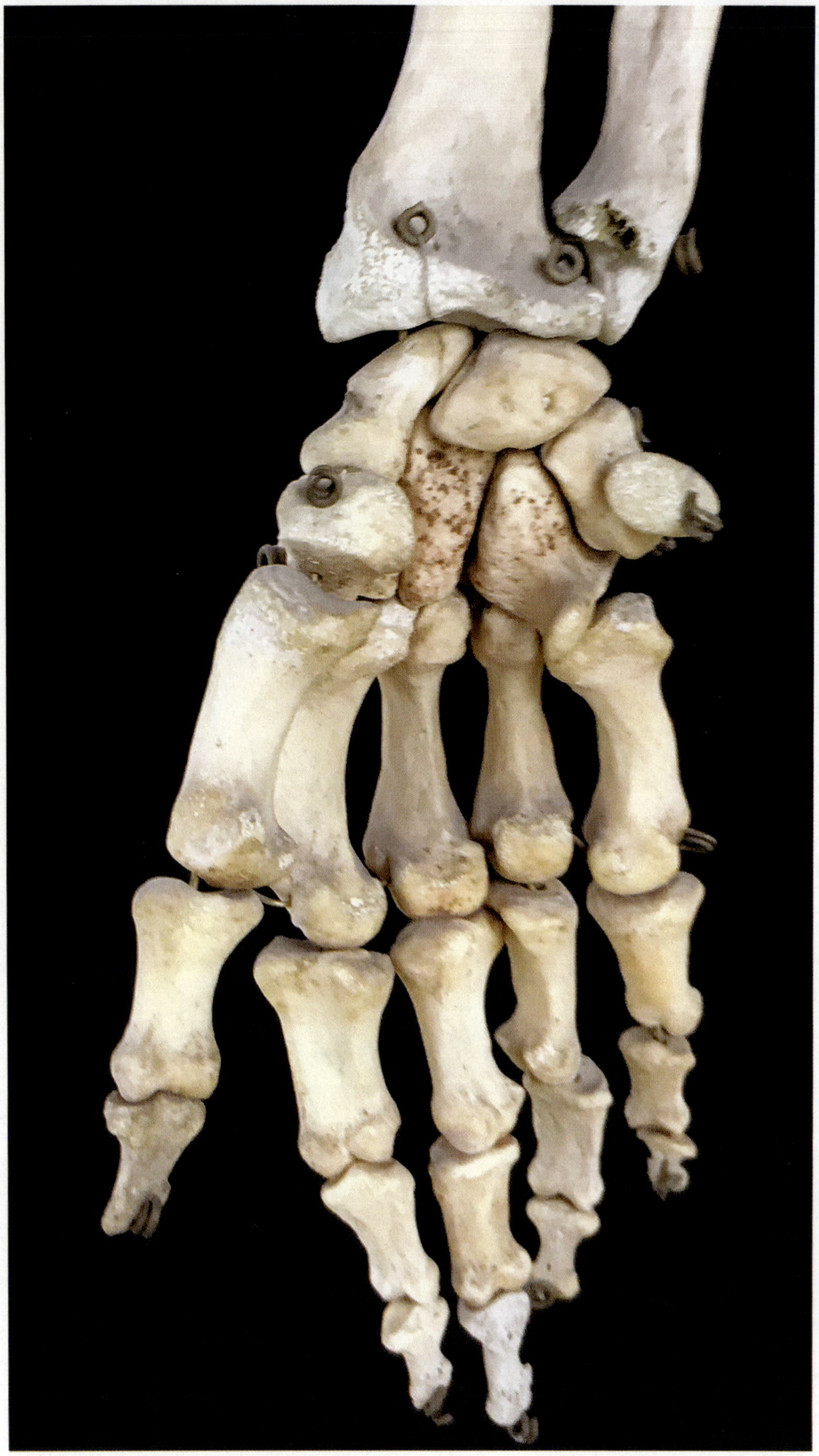

**Figure 149.** Achondroplasia ("chondrodystrophie" on the museum card) in the right hand of a 40-year-old male (MN670). The metacarpals and phalanges are short, robust, and approximately the same length, and the distal epiphyses of the proximal phalanges are irregularly shaped. See Ortner and Putschar (1981); Spranger et al. (2002) for more on achondroplasia.

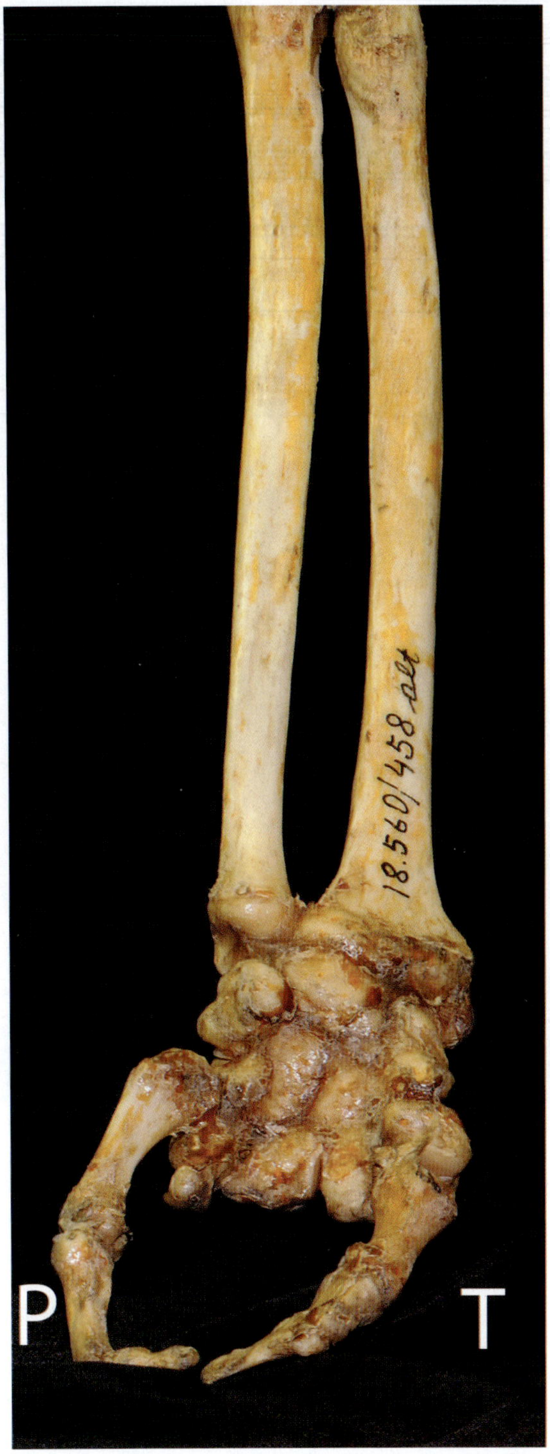


**Figure 150.** Palmar view of ectrodactyly or "lobster claw" of the left hand in a 30-year-old male who died in 1887 (MN18.560/458a). Ectrodactyly is mostly an autosomal dominant inherited defect involving the medial cleft of the hand. The museum card reports "malformation of the hand." This individual is only represented by the hand and arm, but lobster claw may be part of a syndrome consisting of ectrodactyly, ectodermal dysplasia and cleft palate (EEC) (Mumoli 2012).

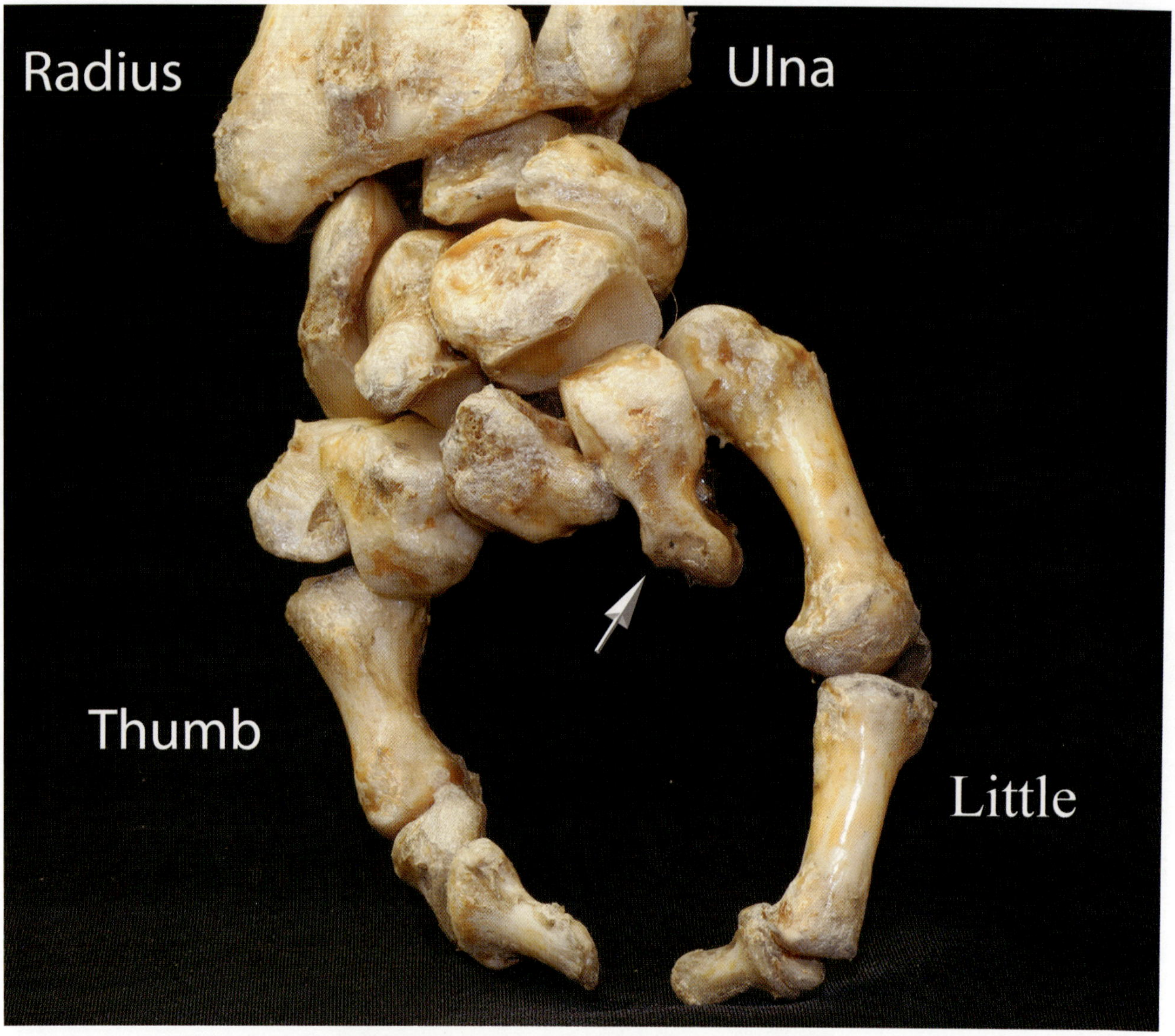


**Figure 151.** Dorsal view of ectrodactyly or "lobster claw" of the left hand of a 30-year-old male who died in 1887 (MN18.560/458a). Note the rudimentary fourth metacarpal (arrow) and absence of metacarpals and phalanges 2 through 4. The museum card reports "malformation of the hand." This individual is only represented by the hand and arm, but lobster claw may be part of a syndrome consisting of ectrodactyly, ectodermal dysplasia and cleft palate (EEC) (Mumoli 2012).

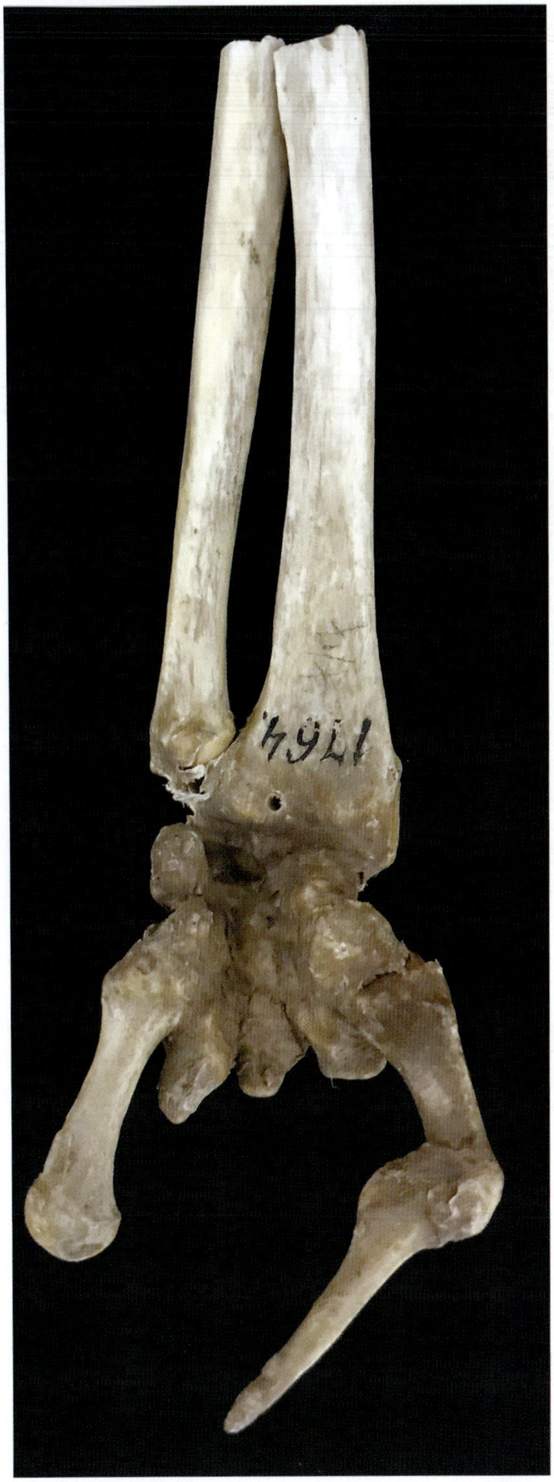

**Figure 152.** Palmar view of ectrodactyly "lobster claw" or "cleft hand" of the left hand in an individual who died in 1860 (MN1764). This hand exhibits medial cleft/missing fingers 2 through 4 and tapering/"penciling" of the thumb. The museum card reports "defective hand." This individual is only represented by the hand and arm, but may have had EEC (Ectrodactyly, Ectodermal Dysplasia-Clefting Syndrome) (Mumoli 2012).

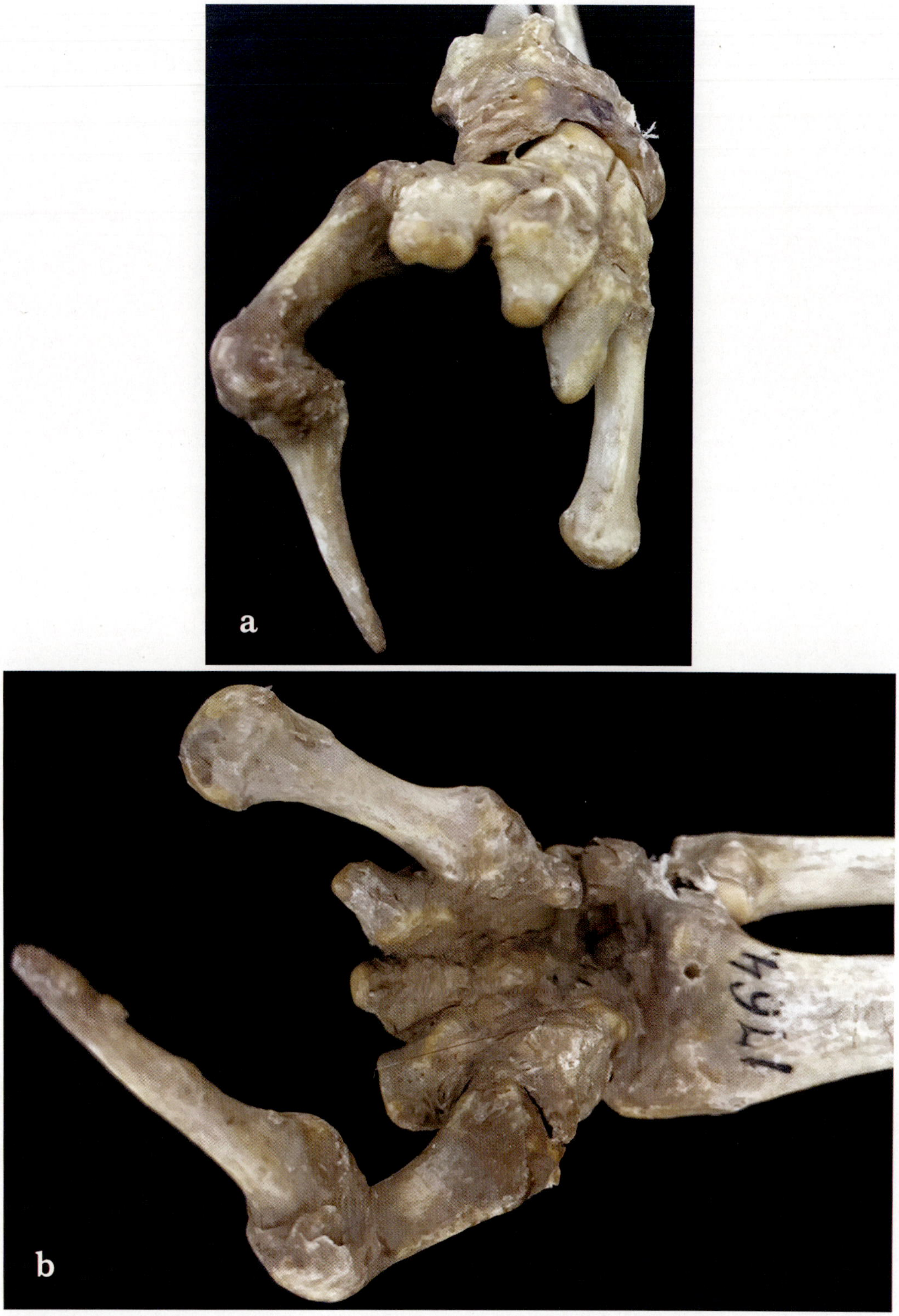


**Figure 153a-b.** Magnified views of ectrodactyly "lobster claw" or "cleft hand" of the left hand of an individual who died in 1860 (MN1764). The museum card reports "defective hand." This individual is only represented by the hand and arm, but may have had a syndrome consisting of Ectrodactyly, Ectodermal Dysplasia-Clefting Syndrome (EEC) (Mumoli 2012).

## Leg and Foot

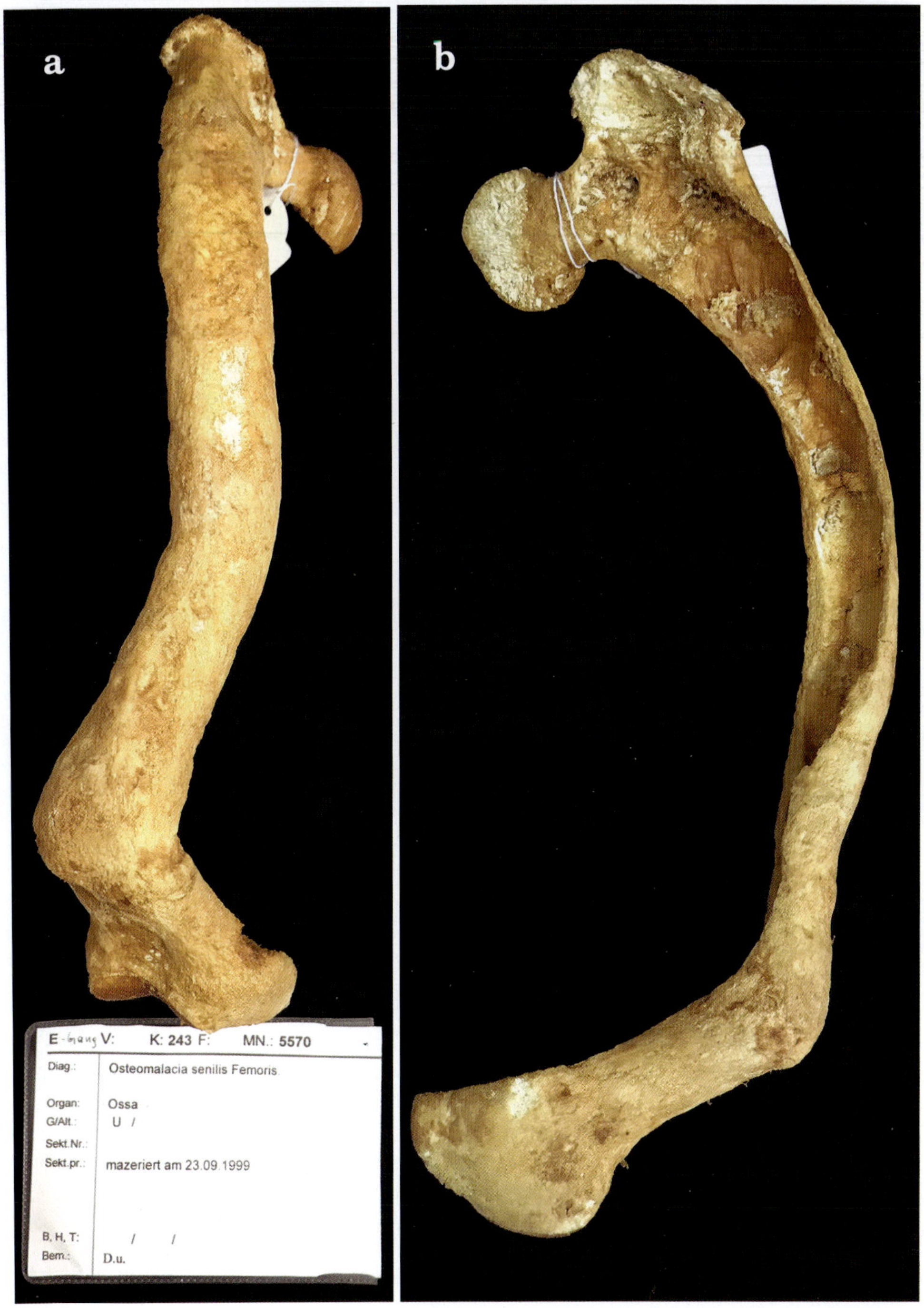


**Figure 154a-b.** (a) Anterior and (b) cross-section of a right femur with senile osteomalacia (vitamin D deficiency) in an individual who died in 1893 (MN5570). (b) Longitudinal cross-section of the femur reveals a thin cortex and absence of medullary cavity and trabeculae. See Geshchickter and Copeland (1936); Snapper (1957) for more on rickets and osteomalacia, and Griffiths (1981) for radiographic features of osteomalacia.

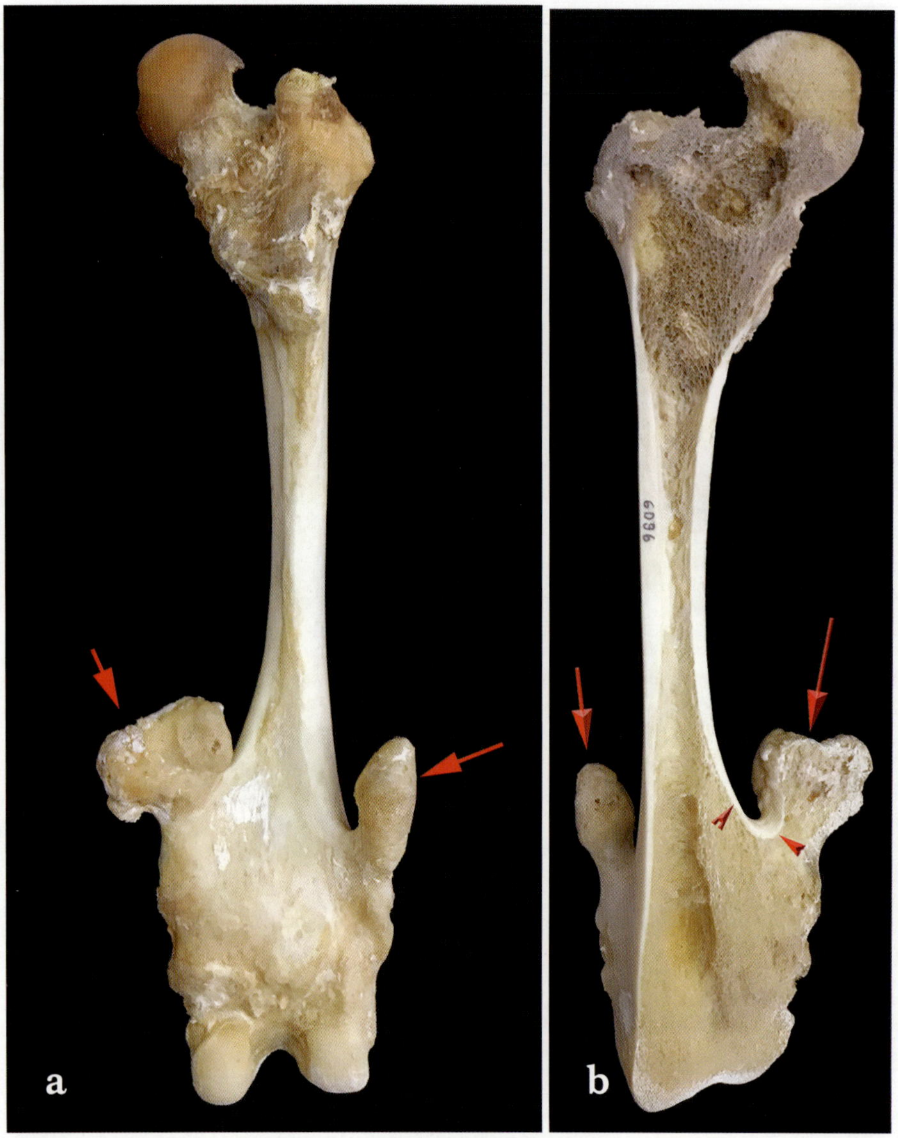


**Figure 155a-b.** (a) Posterior external and (b) anterior internal views of multiple hereditary osteochondromas, or exostoses, in the right femur of a 27-year-old male who died in 1930 (MN6096). Osteochondromas are benign bony projections that typically develop by the early teens, slow during adolescence, and with no new osteochondromas in adults. These cartilage capped projections are typically located near the ends of long bones and are oriented parallel to the long axis of diaphysis. Although most are benign, multiple hereditary exostoses have an increased risk of malignant transformation (Cazjka and DiCaprio 2015).

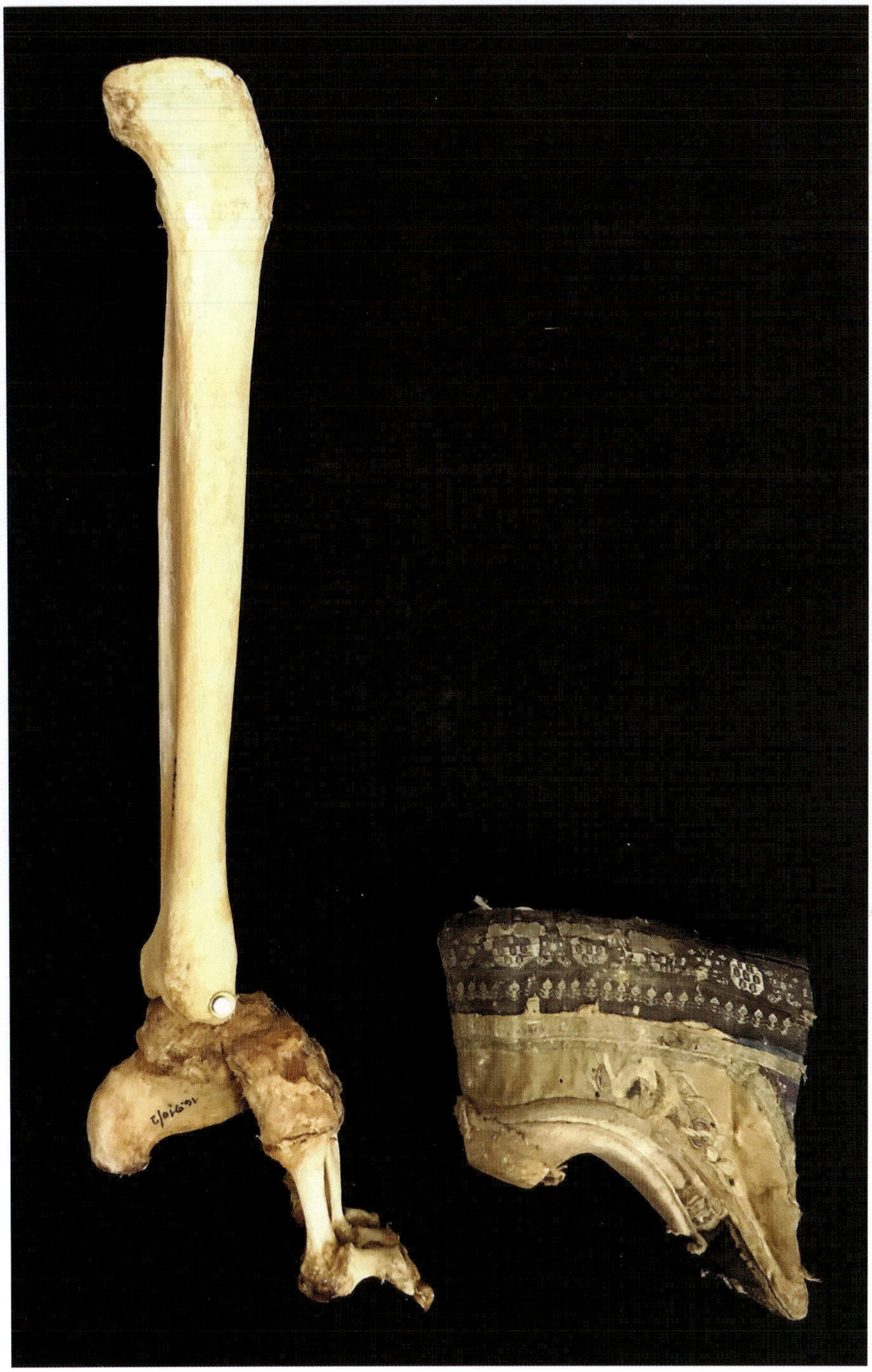

**Figure 156.** Chinese foot binding of the left foot (MN16.910/2) in an adult female and the heavily used silk shoe that she wore (Figure 158). Chinese foot binding was practiced for more than 1,000 years starting in the late 13th century (Foreman 2015; Mann et al. 1990; Richardson 2009). See Ortner and Putschar (1981) for an example of Chinese foot binding and Richardson (2009) for a radiological case report of Chinese foot binding.

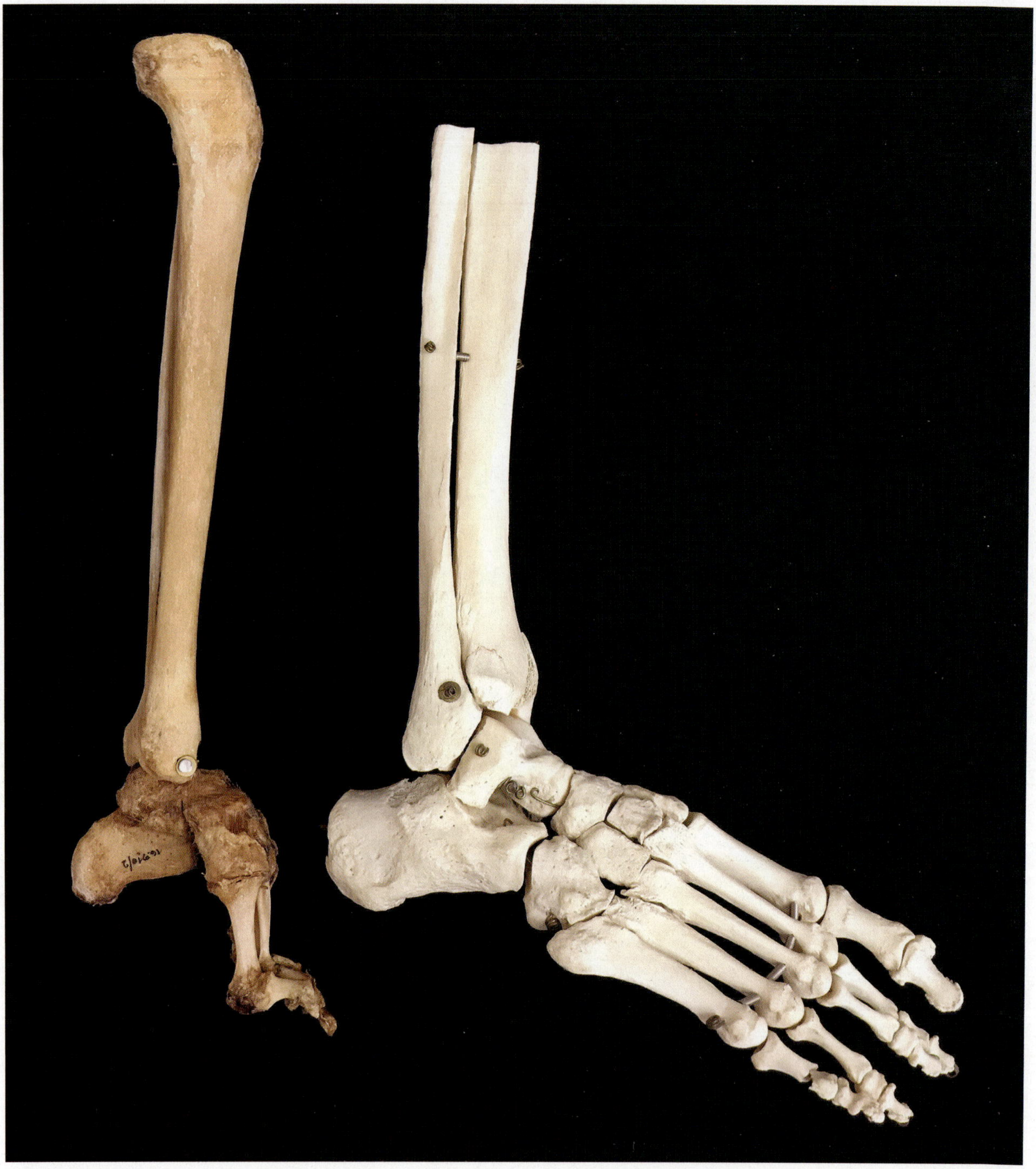

**Figure 157.** Chinese foot binding ("Lotus foot") of the left foot (MN16.910/2) in an adult female compared to a typically sized right foot. The ideal size of a Chinese bound foot in the 19th century was no more than 3 inches (7-8 cm) long and shaped like a crescent moon. Fashion, modesty, status, marriage, and possibly erotica were factors in the development and persistence of bound feet in Chinese females. The practice usually began between 5-8 years of age, often resulting in infection (gangrene) and even death. This painful practice that lasted more than 1,000 years was outlawed in China in 1912. See Mann et al. (1990); Ortner and Putschar (1981) for examples of Chinese foot binding.

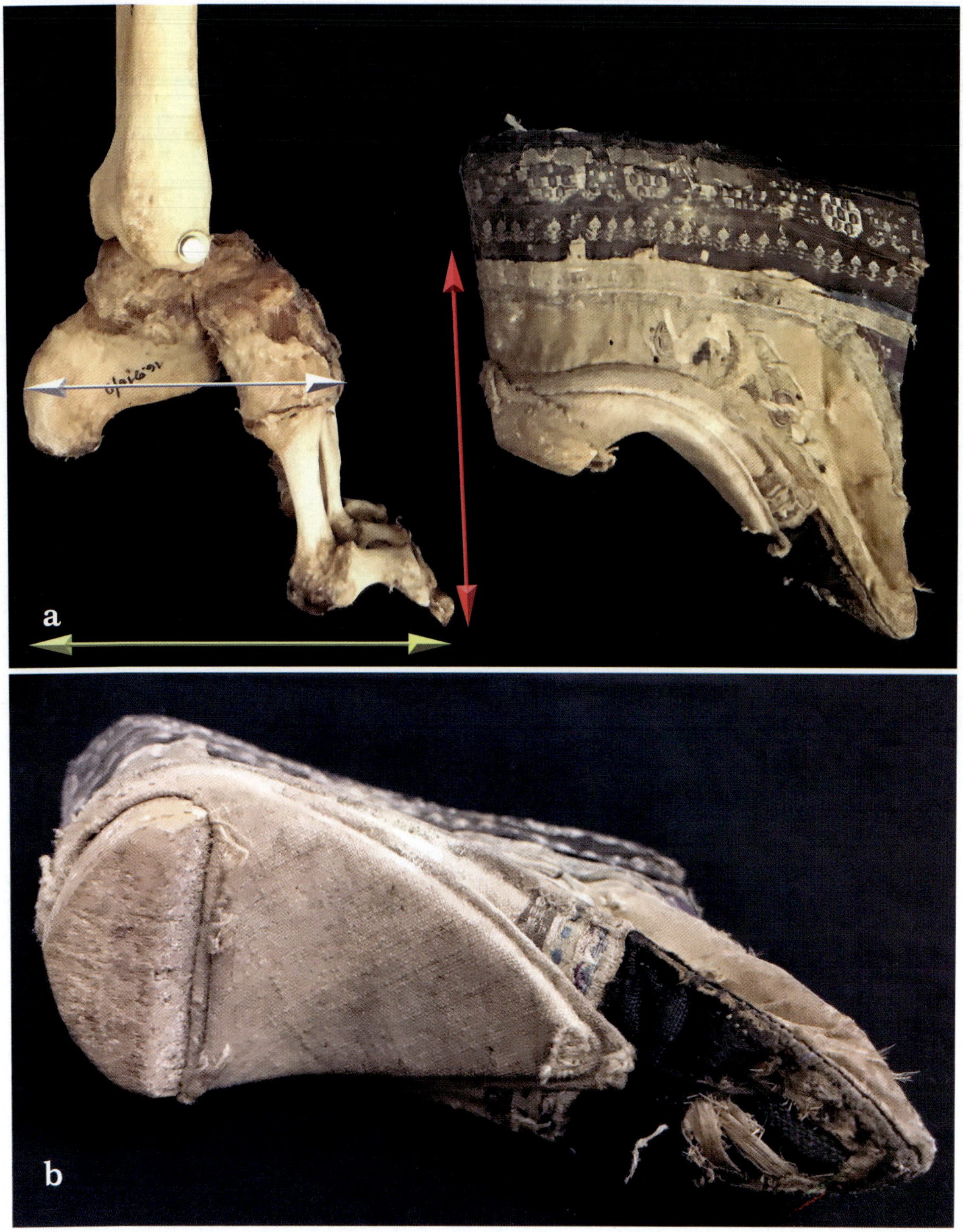


**Figure 158a-b.** (a) Example of Chinese foot binding of the left foot of an adult female (MN16.910/2) and the shoe this female wore. (a) The length from the heel to the cuneiforms (white arrow) is 8 cm, the height of the talar dome to the inferior portion of the phalanges (red arrow) is 9.3 cm and the length of the foot from the heel to the toes (yellow arrow) is 10.7 cm. (b) Magnified view of the show showing what appears to be years' of walking and/or standing in this hand-made shoe. See Mann et al. (1990); Ortner and Putschar (1981).

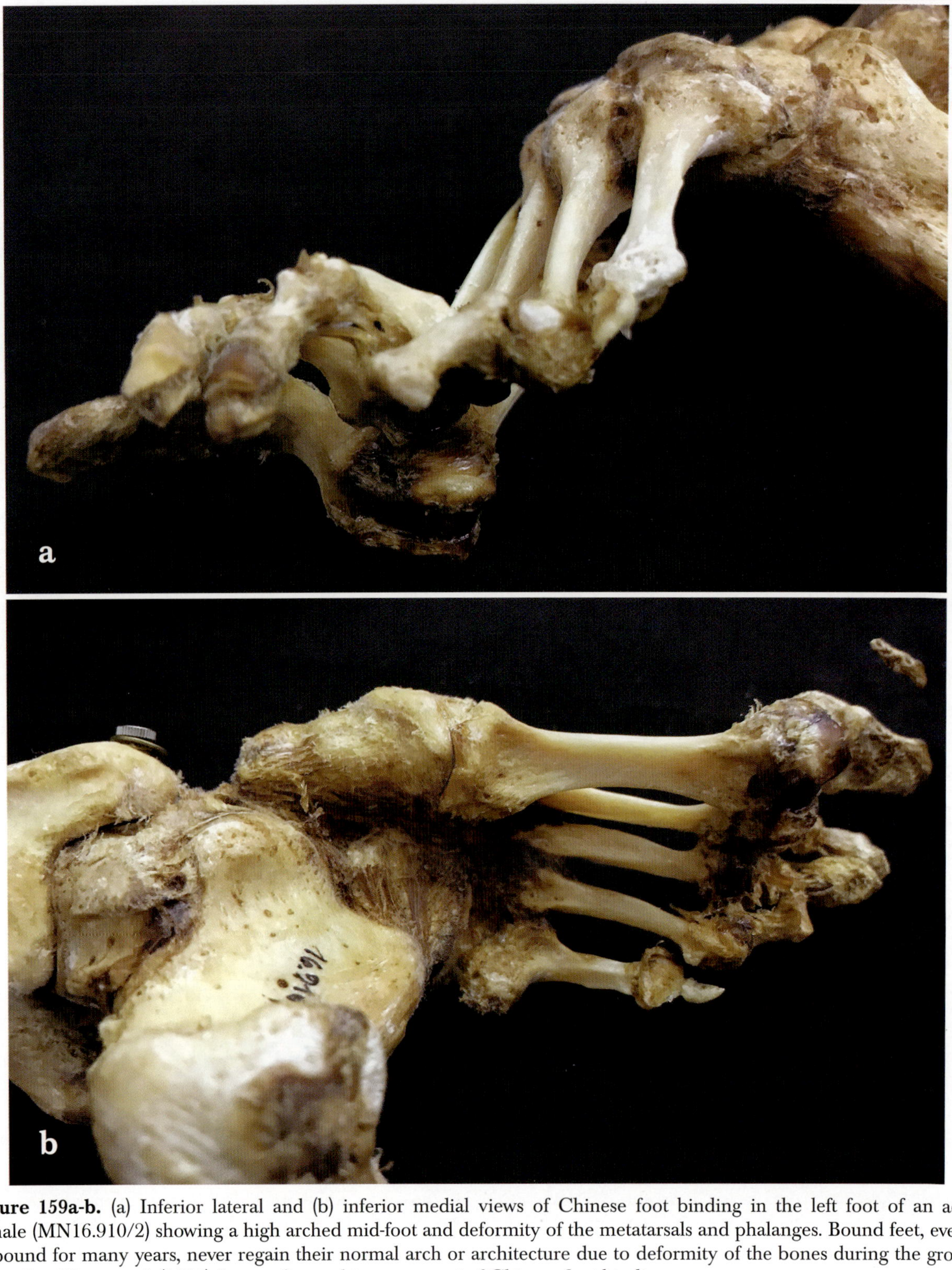


**Figure 159a-b.** (a) Inferior lateral and (b) inferior medial views of Chinese foot binding in the left foot of an adult female (MN16.910/2) showing a high arched mid-foot and deformity of the metatarsals and phalanges. Bound feet, even if unbound for many years, never regain their normal arch or architecture due to deformity of the bones during the growth years. See Mann et al. (1990) for a radiographic case report of Chinese foot binding.

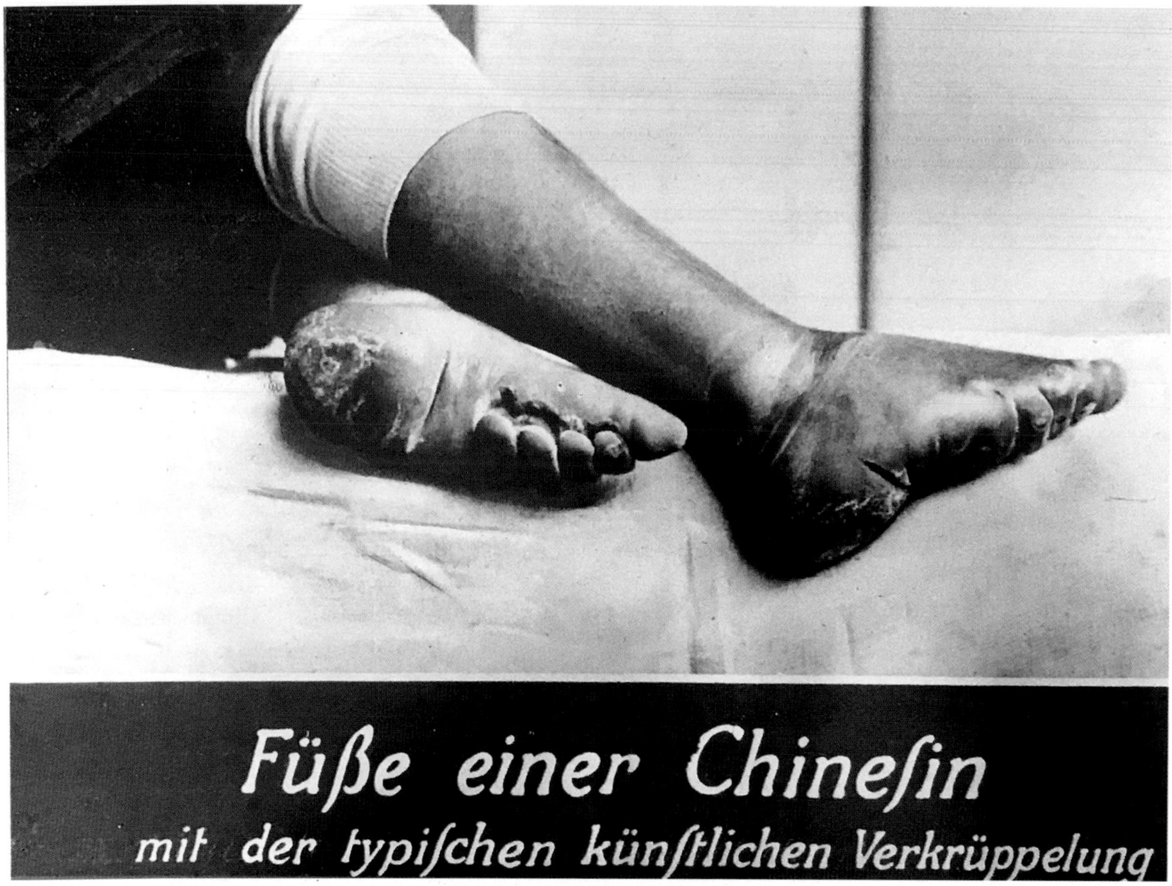


**Figure 160.** An example of Chinese foot binding in an adult female (Narrenturm).

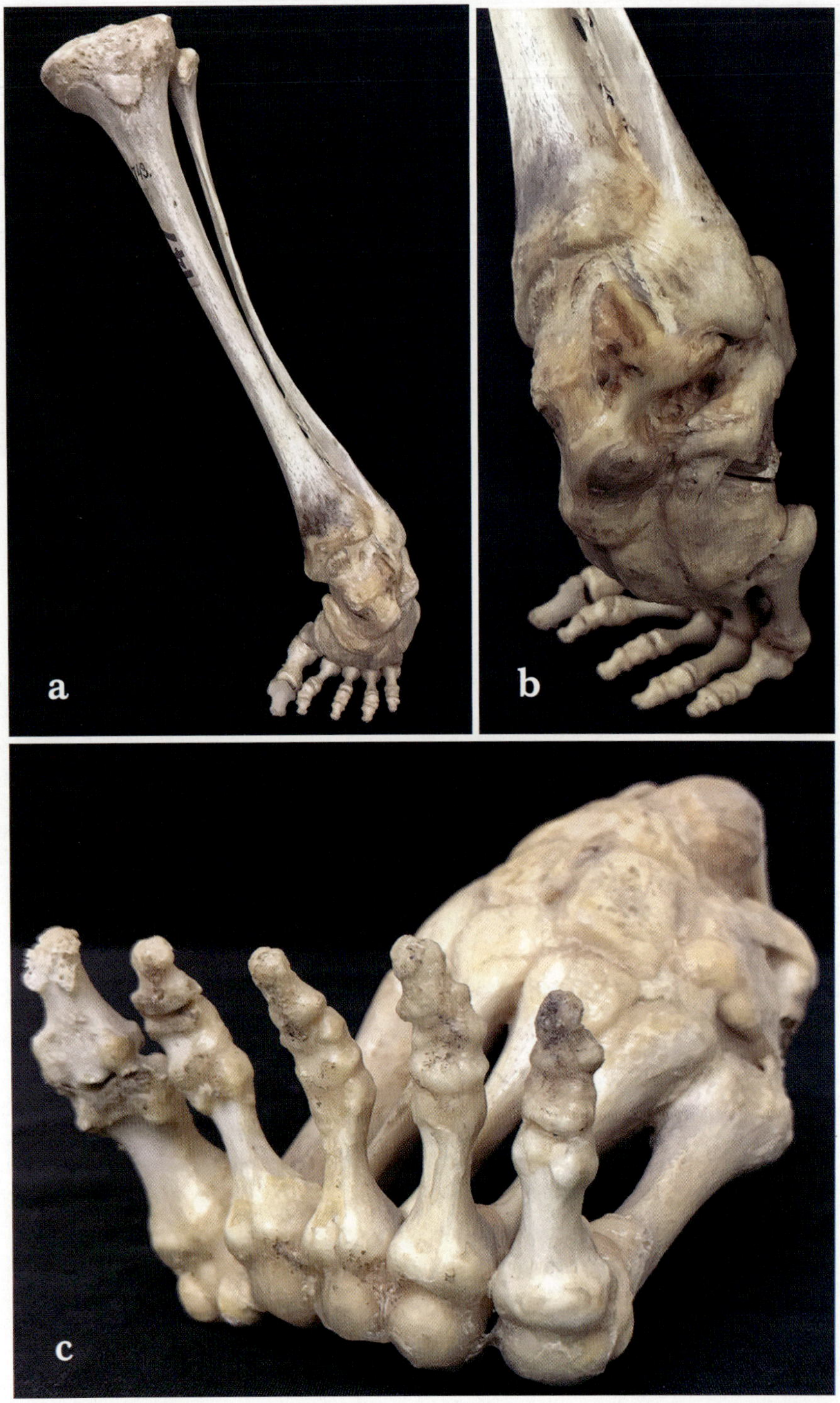


**Figure 161a-c.** (a) Anterior, (b) oblique and (c) distal views of left clubfoot (inversion) ("dropfoot" written on the museum card which is a general meaning for having difficulty lifting the front of the foot) resulting in extreme dorsiflexion of the toes ("toe walking") in an approximately 13 to 18-year-old individual (based on epiphyseal union of the proximal tibia) who died of unknown causes in 1871 (MN21.055/1743). Note the position of the tibia on the talus and that the distal left fibula has developed to fit the lateral contour of the tibia. See Mann et al. (1990) for a radiographic case report of foot binding.

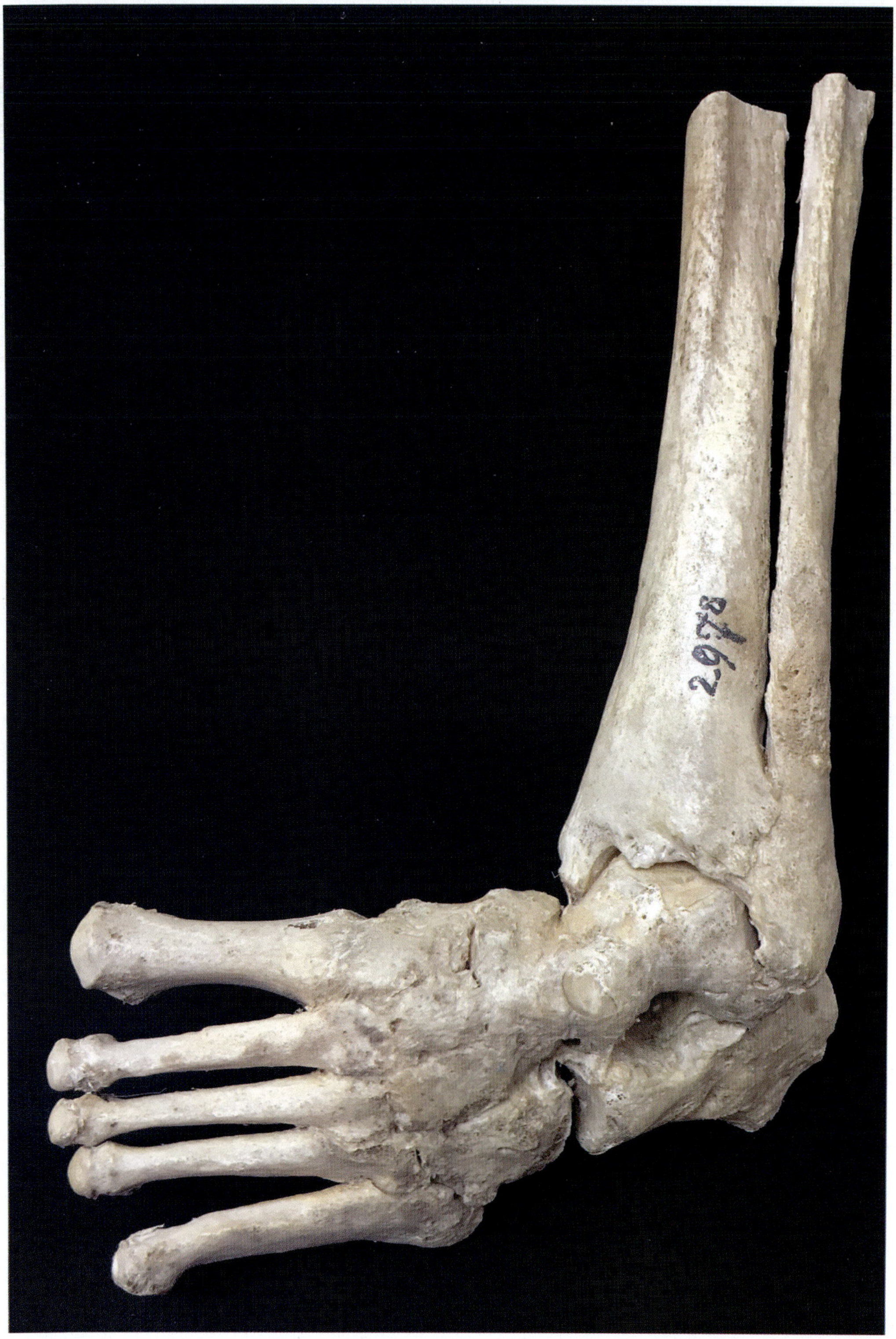


**Figure 162.** Clubfoot (pes valgus) resulting in inversion of the left foot of a 64-year-old female who died in 1876 (MN32.055/2978). See *Campbell's Operative Orthopedics* (1956); Mann et al. (1990); Ortner and Putschar (1981) for early examples and descriptions of clubfoot.

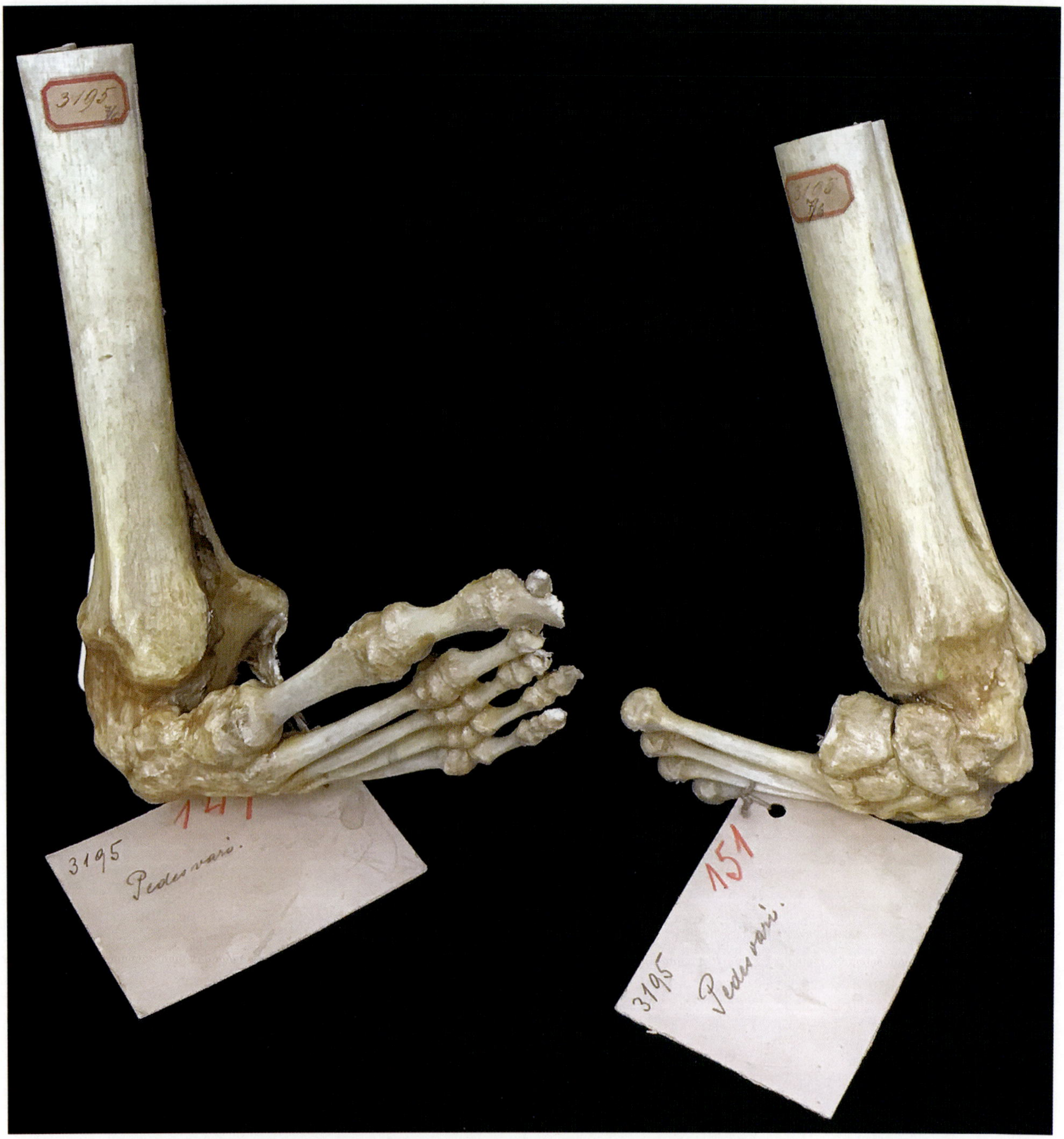


**Figure 163.** Bilateral clubfoot ("pedes vari" on the card) in a 46-year-old male who died in 1877 (MN3195). See Ortner and Putschar (1981) and *Campbell's Operative Orthopaedics* (1956) for more on clubfoot deformity.

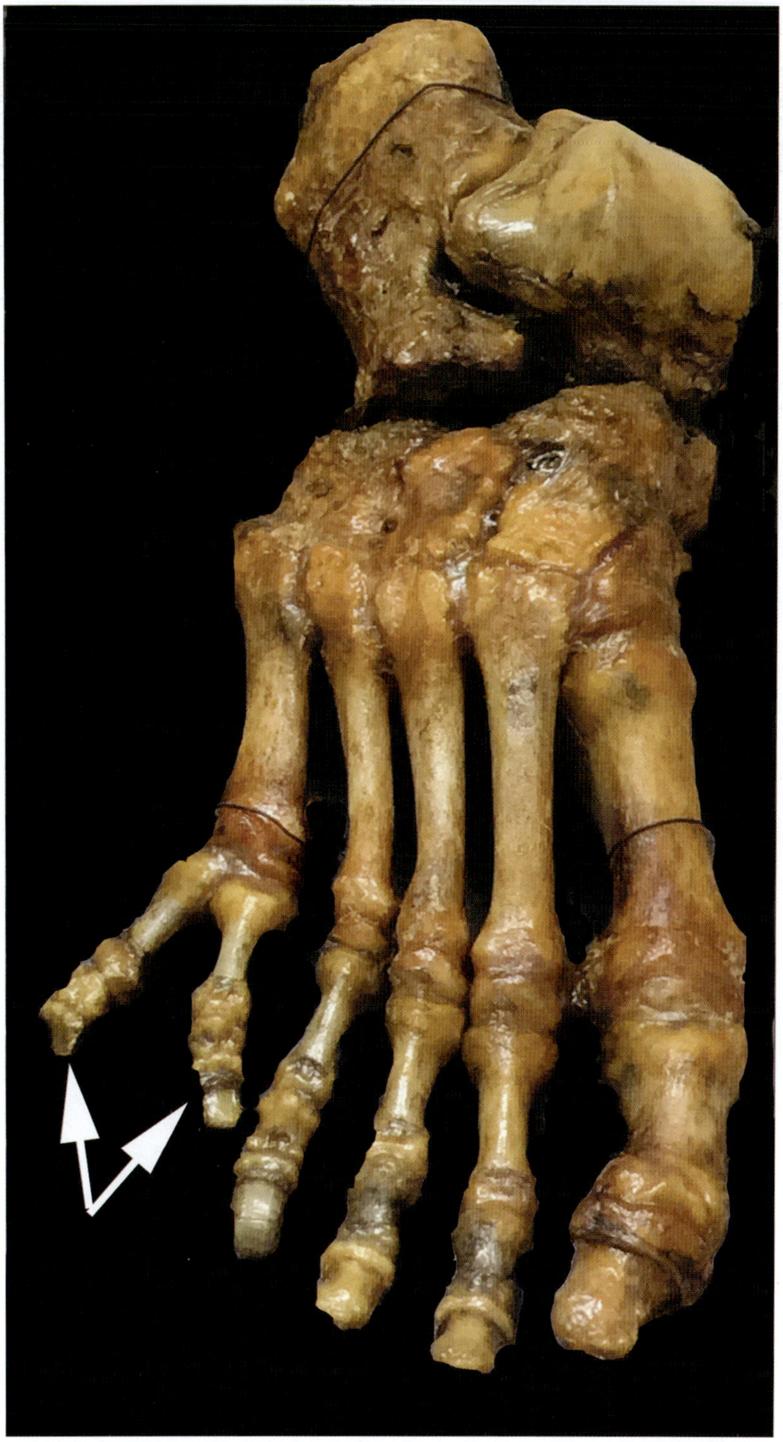

**Figure 164.** Polydactyly is a common malformation that often runs in families and is due to genetic mutations or environmental causes (Bellovits 2003). This individual exhibits duplication of the proximal, middle and distal phalanges (arrows) (Type IV, Wassel Classification; Manske et al. 2017) (MN2866b). Note the unusually thick diaphysis of the fifth metatarsal. Ortner and Putschar (1981; p. 363) present this example of polydactyly. See *Campbell's Operative Orthopaedics* (1956) for more on polydactyly.

# Chapter 2

# INFECTIOUS DISEASE

## Skull

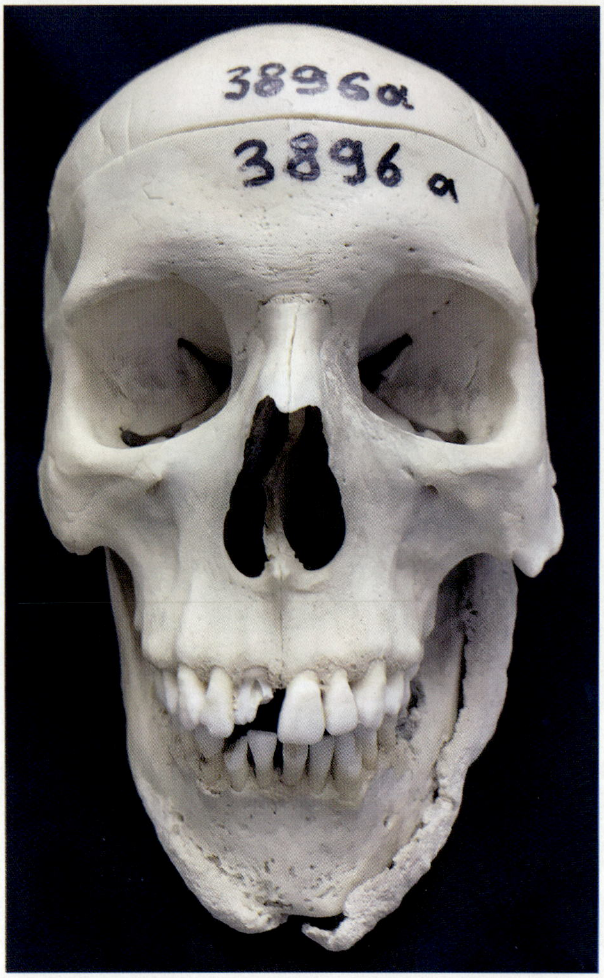


**Figure 165.** "Phossy jaw" or necrosis of the mandible in an unknown individual who died in 1892 (MN3896a). "Phossy jaw," an occupational hazard, resulted from chronic exposure to white or yellow phosphorus in matchstick workers during the 19th and early 20th centuries. The reported mortality rate of individuals with "phossy jaw" osteonecrosis between 1858 and 1906 was 20% (Marx 2008). The use of phosphorus to produce a "strike anywhere effect" on matchstick tips ceased after the Industrial Revolution (Pollock et al. 2015). The involucrum or reactive bone in this individual is loosely attached to the mandible. See Ortner and Putschar (1981) for more on phosphorus necrosis.

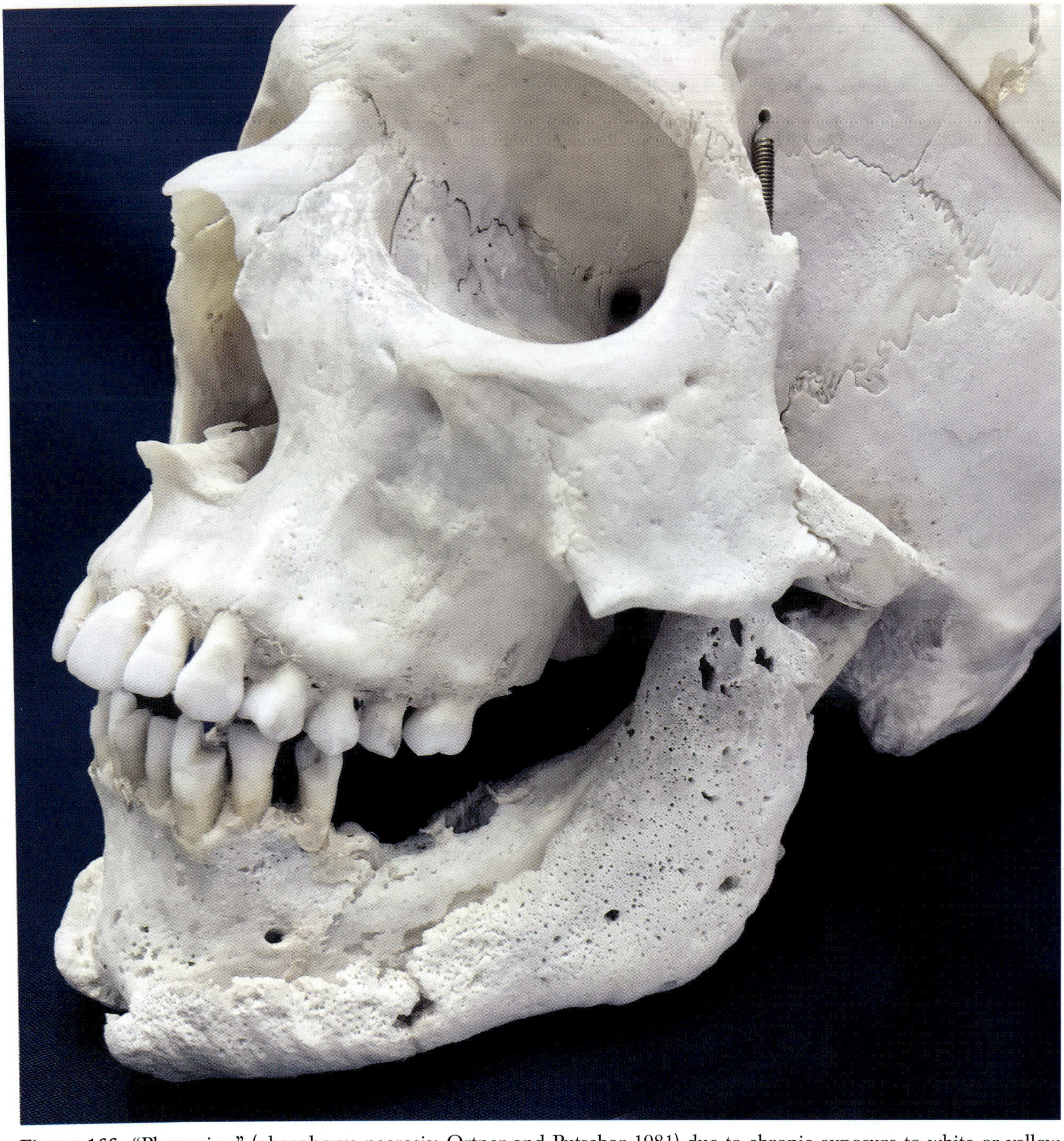

**Figure 166.** "Phossy jaw" (phosphorus necrosis; Ortner and Putschar 1981) due to chronic exposure to white or yellow phosphorus in the making of matchsticks in an individual who died in 1892 (MN3896a). Long-term exposure to the fumes of cooked/heated phosphorus resulted in necrosis of the mandible and formation of a porous and weakly formed involucrum that is loosely attached to the mandible.

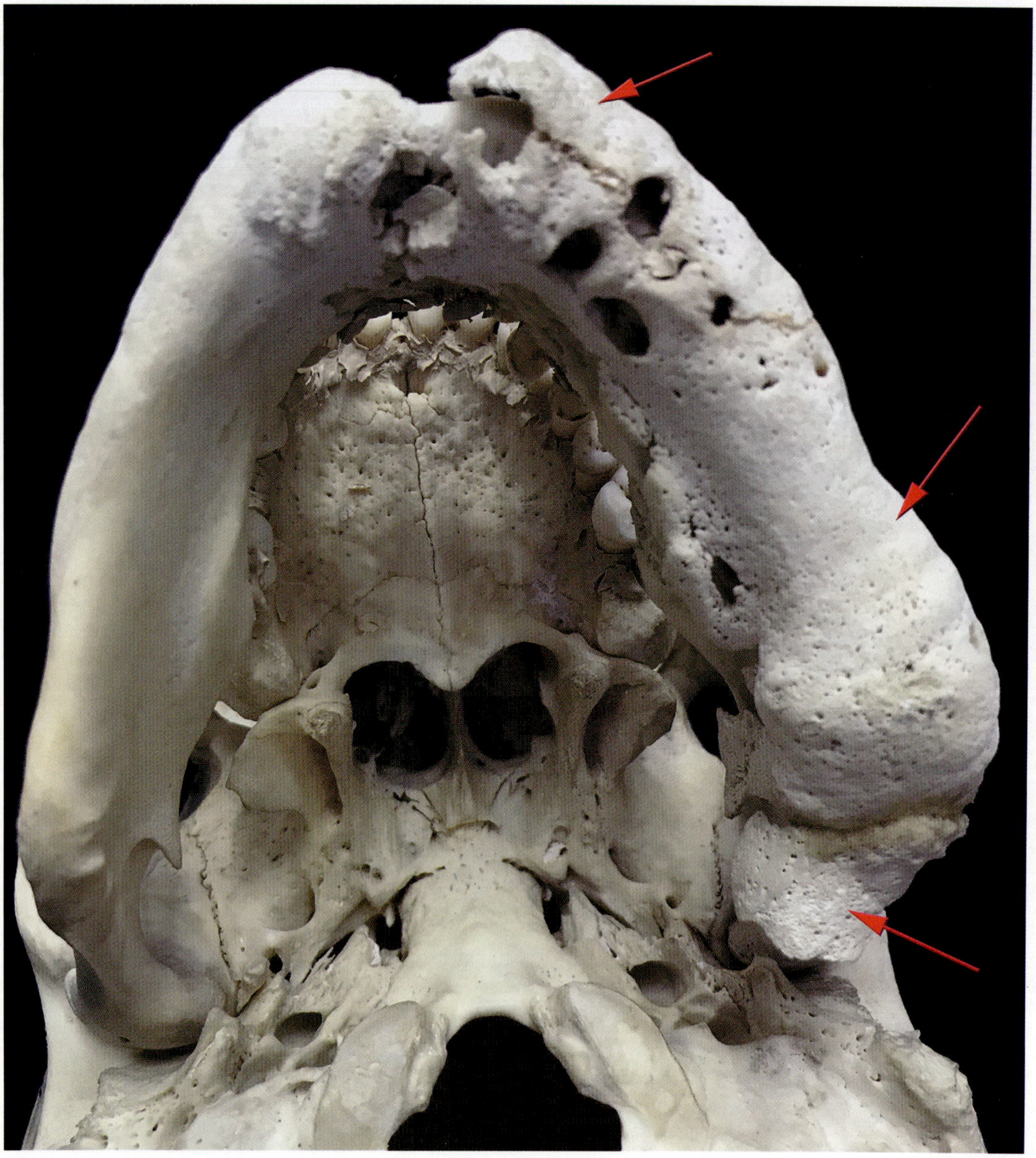

**Figure 167.** "Phossy jaw" due to chronic exposure to fumes produced by cooking or heating white or yellow phosphorus in the making of "strike anywhere" matchsticks in an individual who died in 1892 (MN3896a). Note the involucrum loosely adhering to and encasing the left lingual (arrow) and buccal half of the mandible.

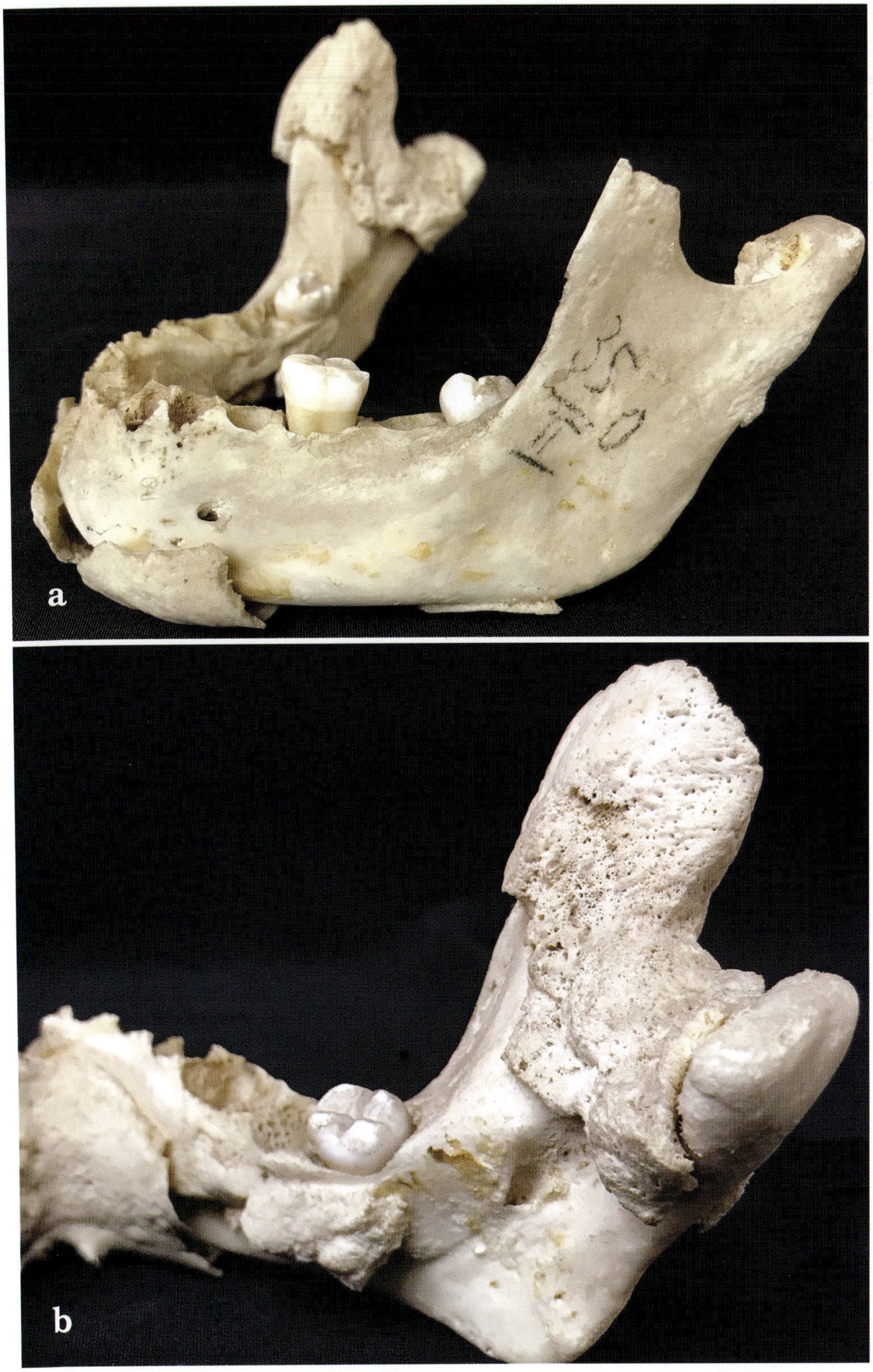


**Figure 168a-b.** (a) Left lateral and (b) posterior oblique views of "Phossy jaw" (listed on the accession card as "periostitis ossificans") due to exposure to white or yellow phosphorus fumes in an individual who died in 1880 (MN21.055/3512). Note the loosely attached bone (involucrum) over much of the mandible that can easily be removed by handling.

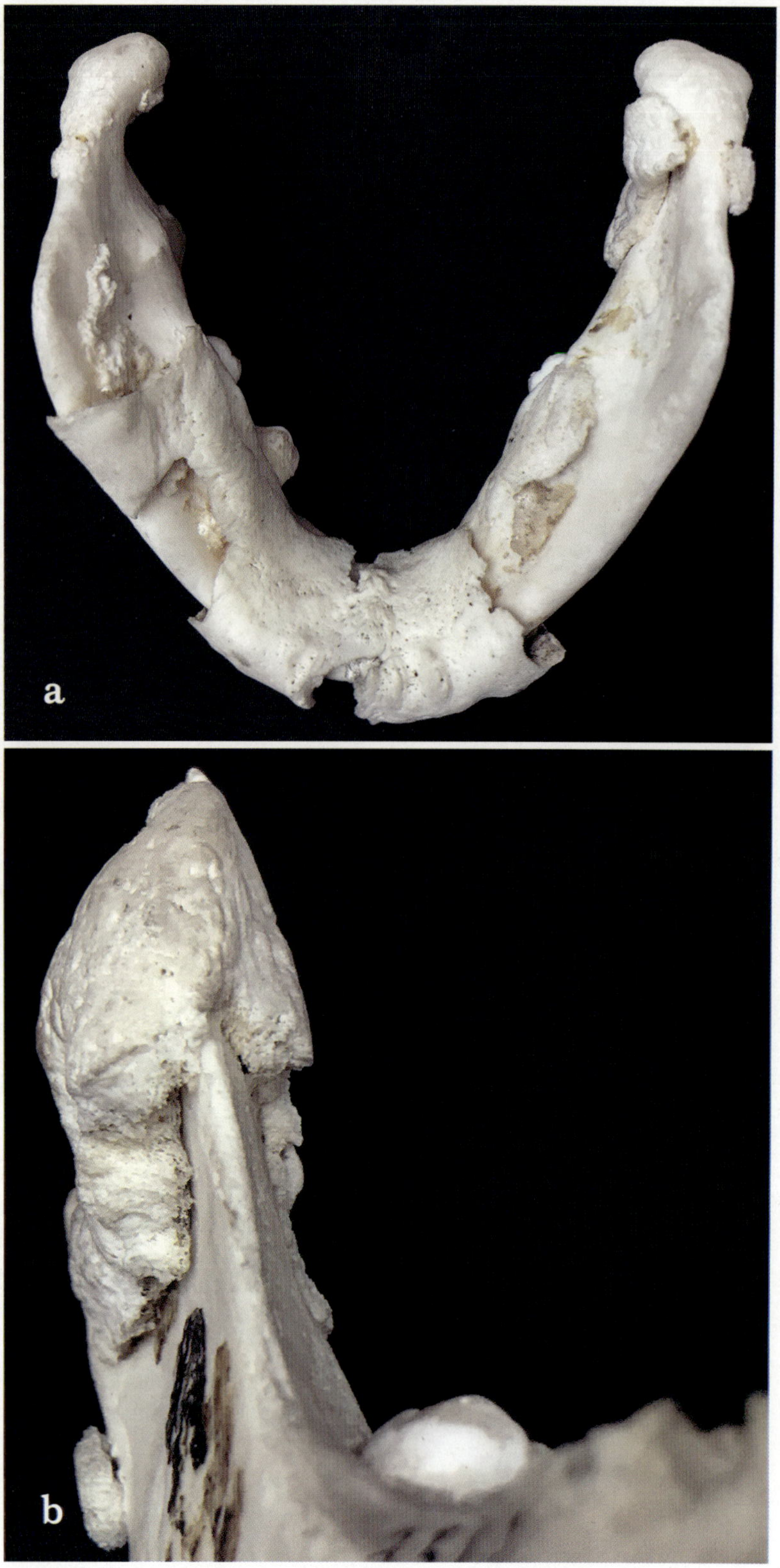


**Figure 169a-b.** (a) Inferior and (b) magnified anterior views of "Phossy jaw" listed on the museum card as "periostitis ossificans" in the mandible of an individual who died in 1880 (MN21.055/3512).

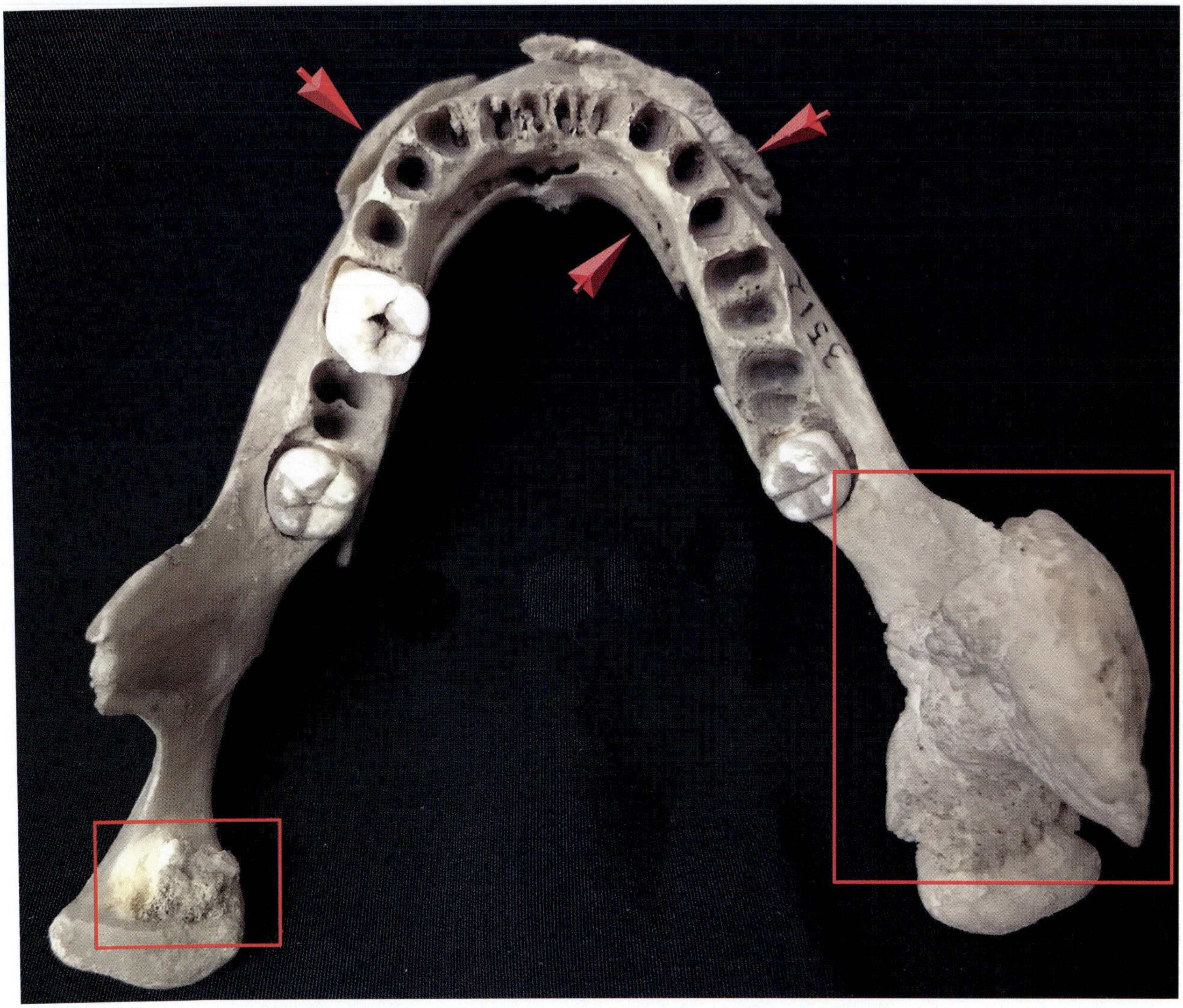

**Figure 170.** "Phossy jaw" listed on the museum card as "periostitis ossificans" in an individual who died in 1880 (MN21.055/3512). Note the loosely attached bone (involucrum; arrows and boxes) covering much of the mandible and that can be easily removed by handling.

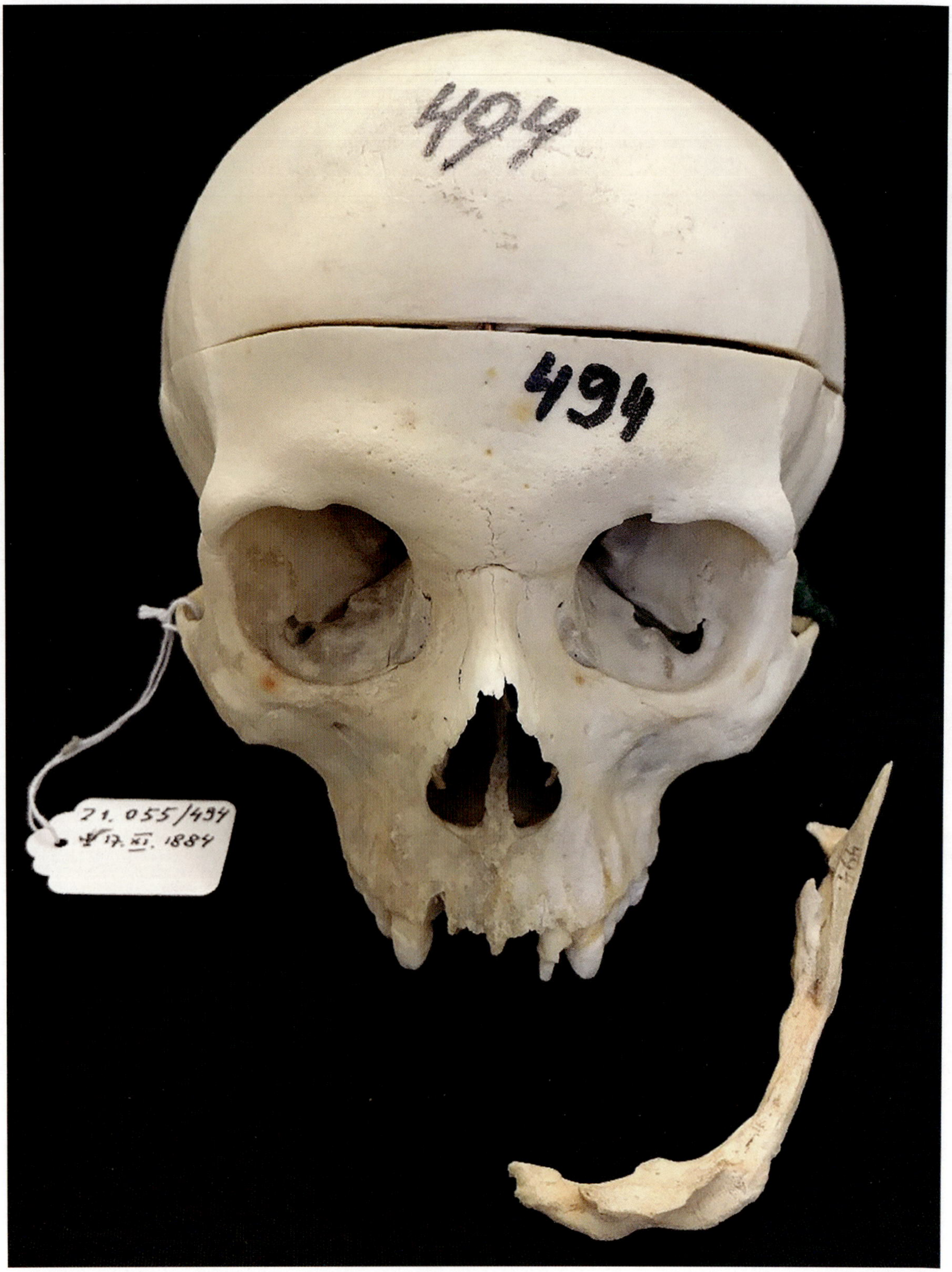


**Figure 171.** Atrophied mandible in an individual listed as having "phossy jaw" (necrosis phosphorus) and who died in 1884 (MN21.055/494). Long-term exposure to the fumes of heated white or yellow phosphorus often produced osteonecrosis (death) of the mandible.

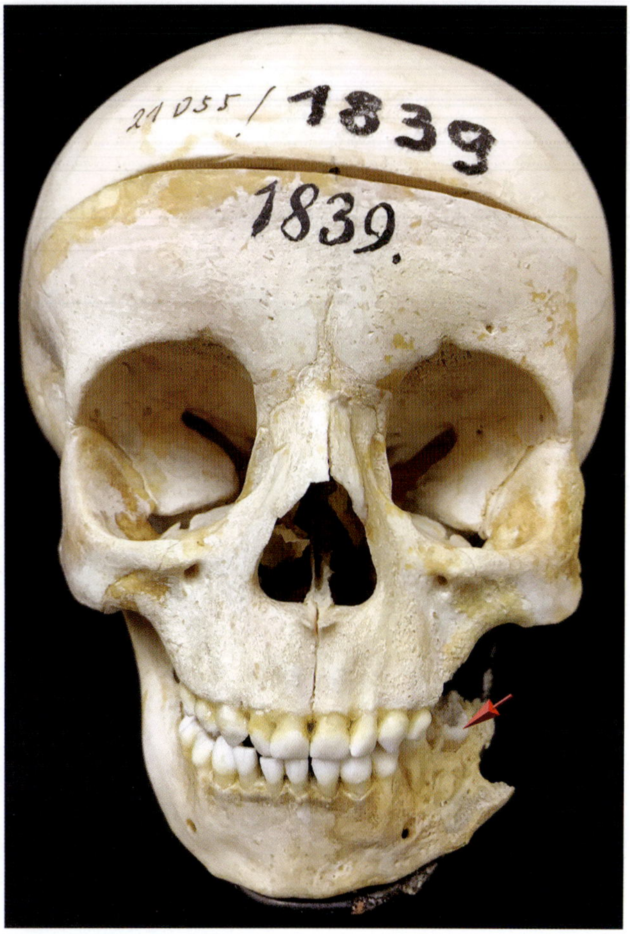


**Figure 172.** Four-year-old male who died in 1871 listed as "Caries, cause unknown" in the left mandible (MN21.055/1839) possibly due to congenital syphilis (authors' opinion). The distal half of the corpus of the left half of the mandible and entire ramus have been destroyed, as well as the vomer, nasal septum, turbinates, and much of the sphenoid. There are scattered patches of periostitis and pitting on the maxillae and left greater wing of the sphenoid. The maxillary central incisors do not exhibit Hutchinson's ("notched") incisors and the first permanent molars are not "mulberry molars," features often seen in congenital syphilis (Hutchinson 1861). The left upper and lower first (arrow) permanent molars, however, exhibit an unusual cusp pattern (Figures 173 and 174) known as Fournier's molars (Iannou et al. 2016) often present in congenital syphilis. While congenital syphilis, "the great imitator," is a possibility in this child, the differential diagnosis would include bone metastases, multiple myeloma, osteomyelitis, lymphoma, tuberculosis, histiocytosis, and leukemia (Petroulia et al. 2020) as well as environmental, genetic or systemic variables. In a study of 20 infants with congenital osseous syphilis, only one had calvarial lesions (Levin 1970). The pattern, morphology and distribution of the periosteal lesions in this skull and significant bone destruction in the mandible suggest against congenital syphilis, in contrast to the Fournier's molars that favor congenital syphilis, but that have been noted in at least one 7-year-old boy with "nonsyphilitic mulberry molars" (Koneru et al. 2018). See Cooper and Sanchez (2018) for a list of osseous features associated with congenital syphilis.

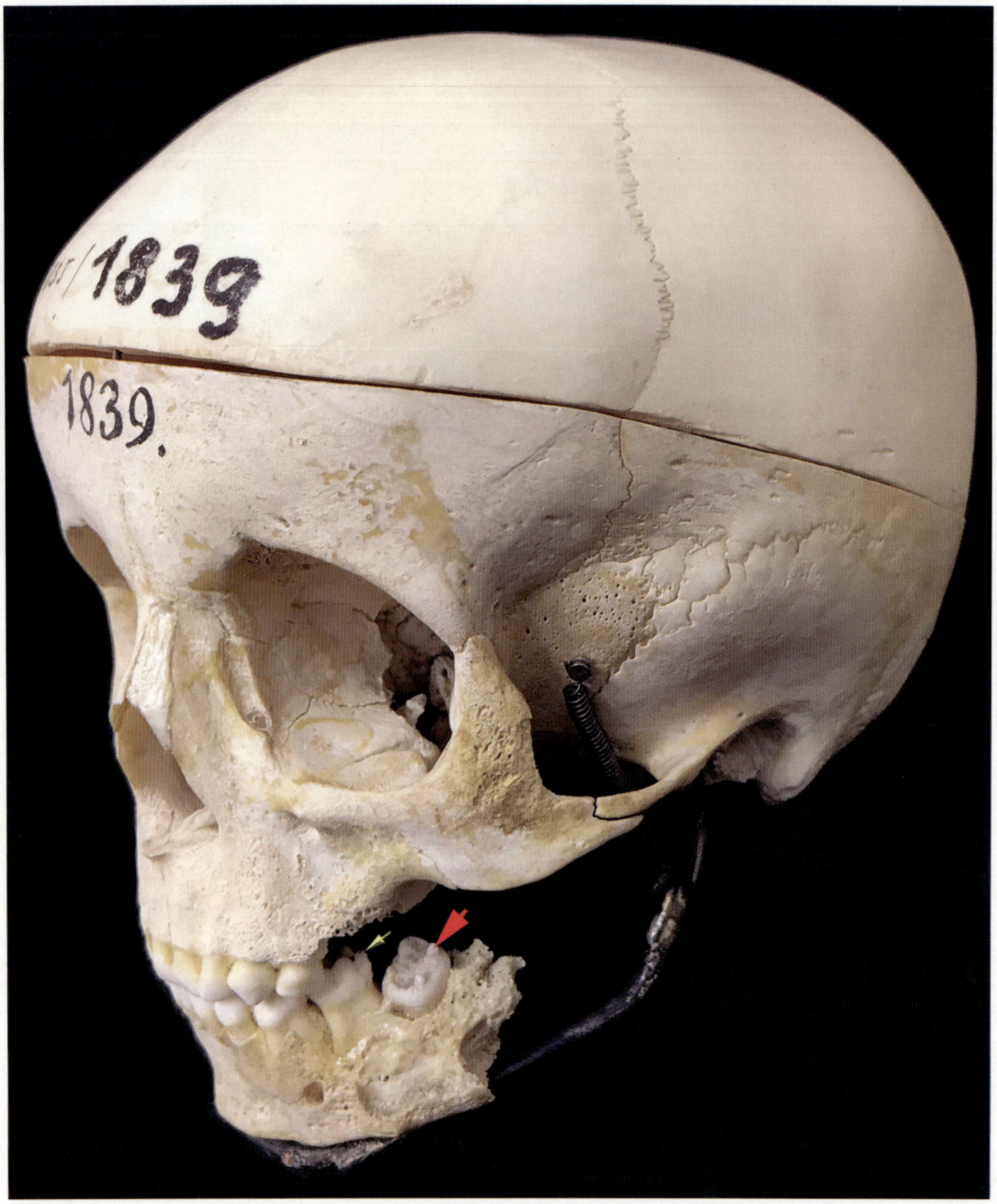


**Figure 173.** A four-year-old male who died in 1871 listed on the museum card as having "Caries, cause unknown" in the left mandible and a carious lesion of the left maxilla possibly due to congenital syphilis or other nonspecific infection (MN21.055/1839). A portion of the distal corpus of the left half of the mandible and entire ramus have been destroyed, as well as the vomer and much of the sphenoid, with scattered patches of periostitis on the left maxilla and left greater wing of the sphenoid. Note the unusual "cusp peaks" on the deciduous second (yellow arrow) and permanent first (red arrow) molars consistent with Fournier's molar (Ioannou et al. 2016) and often present in congenital syphilis. Treatment with Mercury, however, could result in non-congenital deformities (Yamunadevi et al. 2021). See Cooper and Sanchez (2018) for a list of osseous features associated with congenital syphilis and Koneru et al. (2018) for a rare case of a 7-year-old boy with nonsyphilitic "mulberry molars."

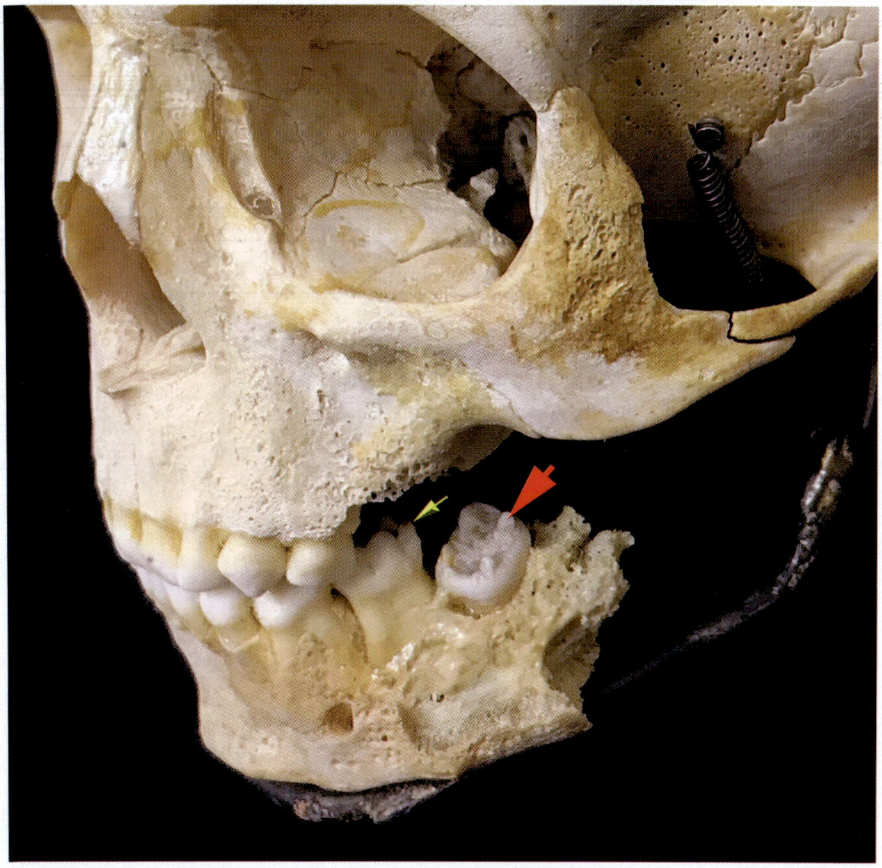

**Figure 174.** Magnified view of a four-year-old male who died in 1871 listed on the museum card as having "Caries, cause unknown" in the left mandible and a carious lesion of the left maxilla possibly due to congenital syphilis or other nonspecific infection (MN21.055/1839). A portion of the distal corpus of the left half of the mandible and entire ramus have been destroyed, as well as the vomer and much of the sphenoid, with scattered patches of periostitis on the left maxilla and left greater wing of the sphenoid. Note the unusual "cusp peaks" on the deciduous second (yellow arrow) and permanent first (red arrow) molars consistent with Fournier's molar (Ioannou et al. 2016) and often found in association with congenital syphilis or from treatment with Mercury (Yamunadevi et al. 2021). See Cooper and Sanchez (2018) for a list of osseous features associated with congenital syphilis and Koneru et al. (2018) for a rare case of a 7-year-old boy with nonsyphilitic "mulberry molars."

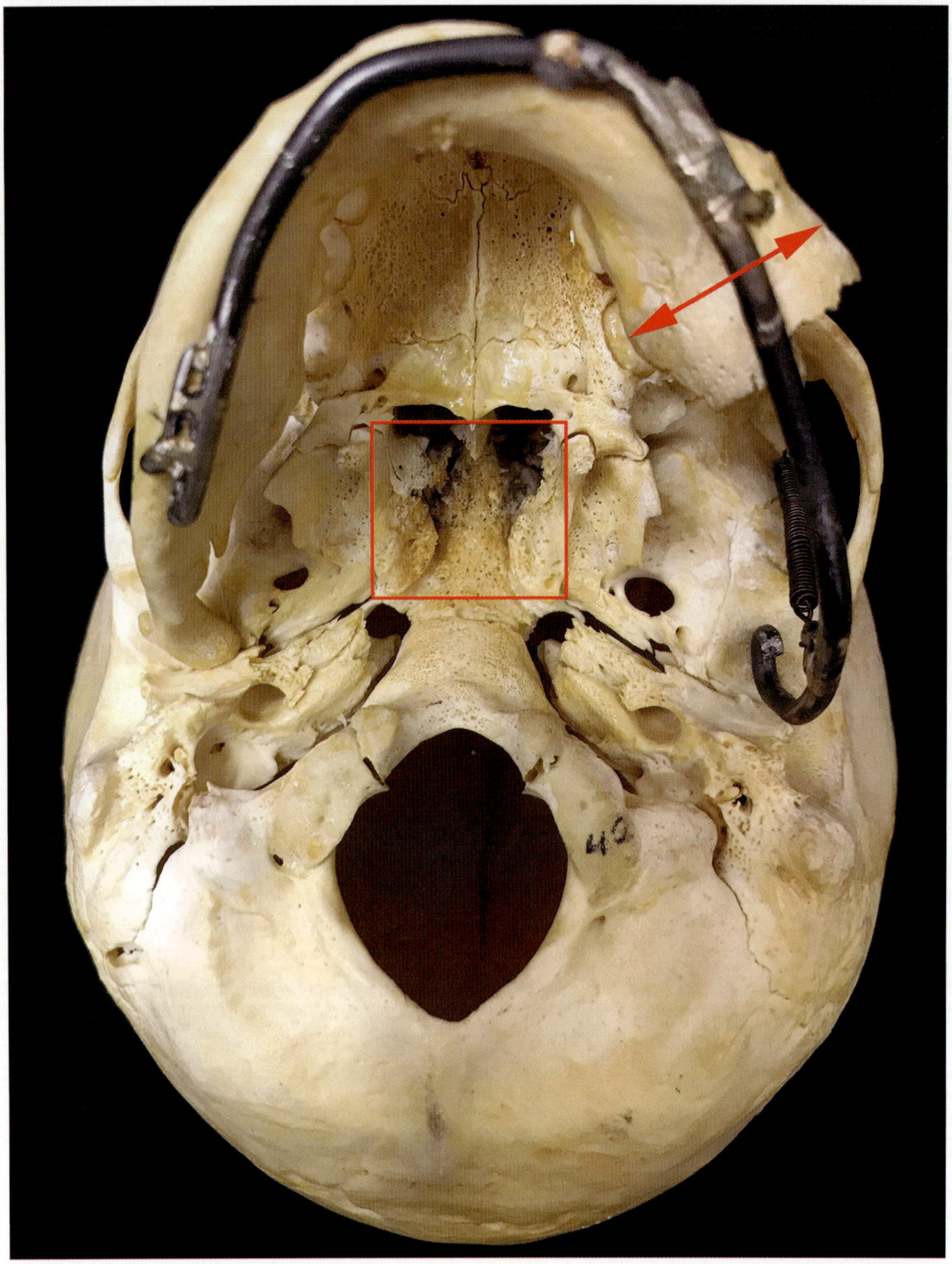

**Figure 175.** Inferior view of cranium of a four-year-old male who died in 1871 with "caries, cause unknown" resulting in destruction of the left half of the mandible, possibly due to congenital syphilis, other nonspecific infection or hypoplasia and crown defects from treatment with Mercury (Yamunadevi et al. 2021) (MN21.055/1839). A portion of the left posterior mandible has both resorbed and thickened (corpus; double arrow) and the entire ramus have been destroyed, as well as the vomer and much of the sphenoid (square). See Cooper and Sanchez (2018) for a list of osseous features associated with congenital syphilis and Koneru et al. (2018) for a rare case of a 7-year-old boy with nonsyphilitic "mulberry molars" whose mother tested negative for syphilis. Treatment with Mercury might also produce these unusual molars.

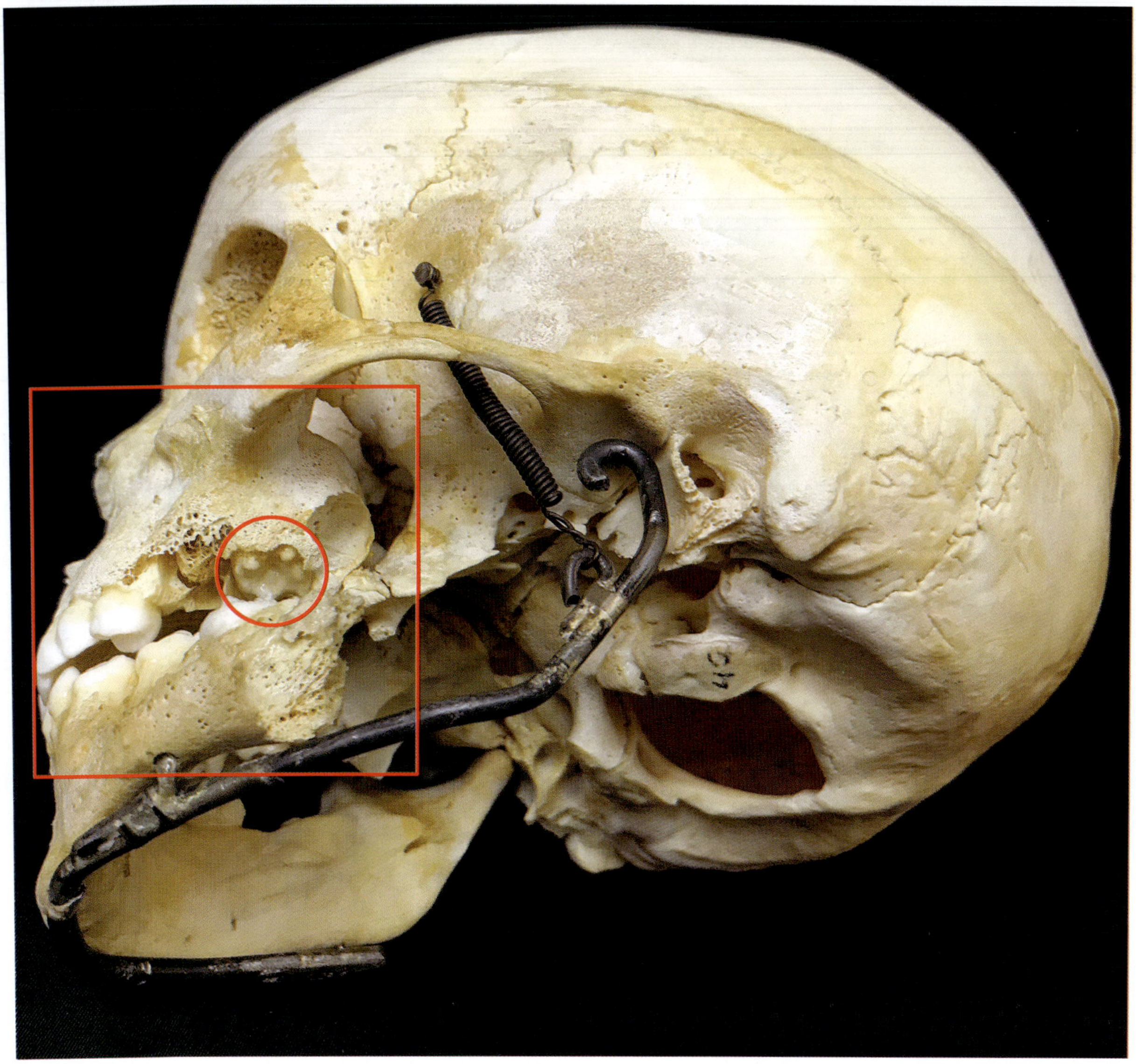

**Figure 176.** A four-year-old male who died in 1871 listed as having "Caries, cause unknown" resulting in destruction of the left half of the mandible possibly due to congenital syphilis or other nonspecific infection (MN21.055/1839). A portion of the corpus of the posterior left mandible and entire ramus were destroyed. The prominent "peaked" cusps on the maxillary permanent first molar (circle) indicate long-term developmental stress affecting development of the molars sometime between 18 weeks in utero to 10 months (see Nelson and Ash 2010), possibly due to congenital syphilis (Ioannou et al. 2016), other nonspecific infection, or treatment with Mercury (Yamunadevi et al. 2021). See Cooper and Sanchez (2018) for a list of osseous features associated with congenital syphilis and Koneru et al. (2018) for a rare case of a 7-year-old boy with non-syphilitic "mulberry molars." See Figure 177 for an indicator of developmental stress in the orbits.

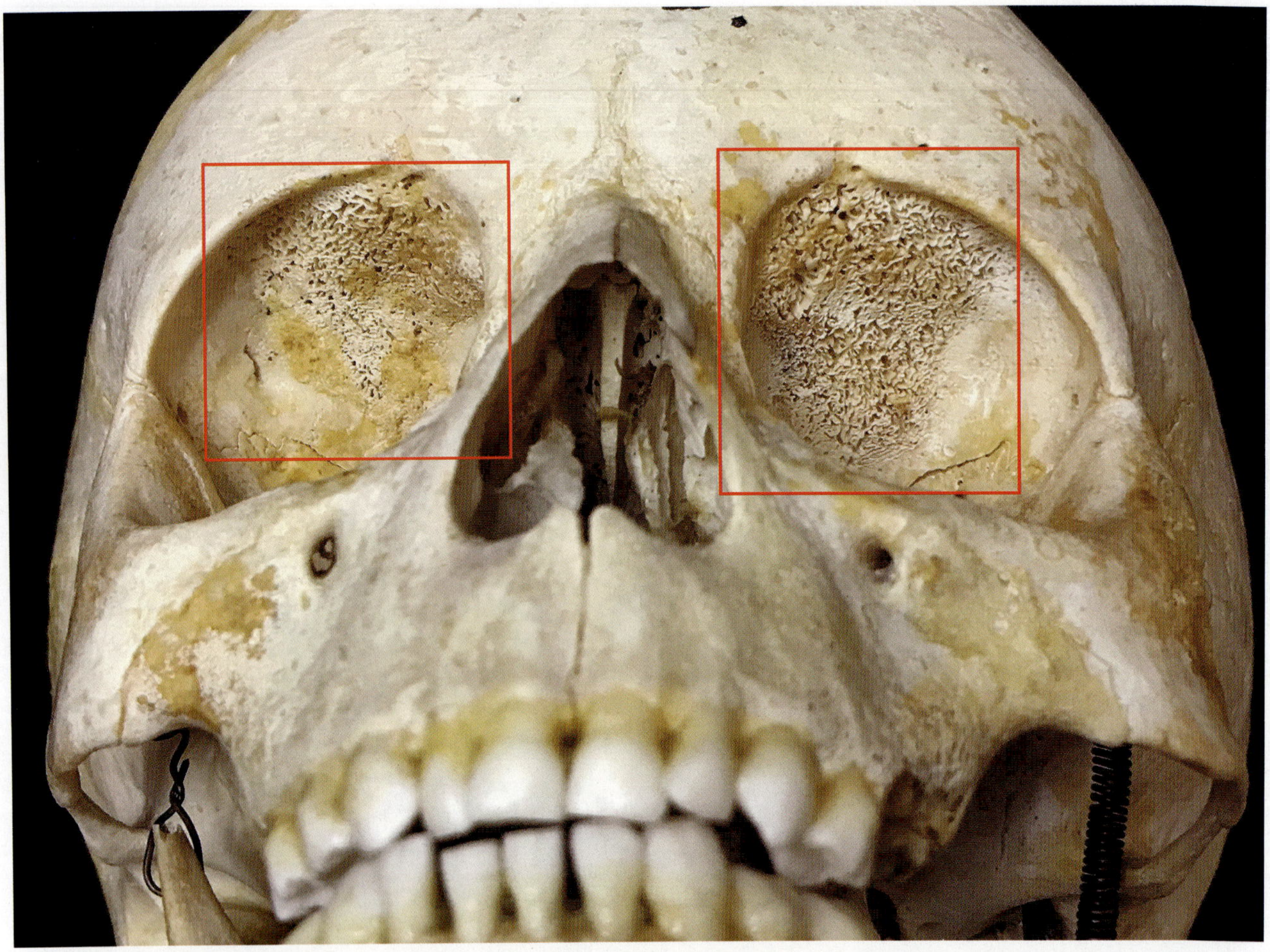

**Figure 177.** Four-year-old male who died in 1871 listed as having "caries, cause unknown" resulting in destruction of the left maxilla and mandible, possibly due to congenital syphilis or other nonspecific infection (MN21.055/1839). The orbits exhibit coral-like thickening of the plates (boxes) consistent with developmental growth disturbances such as iron-deficiency anemia, nutritional deficiency leading to subperiosteal bleeding as in scurvy, and other forms of physiological stress during the growth years (see Walker et al. 2009). The severity of expression of cribra orbitalia can range from a few pits in one or both superior plates, to 3-dimensional coral-like growths as in this child. Cribra orbitalia is often found in individuals with pitting and thickening of the outer table known as porotic hyperostosis (Mann and Hunt 2005; Walker et al. 2009).

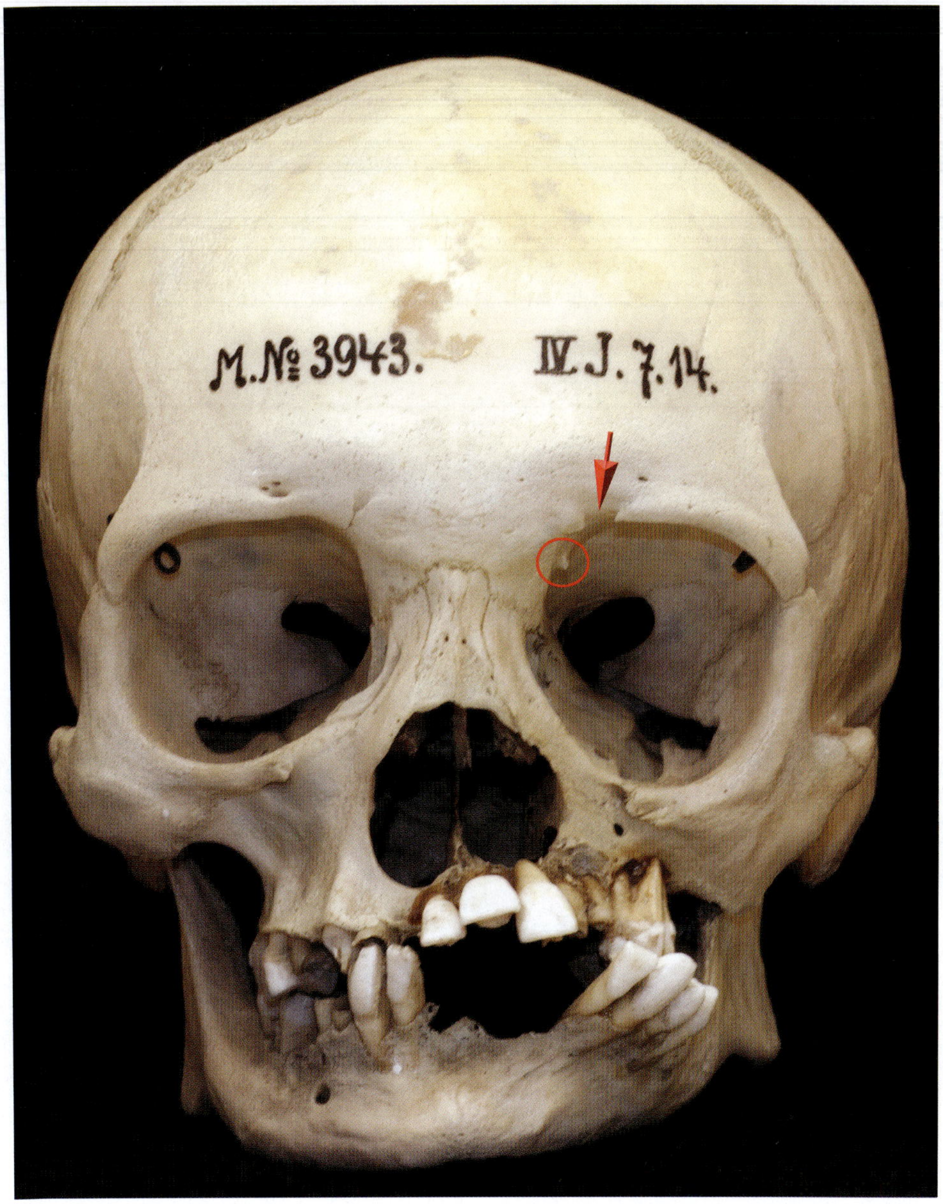


**Figure 178.** Anterior view showing maldevelopment and misalignment of the maxilla and mandible resulting in alveolar prognathism in an adult female who suffered from "noma as a child" (on the museum card) and died before 1869 (MN3943). Noma disease (cancrum oris; Maley et al. 2015) is a type of gangrene that destroys the tissues of the mouth and other areas and often occurs in malnourished children where cleanliness and sanitation are lacking (Bolivar et al. 2012; National Library of Medicine 2022; Srour et al. 2017). Note that several of the anterior maxillary teeth protrude into the floor of the nasal aperture and there is no mental foramen in the left portion of the mandible, likely due to resorption of bone in that area. Although asymmetry of the supraorbital notch and foramen is a common anatomical variant, the left orbit exhibits a supraorbital notch (arrow) and the right orbit a supraorbital foramen and there is calcification of the left trochlear spur (circle), both common anatomical variants.

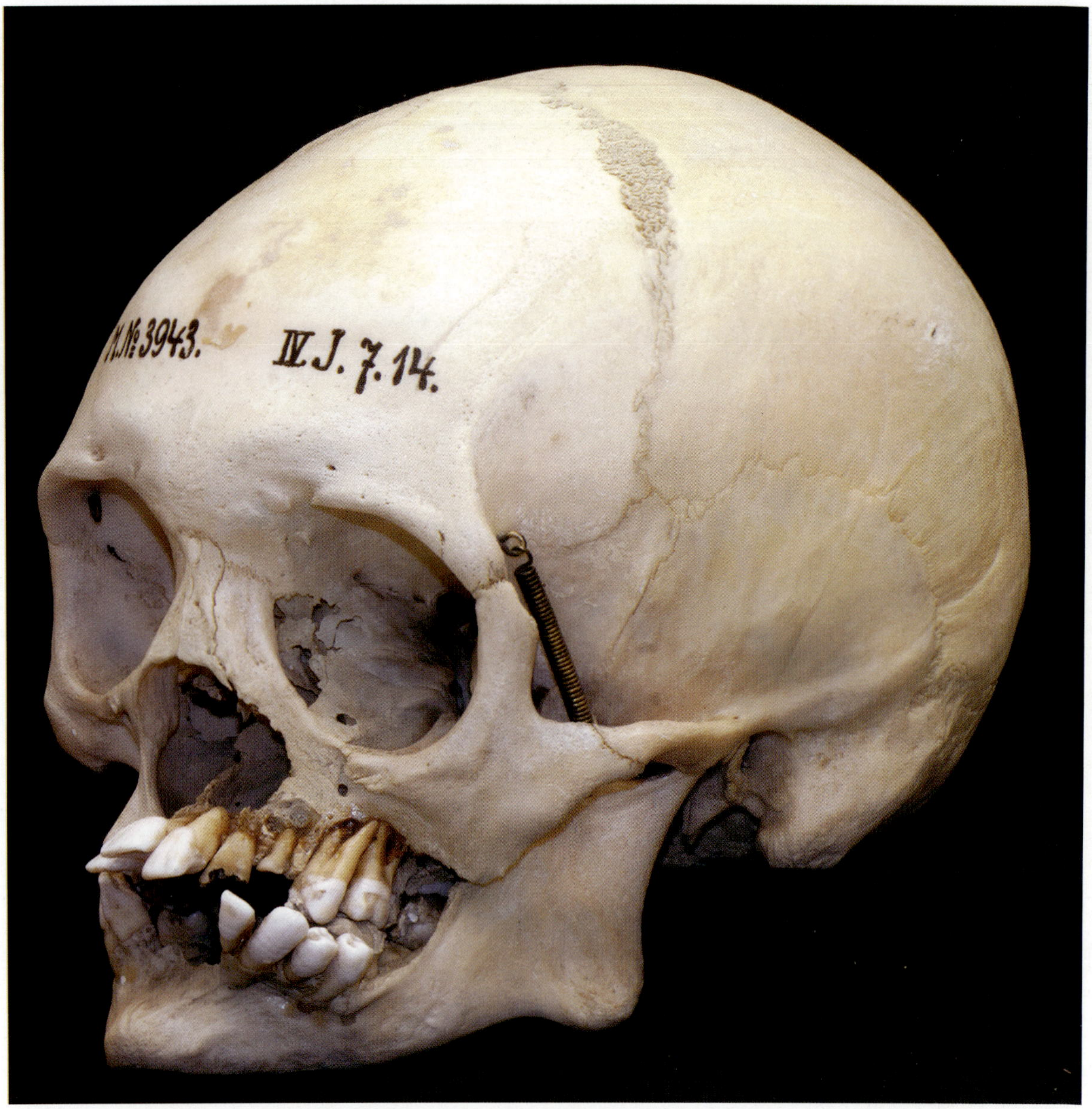


**Figure 179.** Left oblique view of maldevelopment and misalignment of the maxilla and mandible resulting in alveolar prognathism in an adult female who suffered from "noma as a child" (on the museum card) and died before 1869 (MN3943). Noma disease (cancrum oris; Maley et al. 2015) is a type of gangrene that destroys the tissues of the mouth and other areas and often occurs in malnourished children where cleanliness and sanitation are lacking (Library of Medicine 2022; National Bolivar et al. 2012; Srour et al. 2017). Note that several of the anterior maxillary teeth protrude into the floor of the nasal aperture and there is no mental foramen in the left portion of the mandible, likely due to resorption of bone in that area.

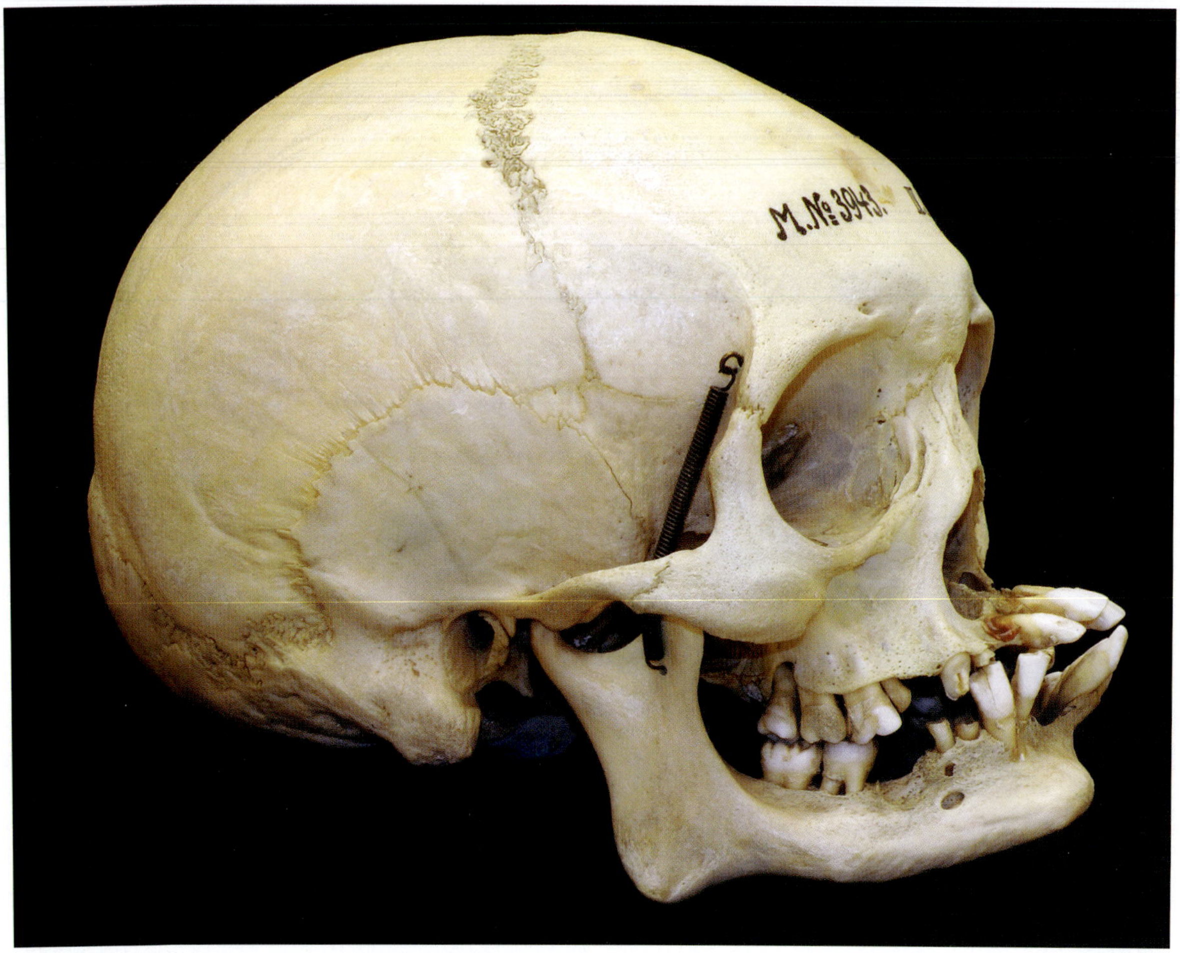


**Figure 180.** Right oblique view showing maldevelopment and misalignment of the maxilla and mandible resulting in alveolar prognathism in an adult female who suffered from "noma as a child" and died before 1869 (MN3943). Noma disease (cancrum oris; Maley et al. 2015) is a type of gangrene that destroys the tissues of the mouth and other areas and often occurs in malnourished children where cleanliness and sanitation are lacking (Bolivar et al. 2012; National Library of Medicine 2022; Srour et al. 2017). Note that several of the anterior maxillary teeth protrude into the floor of the nasal aperture and there is no mental foramen in the left portion of the mandible, likely due to resorption of bone in that area.

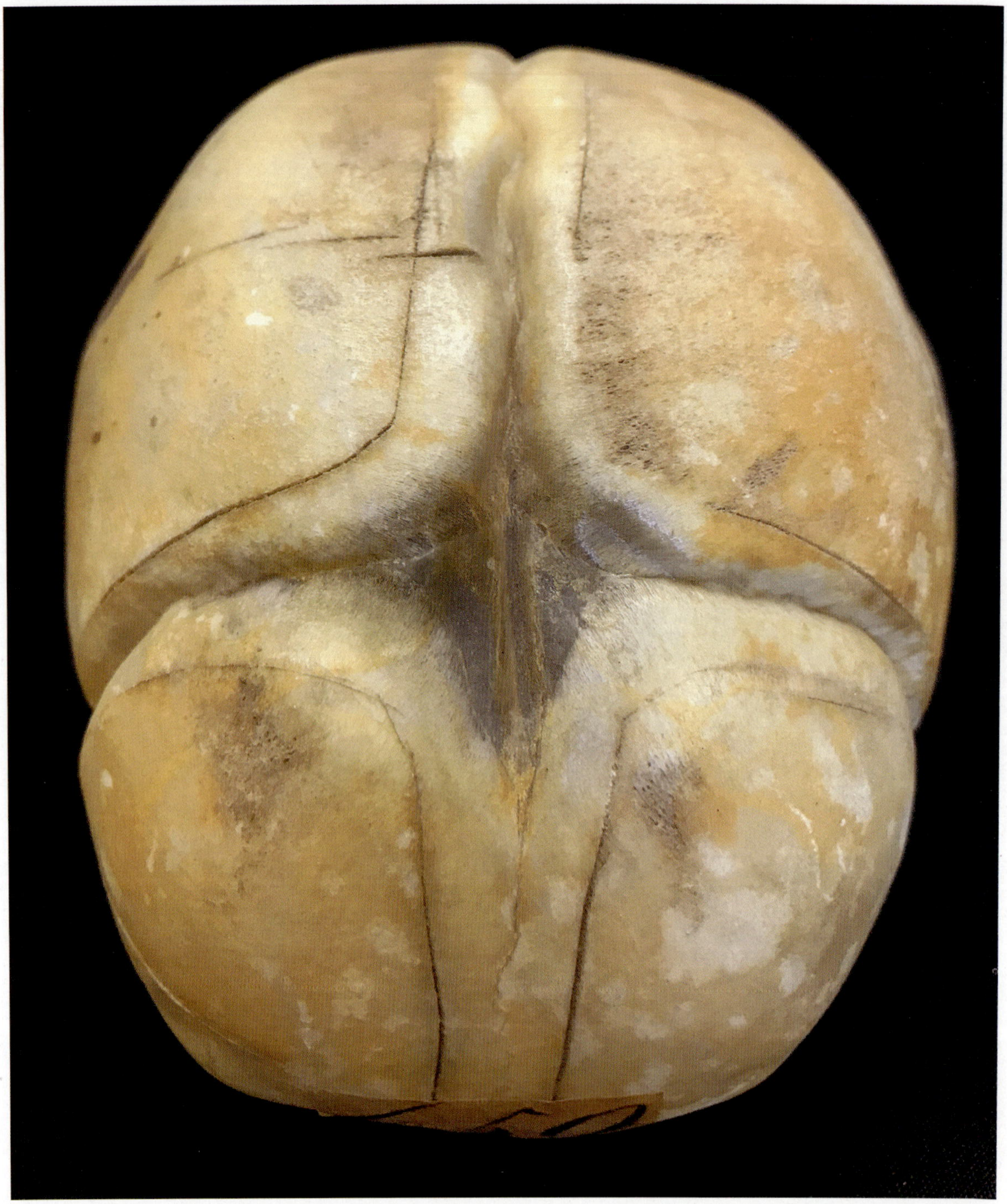

**Figure 181.** Rachitis (rickets) due to deficiency of vitamin D in a 7 ½-month-old male who died in 1863 (MN21.055/659). Note the prominently curved sutures that give the skull a "hot-cross bun" appearance present in some children with rickets and congenital syphilis (Power 1914). The diamond shaped translucency is the dura mater.

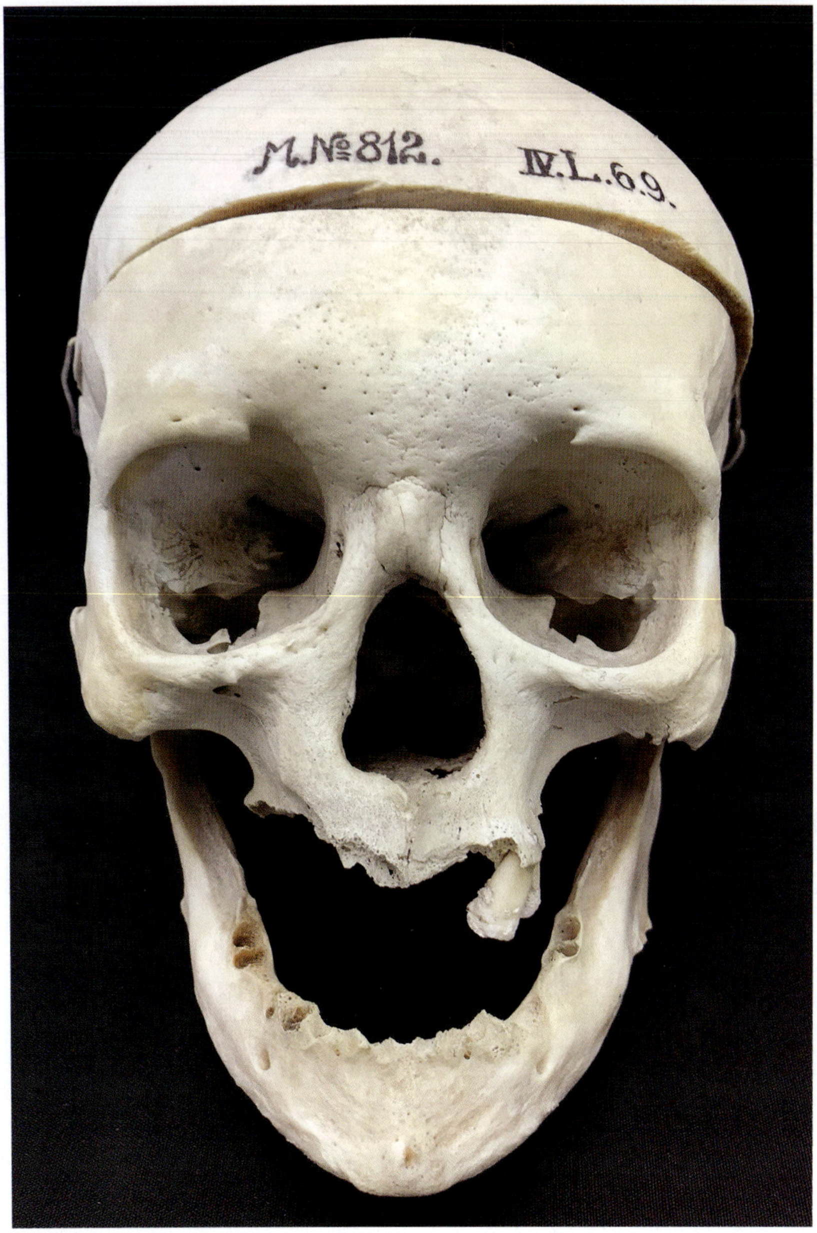


**Figure 182.** Tertiary or late stage syphilis ("Lues cranii" on the museum card) resulting in erosion of the hard palate, nasal aperture with loss of the nasal bones, septum, and turbinates in a 48-year-old female who died in 1825 (MN812). While the actual incidence of osseous syphilis is unknown, several studies from the early 1900s indicate that it is not a common occurrence in people with syphilis. Turner (1930), in Baltimore, Maryland, noted osseous lesions in only 8.8% of 10,000 patients with late stage syphilis, while Buchman and Lieberman (1941) found osseous lesions in 5% (119) of 2,400 cases of syphilis of all stages. Osteomyelitis resulting in the deposition of bone and evidence of infection along the anterior surface of the tibia resulting in anteriorly curved tibia known as "sabre shin" may also accompany cranial lesions due to venereal syphilis (Manchester 1983). See Aufderheide and Rodrigues-Martin (1998); Griffiths (1981); Hackett (1963, 1976); Moller-Christensen (1961, 1967, 1978); Ortner (2003); Ortner and Putschar (1981); Steinbock (1976) for more on syphilis and the effects of leprosy in the cranium.

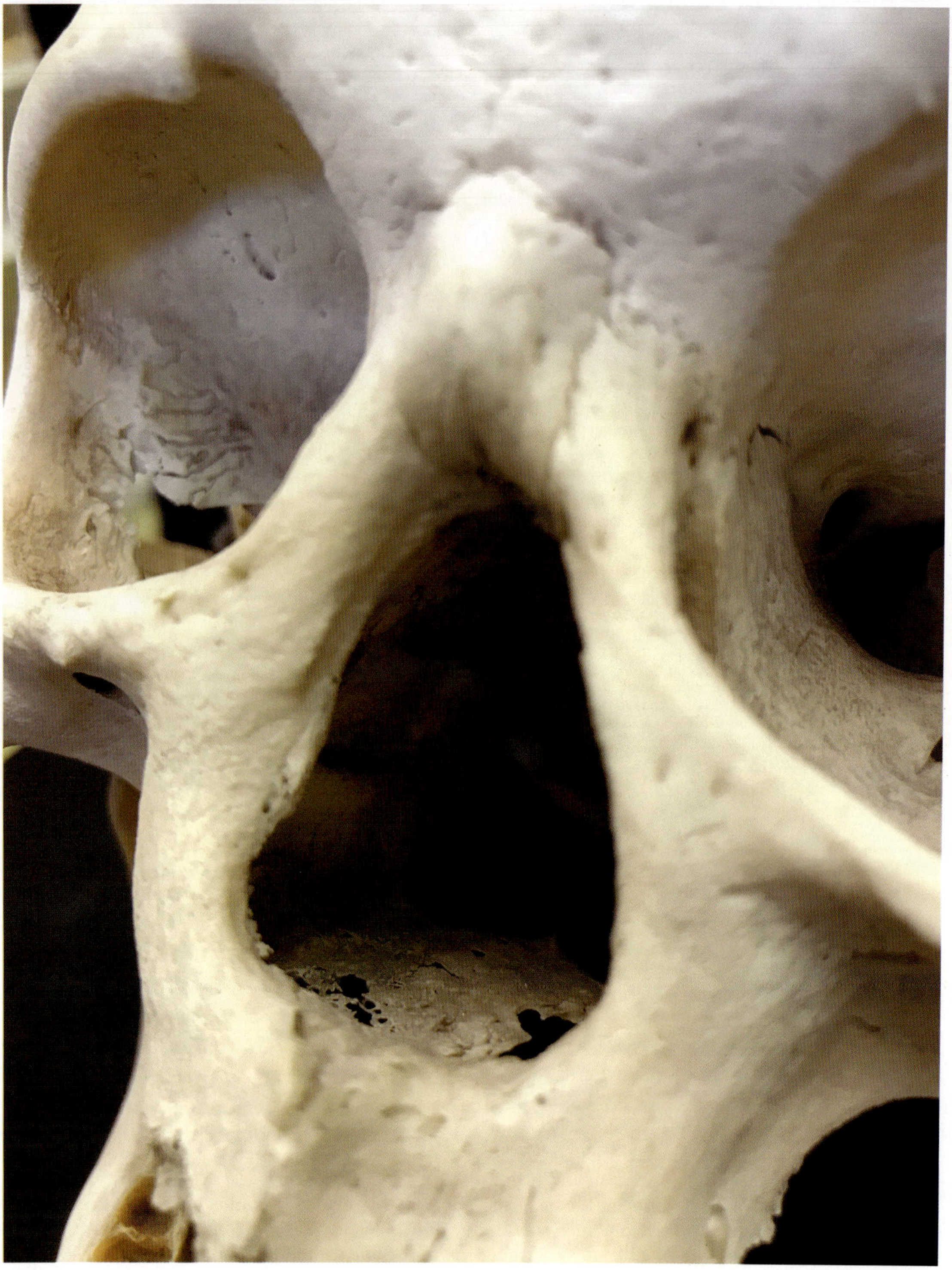

**Figure 183.** Tertiary syphilis resulting in erosion of the nasal aperture with loss of the nasal bones, septum, and turbinate bones in a 48-year-old female (MN812). Note perforations in the floor of the nasal cavity. See Aufderheide and Rodrigues-Martin (1998); Griffiths (1981); Manchester (1983); Moller-Christensen (1961, 1967, 1978); Ortner (2003); Ortner and Putschar (1981); Steinbock (1976) for more on syphilis and the effects of leprosy in the cranium.

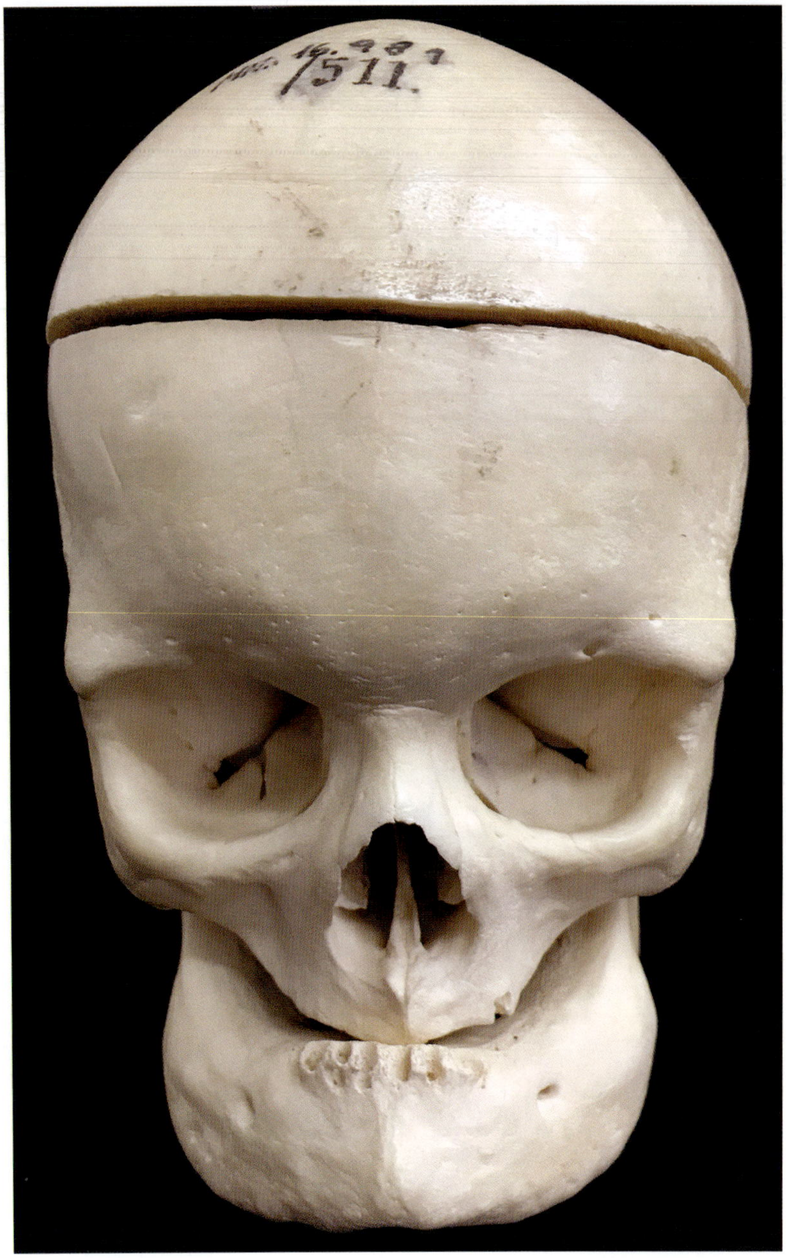

**Figure 184.** Paget's disease (osteitis deformans) of the skull is a non-inflammatory metabolic disorder of the skeleton believed to be the result of a slow virus resulting in excessive osteoclastic bone resorption and compensatory osteoblastic activity (Theroderou et al. 2011; Valenzuela and Pietschmann 2017). The skull in this 56-year-old female who died in 1919 is greatly thickened, most noticeably the mandible which is misshapen and exhibits an unusual contour of the corpus (MN16.981/511). All of the sutures are obliterated and the skull is unusually heavy. See Aufderheide and Rodrigues-Martin (1998); Griffiths (1981); Ortner and Putschar (1981) for more on Leontiasis ossea.

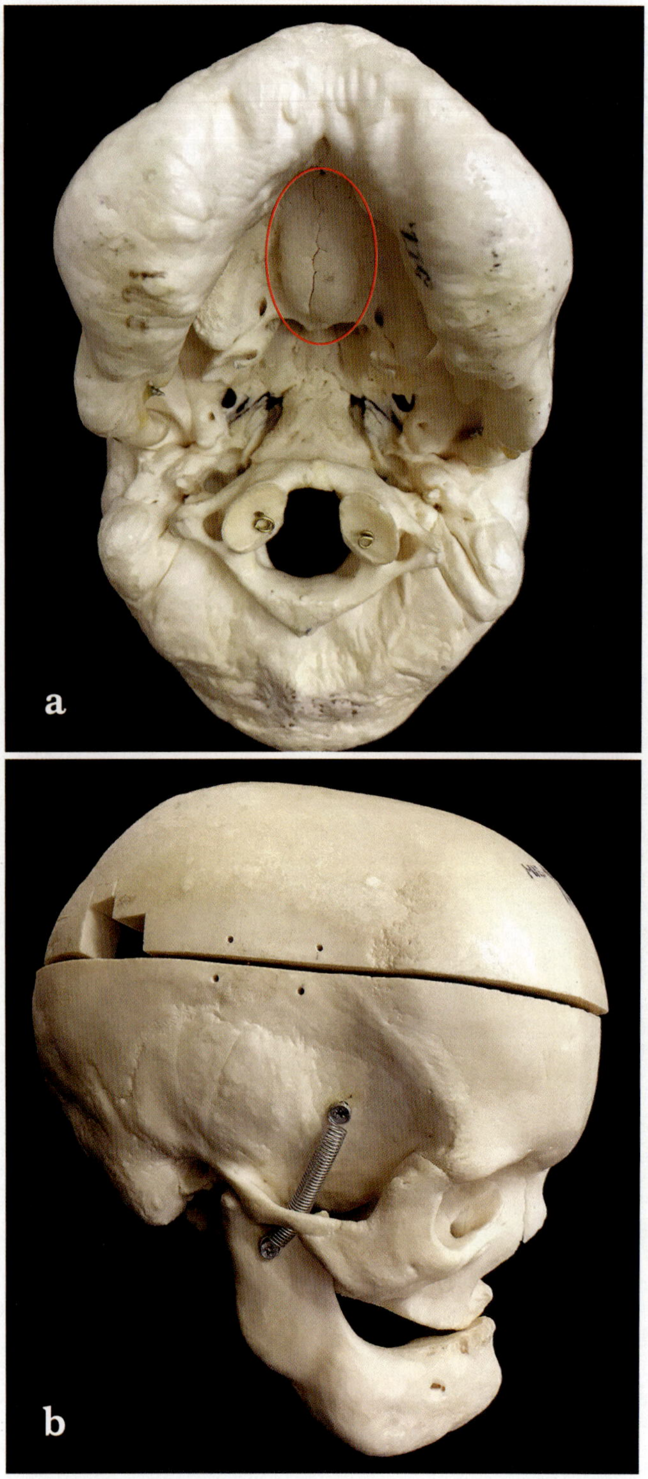


**Figure 185a-b.** Inferior (a) and right lateral (b) views of Paget's skull in a 56-year-old female who died in 1919, showing a greatly enlarged and misshapen mandible primarily affecting the mandibular corpus and not the ramus (MN16.981/511). (a) Note the large torus palatinus (oval). All of the sutures are obliterated. See Aufderheide and Rodrigues-Martin (1998); Griffiths (1981); Ortner and Putschar (1981) for more on the radiographic features of Paget's disease.

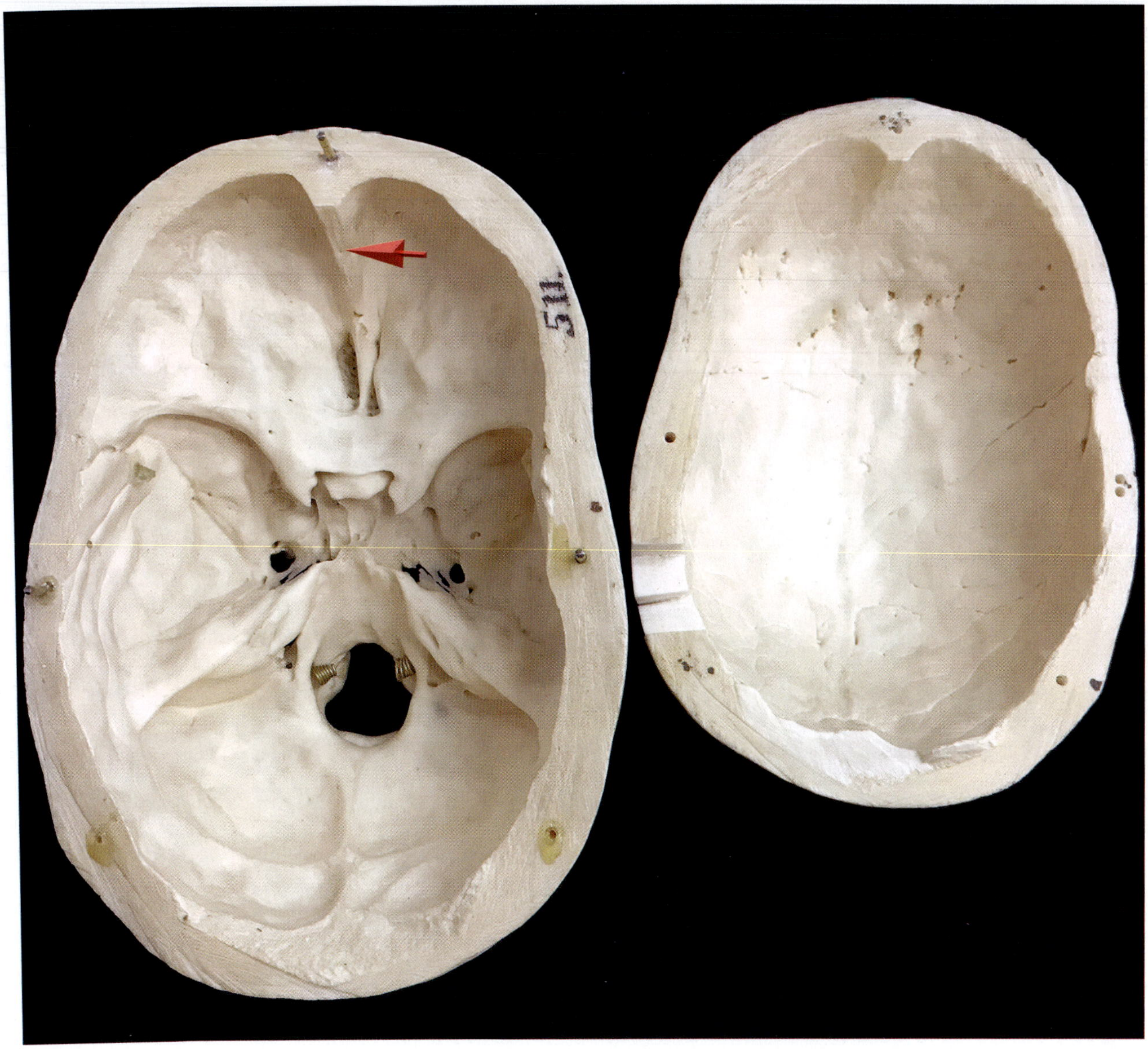


**Figure 186.** Endocranial view of Paget's skull in a 56-year-old female who died in 1919, showing thickening of the vault and frontal crest (arrow) (MN16.981/511). All of the sutures are obliterated endocranially and ectocranially. See Griffiths (1981); Ortner and Putschar (1981) for more on the radiographic features of Paget's disease.

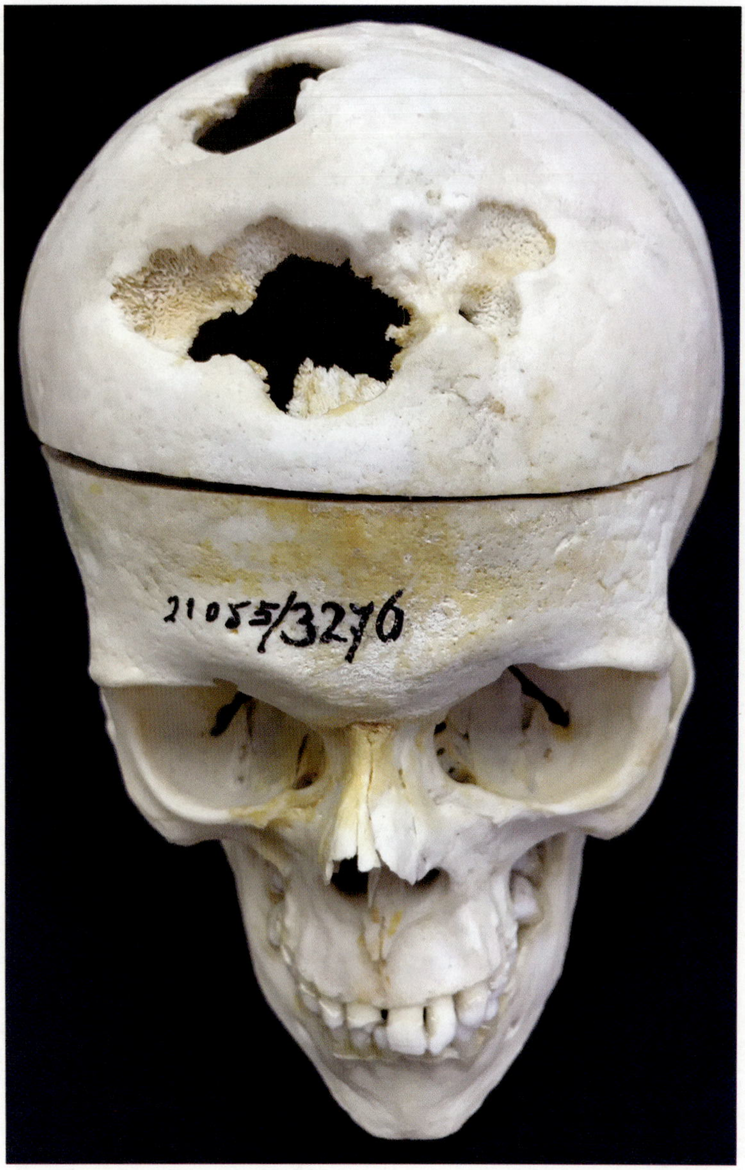


**Figure 187.** Tertiary/late stage syphilis of the frontal (most commonly) and right parietal bones resulting in large, irregular lytic lesions with sclerotic margins in an adult male (MN21.055/3276). Osseous lesions, often referred to as caries sicca, are usually not present in early and secondary syphilis, but appear in about one-third of untreated individuals with the disease 10 to 30 years after initial infection (Huang 2007). Cranial lesions usually begin in the periosteal or diploic areas, often sparing the inner table (Ortner 2003). Note the tripartite nasal bones, an extremely rare condition. See Aufderheide and Rodrigues-Martin (1998); Manchester (1983); Ortner and Putschar (1981); Steinbock (1976) for more on treponematosis and syphilis in bone.

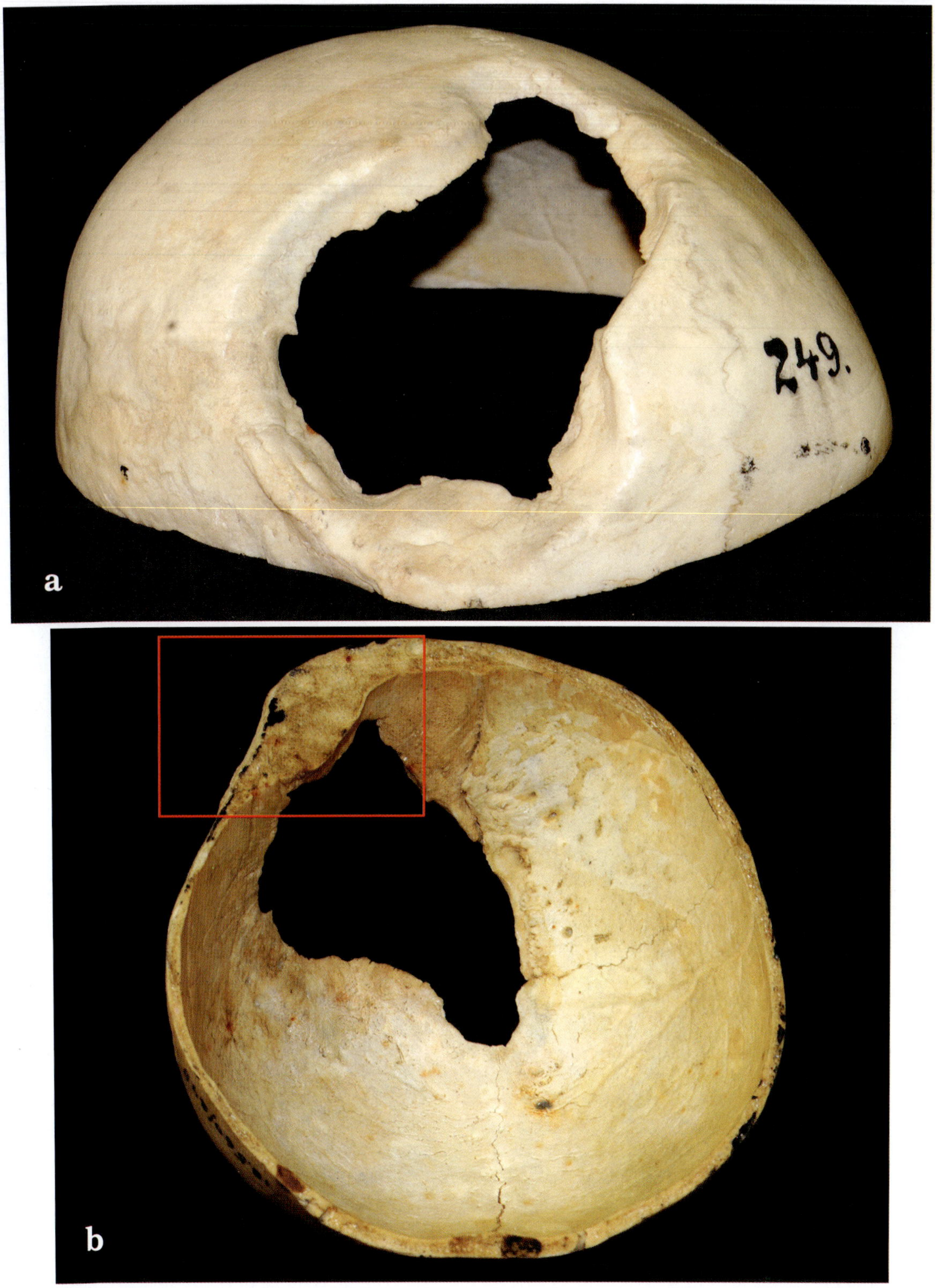


**Figure 188a-b.** (a) Right oblique and (b) endocranial views of tertiary syphilis in the cranium of an adult male who died in 1930, resulting in (b) sandwich-like thickening (rectangle) between the inner and outer tables of the frontal bone (MN18.560/249). See Manchester (1983); Ortner and Putschar (1981); Steinbock (1976) for more on treponematosis and syphilis in bone.

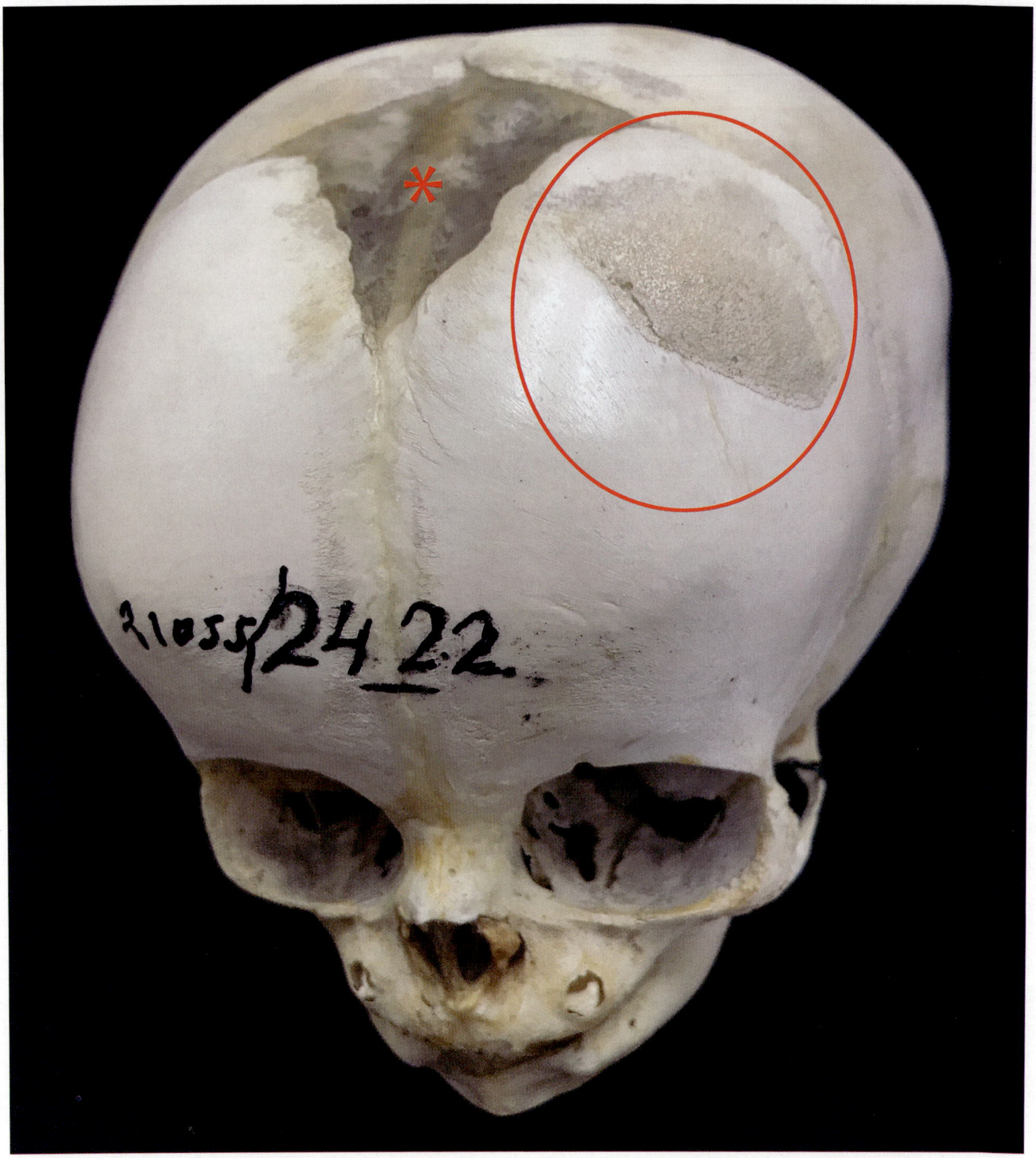


**Figure 189.** An 8-month-old female with a well-demarcated lesion (proliferative periostitis; oval) on the outer table of the left frontal bone, possibly the result of congenital syphilis (MN21.055/2422). Note the normal features of the anterior fontanelle (*). See Aufderheide and Rodrigues-Martin (1998); Ortner and Putschar (1981); Steinbock (1976) for more on treponematosis and syphilis in bone and Griffiths (1981) on the radiographic features of acquired and congenital syphilis.

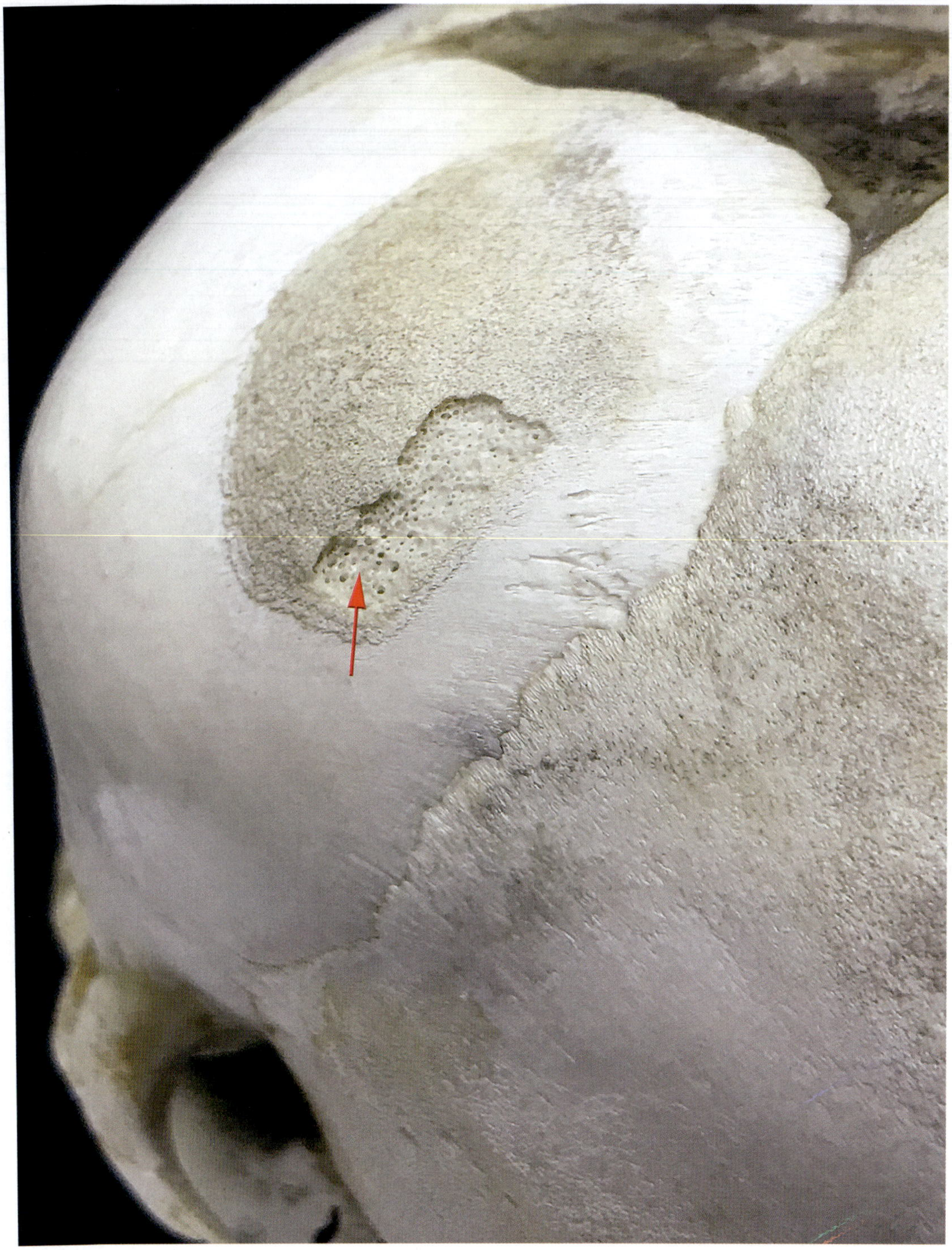

**Figure 190.** A well-demarcated lesion in the left frontal bone of an 8-month-old female, possibly the result of congenital syphilis (MN21.055/2422). Note that this lesion, proliferative periostitis that is rarely seen in dry bone, is fragile and easily separates from the outer table of the cranium by handling, revealing porous bone beneath it (arrow). See Aufderheide and Rodrigues-Martin (1998); Griffiths (1981); Ortner and Putschar (1981); Steinbock (1976) for more on treponematosis and syphilis in bone.

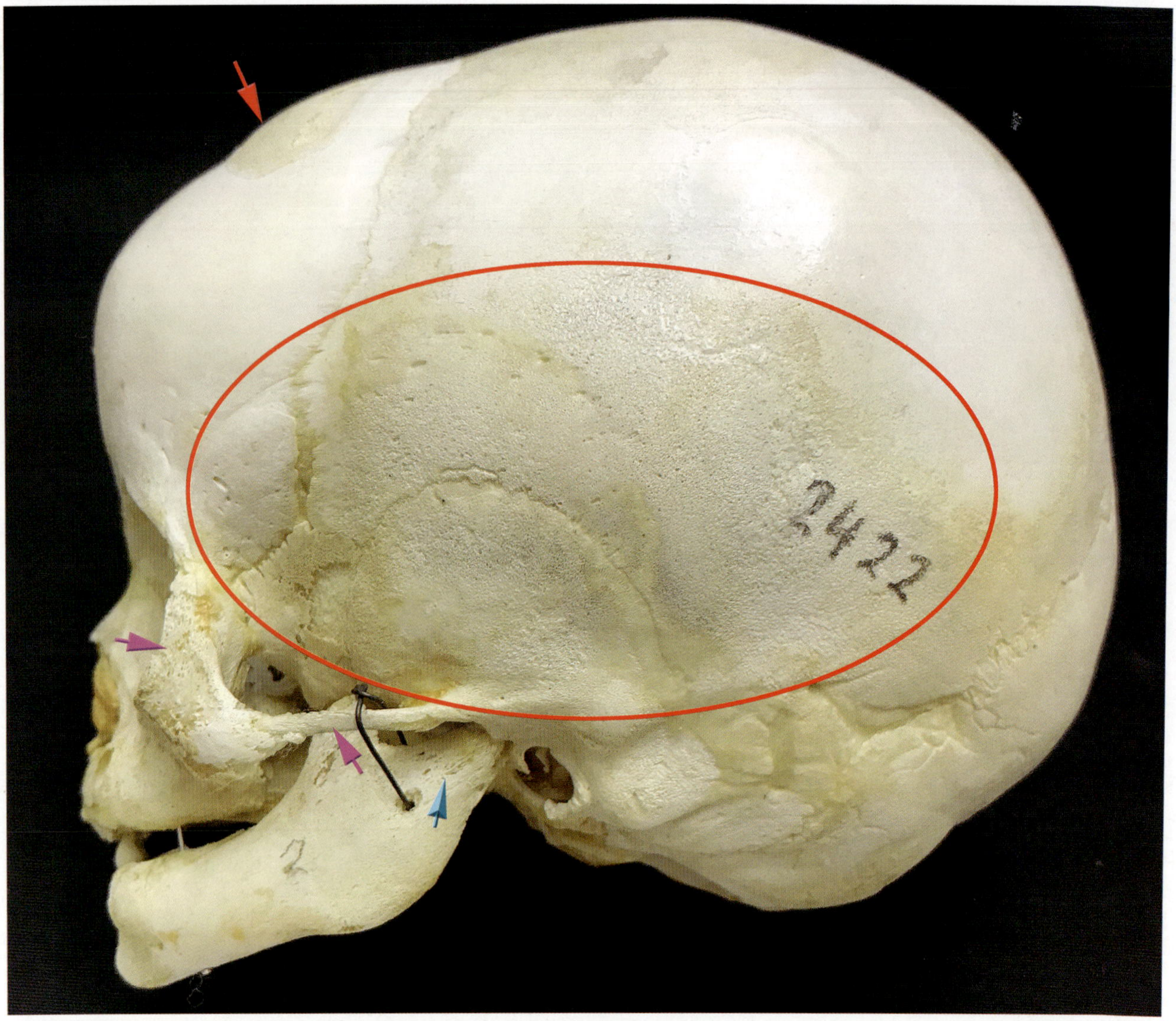


**Figure 191.** Proliferative periostitis of the frontal bone (see Figures 192-194) and porous reactive bone consistent with congenital syphilis over much of the left cranium (oval), left malar and zygomatic arch (purple arrows) and left mandibular ramus (blue arrow) (MN21.055 2422). See Aufderheide and Rodrigues-Martin (1998); Griffiths (1981); Ortner and Putschar (1981); Steinbock (1976) for more on treponematosis and syphilis in bone.

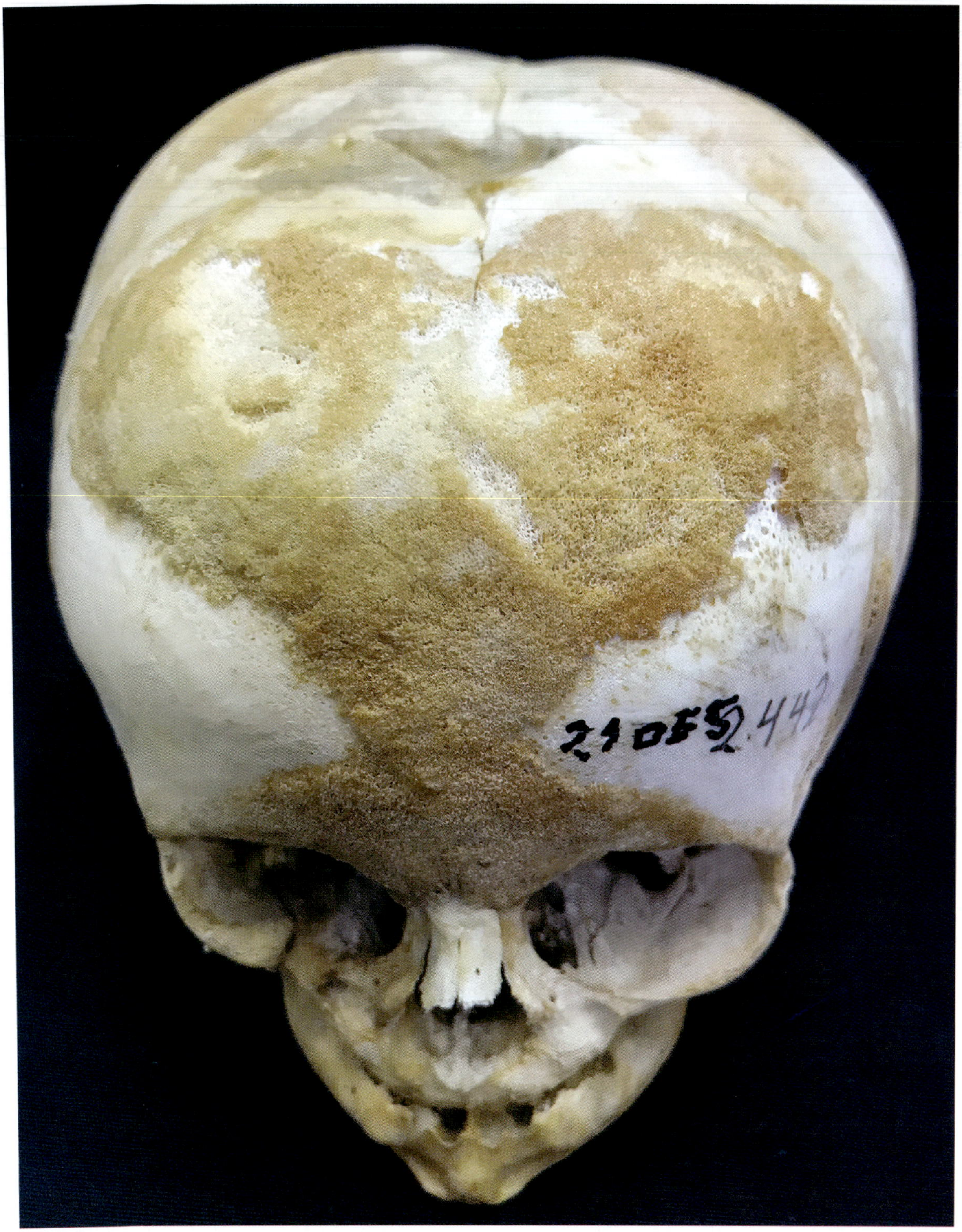

**Figure 192.** Superior-anterior view of unknown disease, more consistent with proliferative periostitis than neoplasia, with a rare "mossy-like" appearance in dry bone resulting in an osteoblastic response in the cranium of a 10-month-old female, possibly due to congenital syphilis (MN21.055/2442). See Aufderheide and Rodrigues-Martin (1998); Ortner and Putschar (1981); Steinbock (1976) for more on treponematosis and syphilis in bone.

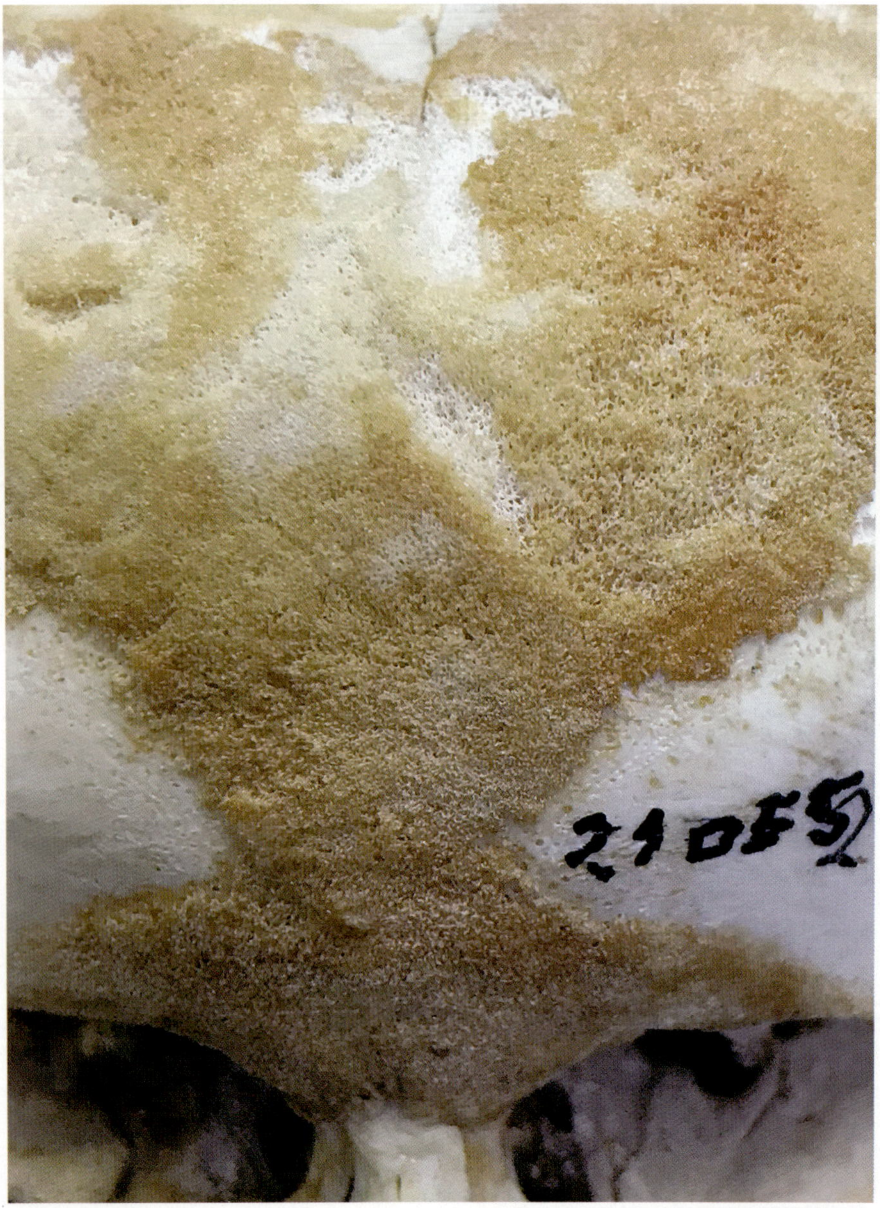

**Figure 193.** Unknown disease, more likely infection than neoplasia, resulting in a rare "mossy-like" proliferative periostitis (osteoblastic response) in dry bone, in the cranium of a 10-month-old female that is possibly due to congenital syphilis (MN21.055/2442). See Aufderheide and Rodrigues-Martin (1998); Ortner and Putschar (1981); Steinbock (1976) for more on treponematosis and syphilis in bone.

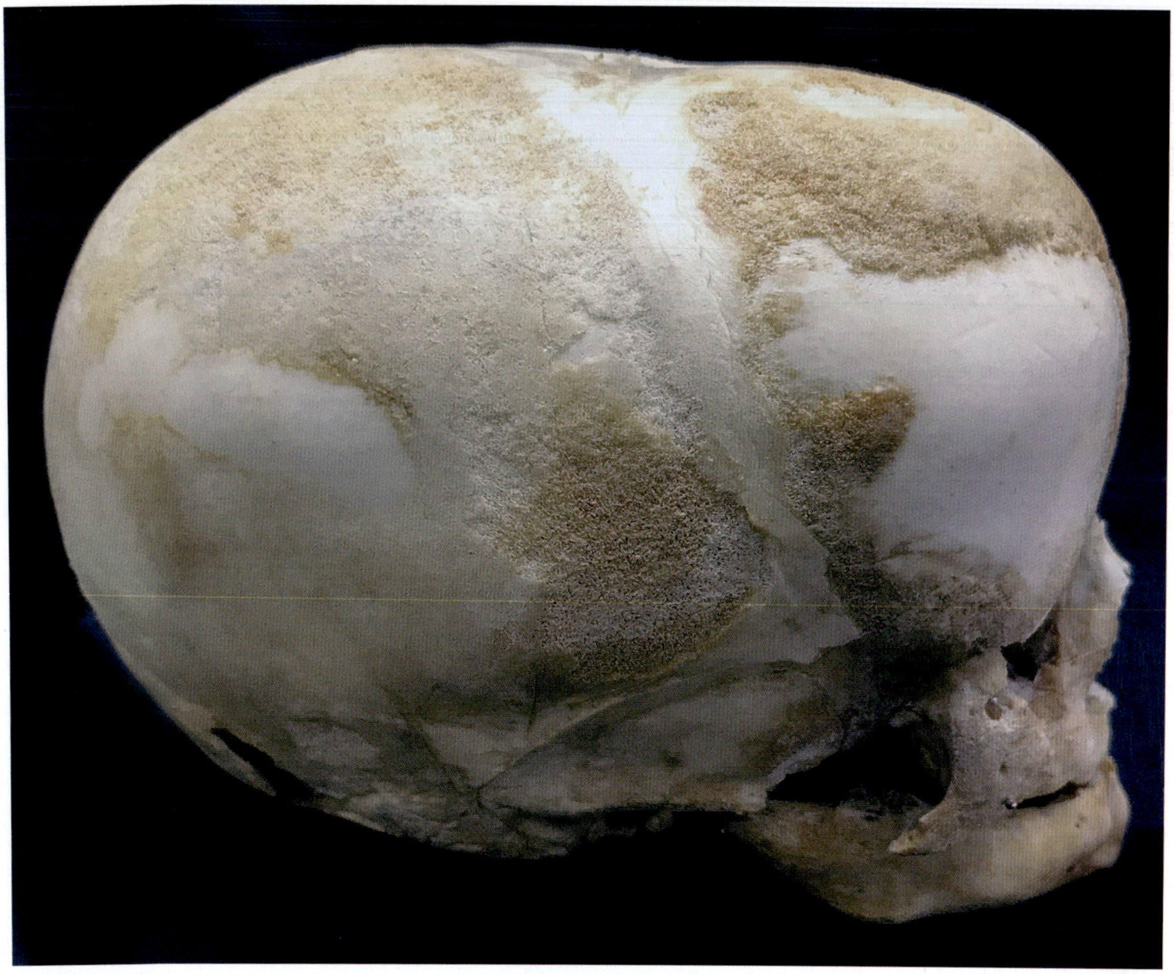

**Figure 194.** Right lateral view of an unknown disease, more likely infection than neoplasia, resulting in a rare "mossy-like" proliferative periostitis (osteoblastic response) in the cranium of a 10-month-old female, possibly due to congenital syphilis (MN21.055/2442). See Aufderheide and Rodrigues-Martin (1998); Ortner and Putschar (1981); Steinbock (1976) for more on treponematosis and syphilis in bone.

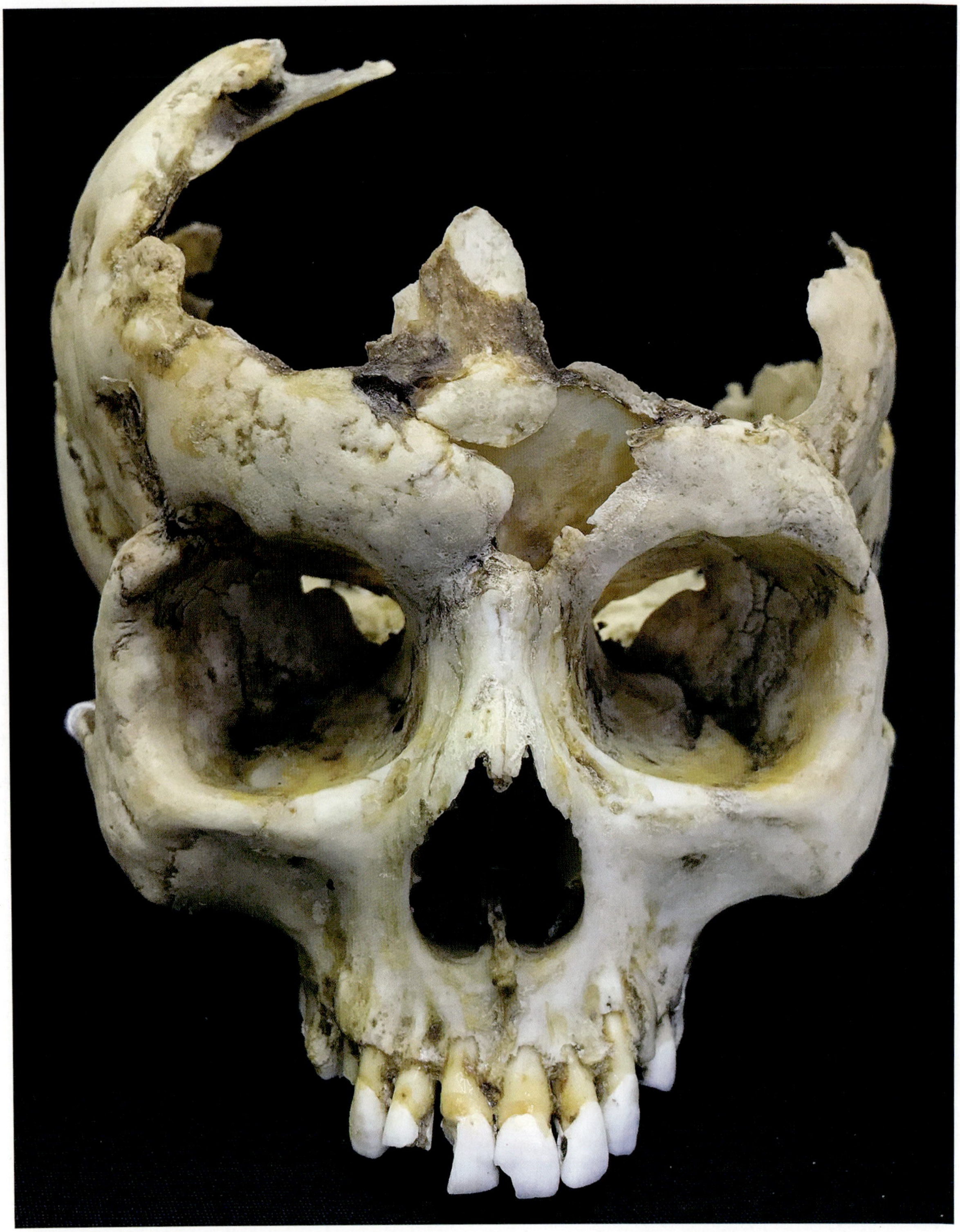

**Figure 195.** Tertiary syphilis ("lues") in an adult male who died in 1871 (MN21.055/2174). The infection eroded the calvarium and dental alveoli (tooth sockets) and resulted in thickened zygomas (malars). See Armelagos et al. (2012); Ortner and Putschar (1981); Steinbock (1976) for more on syphilis in bone. Characteristic cranial lesions (caries sicca) in acquired syphilis such as this consist of nodes, cavitations and stellate (radial) scarring (Hackett 1963, 1976).

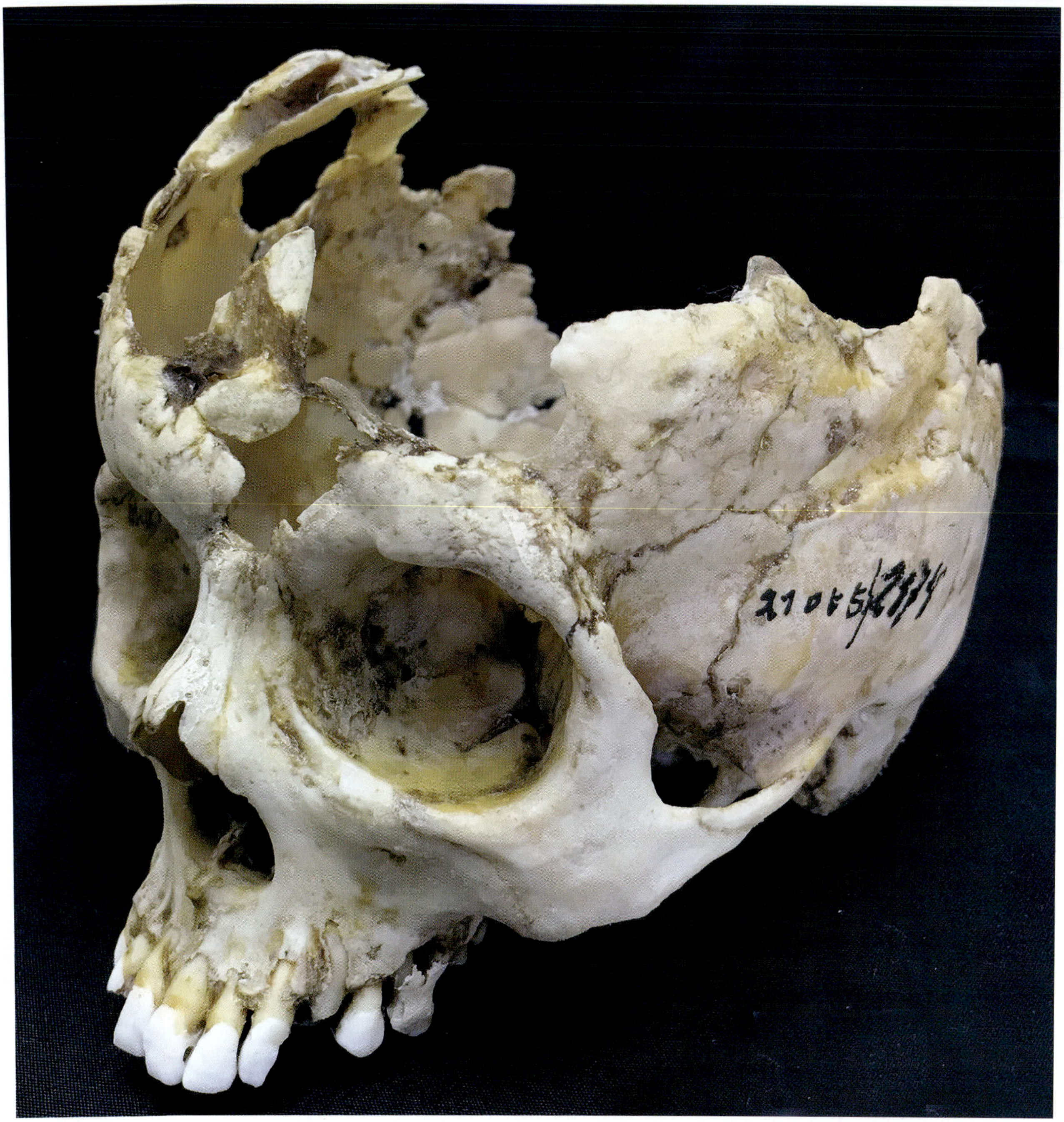

**Figure 196.** Tertiary syphilis ("lues") resulting in caries sicca (Betsinger and Smith 2019; Turk 1995; Virchow 1858) of the skull in an adult male who died in 1871 (MN21.055/2174). The infection eroded most of the calvarium and maxillary dental alveoli and thickened zygomas (malars). See Armelagos et al. (2012); Aufderheide and Rodrigues-Martin (1998); Hackett (1963, 1976); Manchester (1983); Ortner and Putschar (1981); Steinbock (1976); Walker et al. (2015) for more on syphilis in bone.

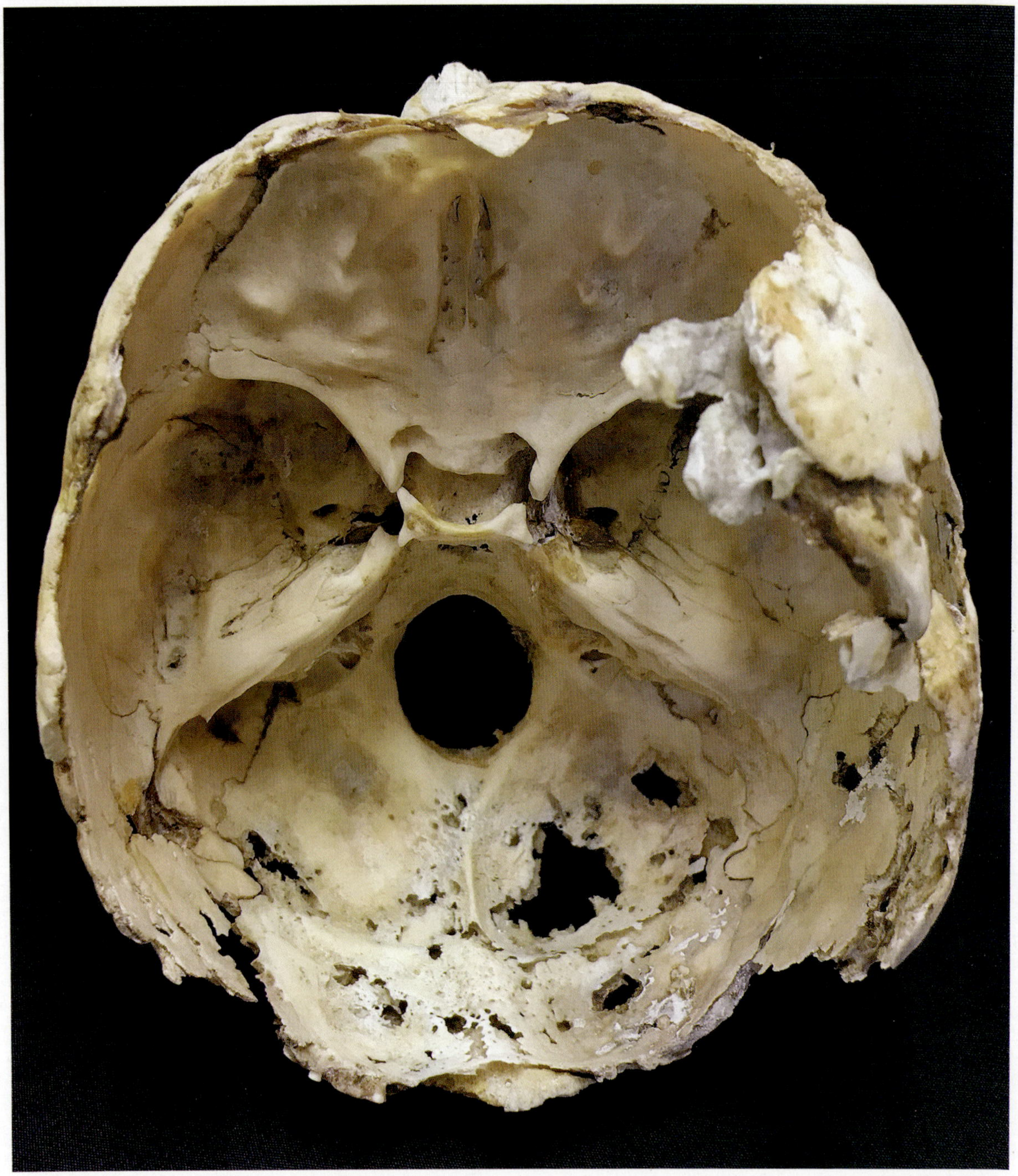

**Figure 197.** Superior endocranial view of tertiary syphilis ("lues") in an adult male who died in 1871 (MN21.055/2174). The infection eroded nearly all of the calvarium, including the cranial base leaving large lytic lesions with irregular margins sometimes referred to as caries sicca (Betsinger and Smith 2019; Turk 1995; Virchow 1858; Walker et al. 2015). See Armelagos et al. (2012); Aufderheide and Rodrigues-Martin (1998); Hackett (1963, 1976); Manchester (1983); Ortner and Putschar (1981); Steinbock (1976) for more on syphilis in bone.

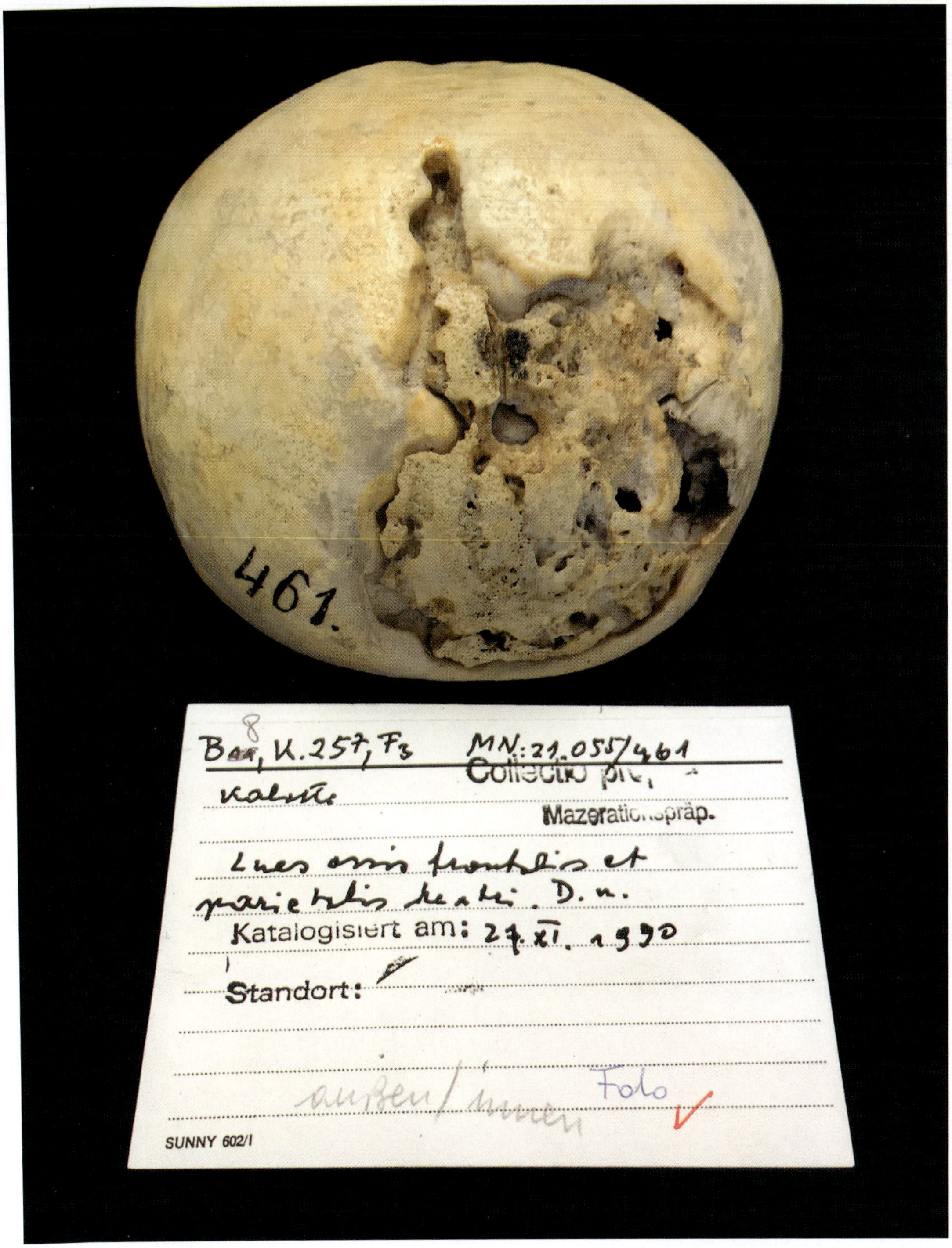


**Figure 198.** Anterior-superior view of cranium and the museum record for a skull with tertiary syphilis ("lues" on the museum card) mostly affecting the left frontal and left parietal bones in an adult who died in 1885 (MN21.055/461). See Hackett (1963, 1976); Manchester (1983); Ortner and Putschar (1981); Steinbock (1976) for more on syphilis in bone. Characteristic cranial lesions such as these due to syphilis are sometimes called caries sicca (Armelagos et al. 2012; Betsinger and Smith 2019; Turk 1995; Virchow 1858; Walker et al. 2015).

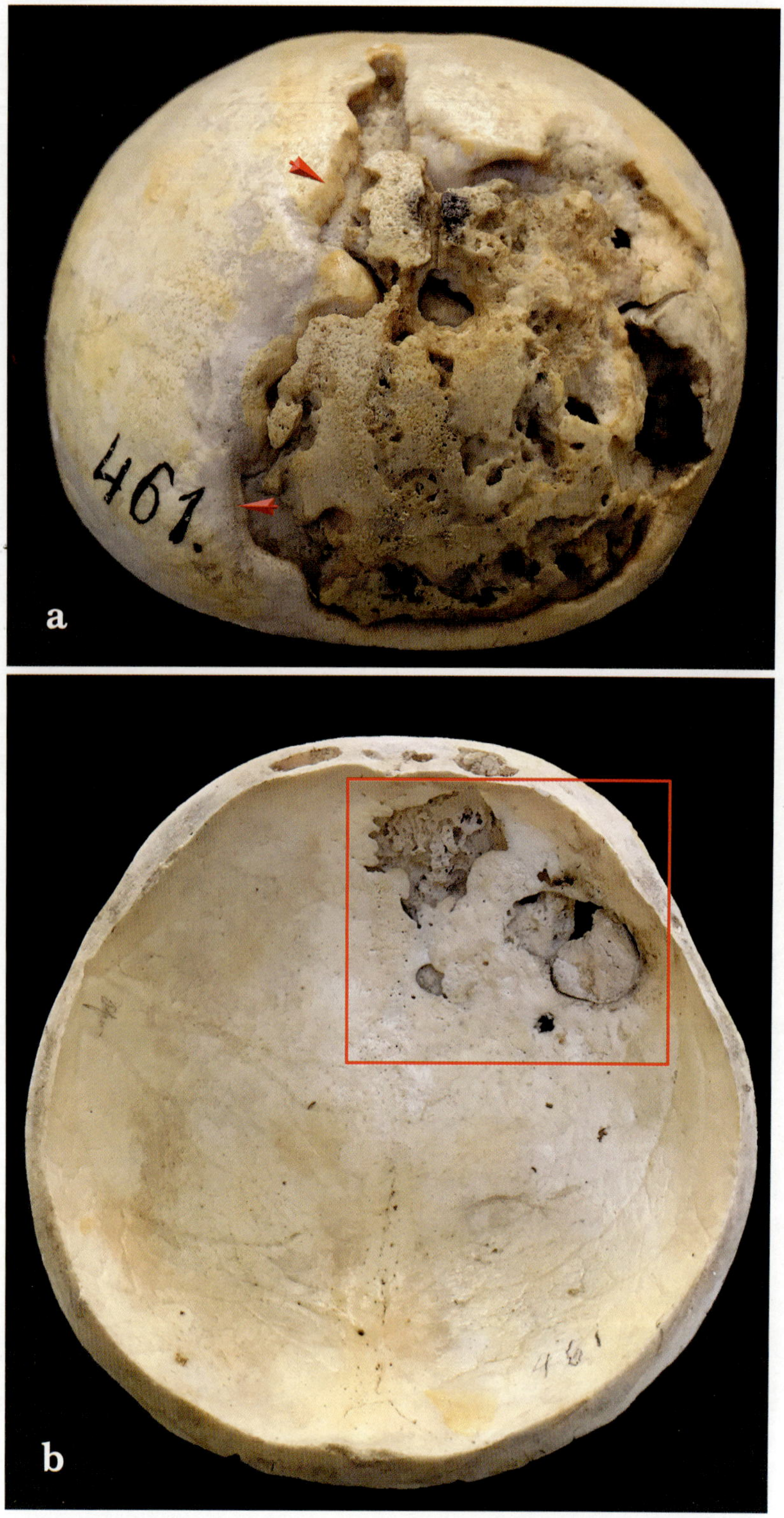


**Figure 199a-b.** (a) Ectocranial and (b) endocranial comparison of bone destruction in the frontal bone of an adult who died in 1871 as a result of tertiary syphilis ("lues") (MN21.055/461). Note the outer vault exhibits much more destruction than endocranially and with smooth, sloping borders. The endocranium shows several well-defined areas of bone sequestration (necrotic/dead bone). See Hackett (1963, 1976); Ortner and Putschar (1981) for more on syphilis in bone.

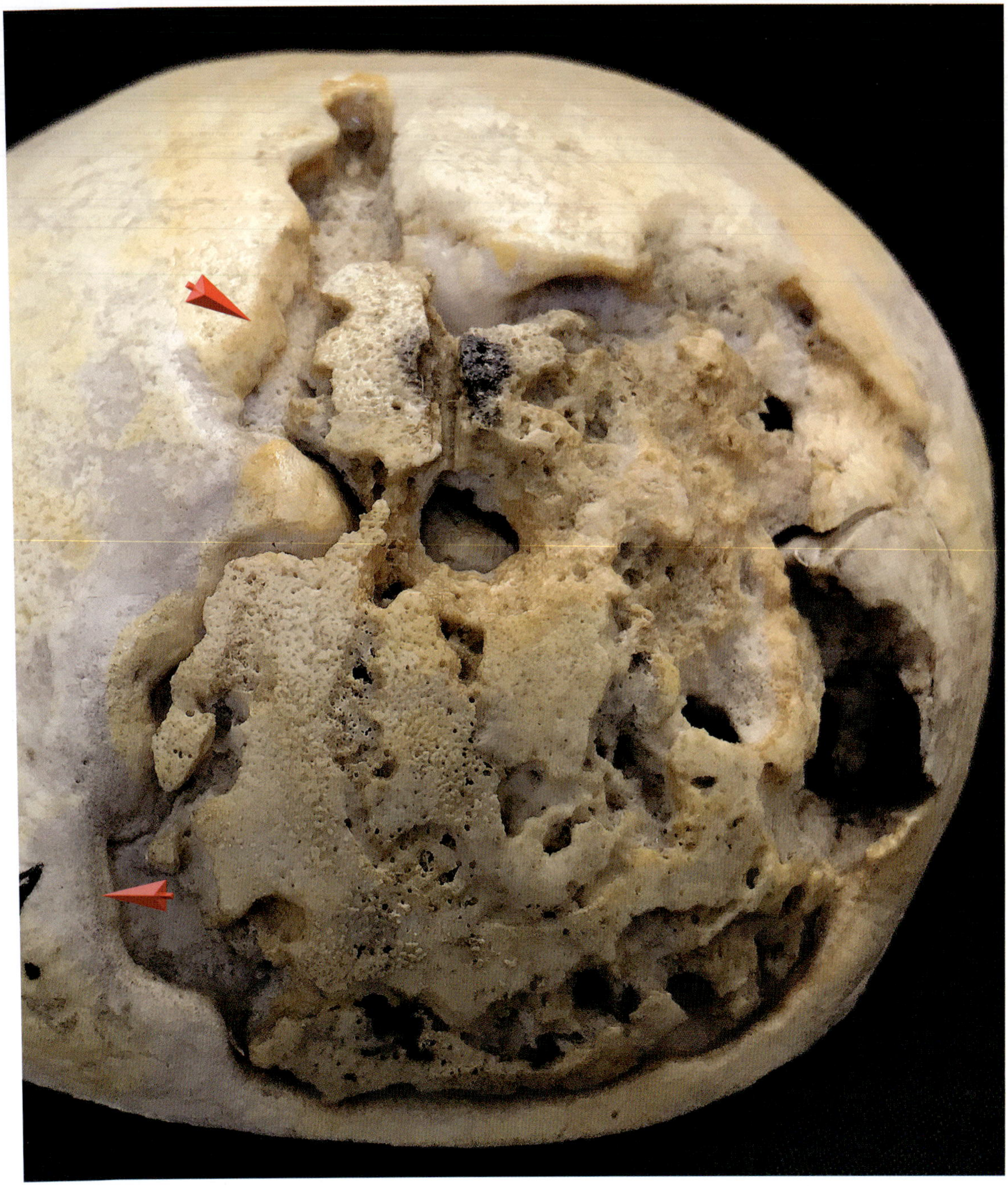

**Figure 200.** Magnified anterior view of cranium with tertiary syphilis (chronic osteomyelitis) mostly affecting the left frontal and left parietal bones in an adult who died in 1885 (MN21.055/461). The central sequestrum or island of necrotic bone has been separated from the rest of the skull by unhealthy granulation tissue/pus. Note the smooth, rounded margins (arrows) of sclerotic bone (involucrum) along the unaffected areas of the frontal bone, lesions known as caries sicca (Virchow 1858) and characteristic of tertiary syphilis. See Armelagos et al. (2012); Aufderheide and Rodrigues-Martin (1998); Betsinger and Smith (2019); Hackett (1963, 1976); Manchester (1983); Ortner and Putschar (1981); Steinbock (1976); Walker et al. (2015) for more on syphilis in bone.

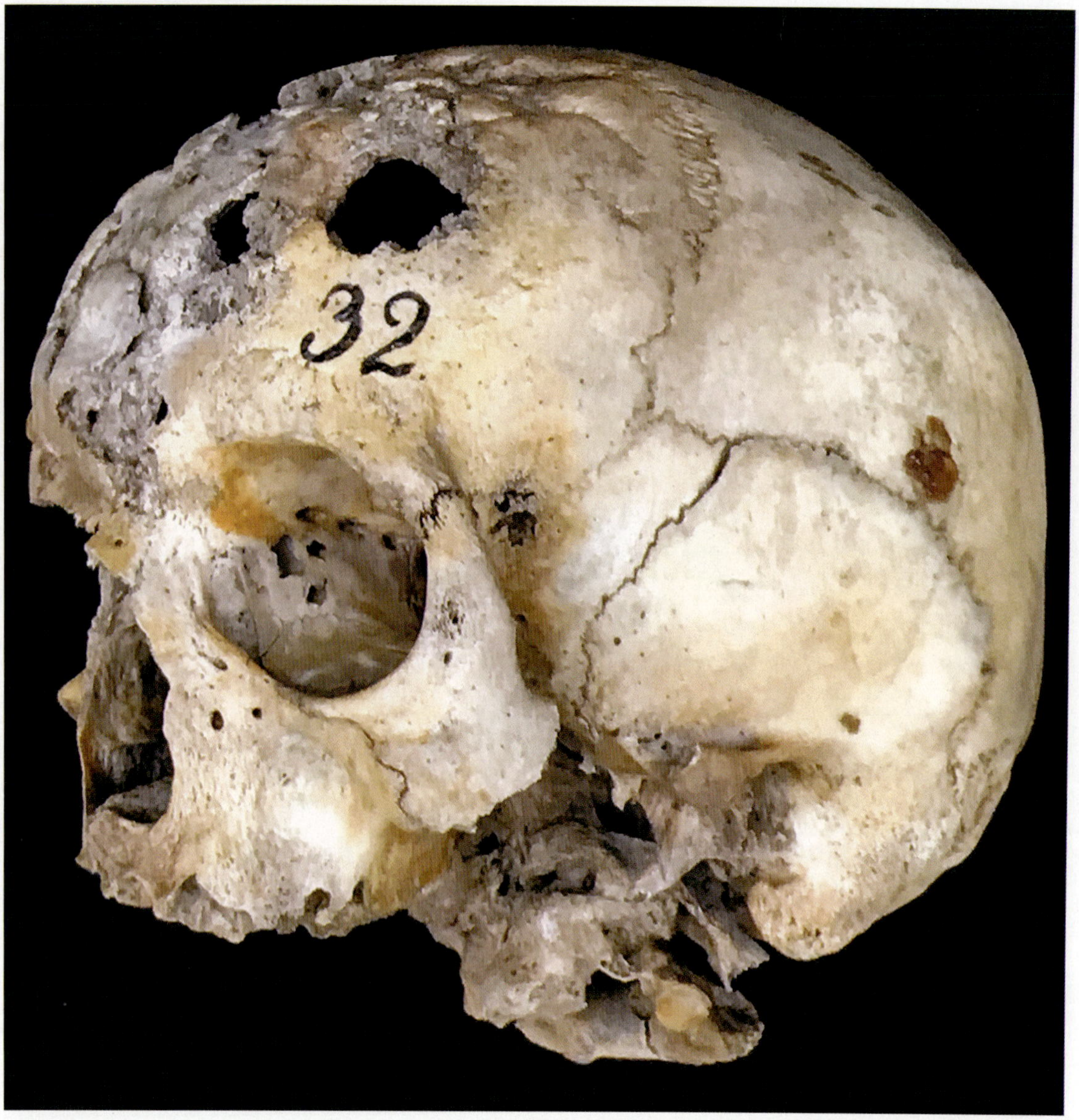


**Figure 201.** Tertiary syphilis affecting much of the cranium, other than the left temporal bone, in an adult (MN21.055/32). The lesions originated in the outer table beneath the periosteum and spread inward destroying the diploe and inner table. Several of the lytic lesions perforated the inner table. See Armelagos et al. (2012); Aufderheide and Rodrigues-Martin (1998); Betsinger and Smith (2019); Hackett (1963, 1976); Manchester (1983); Ortner and Putschar (1981); Steinbock (1976); Walker et al. (2015) for more on syphilis in bone.

**Figure 202.** Tertiary syphilis affecting much of the cranium, other than the left temporal bone, in an adult (MN21.055/32). The infection resulted in several islands (*) of necrotic bone (sequestra) surrounded by irregular margins of healthy bone, some of which in the frontal bone have raised, sclerotic margins. These lesions correspond to "button sequestra" radiologically. Based on the degree of healing, it is likely that infection began in the frontal bone followed by the parietal bones. See Armelagos et al. (2012); Aufderheide and Rodrigues-Martin (1998); Betsinger and Smith (2019); Hackett (1963, 1976); Manchester (1983); Ortner and Putschar (1981); Steinbock (1976); Walker et al. (2015) for more on syphilis in bone.

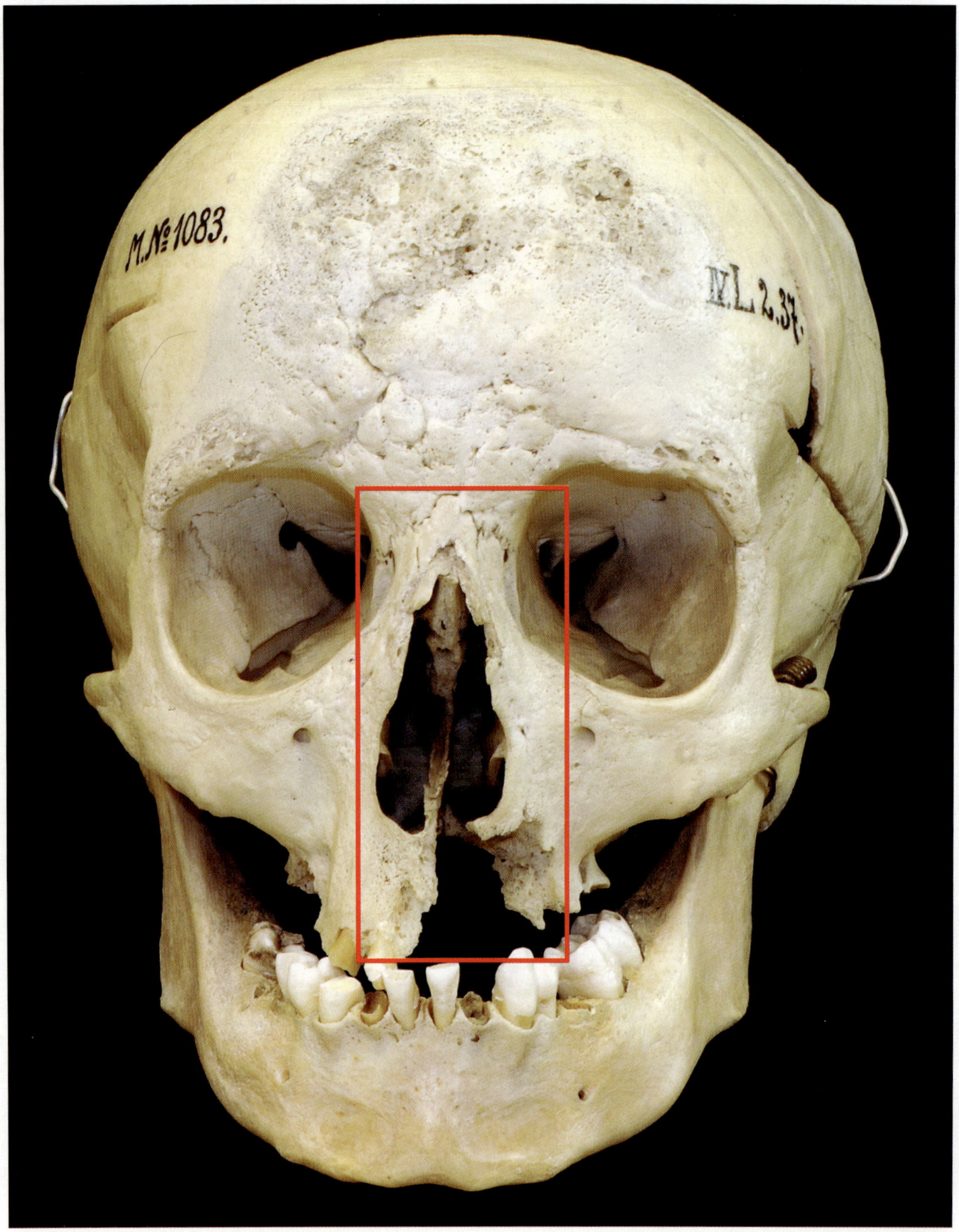


**Figure 203.** Tertiary syphilis affecting the vault and facial region in a 27-year-old female who died in 1828 (MN1083). The infection resulted in perforation of the nasal floor, maxilla, and hard palate (rectangle), antemortem loss of several maxillary teeth, and nearly complete resorption of the nasal bones. It doesn't appear as though the infection in the frontal bone perforated the inner table. See Armelagos et al. (2012); Aufderheide and Rodrigues-Martin (1998); Hackett (1963, 1976); Mann and Hunt (2005); Ortner and Putschar (1981); Steinbock (1976); Walker et al. (2015) for more on syphilis in bone.

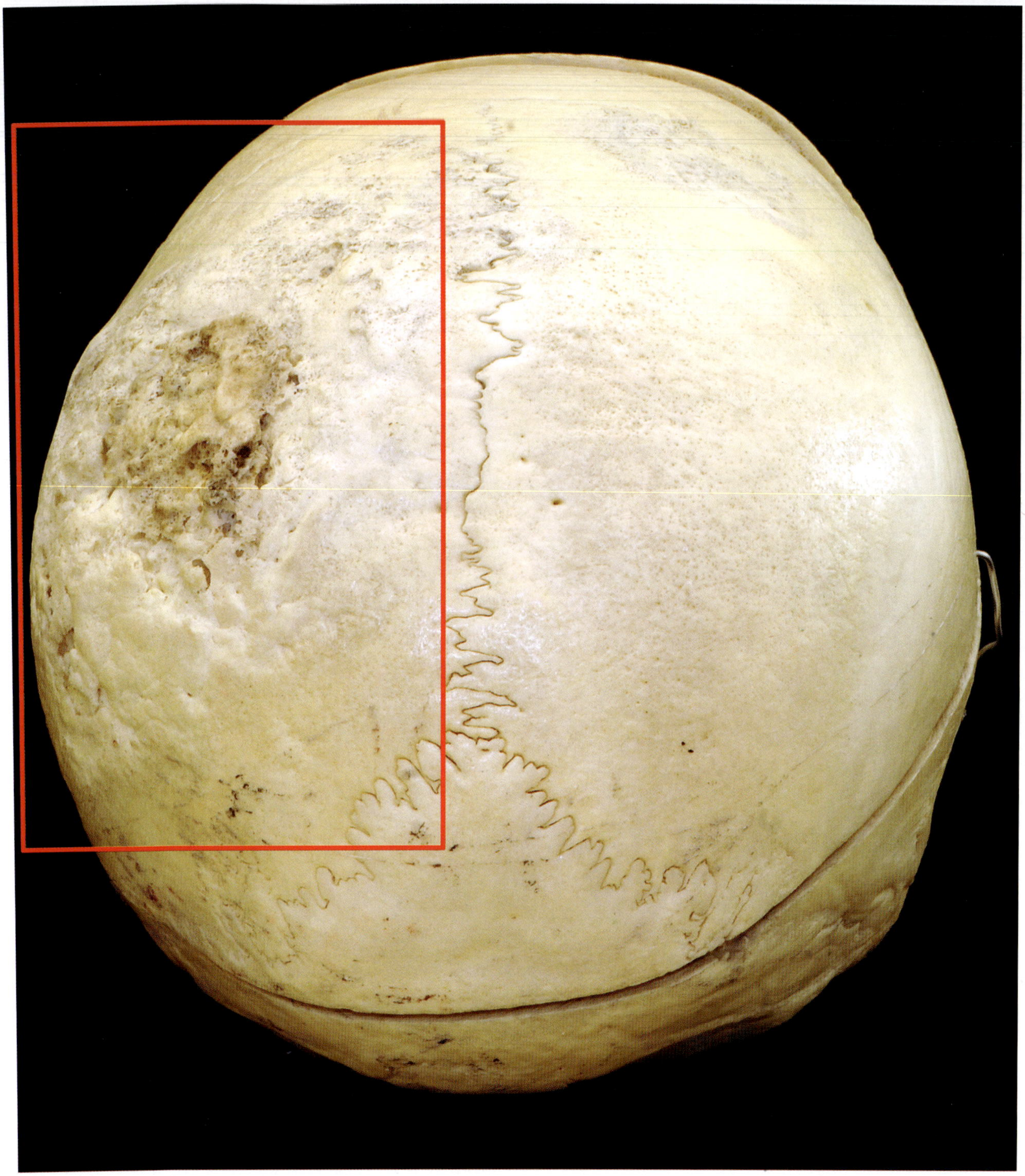

**Figure 204.** Superior view showing tertiary syphilis affecting the left parietal bone (rectangle) in a 27-year-old female who died in 1828 (MN1083). Infection resulted in a crater-like lesion involving the outer table and diploe. See Armelalgos et al. (2012); Aufderheide and Rodrigues-Martin (1998); Hackett (1963, 1976); Manchester (1983); Ortner and Putschar (1981); Steinbock (1976); Walker et al. (2015) for more on syphilis in bone.

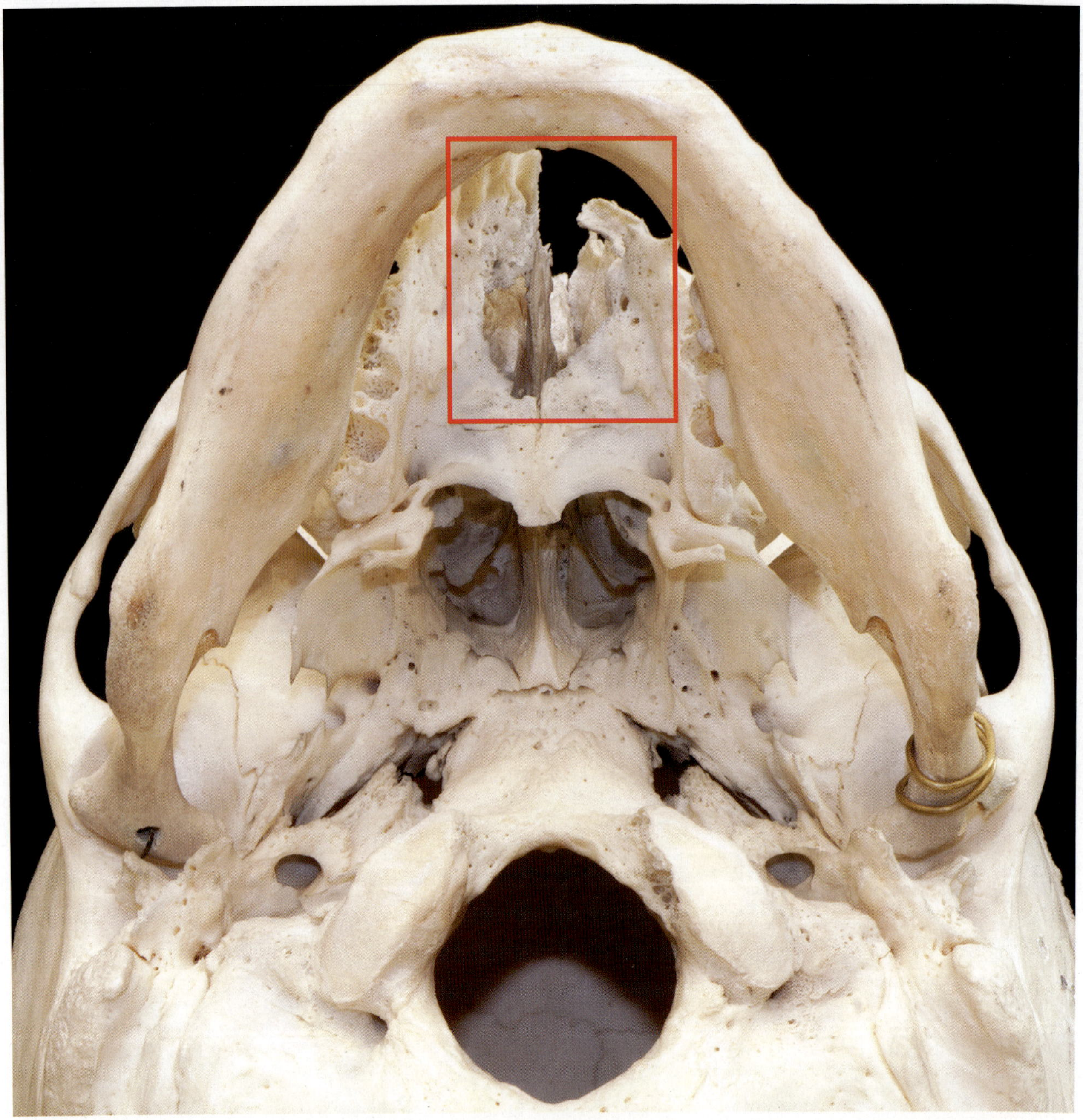

**Figure 205.** Inferior view of tertiary syphilis affecting the cranial vault and facial region in a 27-year-old female who died in 1828 (MN1083). The infection resulted in perforation and resorption of the nasal floor and nasal bones and much of the hard palate of the maxilla (rectangle) with antemortem loss of several teeth. See Armelagos et al. (2012); Aufderheide and Rodrigues-Martin (1998); Hackett (1963, 1976); Manchester (1983); Mann and Hunt (2005); Ortner and Putschar (1981); Steinbock (1976); Walker et al. (2015) for more on syphilis in bone.

**Figure 206.** Ectocranial vault of an adult who died of "gumma in the lung" (tertiary syphilis) in 1926 (MN18.560/21). The vault exhibits numerous large cratered lesions that did not perforate the inner table and a "melted-wax" rim of reactive bone in some areas. Each crater would have had a central island of necrotic/devascularized bone (sequestrum) that the body resorbed or, more likely, were passed through one or more draining tracts (cloacae) in the scalp. See Aufderheide and Rodrigues-Martin (1998); Hackett (1963, 1976); Manchester (1983); Mann and Hunt (2005); Ortner and Putschar (1981); Steinbock (1976); Walker et al. (2015) for more on syphilis in bone and its history.

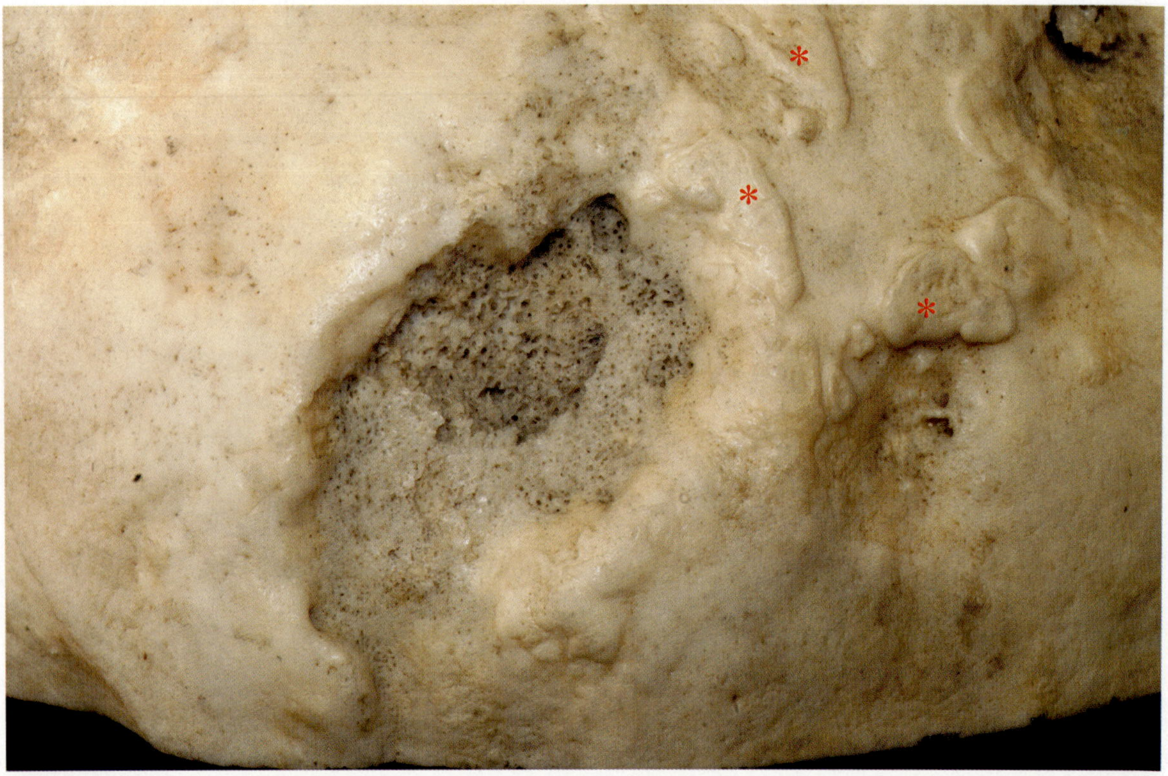

**Figure 207.** Magnified ectocranial view of the vault of an adult with tertiary syphilis ("gumma in the lung") who died in 1926 (MN18.560/21). The vault exhibits numerous large crater-like lesions that did not perforate the inner table and areas of raised bone resembling "melted-wax" (*) surrounding the craters. See Figure 208 for the endocranium. See Aufderheide and Rodrigues-Martin (1998); Hackett (1963, 1976); Manchester (1983); Mann and Hunt (2005); Ortner and Putschar (1981); Steinbock (1976) for more on syphilis and its history in bone.

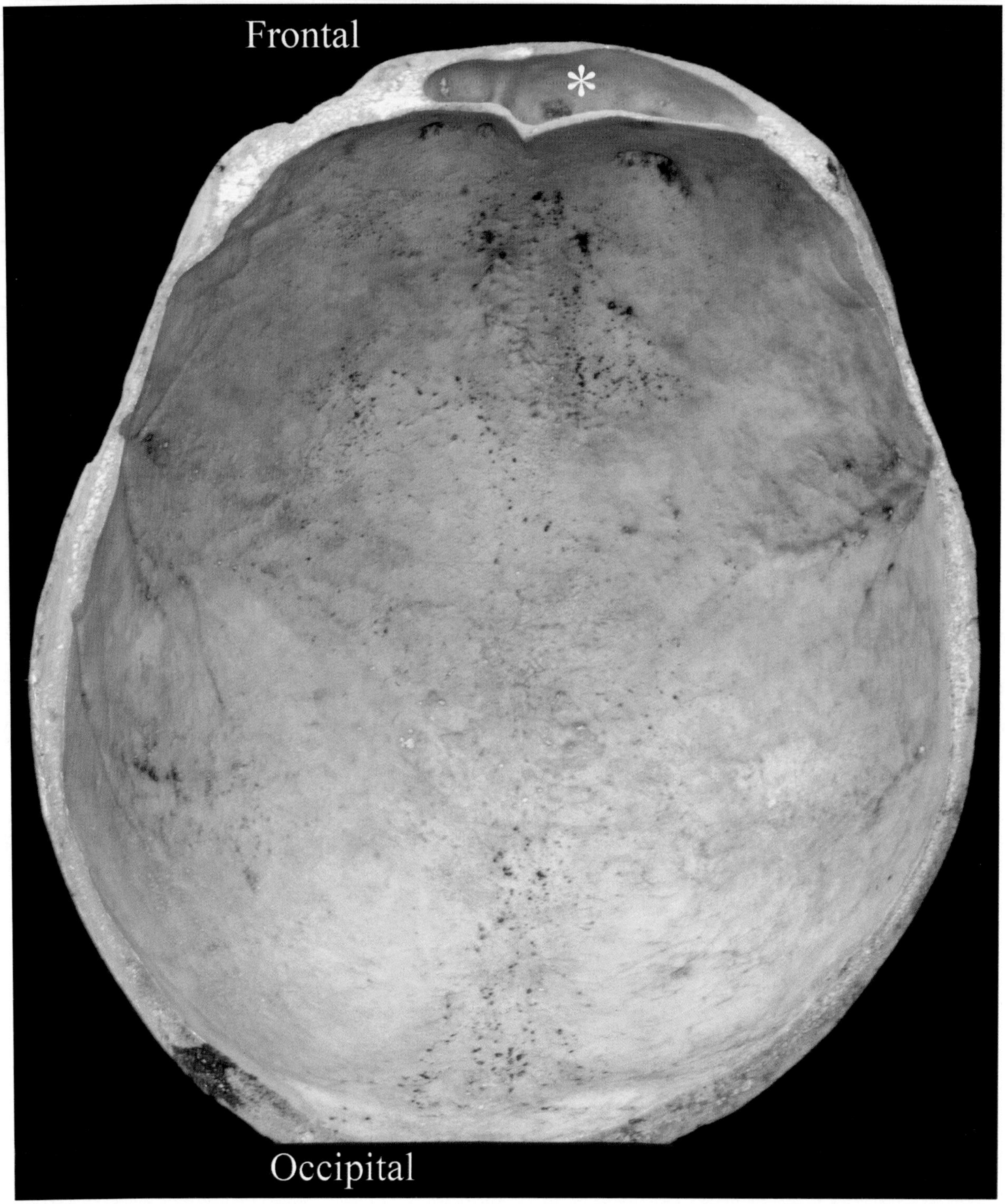


**Figure 208.** Endocranial vault of an adult with tertiary syphilis ("gumma in the lung") who died in 1926 (MN18.560/21). The outer vault (see Figure 207) exhibits numerous large crater-like lesions, while the endocranial surface shows only mild thickening and surface pitting, primarily in the frontal bone (see Figures 206 and 207). The frontal sinuses (*) are grossly unaffected. See Aufderheide and Rodrigues-Martin (1998); Hackett (1963, 1976); Manchester (1983); Ortner and Putschar (1981); Steinbock (1976) for more on syphilis in bone.

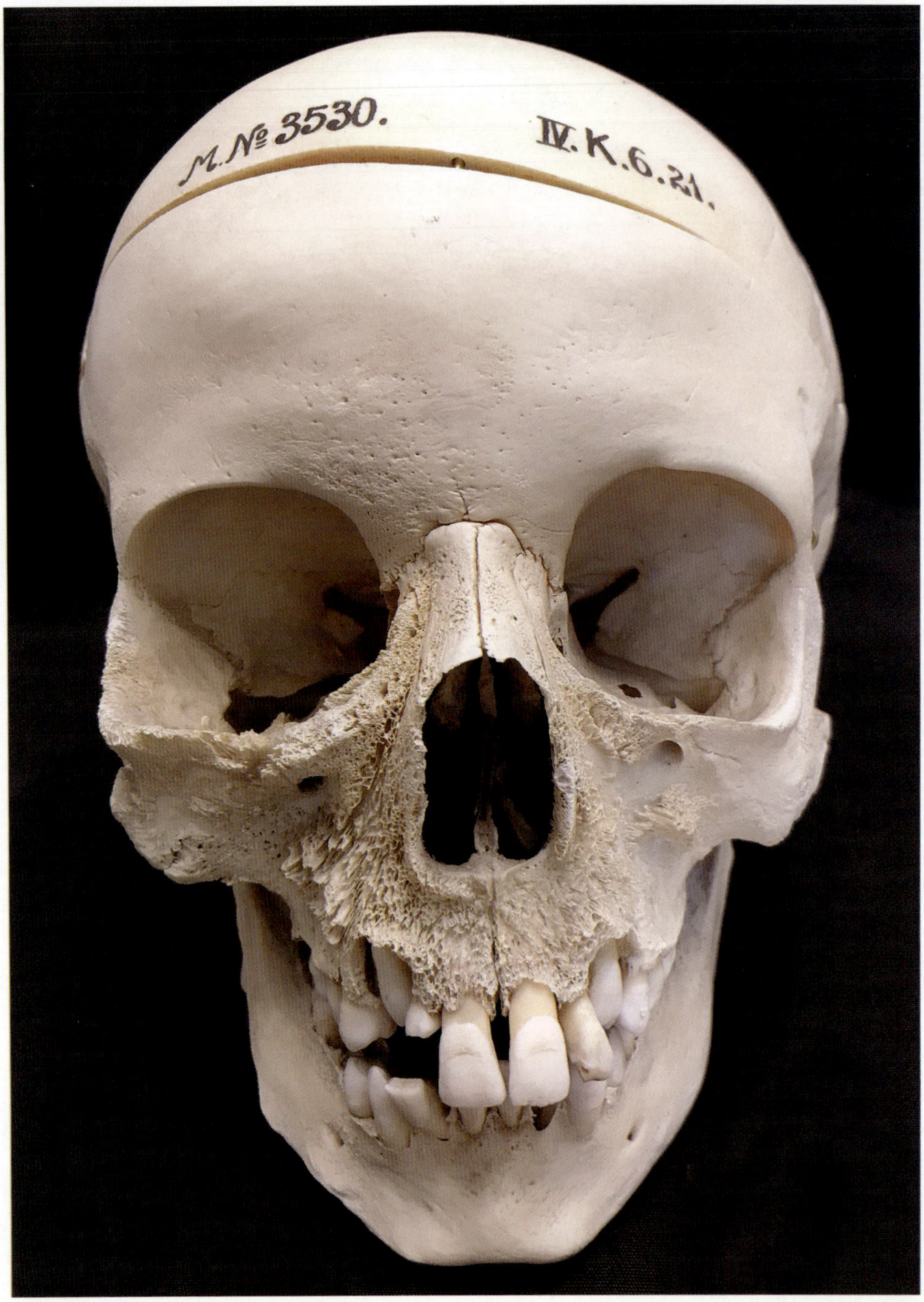


**Figure 209.** Tuberculous osteomyelitis (Lupus vulgaris), a rare form of Mycobacterium tuberculosis in the facial region (Bansal et al. 2015; Rodriguez and Carnelio 2005; Steinbock 1976) in a 16-year-old male who died in 1876 (MN21.055/3530). Note the raised "coral-like" spicules of reactive bone affecting the maxilla, right orbit, and nasal bones and bilateral periapical lesions affecting the canine teeth. See Aufderheide and Rodrigues-Martin (1998); Buikstra (1981); Buzic and Giuffra (2020); Karg and Somvanshi (2011); Manchester (1983); Ortner (2003); Ortner and Putschar (1981); Steyn and Buskes (2016) for more on the skeletal expression and antiquity of tuberculosis.

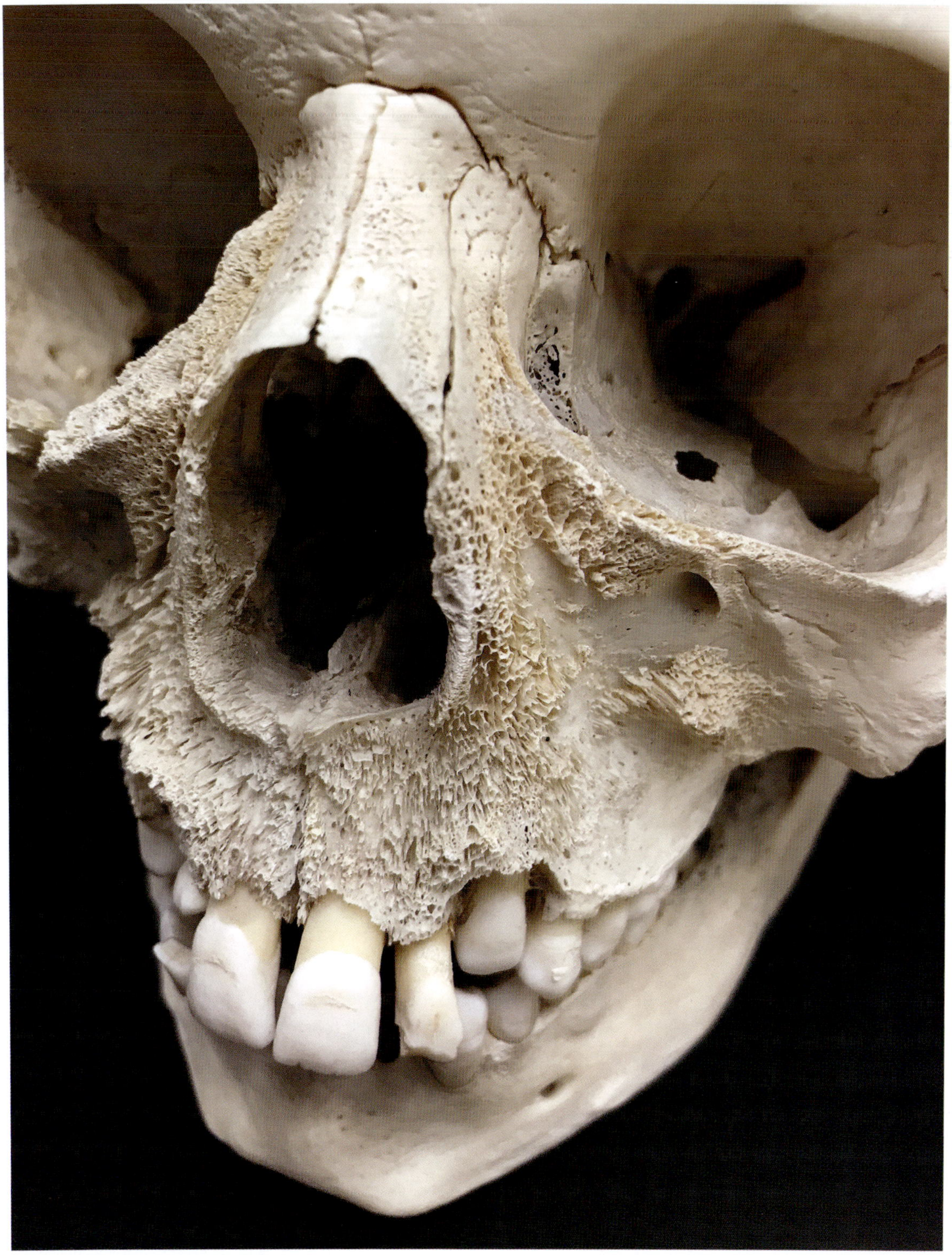

**Figure 210.** Magnified view of raised lesions due to tuberculous osteomyelitis (Lupus vulgaris) of the facial region in a 16-year-old male (MN21.055/3530). See Buikstra (1981); Buzic and Giuffra (2020); Garg and Somvanshi (2011); Geshchickter and Copeland (1936); Ortner (2003); Ortner and Putschar (1981); Steinbock (1976) for more on tuberculosis in bone.

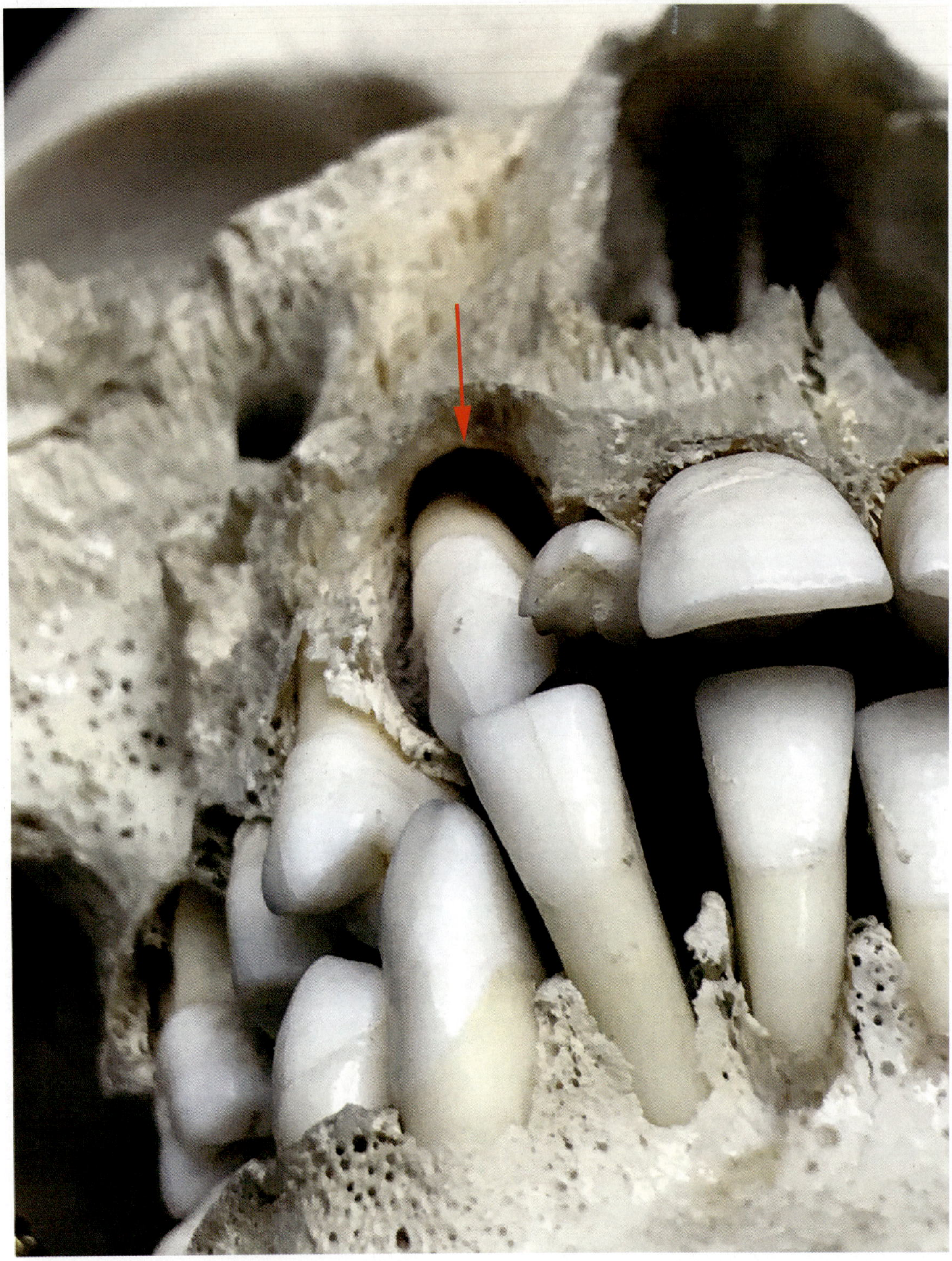

**Figure 211.** Anterior-inferior magnified view of tuberculous osteomyelitis (Lupus vulgaris) of the maxilla and mandible with advanced resorption, likely a periapical lesion, of the maxillary canine tooth (arrow) in a 16-year-old male; a similar lesion is present in the maxillary left canine (MN21.055/3530). See Buikstra (1981); Geshchickter and Copeland (1936); Ortner and Putschar (1981); Steinbock (1976) for more on tuberculosis in bone.

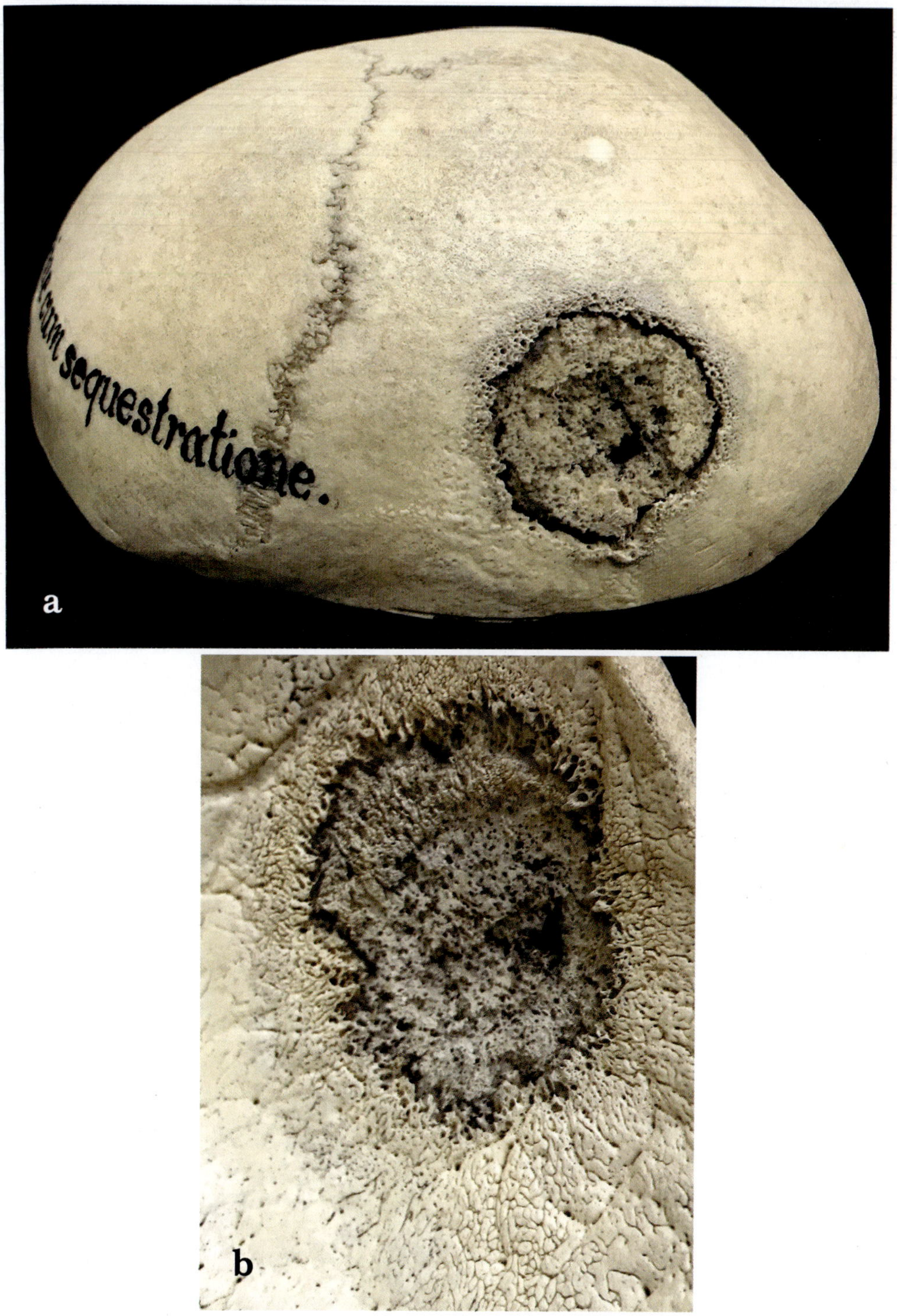


**Figure 212a-b.** (a) Left lateral and (b) endocranial views of chronic osteomyelitis ("Osteomyelitis calvariea cum sequestratione" on the museum card) in an individual who died in 1928, possibly due to tuberculosis with a large area of necrotic bone (sequestrum) that has been "walled off" from the surrounding living bone (MN21.055/5747). Note the globular or "bubble-like" pattern of bone known as serpens endocrania symmetirca (SES) surrounding the sequestrum endocranially (Hershkovitz et al. 2002). See Buikstra (1981); Geshchickter and Copeland (1936); Ortner (2003); Ortner and Putschar (1981) for more on tuberculosis in bone.

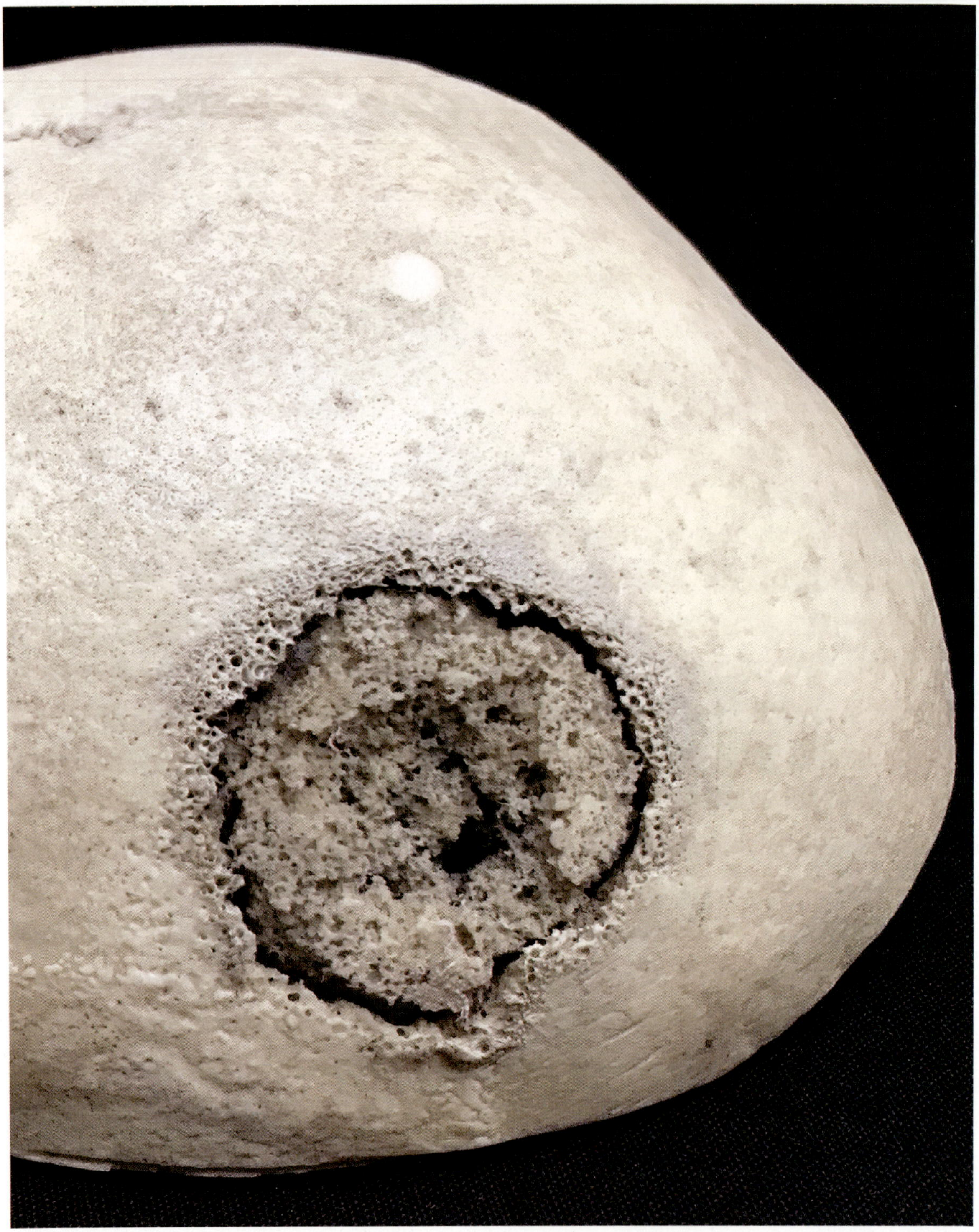

**Figure 213.** Magnified ectocranial view of the large lesion (chronic osteomyelitis) in an unknown individual who died in 1928, possibly Mycobacterium tuberculosis, with a central sequestrum in the left parietal bone (MN21.055/5747). Note the raised margin of porous reactive bone surrounding the lesion. See Buikstra (1981); Buzic and Giuffra (2020); Christopher (1936); Garg and Somvanshi (2011); Geshchickter and Copeland (1936); Ortner (2003); Ortner and Putschar (1981); Steinbock (1976) for more on tuberculosis in bone.

**Figure 214.** Magnified endocranial view of the osteomyelitis (possibly tuberculosis) lesion and sequestrum in an unknown individual who died in 1928 (MN MN21.055/5747). Note the irregular, spiked lesions bordering the central sequestrum (island necrotic bone) and the globular (rectangle) inner table referred to as serpens endocrania symmetrica (SES) (Hershkovitz et al. 1999, 2002; Lewis 2004).

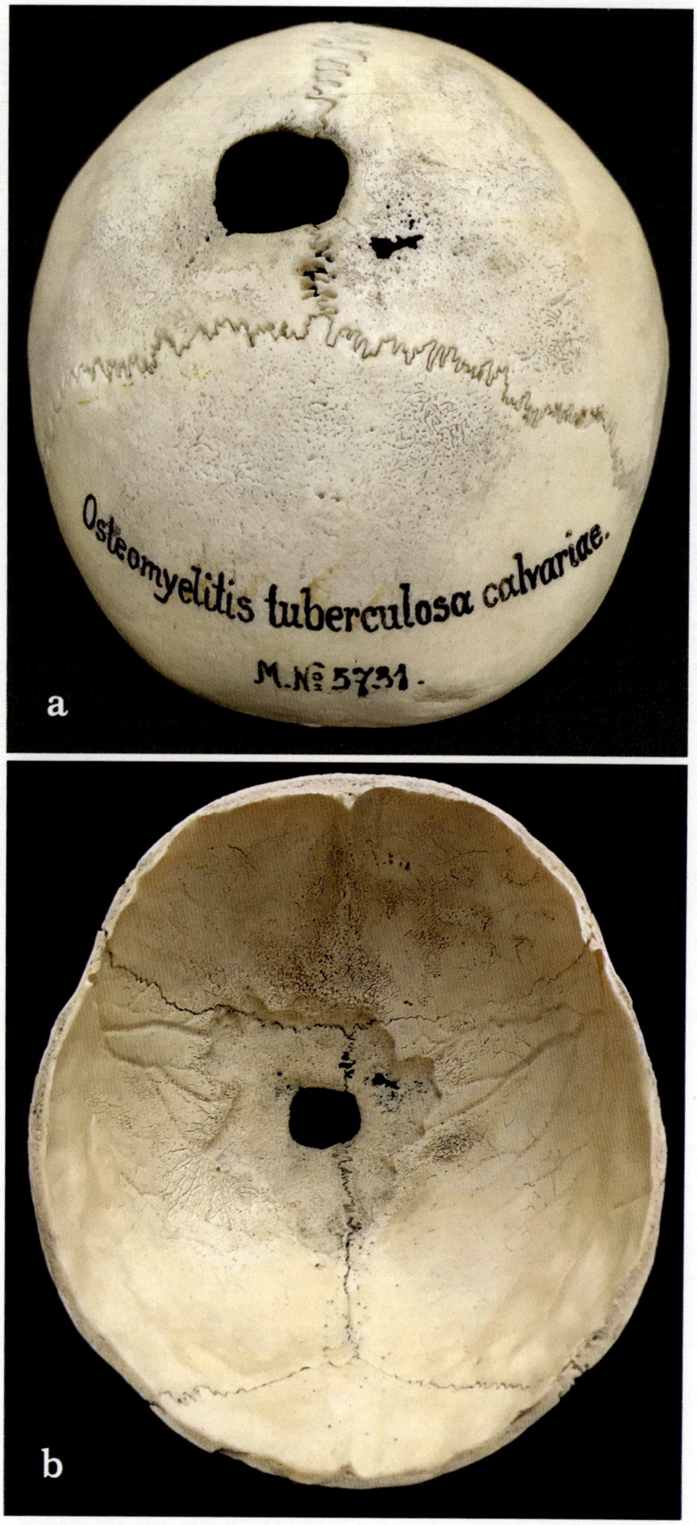


**Figure 215a-b.** (a) Ectocranial and (b) endocranial views of Mycobacterium tuberculosis of the cranium in an individual who died before 1843, resulting in a large perforating lesion near vertex that is surrounded by serpentine/maze-like bone, serpens endocrania symmetrica (SES), often associated with tuberculosis and intrathoracic disease (Hershkovitz et al. 1999, 2002) (MN5731).

**Figure 216.** Magnified view of a tuberculous lesion in an unknown individual who died before 1843, that perforated and eroded the inner and outer tables (MN5731). Note scattered patches of serpens endocrania symmetrica (SES) associated with hypertrophic osteoarthropathy (pulmonary disease), infection, and tuberculosis (see Hershkovitz et al. 1999, 2002). SES has been shown to affect the superficial layer of the endocranium without involving the diploe. See Aufderheide and Rodrigues-Martin (1998); Buikstra (1981); Buzic and Giuffra (2020); Christopher (1936); Garg and Somvanshi (2011); Geschickter and Copeland (1936); Lewis (2004); Ortner (2003); Ortner and Putschar (1981); Steinbock (1976) for more on tuberculosis in bone and SES.

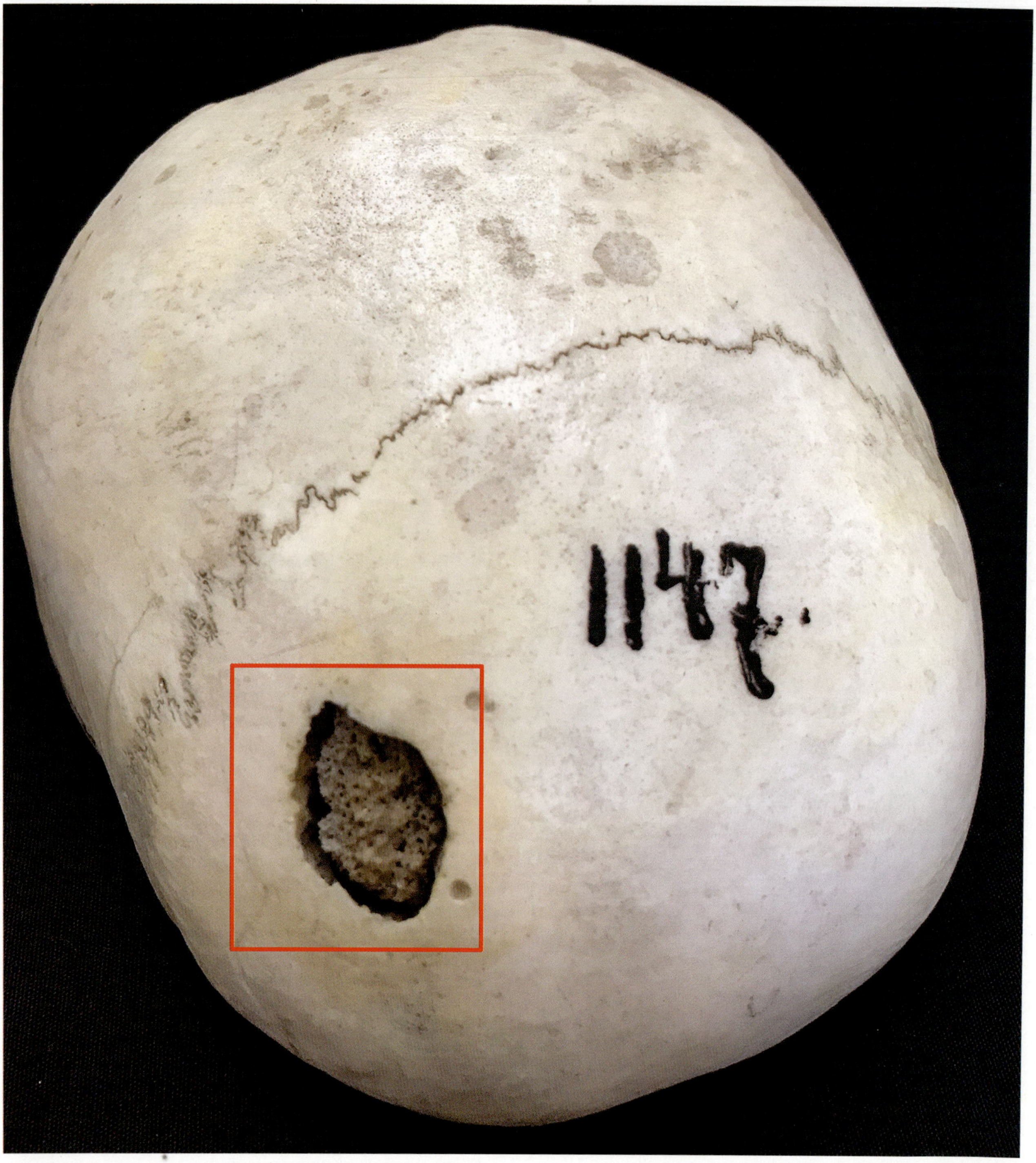


**Figure 217.** Ectocranial view of a large solitary sequestrum (square), possibly due to tuberculosis or syphilis, in the right half of the frontal bone (MN1147). The accession card reports spina bifida, spondylitis and scoliosis in this individual who died in 1829. See Aufderheide and Rodrigues-Martin (1998); Garg and Somvanshi (2011) for more on tuberculosis.

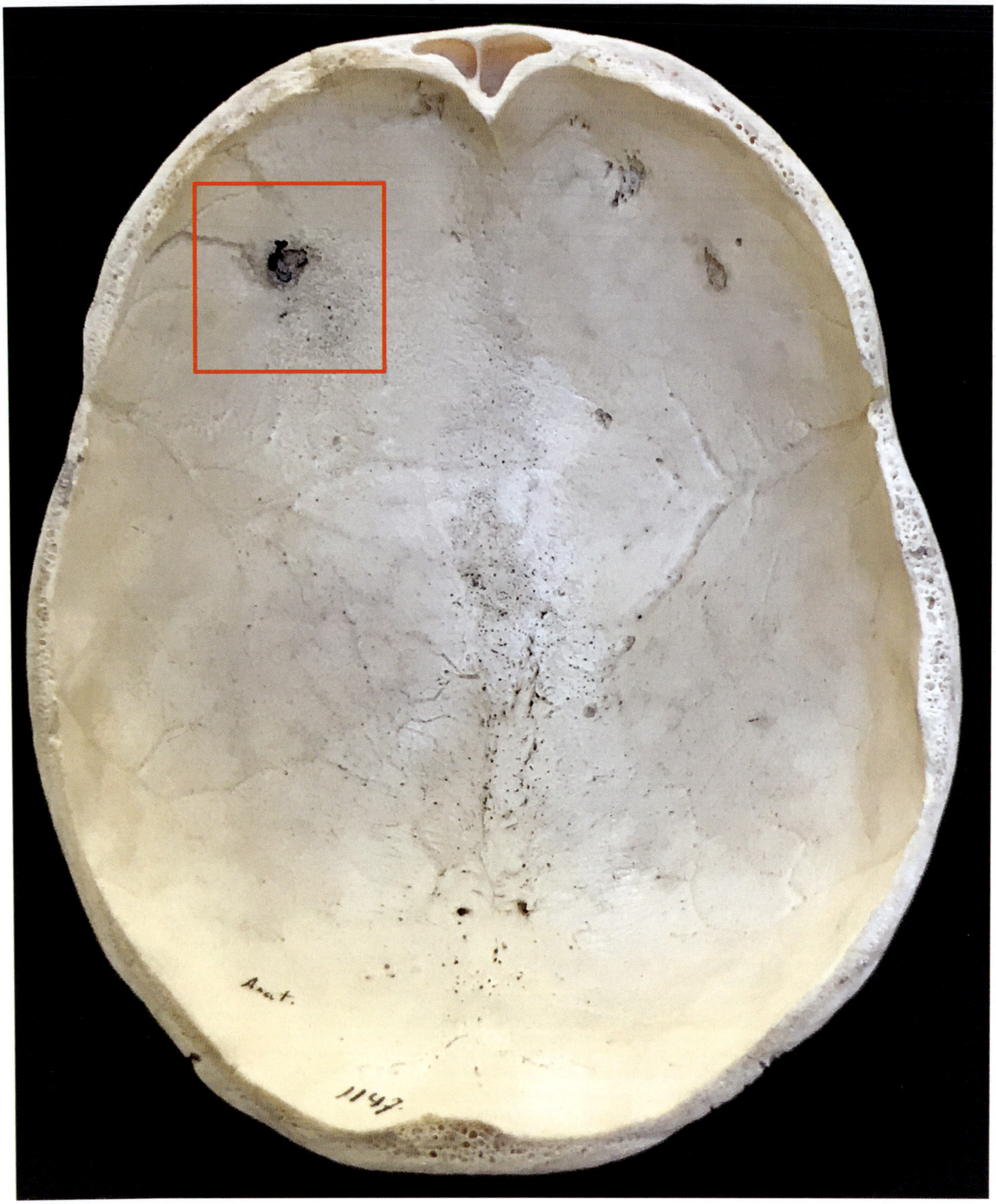


**Figure 218.** Endocranial view of a large solitary sequestrum (square), possibly due to tuberculosis, in the right half of the frontal bone (MN1147). Note that the lesion is much larger in the outer table of the skull (Figure 217). See Buikstra (1981); Christopher (1936); Geshchickter and Copeland (1936); Ortner and Putschar (1981); Steinbock (1976) for tuberculosis in bone.

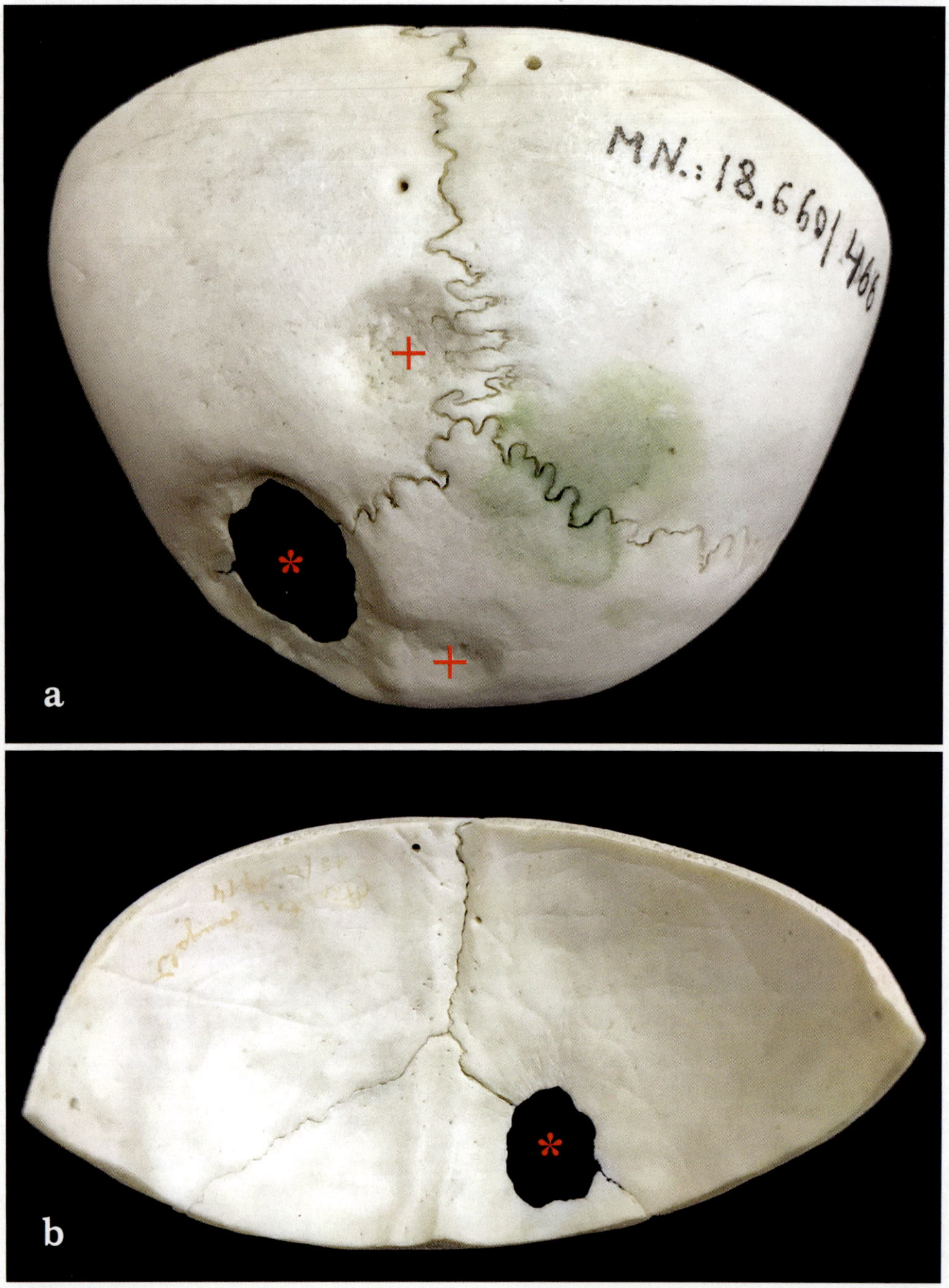


**Figure 219a-b.** (a) Ectocranial and (b) endocranial views of a large destructive tubercular lesion (*) ("Tuberkulose" on the museum card) along the lambdoidal suture involving the left parietal and occipital bones. There are two crater-like lesions (+) in the left parietal and occipital bones that exhibit healing. Based on the thin cranial bones and open sutures this appears to be a child or young adult (MN18.660/456). If these lesions are indeed due to tuberculosis, they are atypical in that they do not exhibit any apparent evidence of infection in the form of pitting or reactive bone in or around the three lesions. See Buikstra (1981); Buzic and Giuffra (2020); Christopher (1936); Garg and Somvanshi (2011); Geshchickter and Copeland (1936); Mann and Hunt (2005); Ortner (2003); Ortner and Putschar (1981); Steinbock (1976) for more on tuberculosis in bone.

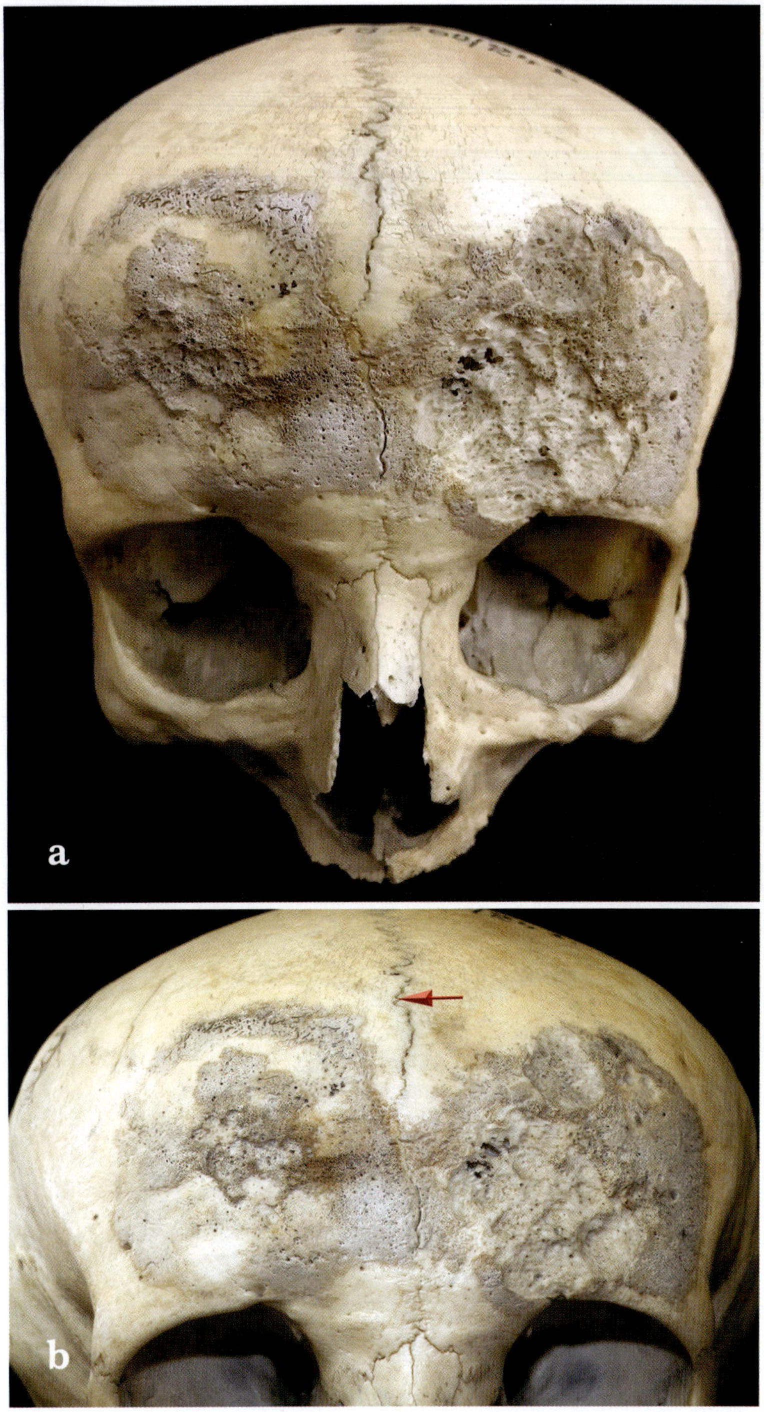


**Figure 220a-b.** (a-b) Frontal views of tertiary syphilis ("Usura ossis frontalis" on the museum card) confined to the frontal bone in a 40-year-old male who died of syphilis in 1930 (MN18.560/241). Syphilis is typically a combined osteoblastic and osteoclastic process that can affect any bone at any age (Christopher 1936). Bilateral and symmetrical osteoblastic (proliferative periostitis) and osteoclastic (lytic) lesions in this individual involve much of the frontal bone immediately above the orbits. Note the open metopic suture (arrow), a normal anatomical variant.

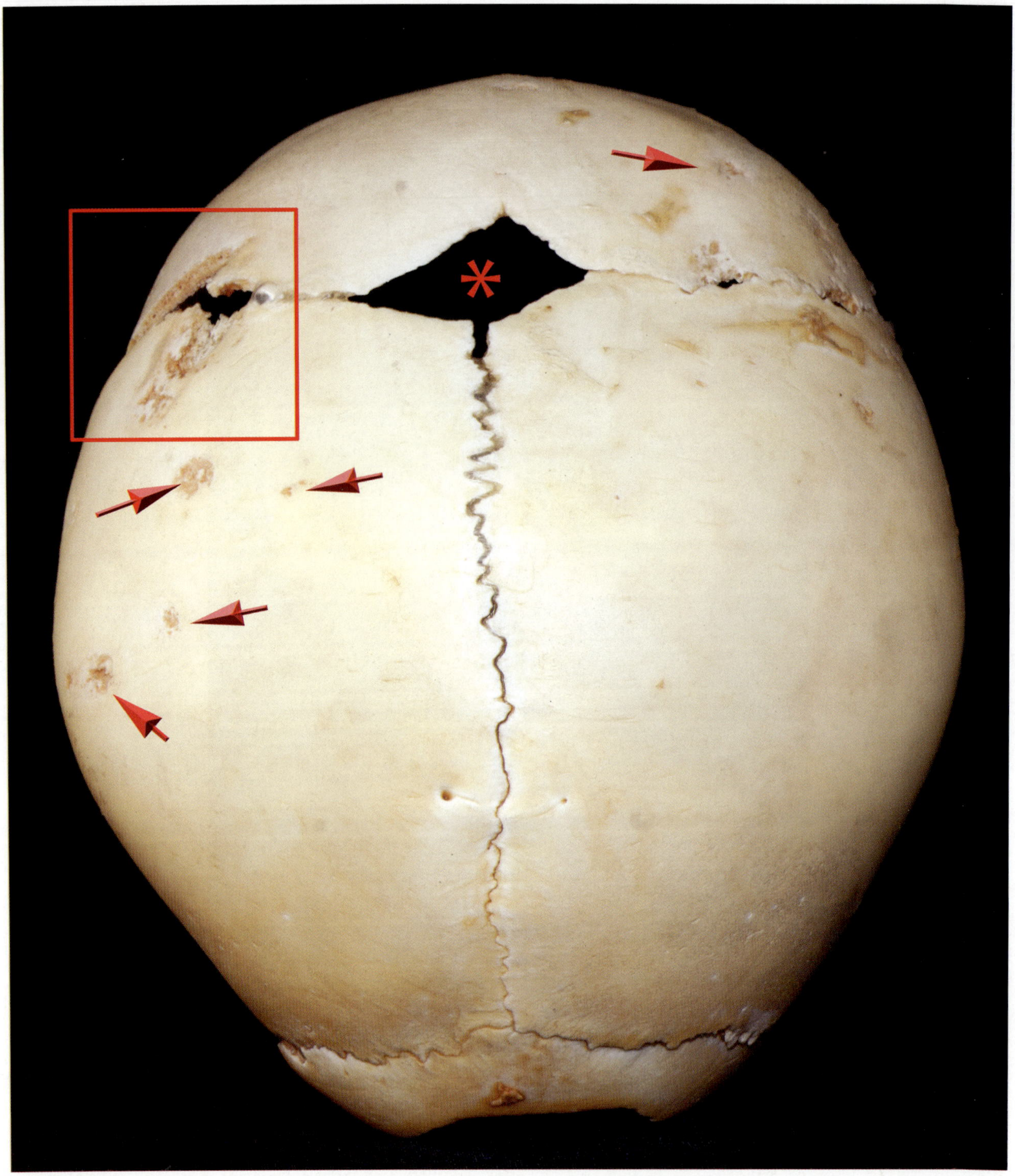

**Figure 221.** Ectocranial view of cranial tuberculosis showing numerous lesions (arrows and square) and partially closed anterior fontanelle (*) in a 13-month-old male (MN10.178). The endocranium exhibits a serpentine maize-like (SES; Hershkovitz et al. 2002) pattern often accompanying (Mycobacterium) tuberculosis and other intrathoracic diseases. See Buikstra (1981); Christopher (1936); Garg and Somvanshi (2011); Geshchickter and Copeland (1936); Ortner (2003); Ortner and Putschar (1981); Steinbock (1976) for more on tuberculosis in bone.

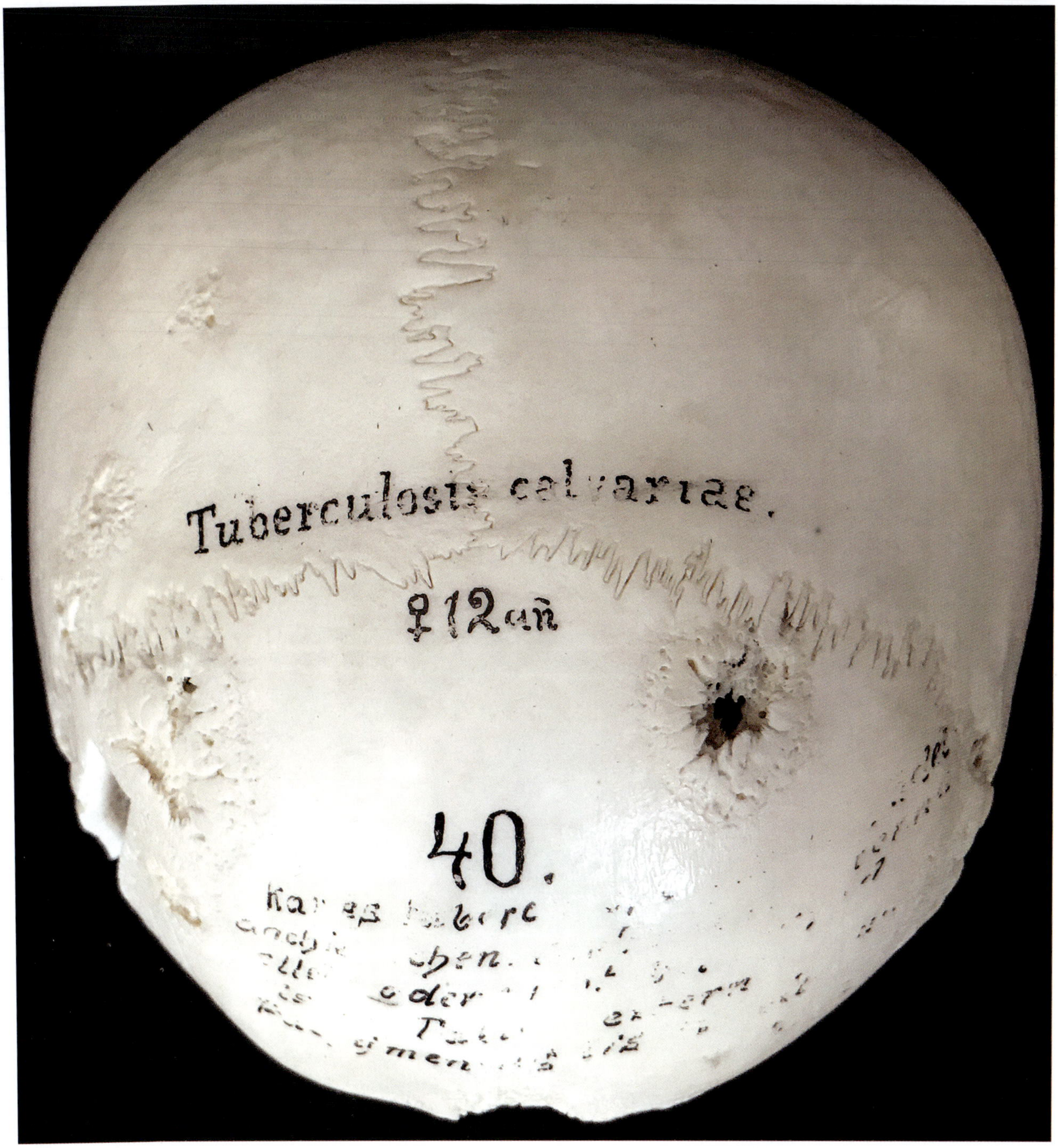


**Figure 222.** Mycobacterium tuberculosis (TB) in the calvarium in a 12-year-old female (MN50). Several lesions with raised irregular margins have perforated the frontal bone. See Aufderheide and Rodrigues-Martin (1998); Buikstra (1981); Christopher (1936); Garg and Somvanshi (2011); Geshchickter and Copeland (1936); Ortner (2003); Ortner and Putschar (1981); Steinbock (1976) for more on tuberculosis in bone.

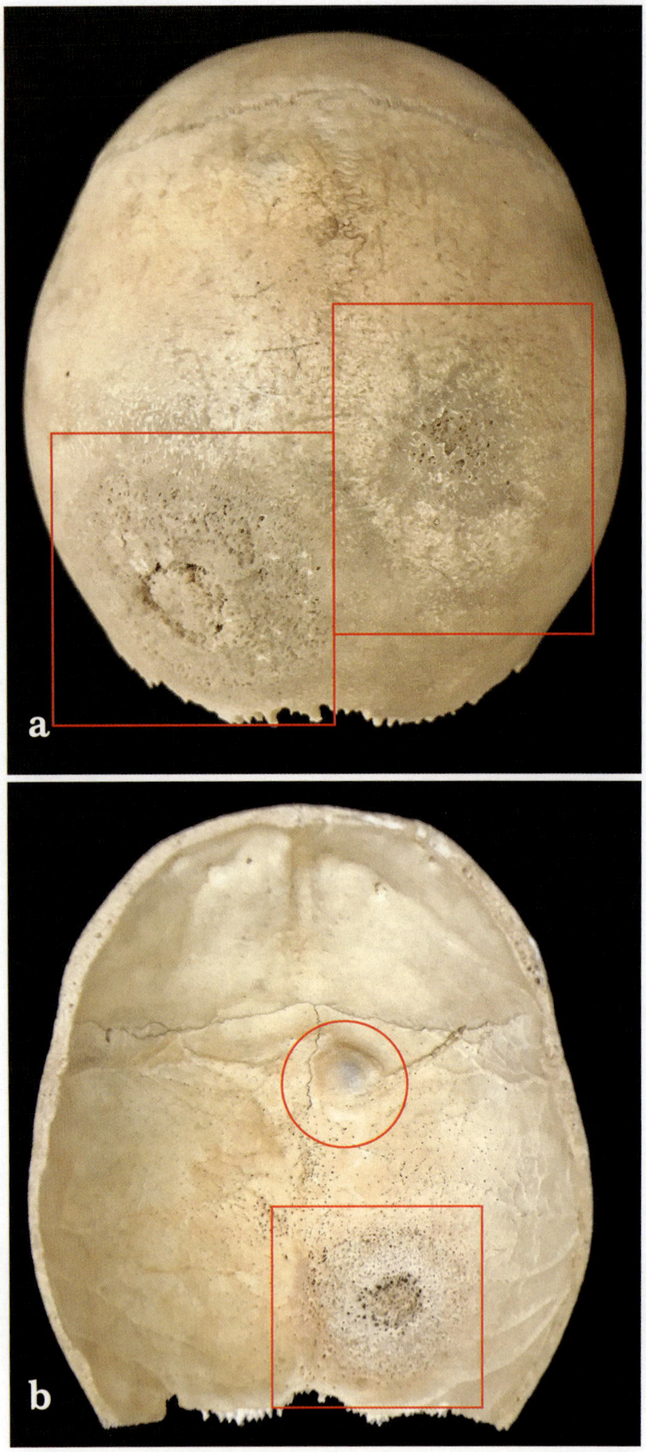


**Figure 223a-b.** (a) Ectocranial and (b) endocranial views of primarily lytic lesions (square/rectangle) due to tuberculosis and a crater-like depression (circle) caused by contact pressure erosion by clusters of arachnoid granulations in the skull of a 15-year-old male who died in 1863 (MN21055.640). See Buikstra (1981); Christopher (1936); Geshchickter and Copeland (1936); Ortner (2003); Ortner and Putschar (1981); Steinbock (1976); Walker et al. (2015) for more on tuberculosis in bone and Mann et al. (2016) for arachnoid granulations.

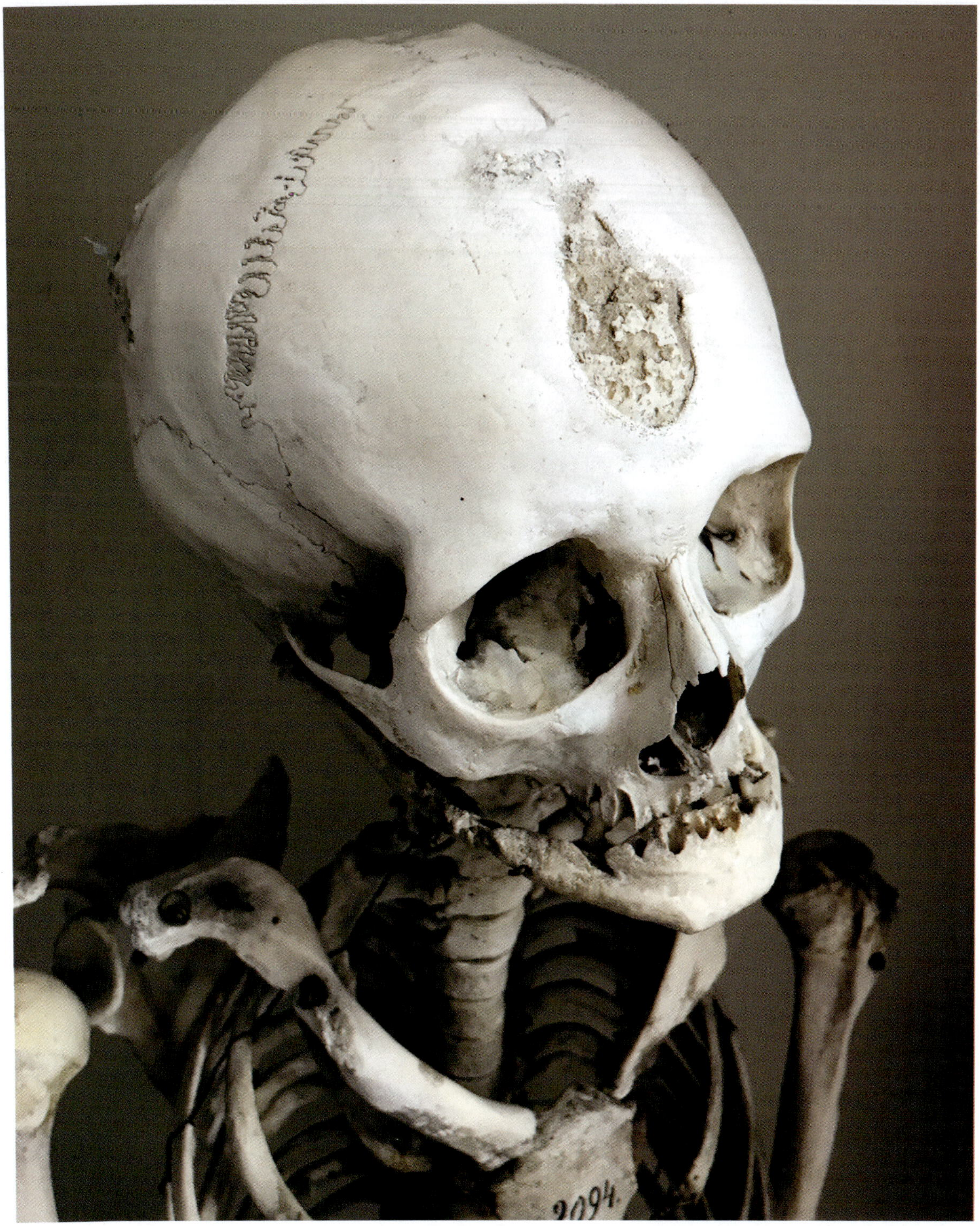


**Figure 224.** Syphilitic lesions ("Lues" on the museum card) of the frontal and parietal bones, mandible, and right clavicle in a 25-year-old female who died in 1869 (MN21.055/2094). See Christopher (1936); Geshchickter and Copeland (1936); Ortner (2003); Ortner and Putschar (1981); Steinbock (1976); Walker et al. (2015) for more on syphilis in bone.

## Spine, Ribs and Pelvis

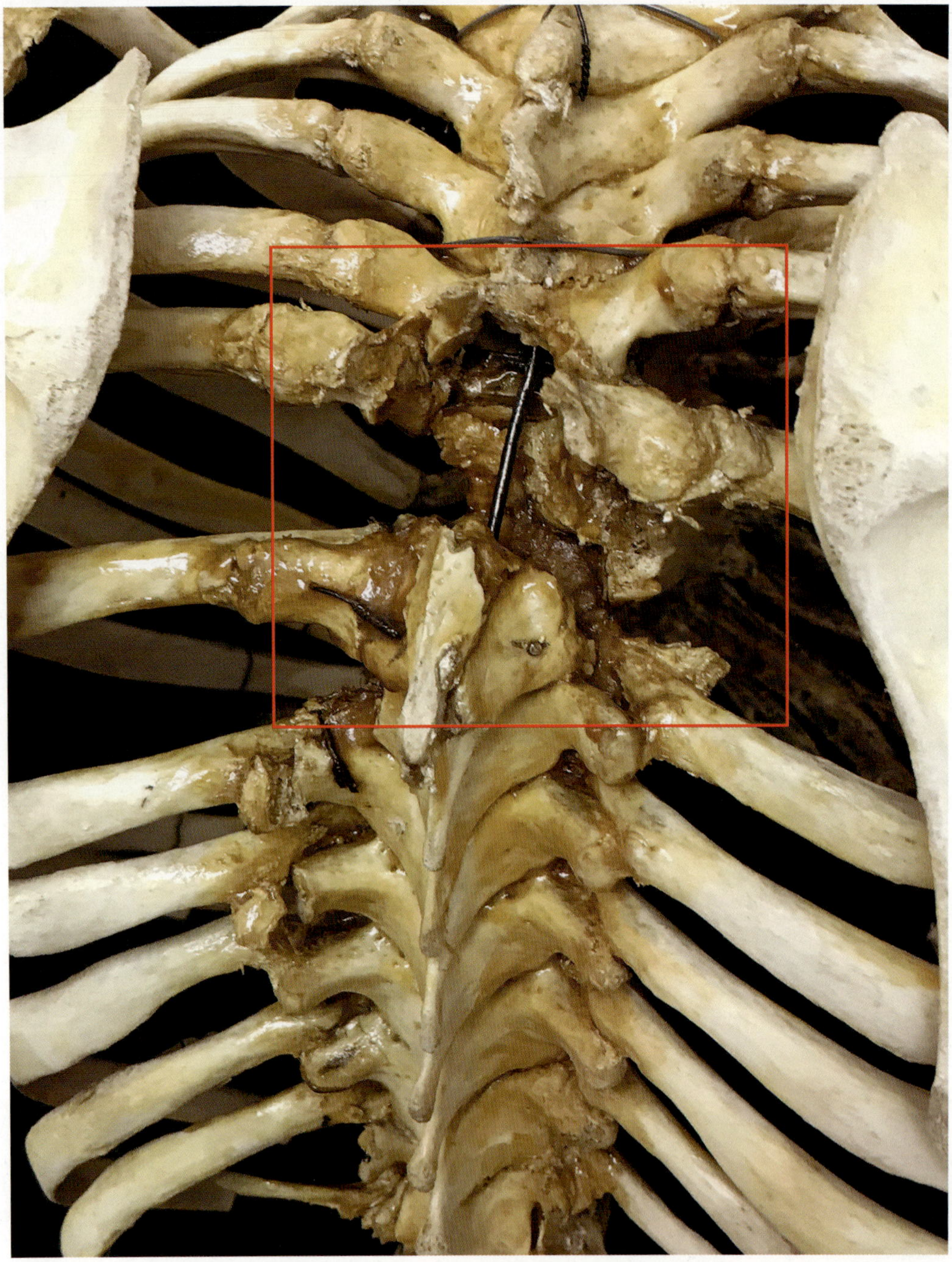

**Figure 225.** Posterior view of the spine of an 18-year-old male with a large defect/lesion in the arch of the thoracic vertebra that resulted in kyphosis (MN3371). "Doesn't look like crushing fracture unless a pathological fracture from a lesion or an infection there. Local destruction and "eaten" bone could have been osteomyelitis, tuberculosis, aneurysmal bone cyst, or an eosinophilic granuloma, are classic in individuals this age that destroy bone and result in pathological fractures" (P. Moroz, personal communication, November 17, 2022). See Lenchik et al. (2004) for more on the classification, diagnosis and treatment of spinal trauma, especially as it relates to osteoporosis.

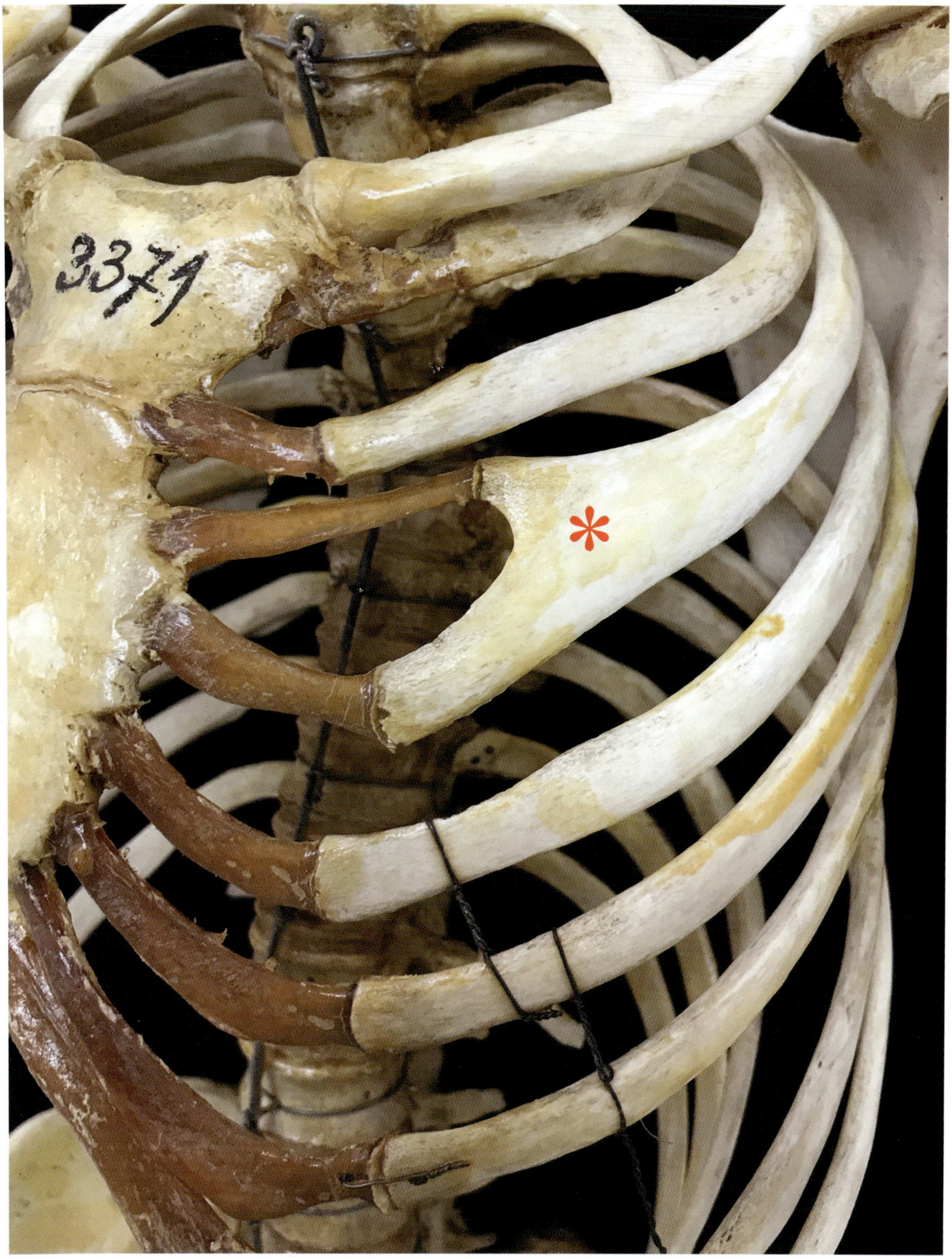


**Figure 226.** Bifurcated left 3rd rib (*) in an 18-year-old male with a large lytic lesion in the arch of the 6th thoracic vertebra that resulted in kyphosis. Rare rib anomalies such as these may be misidentified as trauma and are the result of congenital abnormality of the anterior chest wall (Kryger et al. 2013) (MN3371).

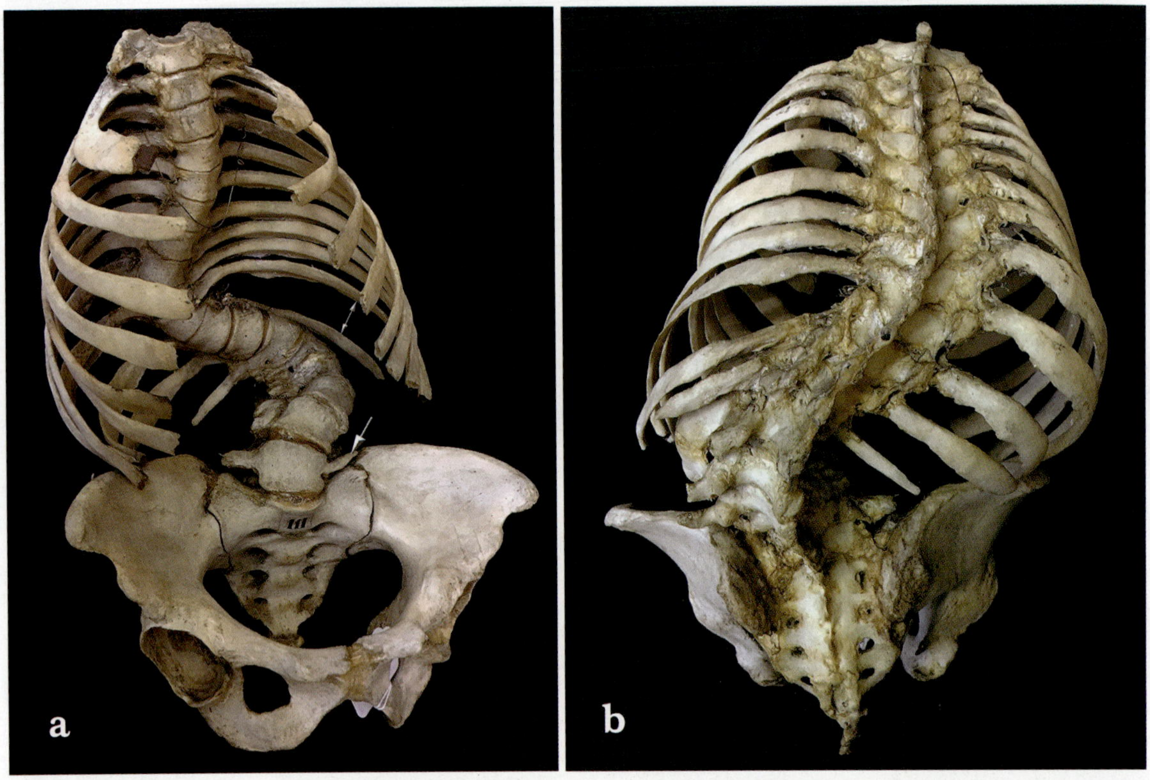


**Figure 227a-b.** (a) Anterior and (b) posterior views of kyphoscoliosis in the spine of a 40-year-old female with a pelvic ring/outlet that measured 18 cm in diameter and who died of tuberculosis in 1892 (MN17.17/111). The kyphoscoliosis in this individual resulted in repositioning of the internal organs causing the left ribs to "open" laterally (double arrow). Spinal curvature also resulted in superior movement of the left transverse process in L5 (arrow). Scoliosis often results in unusual curvature of the anterior/free ends of ribs 11 and 12 where they may come into contact with the pelvis. See *Campbell's Operative Orthopaedics* (1956); Christopher (1936); Garg and Somvonshi (2011) for more on acquired and congenital spinal deformities.

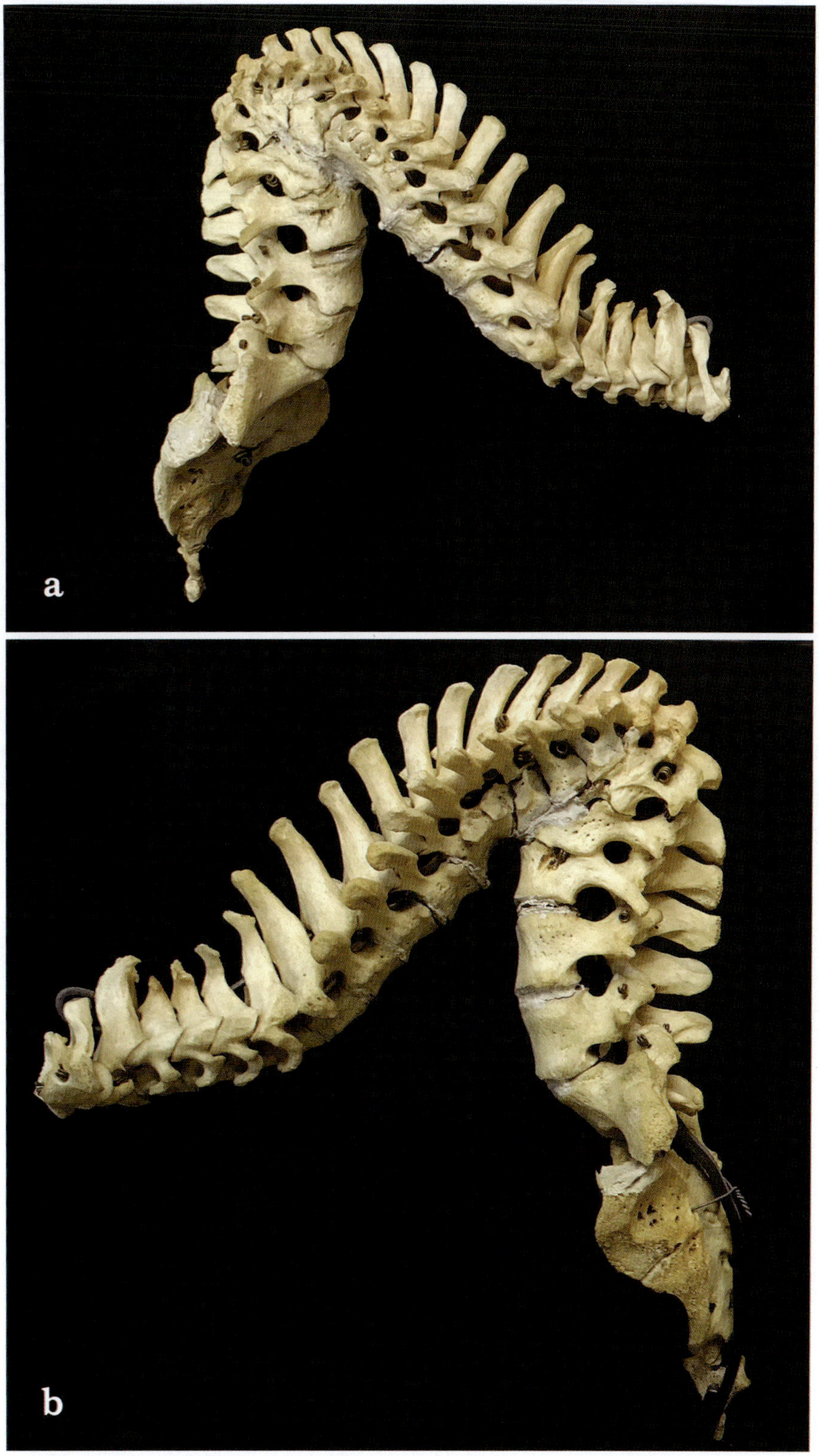


**Figure 228a-b.** (a) Right and (b) left lateral views of thoraco-lumbar region resulting in collapse and angular deformity (gibbus; kyphoscoliosis) of approximately 45° in a subadult who died of tuberculosis in 1825 with both spina bifida and flat foot (MN752).

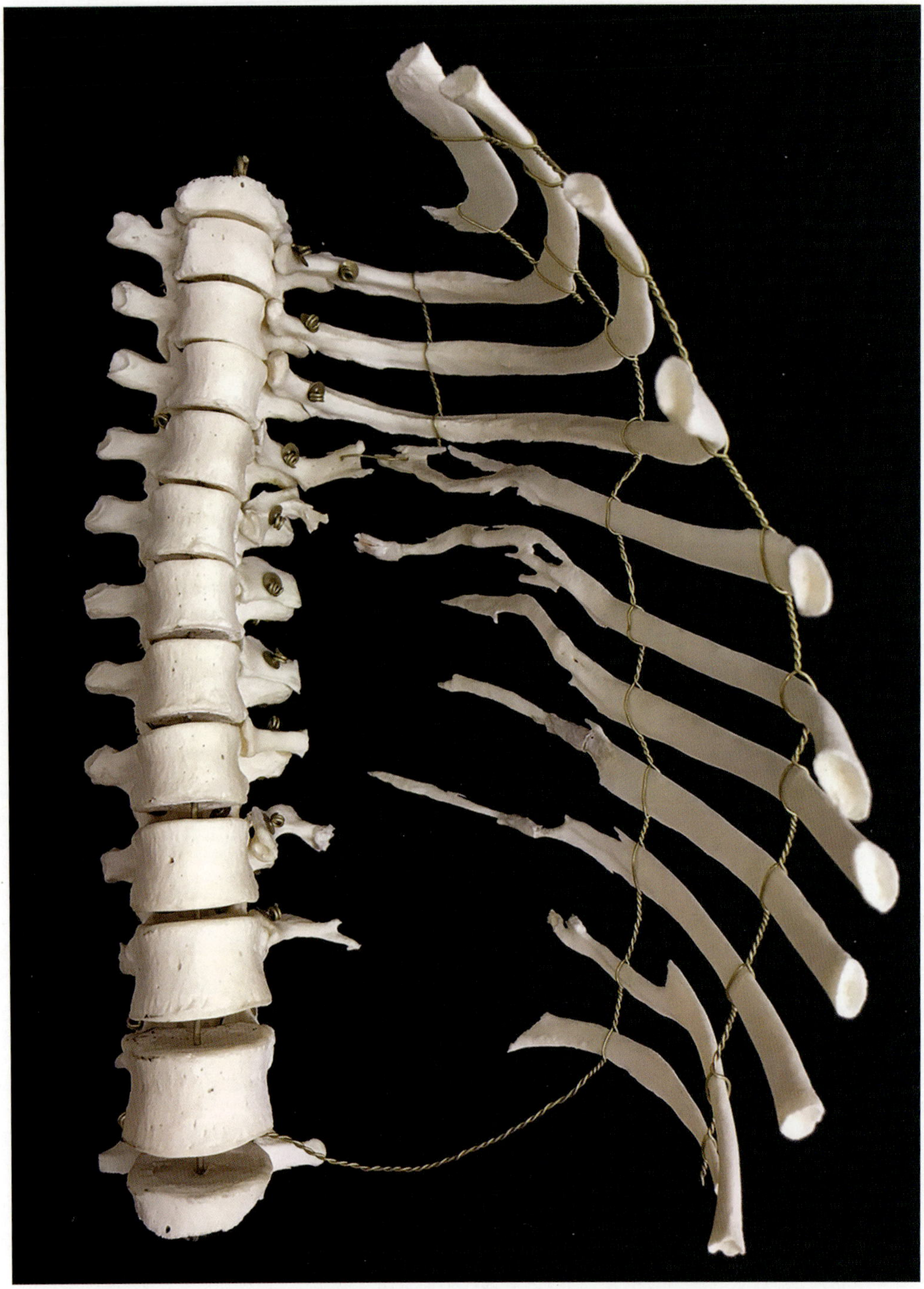

**Figure 229.** Anterior view of thorax following thoracoplasty (surgical removal) of unknown etiology in an adult female who died in 1934 (MN18.660/1309). The bacillus (infection) affected the vertebral ends of the ribs and appears to have spared the vertebrae. Thoracoplasty was often done with the intent of collapsing the lung (compression therapy) "...which could be achieved by creating an artificial pneumothorax or by thoracoplasty to repair the diseased lung. It was postulated that lung collapse would limit the spread of the disease by inducing fibrosis and encapsulating the infected area" (Giller et al. 2020) and Palmar and Berdine (2019; page 71). 2020). See Palfi et al. (2015) for more on tuberculosis in the ribs.

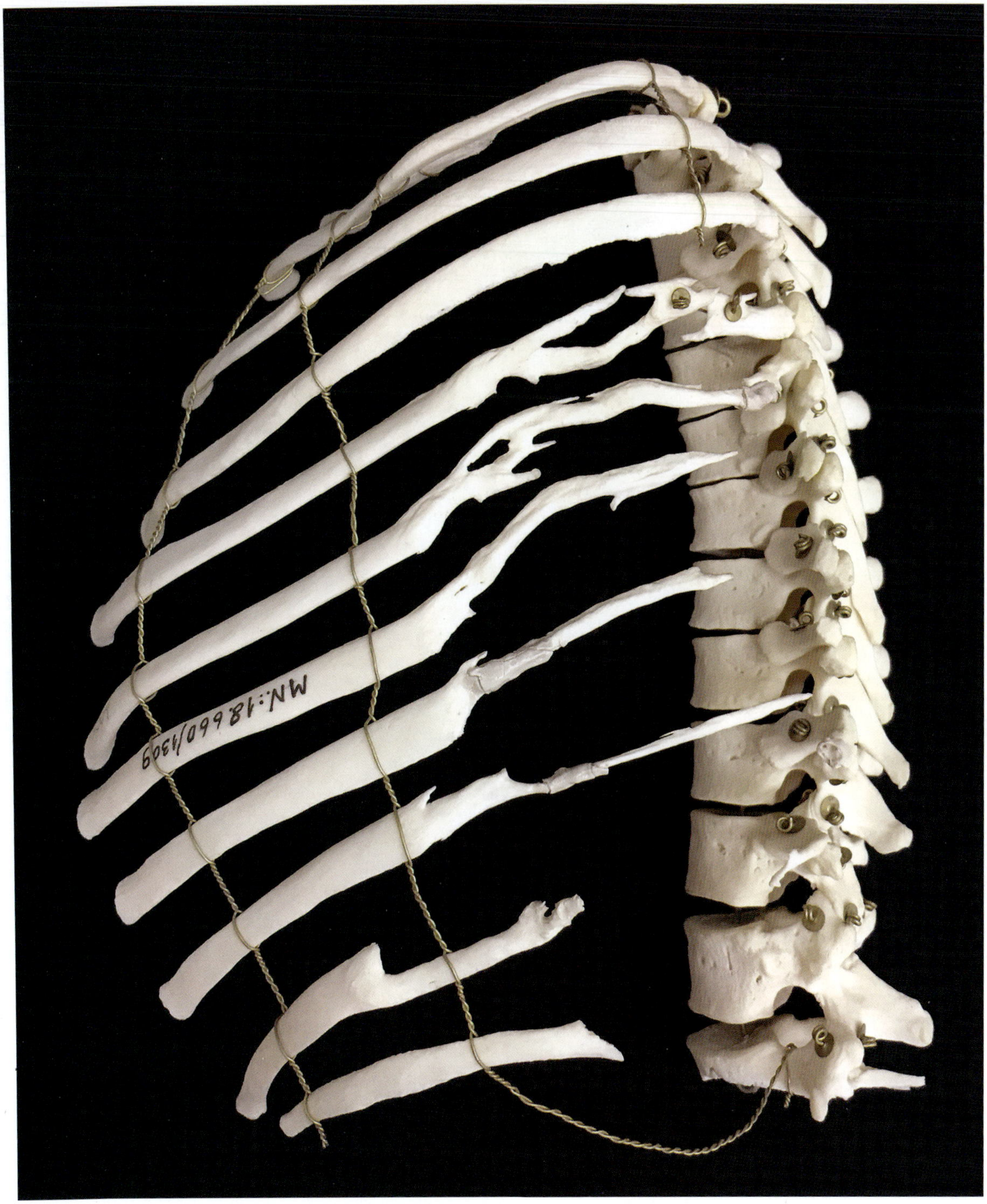


**Figure 230.** Left lateral view of thorax following thoracoplasty of unknown etiology in an adult female who died in 1934 (MN18.660/1309). The bacillus (infection), possibly tuberculosis, affected the vertebral ends of the ribs and appears to have spared the vertebrae. Thoracoplasty is often done with the intent of collapsing the lung (compression therapy) "...which could be achieved by creating an artificial pneumothorax or by thoracoplasty to repair the diseased lung. It was postulated that lung collapse would limit the spread of the disease by inducing fibrosis and encapsulating the infected area" (Giller et al. 2020; Palmar and Berdine 2019; page 71).

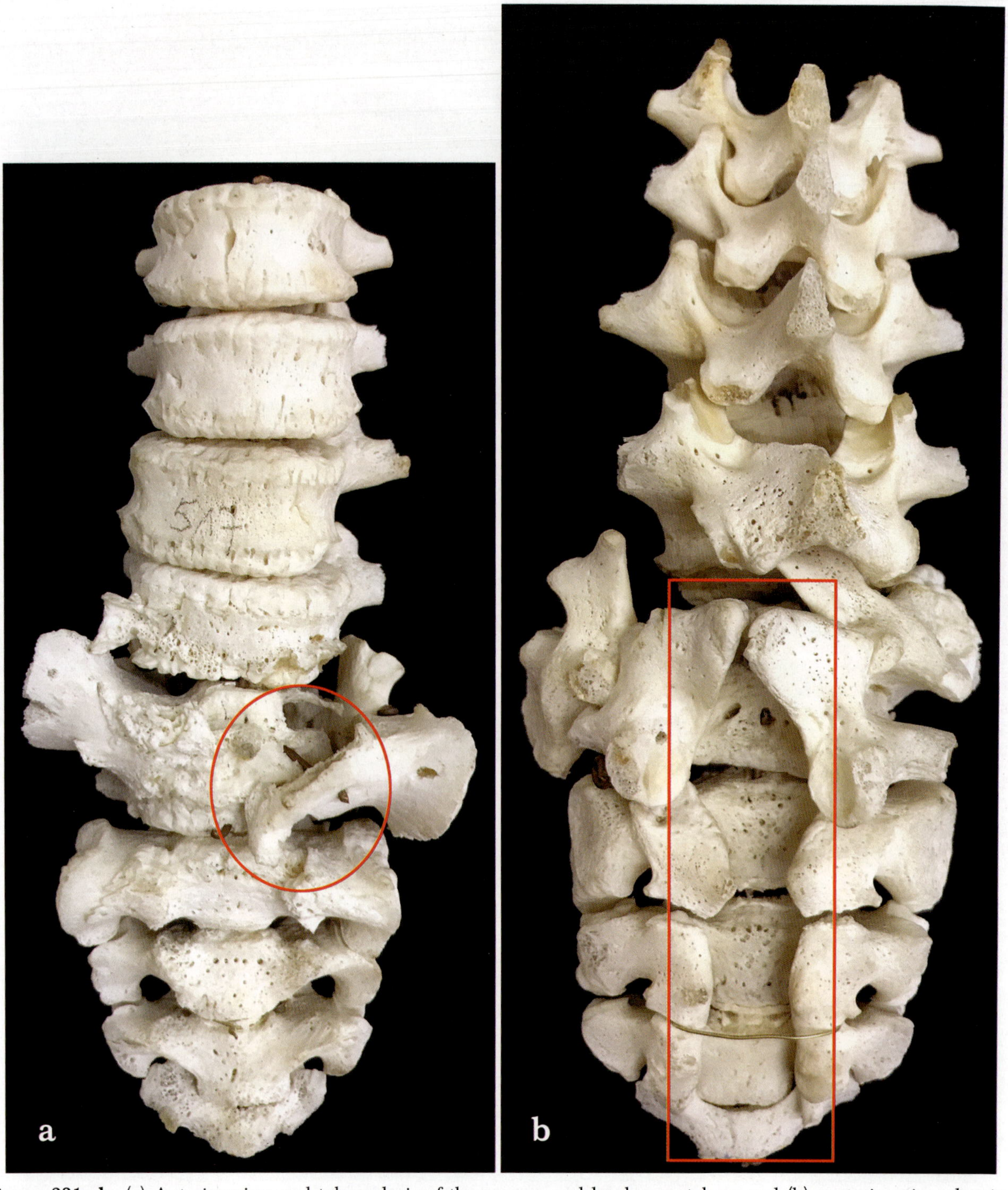


**Figure 231a-b.** (a) Anterior view and tuberculosis of the sacrum and lumbar vertebrae and (b) posterior view showing spina bifida (rectangle) resulting in a wide opening in the sacrum in a 6-year-old male (MN18.660/517). (a) Note destruction of the anterior body of L5 and the sacral segments by tuberculosis that is especially prominent in S1 (oval). See Buikstra (1981); Garg and Ortner (2003); Ortner (2003); Ortner and Putschar (1981); Palfi et al. (2015); Somvanshi (2011); Steinbock (1976) for more on spinal tuberculosis.

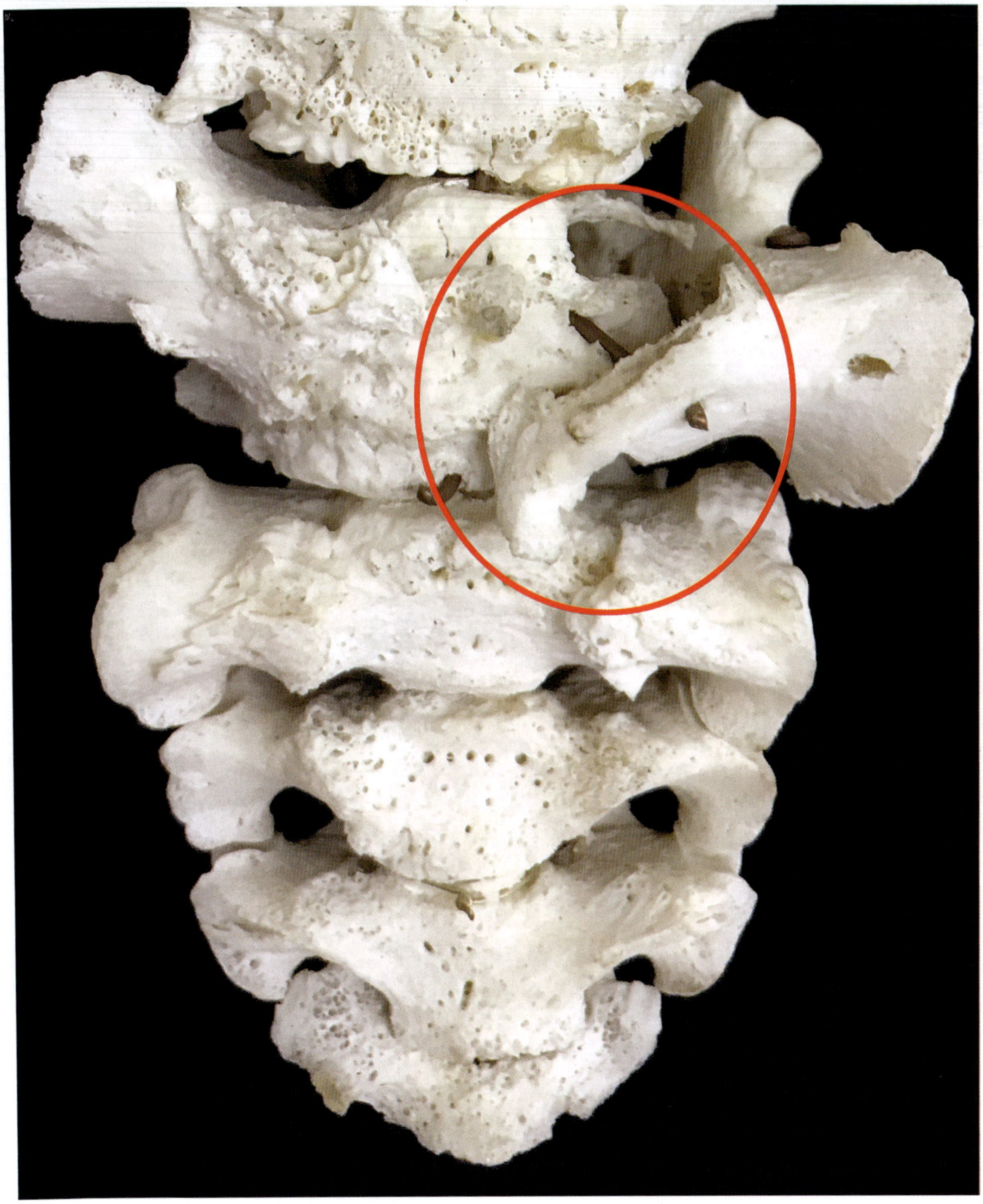

**Figure 232.** Magnified anterior view and tuberculosis of the sacrum and lumbar vertebrae resulting in a wide opening in the posterior sacrum in a 6-year-old male (MN18.660/517). Note destruction of the anterior body of L5 and the sacral segments by tuberculosis that is especially prominent in S1 (oval).

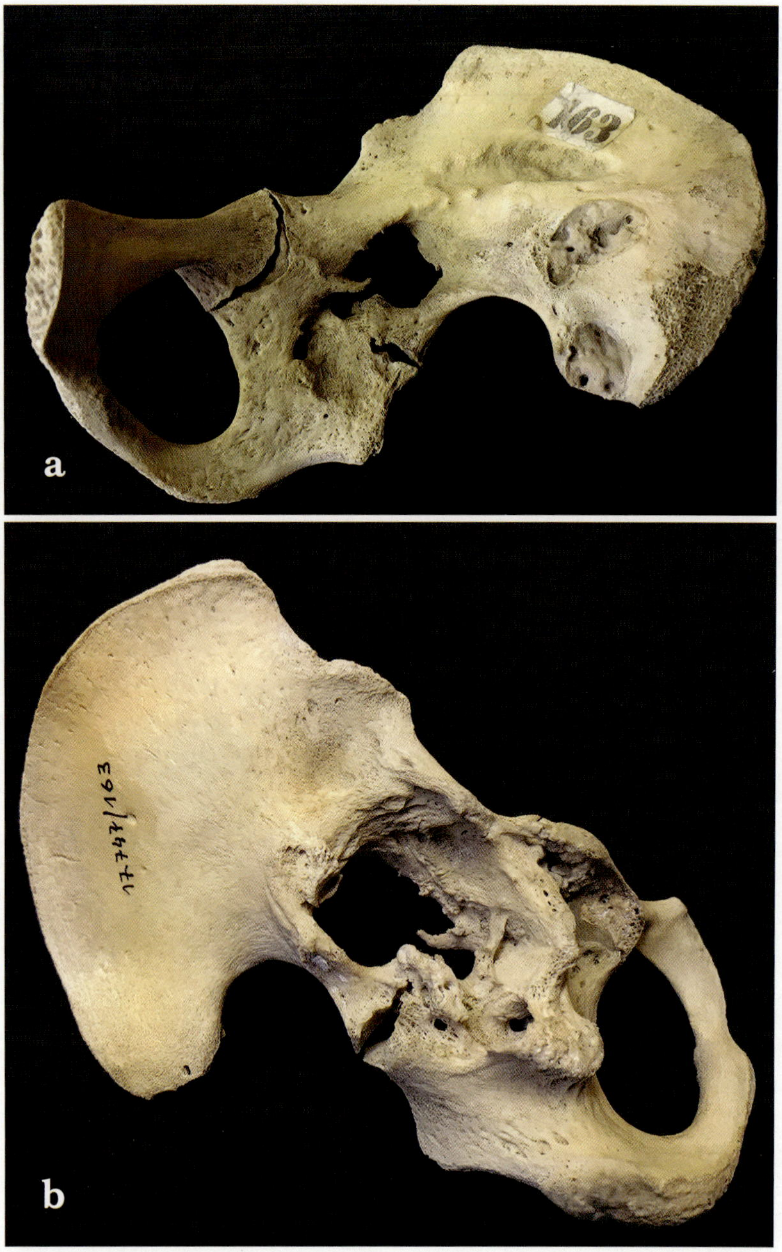


**Figure 233a-b.** (a) Internal and (b) external views of the right innominate with destruction and remodeling of the acetabulum in a juvenile who died of tuberculosis in 1921 (MN17.747/163). See Buikstra (1981); Christopher (1936); Garg and Somvanshi (2011); Saraf and Tuli (2015); Steinbock (1976) for more on tuberculosis of the bones and hip.

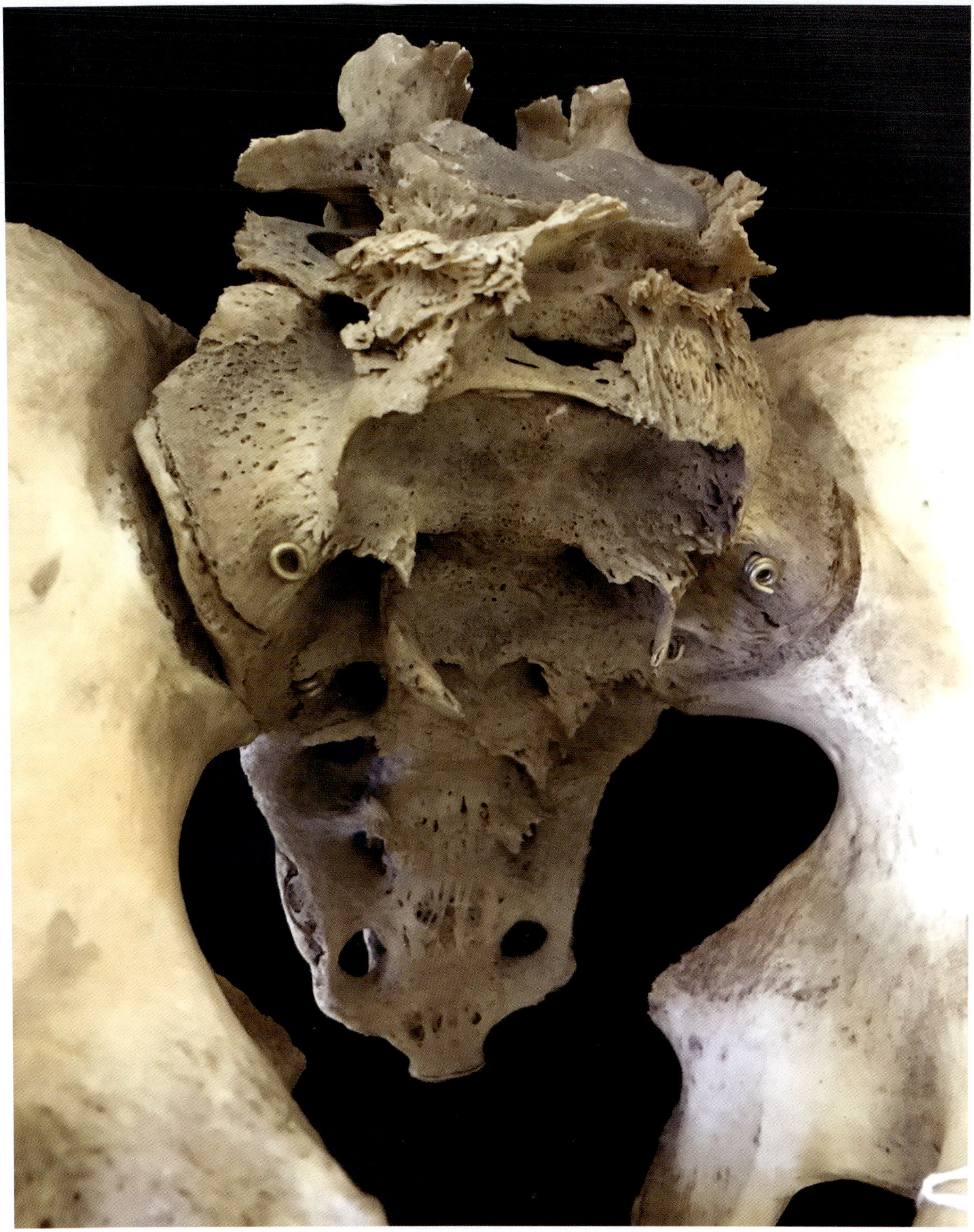

**Figure 234.** Magnified view of tuberculosis of the sacrum and lower lumbar vertebrae resulting in abscesses and reparative bone along the ventral surface of the lower spine and sacrum in an adult female who died in 1928 (MN5655 corrected number). Note the tuberculous infection did not affect the innominate bones. See Buikstra (1981); Christopher (1936); Ortner (2003); Saraf and Tuli (2015); Steinbock (1976) for more on tuberculosis in the joints, including the hip.

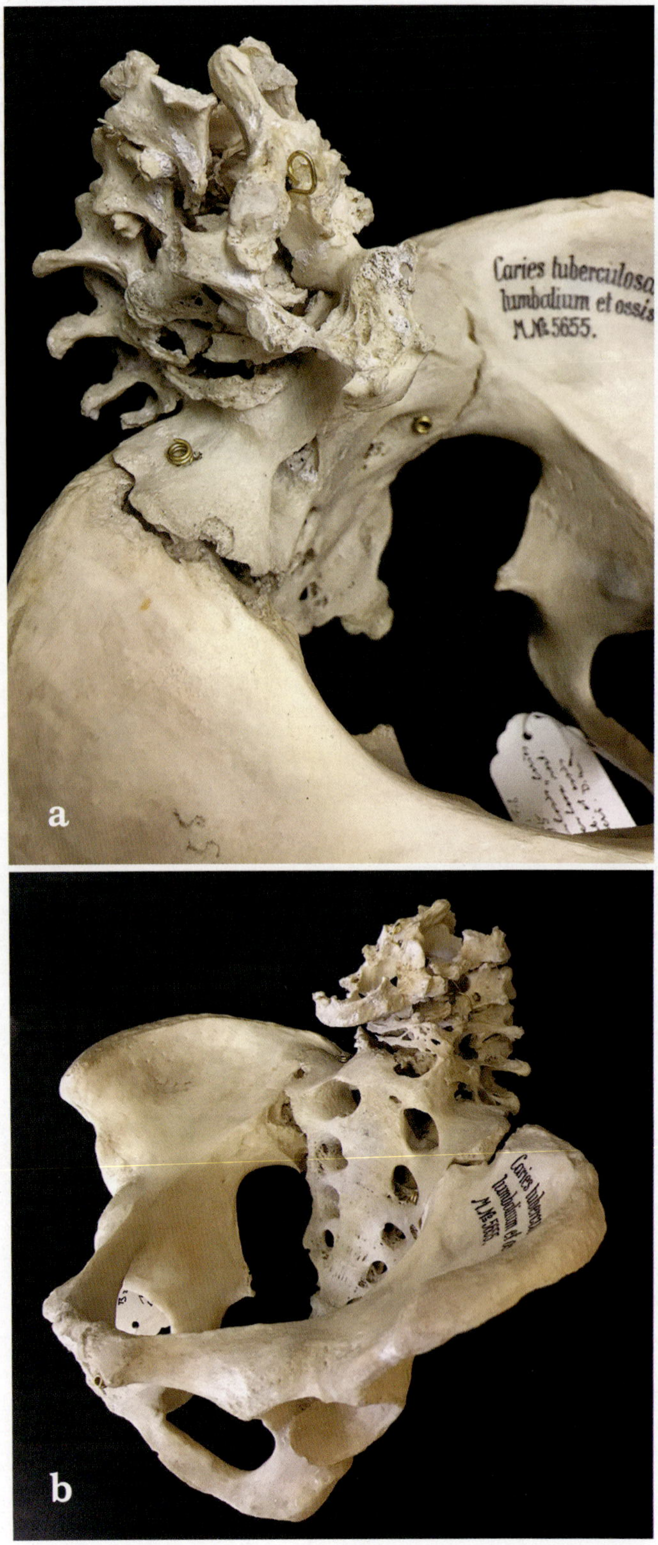


**Figure 235a-b.** (a) Right and (b) left oblique views of lumbar spine and sacrum in an adult female who died of tuberculosis in 1928 (MN5655 corrected number). See Buikstra (1981); Garg and Somvanshi (2011); Saraf and Tuli (2015) for more on tuberculosis in bones.

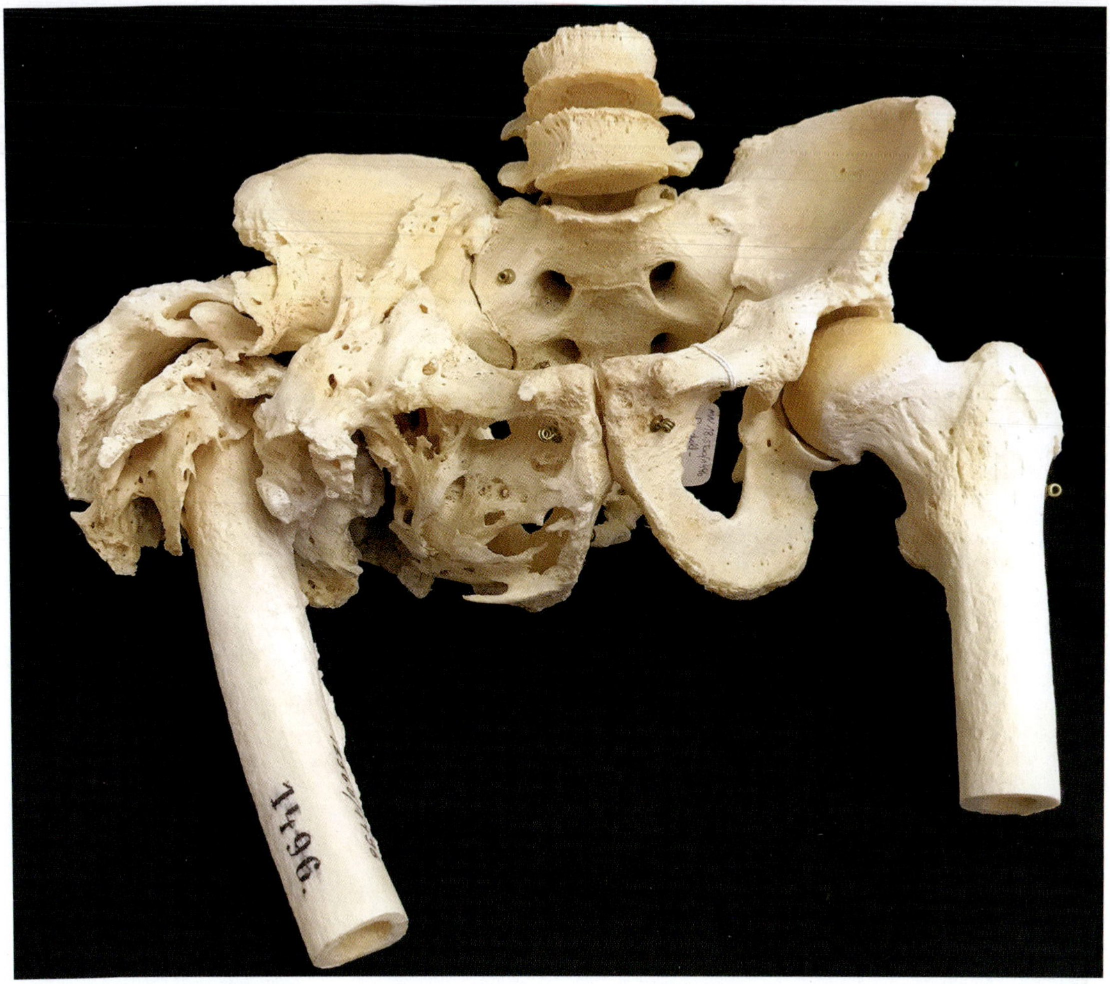


**Figure 236.** Long-standing untreated syphilitic neuropathic arthropathy in the hip (Charcot joint; see Steinbock 1976), a very rare condition of unknown etiology and associated with tabes dorsalis, a manifestation of neurosyphilis (Viens et al. 2010). Spirochetal infection leading to a loss of sensation can result in profound joint destruction many years after initial infection. The femoral head and neck were completely destroyed in this 64-year-old male who died in 1935 (MN18.560/1496). Tertiary or late syphilis can occur anytime two years after the initial infection (Drago et al. 2011; Liang et al. 2018; WHO 2016). This is infection versus a tumor, very unlikely trauma and very unlikely to be some form of dysplasia.

"This is a good case for VINDICATE, a mnemonic used in medicine to establish a differential diagnosis through a process of elimination. In this hip, the fact that it managed to develop some form of extreme joint makes me think of infection. A tumor that is this aggressive would probably not have allowed for a pseudojoint to develop. The waves of layers of bone make me lean toward infection more following the scenario that the person got an infection, it healed or his body fought it for a while, then it got really bad again, then he fought it immunologically, then repeated the process. This would be in the pre-antibiotic era as well. I would go with infection over tumor, but actually, you cannot really tell unless you obtain a culture or biopsy" (P. Moroz, personal communication, November 17, 2022).

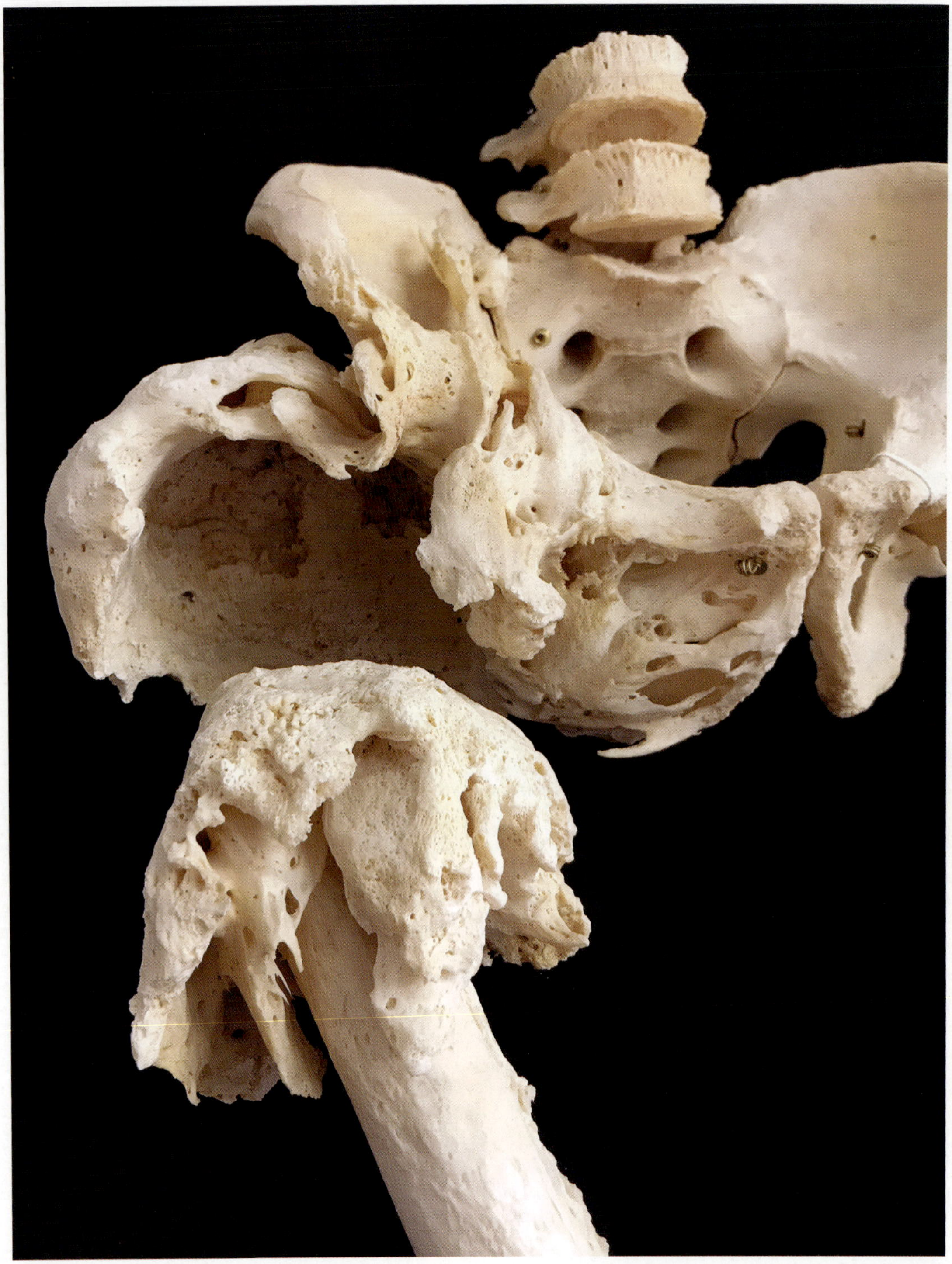

**Figure 237.** Syphilitic ("lues" on the museum card) neuropathic arthropathy (Charcot joint; Aufderheide and Rodrigues-Martin (1998); Steinbock 1976) of the hips with destruction of the right acetabulum, and femoral head and neck in a 64-year-old male who died in 1935 (MN18.560/1496). The femoral head has been removed from the socket in this image to show complete loss/resorption of the femoral head forming what resembles a "palm tree" (P. Moroz personal communication, November 17, 2022) and remodeling and expansion of the acetabulum.

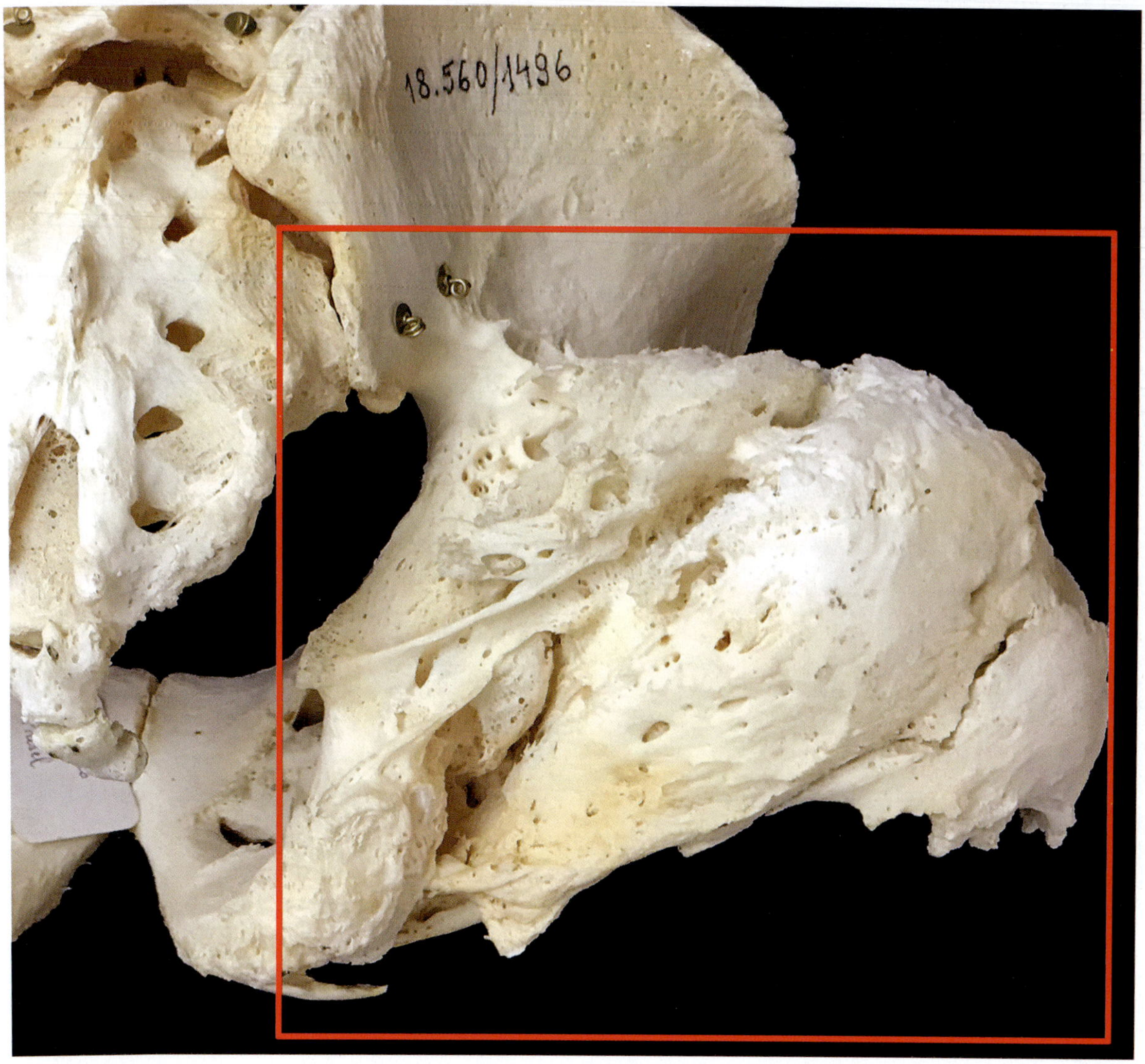


**Figure 238.** Syphilitic ("lues"), neuropathic arthropathy or Charcot joint of the right hip with severe alteration of the acetabulum (rectangle) in a 64-year-old male who died in 1935 (MN18.560/1496). See Aufderheide and Rodrigues-Martin (1998); Cavalli and Predario (2013); Chalmers et al. (2018); Drago et al. (2011); Steinbock (1976); Viens et al. (2010) for more on Charcot arthropathy.

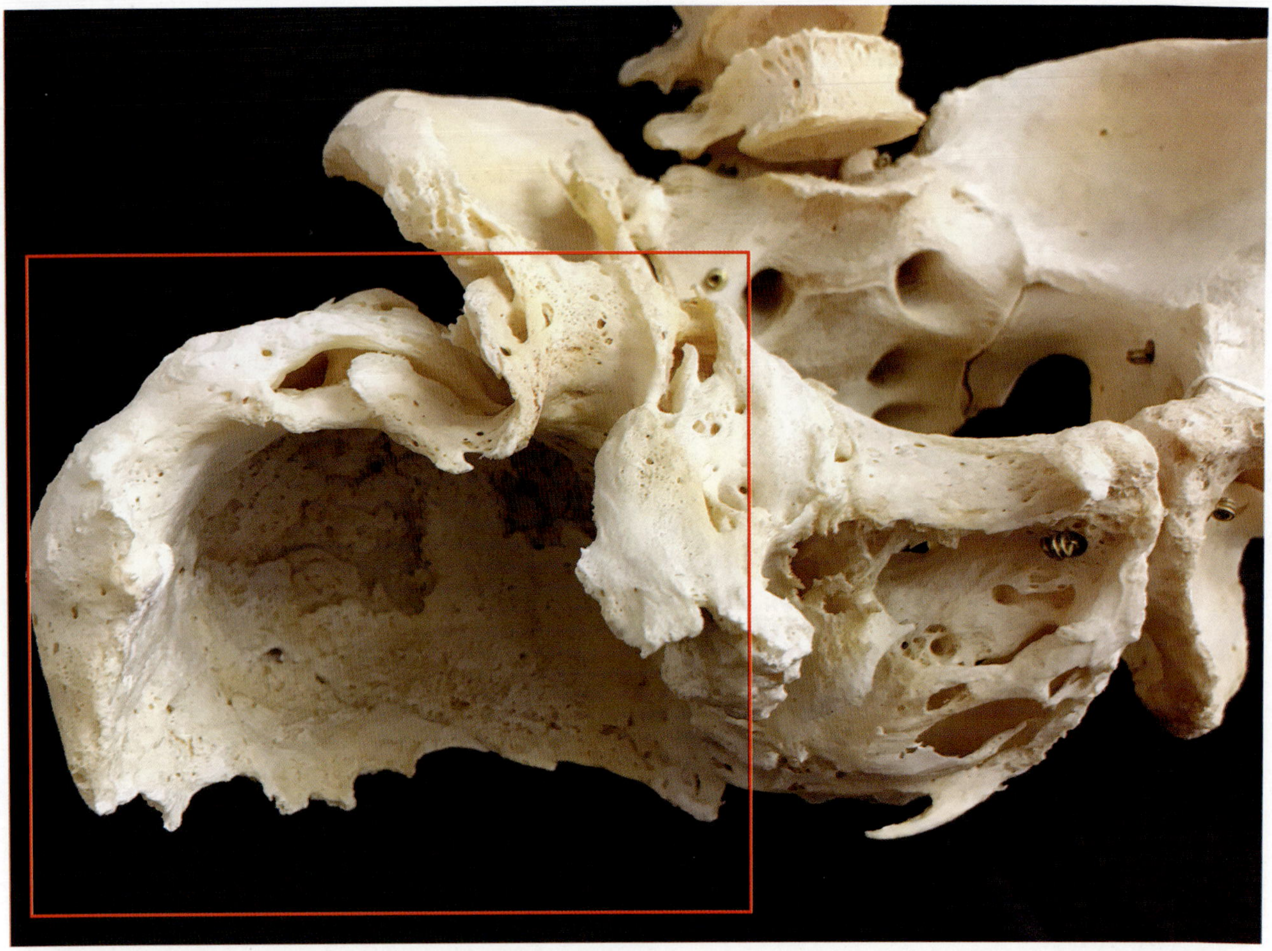

**Figure 239.** Syphilitic neuropathic arthropathy ("lues") of the right hip with destruction of the femoral head and neck and expansion of the acetabulum (rectangle) in a 64-year-old male who died in 1935 (MN18.560/1496). See Aufderheide and Rodrigues-Martin (1998); Cavalli and Predario (2013); Chalmers et al. (2018); Drago et al. (2011); Steinbock (1976); Viens et al. (2010) for more on Charcot arthropathy.

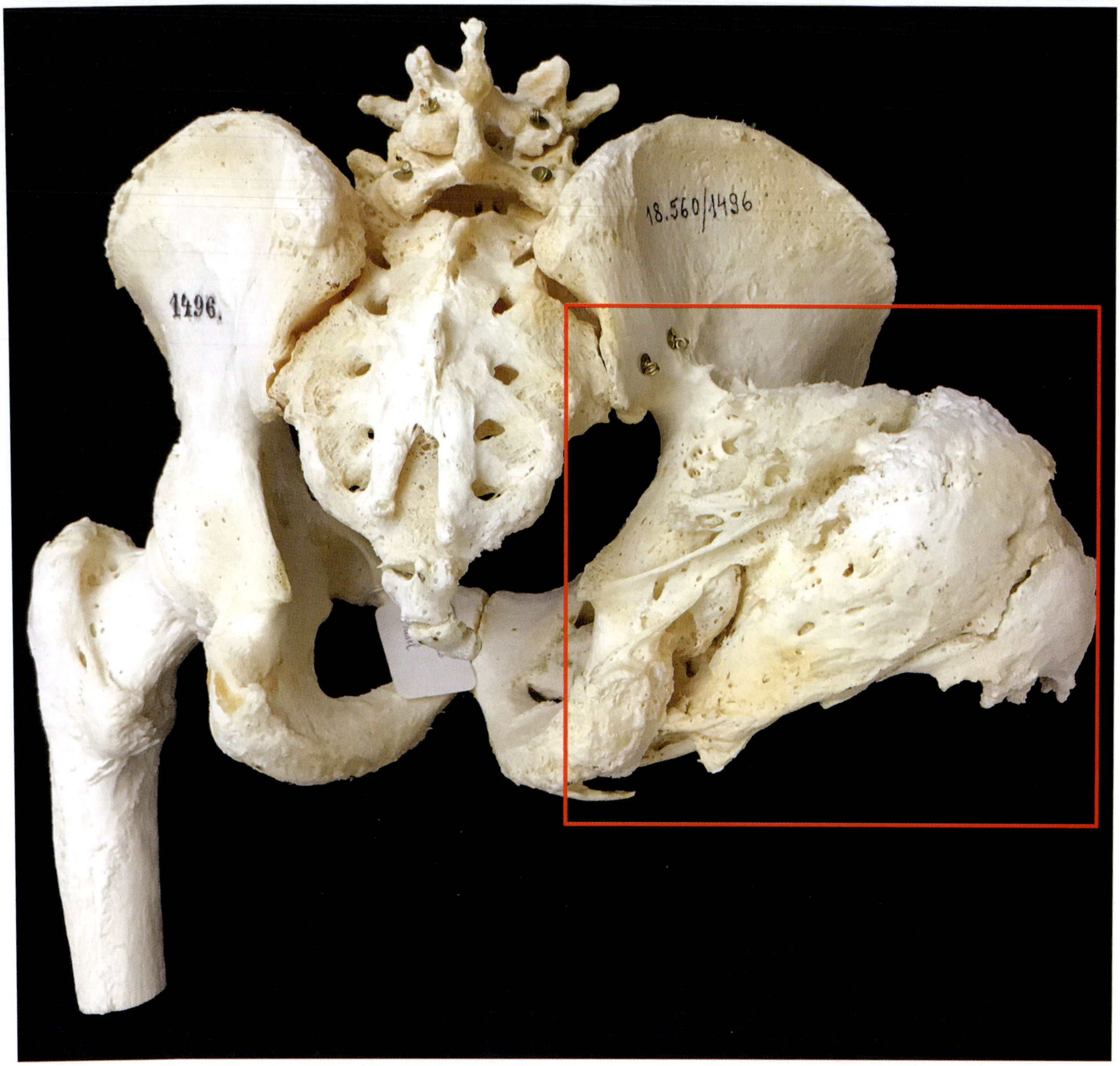


**Figure 240.** Syphilitic neuropathic arthropathy ("lues"), sometimes called Charcot joint of the right hip, with destruction and loss of the femoral head and neck and expansion of the acetabulum (rectangle) in a 64-year-old male who died in 1935 (MN18.560/1496). Neuropathic arthropathy can result in sensory loss of a joint with rapid and progressive mechanical destruction. See Aufderheide and Rodrigues-Martin (1998); Cavalli and Predario (2013); Chalmers et al. (2018); Drago et al. (2011); Steinbock (1976); Viens et al. (2010) for more on Charcot arthropathy.

**Figure 241.** Syphilitic neuropathic arthropathy ("lues") of the right hip resulting in destruction and loss of the femoral head and neck and massive secondary bone growth of the femoral head resembling a palm tree in a 64-year-old male who died in 1935 (MN18.560/1496).

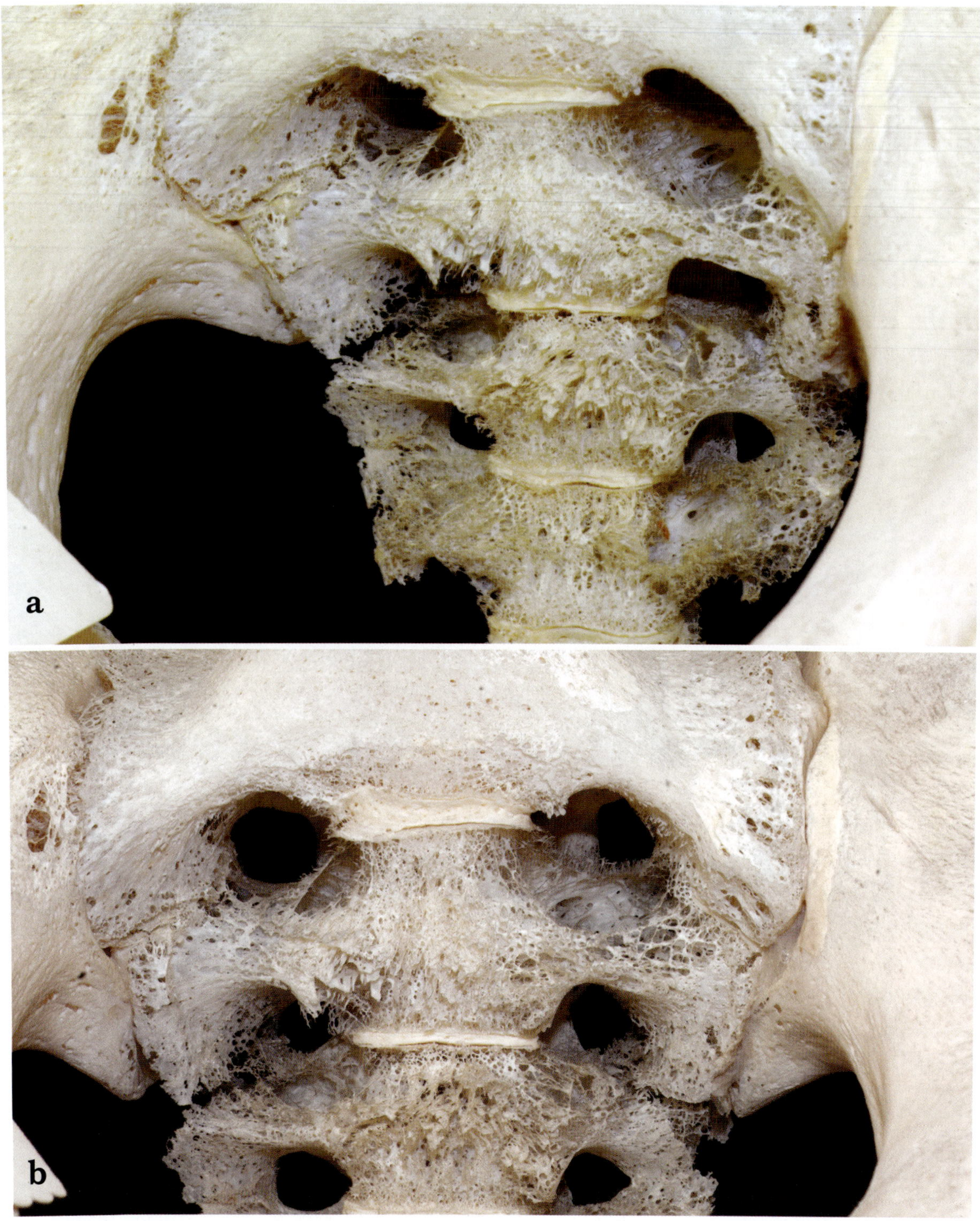


**Figure 242a-b.** Magnified views of "Tuberculosis spondylitis dorsolumbaris" (on the museum card) resulting in destruction of the sacrum and portion of right ilium in a 17-year-old male who died in 1938 (MN18.560/1709). See Aufderheide and Rodrigues-Martin (1998); Buikstra (1981); Christopher (1936); Garg and Somvanshi (2011); Ortner (2003); Ortner and Putschar (1981); Steinbock (1976) for more on tuberculosis in the human skeleton.

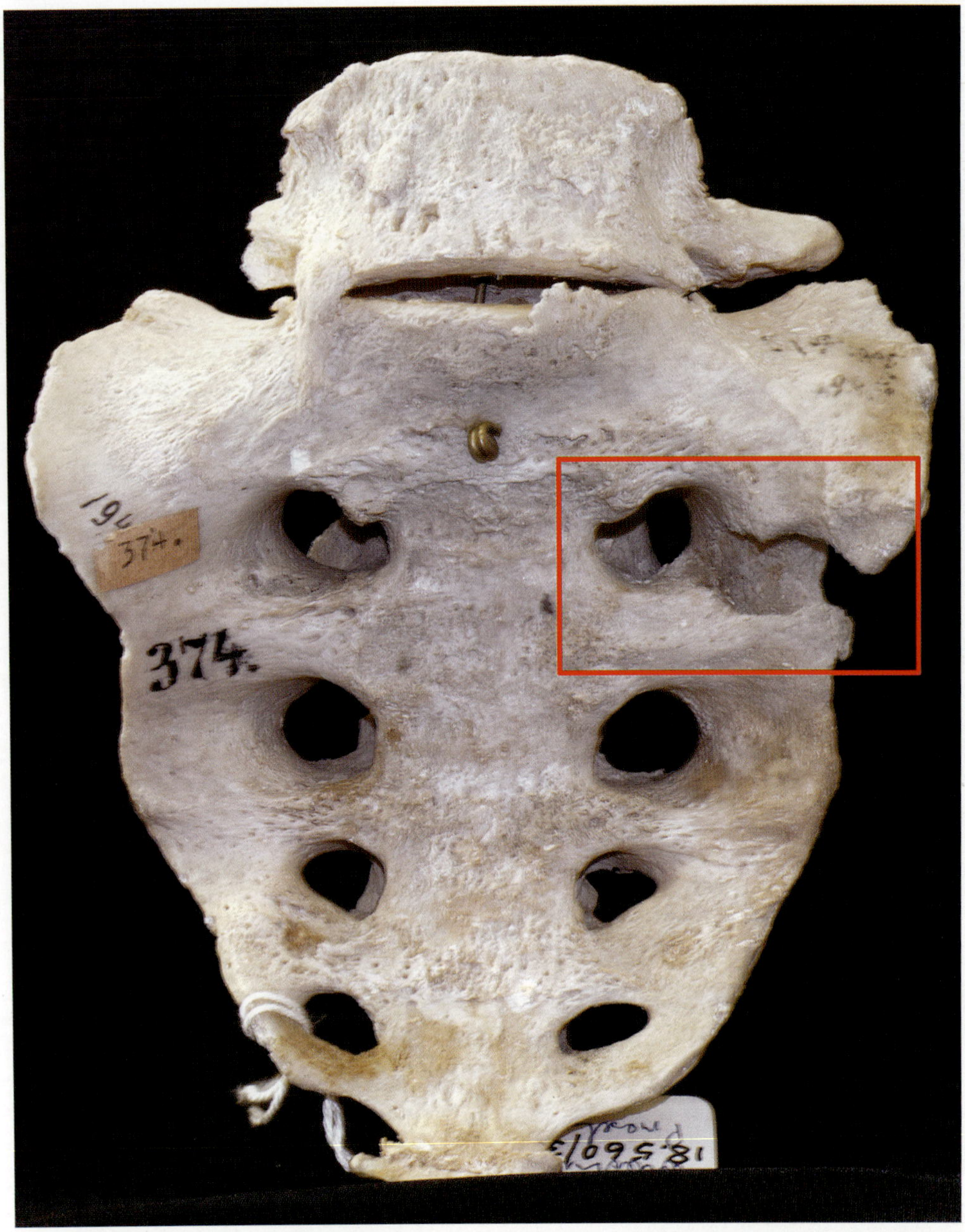


**Figure 243.** Tuberculosis resulted in partial destruction of the left first and second sacral segments (rectangle) in a 53-year-old female who died in 1905 (MN18.560/374). The smooth and rounded margins of the defect indicate that it occurred many years before death. See Buikstra (1981); Christopher (1936); Ortner (2003); Ortner and Putschar (1981); Steinbock (1976) for more on tuberculosis in the human skeleton.

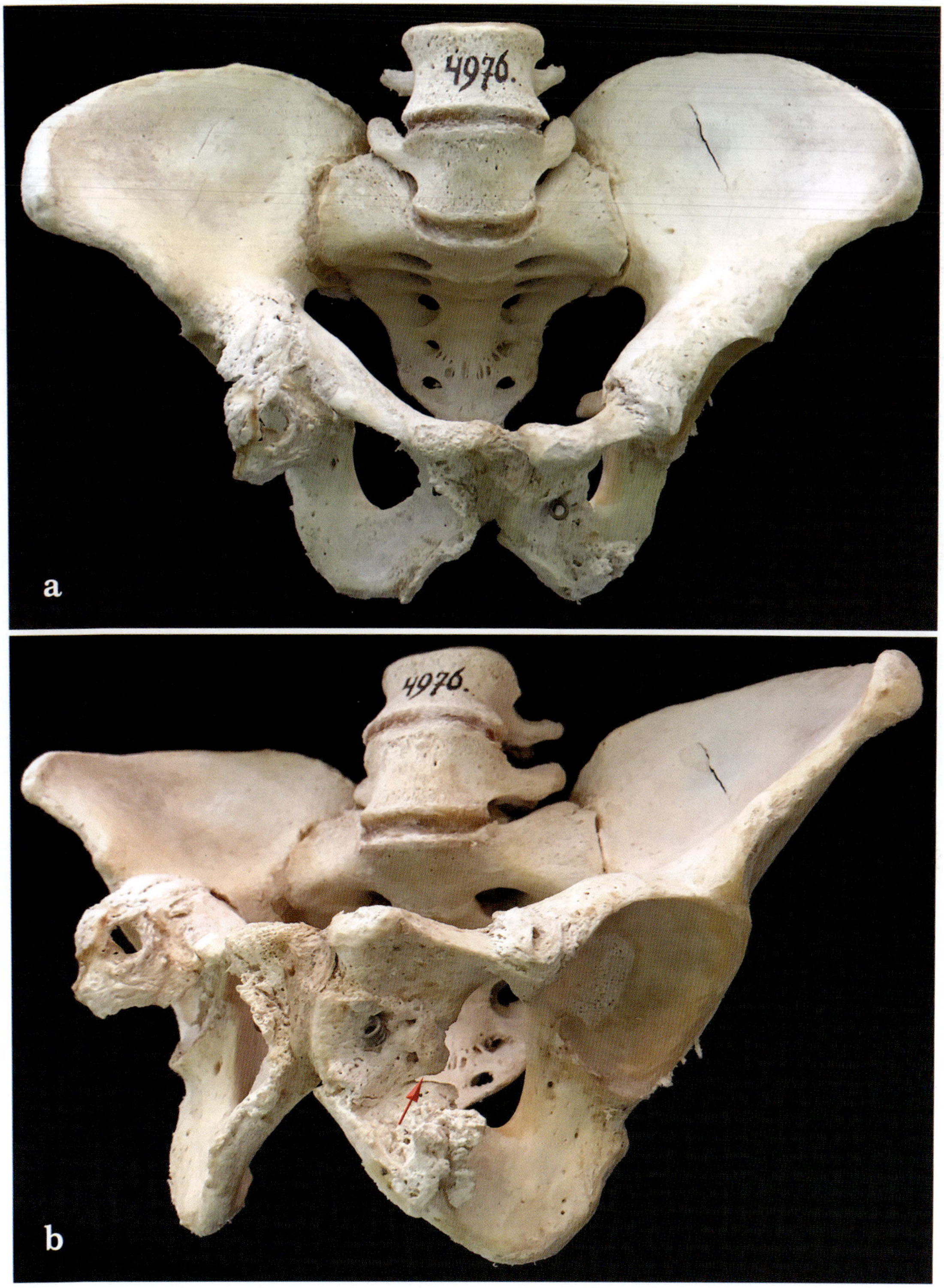


**Figure 244a-b.** (a) Anterior and (b) left anterior oblique views of tuberculosis of the pelvis with possible healed fracture (arrow) of the left ischiopubic ramus in a 42-year-old female who died in 1903 (MN21.055/4976). See Buikstra (1981); Christopher (1936); Garg and Somvanshi (2011); Saraf and Tuli (2015); Steinbock (1976) for more on tuberculosis of the pelvis.

## Arm and Hand

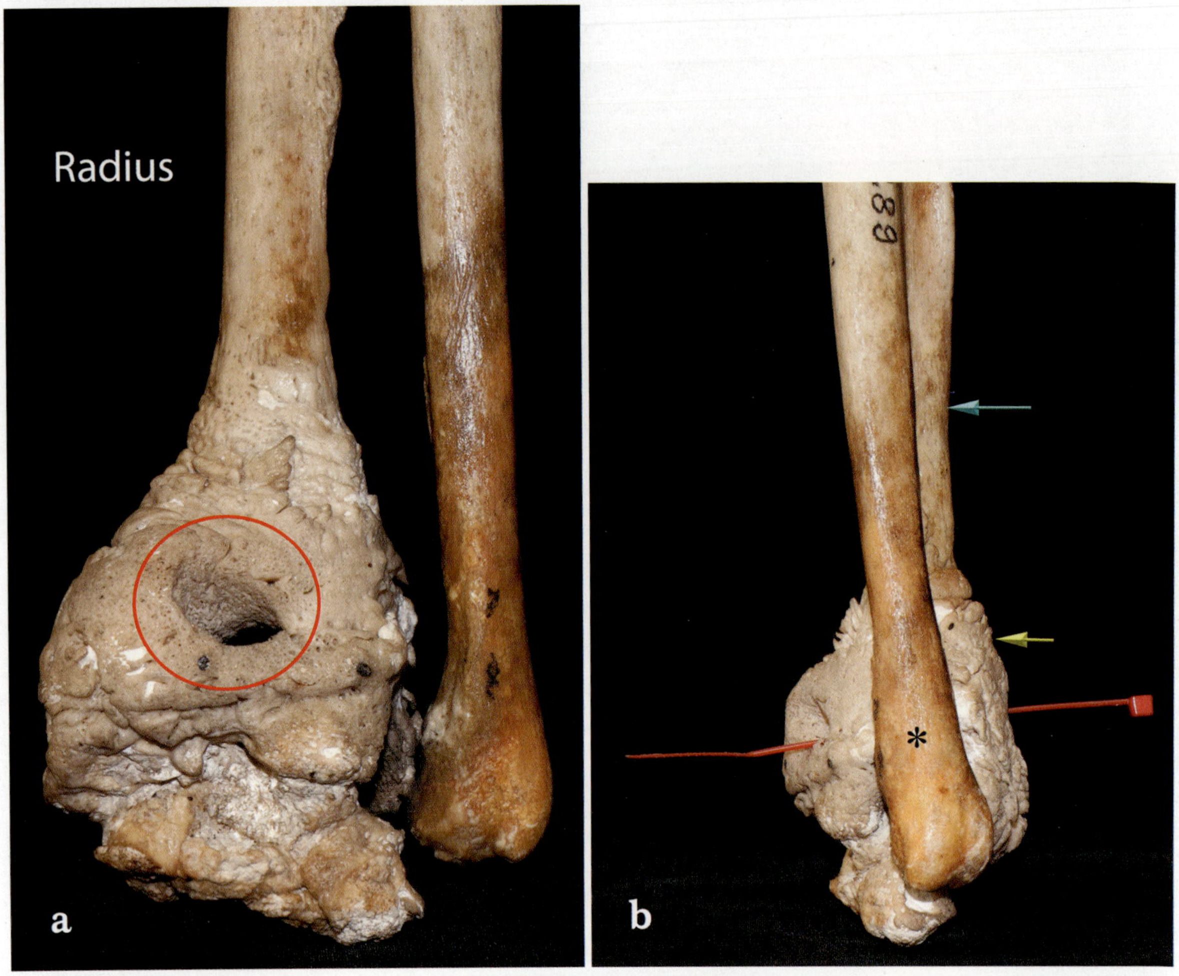


**Figure 245a-b.** (a) Supinated and (b) pronated views of osteomyelitis of unknown origin (possibly tuberculosis) in the right radius and ulna of a 23-year-old male who died in 1880 (MN18.560/289) (blue arrow=radius, yellow arrow=osteomyelitis and *=ulna). Note the cloaca (a:circle and b:red probe) or draining sinus for pus in the distal radius. See Aufderheide and Rodrigues-Martin (1998); Ortner (2003); Ortner and Putschar (1981) for information on infection in dry bone and Desimpel and Posadzy (2017) and Lee et al. (2016) for radiographic examination and features.

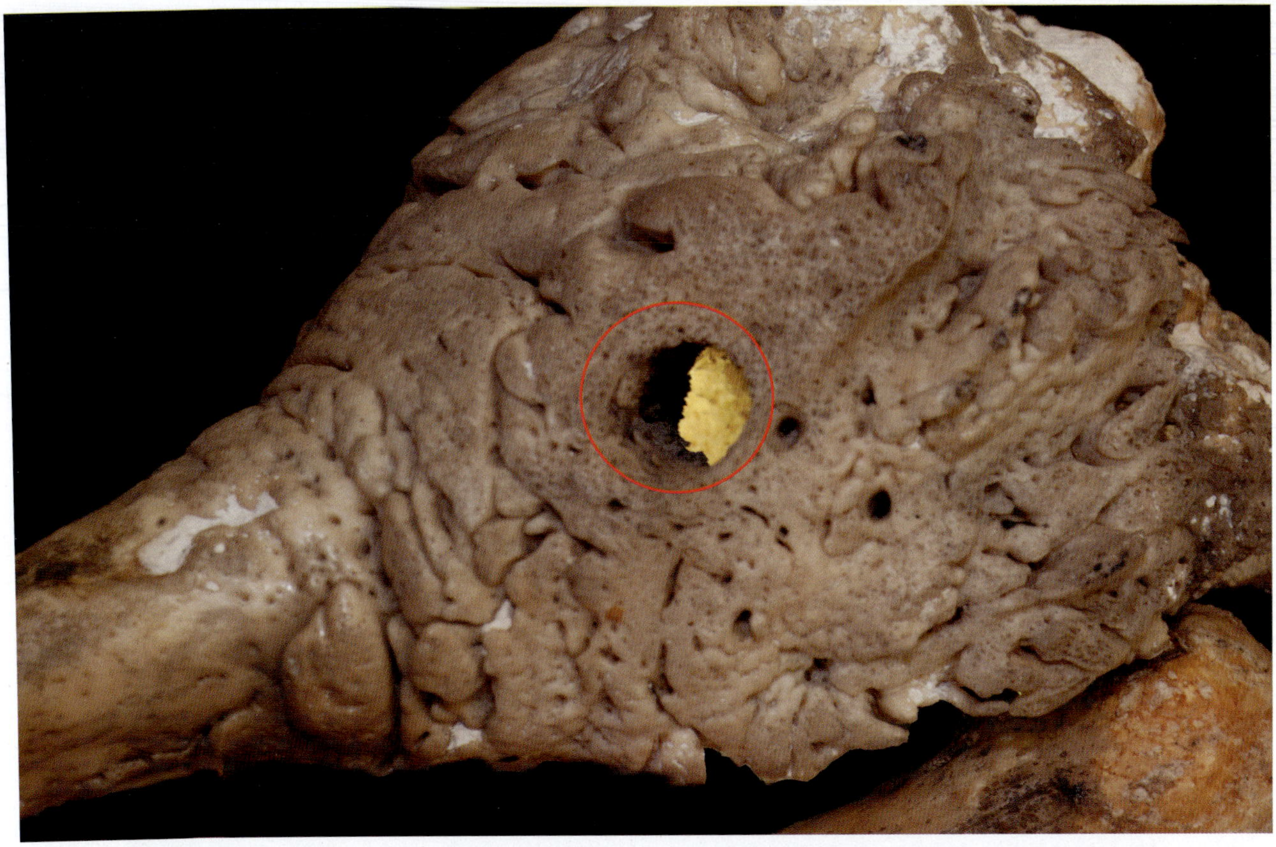

**Figure 246.** Transillumination of right distal radius showing a cloaca (draining sinus; red circle) indicative of chronic osteomyelitis (possibly tuberculosis) in a 23-year-old male who died in 1880 (MN18.560/289). See *Campbell's Operative Orthopaedics* (1956) and Christopher (1936) for more on tuberculosis of joints, including diagnosis and treatment. See Aufderheide and Rodrigues-Martin (1998); Ortner (2003); Ortner and Putschar (1981) for information on infection in dry bone and Desimpel and Posadzy (2017); Lee et al. (2016) for radiographic examination and features.

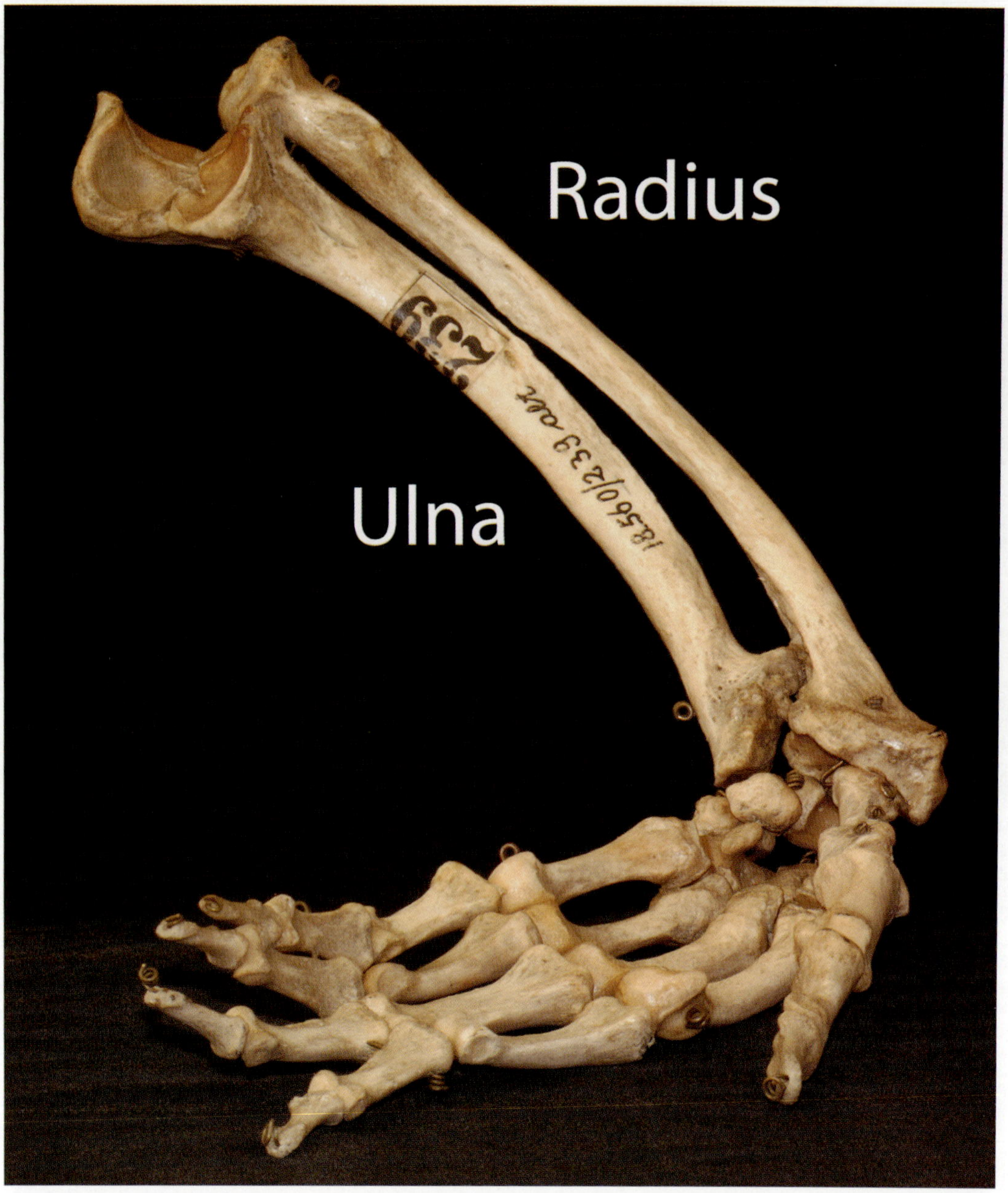


**Figure 247.** Deformation and pseudoarthrosis of the left radius and ulna, probably the result of an accident, in a 25-year-old male who died before 1880 (MN18.560/239a). This individual died of tuberculosis, a slow progressive infection that can affect one (monoarthritis) or several (polyarthritis) joints. See *Campbell's Operative Orthopaedics* (1956) for more on tuberculosis of joints including diagnosis and treatment. See Buikstra (1981); Christopher (1936); Garg and Somvanshi (2011); Geshchickter and Copeland (1936); Ortner (2003); Ortner and Putschar (1981); Steinbock (1976) for more on tuberculosis in bone.

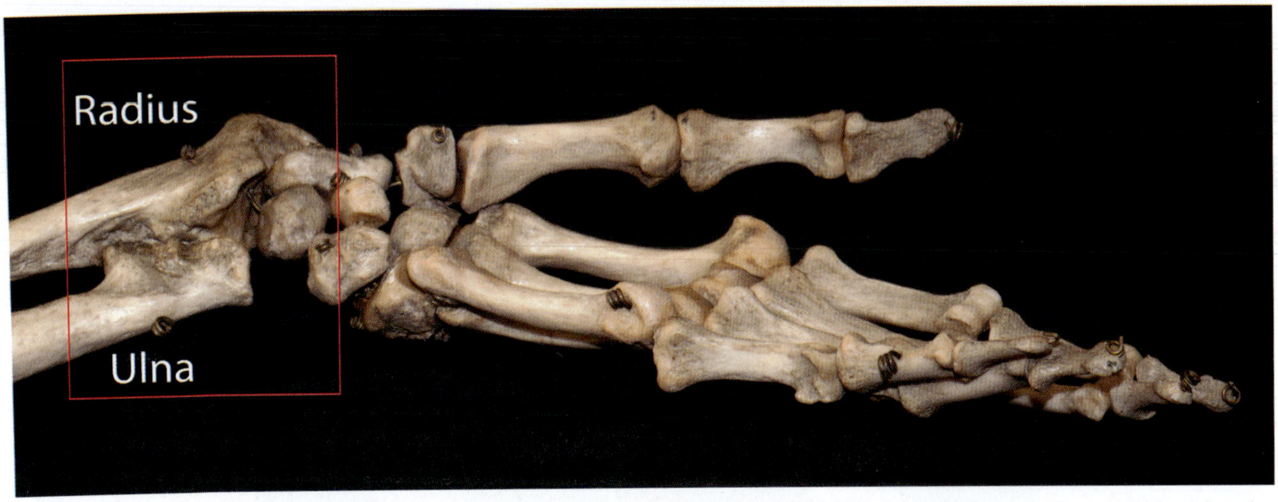


**Figure 248.** Magnified view of the left radius and ulna, probably the result of an accident, in a 25-year-old male who died before 1880 (MN18.560/239a). This individual died of tuberculosis, a slow progressive infection that can affect one (monoarthritis) or several (polyarthritis) joints. See *Campbell's Operative Orthopaedics* (1956) for more on tuberculosis of joints including diagnosis and treatment. See Aufderheide and Rodrigues-Martin (1998); Buikstra (1981); Christopher (1936); Garg and Somvanshi (2011); Geshchickter and Copeland (1936); Ortner (2003); Ortner and Putschar (1981).

## Leg and Foot

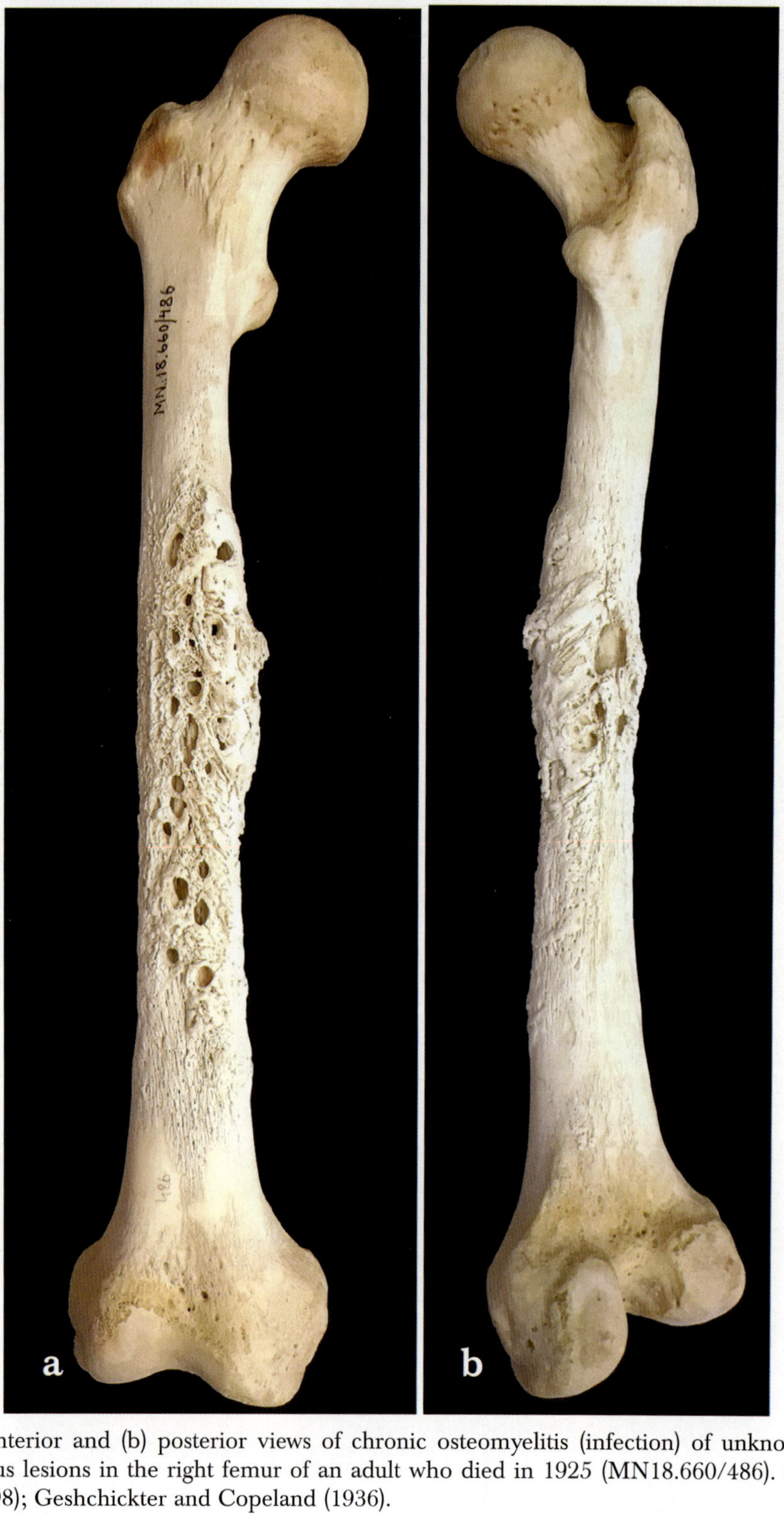


**Figure 249a-b.** (a) Anterior and (b) posterior views of chronic osteomyelitis (infection) of unknown etiology, possibly syphilis, with numerous lesions in the right femur of an adult who died in 1925 (MN18.660/486). See Aufderheide and Rodrigues-Martin (1998); Geshchickter and Copeland (1936).

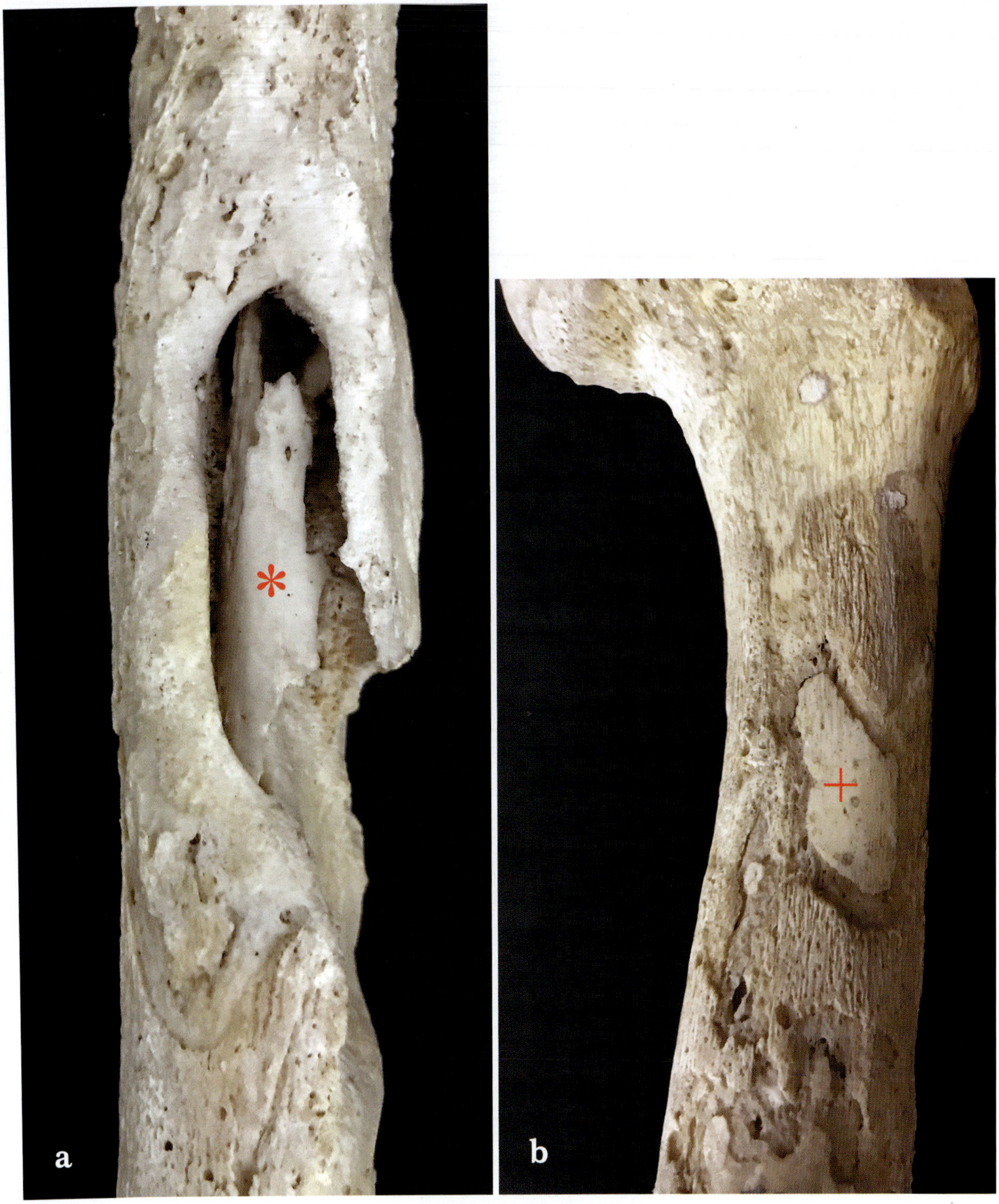

**Figure 250a-b.** (a-b) Magnified anterior views of nonspecific osteomyelitis and sequestrum (*), a segment of the original bone and (b) a small surface sequestrum (+) surrounded by active periostitis with a texture and color resembling tree bark in the left femur of a 49-year-old male who died in 1866 (MN21.055/1351). Chronic osteomyelitis can result in the formation of one or more "islands" of necrotic/dead bone that have been walled-off from the rest of the bone, as well as one or more draining sinuses (cloacae) for pus. See Geshchickter and Copeland (1936).

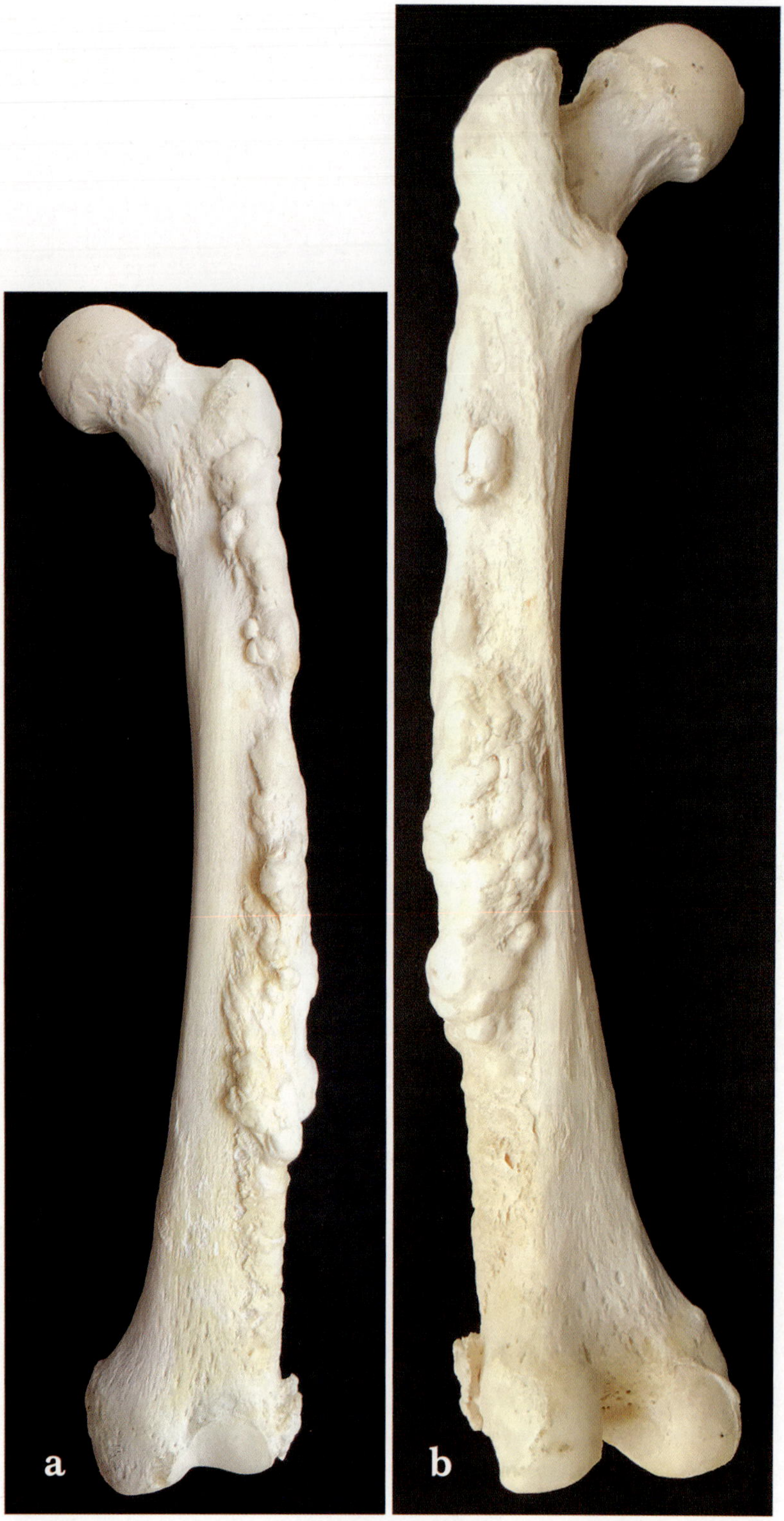


**Figure 251a-b.** (a) Anterior and (b) posterior views of hypertrophic periostitis, possibly hypertrophic osteoarthropathy from lung cancer or other causes, in the left femur of an adult male who died in 1915 (MN16.981/569). Hypertrophic periostitis results in the formation of periosteal new bone on the cortex without underlying lesions (Gaillard and Worsley 2008).

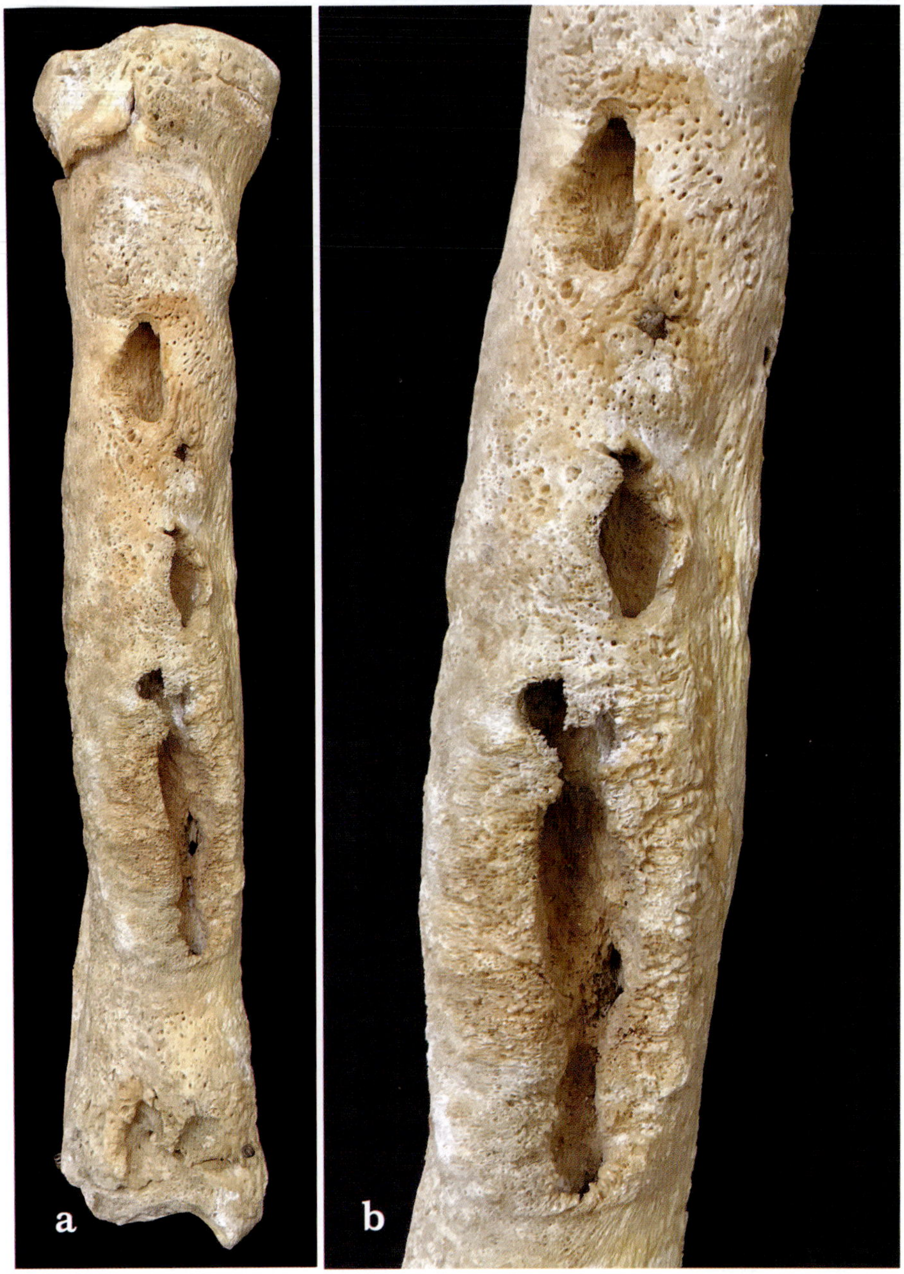


**Figure 252a-b.** (a) Anterior and (b) magnified views of chronic osteomyelitis, a long-term nonspecific bone infection of unknown etiology, with multiple lesions and draining sinuses (cloacae) in the right tibia of a 20-year-old male who died in 1880 (MN21.055/3451). See Aufderheide and Rodrigues-Martin (1998); Christopher (1936) and Geshchickter and Copeland (1936) for more on chronic osteomyelitis.

**Figure 253.** Healed "autoamputation" (stump) of the distal right tibia and fibula due to gangrene in an adult male that occurred in 1880 (MN21.055/3528). Autoamputation is the spontaneous separation of a body part (not self-amputation) due to a variety of diseases and conditions.

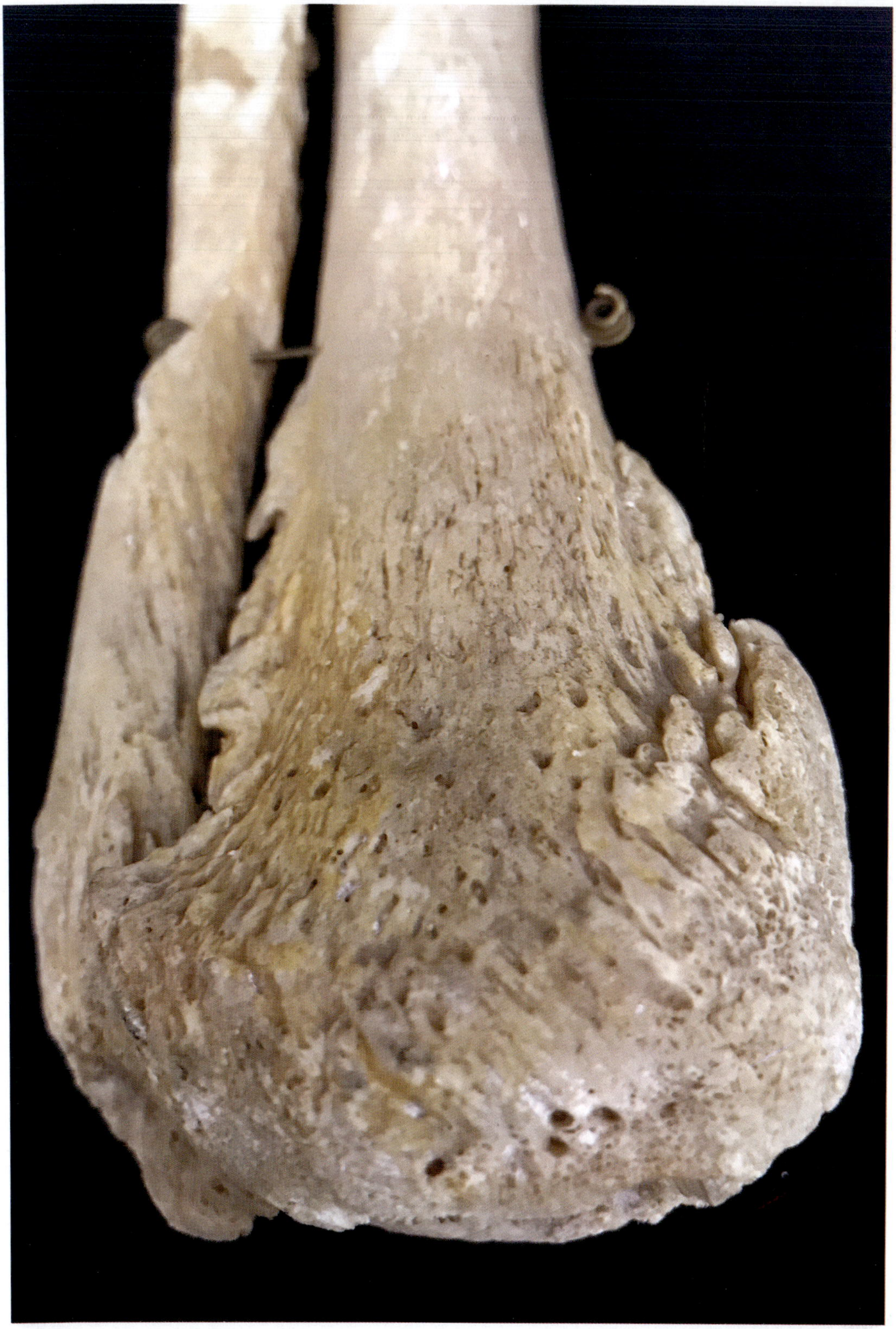

**Figure 254.** Magnified view of healed autoamputation (stump) of the right distal tibia and fibula, just above the ankle, due to gangrene with reported spontaneous separation of the foot from the body in 1880 (MN21.055/3528).

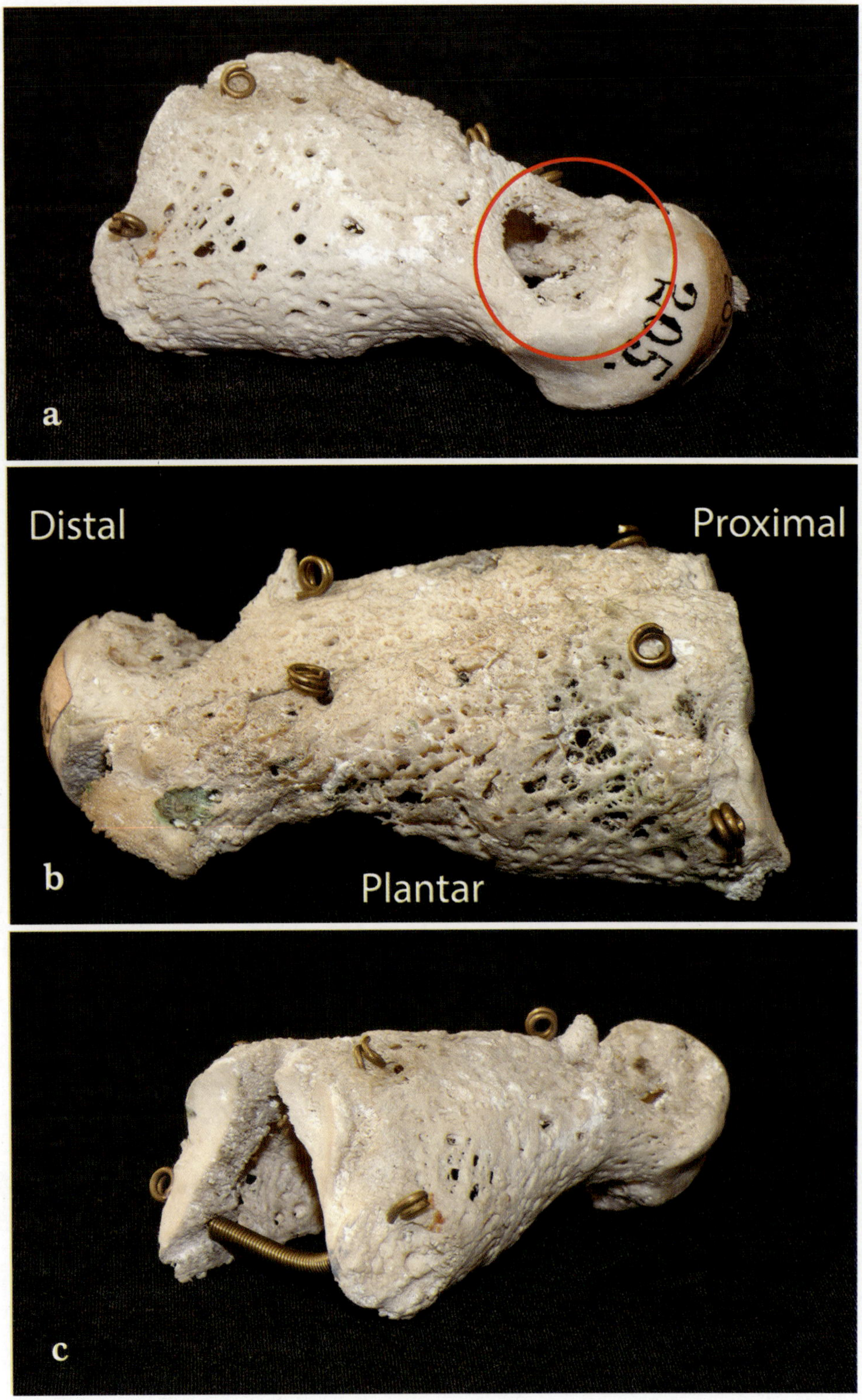


**Figure 255a-c.** (a) Lateral, (b) medial and (c) proximal views of tuberculosis in a right greater hallux (first toe) with an involucrum and (a) cloaca (circle) (first metatarsal) in an individual who died in 1930 (MN18.560/205). See Geshchickter and Copeland (1936).

**Figure 256.** Magnified sagittal view of tuberculosis with sequestrum, cavitation and cloaca (*) of a right greater hallux (first metatarsal) in an individual who died in 1930 (MN18.560/205). No additional information accompanies this individual. See Ortner and Putschar (1981) for more on tuberculosis of the hands and feet and Geshchickter and Copeland (1936) regarding tuberculosis in general. Dhillon and Nagi (2002) reported that involvement of one or more bones in the feet and ankle is an uncommon occurrence in skeletal tuberculosis.

# Chapter 3

# NEOPLASM AND TUMOR

## Skull

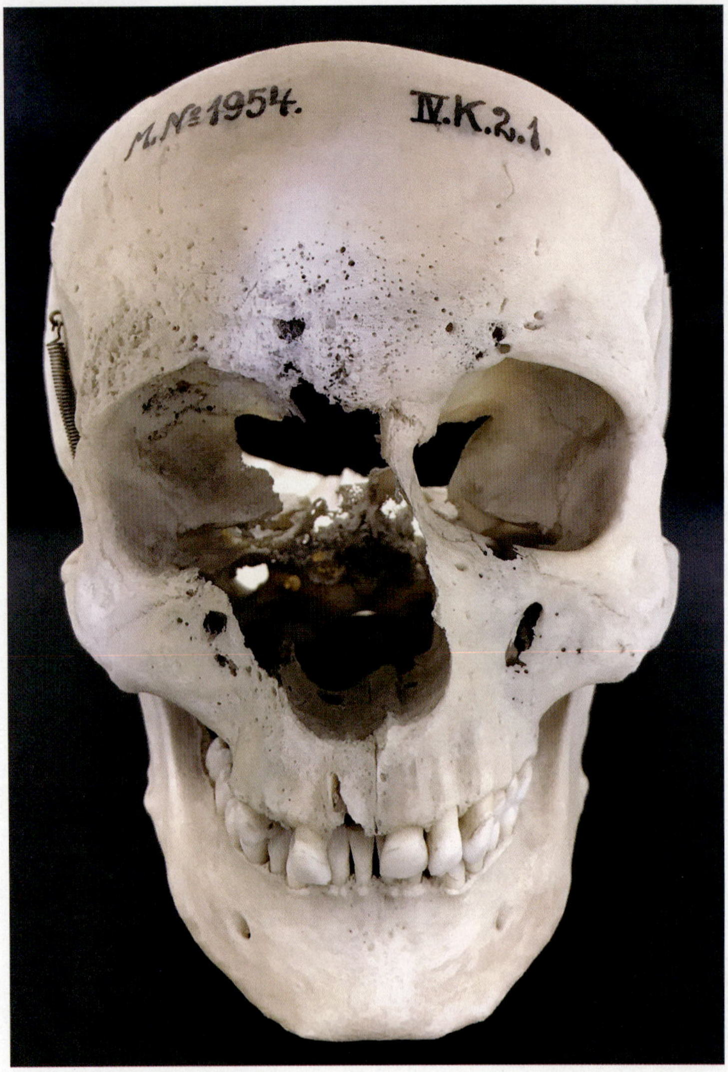


**Figure 257.** Facial carcinoma, "extensive tumor" on the museum card, most likely due to cancer resulting in loss of most of the right orbit, nasal bones, sinuses, infraorbital foramina, frontal bone and cranial base in a 28-year-old male who died in 1840 (MN1954). See Migliorini et al. (2020); Reeder (1993); Steinbock (1976) for more on carcinoma.

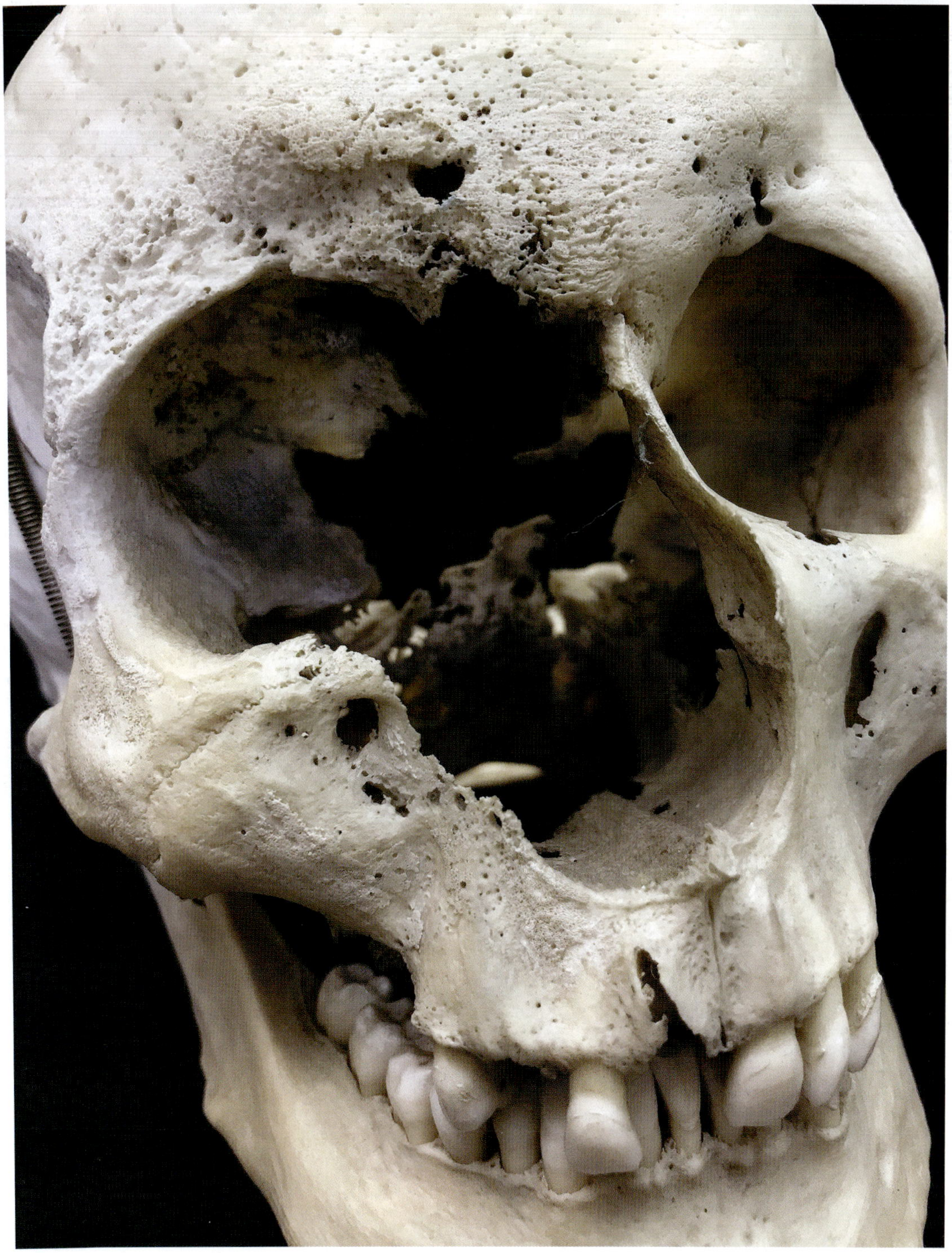

**Figure 258.** Right oblique view of carcinoma, most likely due to cancer, in a 28-year-old male who died in 1840 and that resulted in loss of most of the right orbit, nasal bones, sinuses, infraorbital foramina, and portions of the frontal bone (MN1954). We see the same pattern and degree of bone destruction in contemporary cases of carcinoma. See Migliorini et al. (2020) for metastasis; Steinbock (1976) for more on carcinoma; Reeder (1993) for differentiating carcinoma from other diseases.

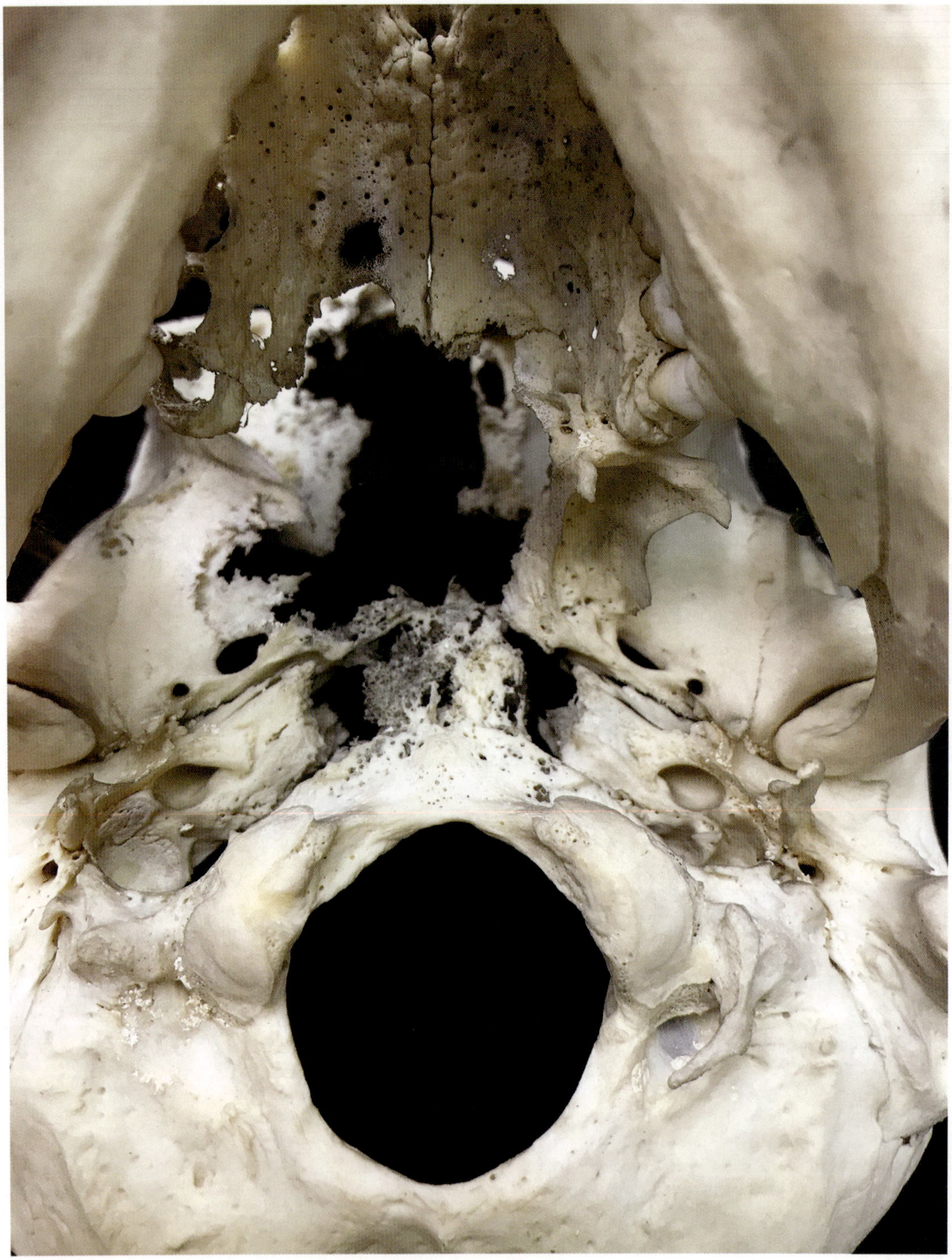

**Figure 259.** Inferior view of the skull of a 28-year-old male who died with facial carcinoma in 1840 (MN1954). Carcinoma resulted in loss of most of the right orbit, nasal bones, sinuses, infraorbital foramina, and portions of the frontal bone. Note the jagged and irregular bony destruction without evidence of remodeling/healing. See Migliorini et al. (2020); Steinbock (1976) for more on carcinoma.

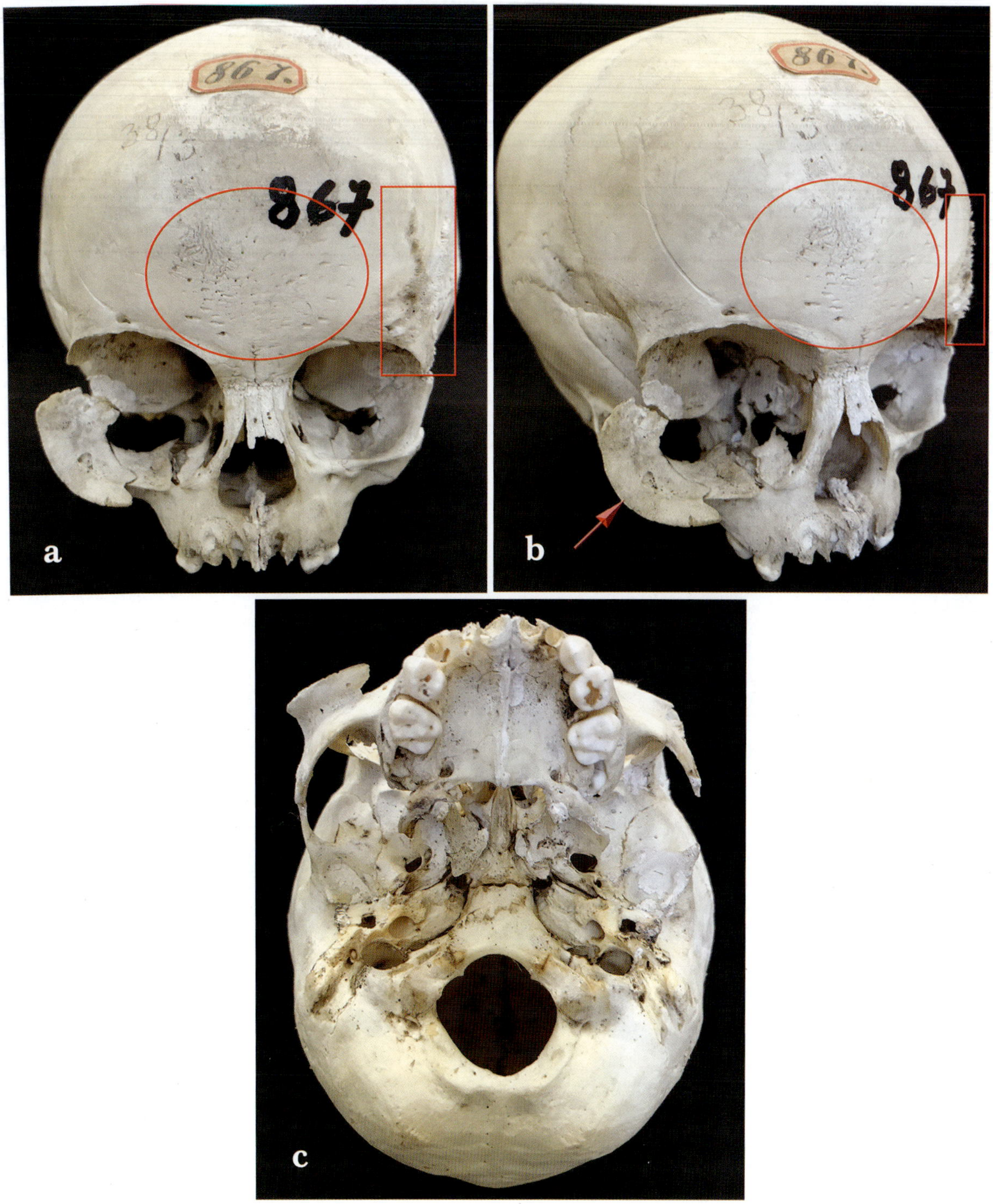


**Figure 260a-c.** (a) Anterior, (b) right oblique and (c) inferior views of a tumor, possibly metastasis secondary to melanoma, of the right orbit (arrow) with hypervascularation of the frontal (a-b; oval) and periosteal new bone (rectangle) on the left frontal bone in a 3-year-old female who died in 1865 (MN21.055/867). Solitary metastases due to malignant skin melanoma to bone or skeletal muscle have been reported to occur in 0.8% of patients and is known for local recurrence, as well as distant metastasis, through lymphatics and the blood stream (Gomez-Leon 2018). See Rose et al. (2017) for more on secondary melanoma within the orbit.

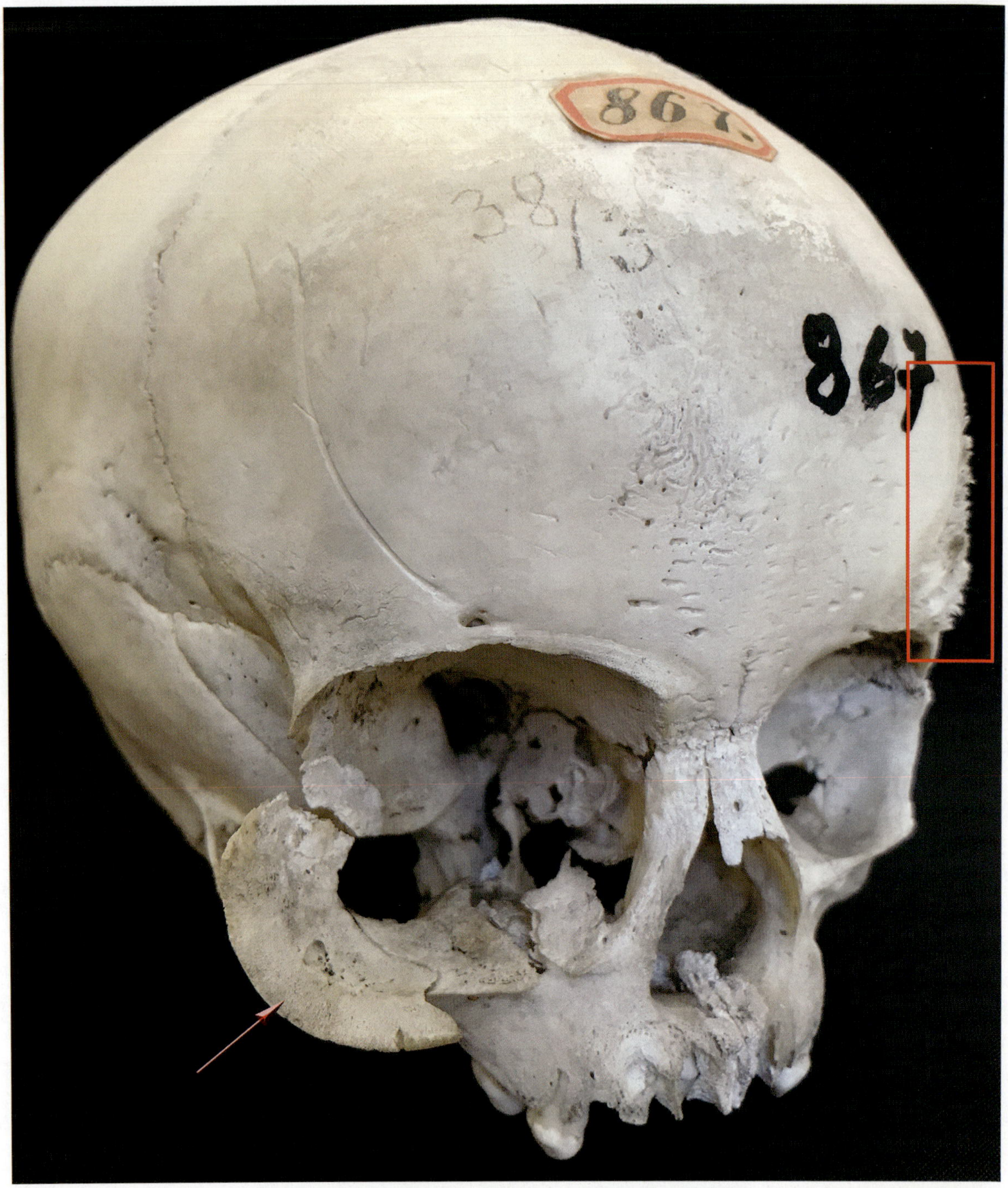


**Figure 261.** Magnified right oblique view of a tumor, possibly melanoma metastasis in a 3-year-old female who died in 1865, resulting in enlargement of the right orbit, especially affecting the orbital surface of the zygomatic (malar; arrow) bone (MN21.055/867). Note the area of hypervascularity in the central portion of the frontal bone and periosteal bone (rectangle). See Gomez-Leon (2018) for more on melanoma metastasis. See Migliorini et al. (2020) for bone metastases in general and Rose et al. (2017) for more on secondary melanoma within the orbit.

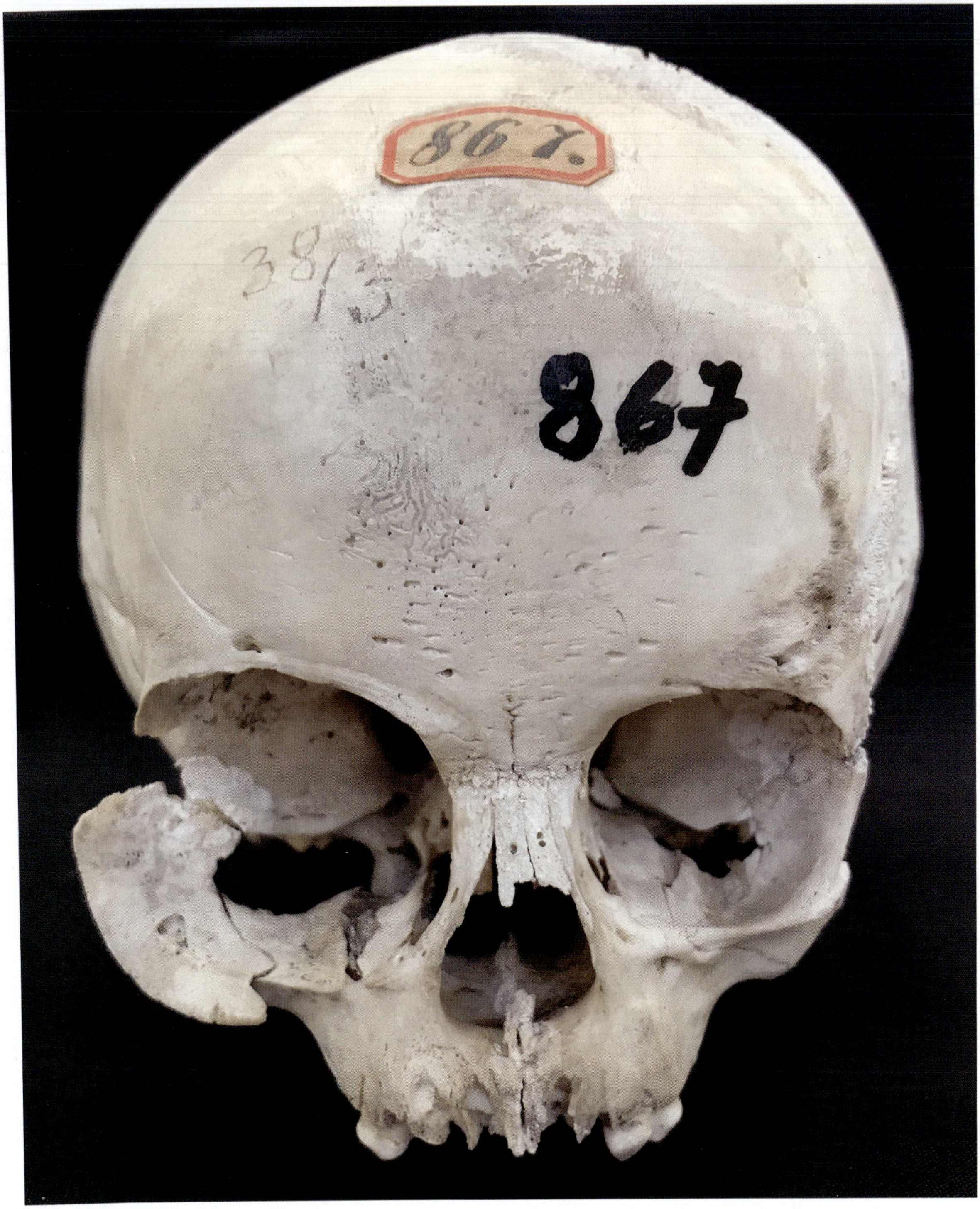


**Figure 262.** Magnified superior-inferior view of tumor, possibly melanoma metastasis, of the right orbit with hypervascularity near the middle of the frontal bone and periosteal new bone along the left frontal bone in a 3-year-old female who died in 1865 (21.055/867). See Rose et al. (2017) for more on secondary melanoma within the orbit.

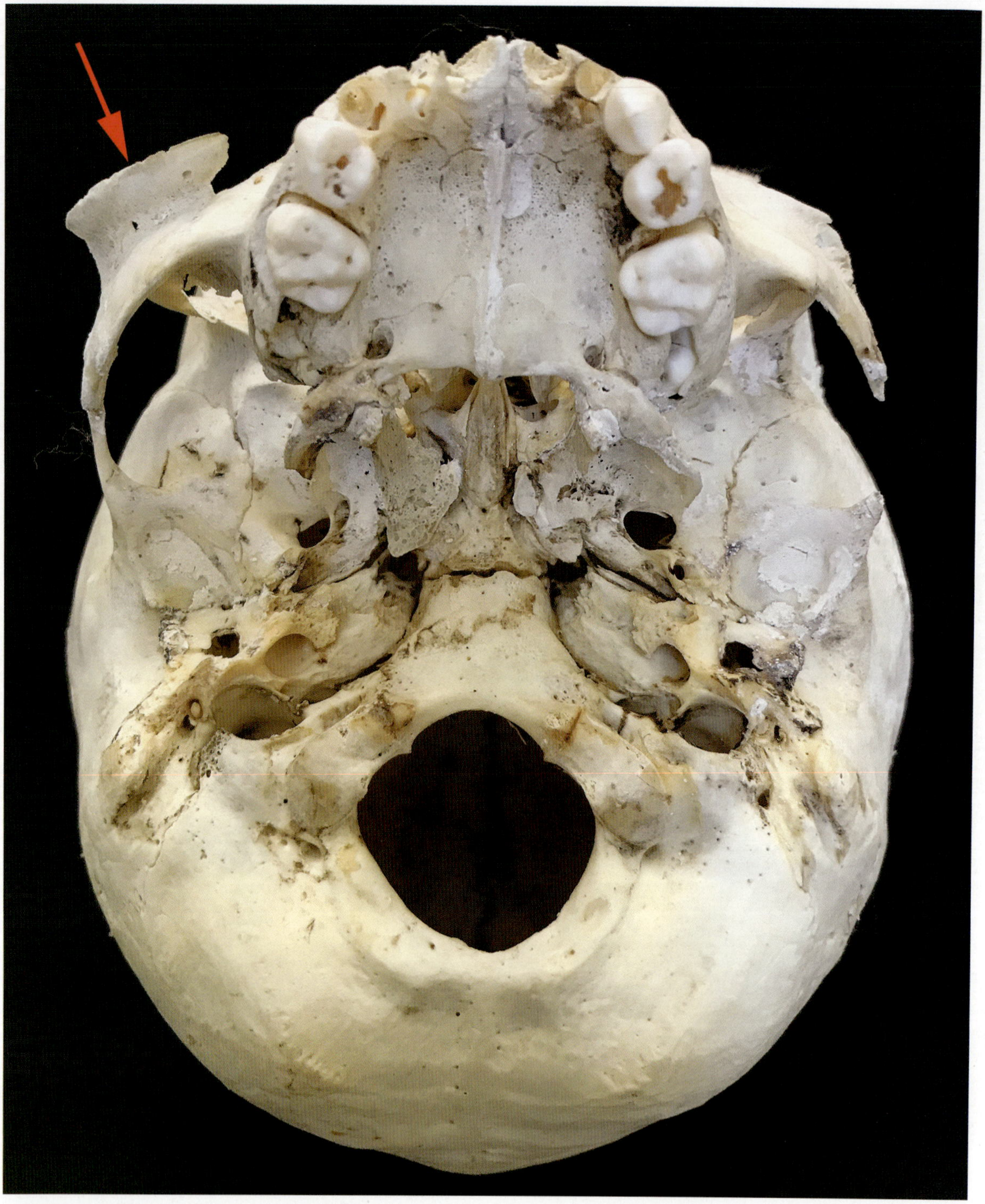

**Figure 263.** Magnified inferior view of tumor, possibly melanoma metastasis, of the right orbit in a 3-year-old female who died in 1865 (MN21.055/867). Note flaring (arrow) of the orbital surface of the zygomatic bone. See Rose et al. (2017) for more on secondary melanoma within the orbit.

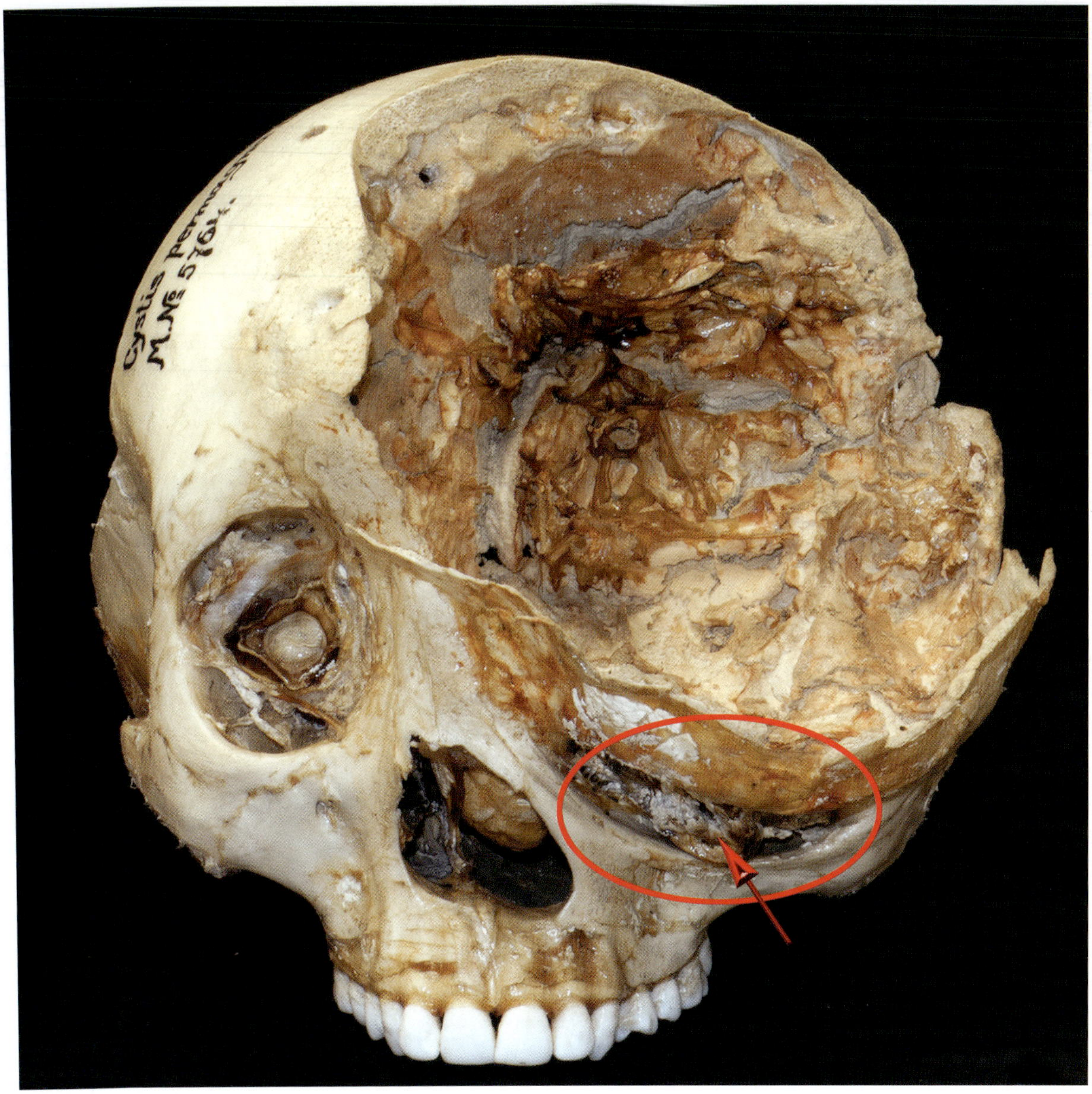

**Figure 264.** Large cystic tumor, possibly hemangioma, involving the left orbit and frontal regions resulting in enlargement and expansion of the left temporal and zygomatic bones and alteration of the nasal bones and aperture in an approximately 14 to 16-year-old male (MN5764). The left eye is overgrown by bone and narrowed (oval) causing the eyeball to bulge anteriorly (arrow). Some soft tissue still adheres to the skull and part of the right eye is present. Hemangiomas are benign vascular tumors that can affect any part of the body, but are often found in the skull, particularly the frontal bone (Yang et al. 2016). Histologically, hemangiomas can be classified into three types consisting of cavernous, capillary, and mixed. See Low and Stokken (2021); Park et al. (2013); Reis et al. (2008) for more on hemangiomas.

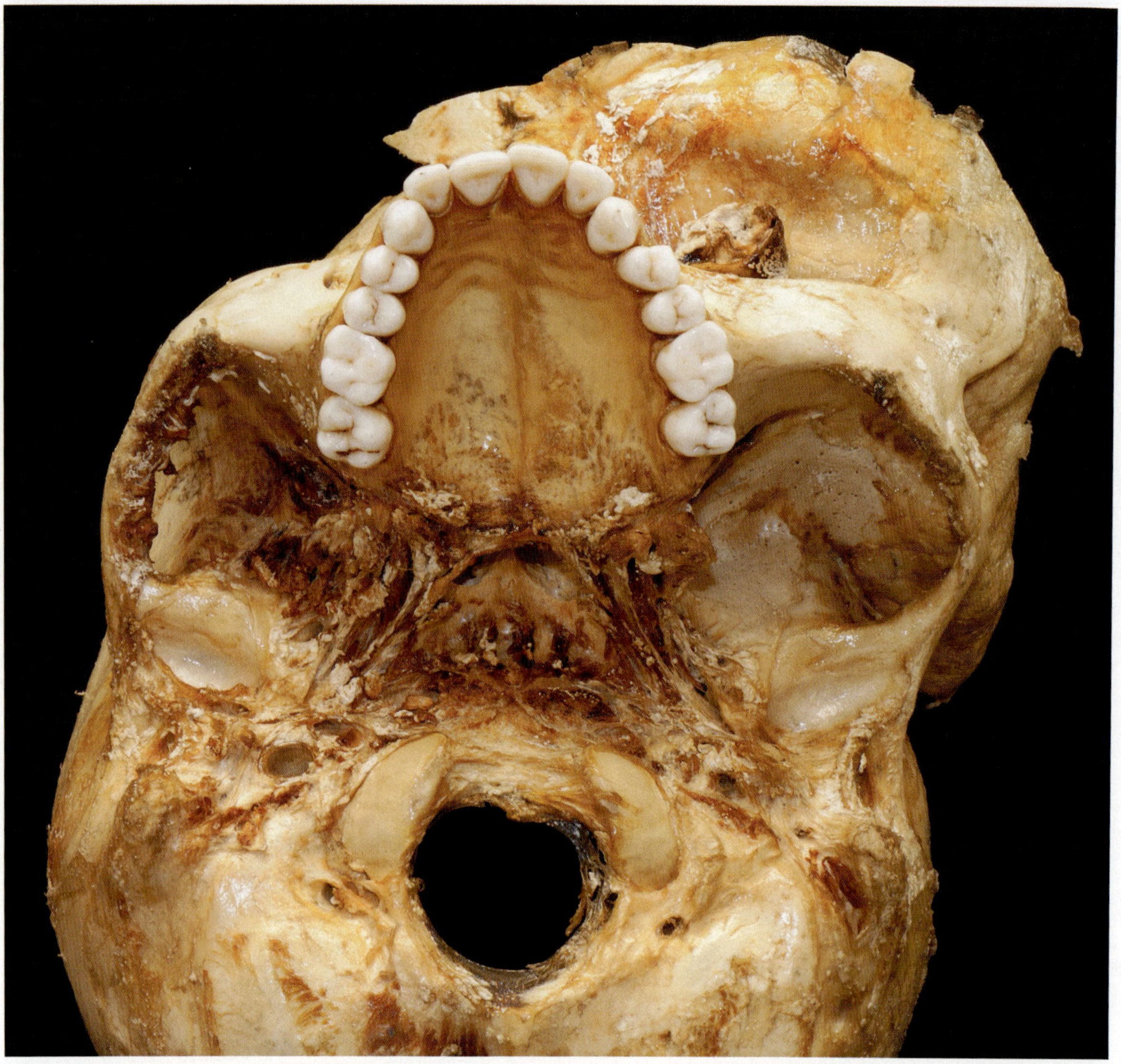

**Figure 265.** Inferior view of large cystic tumor, possibly hemangioma, involving the left orbit and frontal region resulting in enlargement and expansion of the left temporal, sphenoid, and zygomatic bones in an approximately 14 to 16-year-old male (MN5764). The foramen magnum and occipital condyles are asymmetrical. Hemangiomas are benign vascular tumors that can affect any part of the body, but are often found in the skull, particularly the frontal bone (Yang et al. 2016). Histologically, hemangiomas can be classified into three types consisting of cavernous, capillary, and mixed. See Low and Stokken (2021); Park et al. (2013); Reis et al. (2008) for more on hemangiomas.

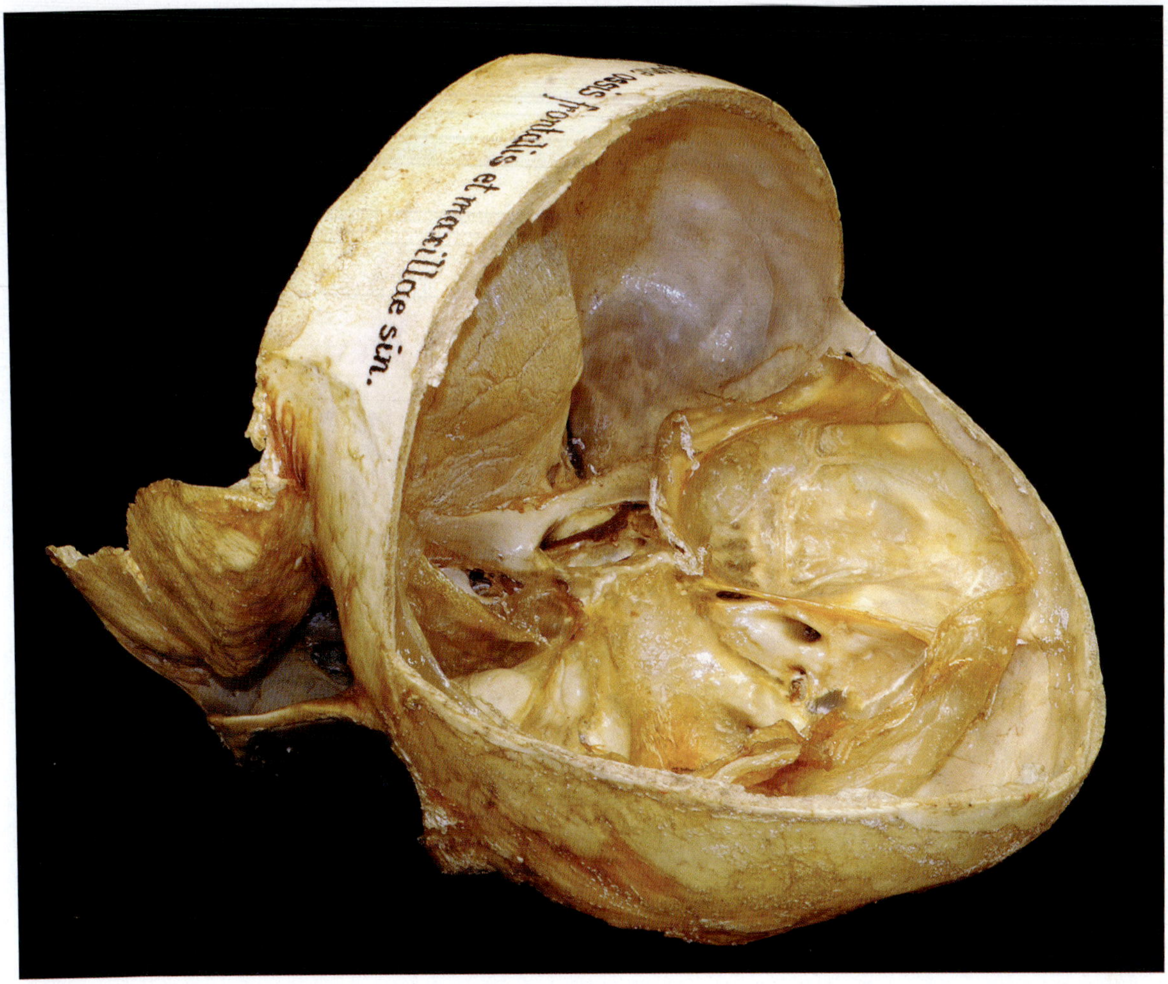


**Figure 266.** Left lateral view of large cystic tumor, possibly hemangioma, involving the left orbit and frontal region resulting in enlargement and expansion of the left temporal, sphenoid, and zygomatic bones in an approximately 14 to 16-year-old male (MN5764). The left orbit has greatly enlarged laterally and posteriorly. Hemangiomas are benign vascular tumors that can affect any part of the body, but are often found in the skull, particularly the frontal bone (Yang et al. 2016). Histologically, hemangiomas can be classified into three types consisting of cavernous, capillary, and mixed. See Low and Stokken (2021); Park et al. (2013), Reis et al. (2008) for more on hemangiomas.

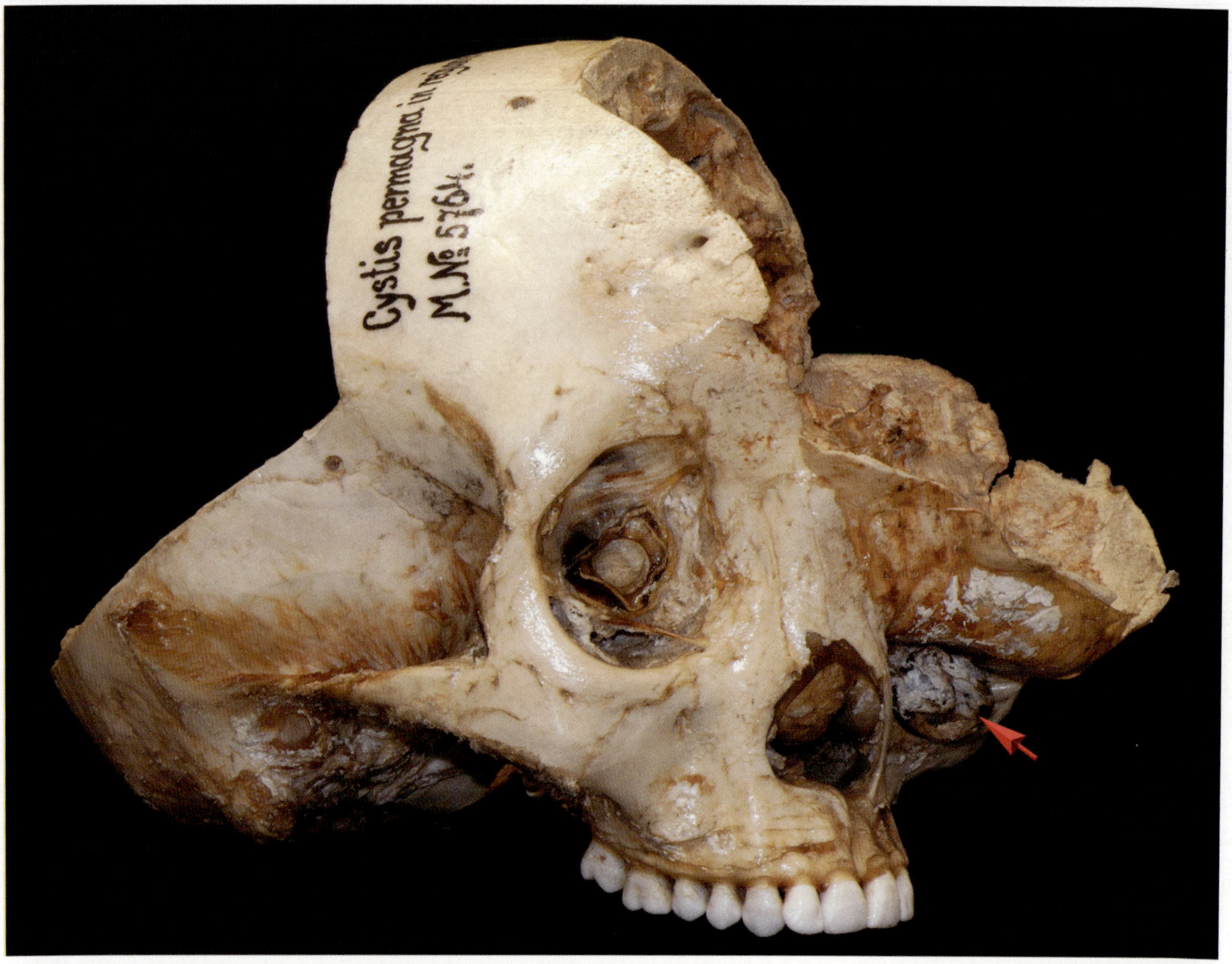


**Figure 267.** Right oblique view of large cystic tumor (possibly hemangioma) involving the left orbit and frontal region resulting in enlargement and expansion of the left temporal, sphenoid, and zygomatic bones in an approximately 14 to 16-year-old male (MN5764). The left orbit has greatly enlarged laterally and posteriorly and narrowed superior-inferior causing the eyeball to bulge (arrow). Hemangiomas are benign vascular tumors that can affect any part of the body, but are often found in the skull, particularly the frontal bone (Yang et al. 2016). Histologically, hemangiomas can be classified into three types consisting of cavernous, capillary, and mixed. See Low and Stokken (2021); Park et al. (2013); Reis et al. (2008) for more on hemangiomas.

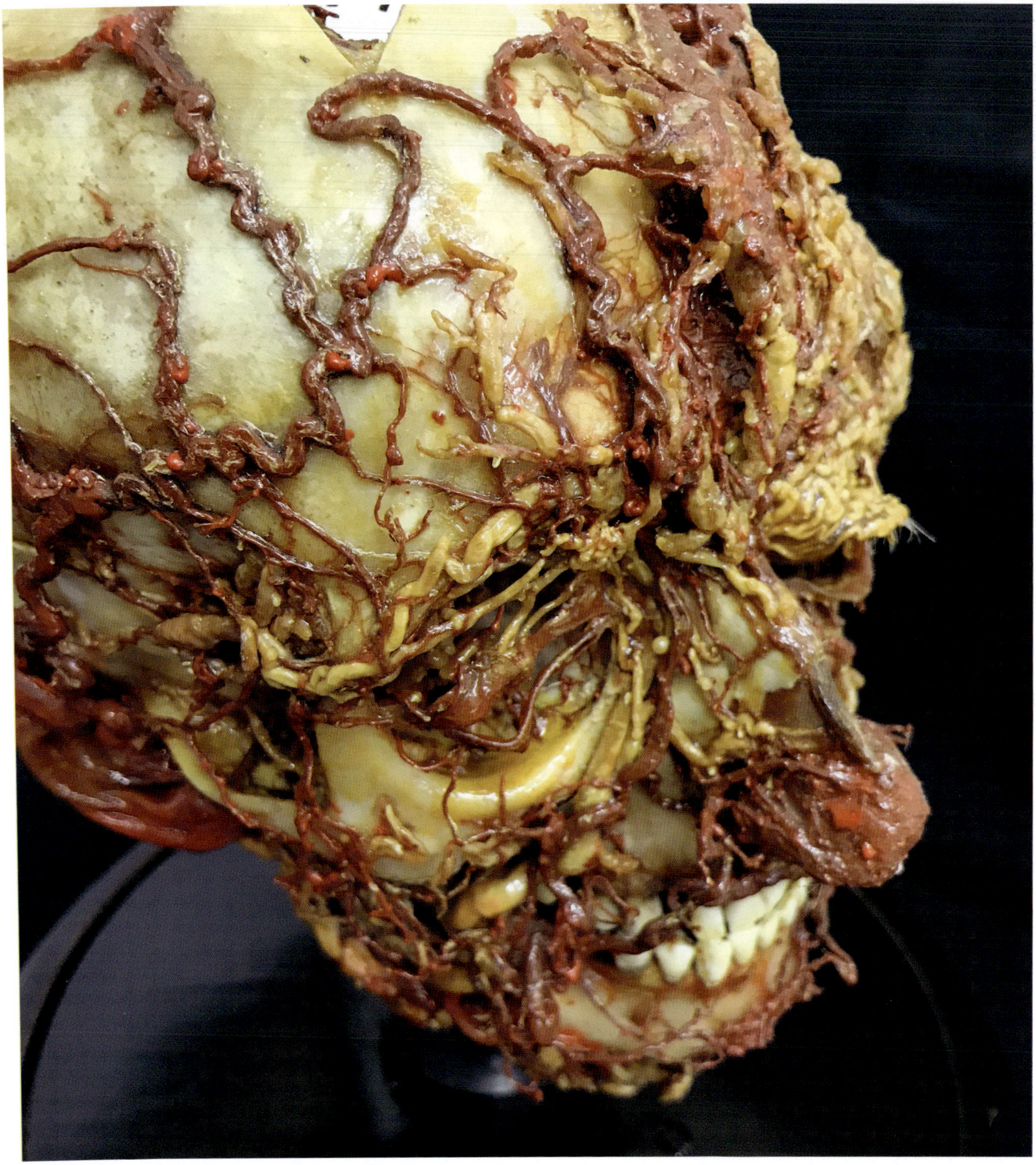

**Figure 268.** Cast corrosion injection method showing arteriovenous aneurysm of the brain ("Angioma arteriale racemosum" on the museum card) in an individual with cancer (MN1902). See Asenjo and Uiberall (1946); Rammos et al. (2016). (Note: Corrosion casting allows the study of anatomical structures by creating a solid model that replicates the space inside or around the tissue or organ. It consists of a three-step process: resin is injected into a structure and allowed to harden (casting). Then the organic material is removed (corrosion). Finally, the cast is cleaned and dried.

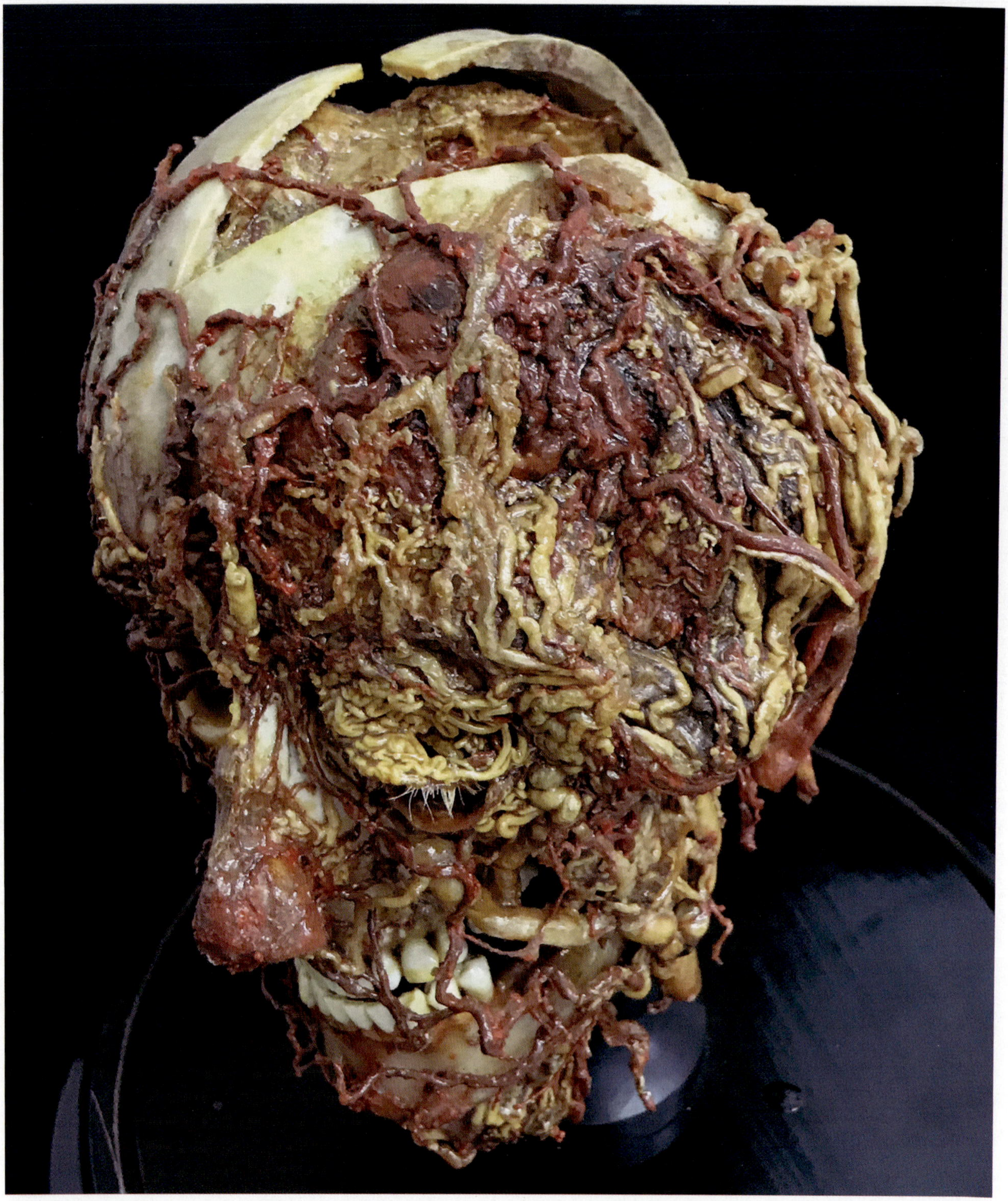

**Figure 269.** Cast corrosion injection method showing arteriovenous aneurysm of the brain and left orbit ("Angioma arteriale racemosum" on the museum card) (MN1902). See Asenjo and Uiberall (1946); Rammos et al. (2016).

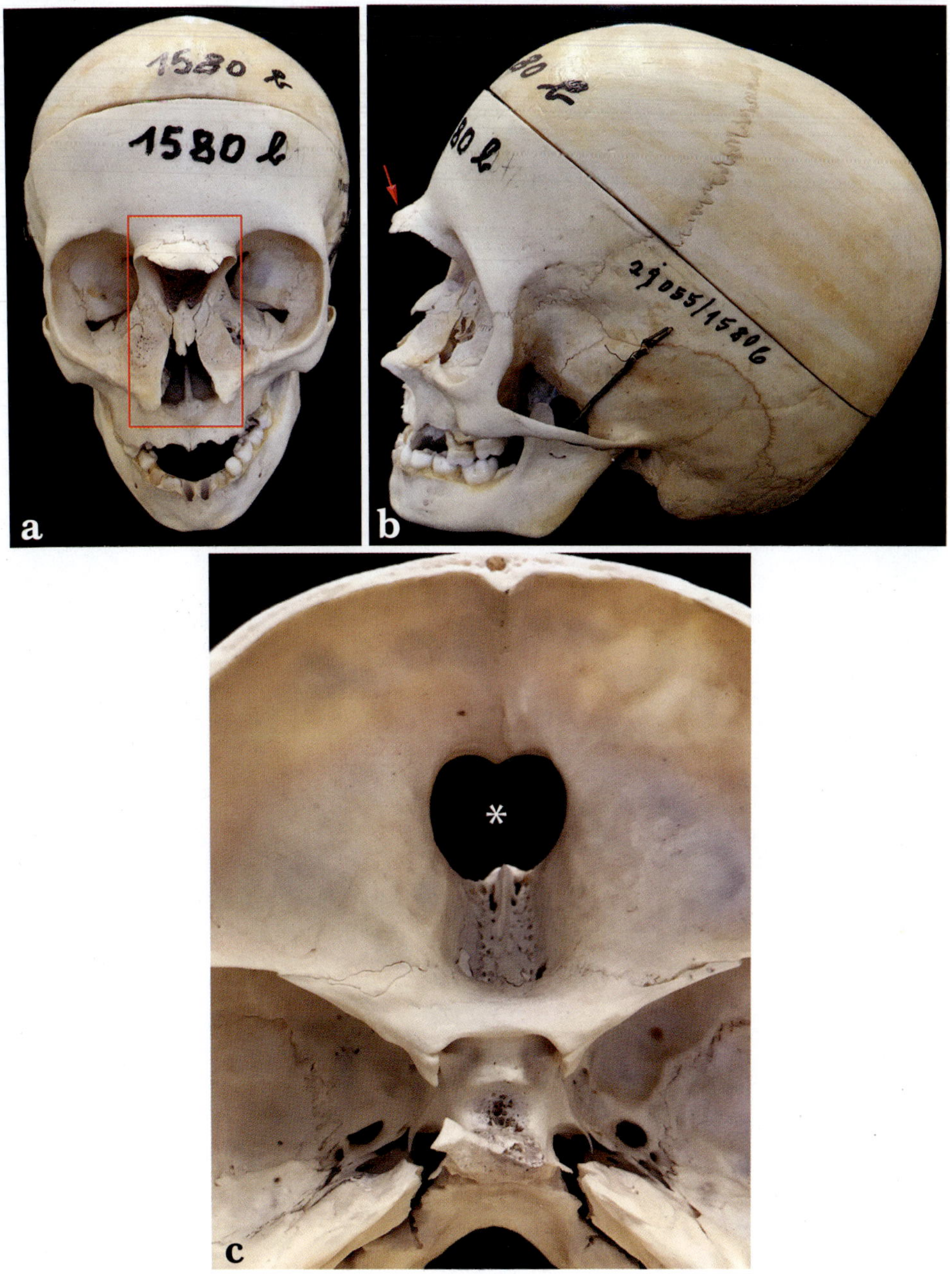


**Figure 270a-c.** (a) Anterior, (b) left lateral showing the shelf-like nasal bones and (c) endocranial views of nasoethmoidal encephalocele defect (*), a neural tube defect resulting in a sac-like protrusion containing brain tissue and spinal fluid, that caused a forward projection of the frontonasal region and hypoplastic nasal bones in a probable subadult female, based on cranial morphology and dental development, who died in 1869 (MN21.055/1580b). Note that this individual, identified on the accession card as having "anterior encephalocele," also exhibits hypertelorism or wide-set orbits. See Dhirawani et al. (2014) for more on the types and classification of encephalocele.

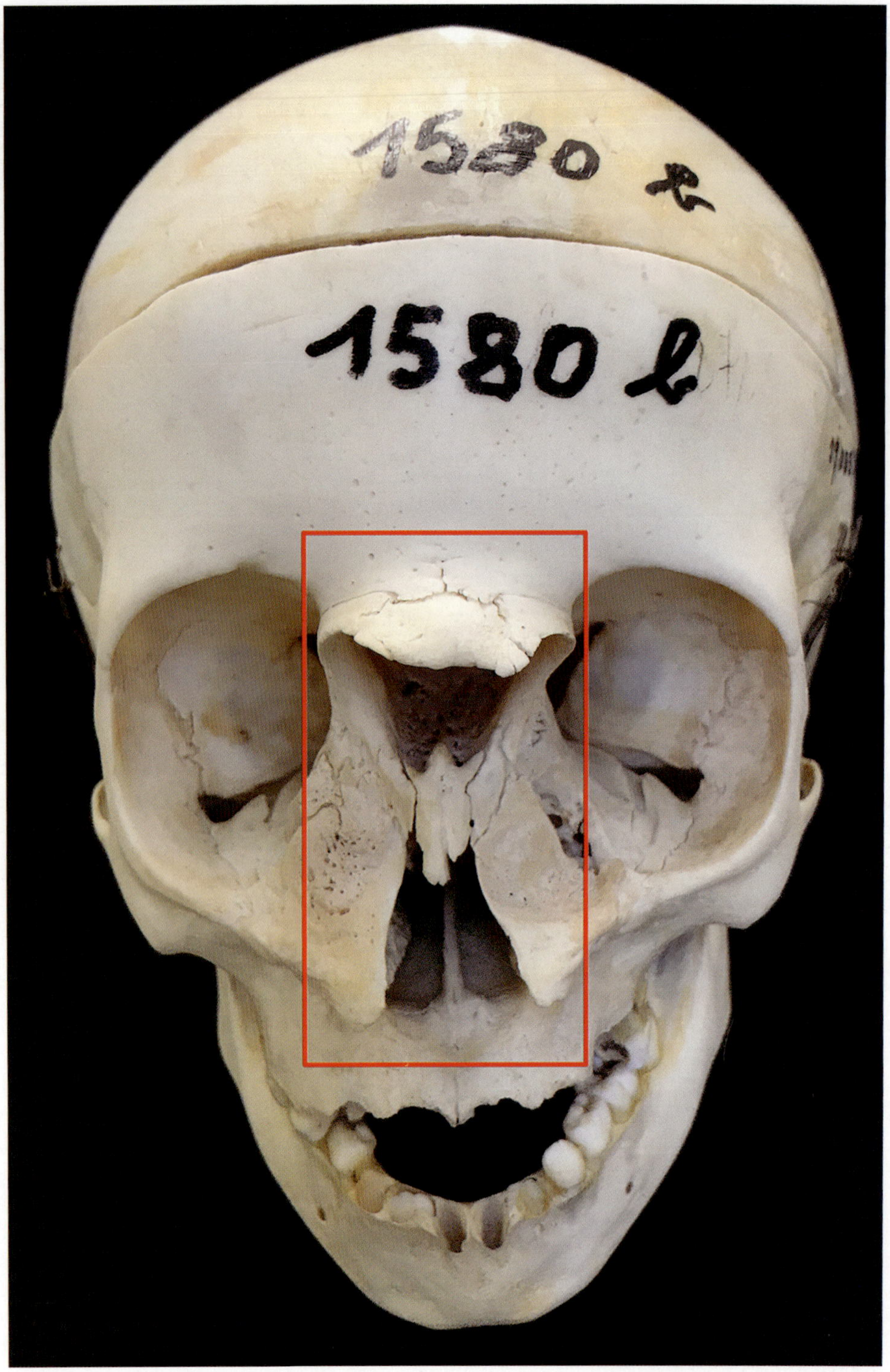


**Figure 271.** Nasoethmoidal encephalocele (rectangle), a neural tube defect resulting in a sac-like protrusion containing brain tissue and spinal fluid, that caused a forward projection of the frontonasal region and hypoplastic nasal bones in a probable subadult female, based on cranial morphology and dental development, who died in 1869 (MN21.055/1580b). Nasoethmoidal encephalocele is the most common of the encephaloceles and, as in this individual, is often accompanied with hyperteleorism (Dhirawani et al. 2014).

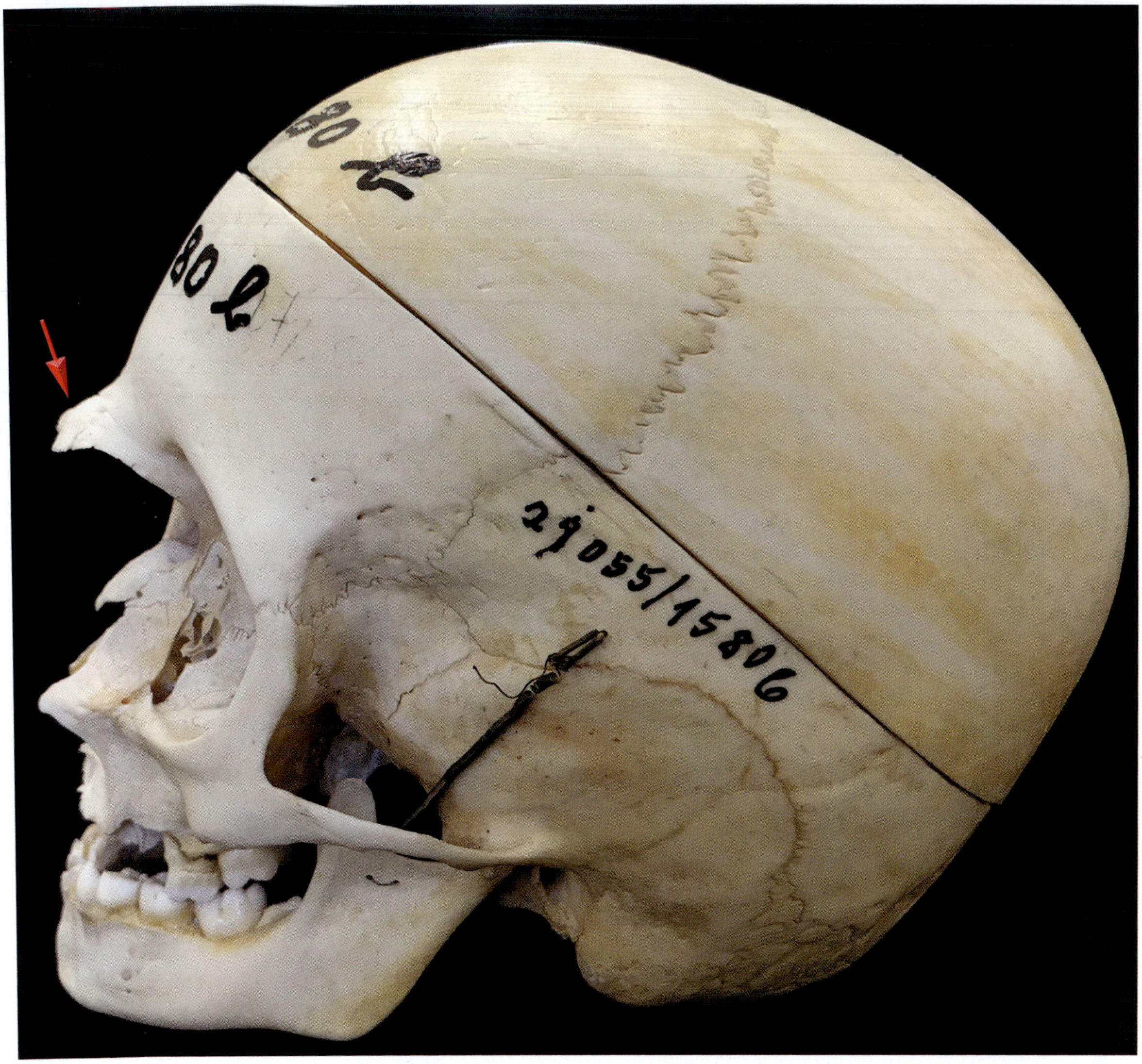

**Figure 272.** Nasoethmoidal encephalocele, a neural tube defect resulting in a sac-like protrusion containing brain tissue and spinal fluid that caused a forward projection of the frontonasal region (arrow), hypertelorism and hypoplastic nasal bones in a probable subadult female, based on cranial morphology and dental development, who died in 1869 (MN21.055/1580b). See Dhirawani et al. (2014) for more on the types and classification of encephaloceles.

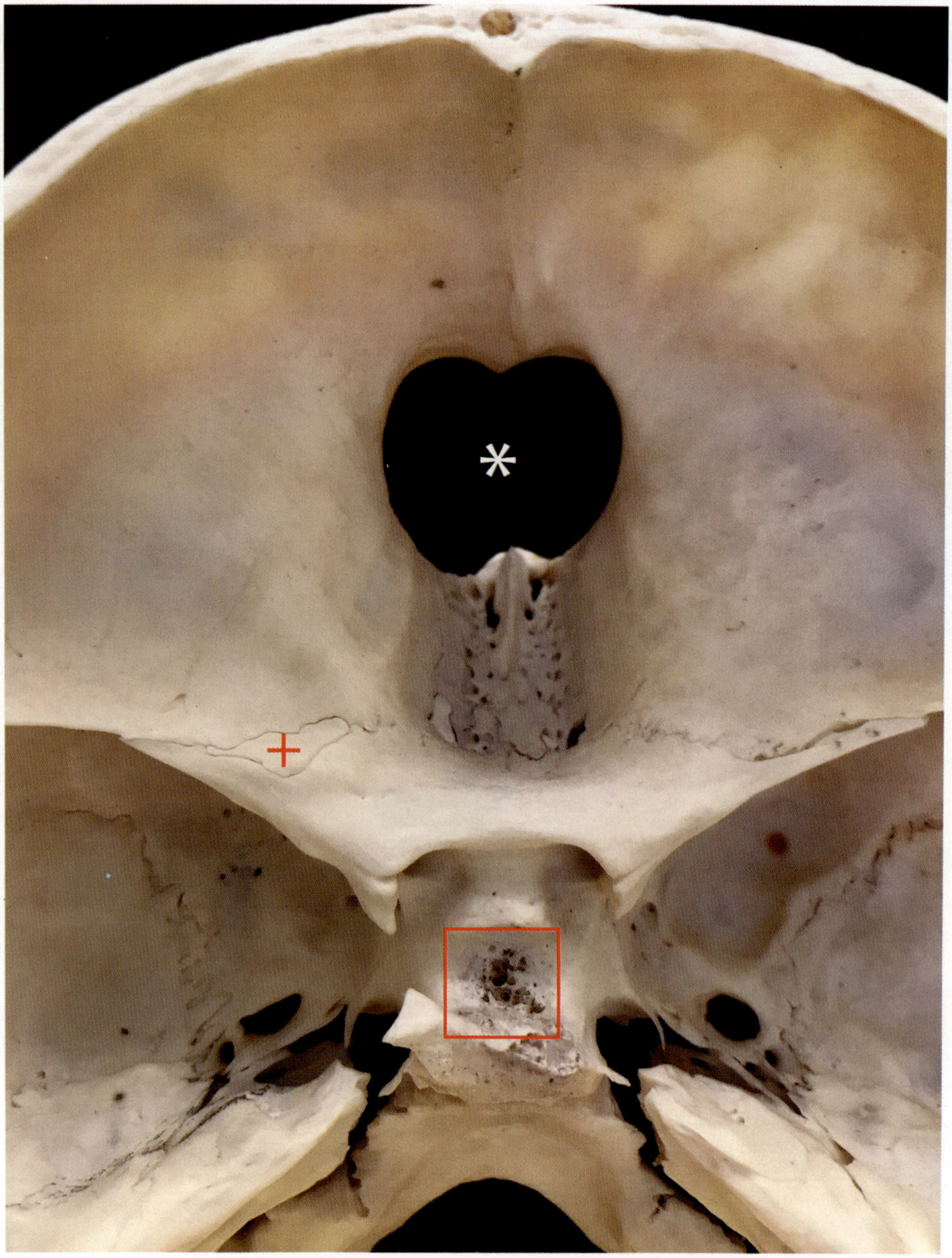


**Figure 273.** Endocranial view showing a nasoethmoidal encephalocele, a neural tube defect resulting in a sac-like protrusion containing brain tissue and spinal fluid, that caused a forward projection of the frontonasal region (*), hypoplastic nasal bones and hypertelorism in a probable subadult female, based on cranial morphology and dental development, who died in 1869 (MN21.055/1580b). Note the lytic lesion (square) of unknown origin in the hypophyseal (pituitary) fossa and the elongated ossicle (+) in the sphenofrontal suture, the latter of which is a common feature rarely reported in the literature.

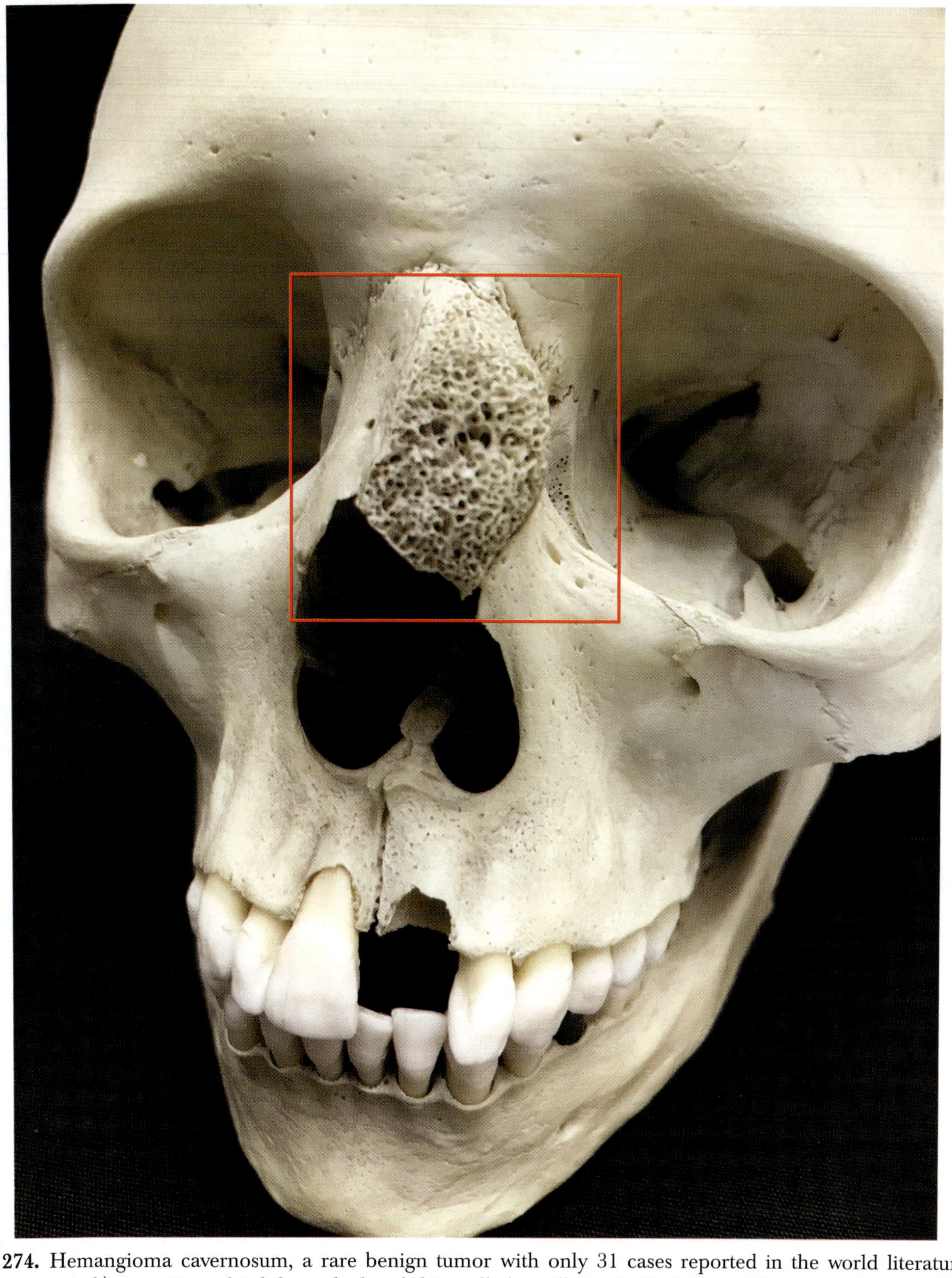

**Figure 274.** Hemangioma cavernosum, a rare benign tumor with only 31 cases reported in the world literature as of 2003 (Layoun et al.) consisting of tightly packed and thin-walled capillaries in the left nasal bone in a 17-year-old female (MN1429).

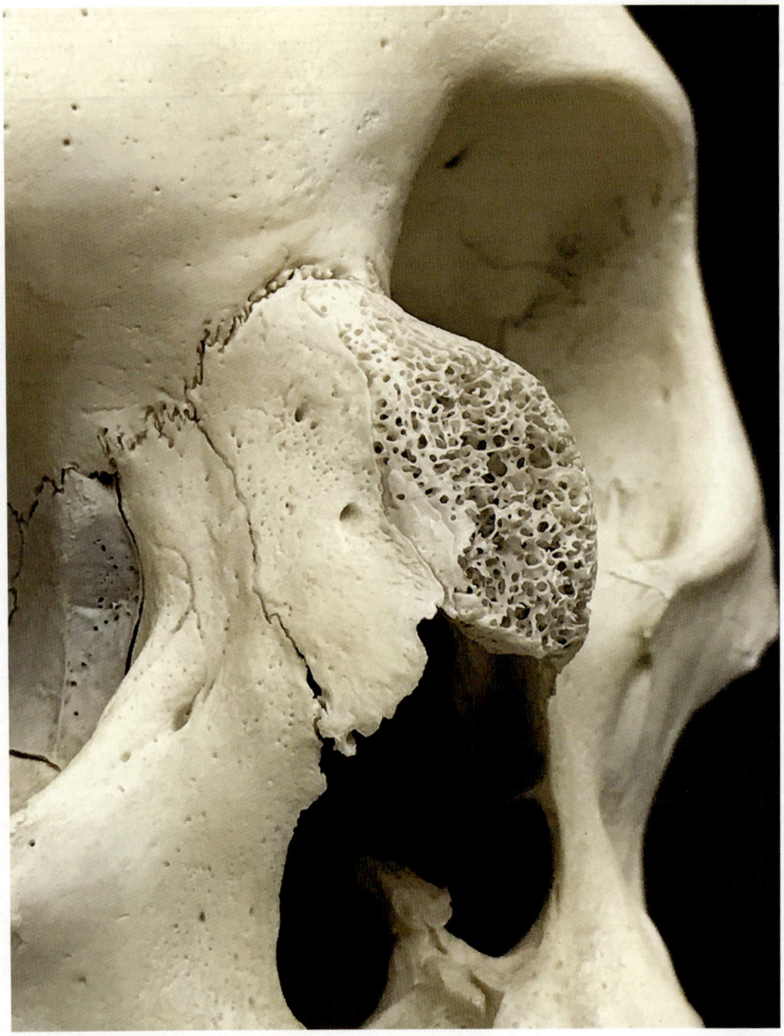

**Figure 275.** Hemangioma cavernosum is a rare benign tumor with only 31 cases reported in the world literature as of 2003 (Layoun et al.) in the left nasal bone of a 17-year-old female (MN1429).

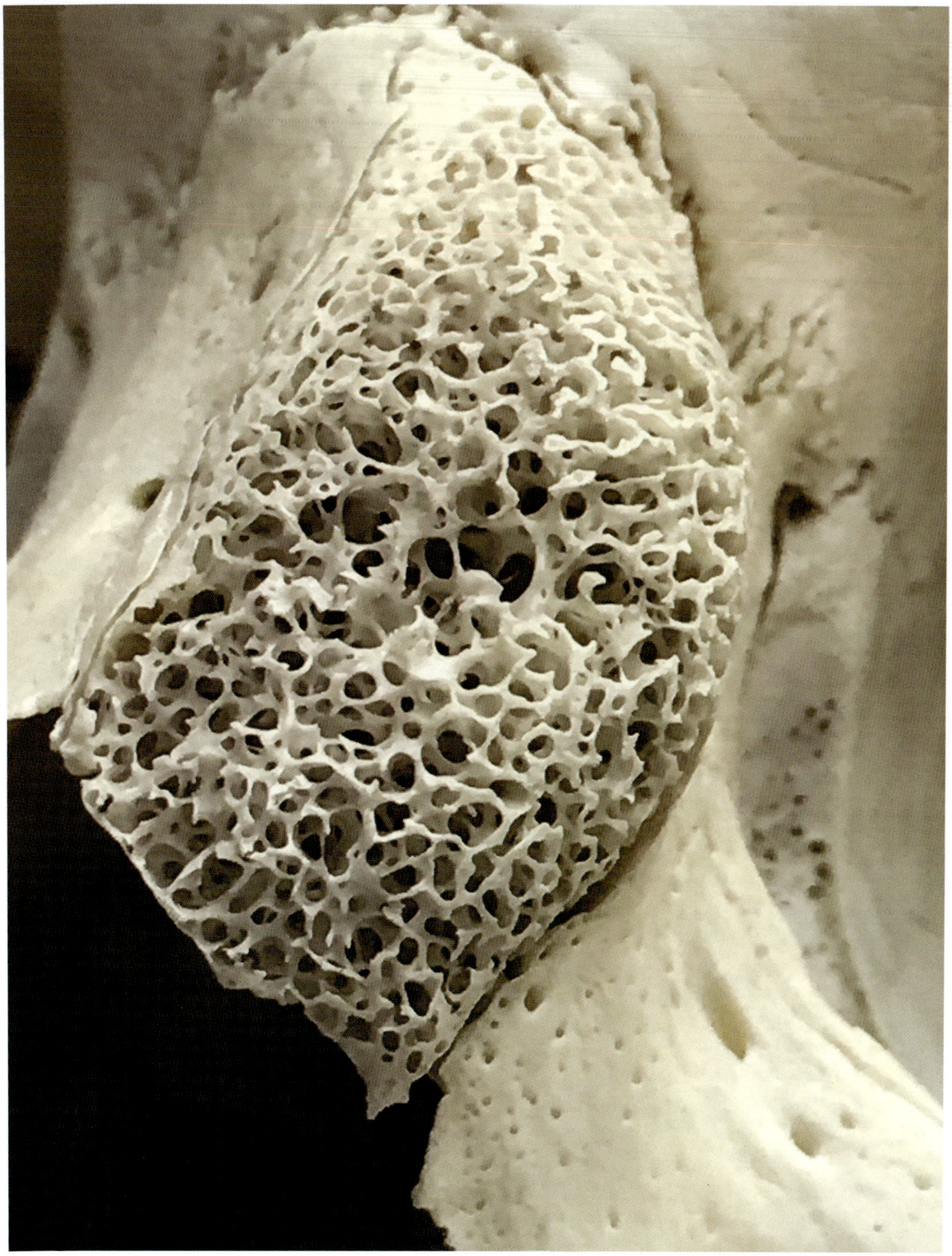

**Figure 276.** Magnified left oblique view of hemangioma cavernosum in the left nasal bone of 17-year-old female (MN1429). Hemangioma cavernosum of the nasal bones is usually a stable tumor of capillaries that can bleed easily, but that can currently be treated with little cosmetic effects (Layoun et al. 2003).

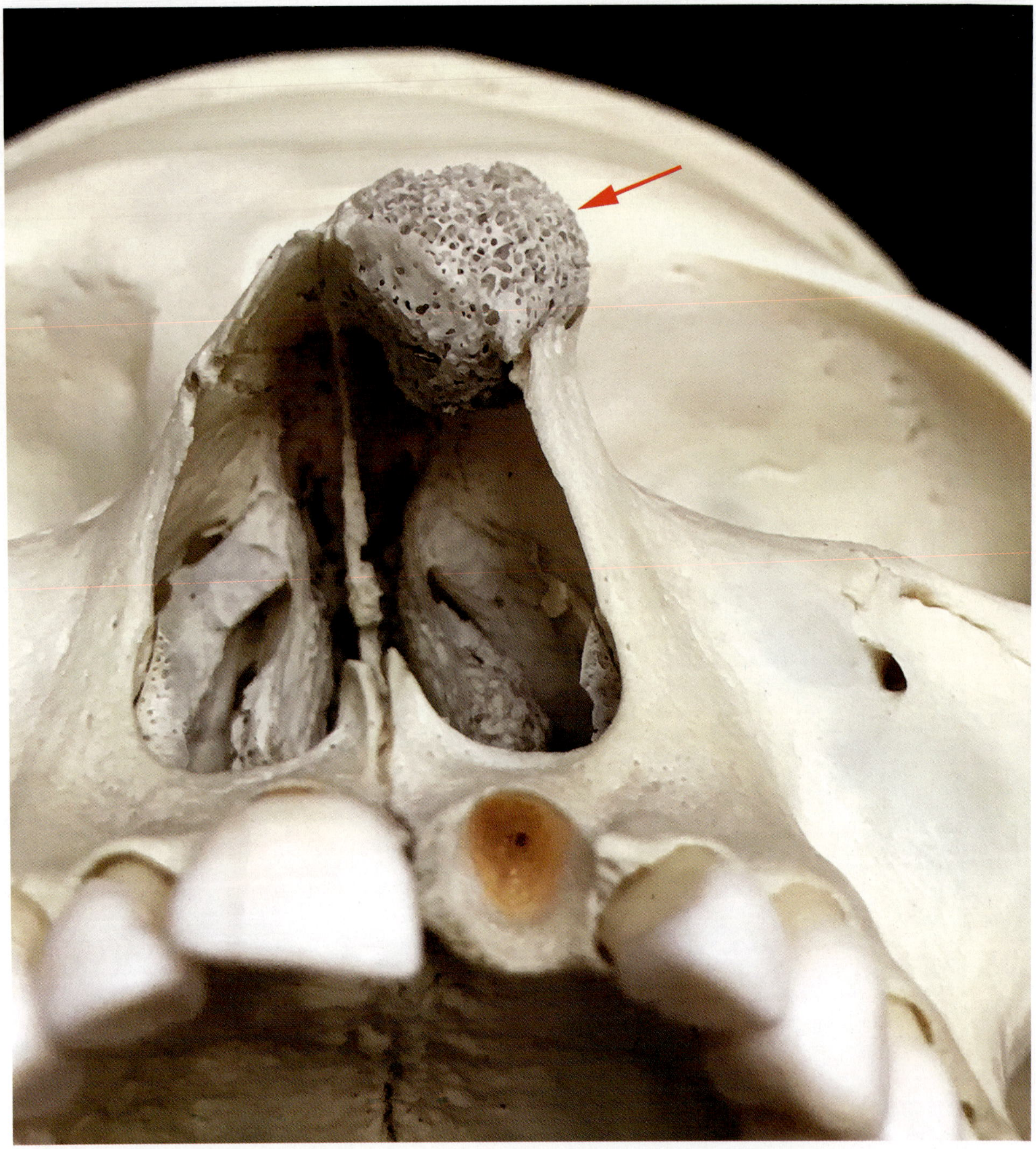

**Figure 277.** Anterior-Inferior view of hemangioma cavernosum of the left nasal bone in a 17-year-old female (MN1429).

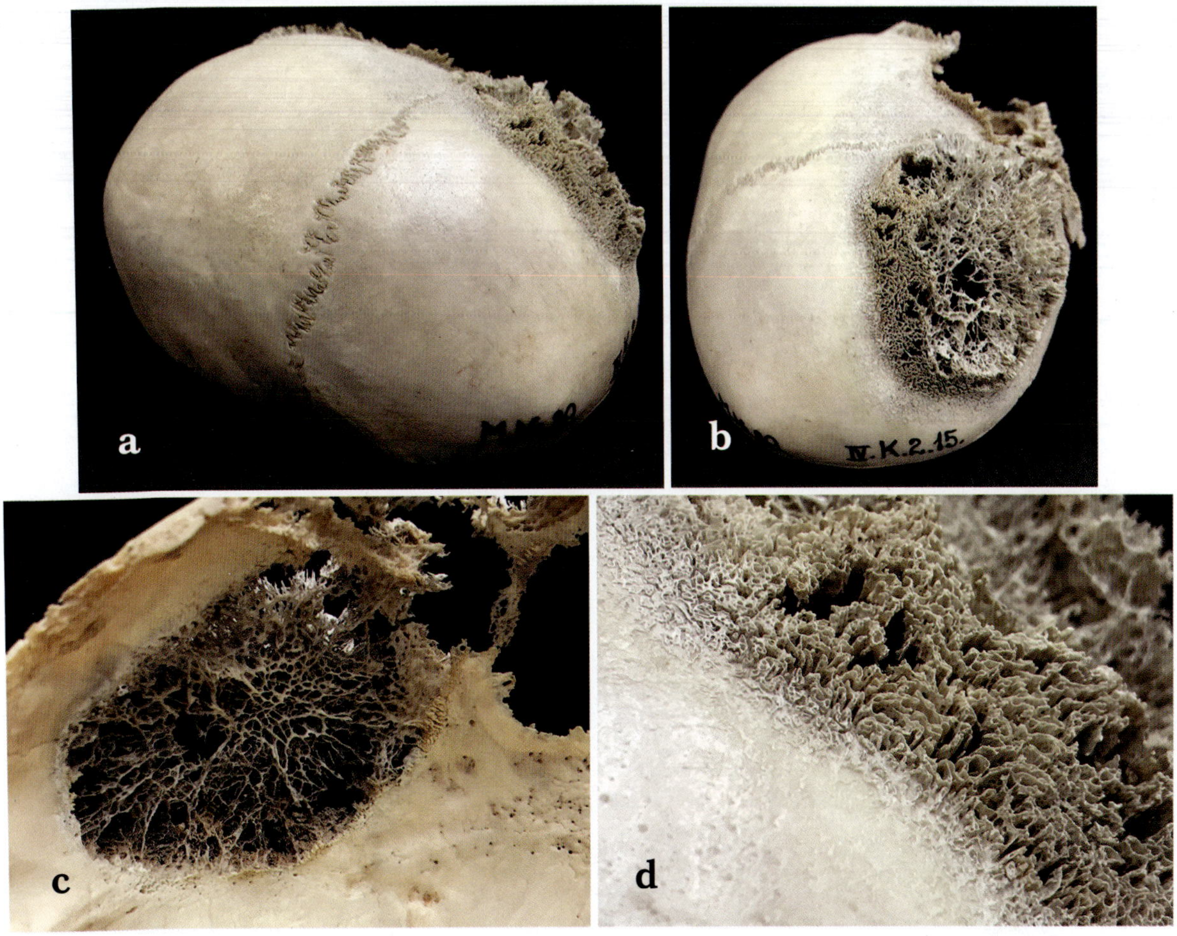


**Figure 278a-d.** Multiple views of metastatic carcinoma of the frontal and parietal bones resulting in a vascular "sunburst" response in a female who died in 1815 (MN39). See Geschickter and Copeland (1936); Migliorini et al. (2020); Rajakulasingam et al. (2020); Reeder (1993); Thomas and Kirsten (1985) for more on metastatic carcinoma and tumors of non-osseous origin.

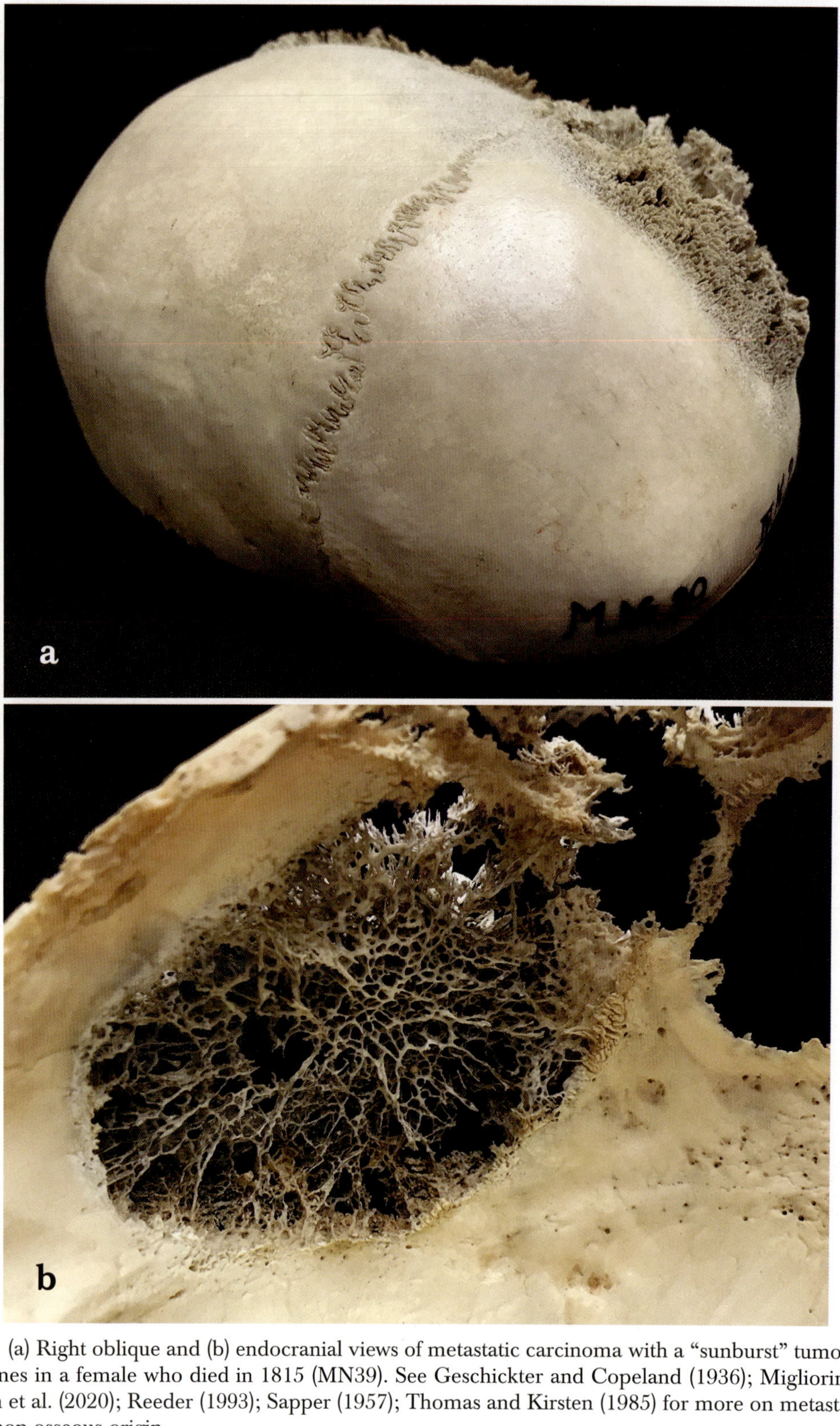


**Figure 279a-b.** (a) Right oblique and (b) endocranial views of metastatic carcinoma with a "sunburst" tumor in the frontal and parietal bones in a female who died in 1815 (MN39). See Geschickter and Copeland (1936); Migliorini et al. (2020); Rajakulasingam et al. (2020); Reeder (1993); Sapper (1957); Thomas and Kirsten (1985) for more on metastatic carcinoma and tumors of non-osseous origin.

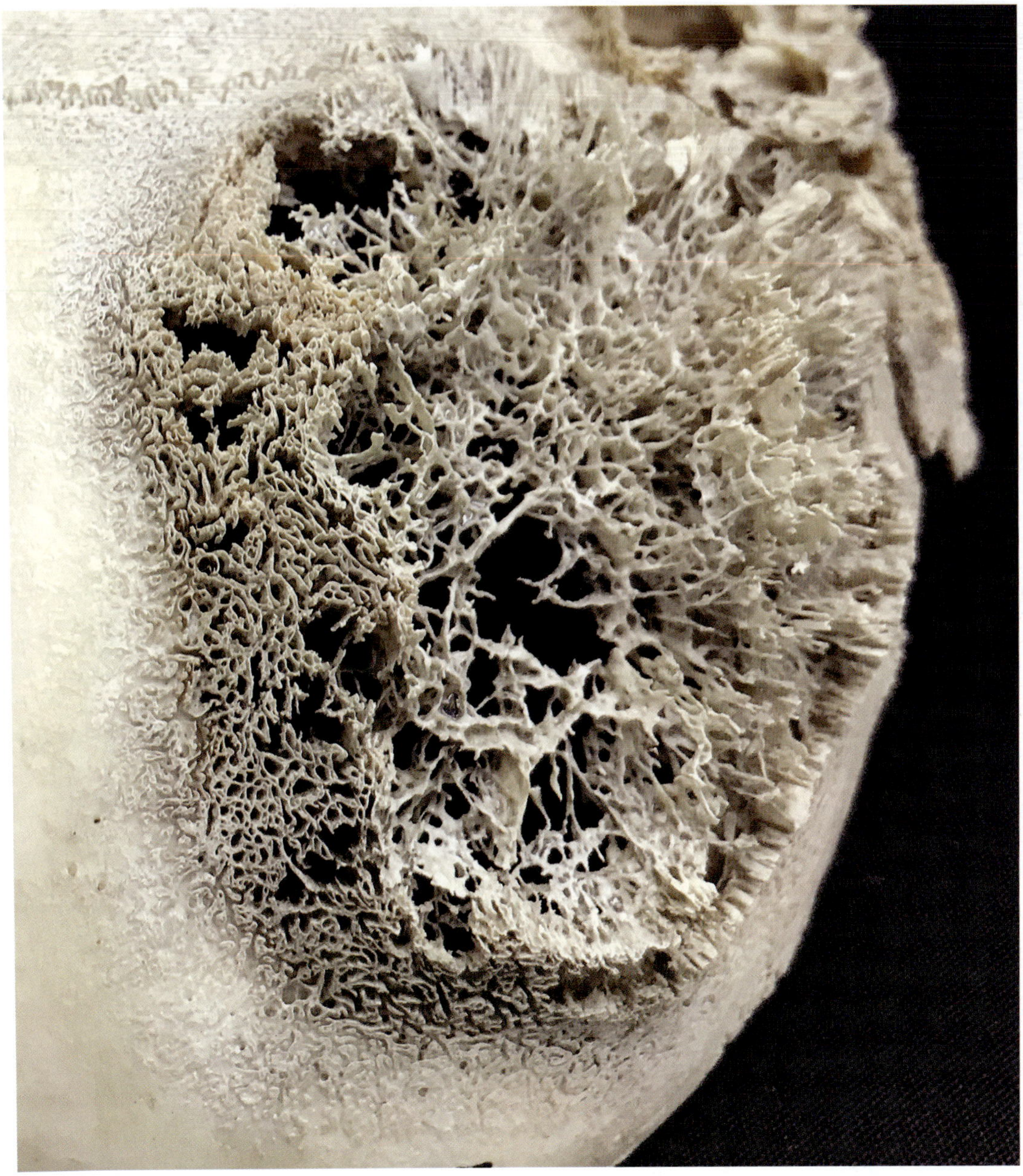

**Figure 280.** Magnified view of metastatic carcinoma of the frontal bone in a female who died in 1815 (MN39). The disease consists of mixed (osteoblastic and osteoclastic) tumors with raised margins that coalesced to erode much of the vault. The large lesion in the frontal bone exhibits an unusual sunburst pattern associated with some metastatic bone tumors (Bloom et al. 1987). See Geschickter and Copeland (1936); Reeder (1993) for more on metastatic carcinoma and tumors of non-osseous origin.

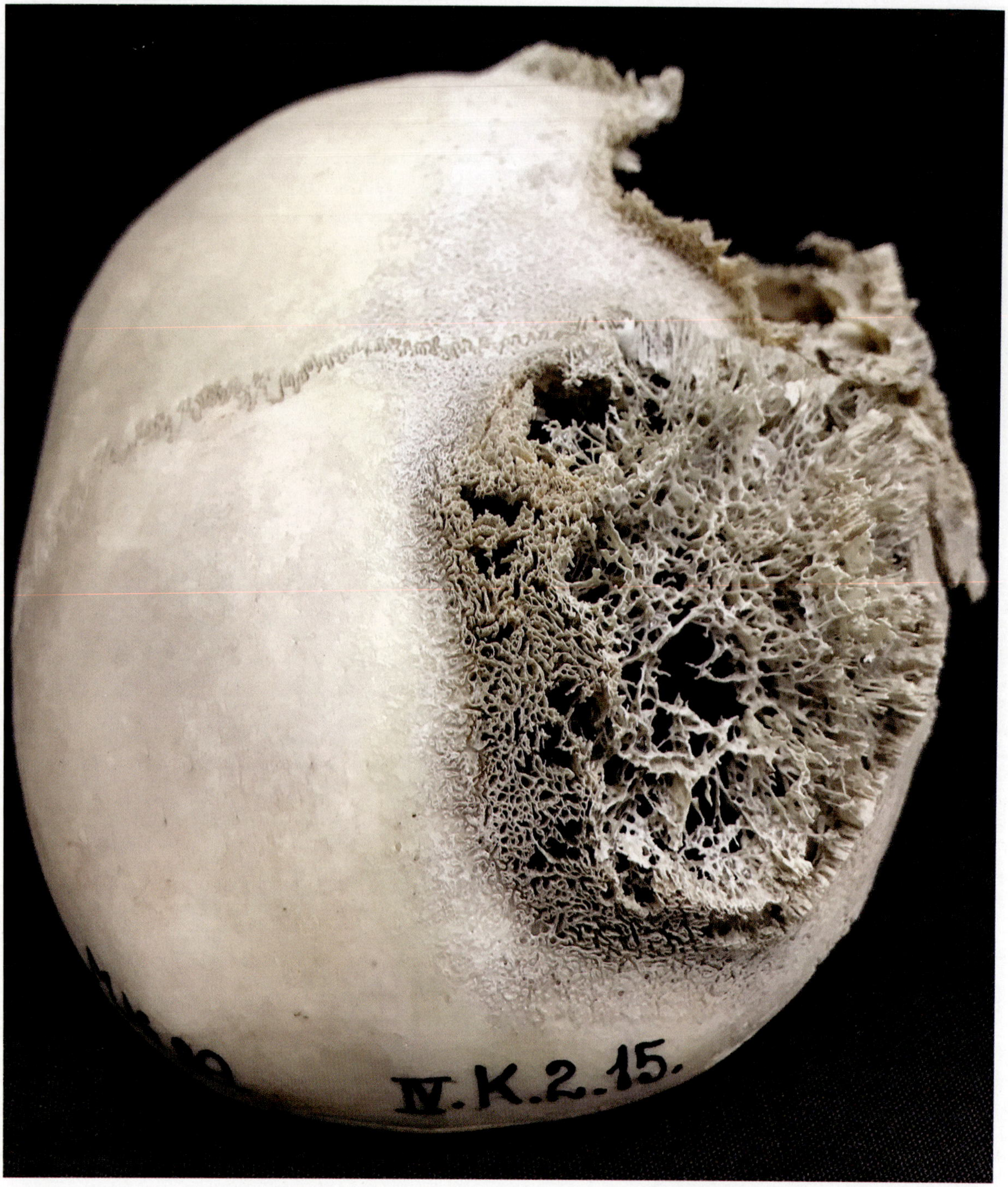


**Figure 281.** Metastatic carcinoma in the frontal and parietal bones in a female who died in 1815 (MN39). The disease consists of mixed (osteoblastic and osteoclastic) tumors with raised margins that coalesced to erode much of the vault. The large lesion in the frontal bone exhibits a "sunburst" pattern sometimes associated with metastatic bone disease (Bloom et al. 1987). See Geschickter and Copeland (1936); Rajakulasingam et al. (2020); Reeder (1993); Sapper (1957); Thomas and Kirsten (1985) for more on metastatic carcinoma and tumors of non-osseous origin. See Steinbock (1976) for more on metastatic carcinoma.

**Figure 282.** Metastatic carcinoma in the frontal and parietal bones in a female who died in 1815 (MN39). The disease consists of mixed (osteoblastic and osteoclastic) lesions/tumors with raised margins that coalesced to erode much of the vault. The large lesion in the frontal bone exhibits a "sunburst" pattern sometimes present in metastatic bone disease (Bloom 1987). See Geschickter and Copeland (1936); Rajakulasingam et al. (2020); Reeder (1993); Sapper (1957); Thomas and Kirsten (1985) for more on metastatic carcinoma and tumors of non-osseous origin.

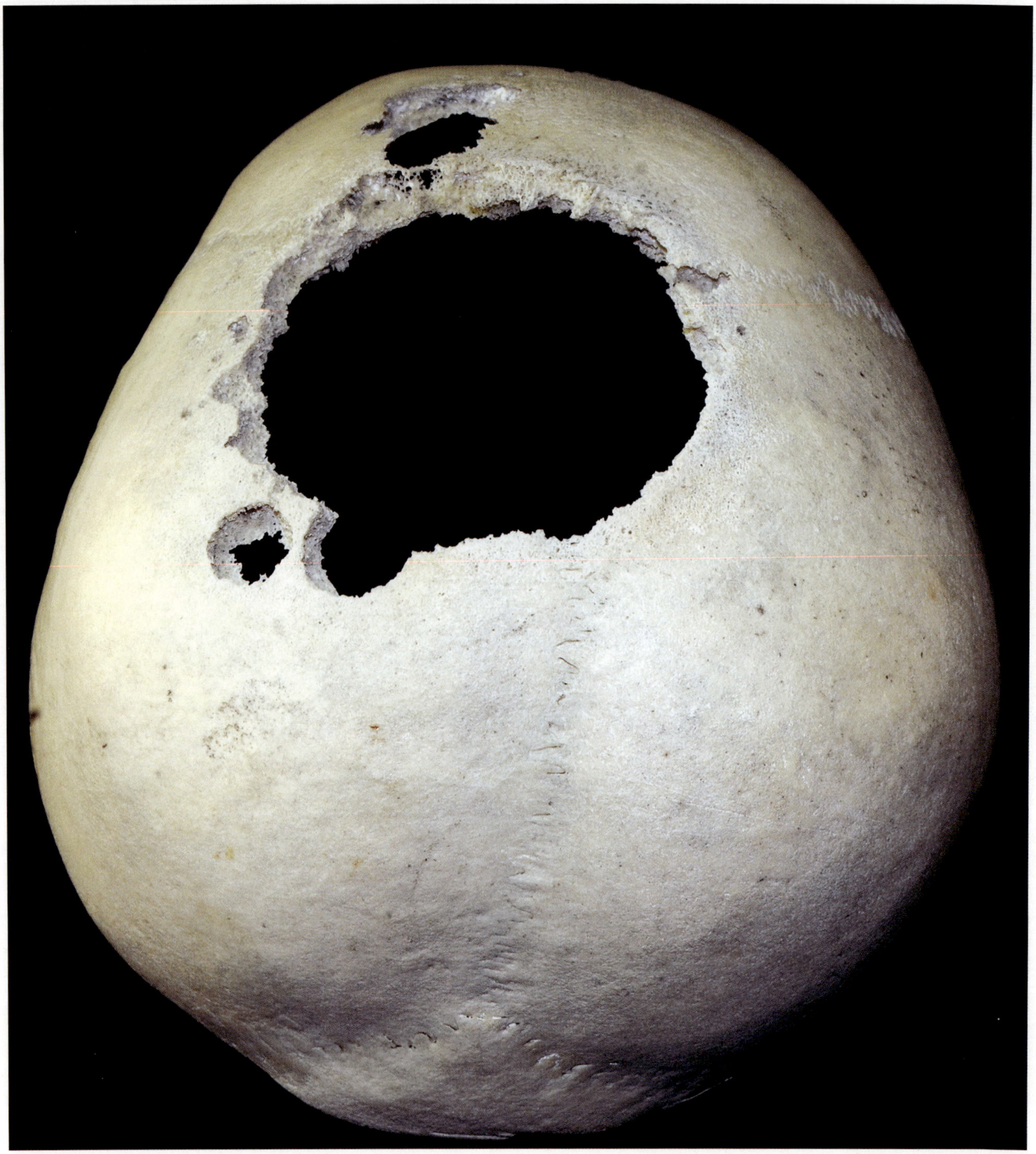

**Figure 283.** Destructive neoplastic tumors ("Destructio neoplastica calvariae" on the museum card) in an unknown adult who died in 1930 (MN18.560/251). Note the scalloped margins and undercut morphology of the defects with little reactive/osteoblastic activity possibly due to carcinoma. See Geschchickter and Copeland (1936); Ortner (2003); Ortner and Putschar (1981); Reeder (1993); Steinbock (1976) for more on bone tumors.

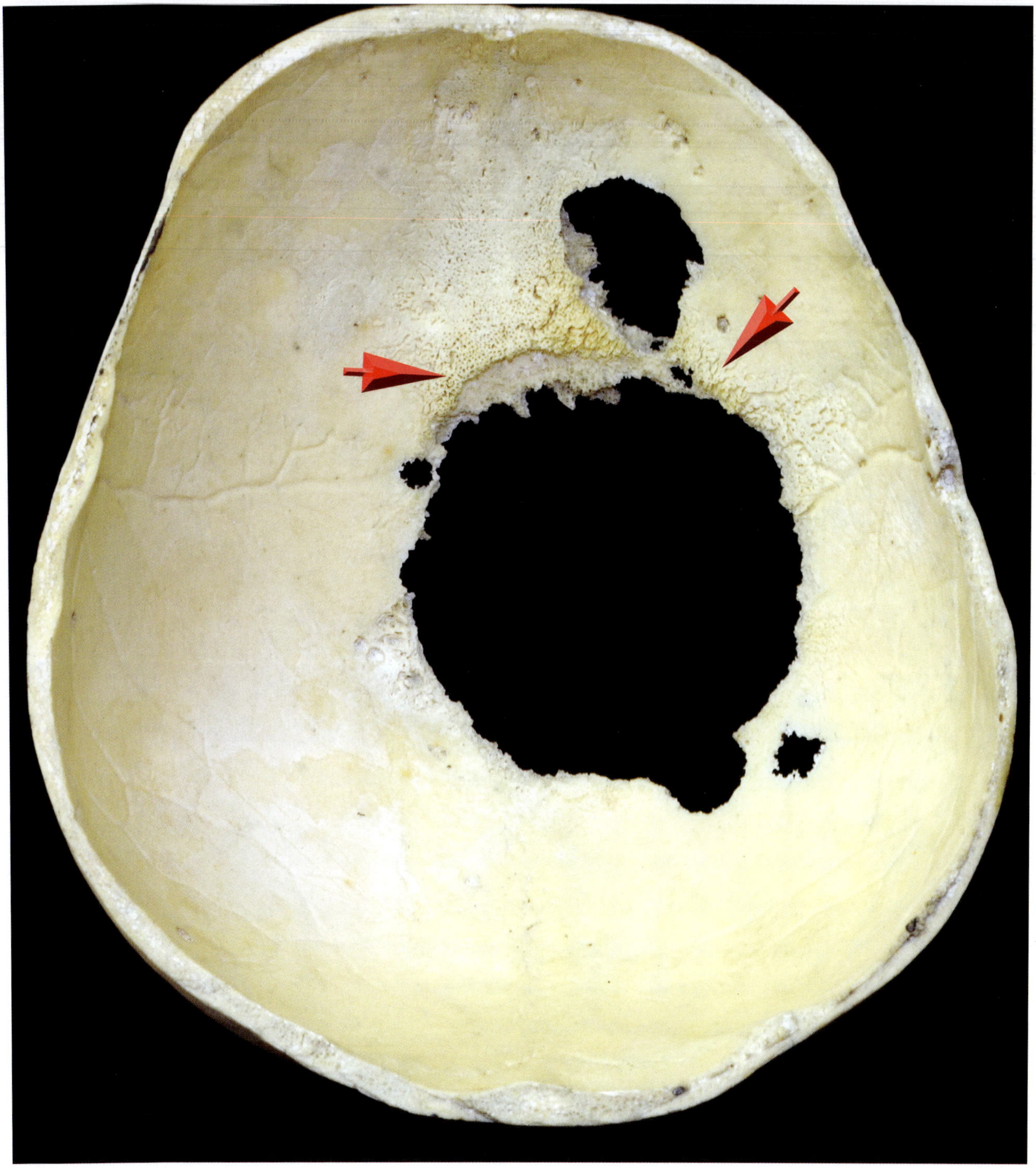

**Figure 284.** Endocranial view of an unknown adult who died in 1930 (MN18.560/251) showing new bone (arrows) growth along the margins of the neoplastic tumors, possibly due to carcinoma. See Aufderheide and Rodrigues-Martin (1998); Geschchickter and Copeland (1936); Ortner (2003); Ortner and Putschar (1981); Reeder (1993); Steinbock (1976) for more on bone tumors.

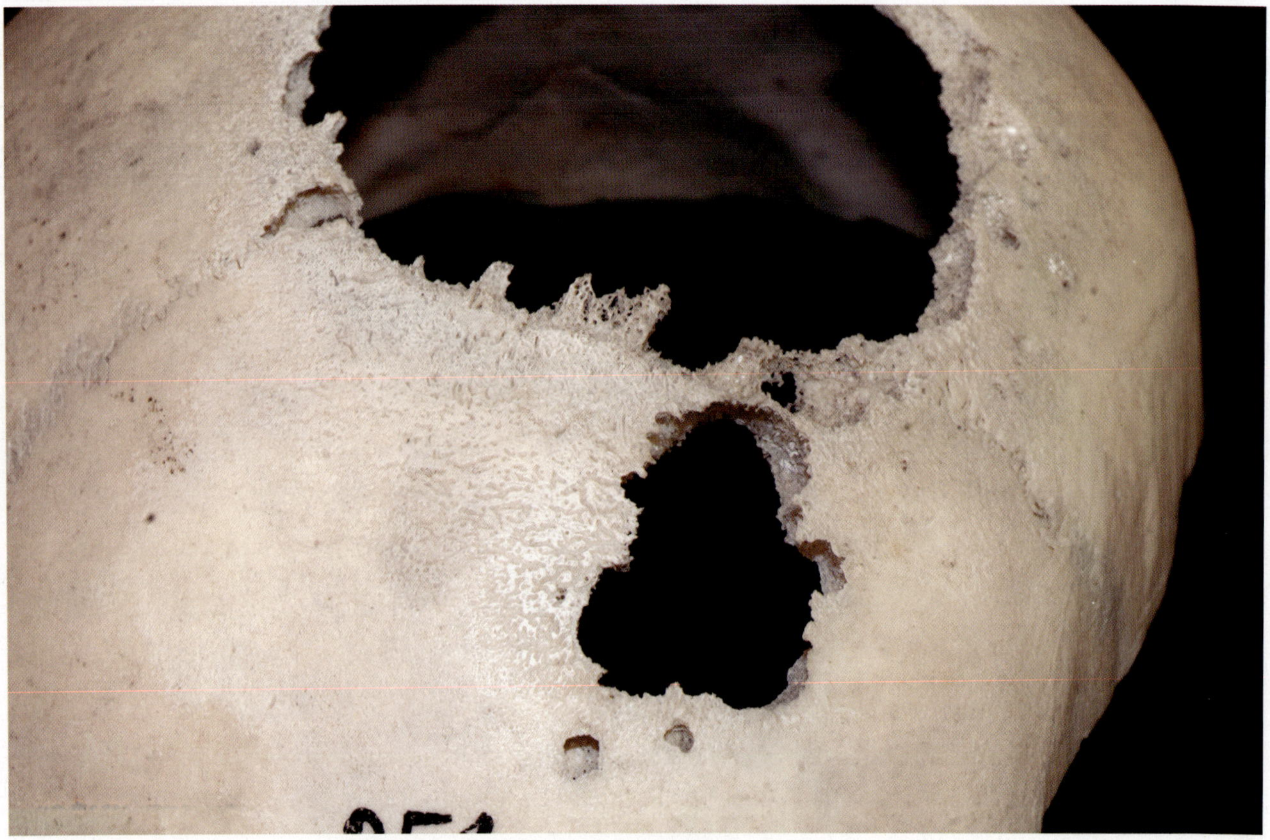

**Figure 285.** Magnified view of the ectocranium of an unknown adult who died in 1930 (MN18.560/251) showing the morphology of the lytic lesions, possibly due to carcinoma. See Geschchickter and Copeland (1936); Ortner (2003); Ortner and Putschar (1981); Reeder (1993); Steinbock (1976) for more on bone tumors.

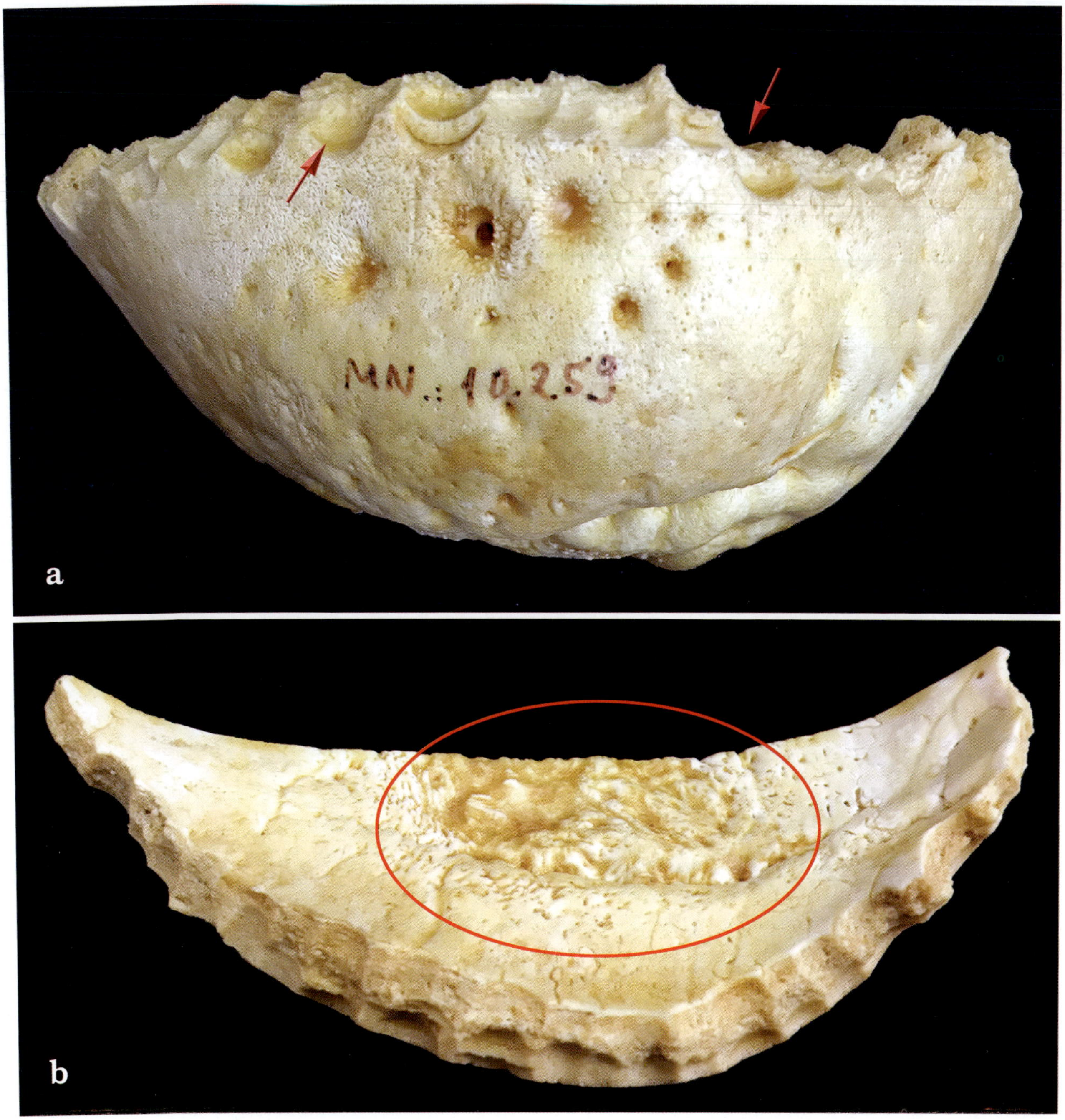


**Figure 286a-b.** (a) Ectocranial and (b) endocranial views of benign osteofibroma (ossifying fibroma) (oval) calvariae in the parietal bones in a 26-year-old male (MN10.259). In (a) note the scalloped margin (arrows) and undulating and pitted outer table and bone measuring 34 mm at its thickest.

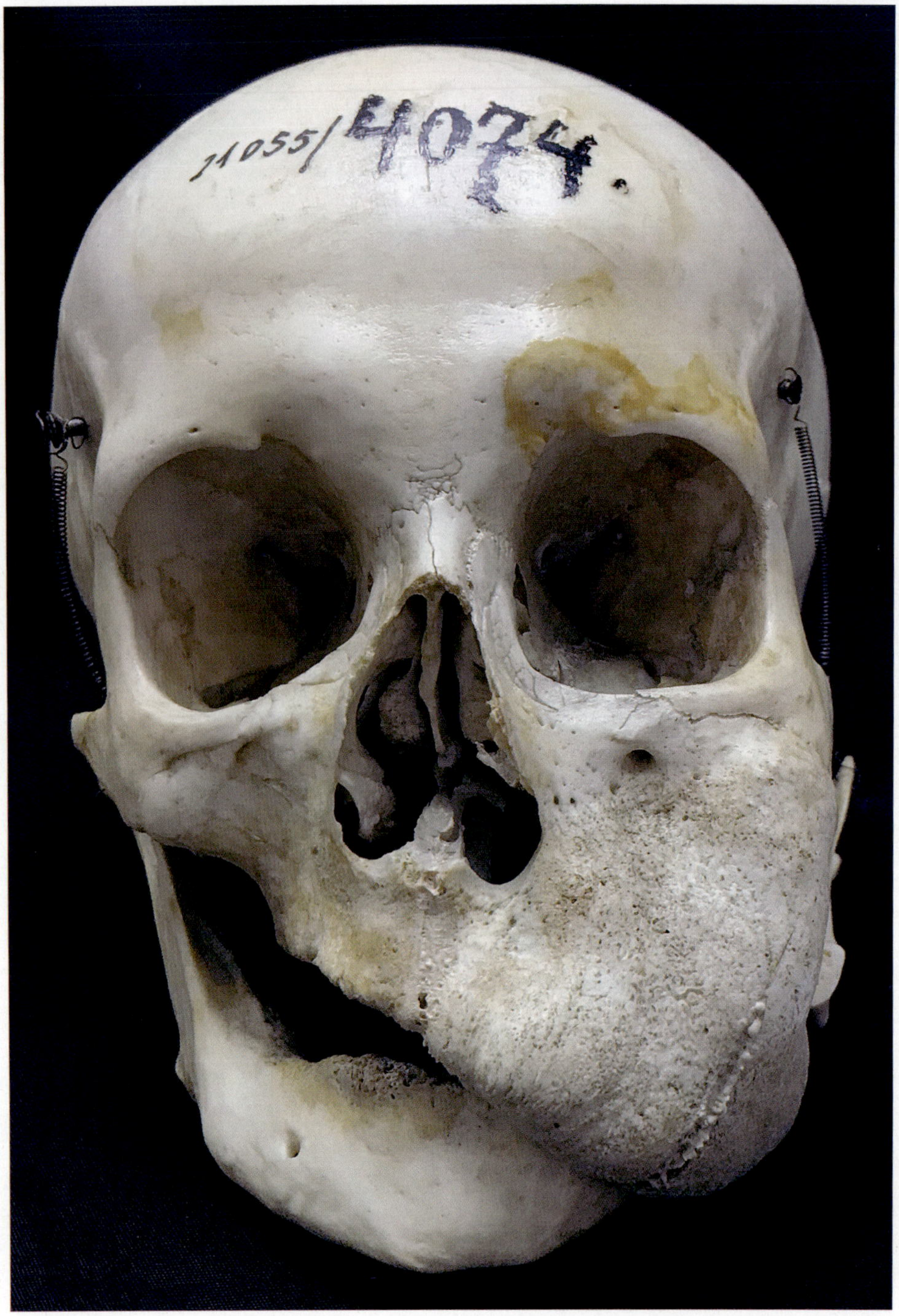


**Figure 287.** Anterior view of a skull with a large, dense hyperostosis (on the museum card) and tumor involving the left maxilla and zygoma in an unknown elderly adult (probable female) who died in 1894 (MN21.055/407). "There are several diseases that could result in a similar condition including Proteus syndrome or possibly fibrous dysplasia. Without histology or knowledge of whether there were other such lesions in the rest of the skeleton, it is hard to say for sure" (K. Thompson, personal communication November 17, 2022.) A section of the maxilla was presumably cut for histological analysis. See Spranger et al. (2002) for more on fibrous dysplasia and Munhoz et al. (2021) for Proteus syndrome.

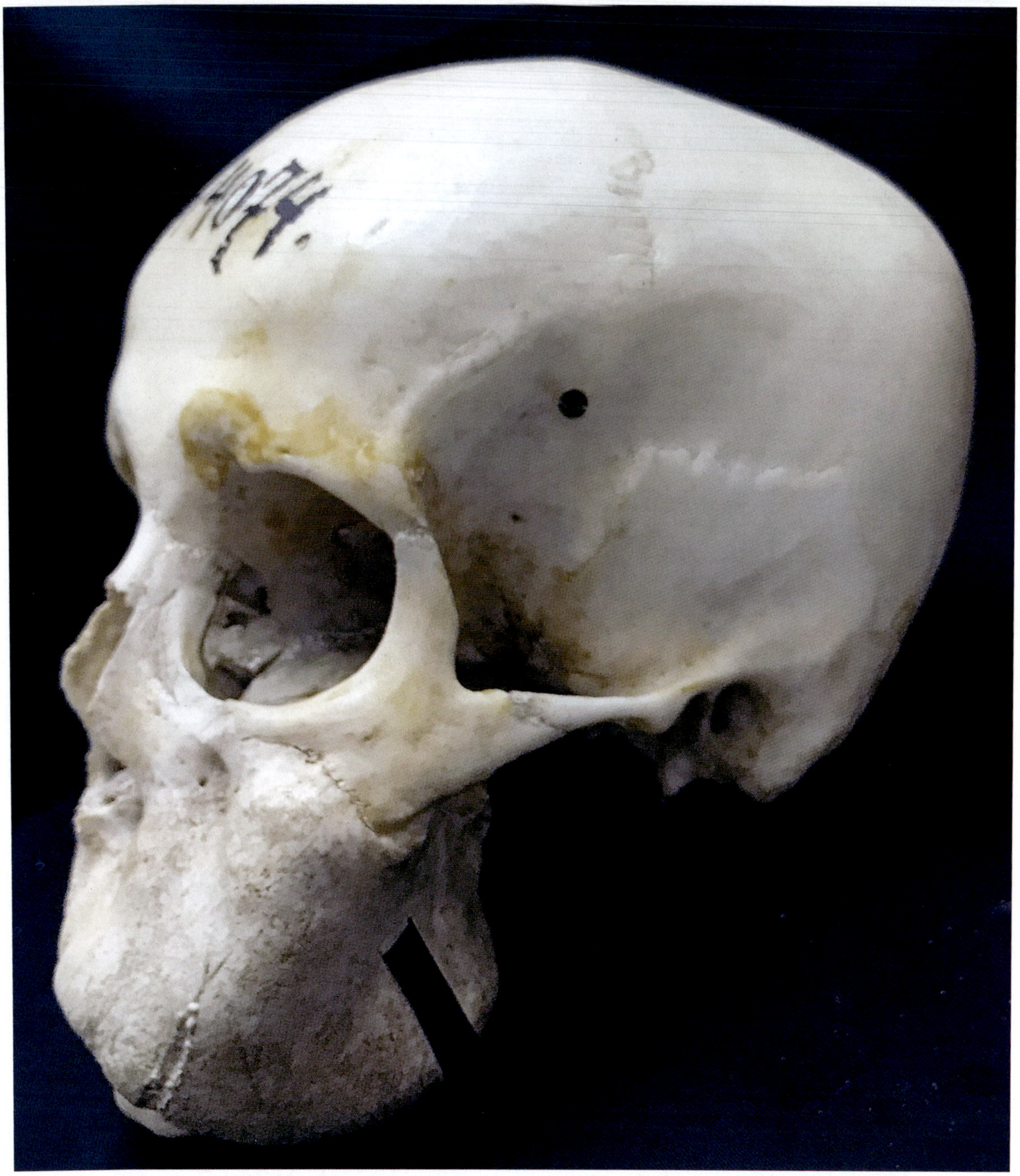


**Figure 288.** Left oblique view of skull with a large, dense hyperostosis (on the museum card) and tumor involving the left maxilla and zygoma in an unknown adult (probable female) who died in 1894 (MN21.055/4074). A section of the maxilla was presumably removed for histologic analysis, but the results were unavailable. See Aufderheide and Rodrigues-Martin (1998); Spranger et al. (2002) for more on fibrous dysplasia and Munhoz et al. (2021) for Proteus syndrome.

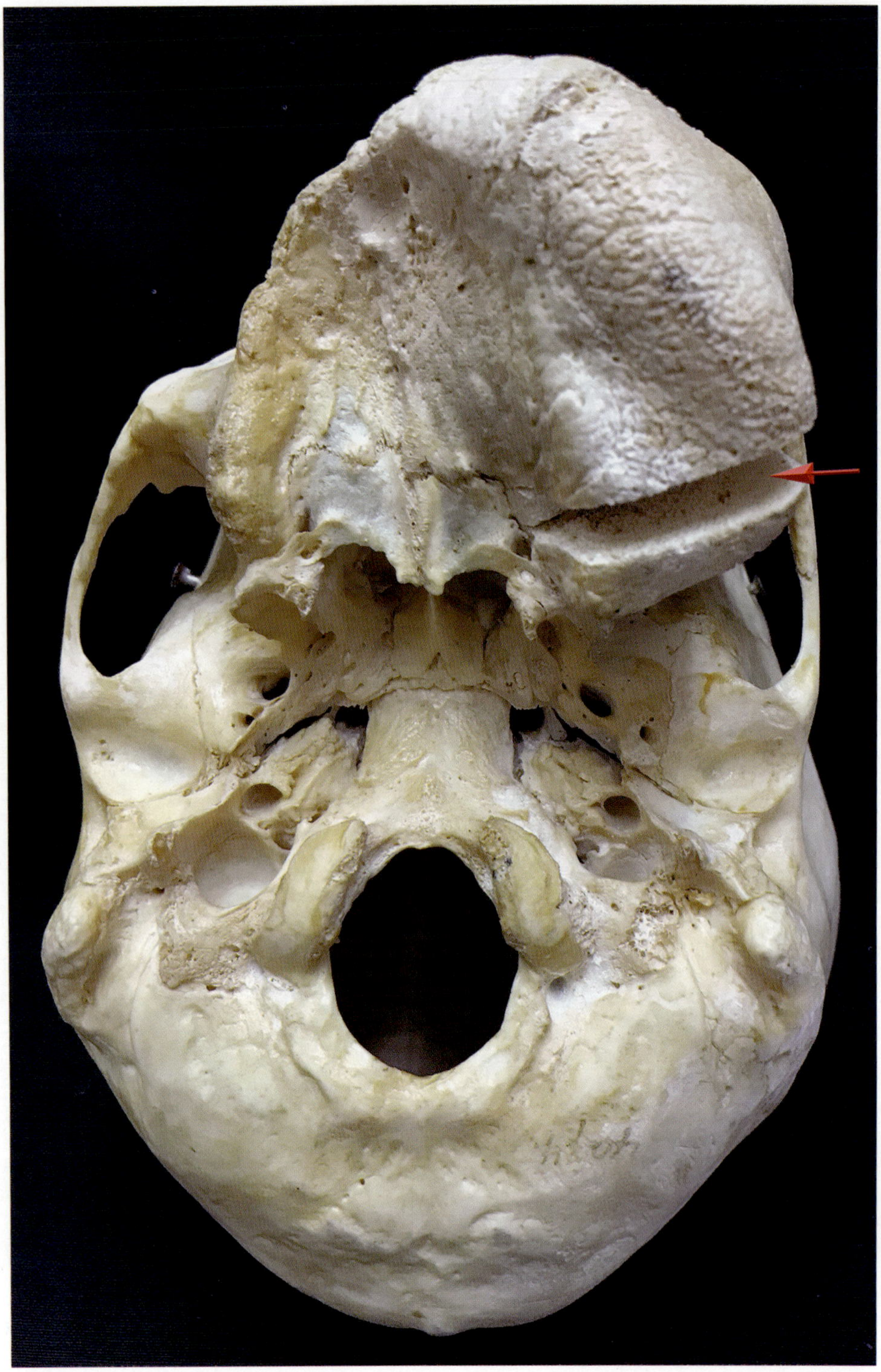

**Figure 289.** Inferior view of skull with large, dense hyperostosis and tumor involving the left maxilla and zygoma in an unknown adult (probable female) who died in 1894 (MN21.055/4074). A portion of the maxilla (arrow) was removed, presumably for histologic analysis. Note that this individual's maxillary right teeth were lost antemortem and remodeled, possibly the result of age or abnormal growth of the left maxilla. See Aufderheide and Rodrigues-Martin (1998); Spranger et al. (2002) for more on fibrous dysplasia and Munhoz et al. (2021) for Proteus syndrome.

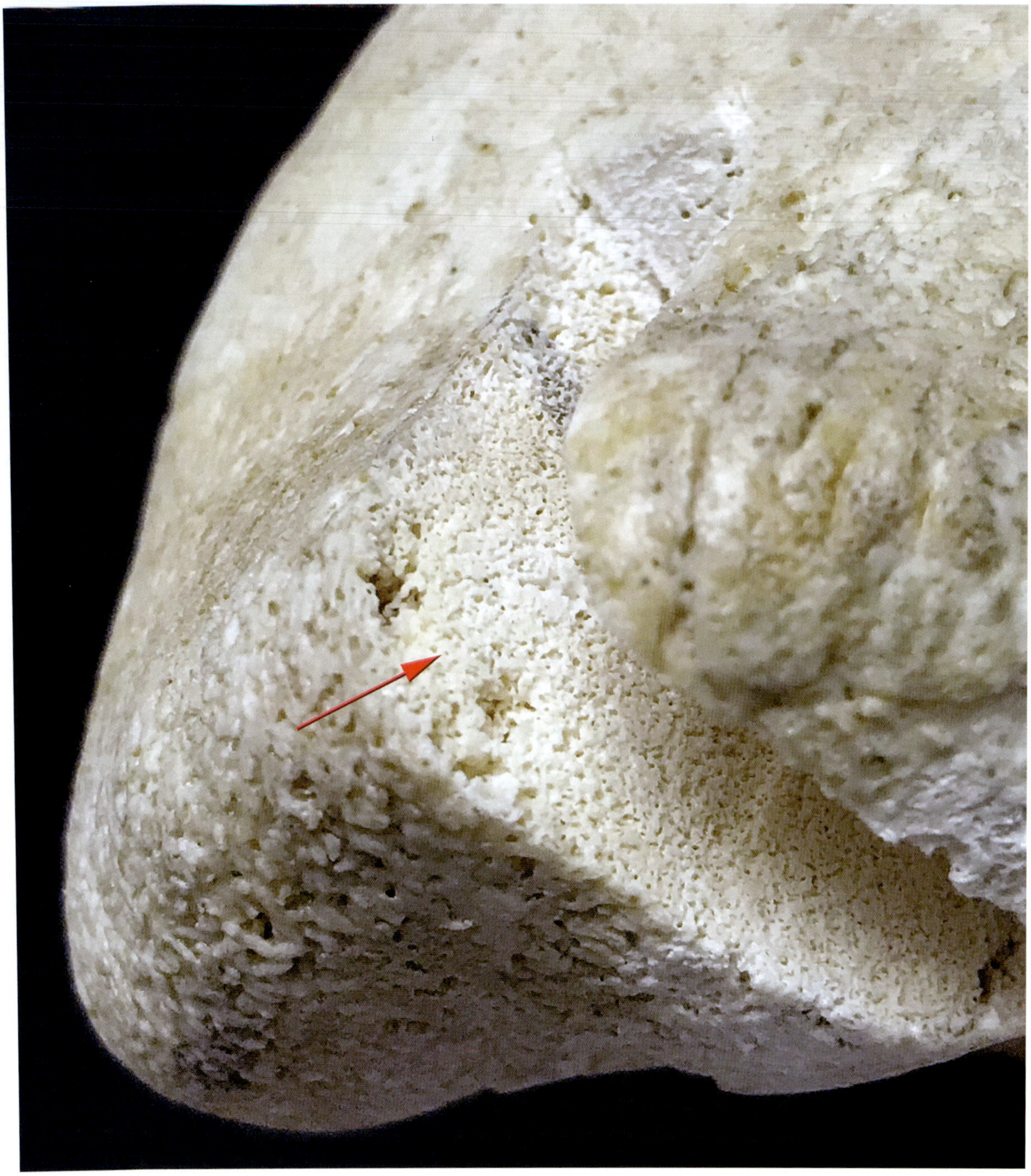

**Figure 290.** Magnified view of area of hyperostotic dense bone and tumor in the facial region that was presumably removed (arrow) for histological examination in an unknown adult (probable female) who died in 1894 (MN21.055/4074). See Aufderheide and Rodrigues-Martin (1998); Spranger et al. (2002) for more on fibrous dysplasia and Munhoz et al. (2021) for Proteus syndrome.

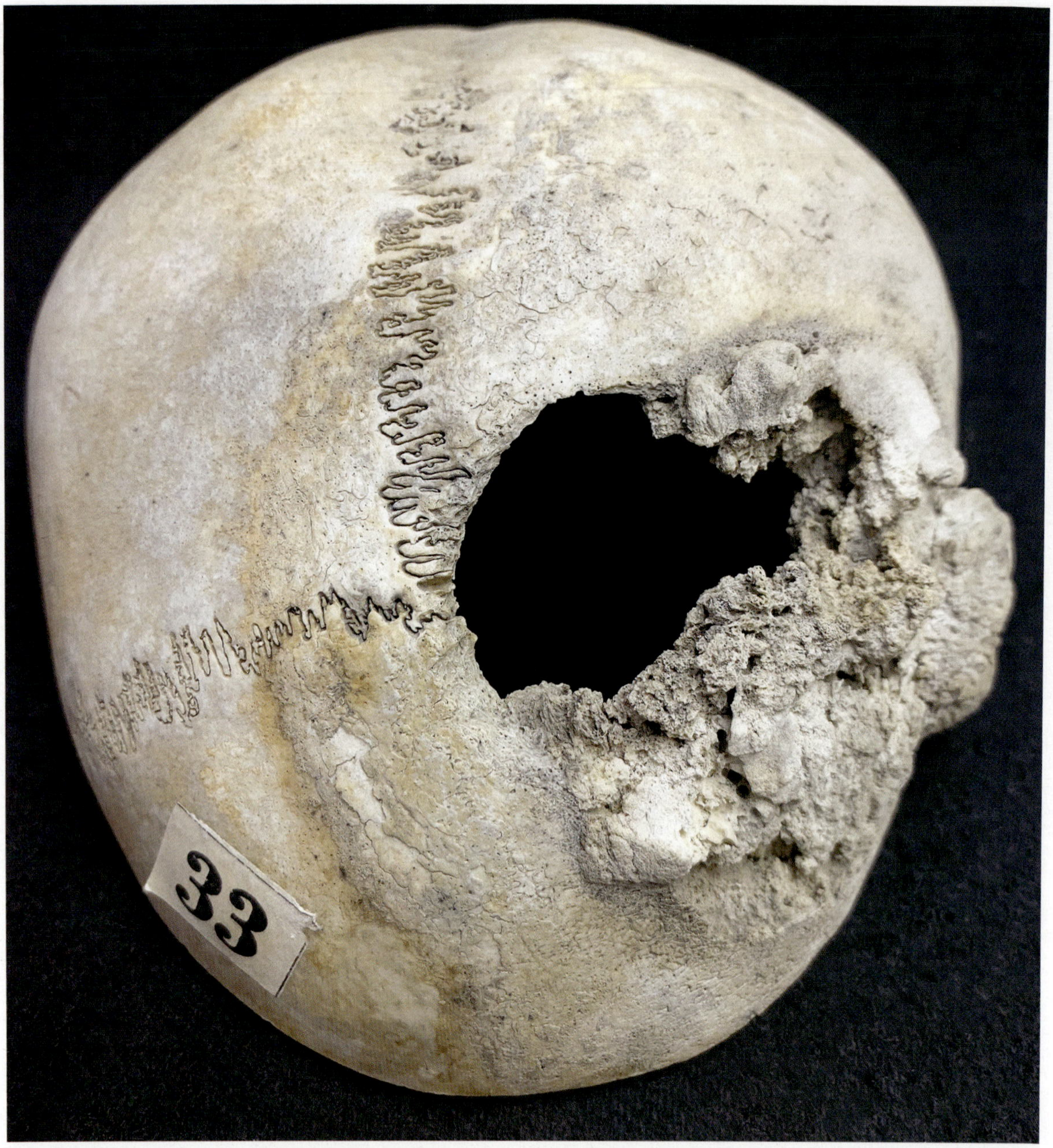


**Figure 291.** Superior view of a "malignant osteoma" (on the museum card) in the left parietal and frontal bones in a 31-year-old male who died in 1868 (MN18.560/33a). Although osteomas are benign tumors (Parvizi and Kim 2010) that may rarely become malignant, identification of these lesions as malignant may have been intended only to describe the osteoblastic growth (osteoma) partially surrounding a large area of (osteoclastic) missing bone.

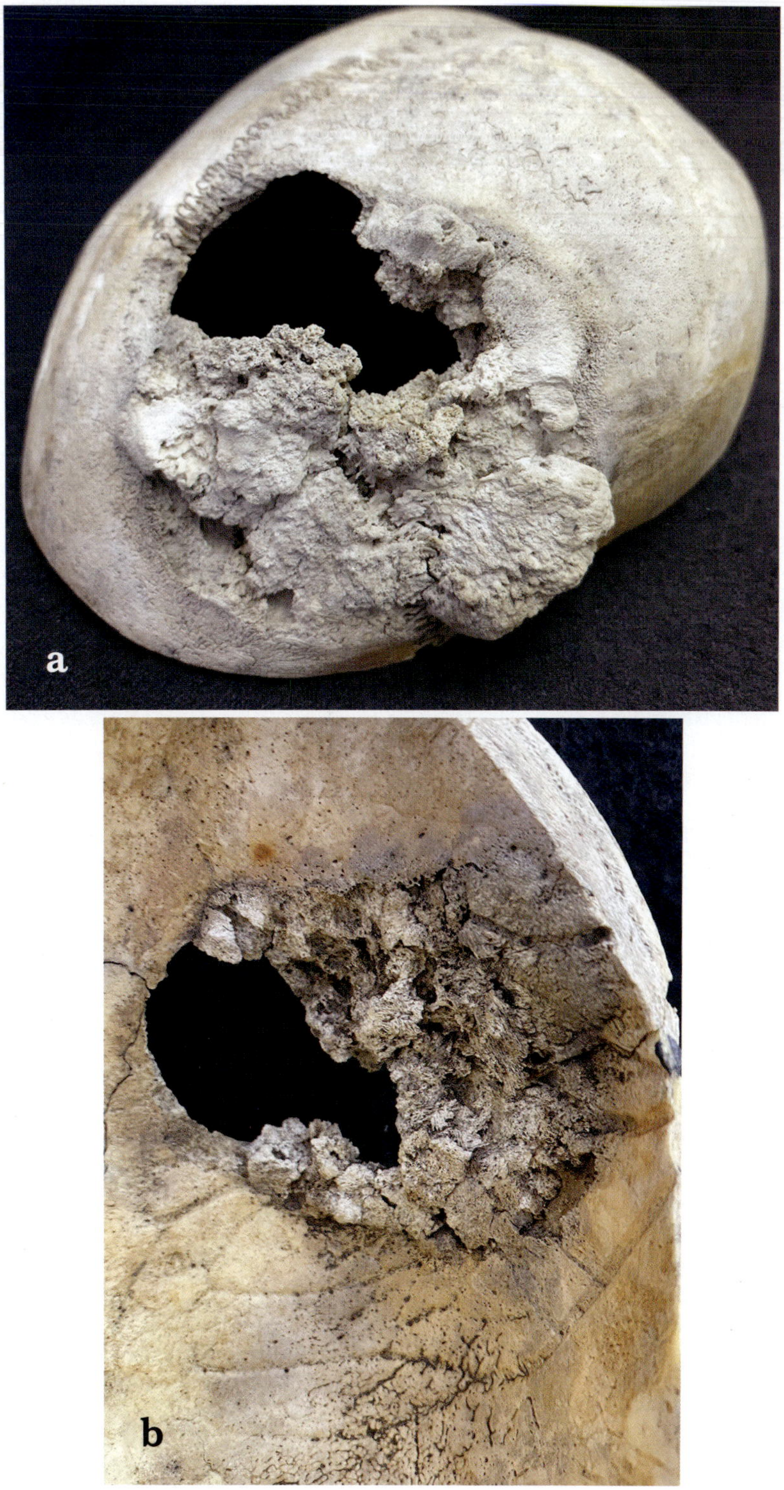


**Figure 292a-b.** (a) Superior oblique and (b) endocranial view of a "malignant osteoma" (on the museum card) in the left parietal and frontal bones in a 31-year-old male who died in 1868 (MN18.560/33a). Note the large bone growths opposite one another along the inner and outer tables. Identification of this tumor as malignant may have been erroneous or used only to describe the osteoblastic and osteoclastic lesions.

**Figure 293.** Left lateral view of a "malignant osteoma" (on the museum card) in the left parietal and frontal bones of a 31-year-old male (MN18.560/33a). Note the clearly defined margins of the tumor and absence of osteoblastic or osteoclastic activity surrounding it.

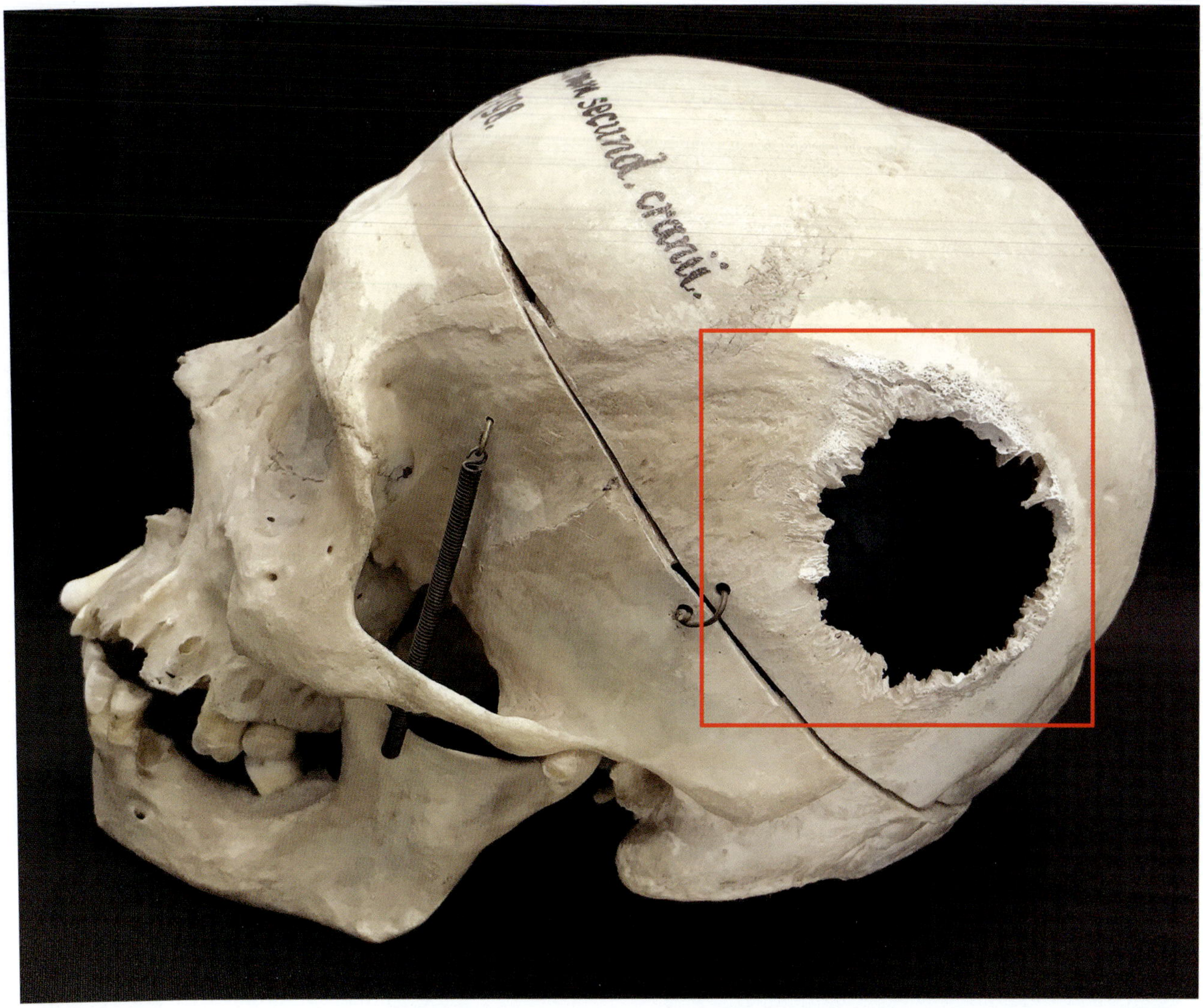


**Figure 294.** Left superior-inferior view of a skull with metastatic carcinoma (secondary/that has spread to the cranium) exhibiting numerous mixed lesions (tumors) in an adult probable male who died in 1828 (MN5798). Note the destruction and loss of a large circular area in the left parietal bone and irregular margin surrounded by new bone (box). There are additional lesions in the right parietal and temporal bones (Figures 295-297). See Aufderheide and Rodrigues-Martin (1998); Rajakulasingam et al. (2020); Reeder (1993); Steinbock (1976) for more on carcinoma and Smith and Mohsin (2013) for its pathophysiology.

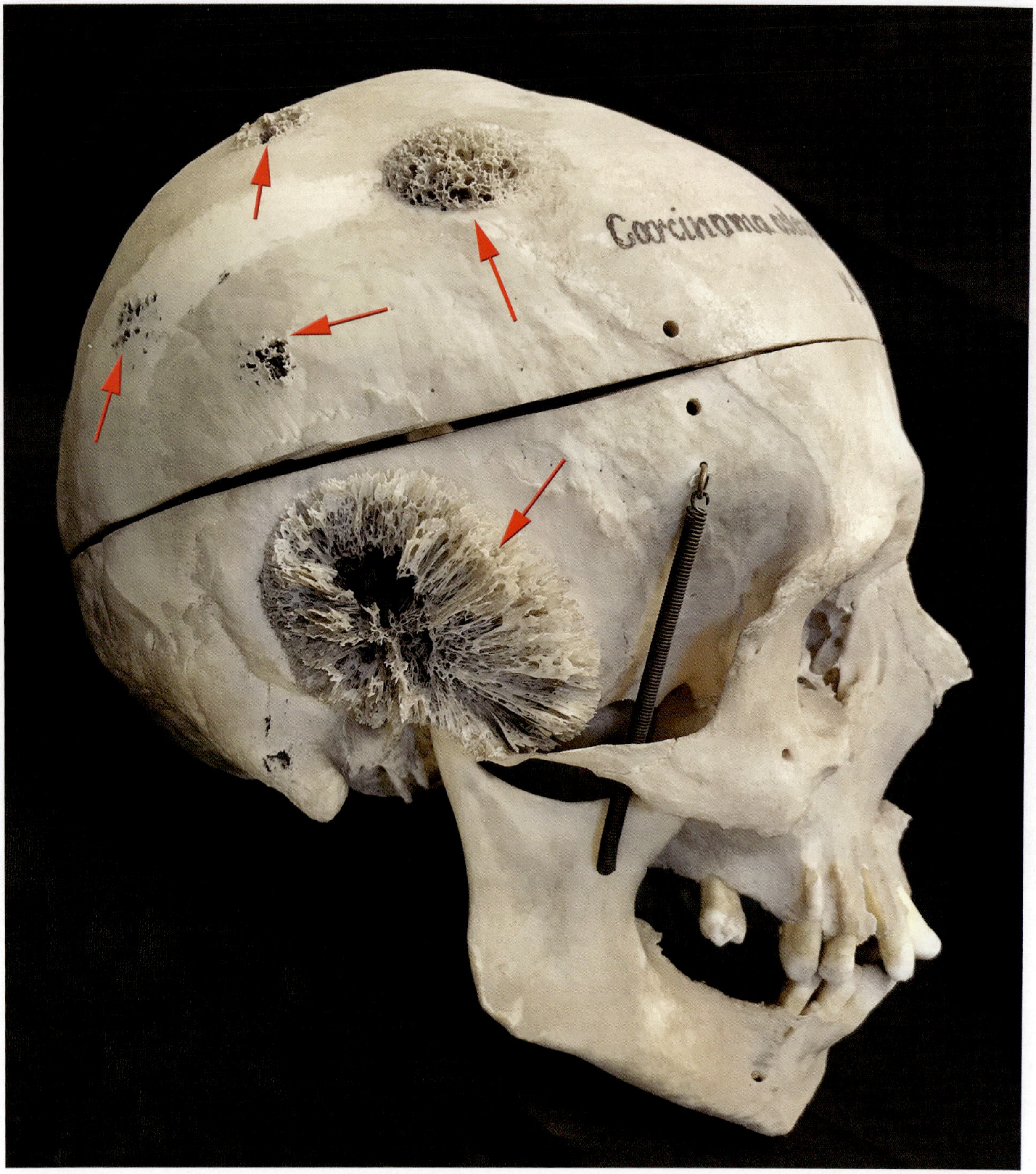


**Figure 295.** Right oblique view of a skull with metastatic carcinoma (secondary/that has spread to the cranium) with numerous lesions (tumors) (arrows), including a large "sunburst" lesion (see Reeder 1993) in the right temporal bone of an adult probable male who died in 1828 (MN5798). The sunburst lesion indicates an aggressive and rapidly growing lesion resulting in the formation of Sharpey's fibers that outpace the response of the periosteum (Lehrer et al. 1970). See Aufderheide and Rodrigues-Martin (1998); Sapper (1957); Steinbock (1976); Thomas and Kirsten (1985) for more on metastatic carcinoma and its distribution in the human skeleton and Smith and Mohsin (2013) for its pathophysiology.

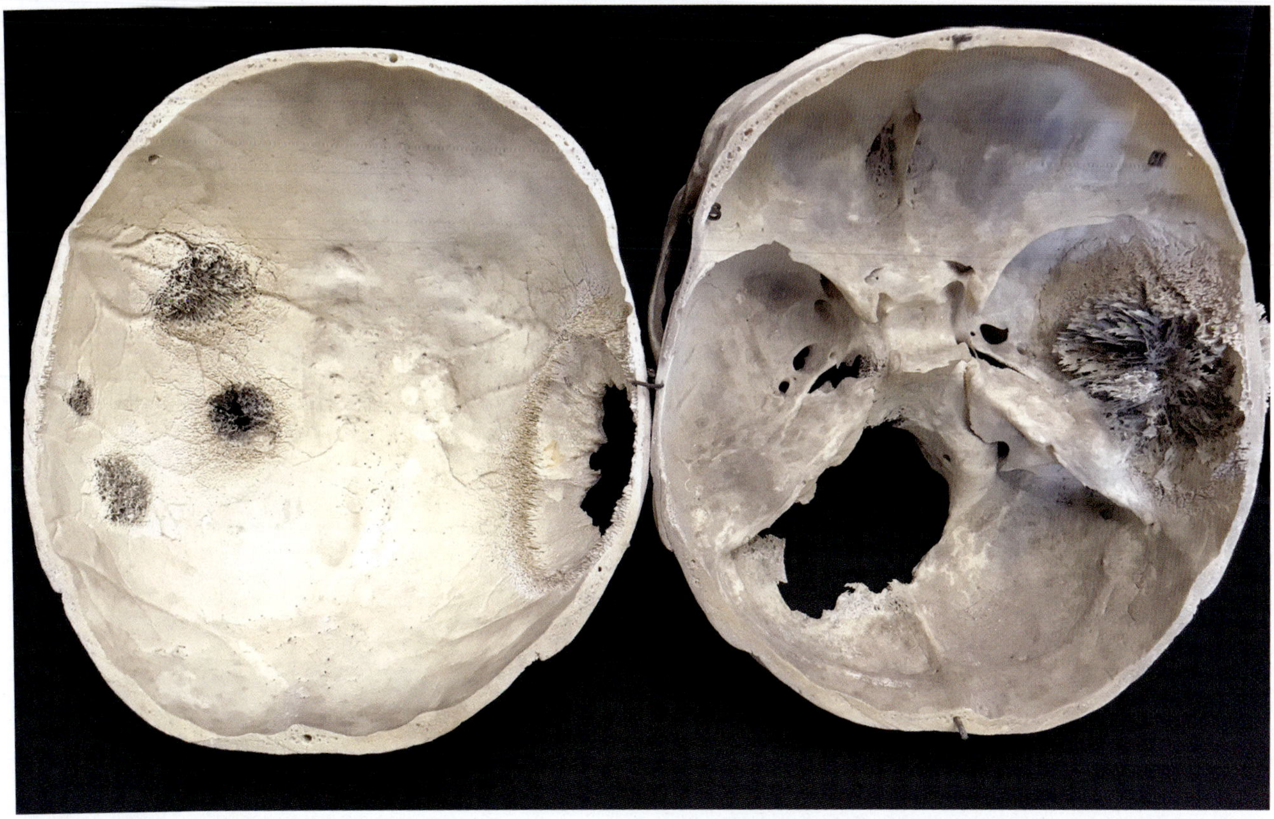

**Figure 296.** Endocranial view of a skull with metastatic carcinoma (secondary/that has spread to the cranium) exhibiting numerous mixed osteoblastic and osteoclastic lesions (tumors), including a large "sunburst" lesion in the right temporal bone in an adult probable male who died in 1828 (MN5798). See Rajakulasingam et al. (2020); Sapper (1957); Thomas and Kirsten (1985); Steinbock (1976) for more on metastatic carcinoma and its distribution in the human skeleton and Smith and Mohsin (2013) for its pathophysiology.

**Figure 297.** Magnified right lateral view of a skull with metastatic carcinoma exhibiting a large "sunburst" lesion (tumor) (Smith and Mohsin 2013) in the right temporal bone in an adult probable male who died in 1828 (MN5798). This "sunburst" pattern is a periosteal reaction in response to a rapidly growing lesion that lays down Sharpey's fibers faster than the periosteum can respond and remodel, resulting in Sharpey's fibers oriented perpendicular to the underlying bone (Lehrer et al. 1970). See Thomas and Kirsten (1985) for more on metastatic carcinoma and its distribution in the human skeleton and Smith and Mohsin (2013) for its pathophysiology.

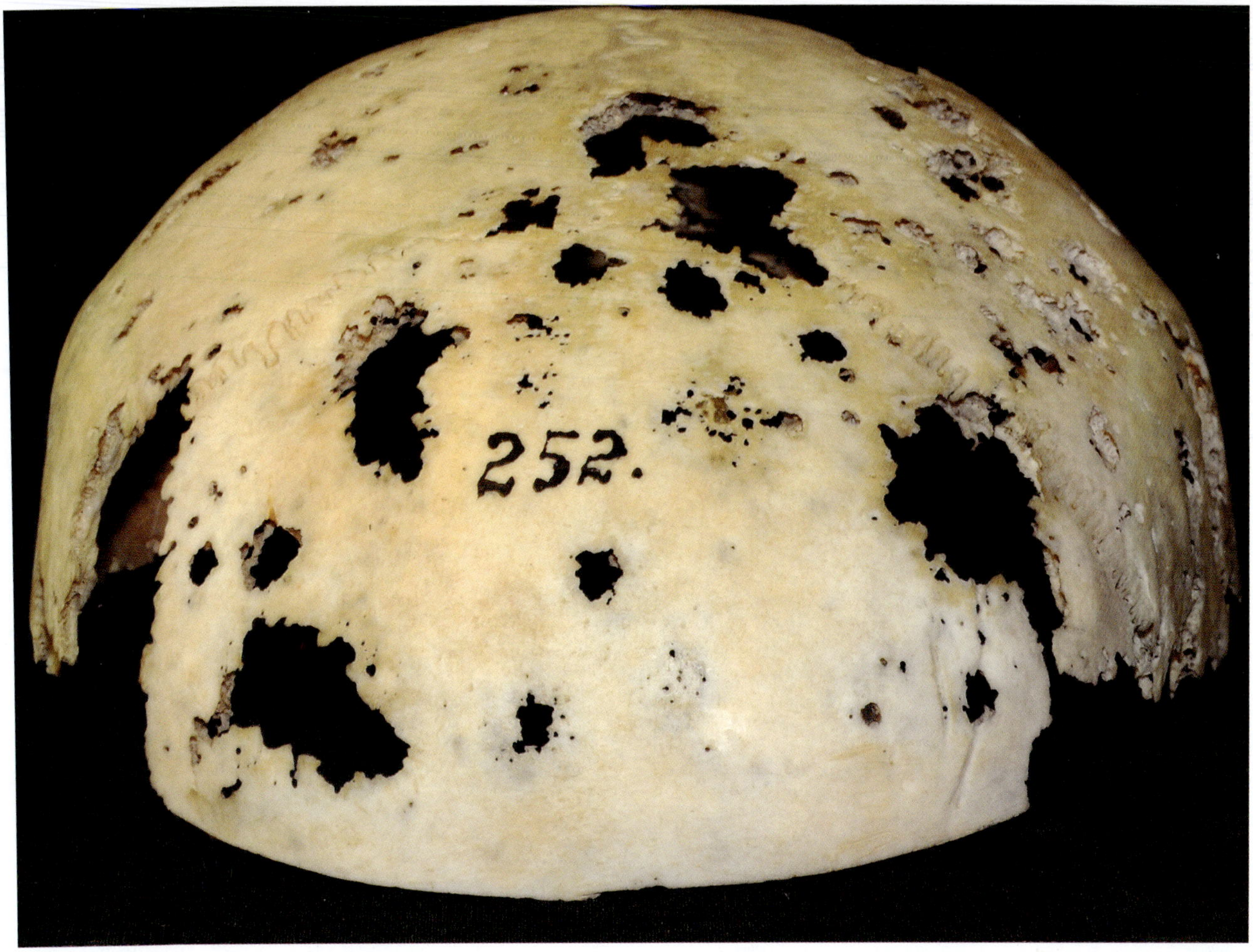


**Figure 298.** Anterior view of a skull with multiple myeloma, a relatively uncommon bone marrow cancer that can affect any bone, but typically the spine, skull, pelvis and ribs, in an adult male who died in 1830 (MN18.560/252). The lytic lesions are scattered throughout the skull and vary in size and shape and exhibit minimal evidence of new bone. Myeloma bone disease (MBD), of which 80% of individuals with multiple myeloma exhibit painful destructive bone lesions, is a complication of multiple myeloma (Hameed et al. 2014). See Aufderheide and Rodrigues-Martin (1998); Ortner (2003); Ortner and Putschar (1981); Rajakulasingam et al. (2020); Reeder (1993); Sapper (1957); Steinbock (1976); Thomas and Kirsten (1985); Watson-Jones (1944) for more on myeloma.

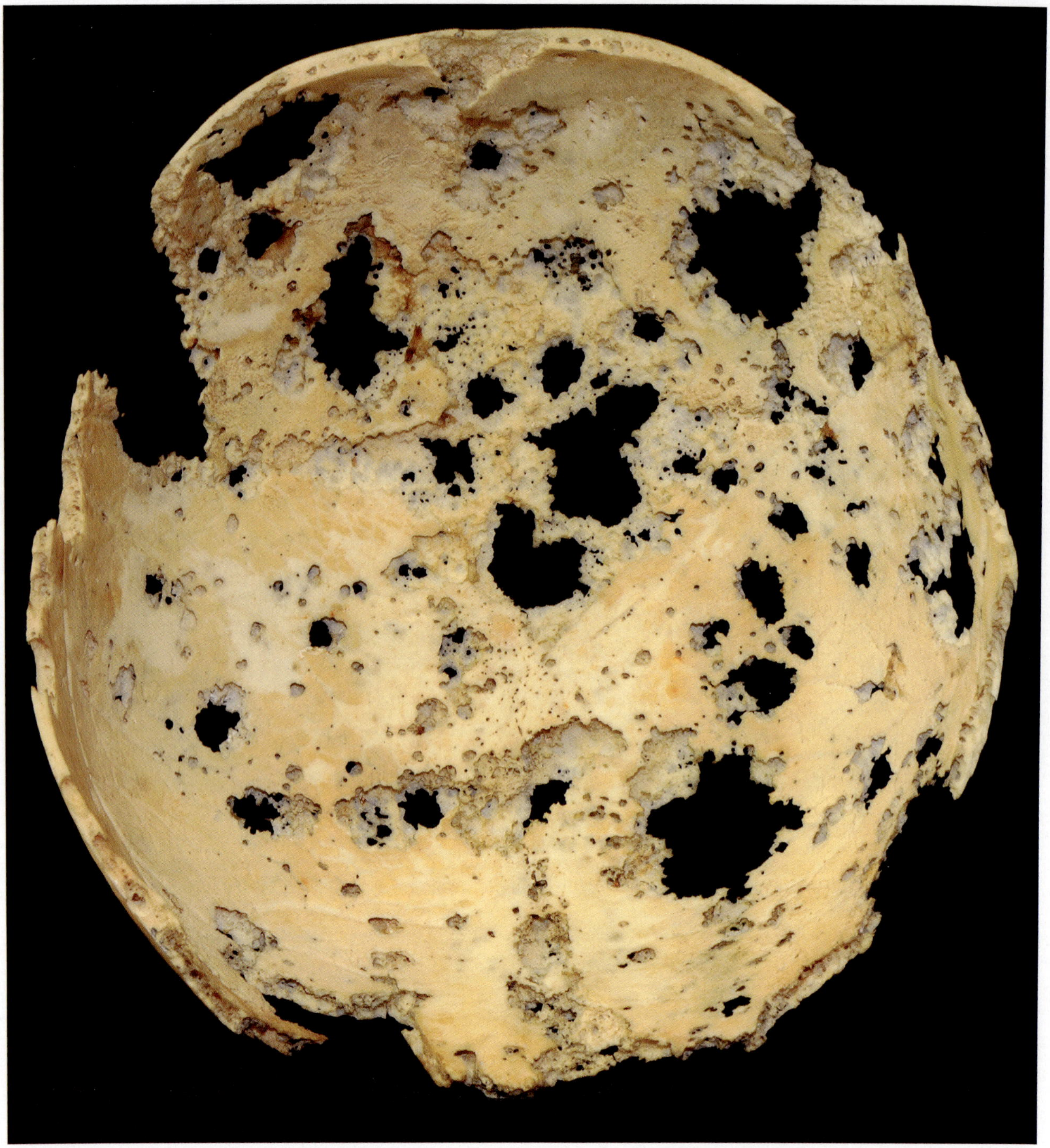

**Figure 299.** Endocranial view of skull with numerous lytic lesions due to multiple myeloma, a relatively rare bone marrow cancer, in an adult male who died in 1830 (MN18.560/252). Myeloma bone disease (MBD) is a painful complication of multiple myeloma (Hameed et al. 2014). See Aufderheide and Rodrigues-Martin (1998); Ortner (2003); Ortner and Putschar (1981); Rajakulasingam et al. (2020); Reeder (1993); Sapper (1957); Steinbock (1976); Thomas and Kirsten (1985); Watson-Jones (1944) for more on myeloma and Hameed et al. (2014) for its pathophysiology.

**Figure 300.** Magnified endocranial view of multiple myeloma, a relatively rare bone marrow cancer, in an adult male who died in 1830 (MN18.560/252). See Aufderheide and Rodrigues-Martin (1998); Hameed et al. (2014); Ortner (2003); Ortner and Putschar (1981); Sapper (1957); Steinbock (1976); Thomas and Kirsten (1985); Watson-Jones (1944) for more on myeloma.

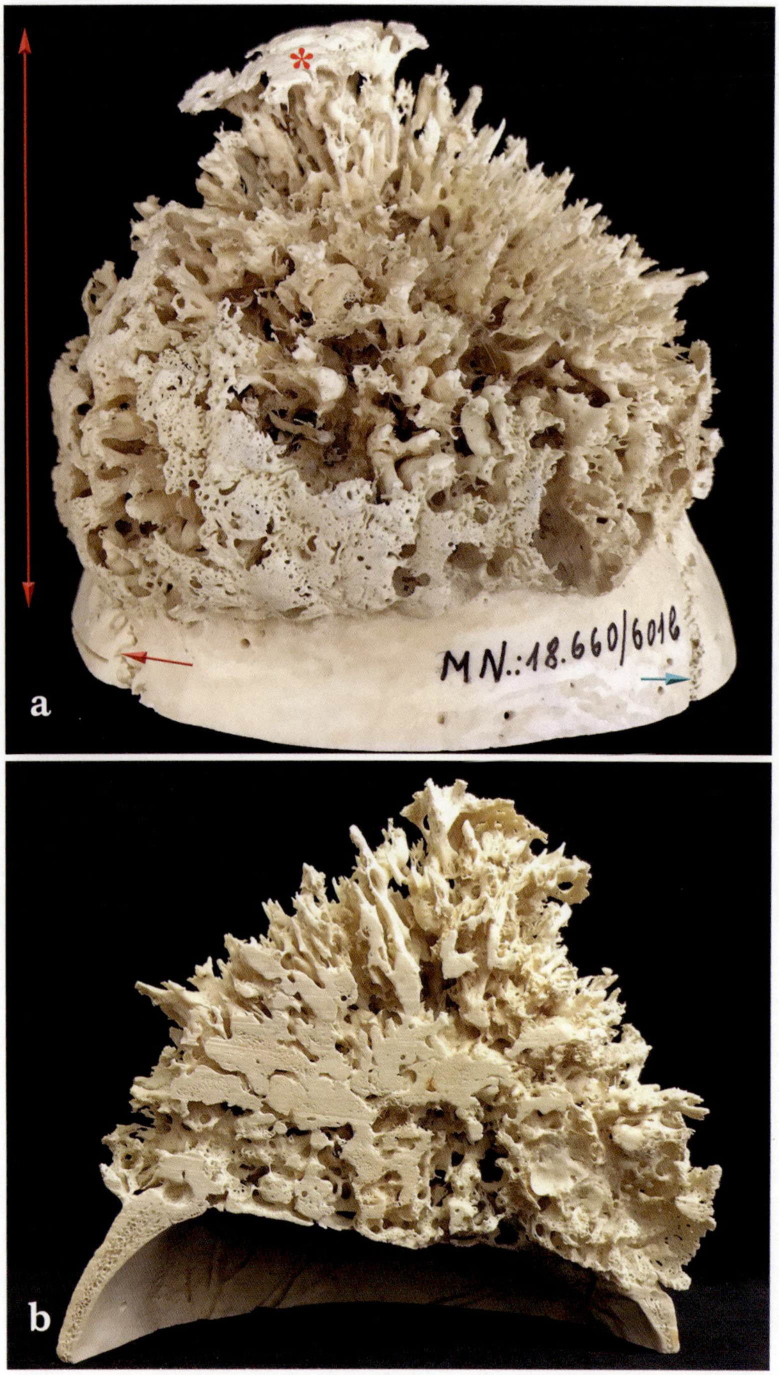


**Figure 301a-b.** (a) Right lateral view of osteosarcoma with a dense, sclerotic sunburst pattern in the right parietal, frontal and occipital bones of a female who died in 1923 (MN18.660/601b). (a) The tumor measures 130 mm (double arrow) inferior-superior and 145 mm anterior-posterior; note the lambdoidal suture (red arrow) and coronal suture (blue arrow) and the sclerotic "cap" of bone (*) at its apex. (b) Sawed internal view of the tumor showing sclerotic bone that has pushed endocranially altering the normal contour of the inner table. See Ortner (2003); Ortner and Putschar (1981); Reeder (1993); Smith-Guzman et al. (2018); Waldron (2009) for more on osteosarcoma in clinical and archaeologically derived bones.

**Figure 302.** Magnified right lateral view of osteosarcoma showing the detail of the coral-like sunburst pattern in the right parietal, frontal and occipital bones of a female who died in 1923 (MN18.660/601b). Note that the morphology of the tumor varies from tubes to curved sheets (*) and tiny, delicate spicules. See Aufderheide and Rodrigues-Martin (1998); Ortner (2003); Ortner and Putschar (1981); Reeder (1993); Smith-Guzman et al. (2018); Waldron (2009) for more on osteosarcoma in clinical and archaeologically derived bones.

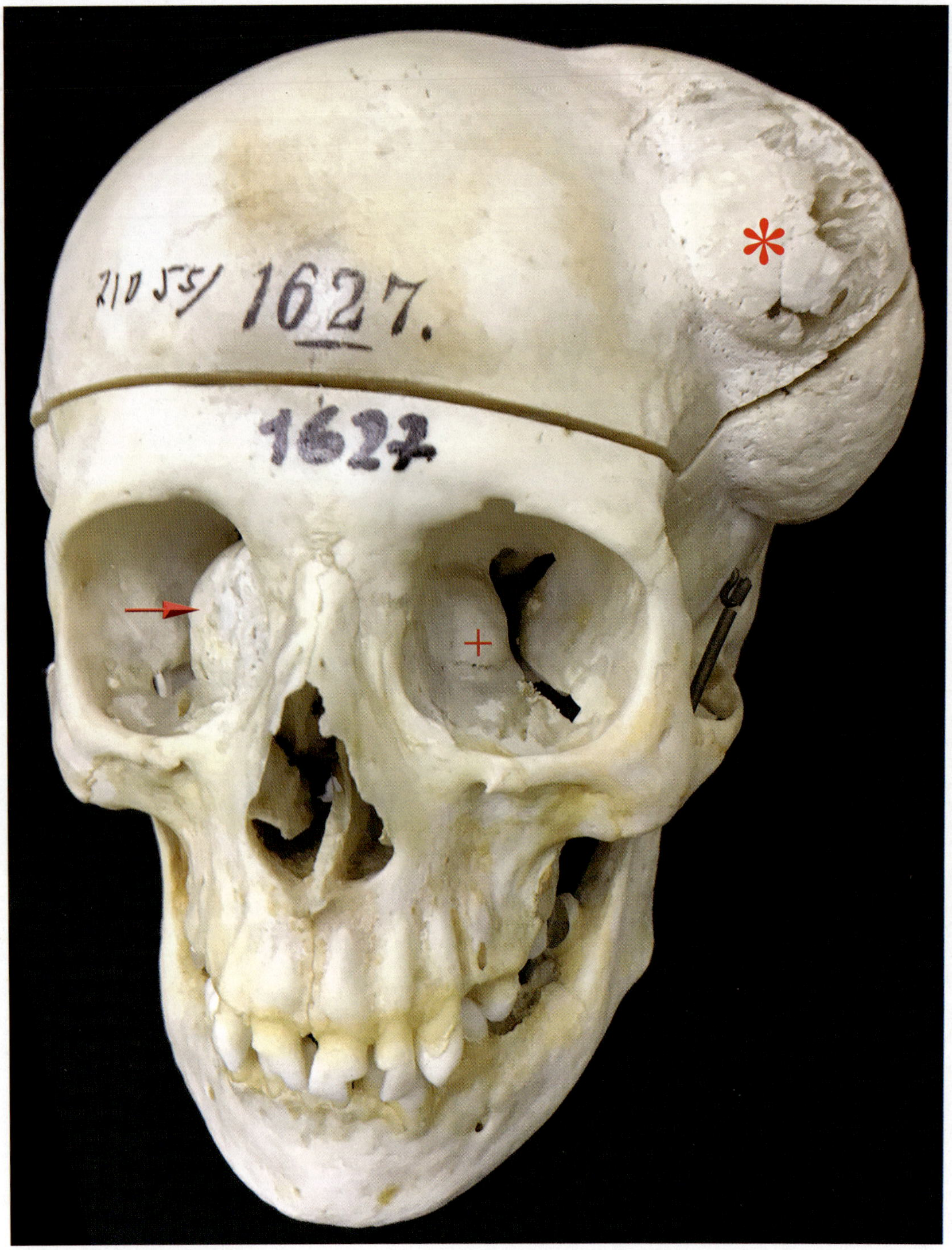


**Figure 303.** Left superior-inferior oblique view of the right orbit (red arrow) and carcinoma gelatinous, also known as mucinous or colloid, in the left parietal and temporal bones (*) in a 45-year-old female with metastatic breast cancer who died in 1870 (MN21.055/1627). The left orbit also exhibits a small dome of expanded bone (+). Mucinous carcinoma refers to a form of malignant cancer of the breast (Ha et al. 2013) and accounts for 2-5% of all breast cancers (Varadharajan et al. 2015).

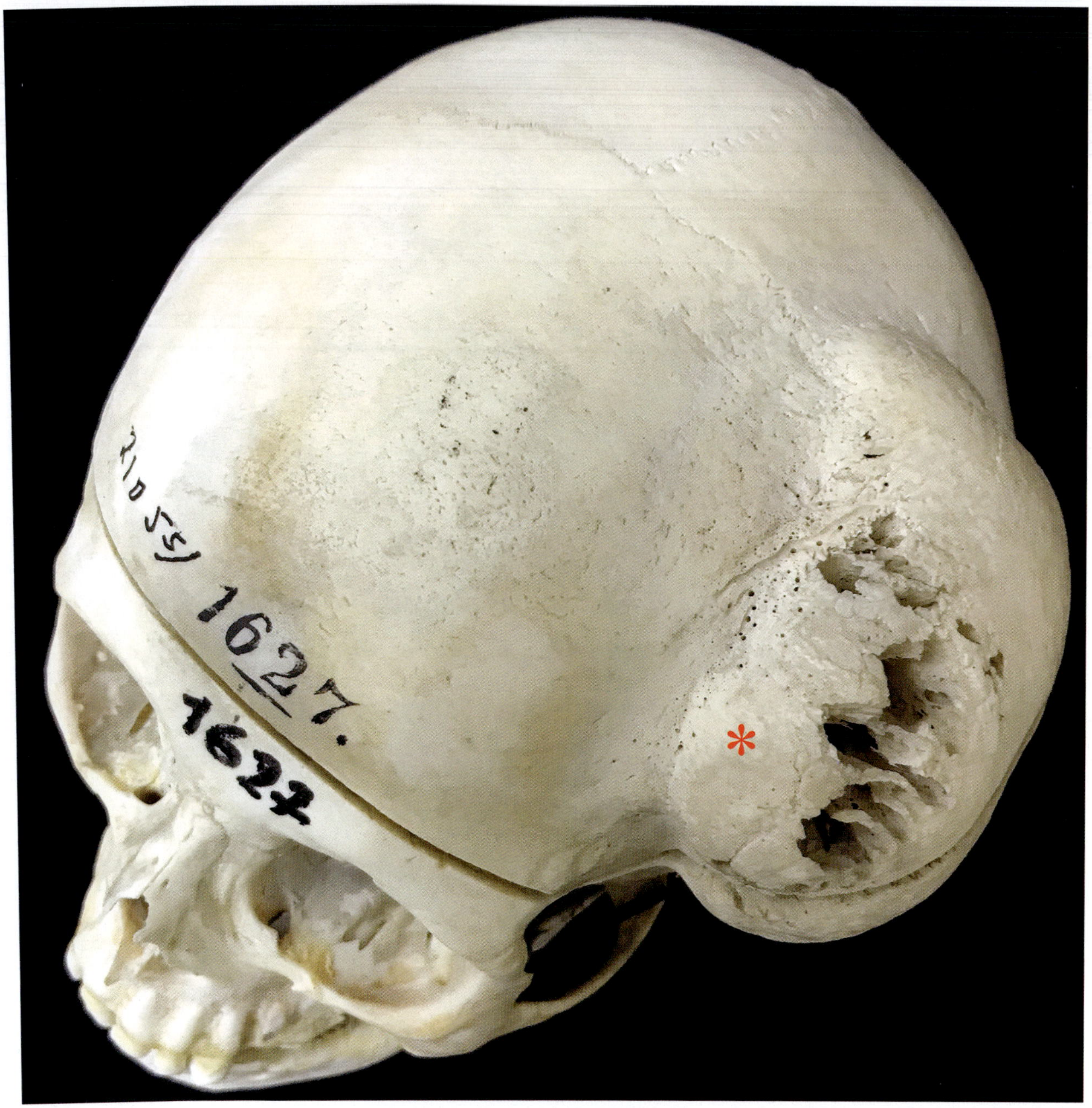


**Figure 304.** Large malignant (mucinous) carcinoma (*) of the left parietal and temporal bones in a 45-year-old female with breast cancer who died in 1870 (MN21.055/1627). Mucinous carcinoma refers to a rare form of malignant cancer of the breast (Ha et al. 2013) that accounts for 2-5% of all breast cancers (Varadharajan et al. 2015).

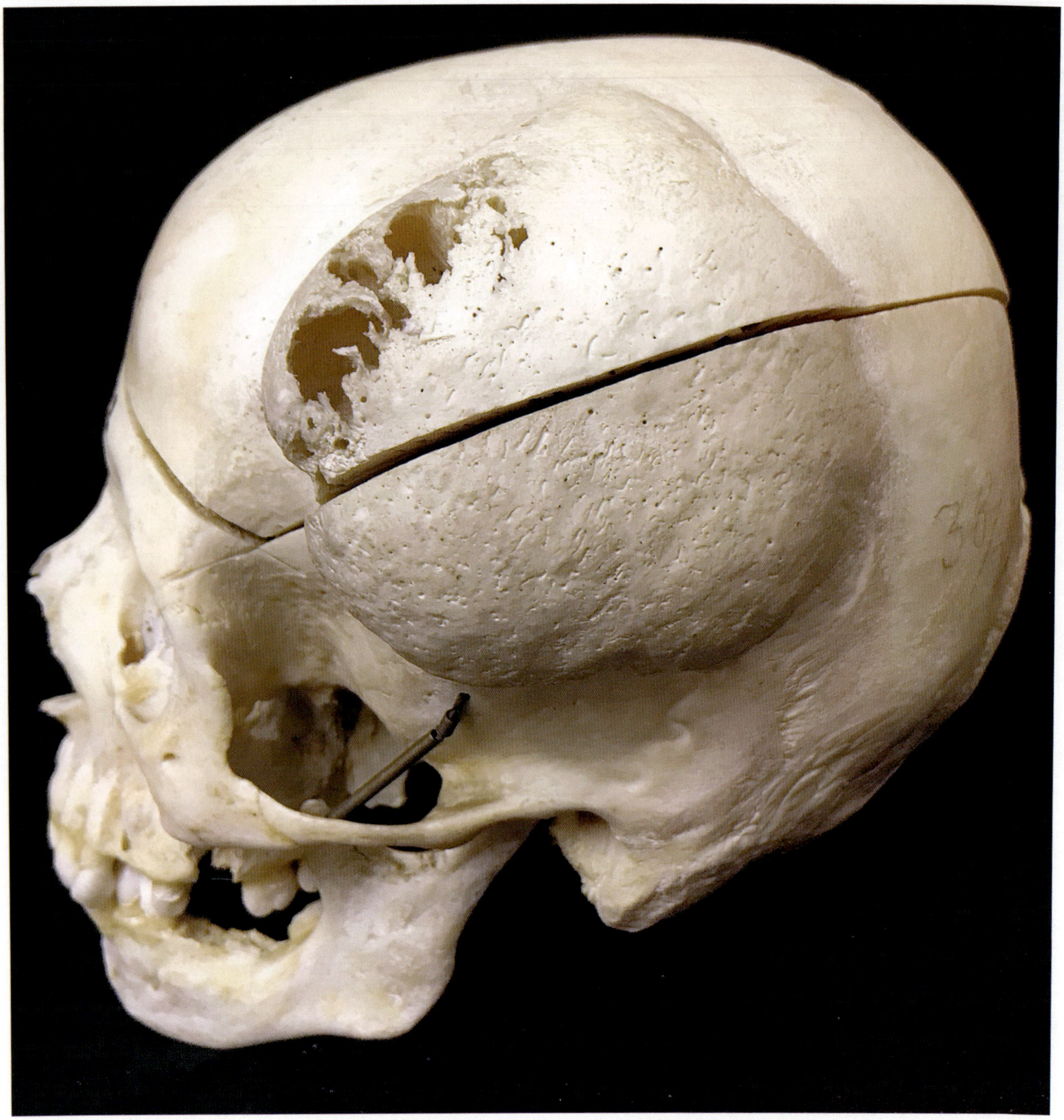

**Figure 305.** Left superior view of a large malignant mucinous tumor in the left parietal and temporal bones of a 45-year-old female with breast cancer who died in 1870 (MN21.055/1627). Mucinous carcinoma refers to a rare type of malignant cancer of the breast (Ha et al. 2013) that accounts for 2-5% of all breast cancers (Varadharajan et al. 2015).

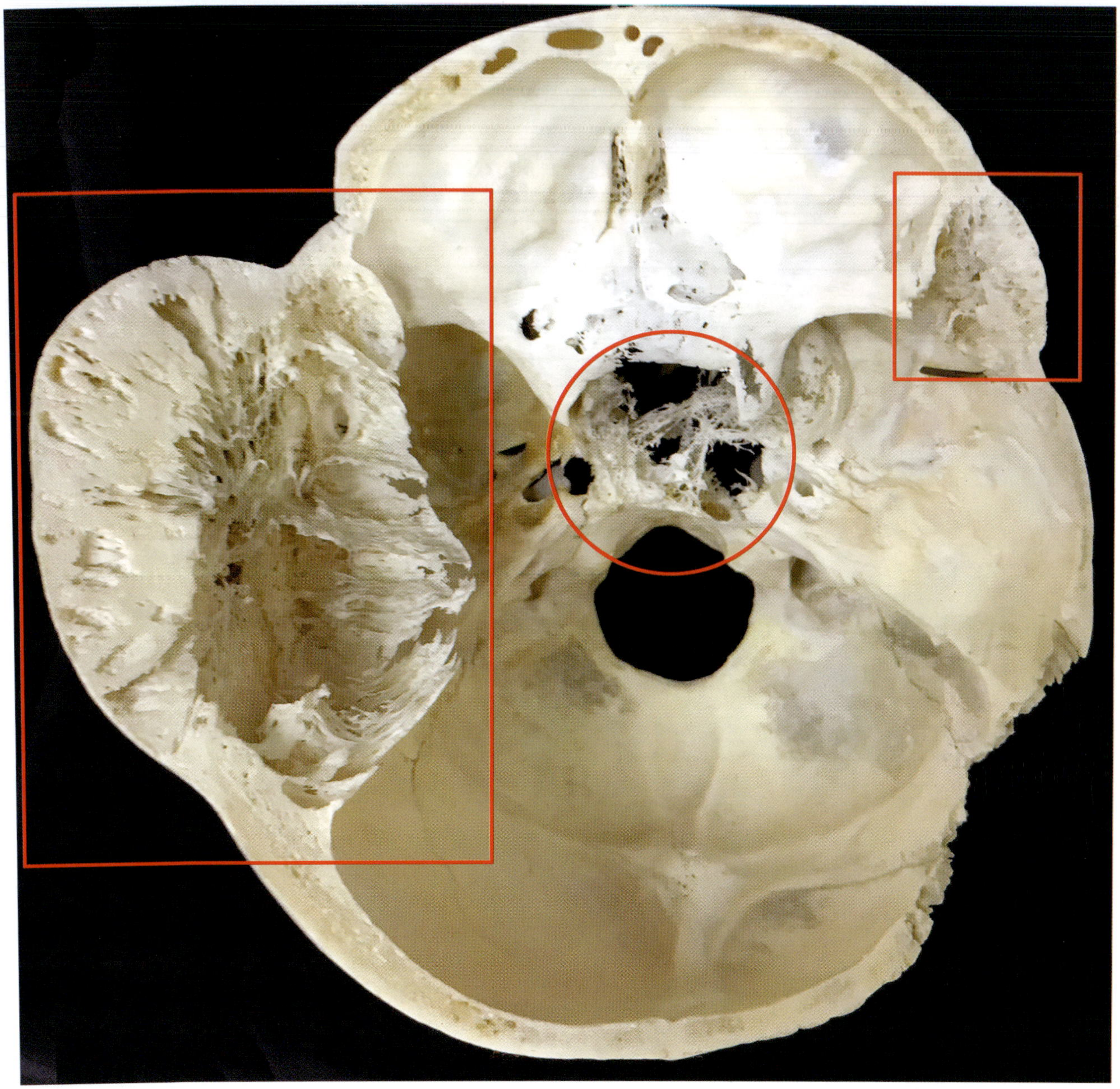

**Figure 306.** Endocranial view of the cranium with malignant cancerous tumors of the right orbit (small square), cranial base (circle) and left parietal and temporal bones (red rectangle) in a 45-year-old female with breast cancer who died in 1870 (MN21.055/1627).

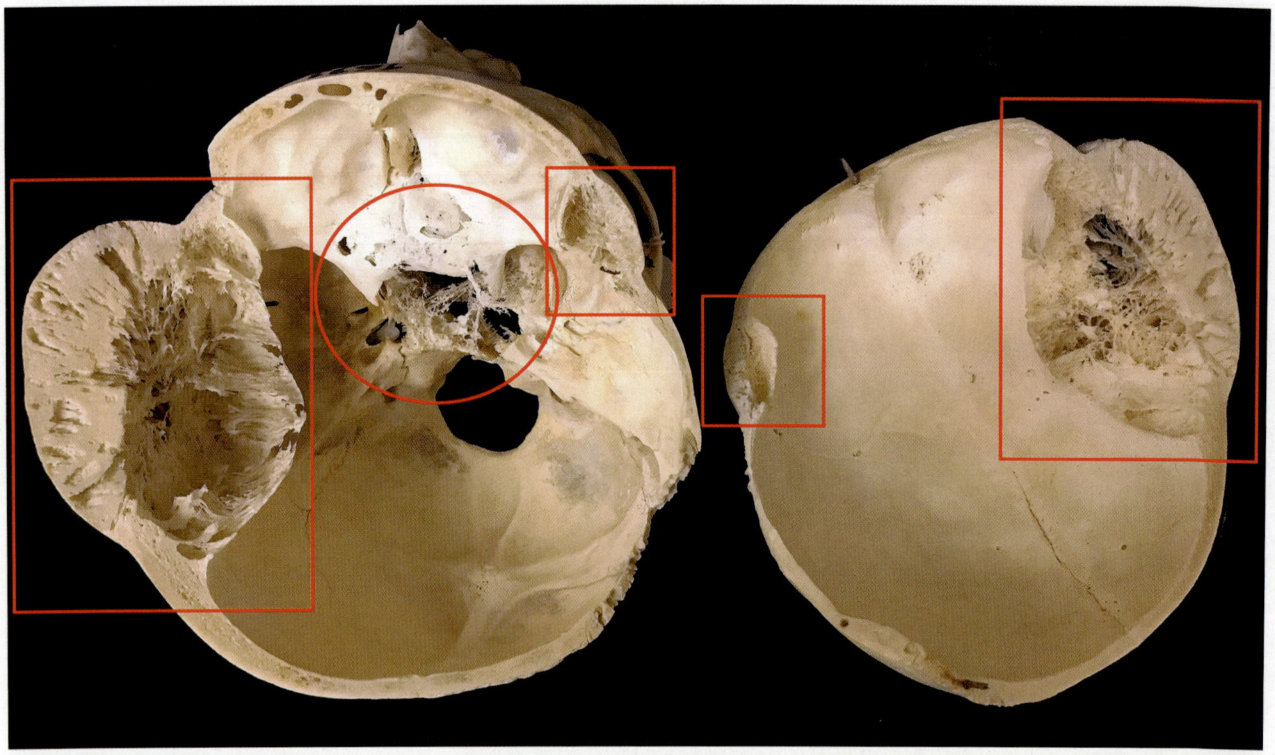

**Figure 307.** Endocranial view of mucinous cancerous tumors in the left parietal and temporal bones (rectangle), cranial base (circle), and right orbit (square) in a 45-year-old female with breast cancer who died in 1870 (MN21.055/1627). Ectocranially, the tumors resulted in the formation of dense/sclerotic bone towards the outer table and migration of the inner table. The lesion in the cranial base resulted in destruction of sella turcica and adjacent bones.

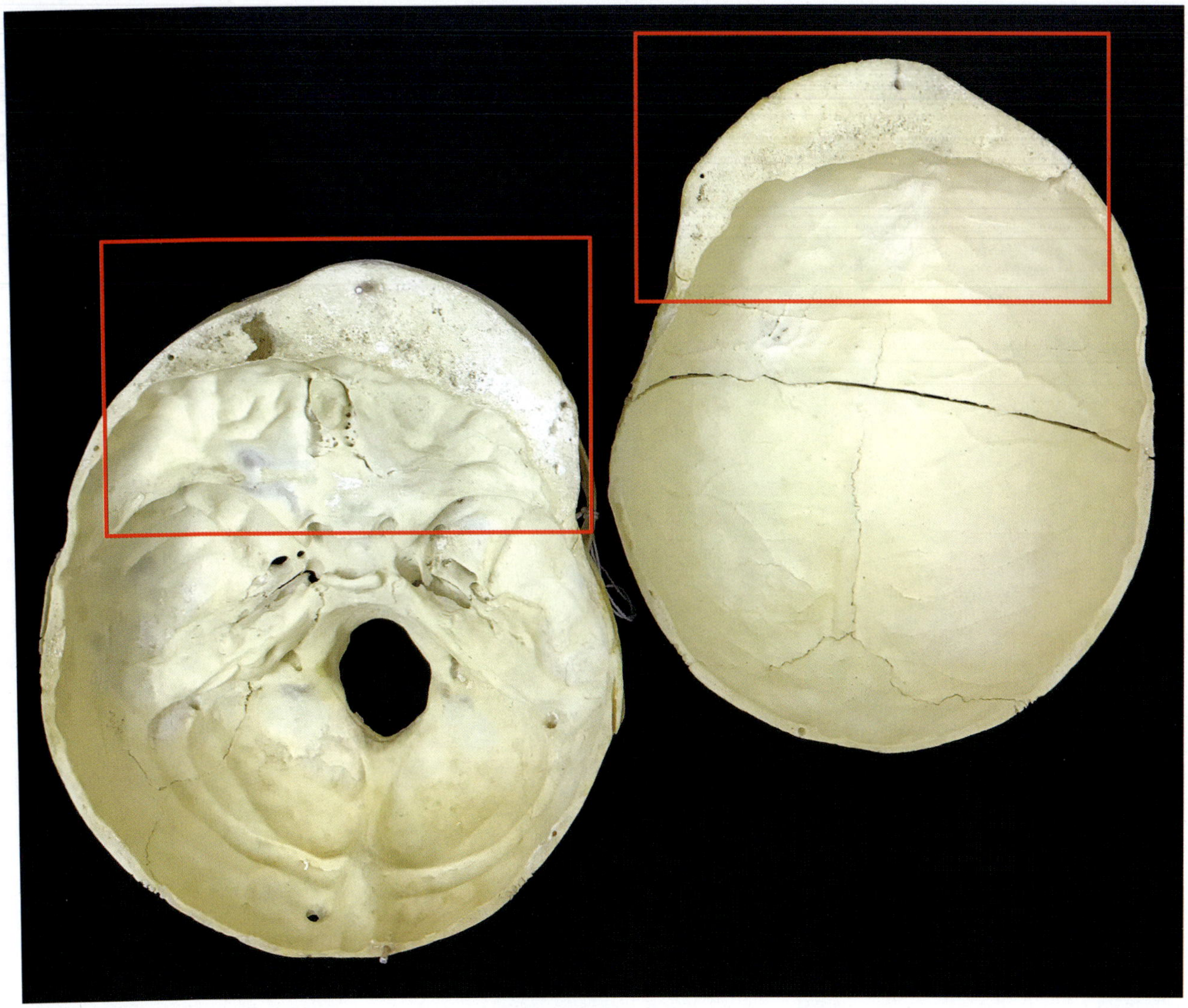

**Figure 308.** Hyperostosis of unknown origin, possibly fibrous dysplasia, but not hyperostosis frontalis internal (HFI), as this growth is asymmetrical and affects the outer table of the skull, in the frontal bone of a 20-year-old male who died in 1875 (MN21.055/2713). See Mann et al. (2016); Raikos et al. (2011); She and Szakacs (2004) for more on HFI.

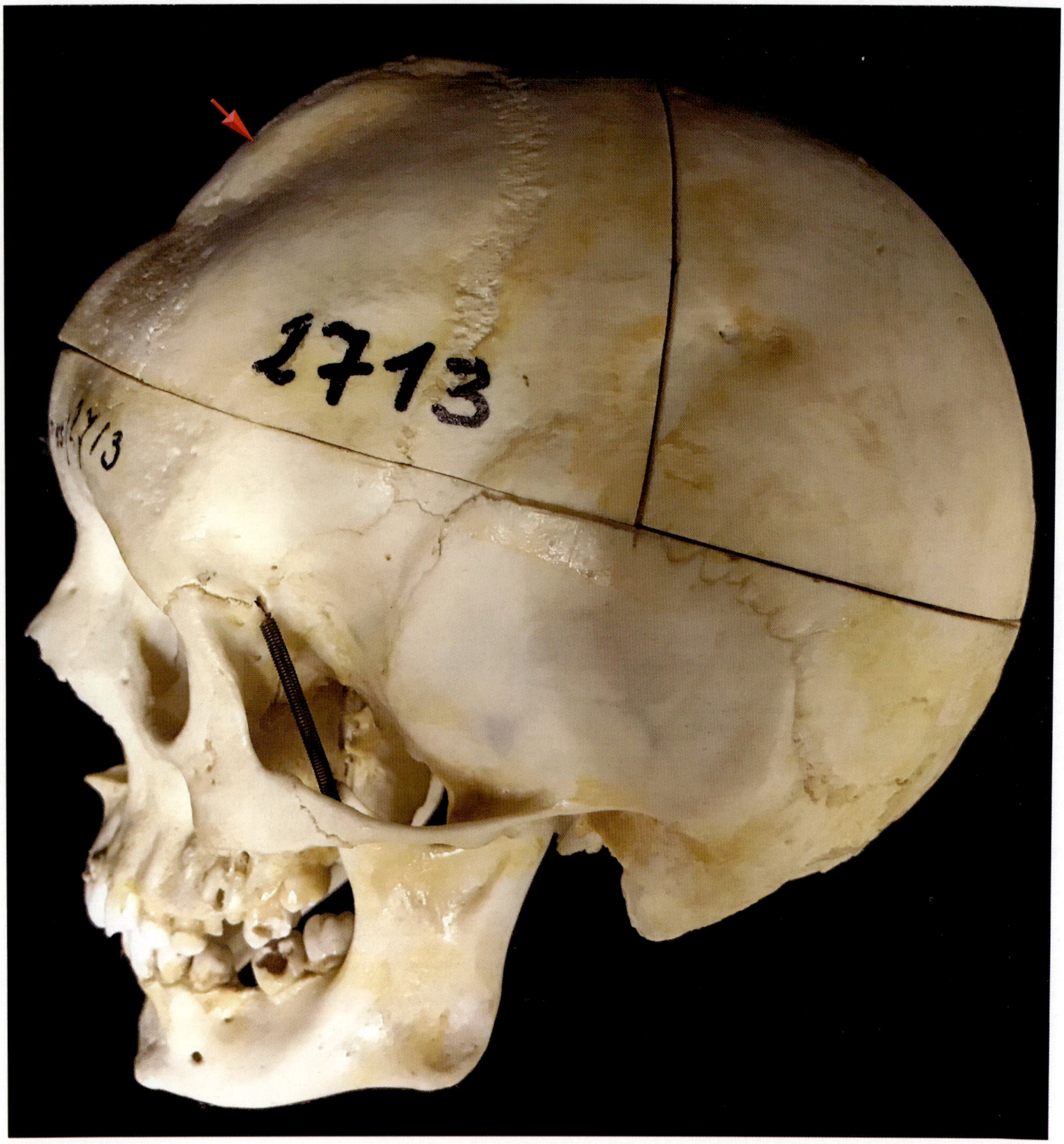


**Figure 309.** Hyperostosis of unknown origin, possibly fibrous dysplasia, but not hyperostosis frontalis interna (HFI) as this lesion is asymmetrical and affects the outer table of the cranium, in the frontal bone of a 20-year-old male who died in 1875 (MN21.055/2713). See Aufderheide and Rodrigues-Martin (1998); Hershkovitz et al. (1999, 2002); Mann et al. (2016); Ortner (2003); Ortner and Putschar (1981); Raikos et al. (2011); She and Szakacs (2004) for more on fibrous dysplasia and HFI.

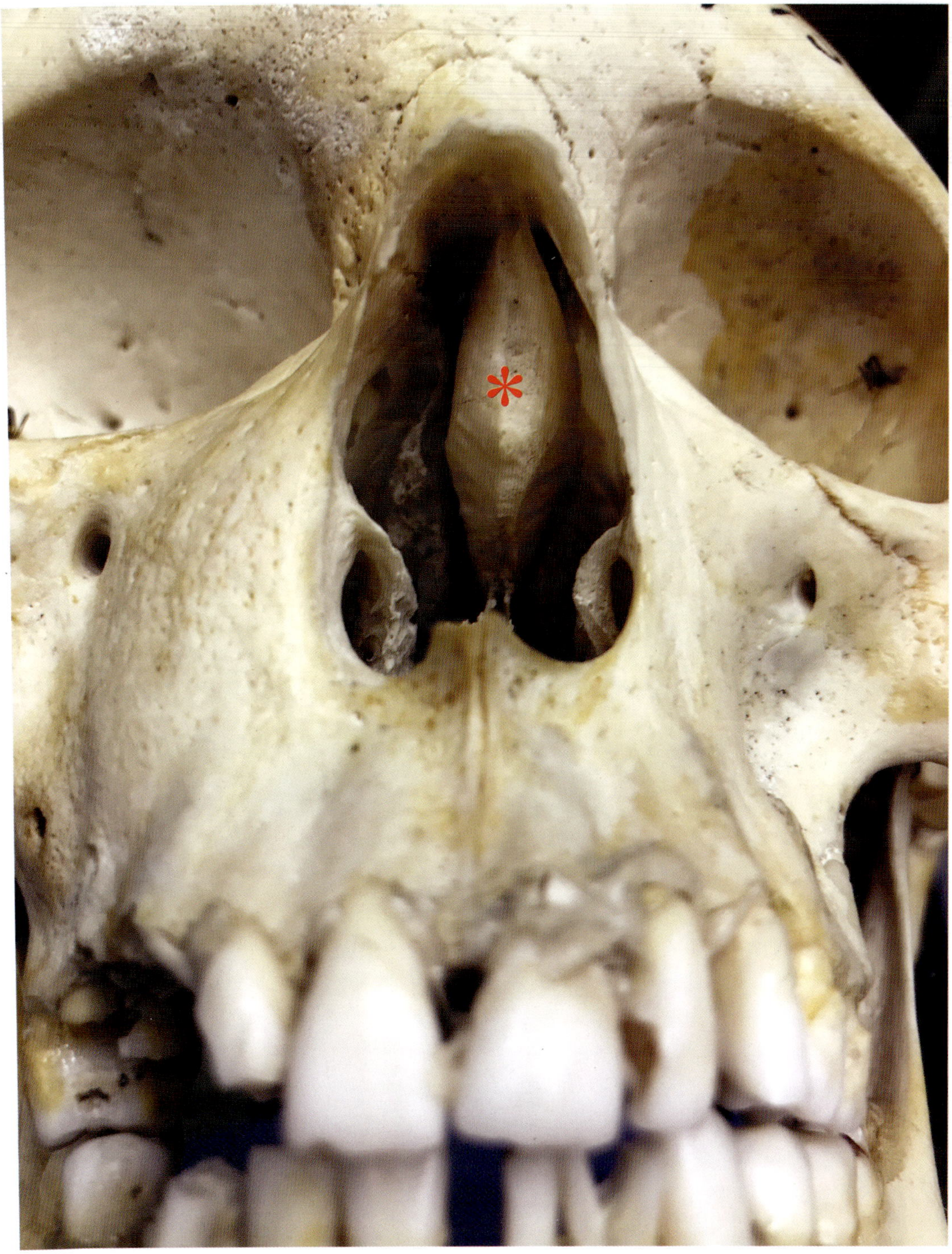

**Figure 310.** Pneumatization of the perpendicular plate of the ethmoid (*) in a 20-year-old male with hyperostosis of the frontal bone who died in 1875 (MN21.055/2713). Although pneumatization of the perpendicular plate of the ethmoid bone can occur in 18% of patients (Lei et al. 2004), it is a rare finding in dry bones and osteological collections. See Koo et al. (2016) for a case report of pneumatization of the perpendicular plate and Nasal Swell Body (septal turbinate).

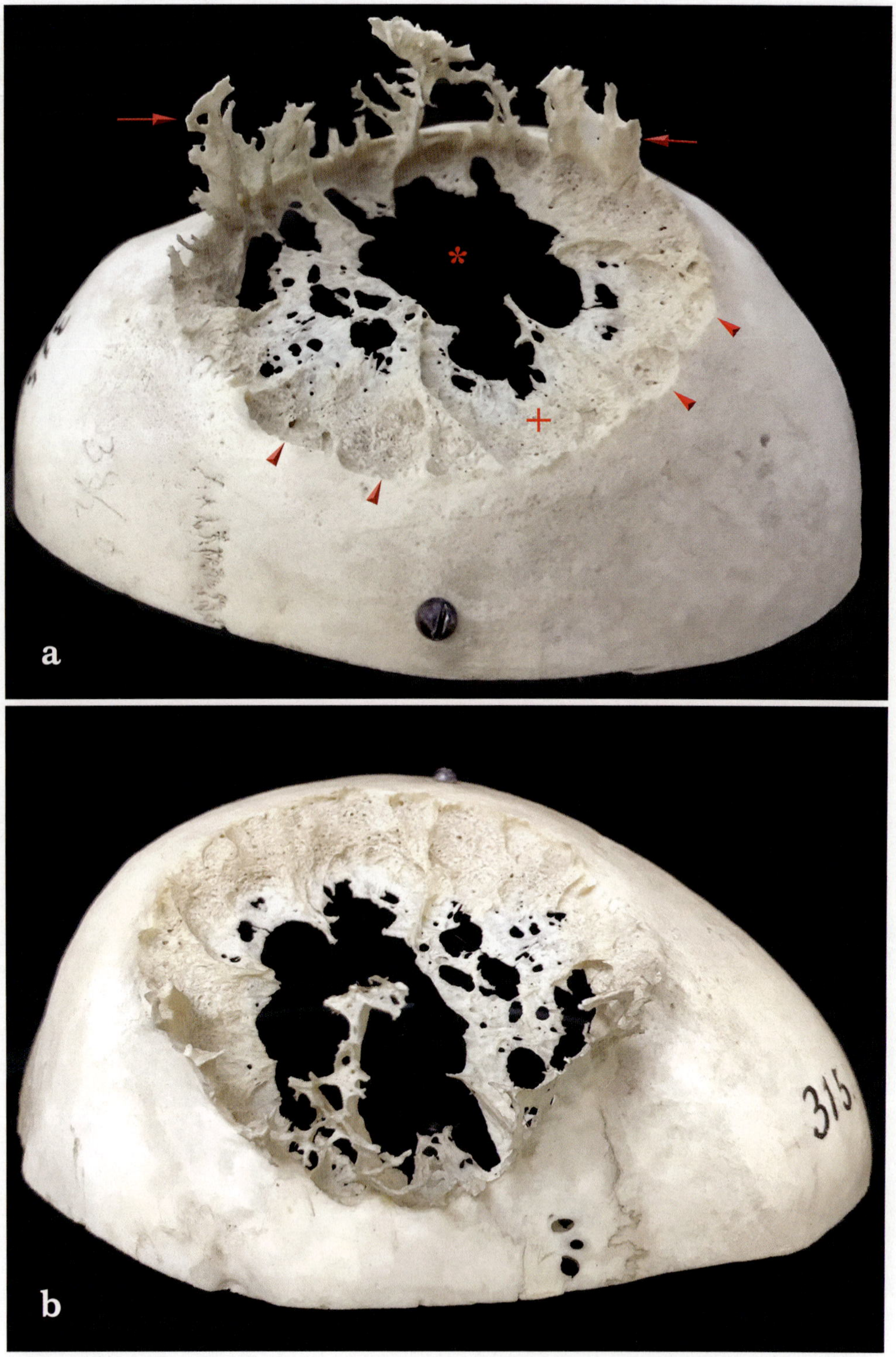


**Figure 311a-b.** (a) Left and (b) right lateral views of a large tumor of unknown origin primarily affecting the parietal bones with encroachment onto the frontal and right temporal bones in an adult (MN21.055/315). (a) Note destruction of bone near the middle of the defect (*), the scalloped margin (arrowheads), vertically oriented new bone along the margin (arrows) and partial retention of the inner table (+).

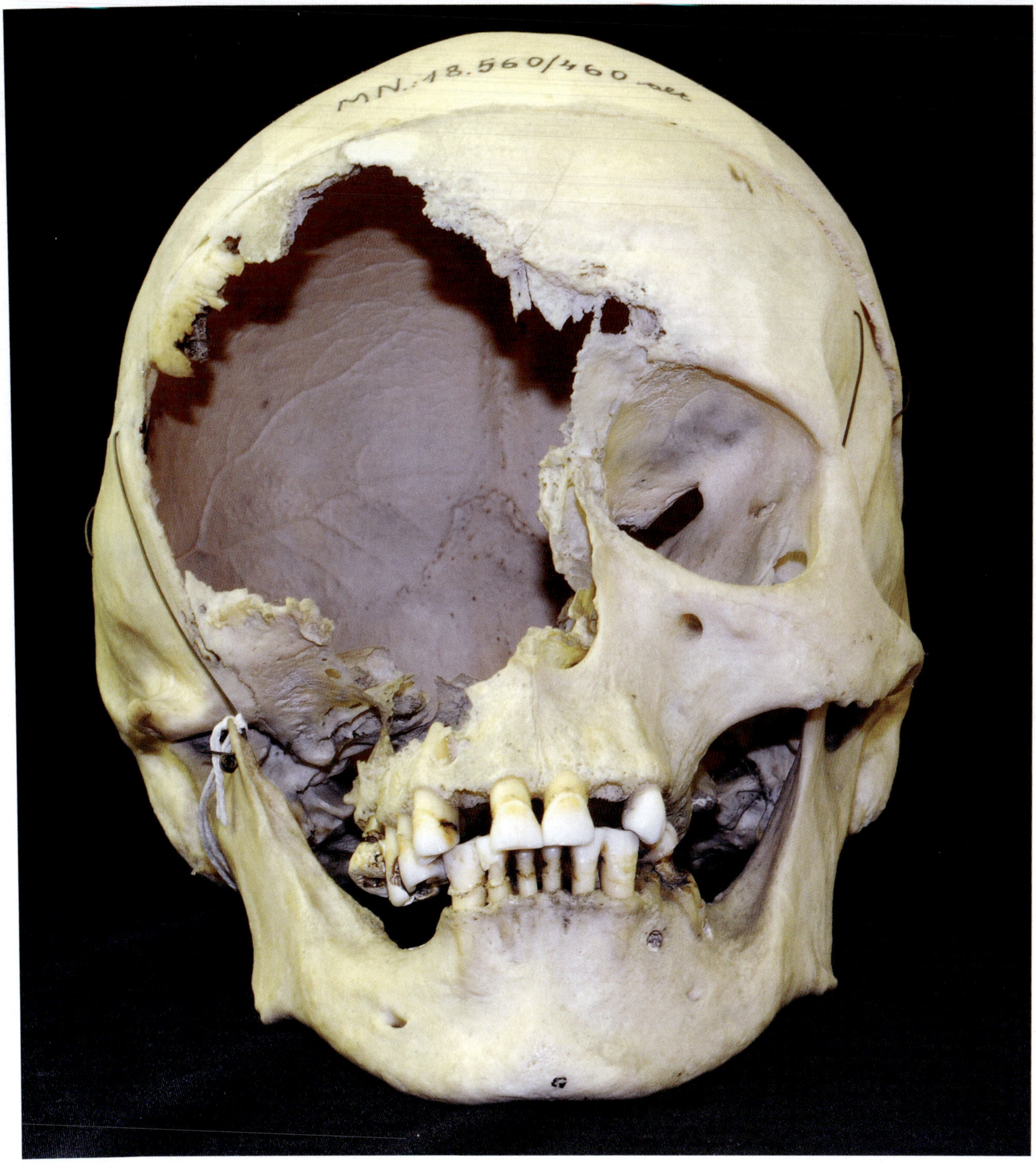


**Figure 312.** Anterior view of epithelial carcinoma showing extensive destruction of the right facial region as well as the frontal and right temporal bones in a 57-year-old male who died in 1887 (MN18.560/460a).

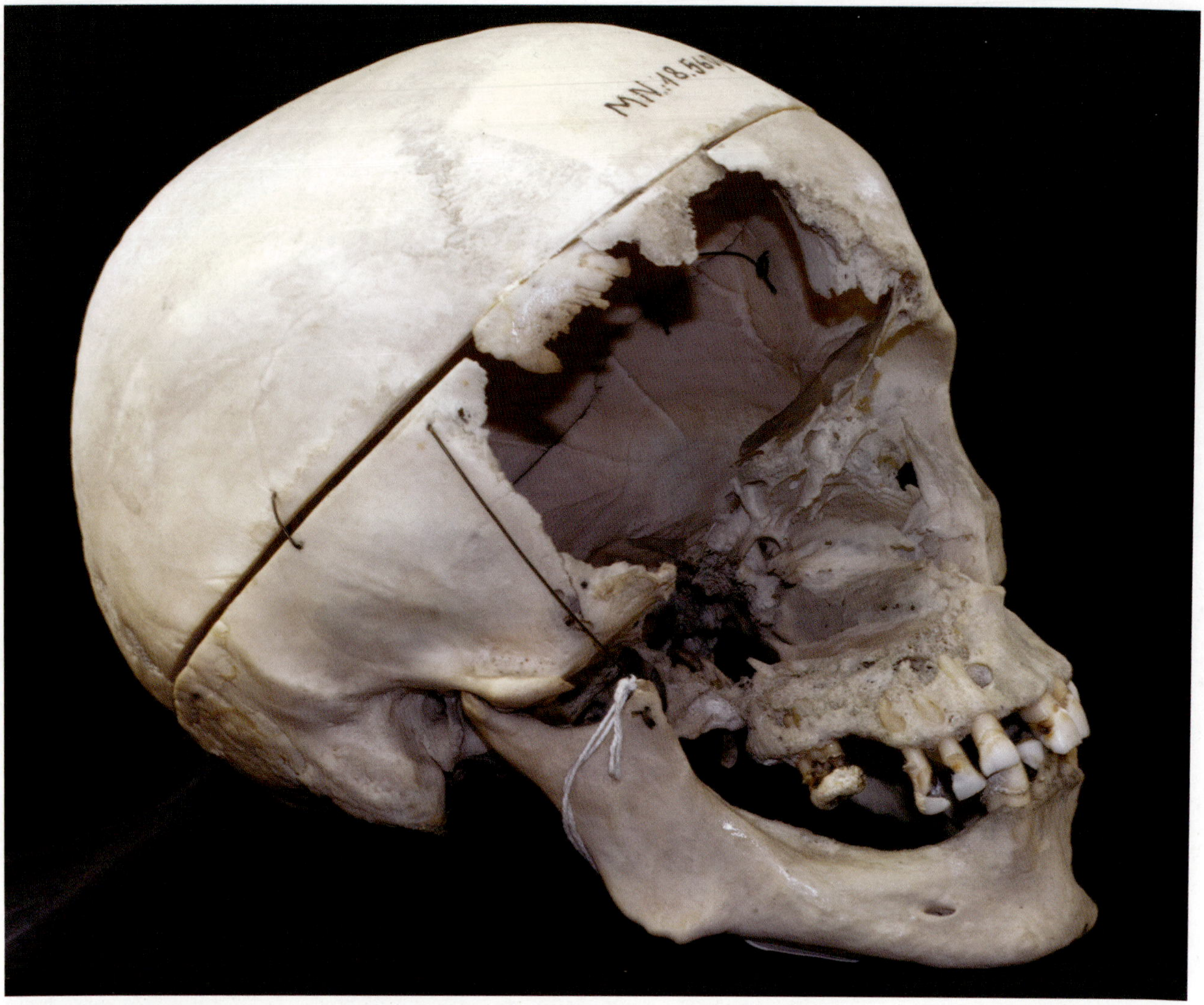

**Figure 313.** Right lateral view of epithelial carcinoma showing extensive destruction of the right facial region as well as the frontal and right temporal bones in a 57-year-old male who died in 1887 (MN18.560/460a).

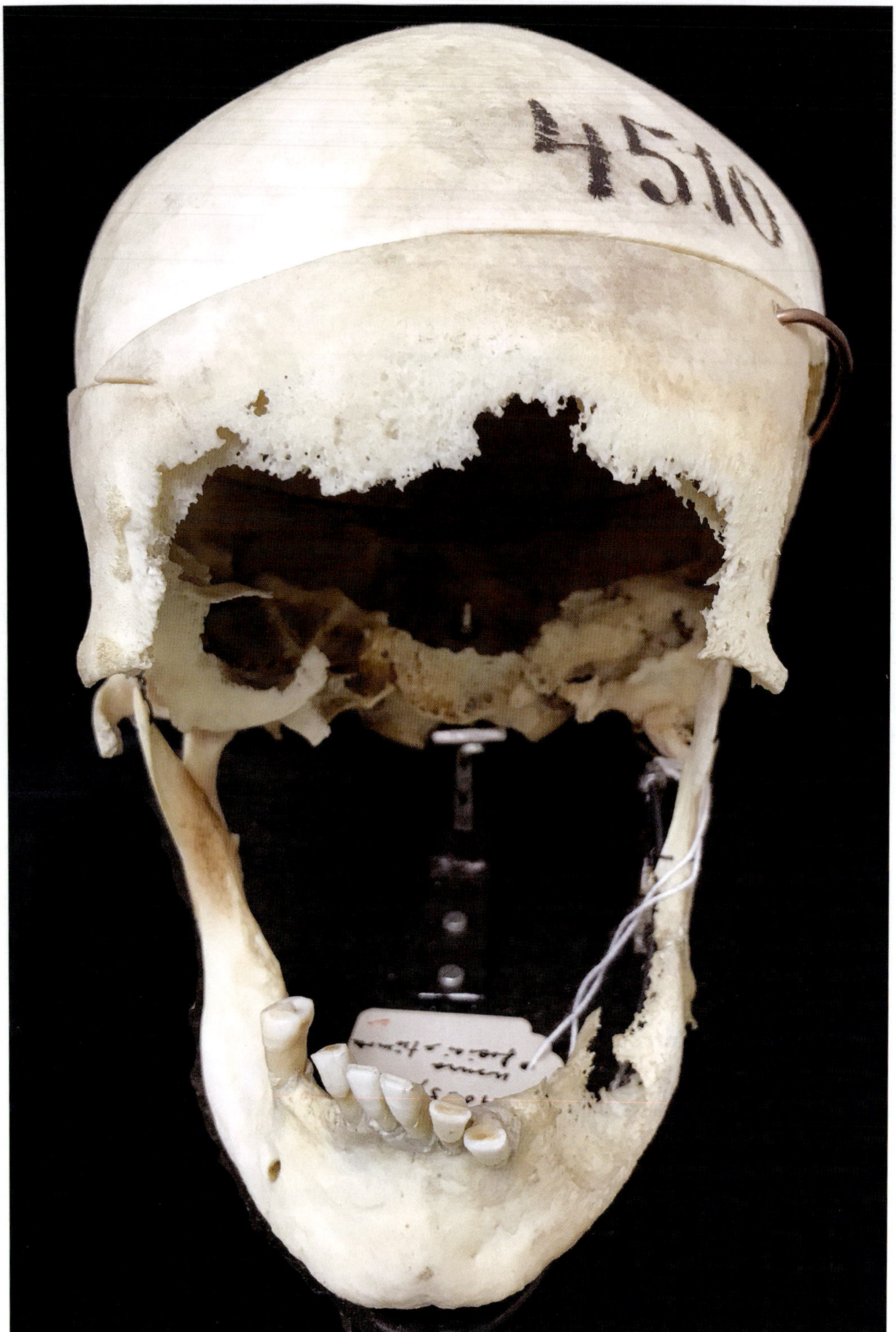


**Figure 314.** Anterior view of skull of an elderly individual of unknown sex showing extensive destruction of the frontal bones and orbits, facial region, cranial base and left half of the mandible of unknown origin (MN21.055/4510) ("Usura facies" on the museum card). The location and extensive area of bone destruction are suggestive of facial carcinoma.

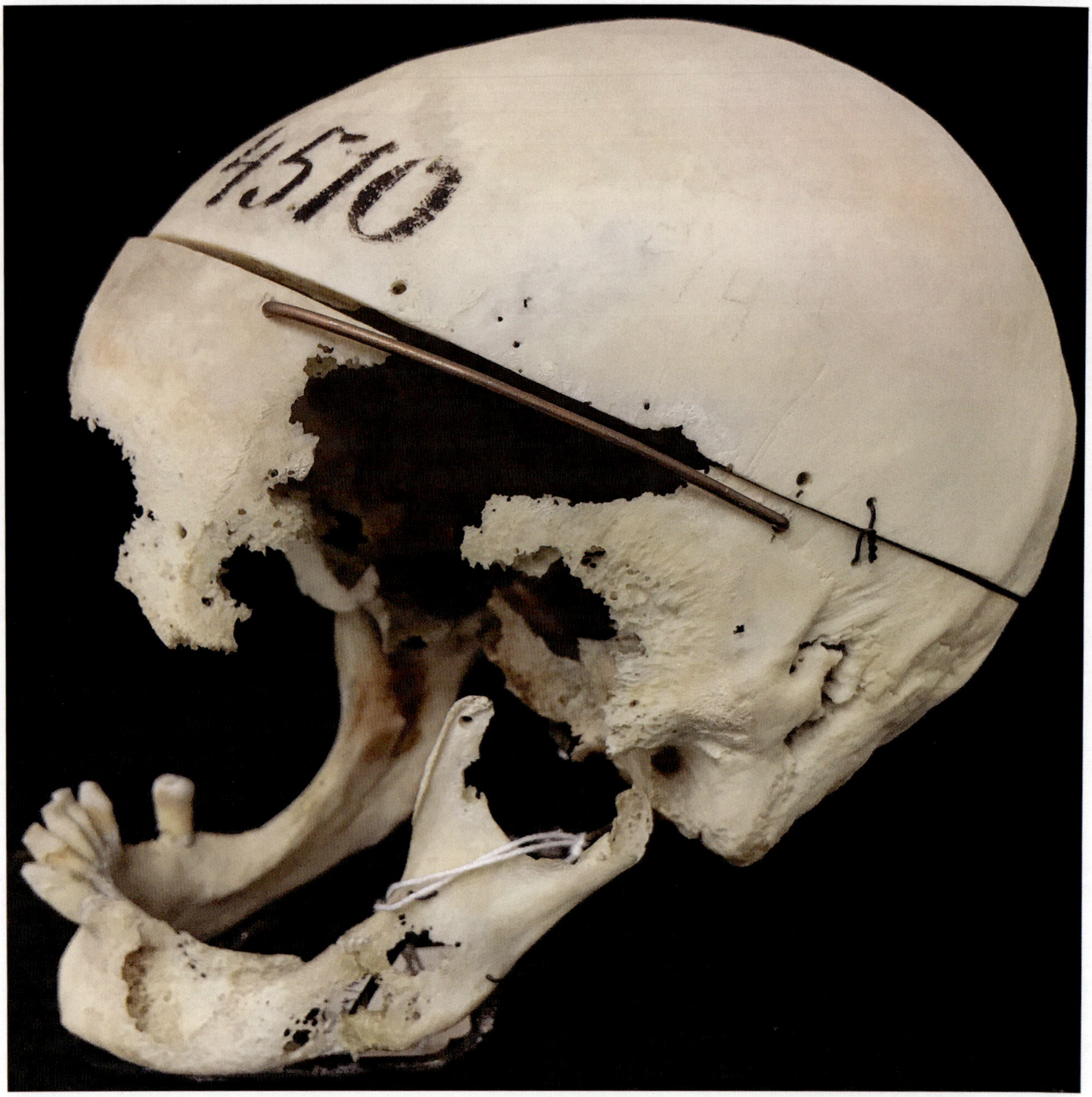


**Figure 315.** Lateral view of skull of an adult showing extensive destruction of the frontal bones and orbits, facial region and left half of the mandible in an elderly individual of unknown sex (MN21.055/4510) ("Usura facies" on the museum card). The location and extensive area of bone destruction are suggestive of facial carcinoma.

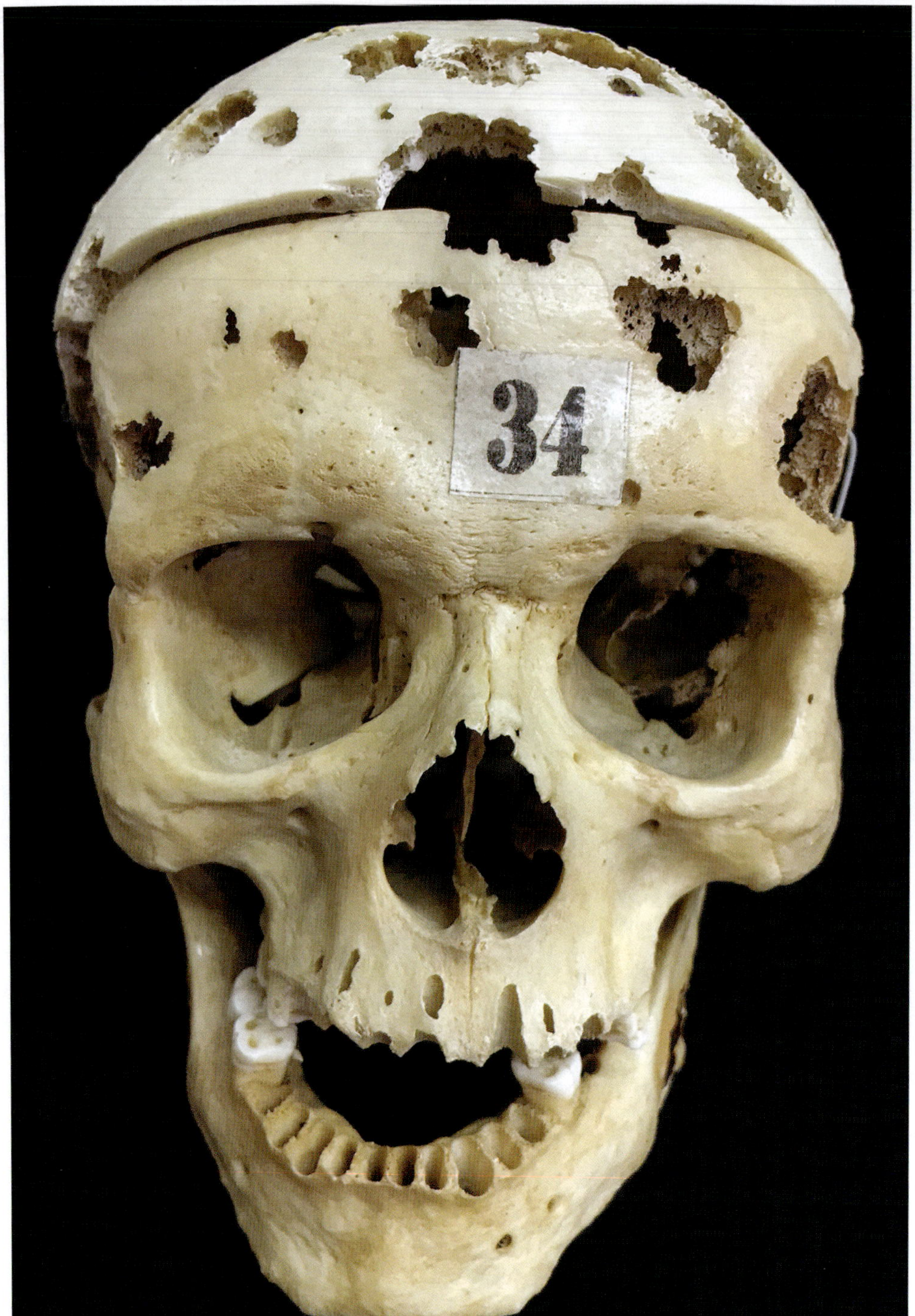


**Figure 316.** Anterior view of a 45-year-old male who died of multiple myeloma with lesions throughout his entire skeleton in 1902 (MN17.747/312). See the Case Report Multiple Myeloma (Chapter 6 in this book) for more about this individual and Watson-Jones (1944) for multiple myeloma in general.

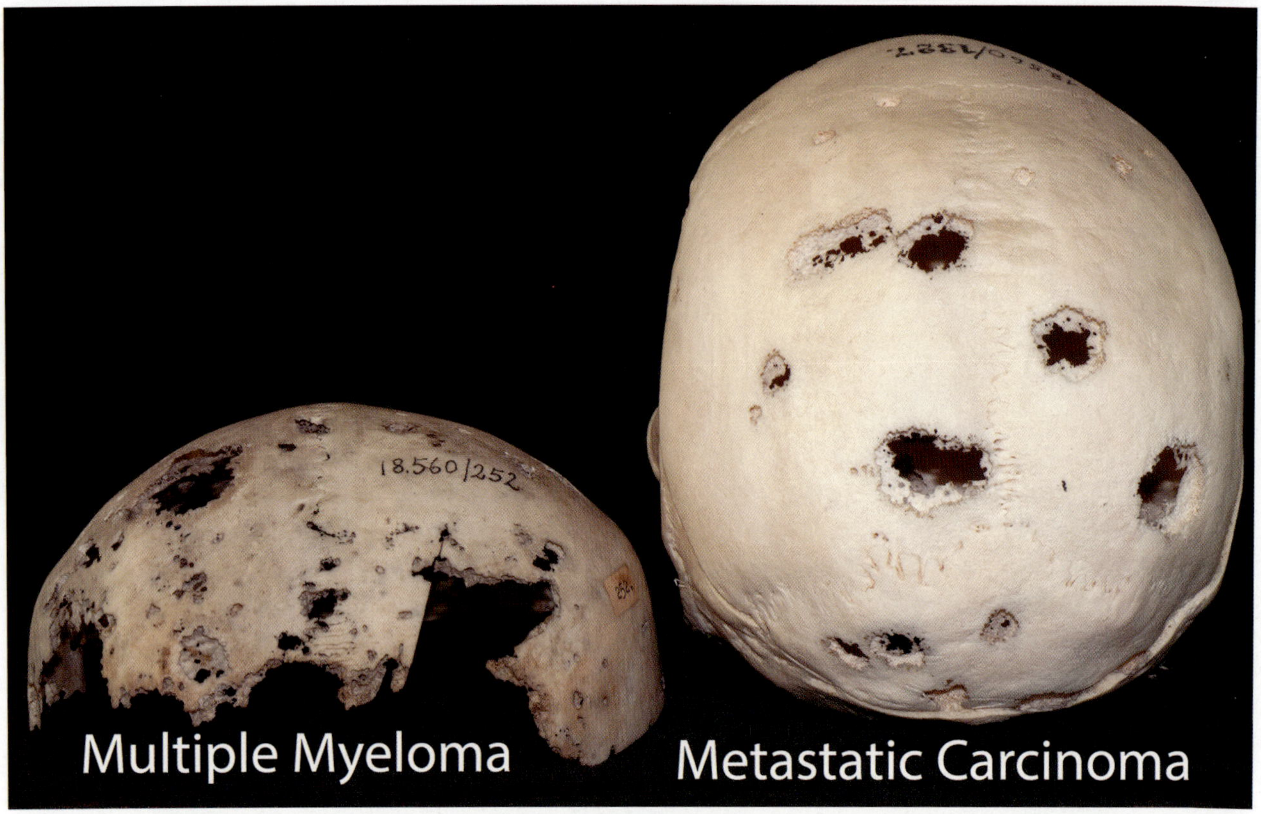


**Figure 317.** Comparison of cranial lesions in multiple myeloma in a male who died in 1930 (MN18.560/252) and metastatic carcinoma in a 60-year-old male secondary to esophageal cancer who died in 1933 (MN18.560/1327). See Rajakulasingam et al. (2020); Rothschild et al. (1998); Sapper (1957); Thomas and Kirsten (1985) for more on distinguishing multiple myeloma and metastatic carcinoma and their distribution in the human skeleton. Macroscopically, there is little difference in the appearance of destructive lesions in multiple myeloma and metastatic bone disease in dry bone. See Biehler-Gomez et al. (2019) for more on comparative macroscopic and radiographic observations of lesions in dry bone.

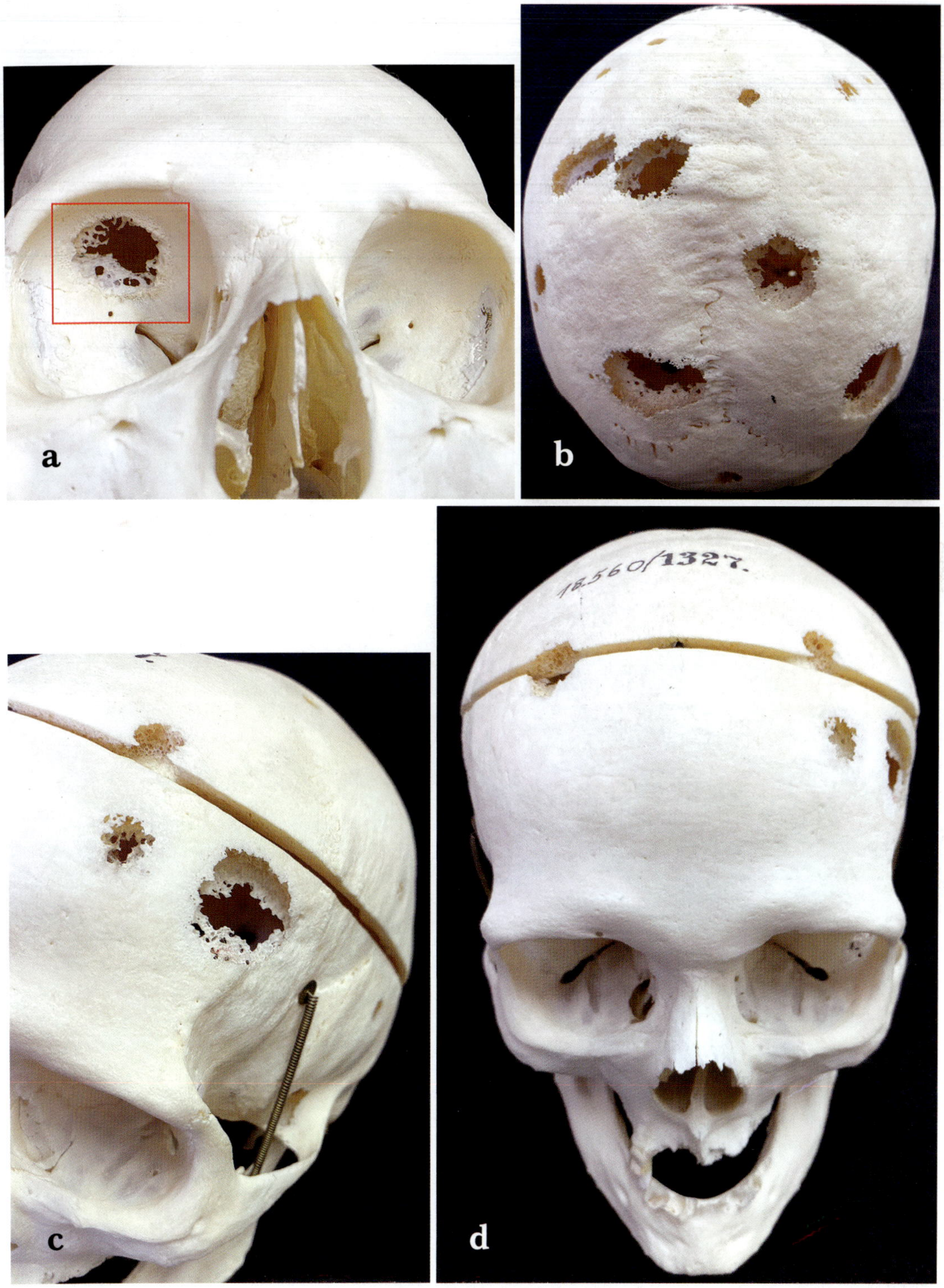


**Figure 318a-d.** (a) Magnified view of right orbit, (b) posterior-superior, (c) left superior-oblique and (d) anterior-superior views of metastatic carcinoma of the skull and right orbital plate secondary to esophageal cancer in a 60-year-old female who died in 1933 (MN18.560/1327). Note the multiple scalloped lesions of varying size with minimal reactive bone surrounding them.

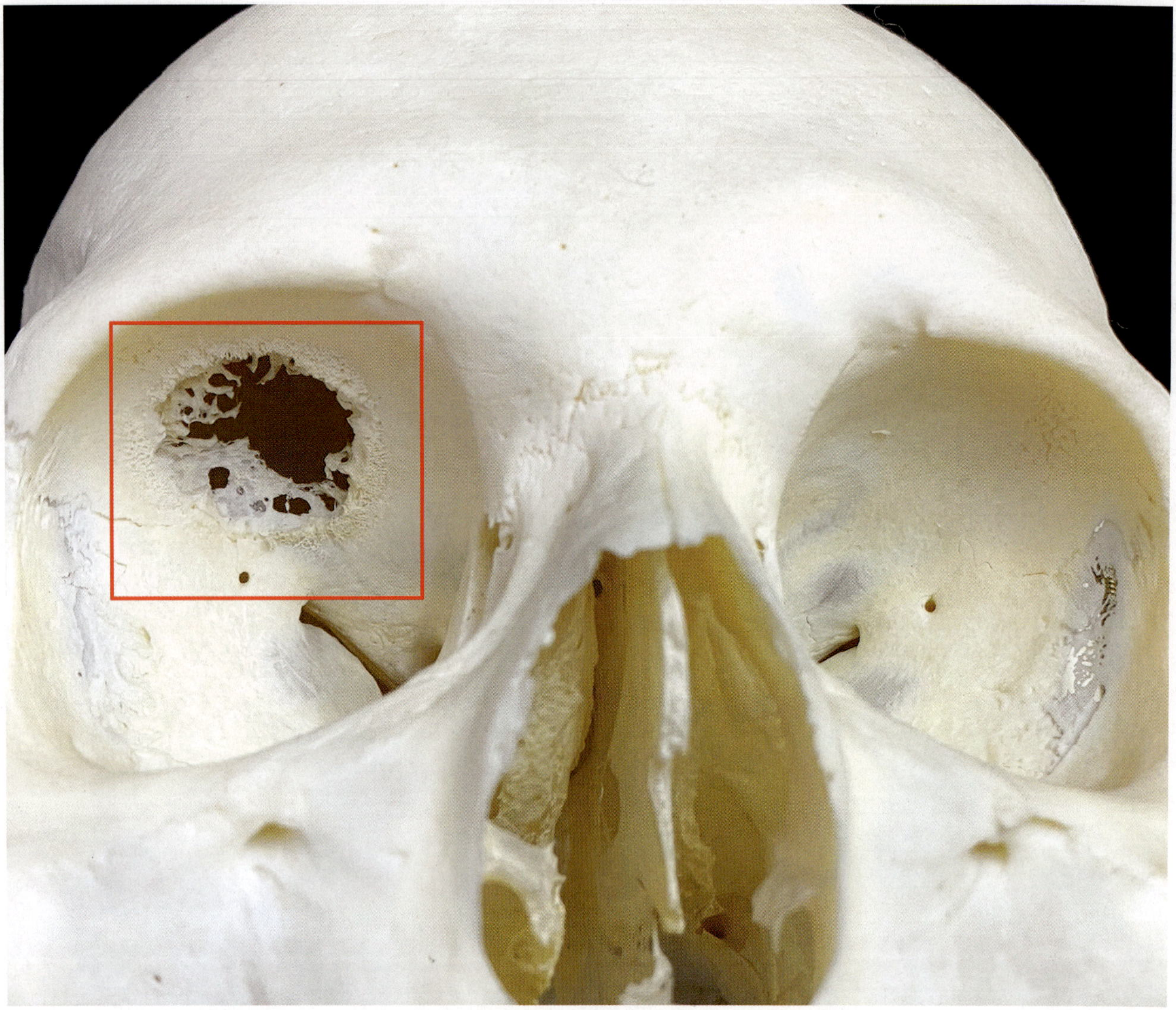

**Figure 319.** Carcinoma of the skull secondary (metastasis) to esophageal cancer in a 60-year-old female who died in 1933 (MN18.560/1327). Note the flame-like projections of the inner table and raised ring of reactive bone encircling this lesion in the superior plate of the right orbit.

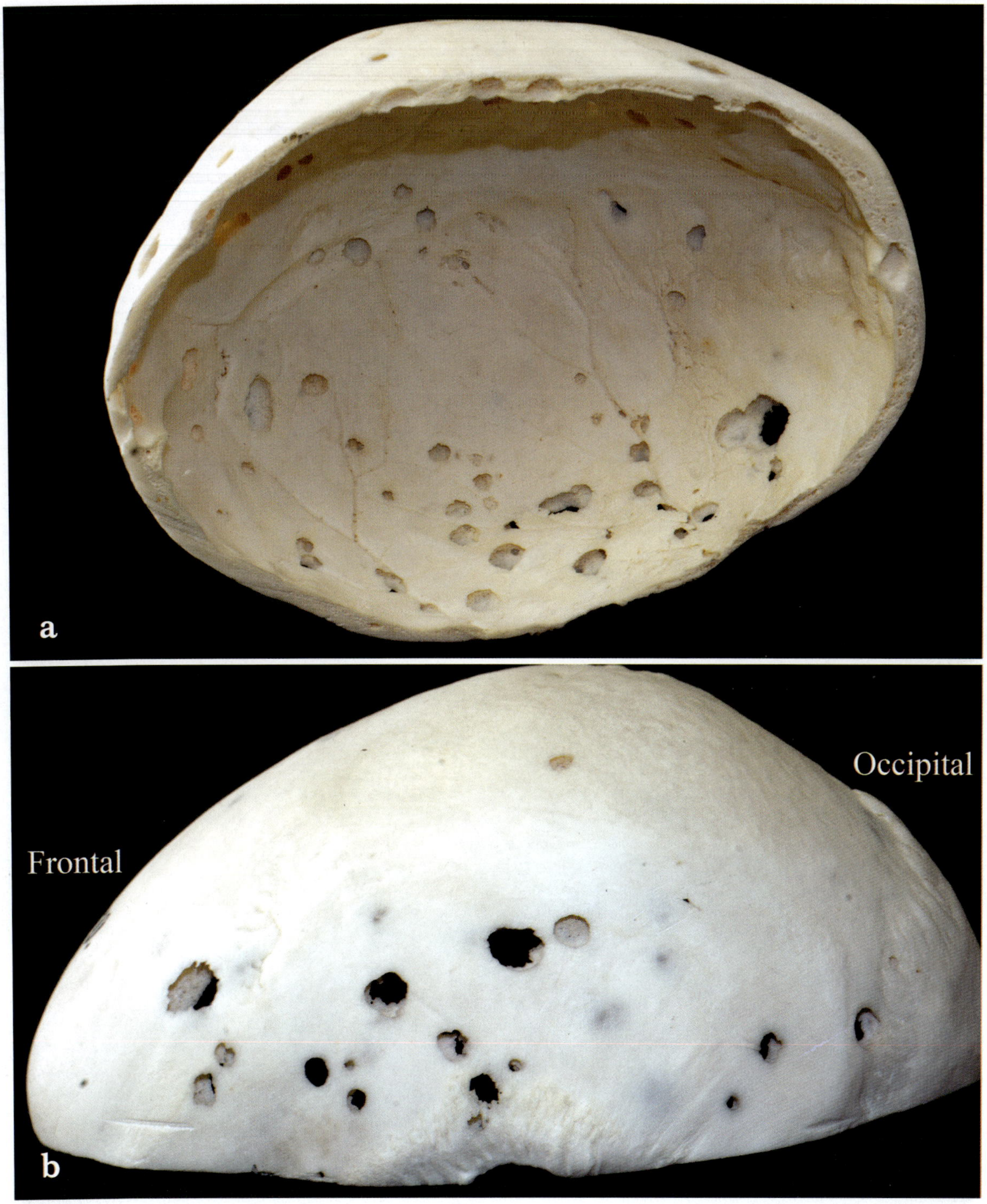


**Figure 320a-b.** (a) Endocranial and (b) left lateral views of hypernephroma ("atypicum secundarium calvariae" on the museum card), also known as renal cell cancer, renal cell adenocarcinoma and renal cell carcinoma, in a 61-year-old female who died in 1935 (MN18.560/1470). Note there are more lesions visible endocranial than ectocranial. See Salapura et al. (2014) for more on hypernephroma and metastasis to bone.

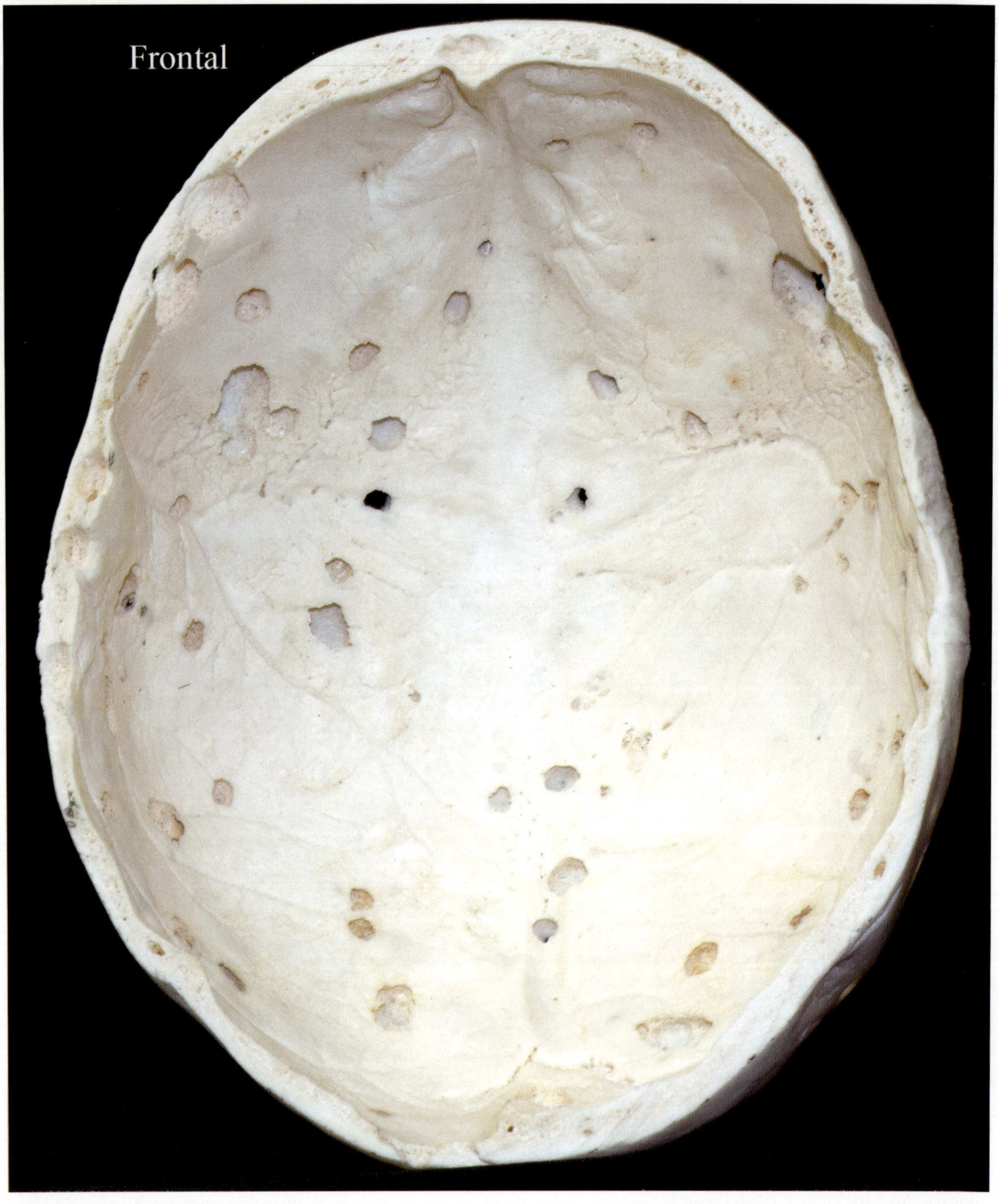


**Figure 321.** Endocranial view of hypernephroma atypicum secundarium calvariae, also known as renal cell cancer and renal cell carcinoma in a 61-year-old female who died in 1935 (MN18.560/1470). See Salapura et al. (2014) for more on hypernephroma and metastasis to bone.

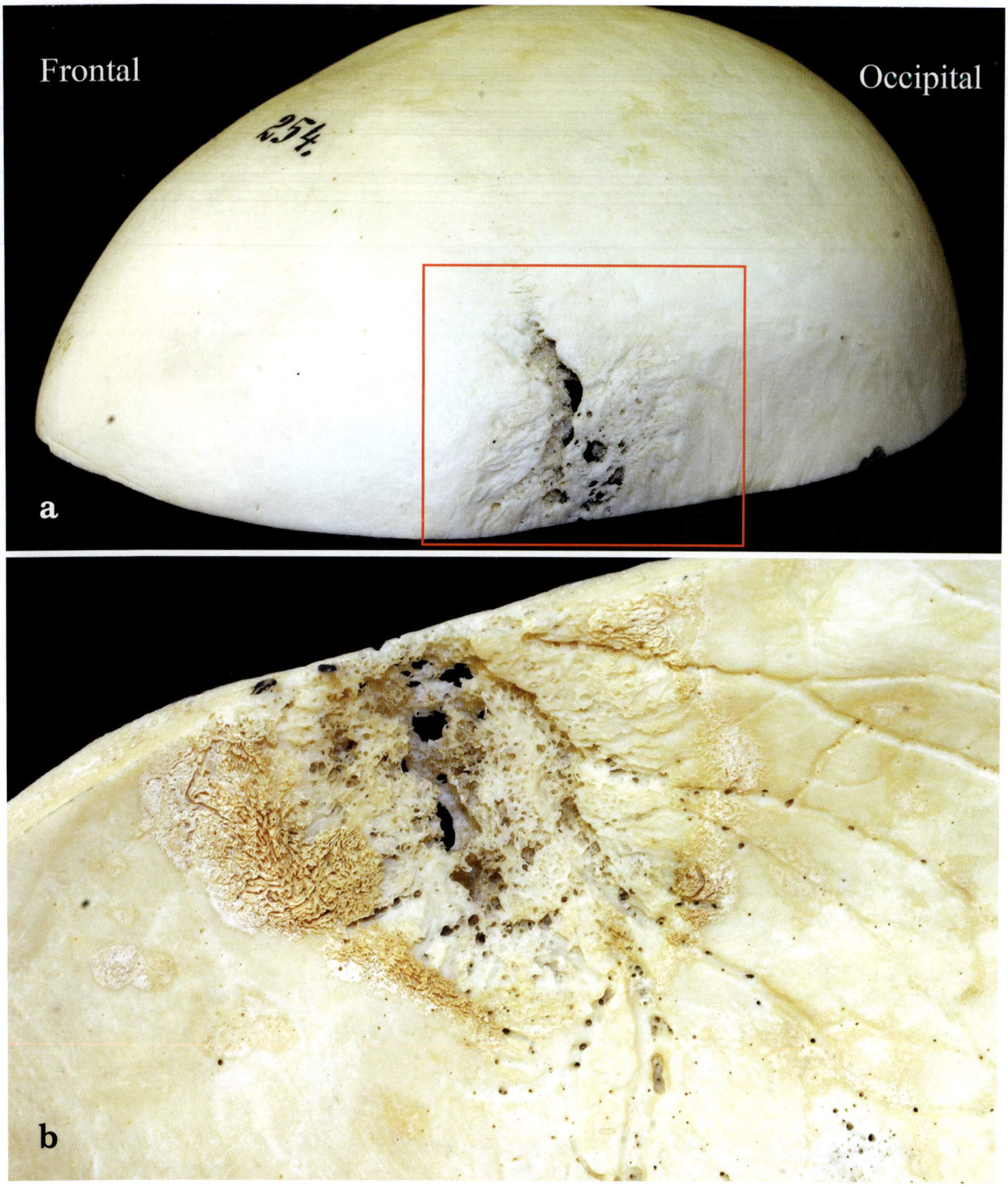


**Figure 322a-b.** (a) Ectocranial and (b) endocranial views of carcinoma bronchi (lung cancer) in a 50-year-old female who died in 1919 showing that the cancer lesions affected both the outer and inner tables (MN18.560/254). Bronchogenic carcinoma, or bronchial carcinoma, is a malignant neoplasm of the lung (Siddiqui et al. 2022). (a) Ectocranial, the lesion (square) appears to be more remodeled than endocranial, the latter of which exhibits woven bone (dark areas) and numerous vascular perforations along the middle meningeal artery grooves. (b) Much of the loosely attached woven bone is missing postmortem.

**Figure 323.** Large button osteoma (arrow), recorded as "exostosis faciei externae" on the museum card, in the left parietal bone of 58-year-old female (MN21.055/1136). Button osteoma, or external exostosis, as written on the museum card, is an ivory-like benign tumor consisting of one or many rounded or dome-like bony growths of varying size on the external surface of the skull. See Chae et al. (2015); Eshed et al. (2002); Mann and Hunt (2005); Ortner and Putschar (1981); Steinbock (1976).

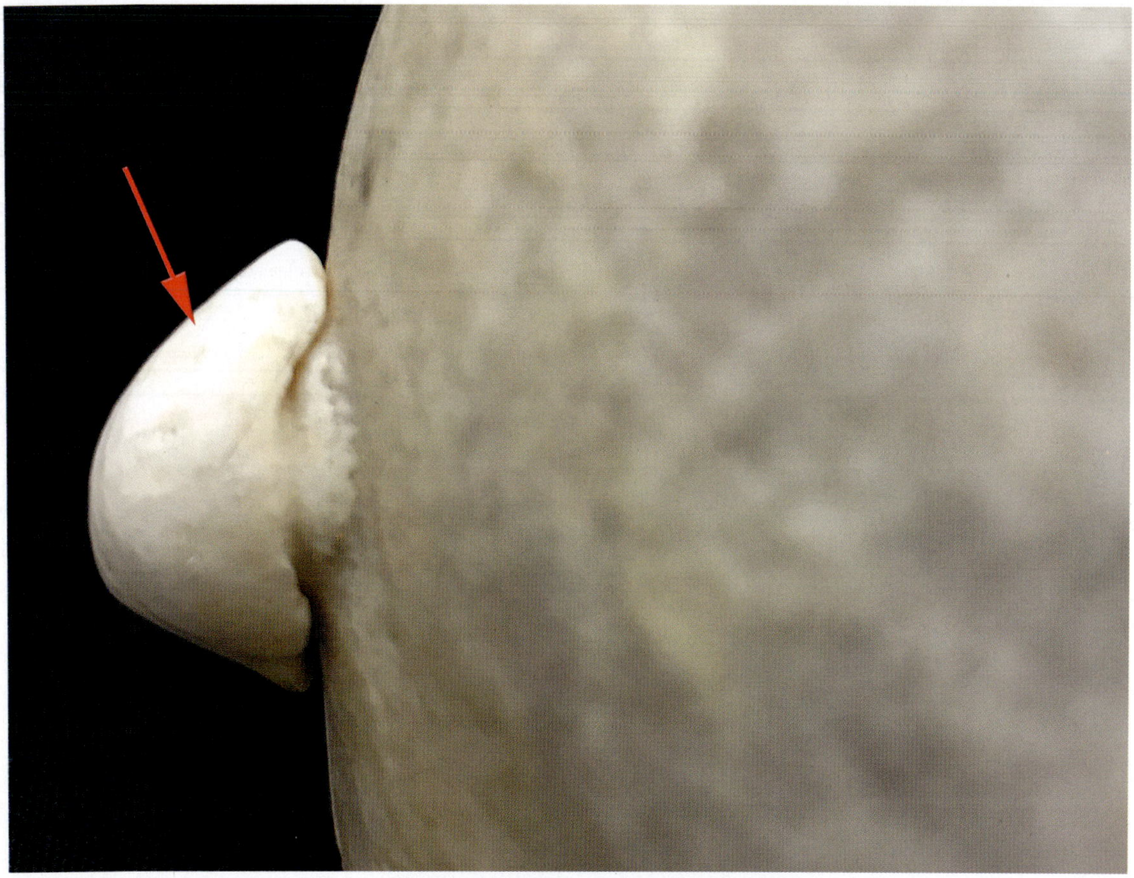

**Figure 324.** Magnified posterior-anterior view of large button osteoma (arrow), recorded as "exostosis faciei externae" on the museum card, in the left parietal bone of a 58-year-old female (MN21.055/1136). Button or "ivory" osteoma is a benign tumor commonly found on the external surface of the skull in all population groups. Note that this osteoma has a mushroom-like appearance with a stalk-like base and a separate margin around its dome. See Chae et al. (2015); Eshed et al. (2002); Mann and Hunt (2005); Ortner and Putschar (1981); Steinbock (1976).

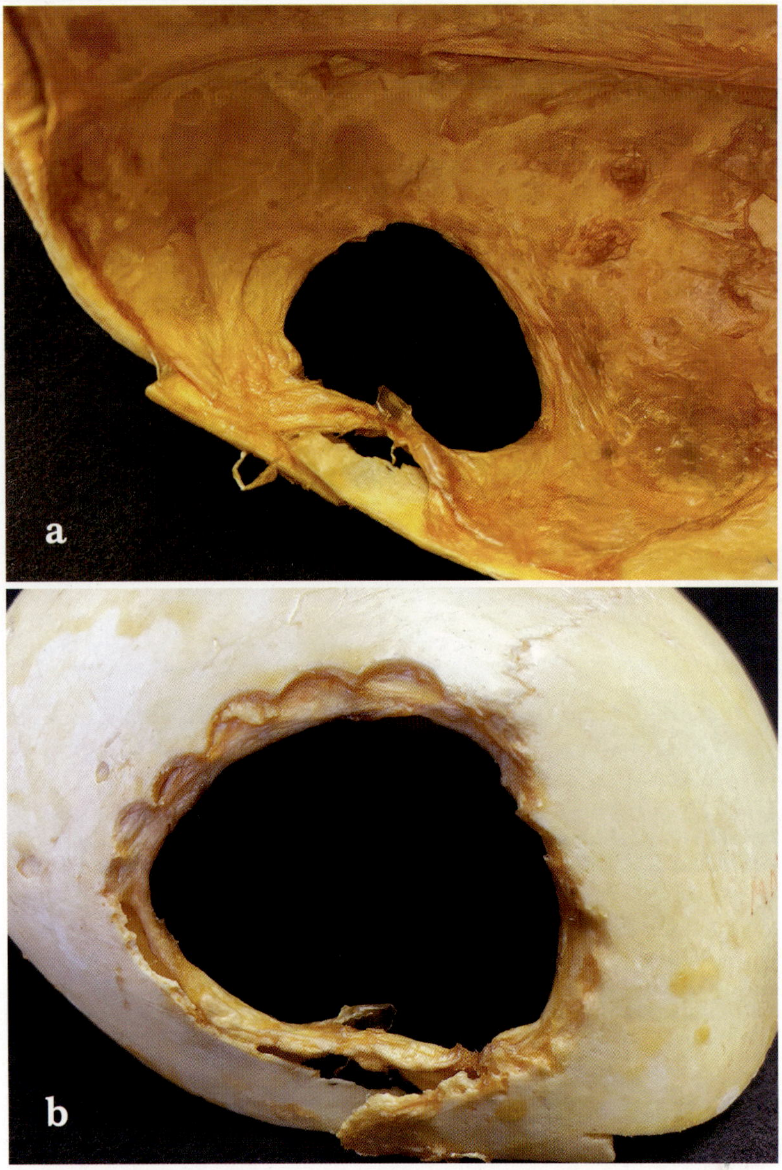


**Figure 325a-b.** (a) Endocranial and (b) ectocranial views of fibroma of the right parietal and frontal bones in a 4-year-old male (MN10.183). Note the uniformly scalloped margins in the ectocranium that could be mistaken for surgical burr holes. See Dumitriu et al. (2014) for more on fibromas.

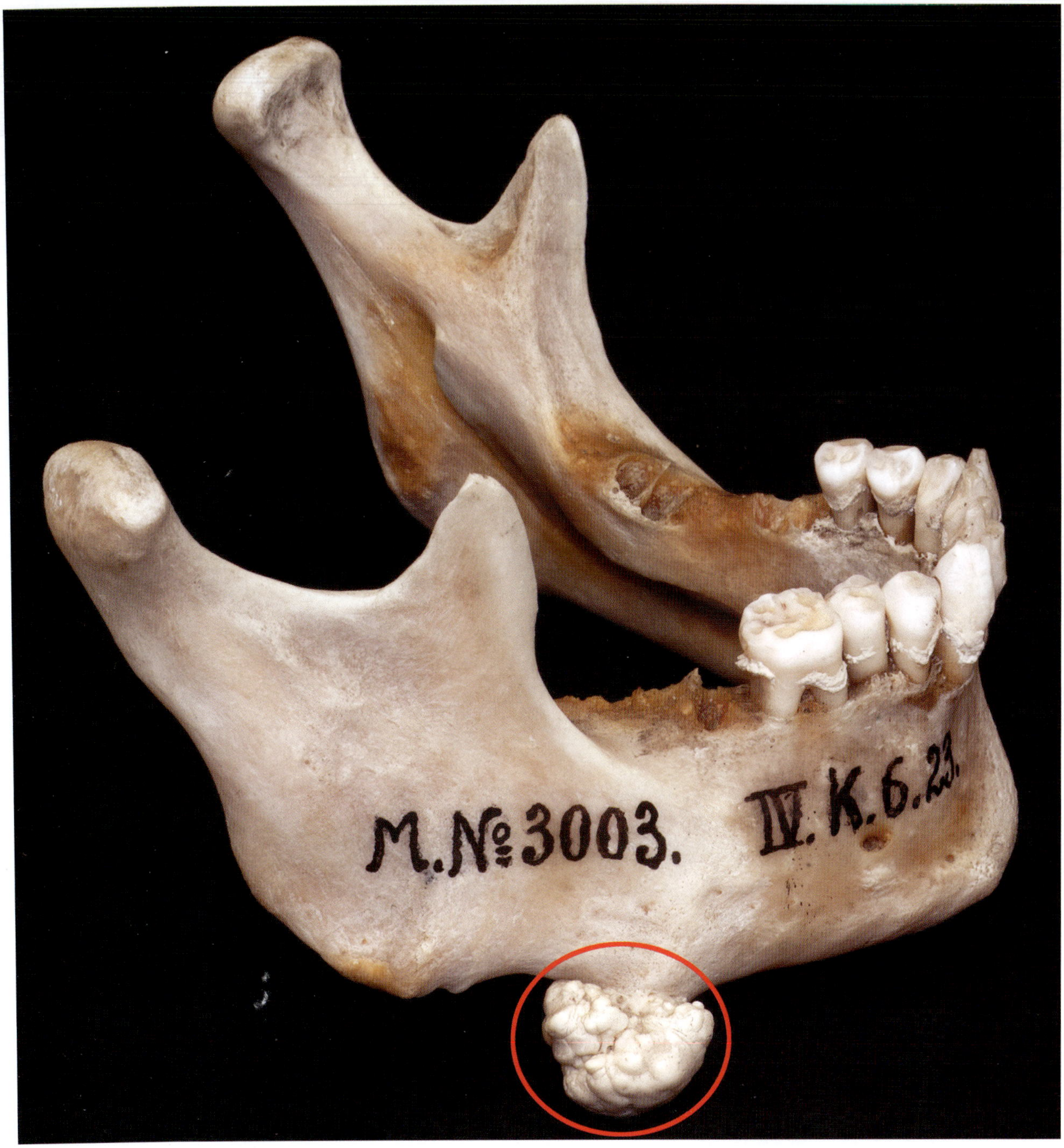


**Figure 326.** Probable osteochondroma/exostosis, a benign tumor (circle), firmly attached along the right inferior border of the mandibular corpus in a 43-year-old male who died in 1866 (MN3003). Osteochondromas such as these are rare. See Geshchickter and Copeland (p. 720; 1936) for a similar example of osteochondroma in the mandible.

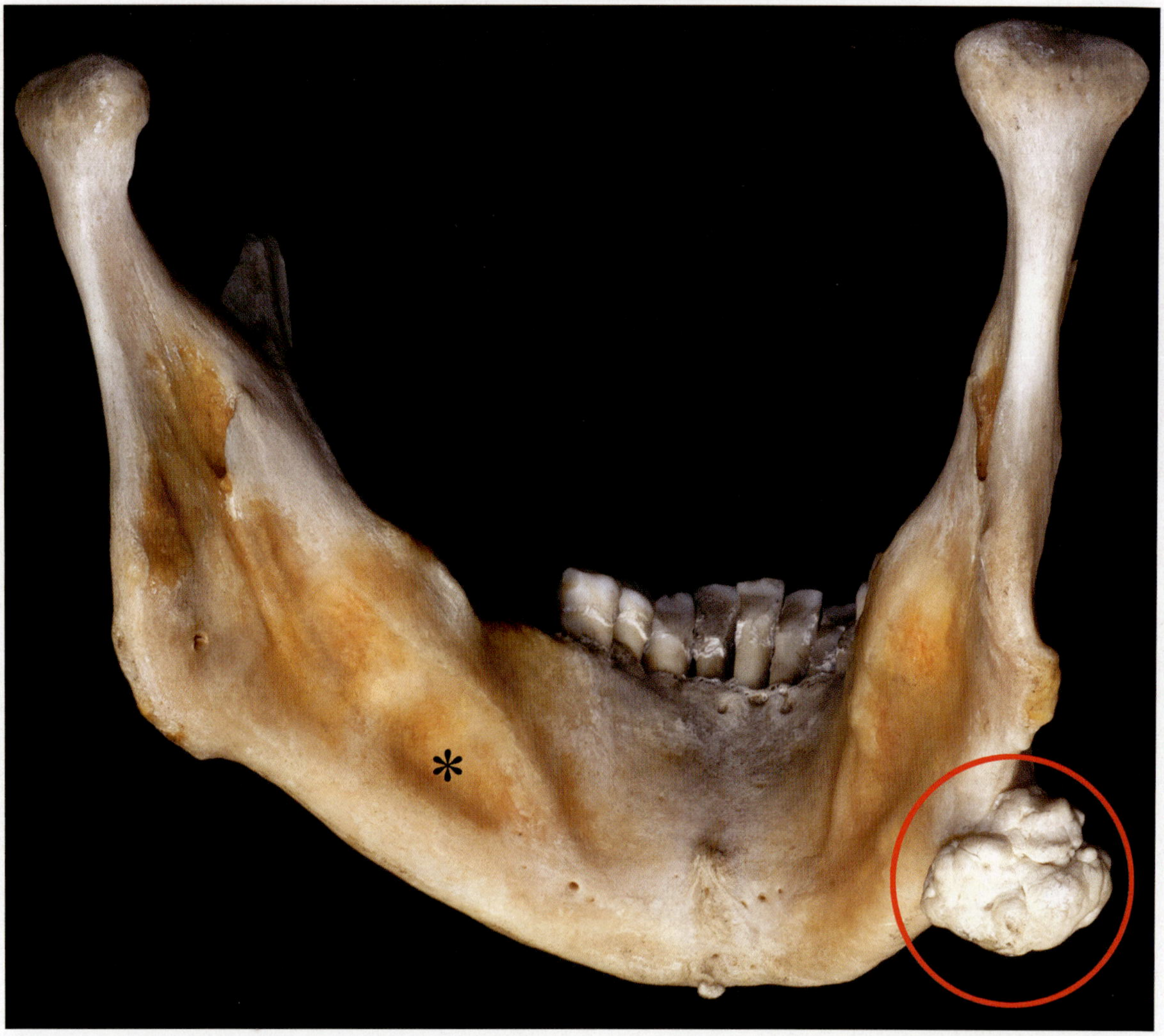


**Figure 327.** Posterior view of globular-shaped probable osteochondroma (circle) firmly attached along the buccal surface of the mandible near its inferior border in a 43-year-old male (MN3003). Osteochondromas, benign tumors, along the inferior border of the mandible are rare. This individual also has a lingual cortical mandibular defect (*) typically from contact with the submandibular salivary gland. Lingual cortical defects, also known as idiopathic bone cavity, Stafne defect, static bone cavity, and salivary gland defect, among others, is a benign and asymptomatic defect in the posterior portion of the mandible that may be unilateral or bilateral, and present overwhelmingly in adult males (Lello and Makek 1985; Mann 2001; Stafne 1942). See Geshchickter and Copeland (1936) for an example of osteoma in the mandible.

## Spine, Ribs and Pelvis

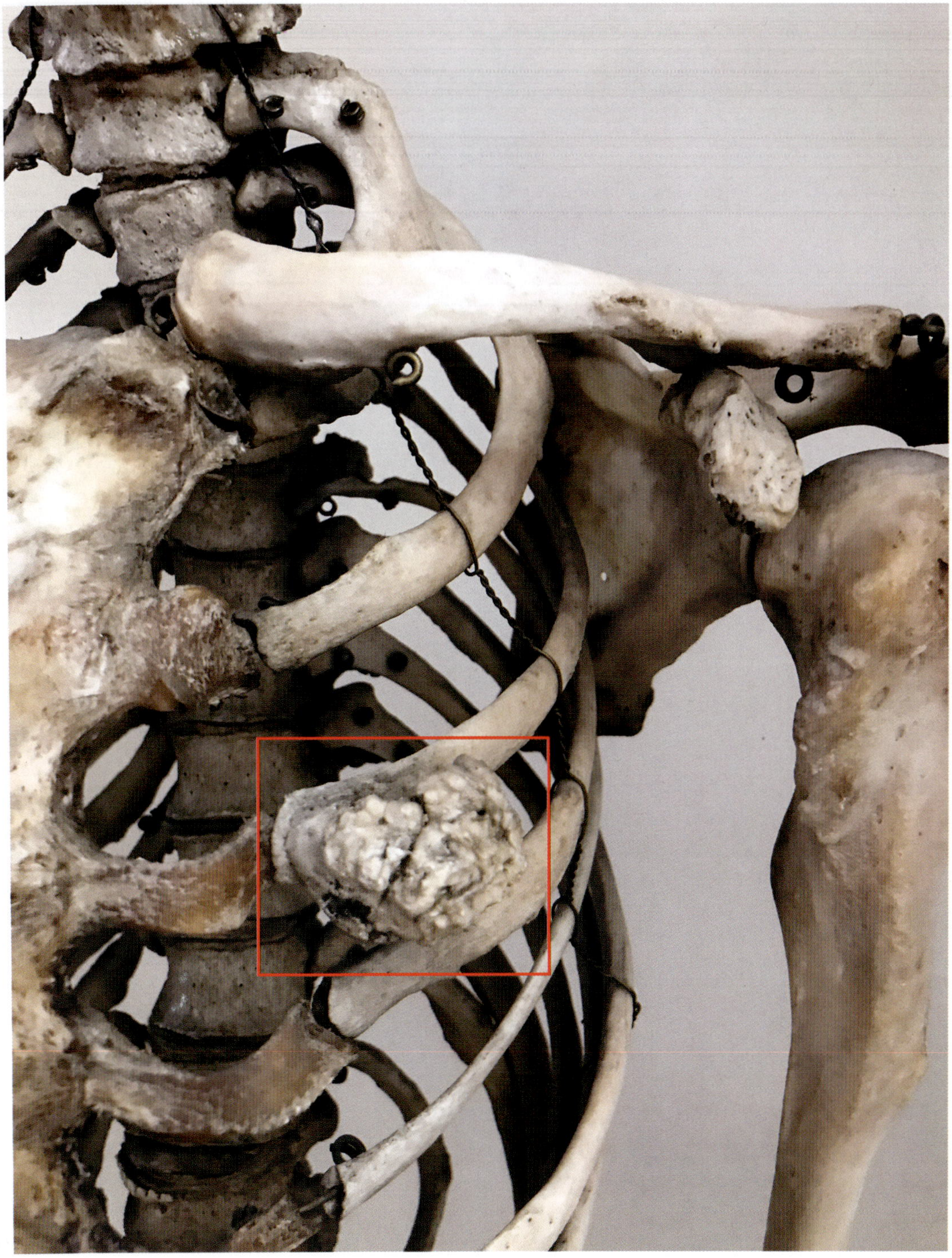

**Figure 328.** Lobulated exostosis on the external surface of left rib 3 near the costochondral junction in a 56-year-old female with hereditary multiple exostoses (HME) of the skeleton (MN605). Lesions in HME have been reported as 70% in distal femur, 70% in proximal tibia, 50% in proximal humerus, and 40% in the ribs (Schmale et al. 1994). See Geshchickter and Copeland (1936) for more on HME and their distribution in the skeleton.

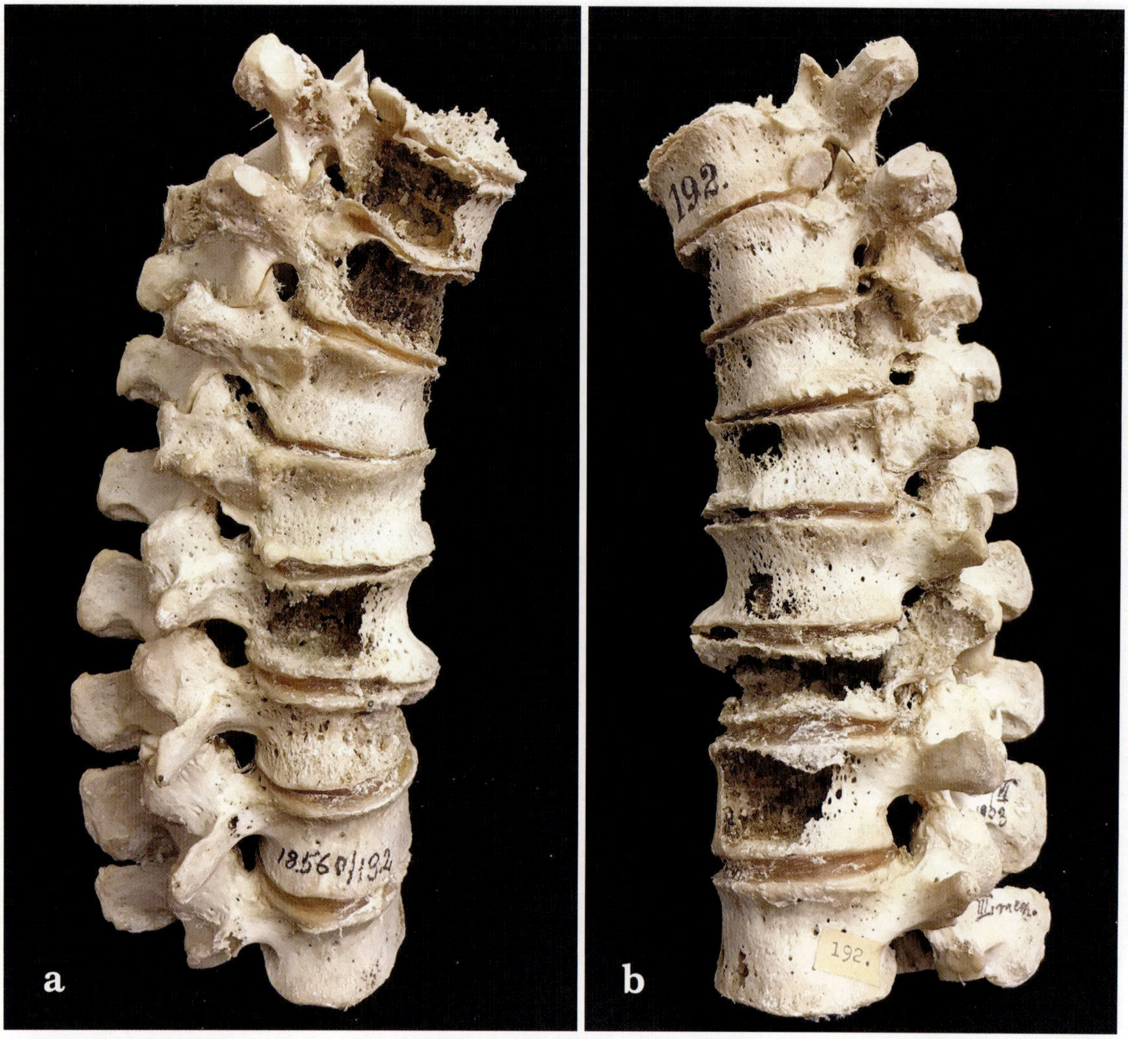


**Figure 329a-b.** (a) Right and (b) left oblique views of destructive lesions in the lower thoracic vertebrae secondary to carcinoma of the stomach in an unknown adult (MN18.560/192). The cancer resulted in lytic (destructive) lesions of the centra leading to collapse and anterior wedging (kyphosis). See Buikstra (1981); Christopher (1936); Steinbock (1976) for more on destructive lesions in the spine.

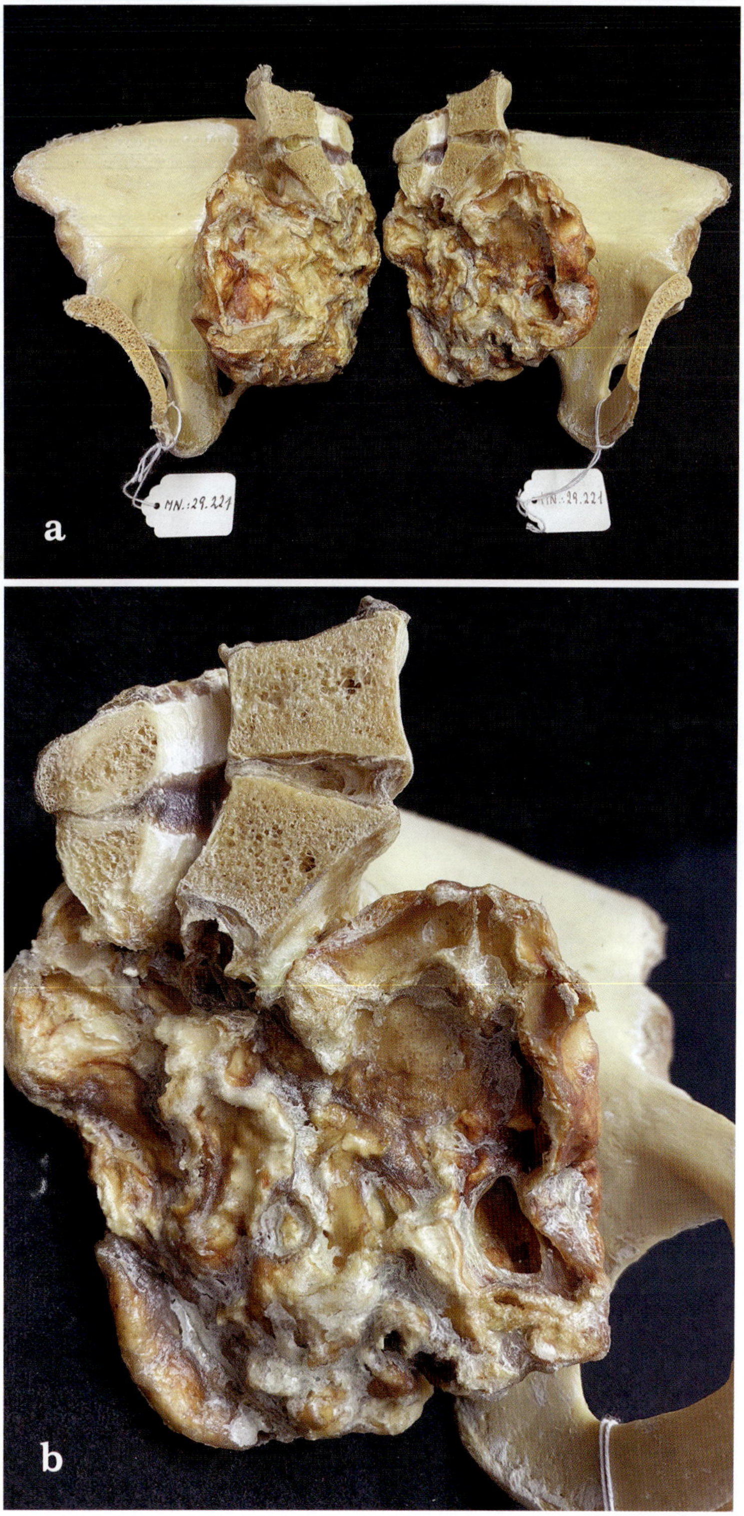


**Figure 330a-b.** (a) Sagitally sectioned lower spine and hips and (b) left innominate with a large sarcoma in a subadult of unknown sex (MN29.221). See Geshchickter and Copeland (1936) for more on sarcoma.

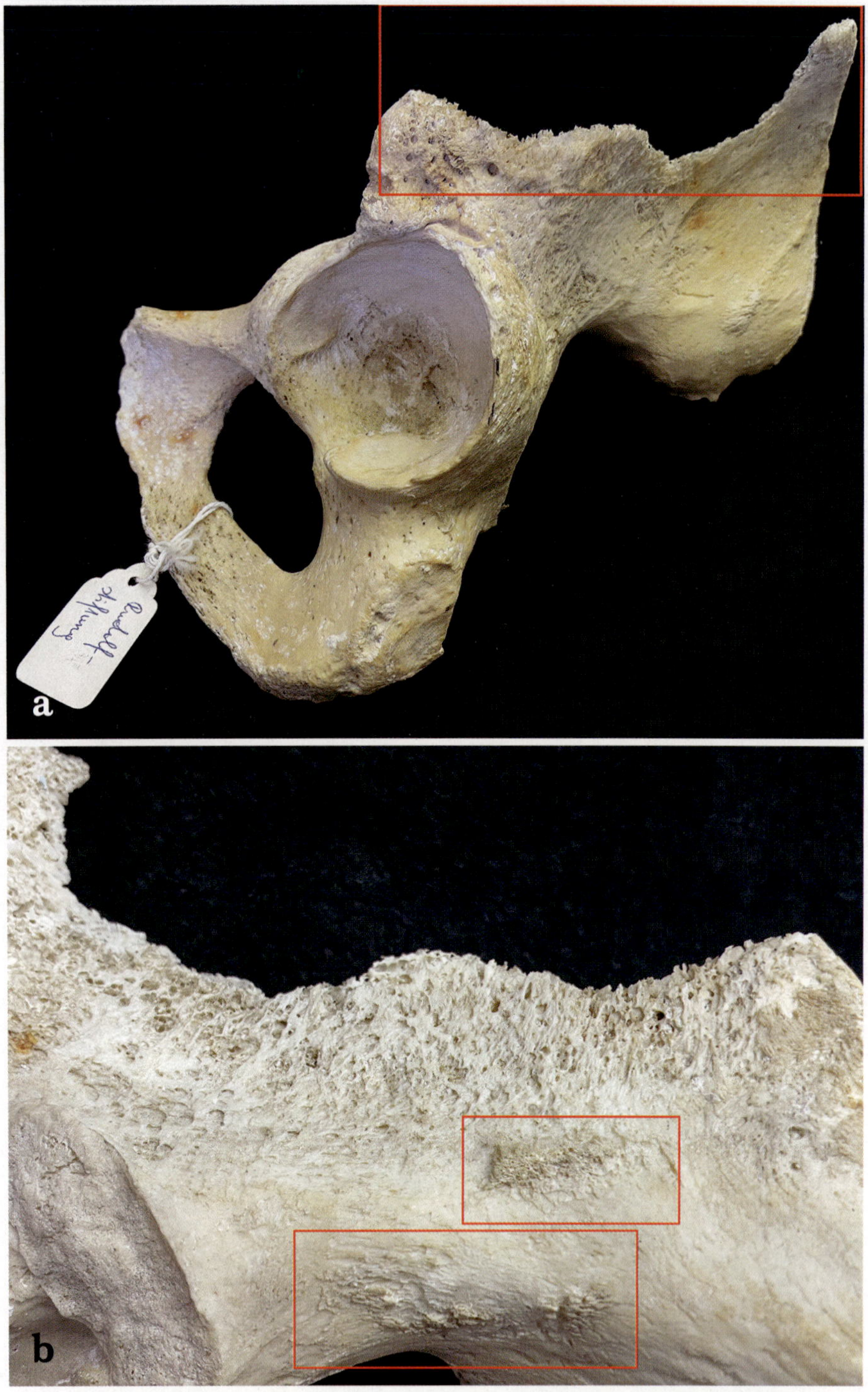


**Figure 331a-b.** (a) External and (b) magnified internal views of left innominate with metastatic carcinoma (rectangles) in a 52-year-old female who died in 1904 (MN18.560/3564). (b) Note the jagged and porous appearance of the bone with slight periosteal remodeling (rectangles).

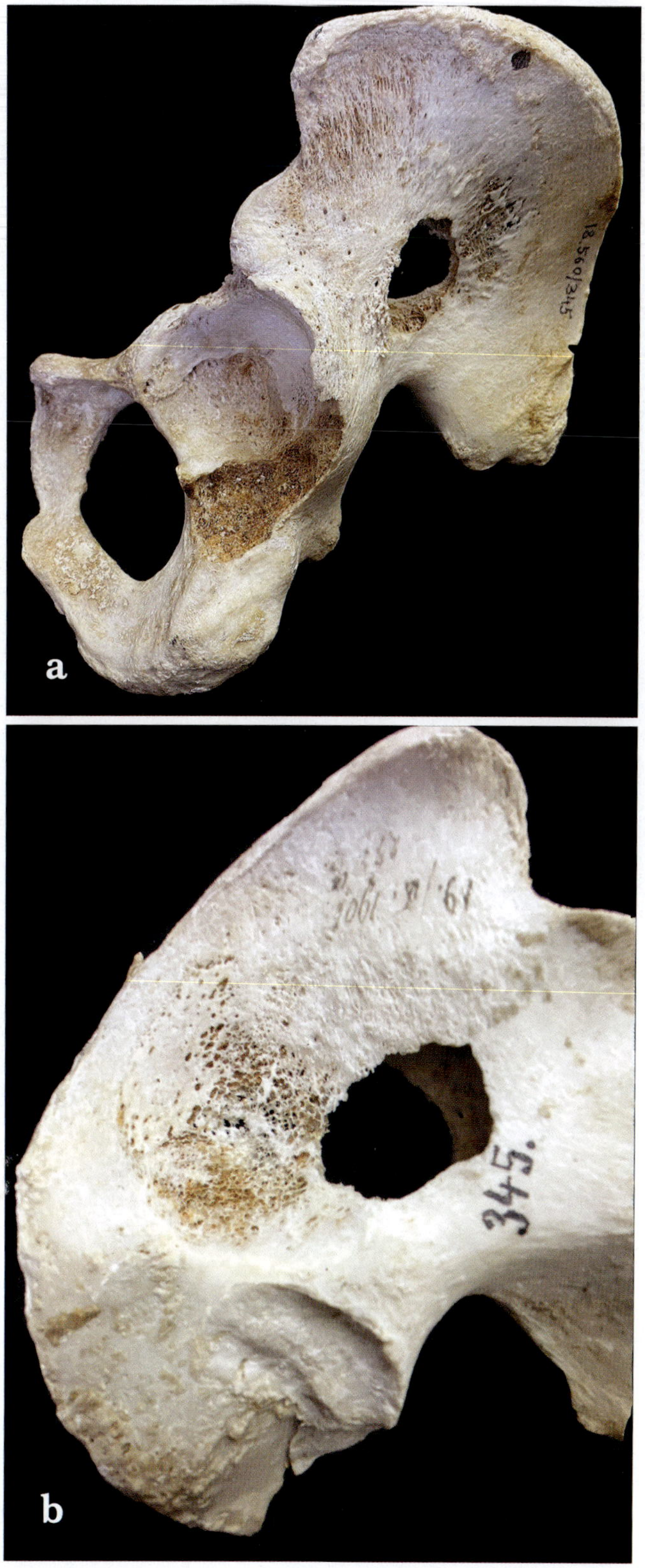


**Figure 332a-b.** (a) External and (b) internal views of left innominate in a 39-year-old male with multiple lesions due to metastatic carcinoma secondary to esophageal cancer who died in 1905 (MN18.560/345).

## Leg

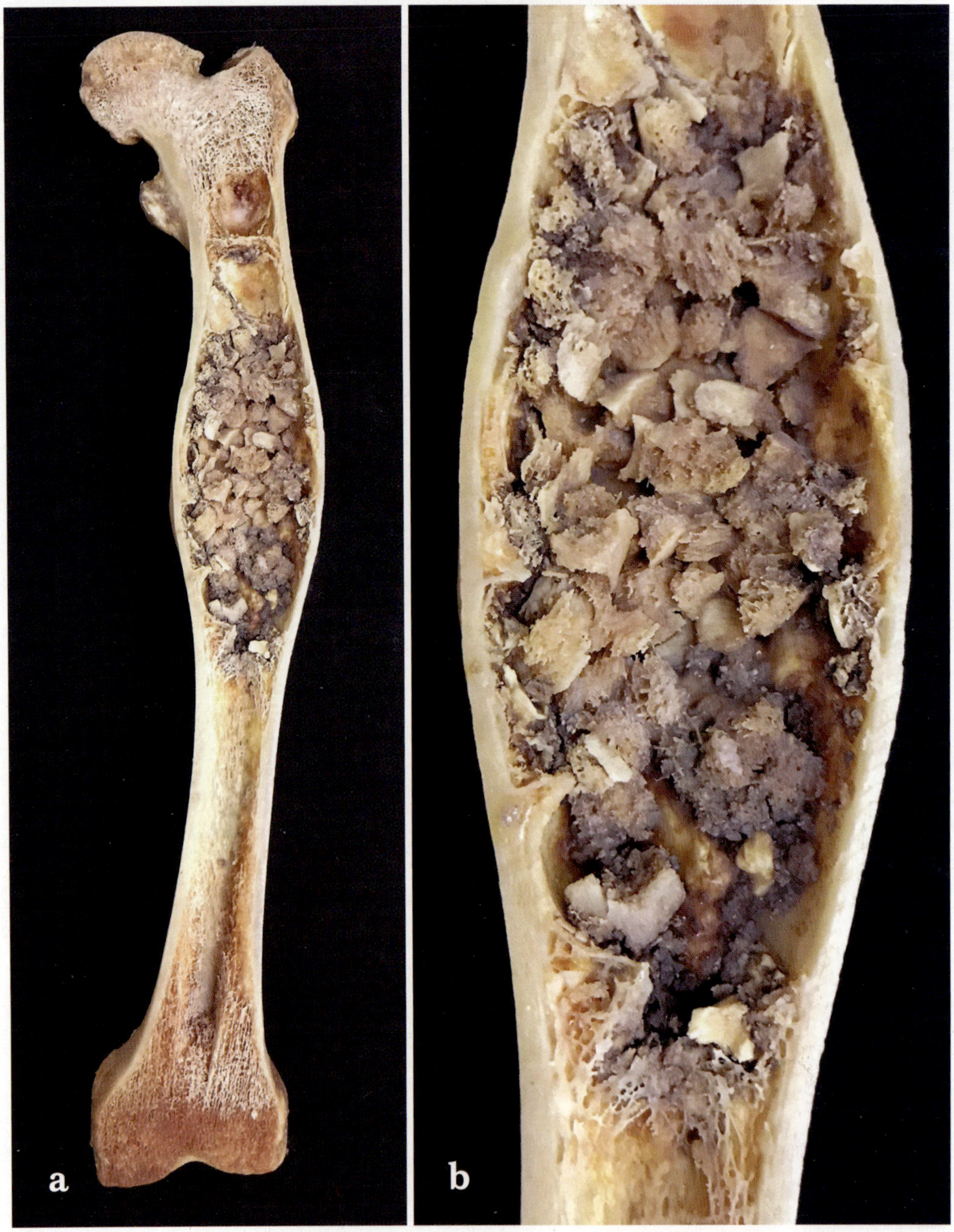


**Figure 333a-b.** (a) Posterior-internal view of monostotic fibrocystic disease ("osteodystrophia fibrosa cystica" on the museum card) with reimplantation of bone material resembling stones in the cystic region in the left femur of a 49-year-old male who died in 1955 and (b) magnified view of reimplantation material (MN5953). This man was injured and surgically treated in 1918, developed problems with his leg and was operated on again in 1955 and died 2 days later. Fibrocystic disease is caused by an overproduction of parathyroid hormone in over active thyroid glands (hyperparathyroidism) resulting in osteoclastic bone loss and replacement by fibrous tissue. See Gask (1934); Krigsten and Phalen (1945); Sing and Yap (2008).

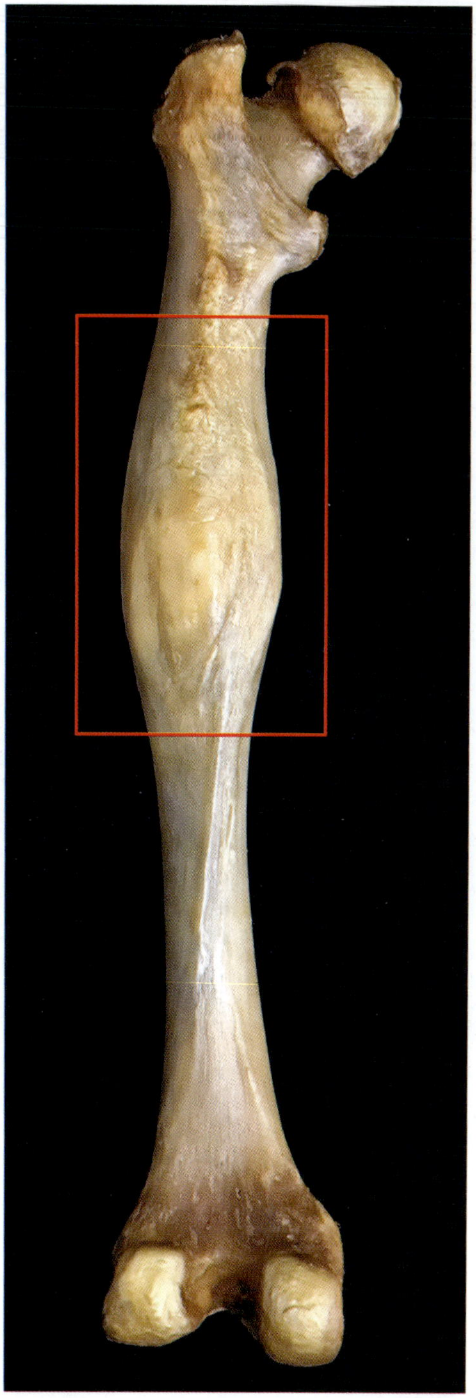

**Figure 334.** Posterior view of monostotic (affecting only one bone) fibrocystic disease with reimplantation of bone material in the cystic region of the left femur of a 49-year-old male injured in 1918 and underwent surgery in 1955 and died 2 days later (MN5953). Fibrocystic disease is caused by an overproduction of parathyroid hormone in over-productive thyroid glands (hyperparathyroidism) resulting in osteoclastic resorption of bone and replacement with fibrous tissue (see Gask 1934; Krigsten and Phalen 1945).

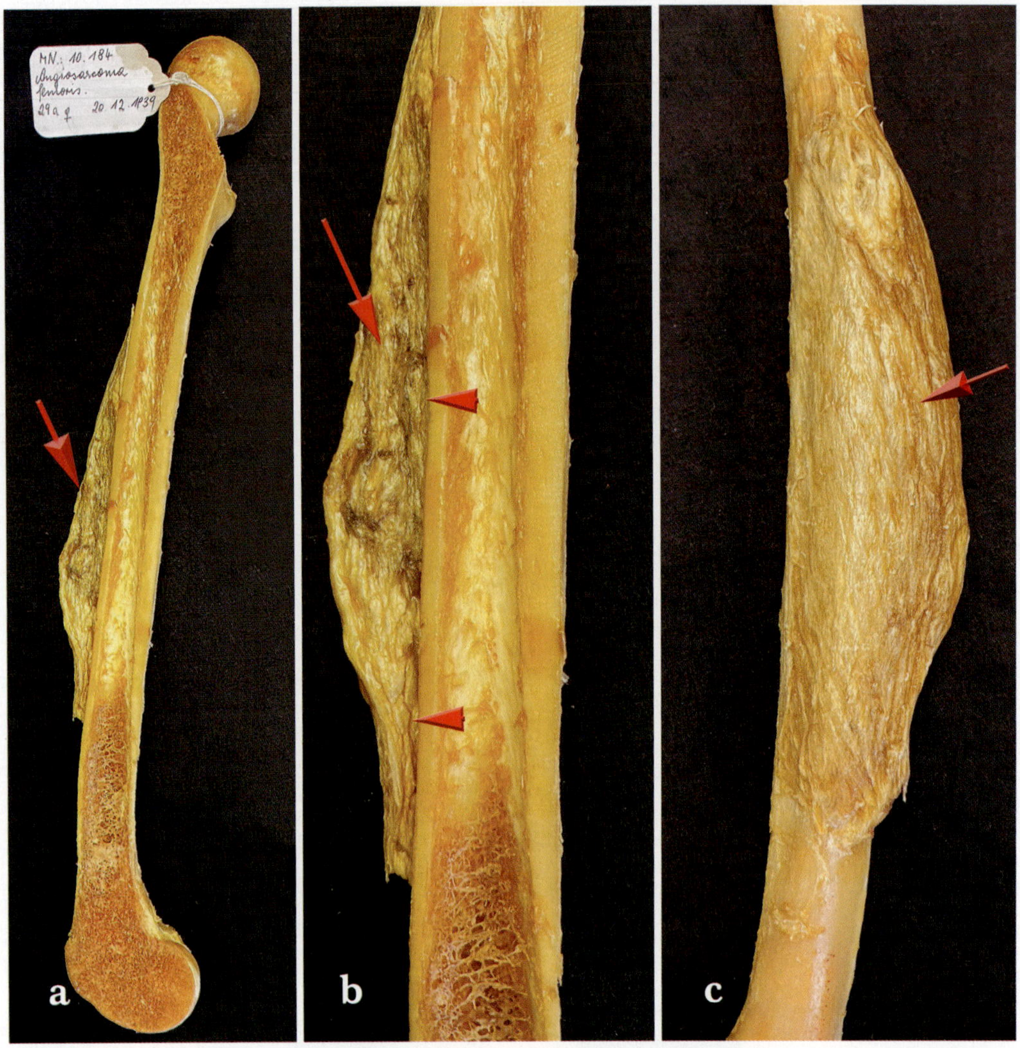


**Figure 335a-c.** (a) Longitudinally sectioned, (b) magnified view of the tumor and (c) medial view of right femur with angiosarcoma in a 29-year-old female who died in 1939. Angiosarcoma is an aggressive malignant tumor of vascular or lymphatic origin (Gaballah et al. 2017). See Mirra (1980); Evans et al. (2003) for more on angiosarcoma.

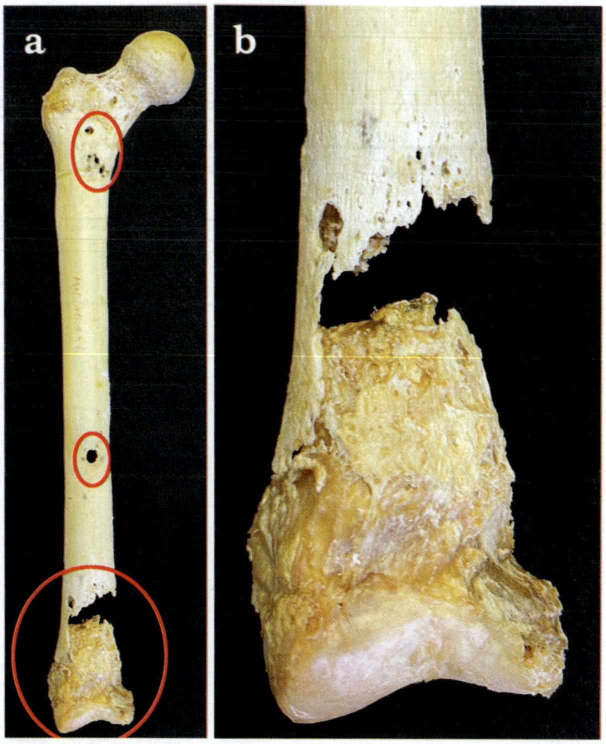


**Figure 336a-b.** (a) Anterior view of multiple destructive lesions and (b) magnified view of the distal right femur in a 29-year-old female with angiosarcoma of bone (MN10.184). Angiosarcoma is an extremely rare malignant tumor of vascular origin that occurs primarily between 50 and 70 years of age (Palmerini et al. 2020).

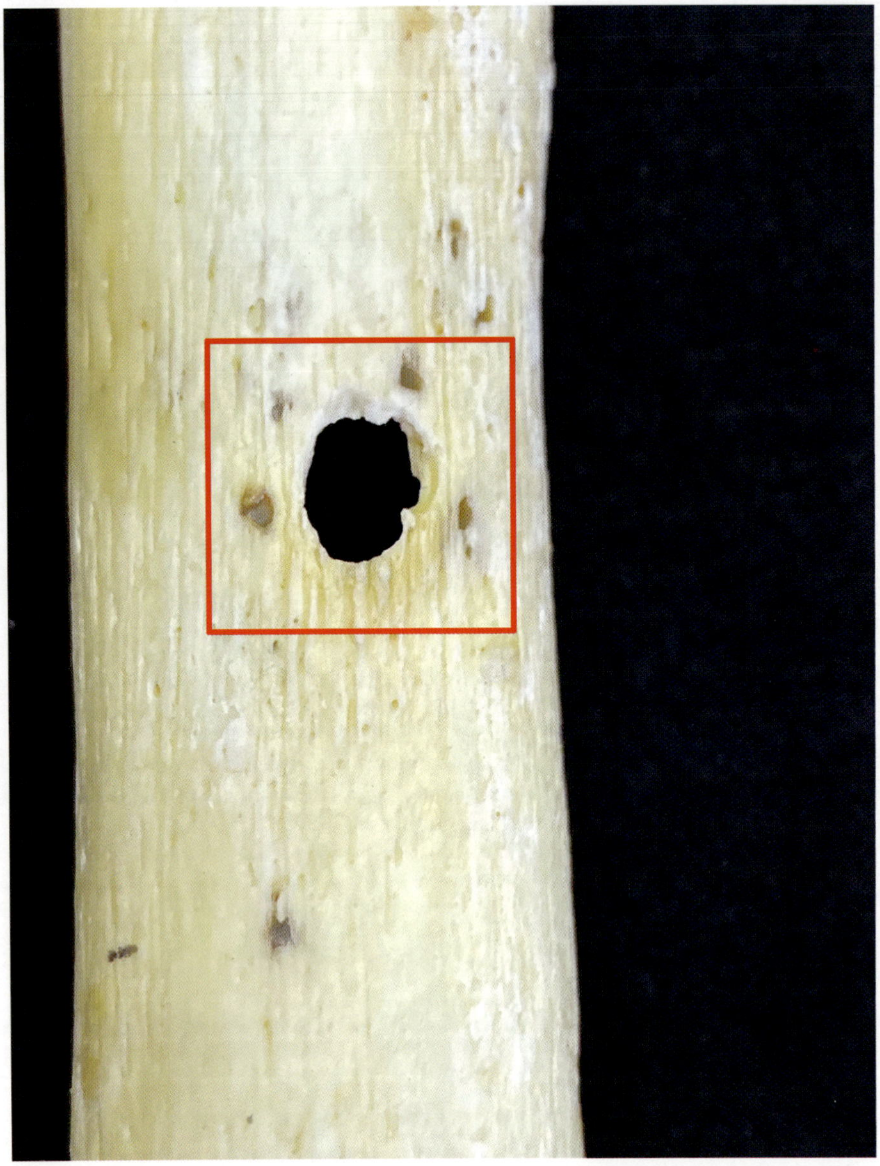

**Figure 337.** Magnified anterior view of a lytic lesion (square) near the midshaft of a right femur in a 29-year-old female with angiosarcoma (MN10.184). Note the presence of several other smaller lytic lesions (dark in color) surrounding the large lytic defect. Angiosarcoma is an extremely rare malignant tumor of vascular origin that occurs primarily between 50 and 70 years of age (Palmerini et al. 2020).

# Chapter 4

# DEGENERATIVE JOINT DISEASE AND OSTEOARTHRITIS

## Spine and Ribs

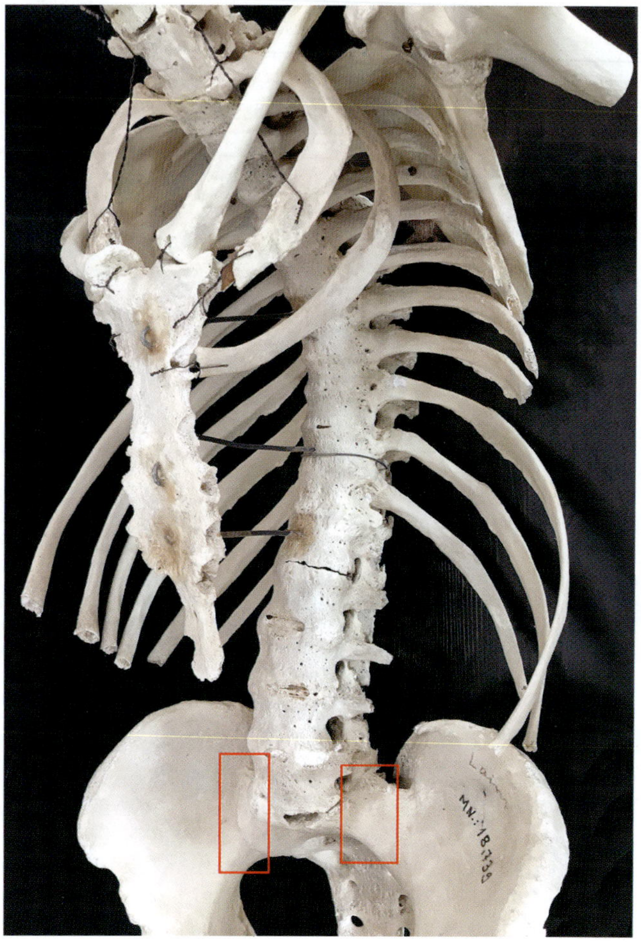


**Figure 338.** Ankylosing spondylitis (AS), a form of arthritis that results in inflammation between the vertebrae and sacroiliac joints, in an unknown individual (innominate bones suggest female) who died before 1974 (MN18.739). The ligaments of the vertebrae have synostosed giving the spine a "bamboo-like" appearance and the sacroiliac joints have fused (rectangles). AS typically differs from other forms of spondylosis in that the spine doesn't exhibit large, bulbous osteophytes joining the vertebral centra. See Mann and Hunt (2005); Ortner (2003, 2012); Ortner and Putschar (1981); Schmorl and Junghanns (1959). AS is also sometimes called Bechterew's disease, Marie-Strumpell disease and rheumatoid spondylitis (Neumann 1967).

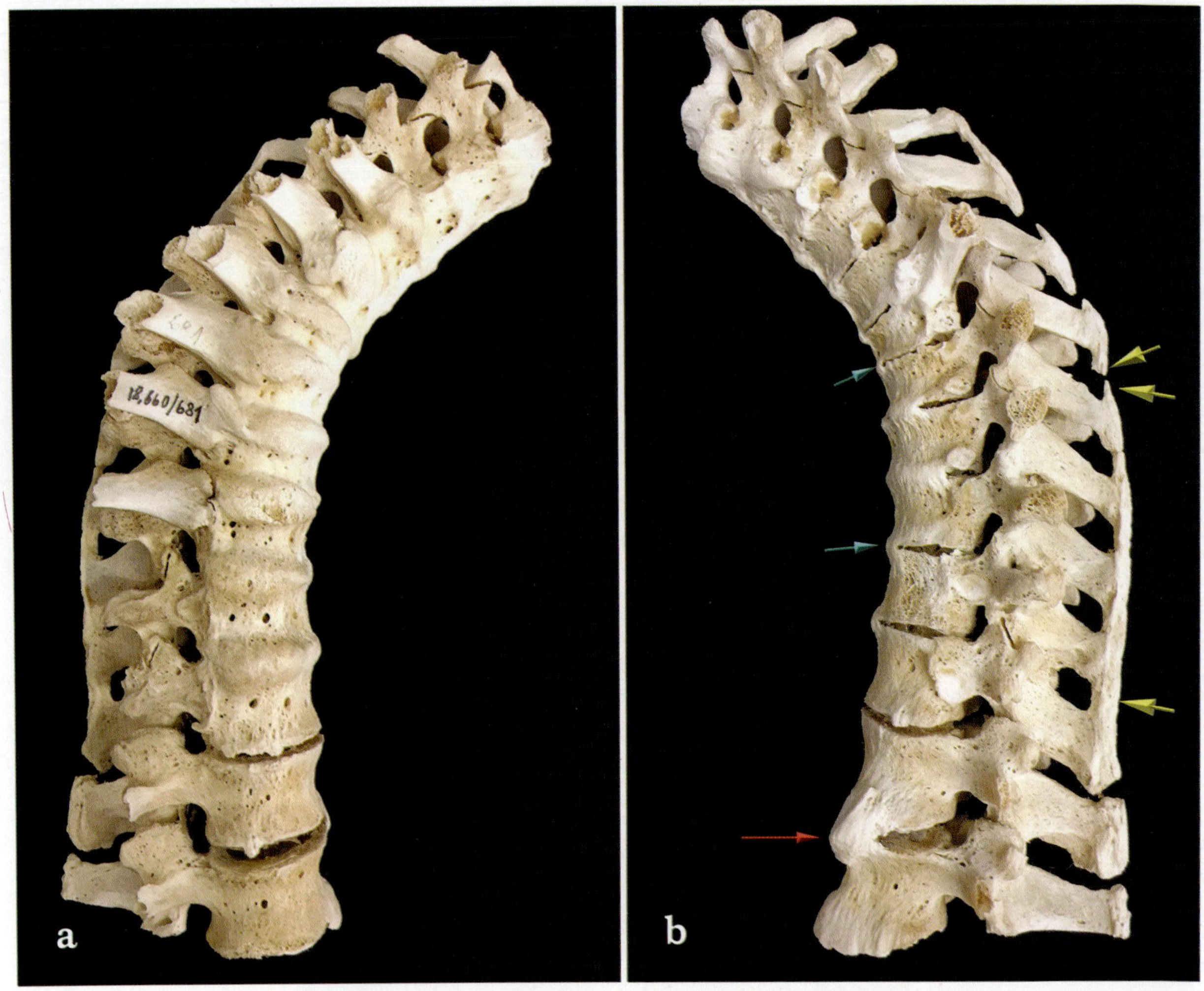


**Figure 339a-b.** (a) Right and (b) left lateral views of chronic ankylosing spondylitis (AS) ("morbus bechterew" on the museum card) in a 74-year-old male who died in 1927 (MN18.660/681). Several vertebrae (blue arrows) exhibit fusion between the vertebrae giving the spine a "bamboo-like" appearance and the lower spine a more bulbous (red arrow) and projecting fusion of the vertebral centra (bodies). Ankylosing spondylitis, often called "bamboo" spine, usually begins to appear in the 30s and results in inflammation of the vertebrae and sacrum resulting in reduction of disc space and fusion between vertebrae and the sacroiliac joints. (b) This condition is also often accompanied by ossification of the supraspinous ligaments (yellow arrows) which further limits mobility of the spine. See Mann and Hunt (2005); Ortner (2003, 2012); Ortner and Putschar (1981); Schmorl and Junghanns (1959).

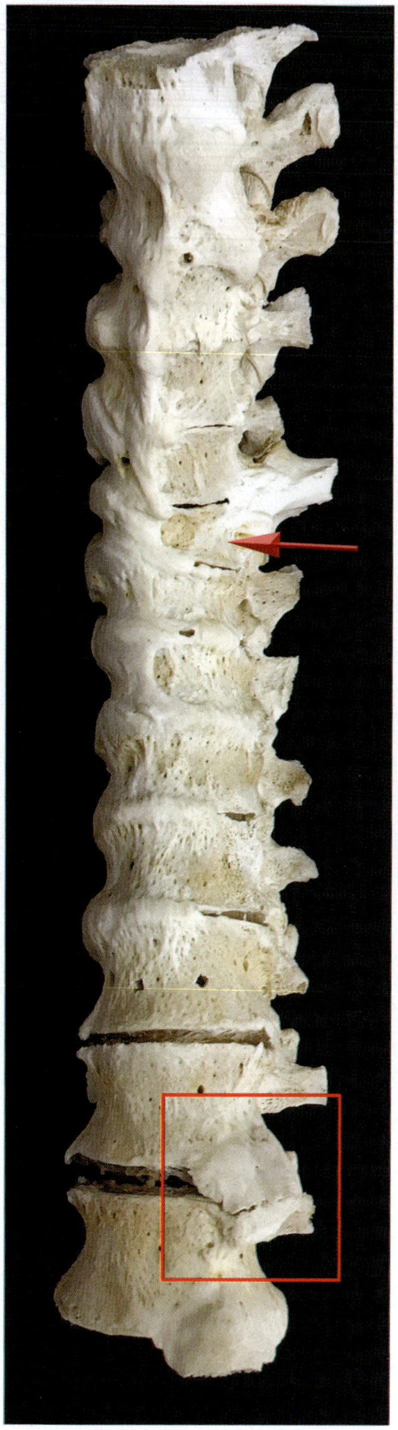

**Figure 340.** Anterior view of chronic ankylosing spondylitis ("bamboo" spine) ("morbus bechterew" on the museum card), compression fracture of a thoracic vertebra (arrow), and large buttressing osteophytes (square) in a 74-year-old male who died in 1927 (MN18.660/681). Compression fracture such as in this individual is typically the result of osteoporosis and often results in a wedge-shaped centrum. See Schmorl and Junghanns (1959) for more on trauma and ankylosing spondylitis and Lenchik et al. (2004) for classification, diagnosis and treatment of osteoporotic vertebral fractures.

**Figure 341.** Left oblique view of ankylosing spondylitis and kyphosis or anterior curvature ("Spondylarthritis ankylopetica" on the museum card) of C2 to L4 vertebrae in an unknown adult who died in 1900 (MN18.560/196). Ankylosing spondylitis, sometimes referred to as "bamboo" spine because of its resemblance to bamboo, is a chronic inflammatory arthritis that often affects the sacrum, spine, and ribs resulting in fusion and limited mobility. See Mann and Hunt (2005); Ortner (2003, 2012); Ortner and Putschar (1981); Schmorl and Junghanns (1959) for more on ankylosing spondylitis and spinal deformities.

## Leg

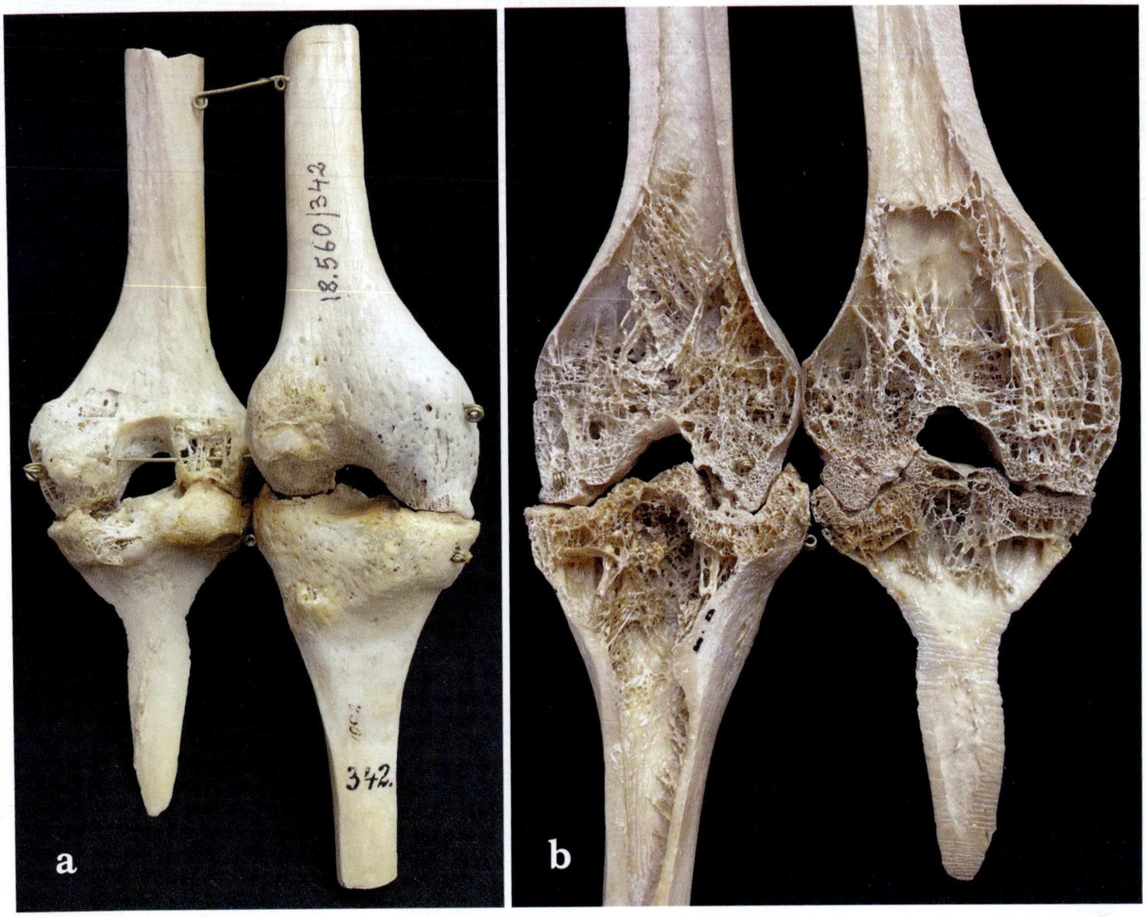


**Figure 342a-b.** (a) External and (b) internal views of a right femur and tibia with extreme osteoporosis due to "Arophie ossium articulationis genu dextri" (on the museum card) in an unknown adult who died in 1930 (MN18.560/342). The femoro-tibial joint has greatly altered resulting in total joint line destruction with surrounding osteoporosis and some bony fusion. Note the altered shape of the joint and extreme loss of trabeculae in both the femur and tibia. This type of primary joint disease would not be infection or degenerative, but arthritis or systemic disease. Juvenile rheumatoid arthritis and hemophilia can result in joint destruction such as this (P. Moroz, personal communication, July 10, 2022).

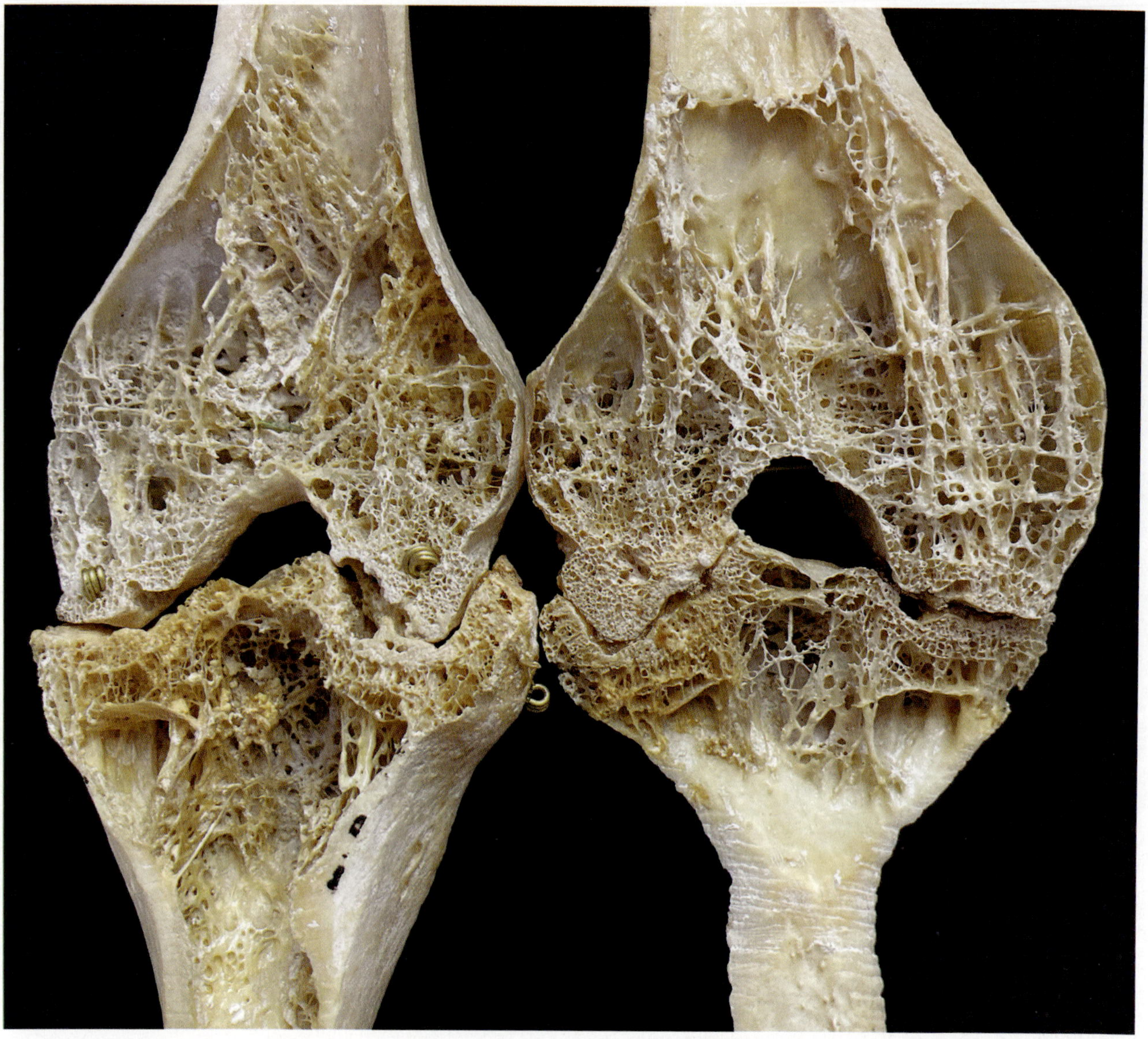

**Figure 343.** Magnified view of a right femur and tibia with extreme osteoporosis due to "Arophie ossium articulationis genu dextri" (on the museum card) in an unknown adult who died in 1930 (MN18.560/342). The femoro-tibial joint has greatly altered resulting in total joint line destruction with surrounding osteoporosis and some boney fusion. Note the altered shape of the joint and extreme loss of trabeculae in both the femur and tibia. This type of primary joint disease would not be infection or degenerative, but arthritis or systemic disease. Juvenile rheumatoid arthritis and hemophilia can result in joint destruction such as this (P. Moroz, personal communication, July 10, 2022).

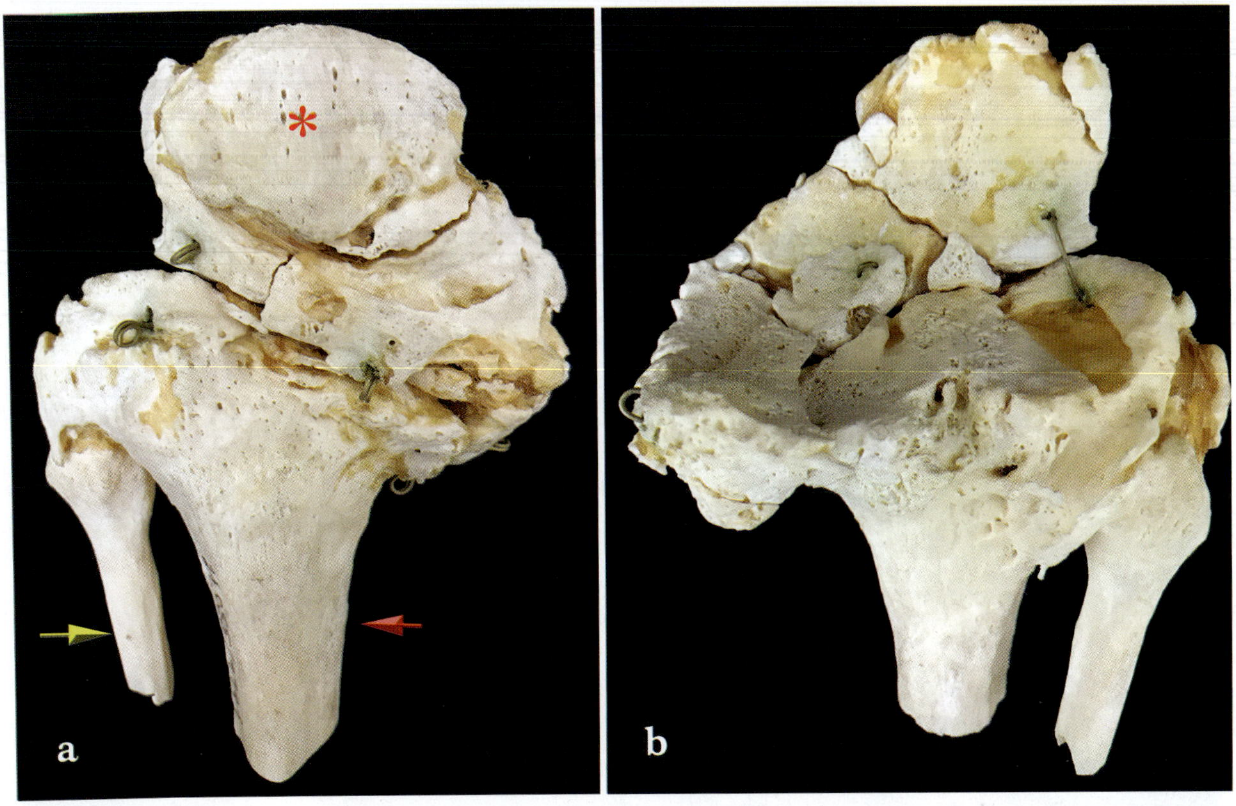


**Figure 344a-b.** (a) Anterior and (b) posterior views of chronic osteoarthritis ("gonarthritis" which means inflammation of the knee joint) of the right knee resulting in destruction, neuropathic arthropathy of the cartilage and production of numerous osteophytes and heterotopic bones in a 75-year-old male who died in 1909 (MN21.055/5593) (*=patella, red arrow=tibia and yellow arrow=fibula). This condition is due to chronic mechanical use of the joint in the absence of the normal sensation of pain, possibly due to neurosyphilis (tabes dorsalis) (Figueiredo et al. 2018). Syphilis was a major cause of neuropathic arthropathy in the 1800s, with diabetes replacing it as a major cause today (Gouveri and Papanas, 2011). Ortner (1981, p. 434) presents this knee as "Charcot's joint of right knee," a term sometimes used to describe this condition in any joint of the body (Ahmad et al., 2020), is usually used only when referring to the foot and ankle. Charcot joint or Charcot arthropathy, a "named disease" (Charcot 1868), is slowly being replaced with "neuropathic arthopathy" (P. Moroz, personal communication, July 12, 2022).

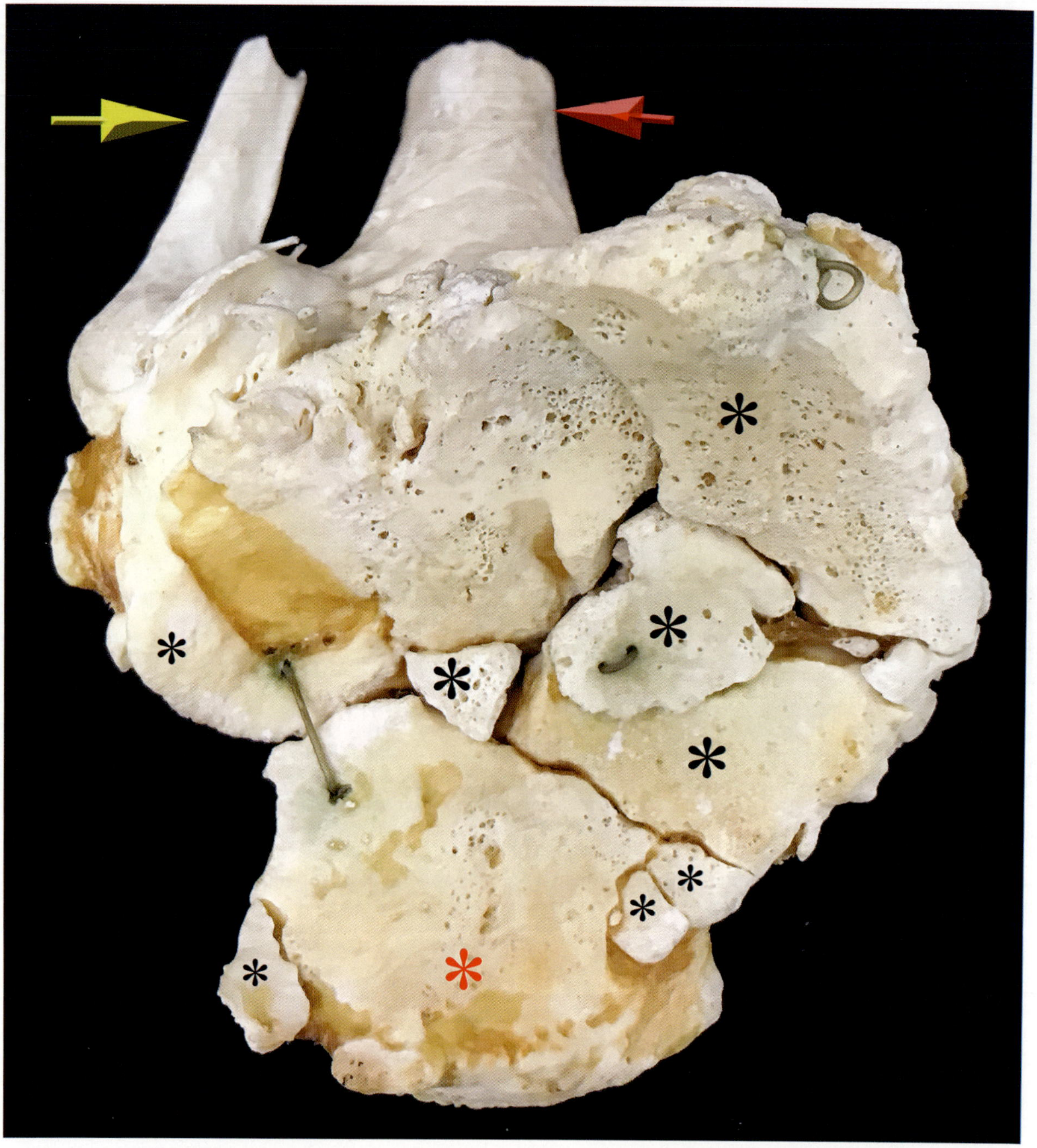

**Figure 345.** Superior view of chronic osteoarthritis ("gonarthritis" which means inflammation of the knee joint) of the right knee resulting in destruction, neuropathic arthropathy, of the cartilage and the production of numerous osteophytes and heterotopic bones (black *), many with articular facets, in a 75-year-old male who died in 1909 (MN21.055/5593) (red *=patella, red arrow=tibia and yellow arrow=fibula). This condition is due to chronic mechanical use of the joint in the absence of the normal sensation of pain, possibly due to neurosyphilis (tabes dorsalis) (Figueiredo et al. 2018). Syphilis was a major cause of neuropathic arthropathy in the 1800s, with diabetes replacing it as a major cause today (Gouveri and Papanas, 2011).

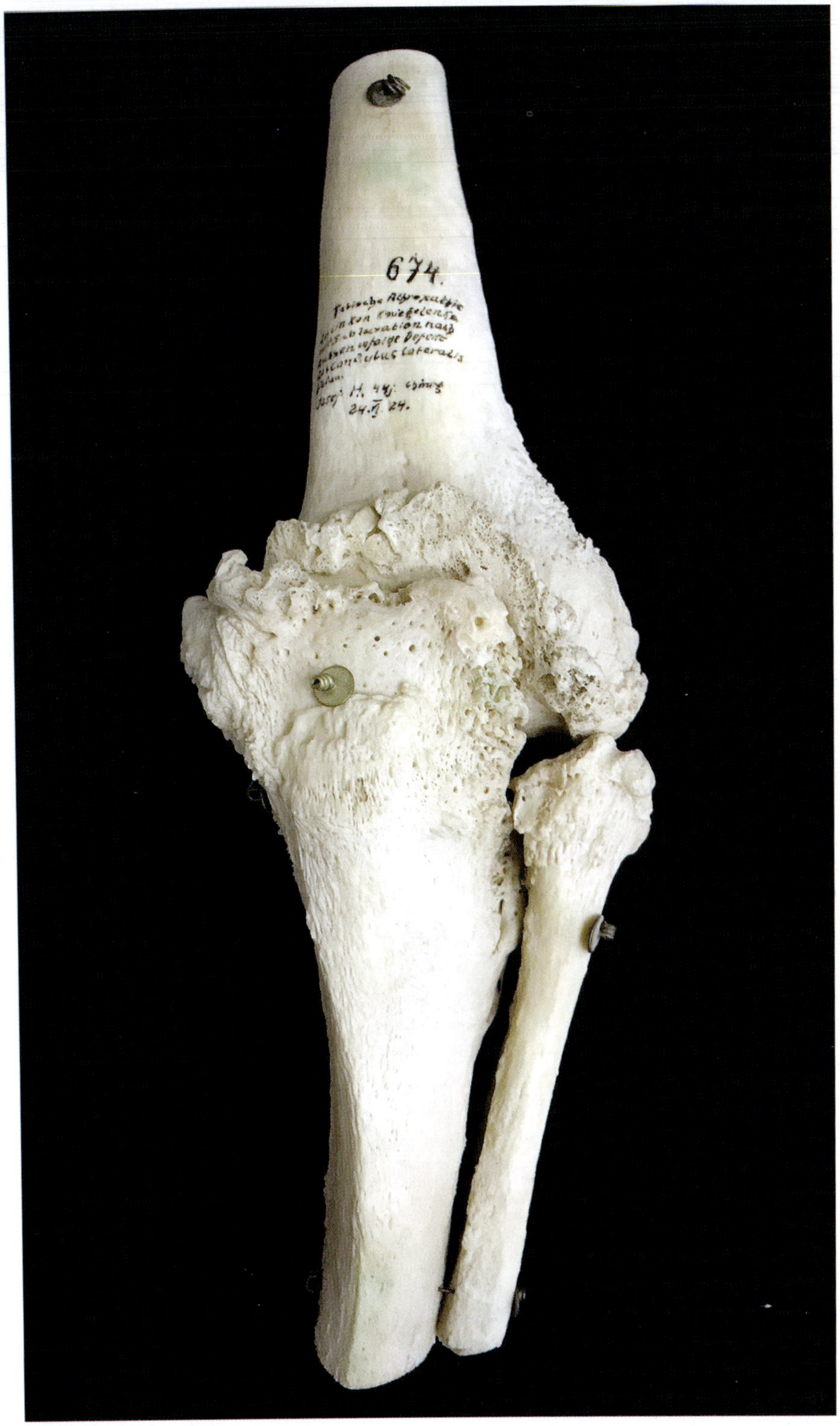


**Figure 346.** Anterior view of a subluxed (dislocated) left knee with destruction of the lateral condyles of the femur and tibia due to osteochondroma in an adult female who died in 1823 (MN674).

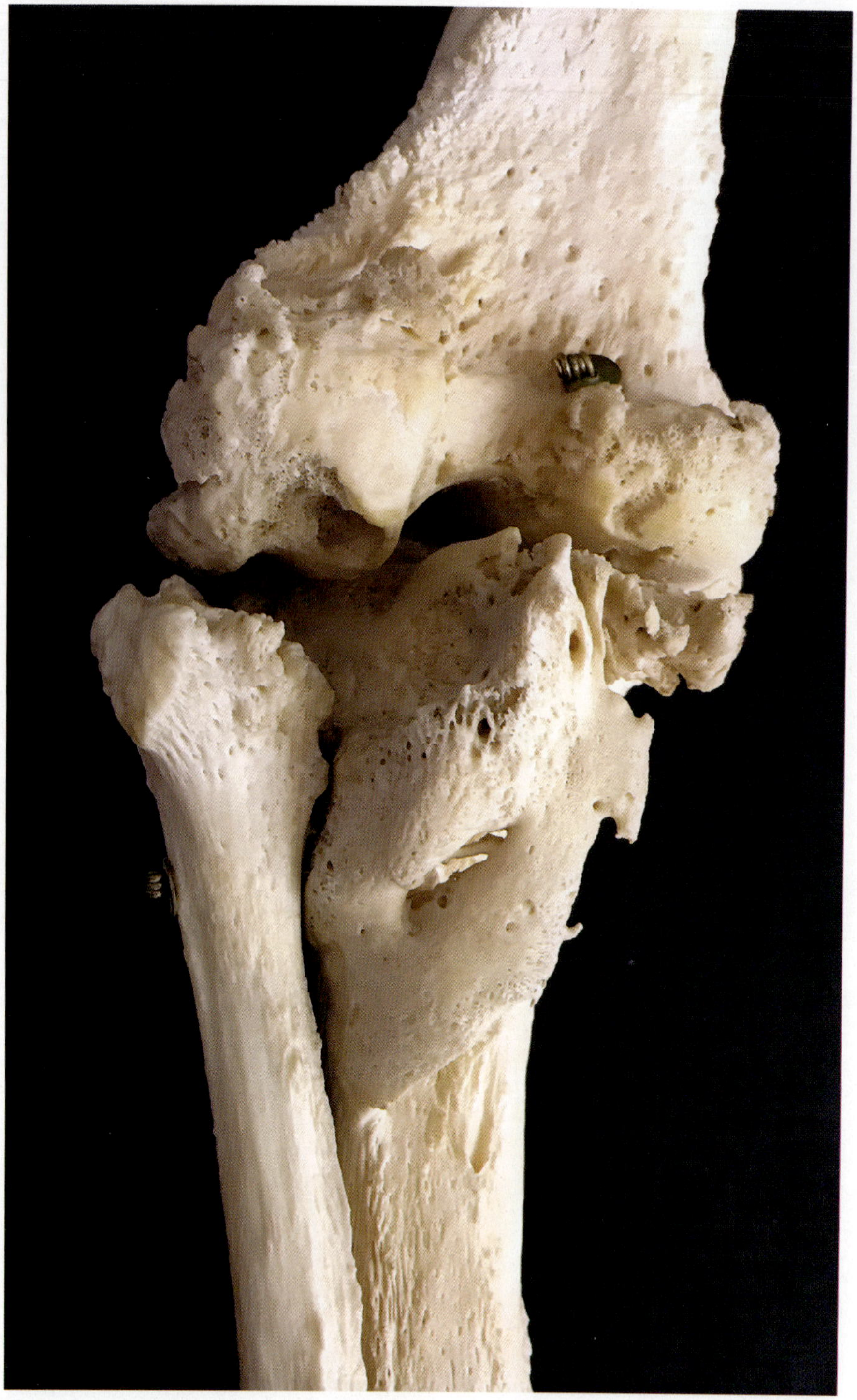

**Figure 347.** Posterior view of a subluxed (dislocated) left knee with destruction of the lateral condyles of the femur and tibia due to osteochondroma in an adult female who died in 1823 (MN674).

# Chapter 5

# TRAUMA AND MEDICAL IMPLANTS

## Skull

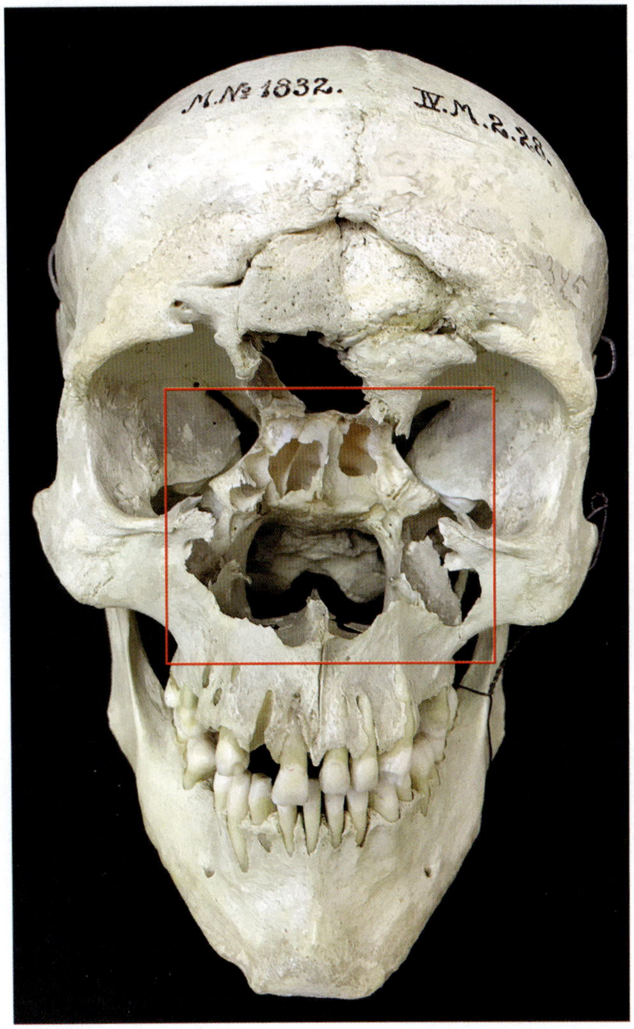


**Figure 348.** Healed blunt force trauma to the frontal bone in a 31-year-old male who was kicked in the head by a horse and died in 1838 (MN1832). The absence of healing and light color of the bone margins suggest that part of the facial region (rectangle) was broken post-mortem. See DiMaio and DiMaio (2001); Kimmerle and Baraybar (2008); Watson-Jones (1944, 1955) for more on skeletal trauma.

**Figure 349.** Healed blunt force trauma (arrows) to the frontal region in a 31-year-old male who was kicked in the head by a horse and died in 1838 (MN1832). The broken bones overlap one another and the broken edges are rounded/smooth indicating remodeling. There is post-mortem breakage of the nasal and maxillary region as indicated by the absence of healing.

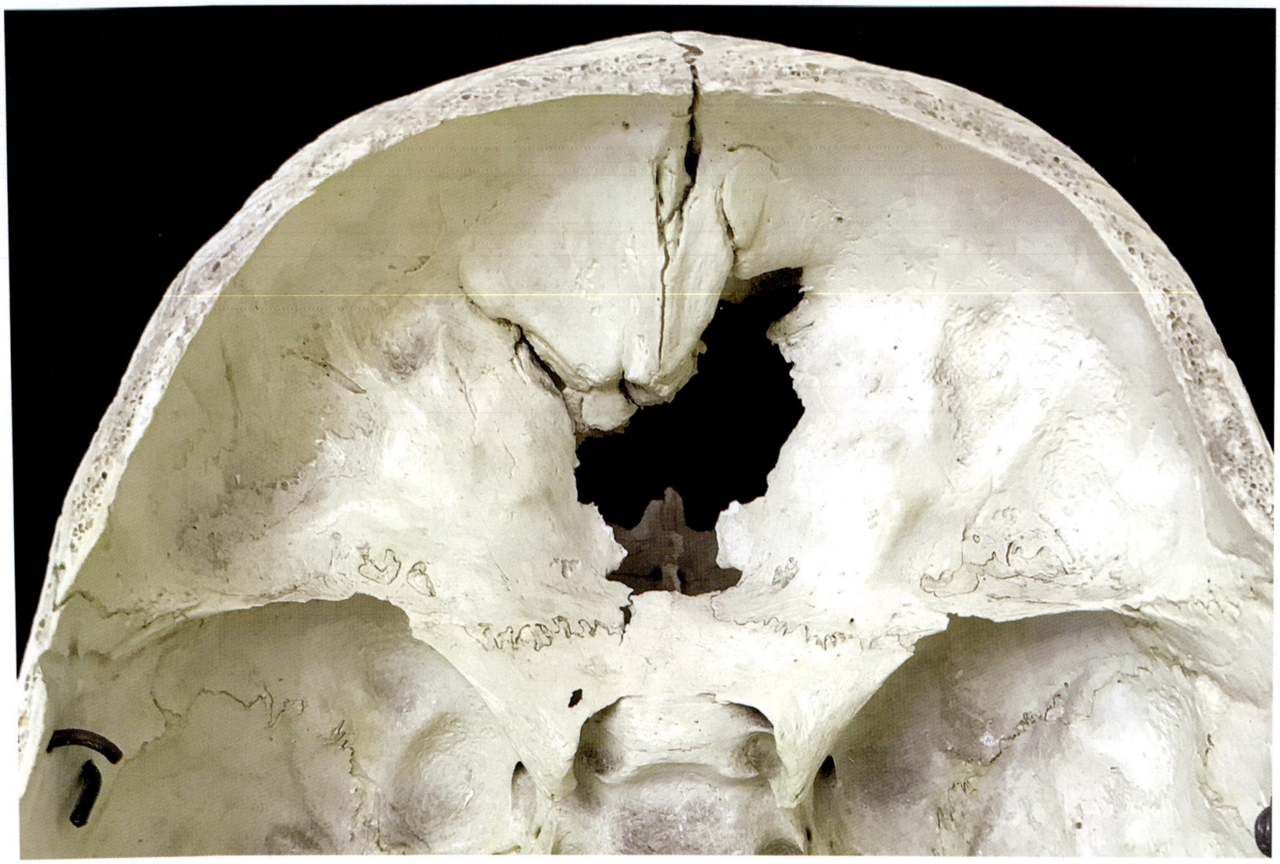

**Figure 350.** Endocranial view of healed blunt force trauma to the frontal region in a 31-year-old male who was kicked in the head by a horse and died in 1838 (MN1832). The margins of the broken bones overlap one another and are rounded and smooth, indicative of healing. See Kimmerle and Baraybar (2008); Watson-Jones (1944); Wedel and Galloway (2014) for more on skeletal trauma.

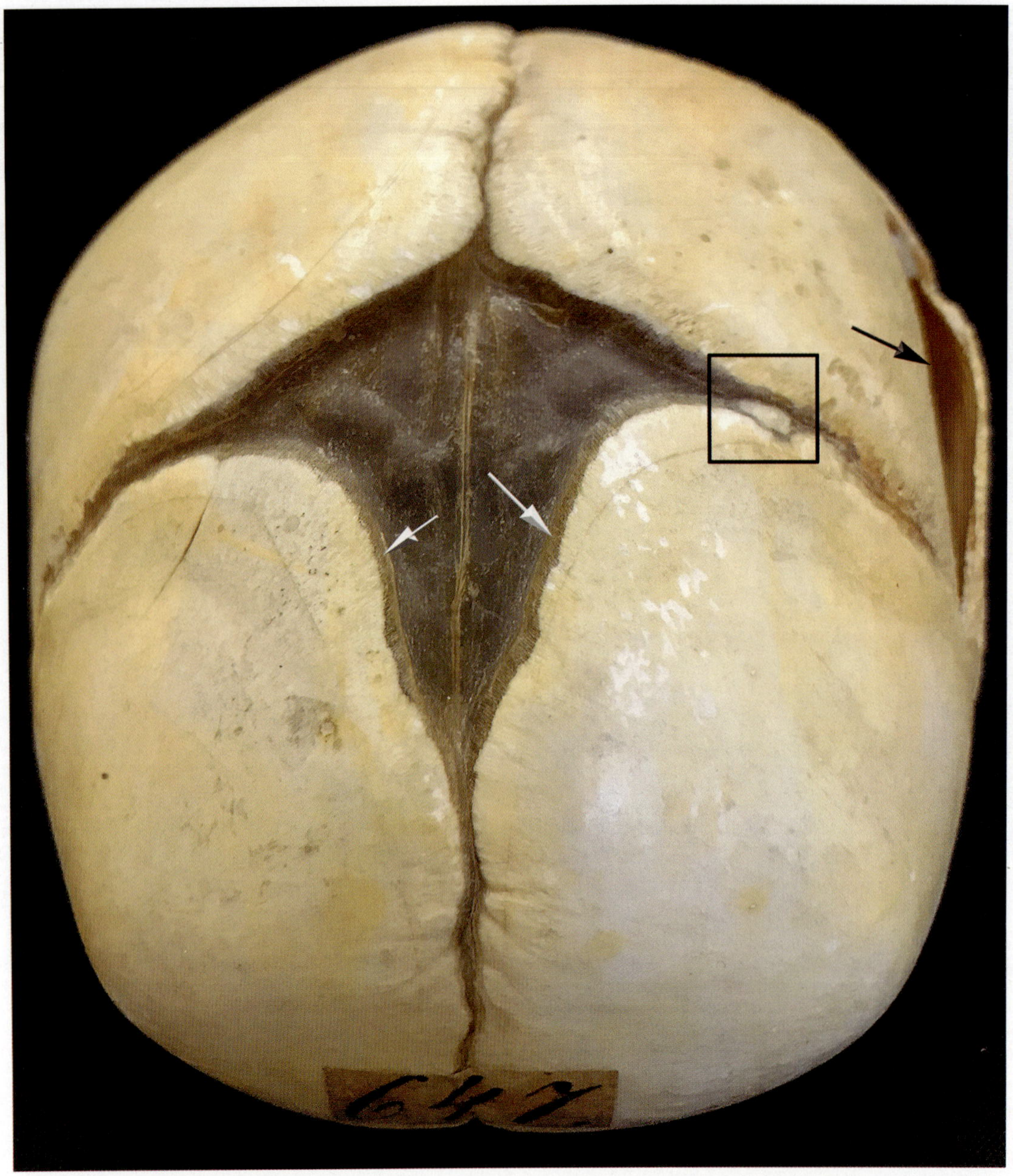


**Figure 351.** Superior view of skull with healing trauma (black arrow) of a juvenile female that was kicked in the head by a horse and died in 1863 (MN21.055/647). Note the open anterior fontanelle, coronal ossicle (square), and development of the interfrontal/metopic suture along its leading edge (white arrows), resulting in growth and increased size of the cranium.

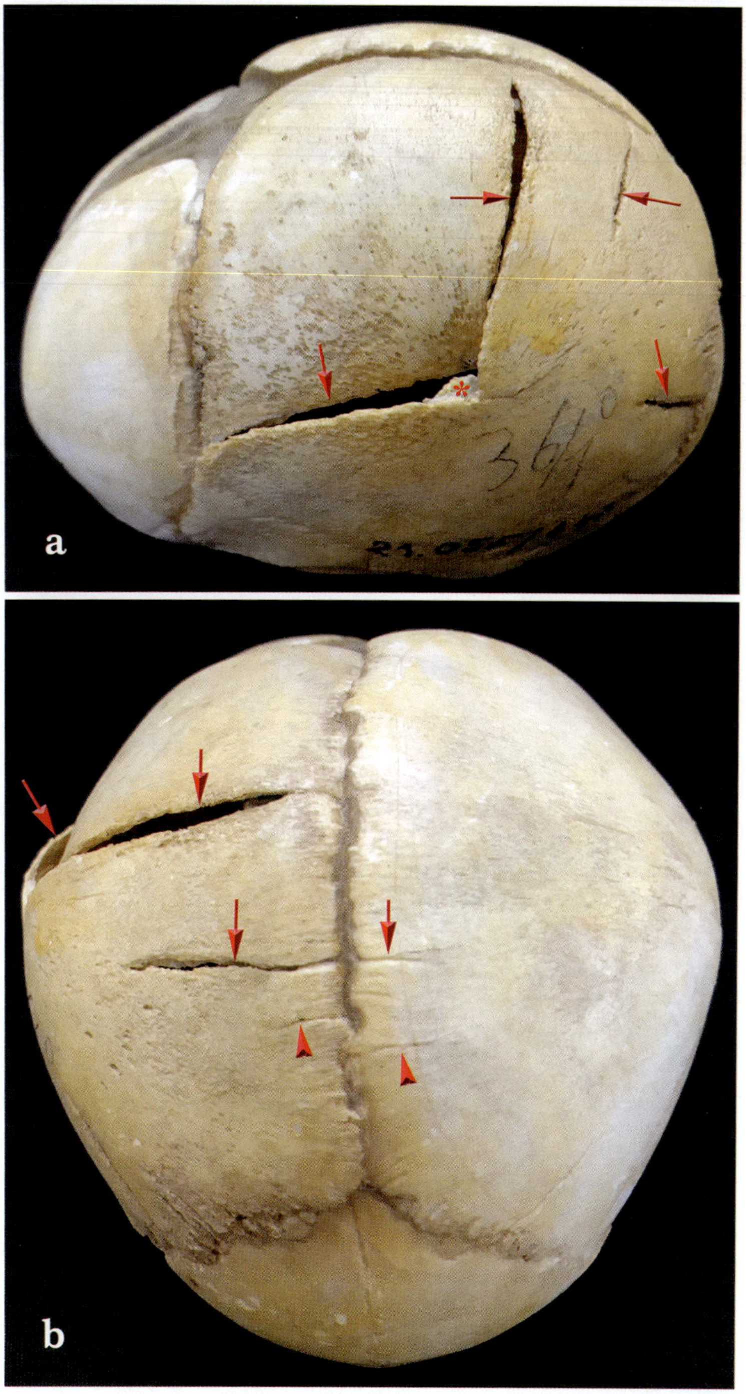


**Figure 352a-b.** (a) Left lateral and (b) superior views of a skull, of a juvenile female that was kicked in the head by a horse and died in 1863 (MN21.055/647). The skull exhibits multiple fractures (arrows), some of which show evidence of remodeling. (a) The site of impact appears to be near the middle of the left parietal bone resulting in diastatic fracture (separation) of the left coronal suture and a small shelf of the inner table (*). See DiMaio and DiMaio (2001); Kimmerle and Baraybar (2008); Watson-Jones (1944); Wedel and Galloway (2014) for more on skeletal trauma.

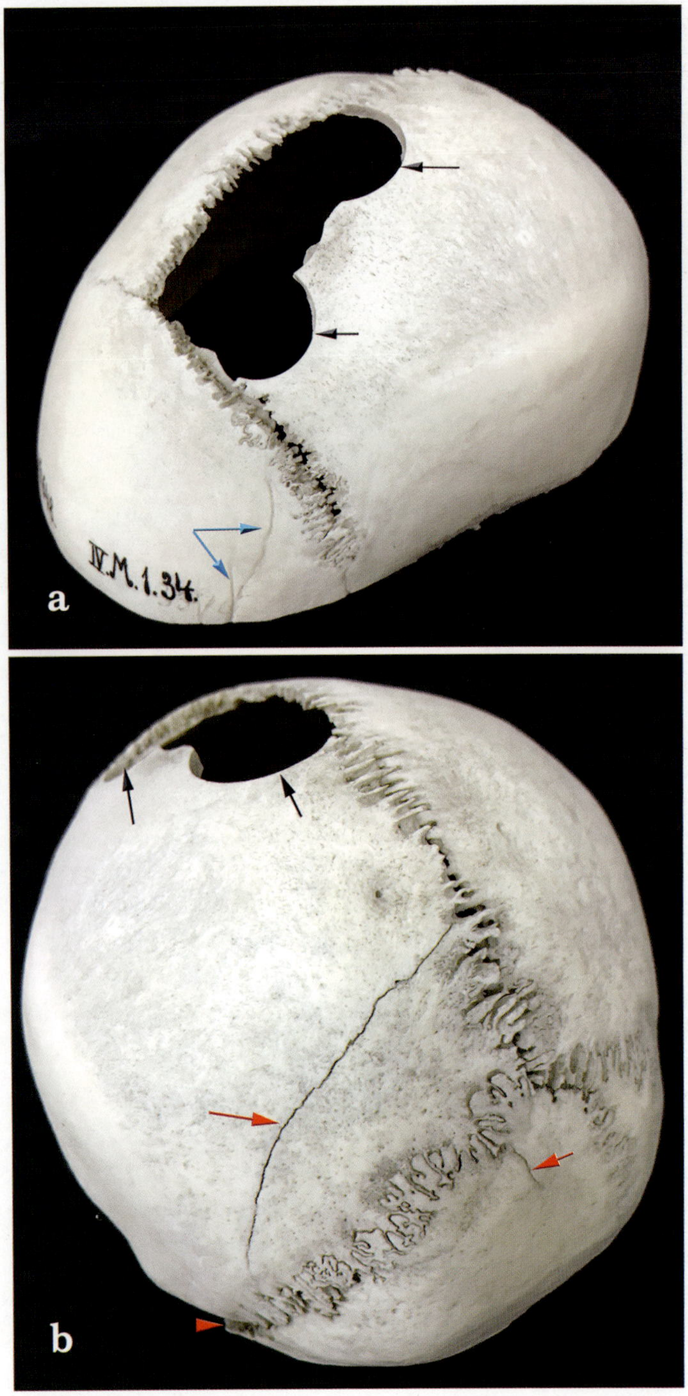


**Figure 353a-b.** Ectocranial views of a female who was kicked in the side/back of the head in the left parietal region by a horse and died in 1822 (MN648). Note the perimortem fractures (red arrows) with no evidence of healing. There are at least two surgical trepanations (black arrows), likely an attempt to relieve cranial pressure and endocranial bleeding. (a) Unrelated to the trauma are accessory grooves (blue arrows) in the left frontal bone. The large fracture is not visible endocranially, but the small fracture is (Figure 354).

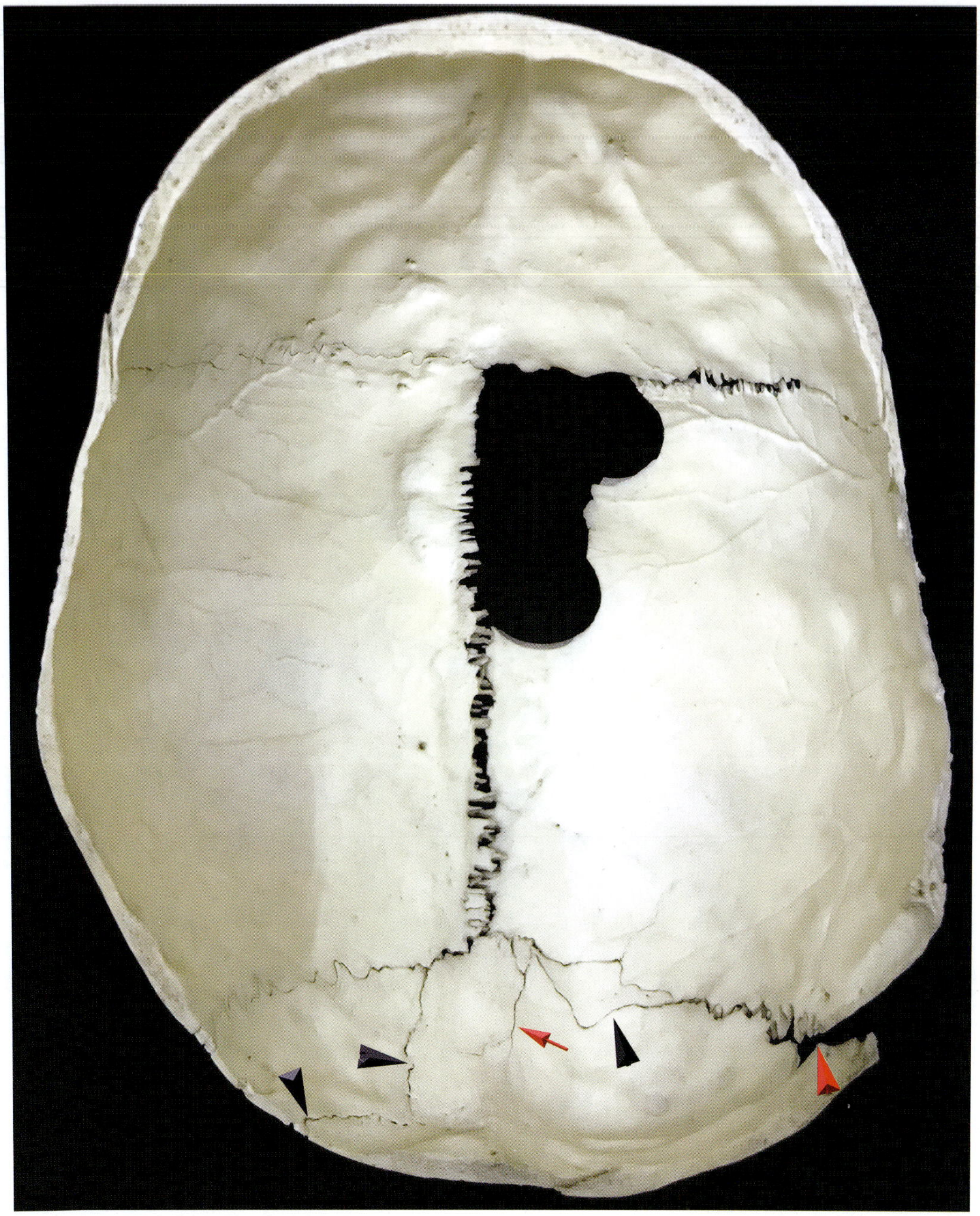

**Figure 354.** Endocranial view of a female kicked in the side of the head by a horse and died in 1822 (MN648). The small fracture (red arrow) is visible both ectocranially and endocranially. There are two accessory bones/ossicles (normal anatomical variants) with sutures (black arrowheads) that could be mistaken for perimortem fractures. Part of the lambdoidal suture opened (red arrowhead; diastatic fracture) as a result of the trauma. See Kimmerle and Baraybar (2008); Watson-Jones (1944); Wedel and Galloway (2014) for more on skeletal trauma.

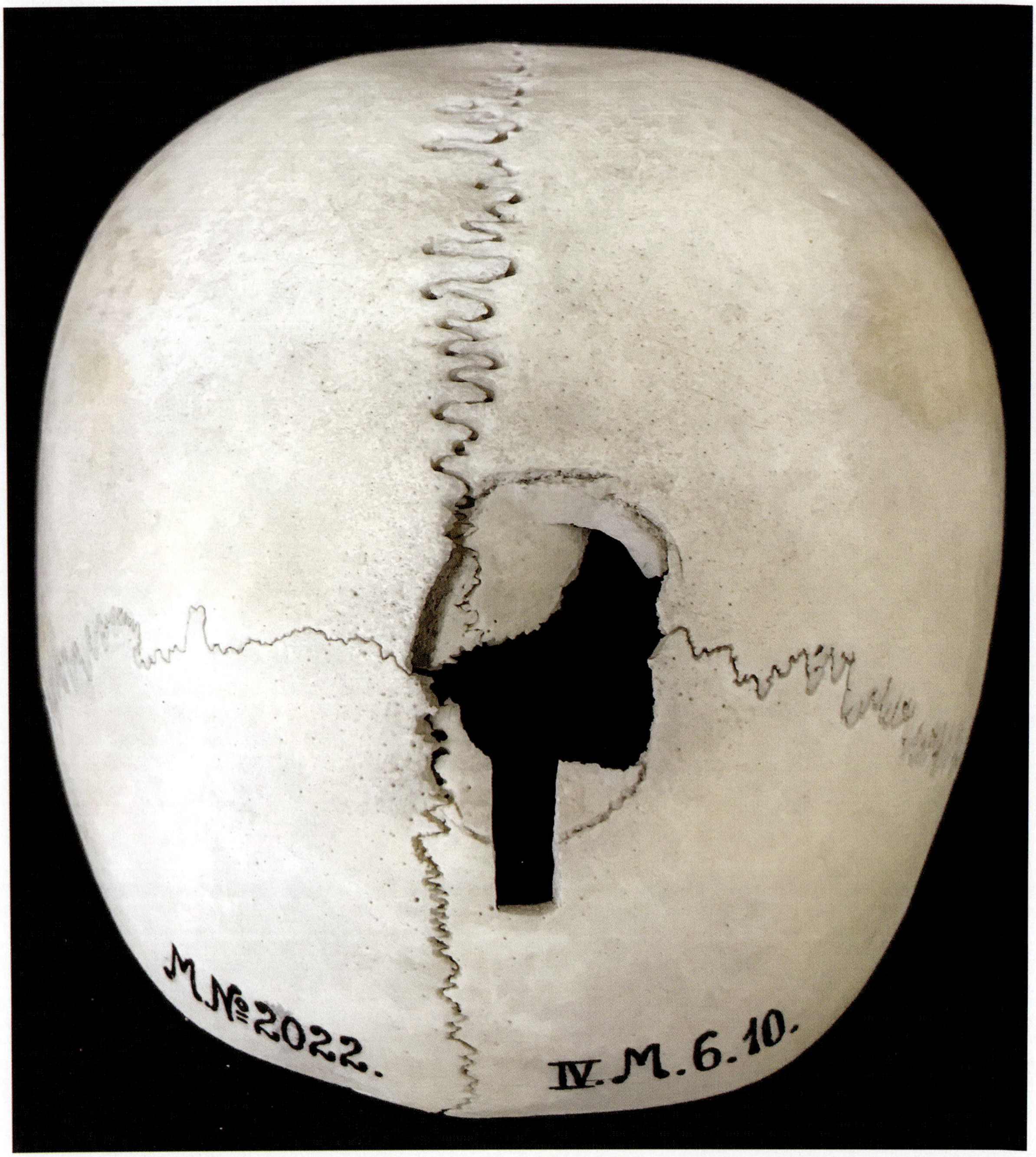


**Figure 355.** Ectocranial view of depressed blunt force trauma to left parietal and frontal bones with no evidence of healing in a 17-year-old male who died in 1843 (MN2022). There appears to be pitting around this trauma that could indicate he survived for a few days; evidence of healing in bone is usually visible about 7-10 days after sustaining trauma (see Barbian and Sledzik 2008 for more on bone healing following cranial trauma). The oval-shaped injury is situated lateral to bregma and involves the coronal, sagittal and open metopic suture. Ortner's (1981; p 58) classic book has a photo of this vault without the rectangular area of sawed piece of bone, indicating that the procedure was done after 1981. See DiMaio and DiMaio (2001); Kimmerle and Baraybar (2008); Watson-Jones (1944); Wedel and Galloway (2014) for more on skeletal trauma.

**Figure 356.** Endocranial view of blunt force trauma to left parietal and frontal bones with no evidence of healing in a 17-year-old male who died in 1843 (MN2022). The rectangular piece of missing bone was sawed from the frontal bone after 1981, presumably for histology. Note beveling (arrow) along the margin and inwardly displaced piece of bone (*) along the metopic suture. The defect is consistent with an oval-shaped implement that struck the decedent to the left of bregma along the left half of the coronal suture.

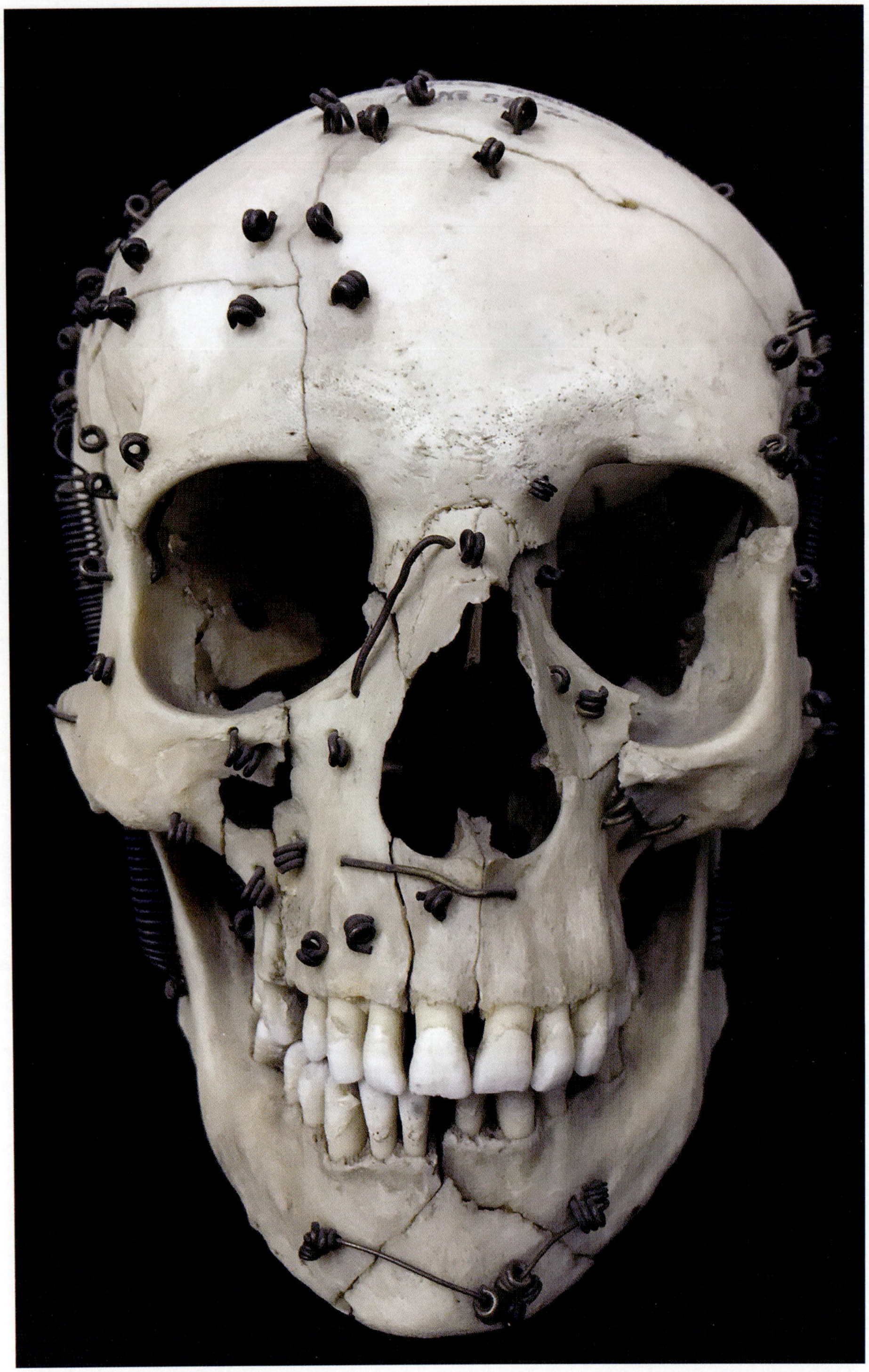

**Figure 357.** Massive blunt force trauma to the head resulting in extensive fracturing of the skull in an unknown adult (MN5752). Note the symphyseal fracture of the mandible and blowout fracture of the left orbit consistent with direct trauma over a large area of the skull and facial region consistent with a fall from a height. As was often done in the 1800s and early 1900s, the skull was reconstructed postmortem with wire through small drill holes.

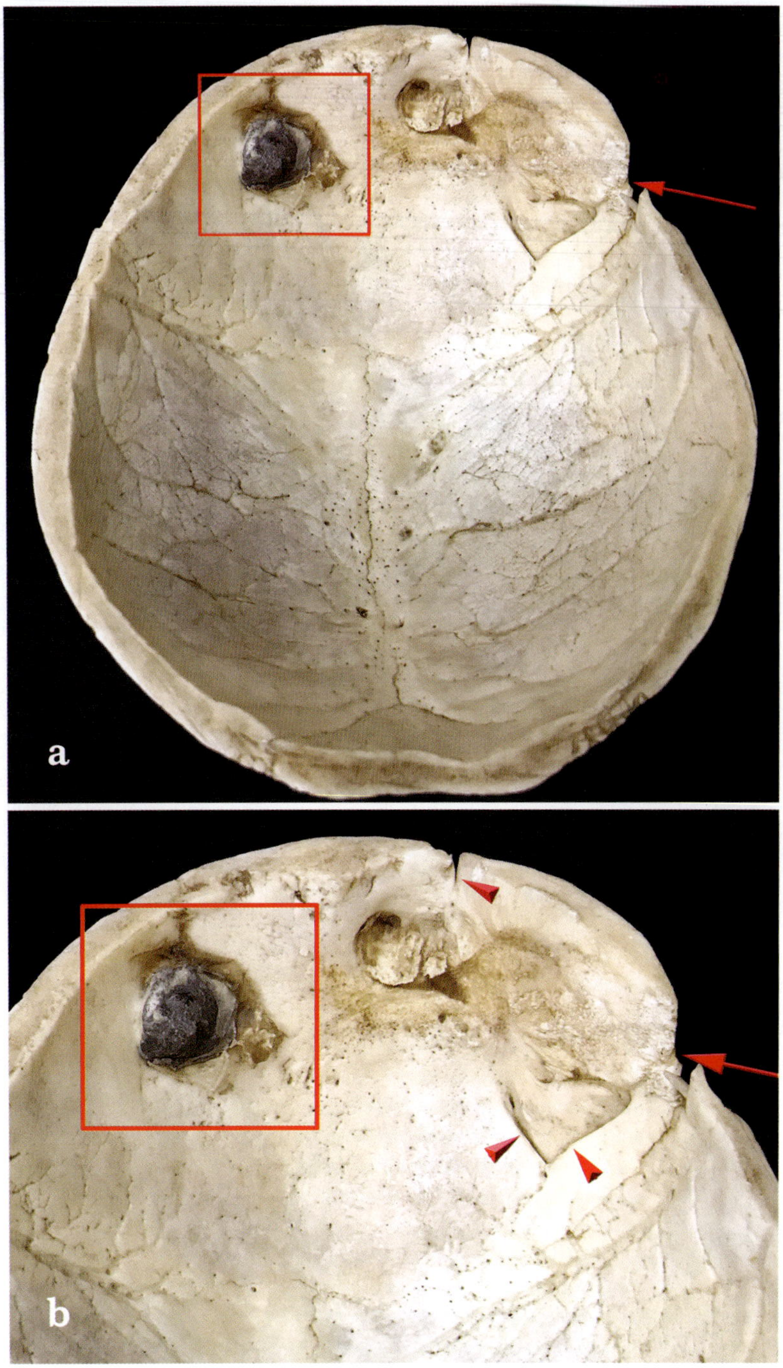


**Figure 360a-b.** (a) Endocranial and (b) magnified views of an embedded musket ball (square) that entered the left frontal bone (arrow) with evidence of healing in an unknown adult male (MN5716). (b) Note the ectocranial fractures (arrowheads) associated with the entry wound. See DiMaio and DiMaio (2001); Kimmerle and Baraybar (2008); Watson-Jones (1944); Wedel and Galloway (2014) for more on gunshot and other types of skeletal trauma.

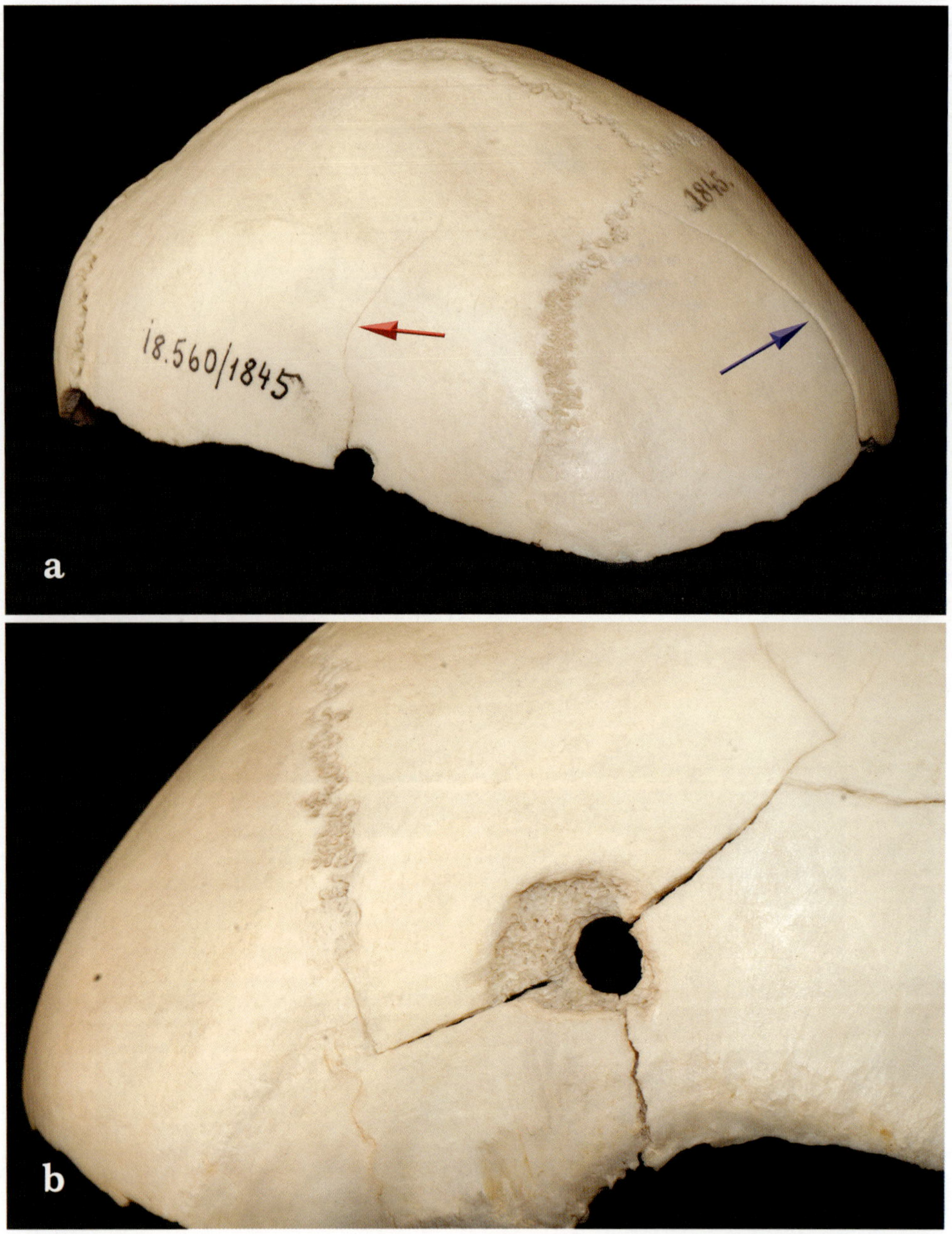


**Figure 361a-b.** (a) Gunshot entry (red arrow) wound in the right parietal bone and an associated vertical radiating fracture in the frontal bone (blue arrow) that (b) exited the left parietal bone in a 20-year-old male who died of a self-inflicted gunshot wound in 1944 (MN18.560/1845). Note that the entry wound (a) is smaller than the exit wound (b) which exhibits external beveling and that both exhibit triradiate fractures. The bullet resulted in separation of much of the cranial vault. See DiMaio and DiMaio (2001); Kimmerle and Baraybar (2008); Otis (1877); Watson-Jones (1944); Wedel and Galloway (2014) for more on gunshot wounds and skeletal trauma.

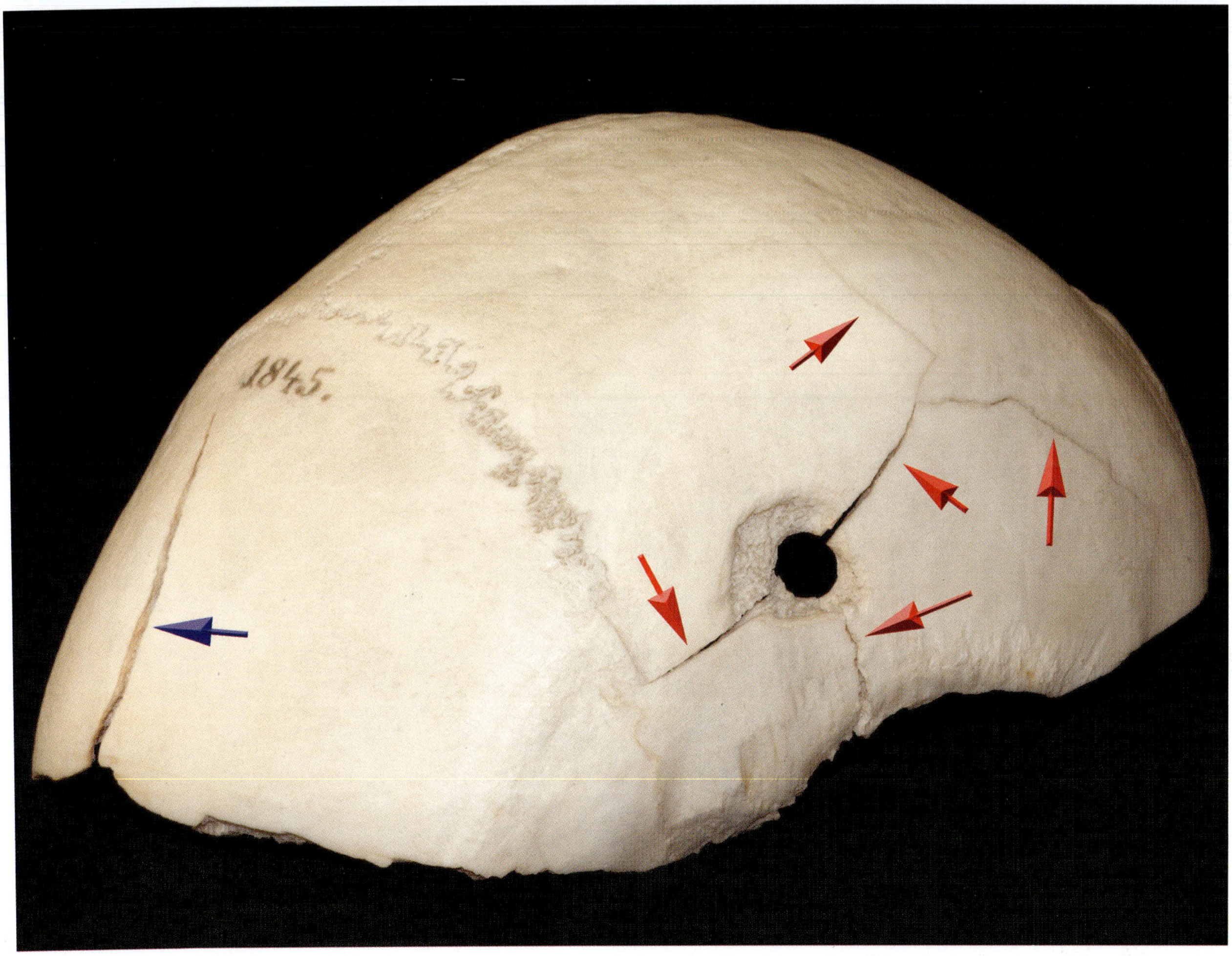


**Figure 362.** Left oblique view of a vertical fracture in the frontal bone (blue arrow) associated with an entry wound in the right parietal bone and multiple exit fractures (red arrows) in the left parietal bone in a 20-year-old male who died of a self-inflicted gunshot wound in 1944 (MN18.560/1845). See DiMaio and DiMaio (2001); Kimmerle and Baraybar (2008); Otis (1877); Watson-Jones (1944); Wedel and Galloway (2014) for more on gunshot wounds and skeletal trauma.

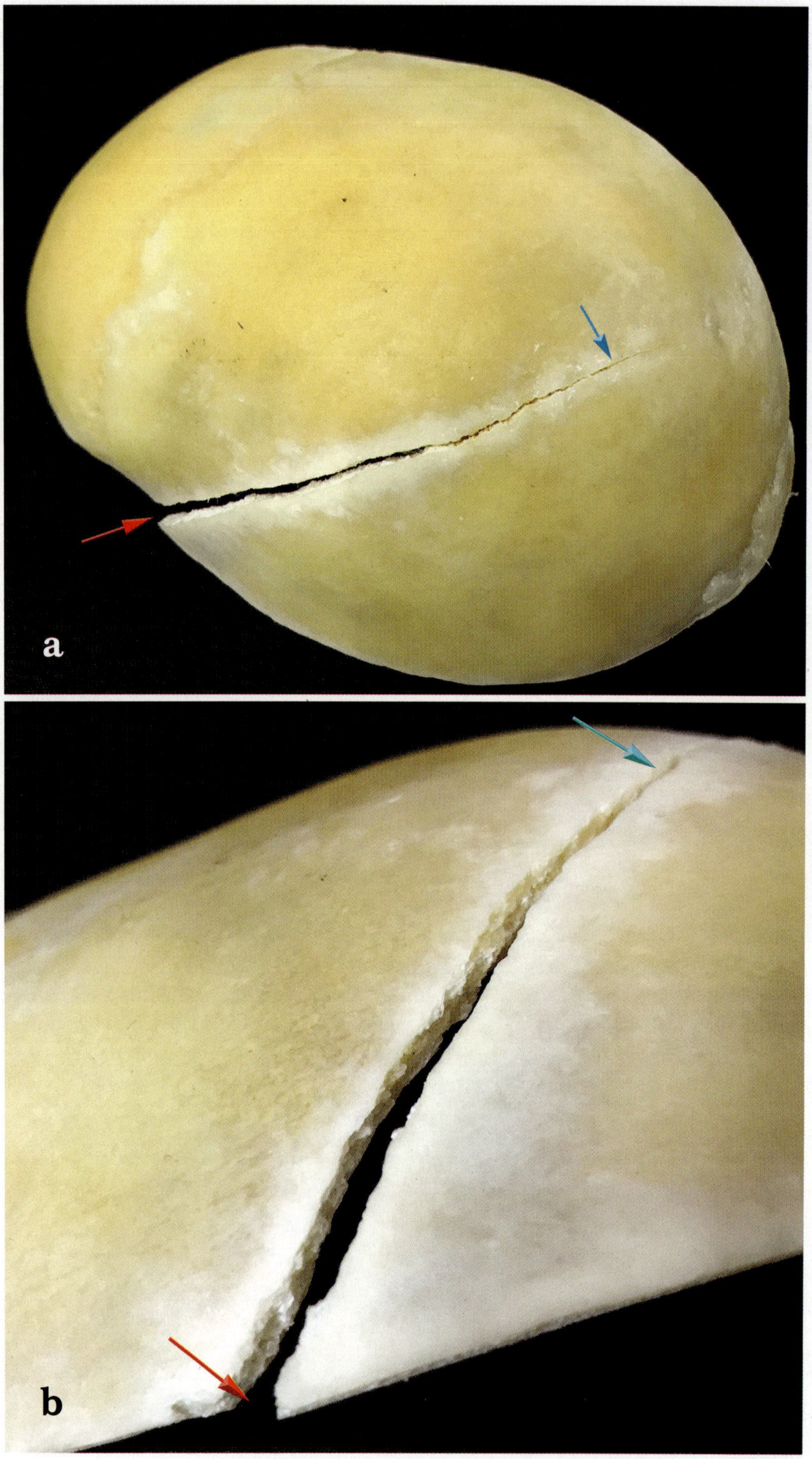


**Figure 363a-b.** (a) Left superior-inferior and (b) left lateral views of blunt force trauma to the left parietal bone in a 45-year-old male who died from a subdural hematoma due to rupture of the middle meningeal artery from a fall on the street in 1915 (MN18.476). (b) Note that the fracture is widest (red arrow) and gradually narrows away (blue arrow) from the point of impact.

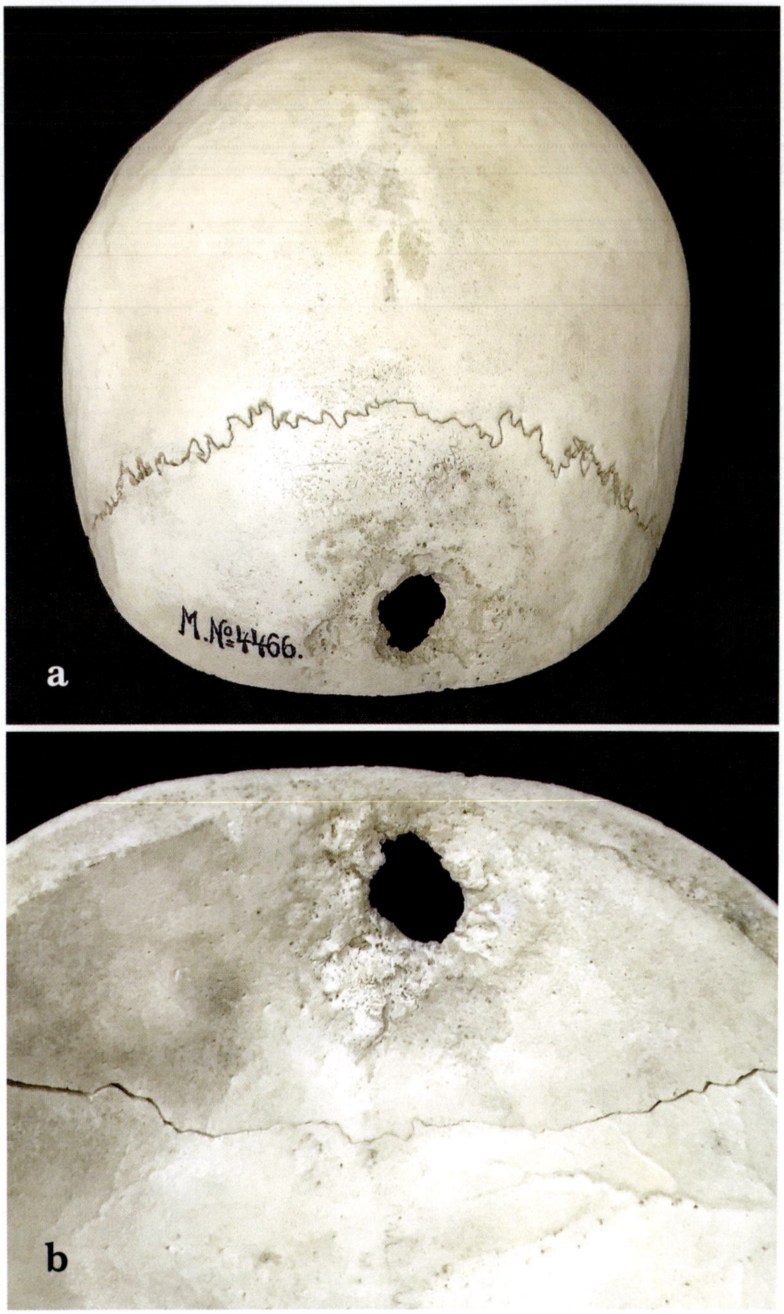


**Figure 364a-b.** (a) Ectocranial and (b) endocranial views of blunt force trauma to the frontal bone in a 30-year-old male who experienced an epileptic seizure and struck the front of his head in 1885 (MN4466). Translated (S. Doll, personal communication, November 14, 2022), the museum card reads "The man had an accident at work a year earlier. While loading his load, he had bumped himself and became unconscious. The next day he was fit for work again. However, he developed epileptic seizures and a bump at the site of the accident injury, from which pieces of bone were removed. Gradually the man became stupid (from the autopsy report 6.2.1885). Cause of death was lues." Note bone remodeling (raised bone and pitting) surrounding the injury indicating that the individual survived the injury and possible infection for several days if not weeks or months. See DiMaio and DiMaio (2001) for skeletal trauma.

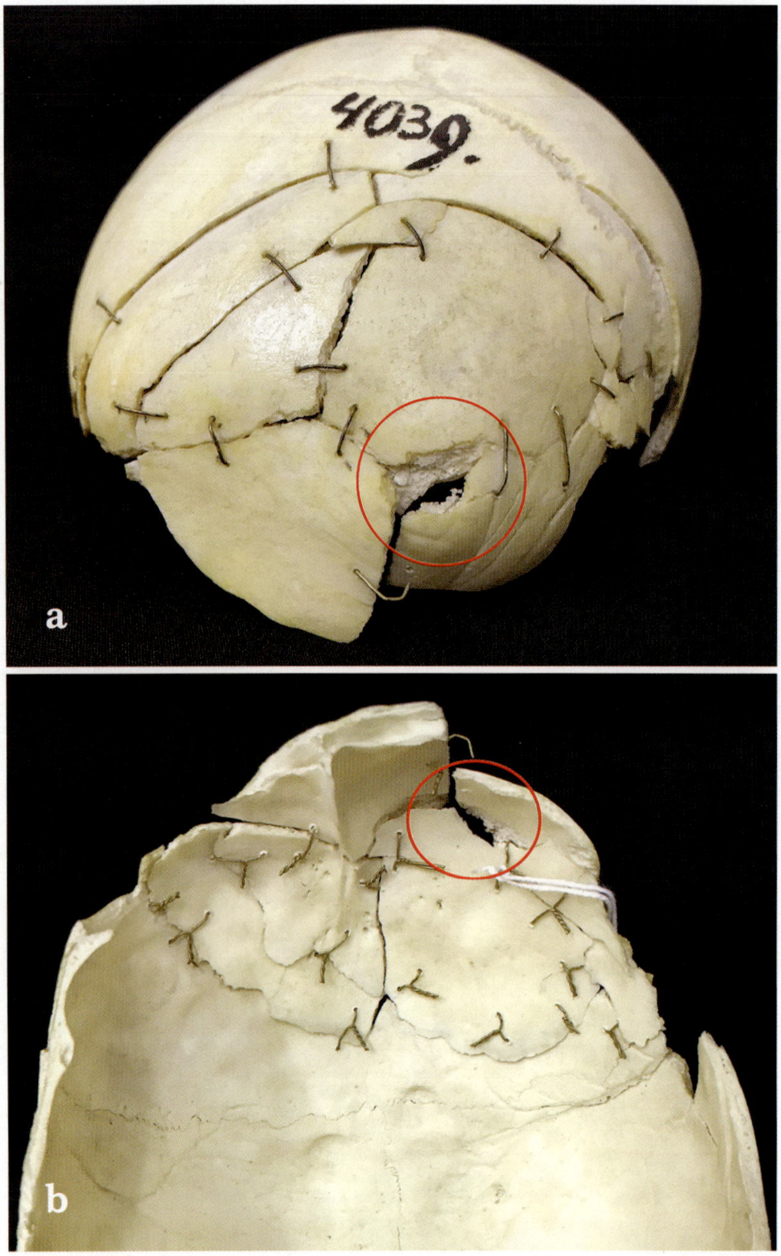


**Figure 365a-b.** (a) Ectocranial and (b) endocranial views of chisel trauma to the frontal bone in a 4-year-old female with no evidence of healing who died before 1869 (MN4039). The site of impact (circle) was the frontal bone, above the left orbit. External beveling such as seen here can occur from trauma. See DiMaio and DiMaio (2001); Kimmerle and Baraybar (2008), Watson-Jones (1944) and Wedel and Galloway (2014) for more on skeletal trauma.

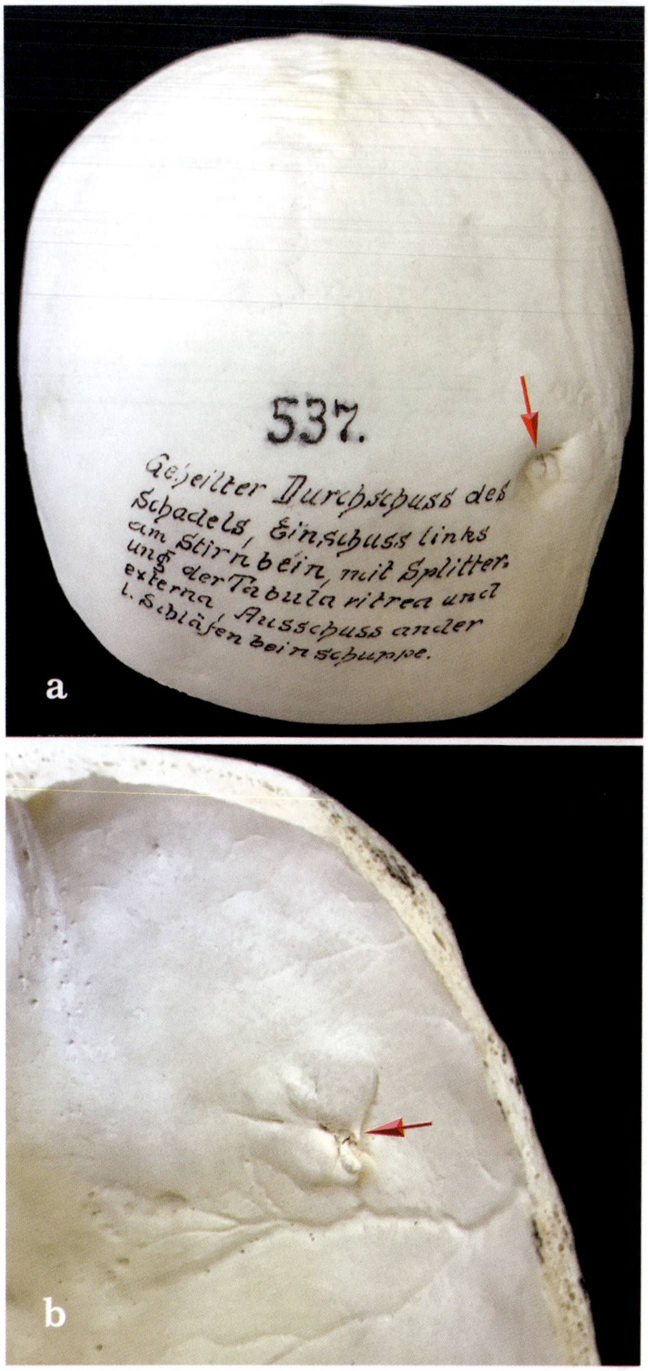


**Figure 366a-b.** (a) External and (b) internal views of a bullet that struck the left frontal bone in an approximately 19-year-old male wounded in World War I (possibly 1917) who survived (b) resulting in internal displacement of several pieces of bone and healing (MN537). See Otis (1877) for more on military gunshot and shrapnel wounds, healing and infection in the human skeleton from the latter half of the 19th century.

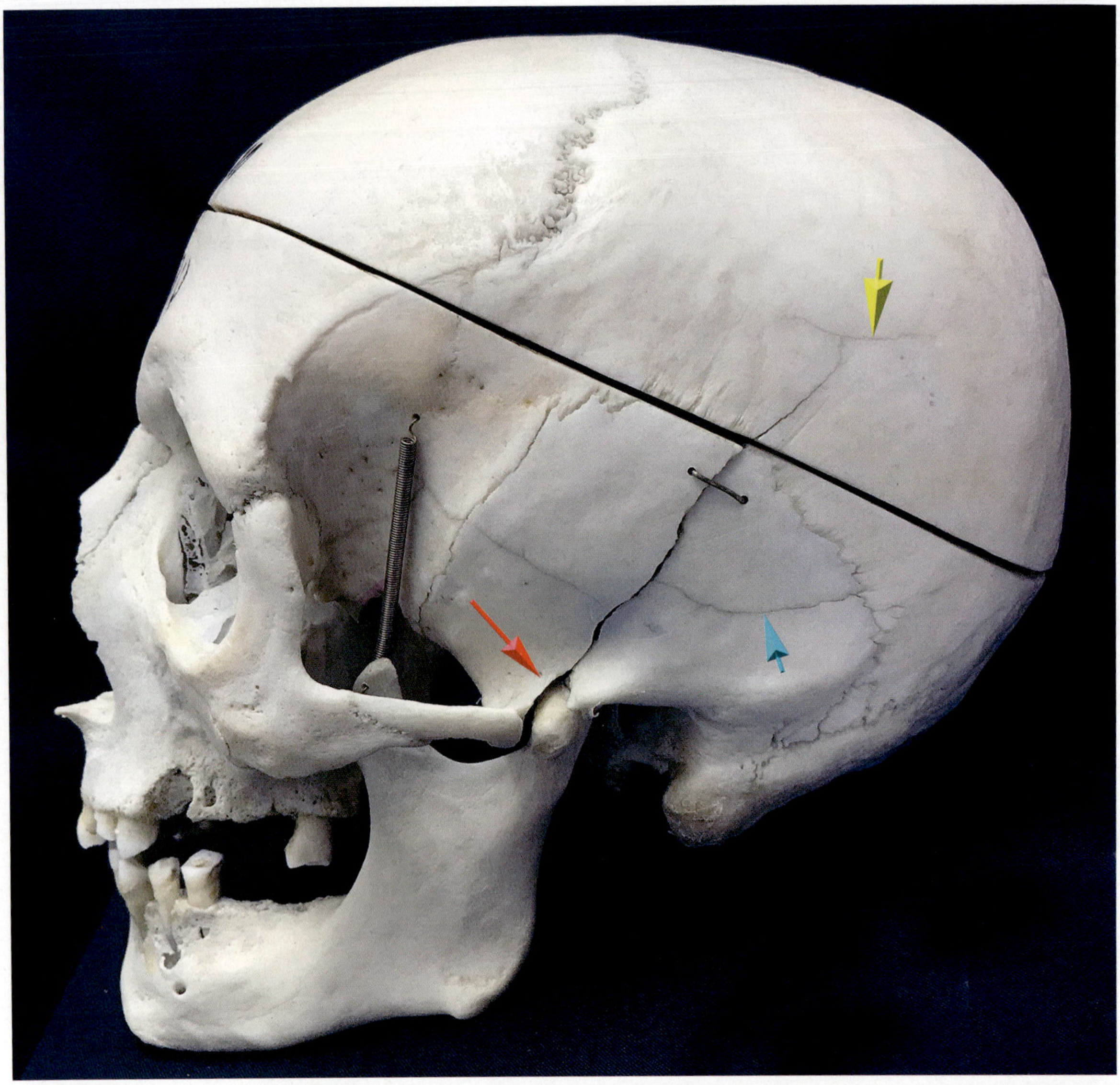

**Figure 367.** Large primary fracture (red arrow) originating at the left temporomandibular joint (TMJ) and that extended into the left parietal bone in an adult male who died in 1888 (MN21.055/2320). Some of the fractures (blue arrow) terminate either at an existing fracture in the left parietal bone (yellow arrow) or suture, both of which illustrate Puppe's rule that a fracture will not cross an existing fracture and often terminates in an open/incompletely fused suture (Amadasi et al. 2022; Viel et al. 2008). The stresses associated with a fracture often travel along an open suture, known as a diastatic fracture, and will sometimes widen it and then continue on in the bone. Note that the fracture in the left temporal and parietal bones is widest at its point of origin in the TMJ and narrows farthest from the point of origin. There is no evidence of bone healing associated with this injury. See DiMaio and DiMaio (2001); Kimmerle and Baraybar (2008), Watson-Jones (1944) and Wedel and Galloway (2014) for more on skeletal trauma.

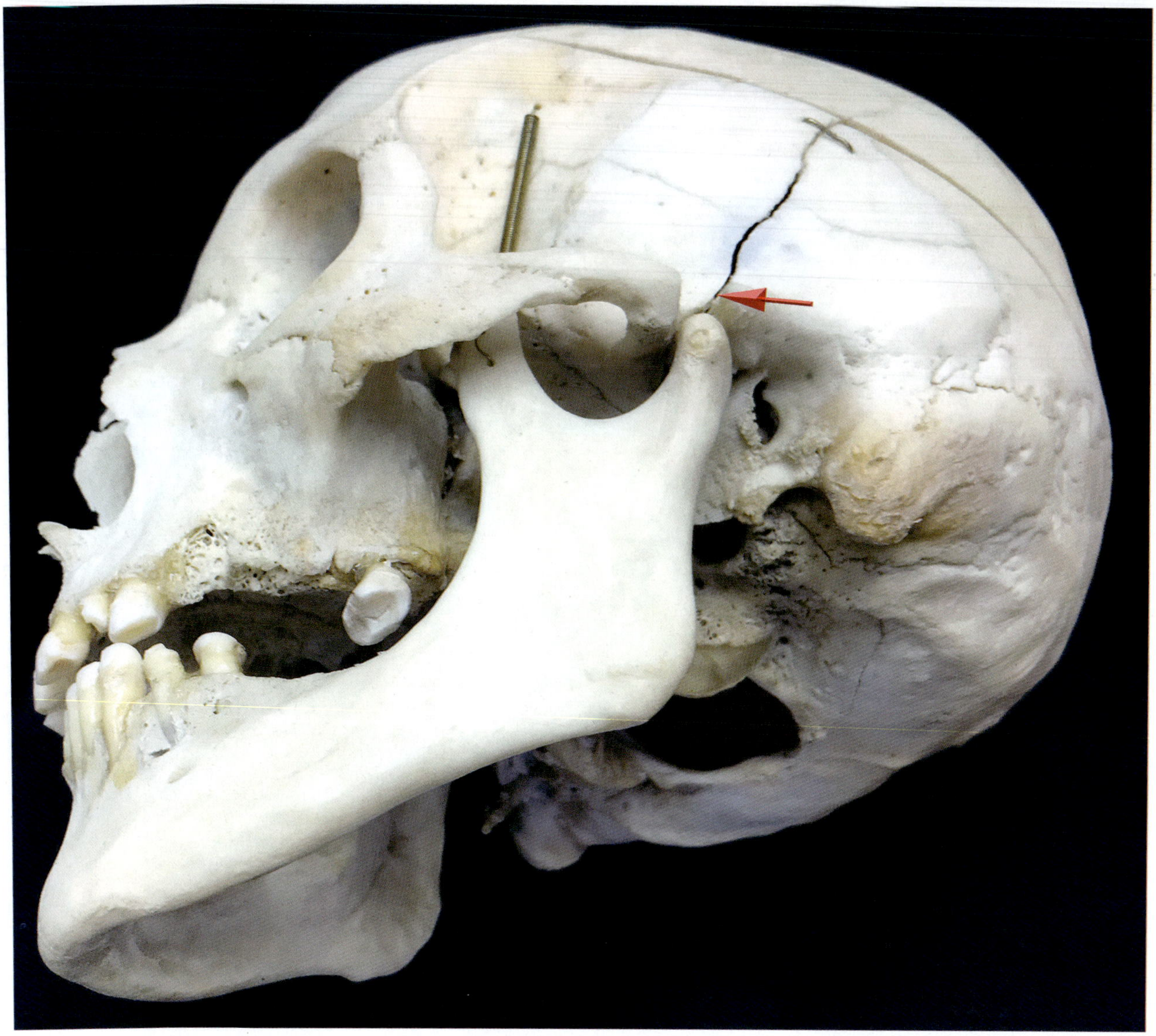

**Figure 368.** Multiple perimortem fractures involving the left temporomandibular joint (red arrow) with several radiating fractures of the left parietal bone in an adult male who died in 1888 (MN21.055/2320). The wide fracture of the left temporomandibular joint (TMJ) and narrowing of the radiating/linear fracture toward the parietal bone indicates that the fracture originated at the TMJ. Fractures are typically widest at their point of impact and narrow distally. The absence of healing/remodeling of the fractures suggests that the individual did not survive the injury or if he did, that it was probably less than 7-10 days (Mann and Hunt 2005). See Barbian and Sledzik (2008) for more on bone healing following cranial trauma and DiMaio and DiMaio (2001); Kimmerle and Baraybar (2008); Watson-Jones (1944); Wedel and Galloway (2014) for more on skeletal trauma.

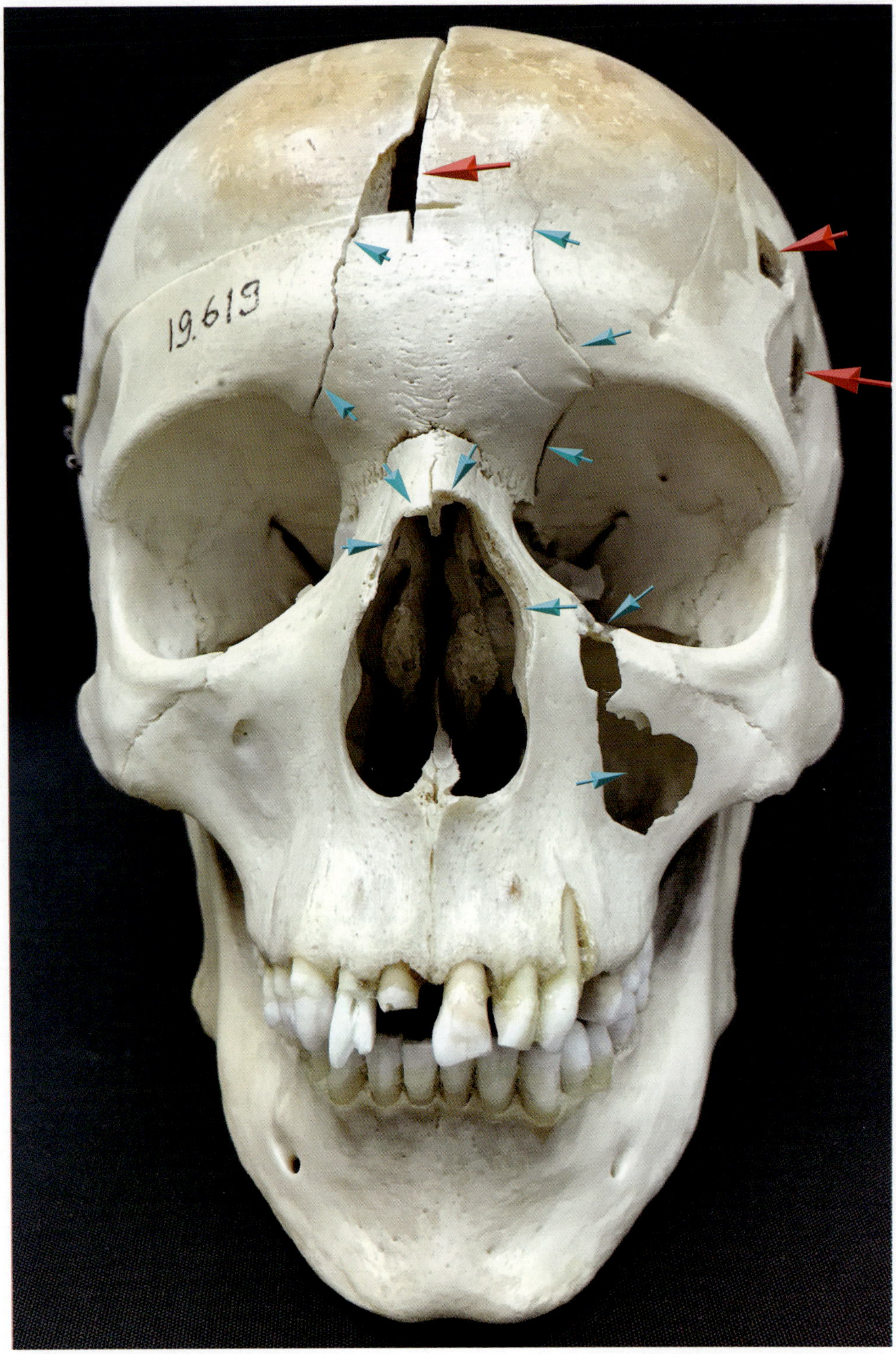


**Figure 369.** Frontal view of a 17-year-old male who sustained at least 11 ice hammer blows (red arrows) to the frontal and left lateral skull (MN19.619). Blue arrows indicate radiating fractures and areas of missing bone due to perimortem trauma. The size and shape of the bony defects in the vault differ based on how deep the ice hammer penetrated the skull. See DiMaio and DiMaio (2001); Kimmerle and Baraybar (2008); Watson-Jones (1944); Wedel and Galloway (2014) for more on skeletal trauma.

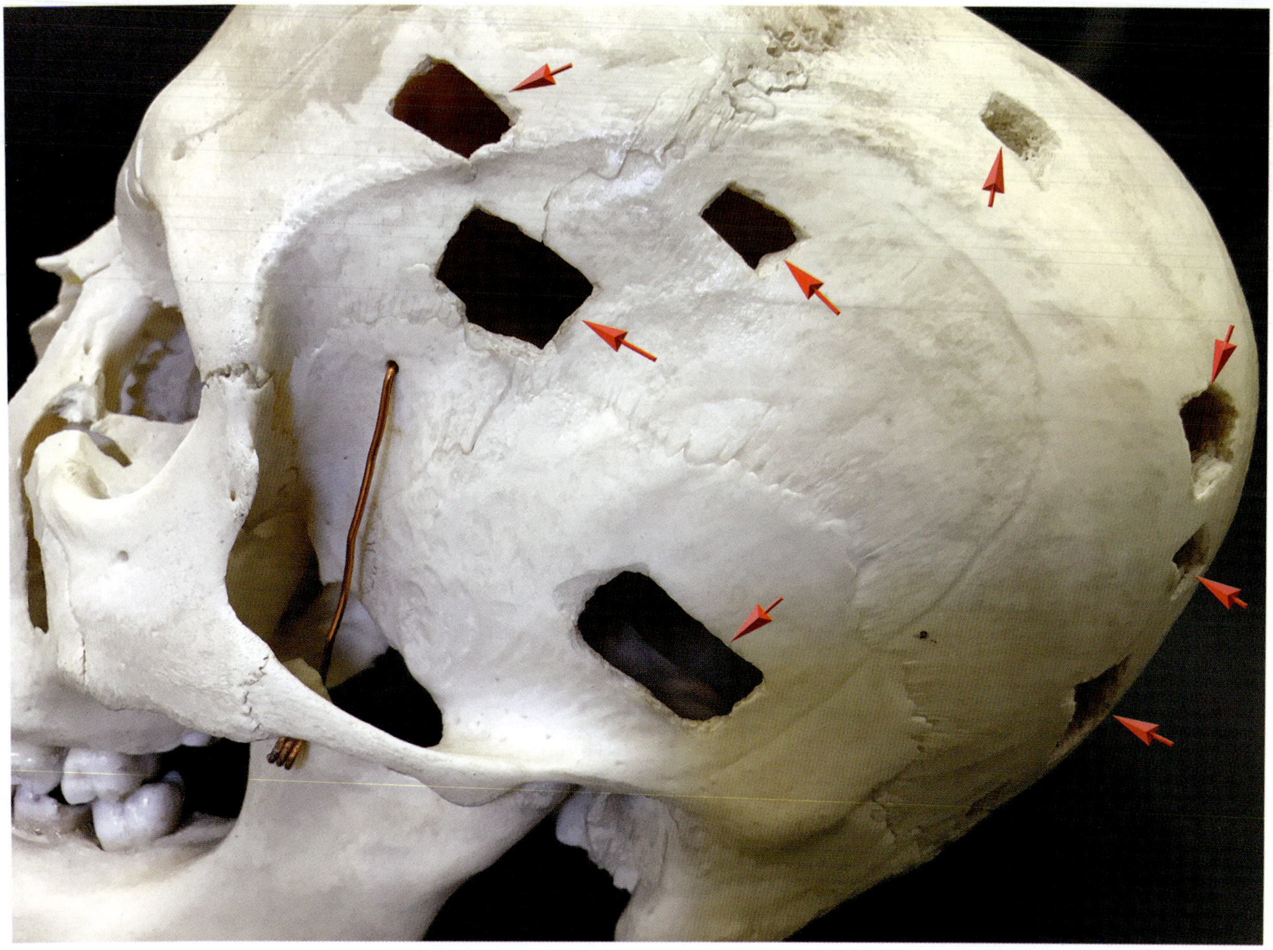

**Figure 370.** Left lateral view of 17-year-old male who sustained at least 11 ice hammer blows (arrows) to the frontal and left lateral skull with no evidence of healing (MN19.619). The size of the bony defects differs based on varying thickness of the vault and the angle of depth that the ice hammer penetrated the skull. Surprisingly, most of the vault injuries are not accompanied by radiating fractures. See DiMaio and DiMaio (2001); Kimmerle and Baraybar (2008); Watson-Jones (1944); Wedel and Galloway (2014) for more on skeletal trauma.

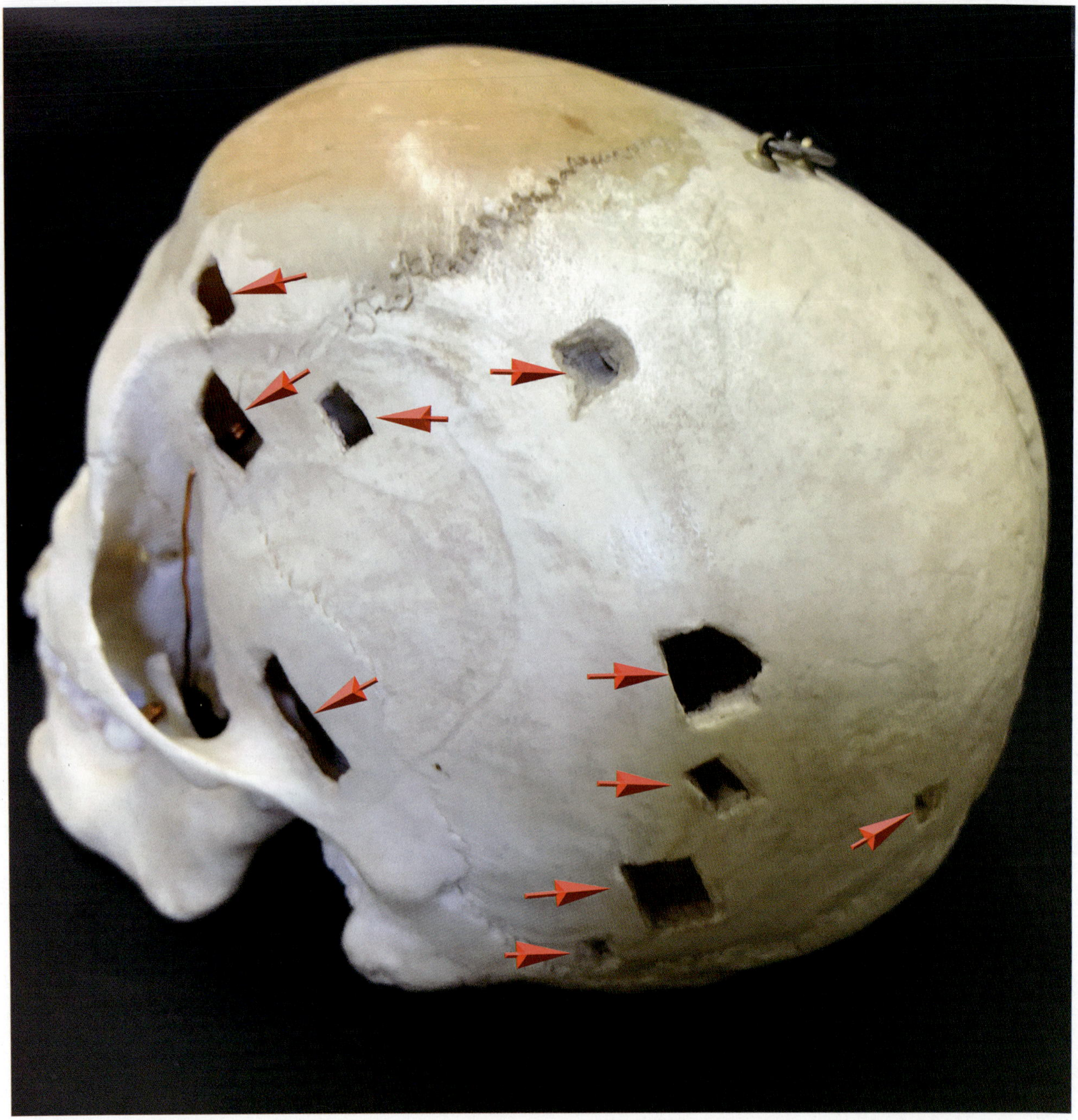

**Figure 371.** Left superior-inferior view of 17-year-old male who sustained at least 11 ice hammer blows (red arrows) to the frontal and left lateral skull with no evidence of healing (MN19.619). The size of the bony defects differs based on varying thickness of the vault and the angle and depth that the ice hammer penetrated the skull. Surprisingly, most of the vault injuries are not accompanied by radiating fractures. See DiMaio and DiMaio (2001); Kimmerle and Baraybar (2008); Watson-Jones (1944) for more on skeletal trauma.

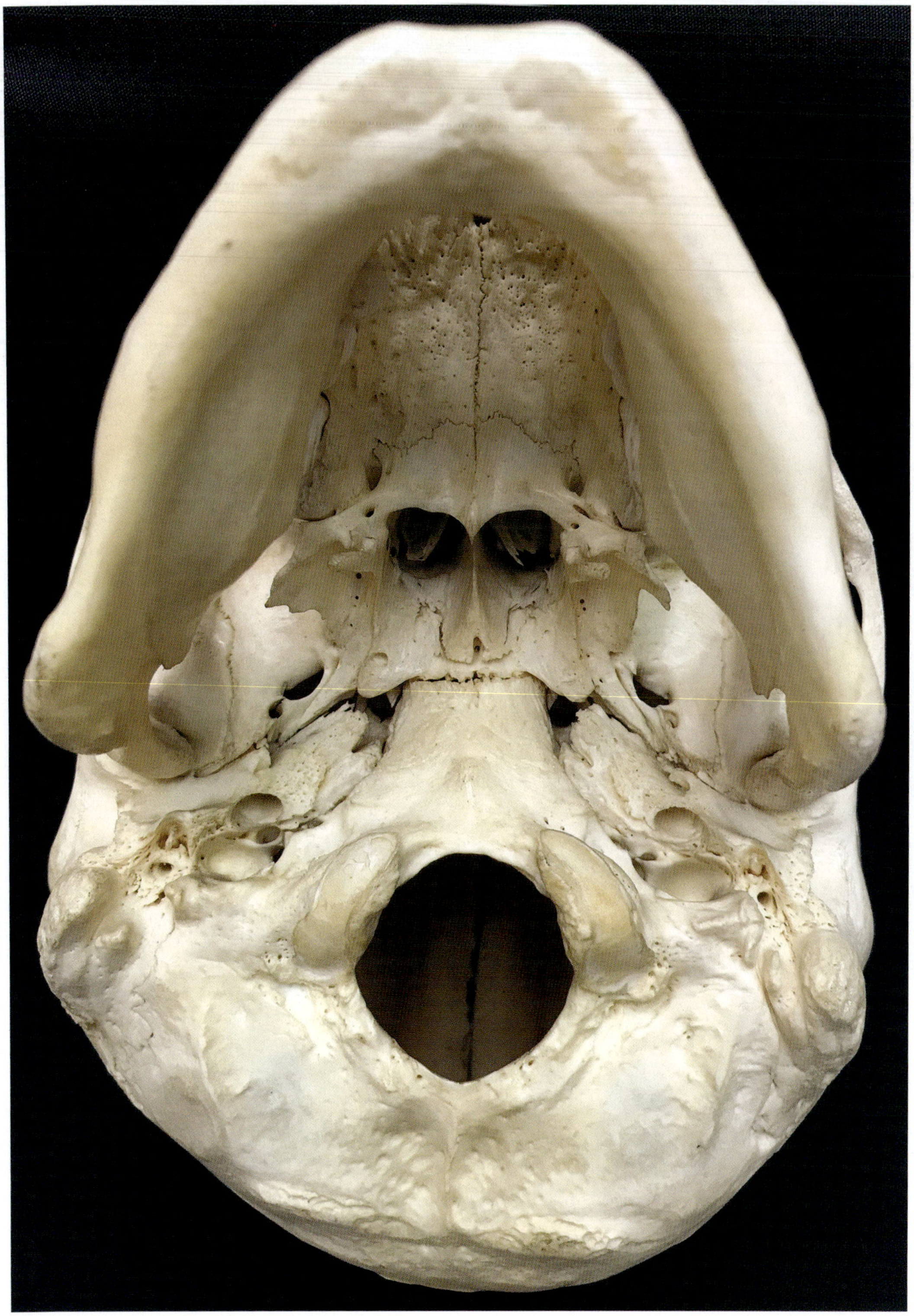

**Figure 372.** Inferior view of 17-year-old male who sustained at least 11 ice hammer blows to the frontal and left lateral skull with no evidence of healing and no visible trauma in the cranial base (MN19.619).

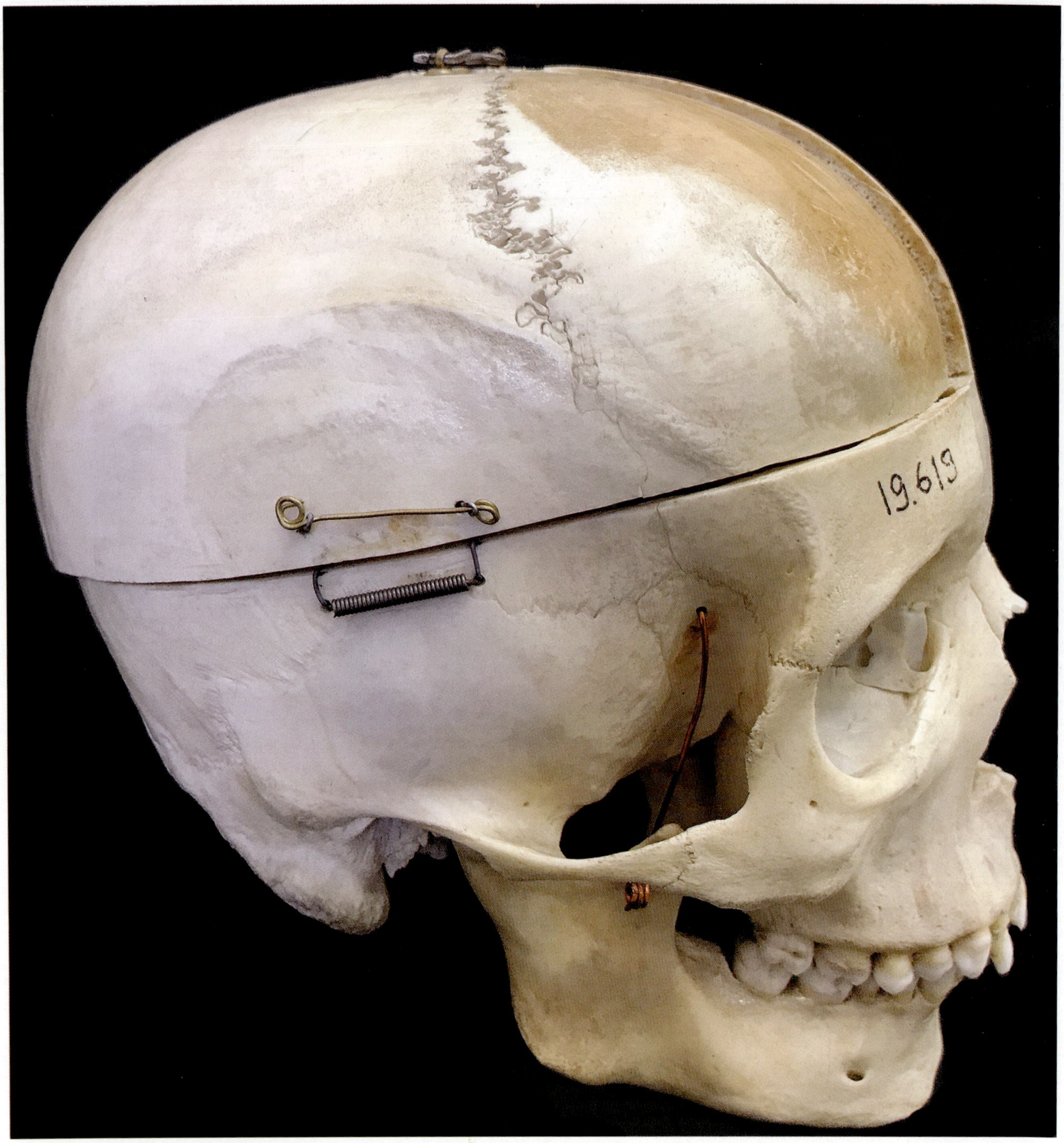


**Figure 373.** Right lateral view of 17-year-old male who sustained at least 11 ice hammer blows to the frontal and left lateral skull with no evidence of healing and no visible evidence of trauma to the right portion of the skull or facial region (MN19.619).

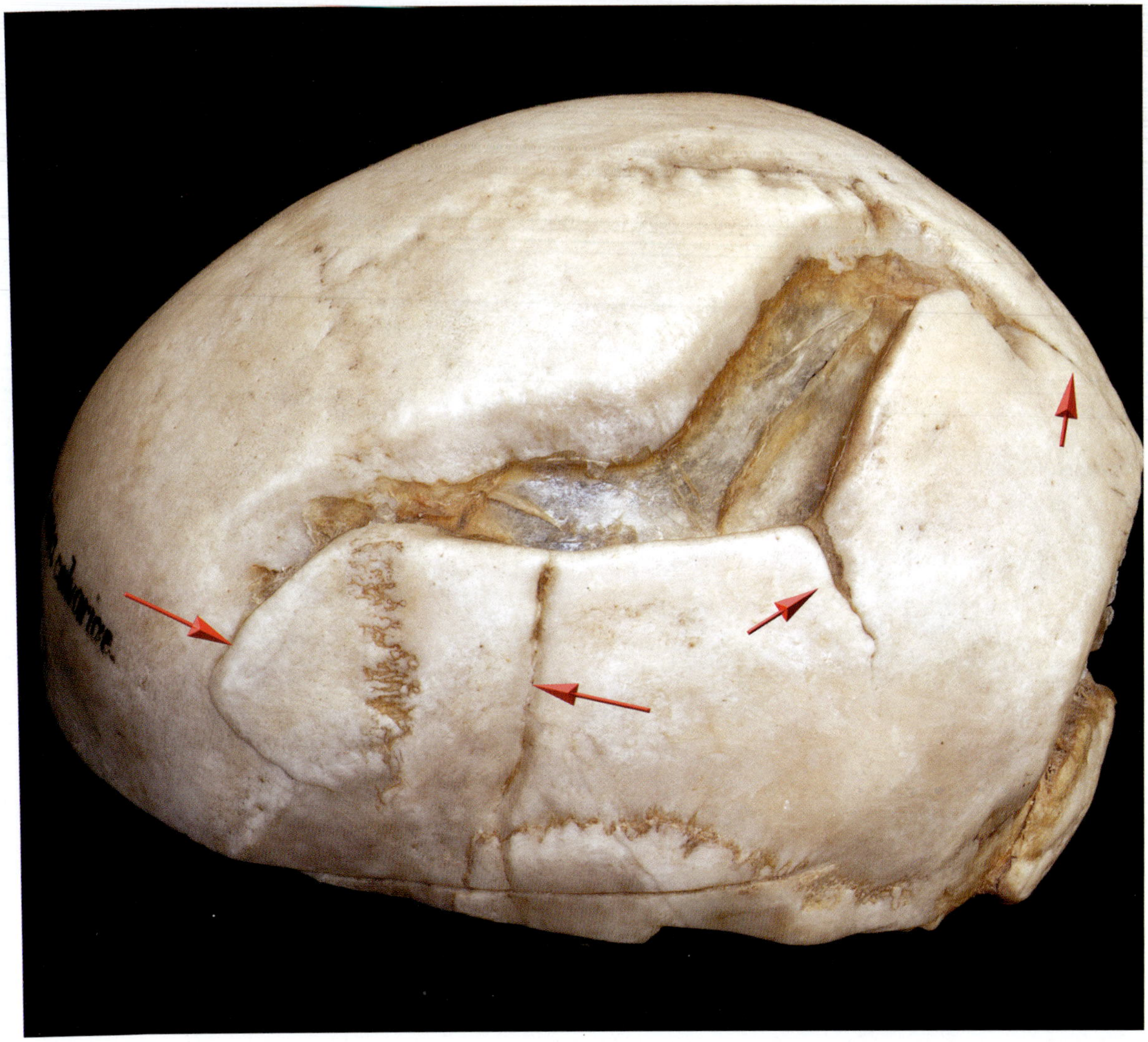

**Figure 374.** Left superior-inferior view of cranium with a healed axe or sword injury in the left parietal and frontal bone with numerous fractures originating from the impact site. The pattern of injury without intersecting fractures suggest a single blow and the smooth and rounded edges of the injury indicate that the individual survived the injury for many months if not years. See Ortner and Putschar (1981) for more on this injury. See DiMaio and DiMaio (2001); Kimmerle and Baraybar (2008); Watson-Jones (1944); Wedel and Galloway (2014) for more on skeletal trauma.

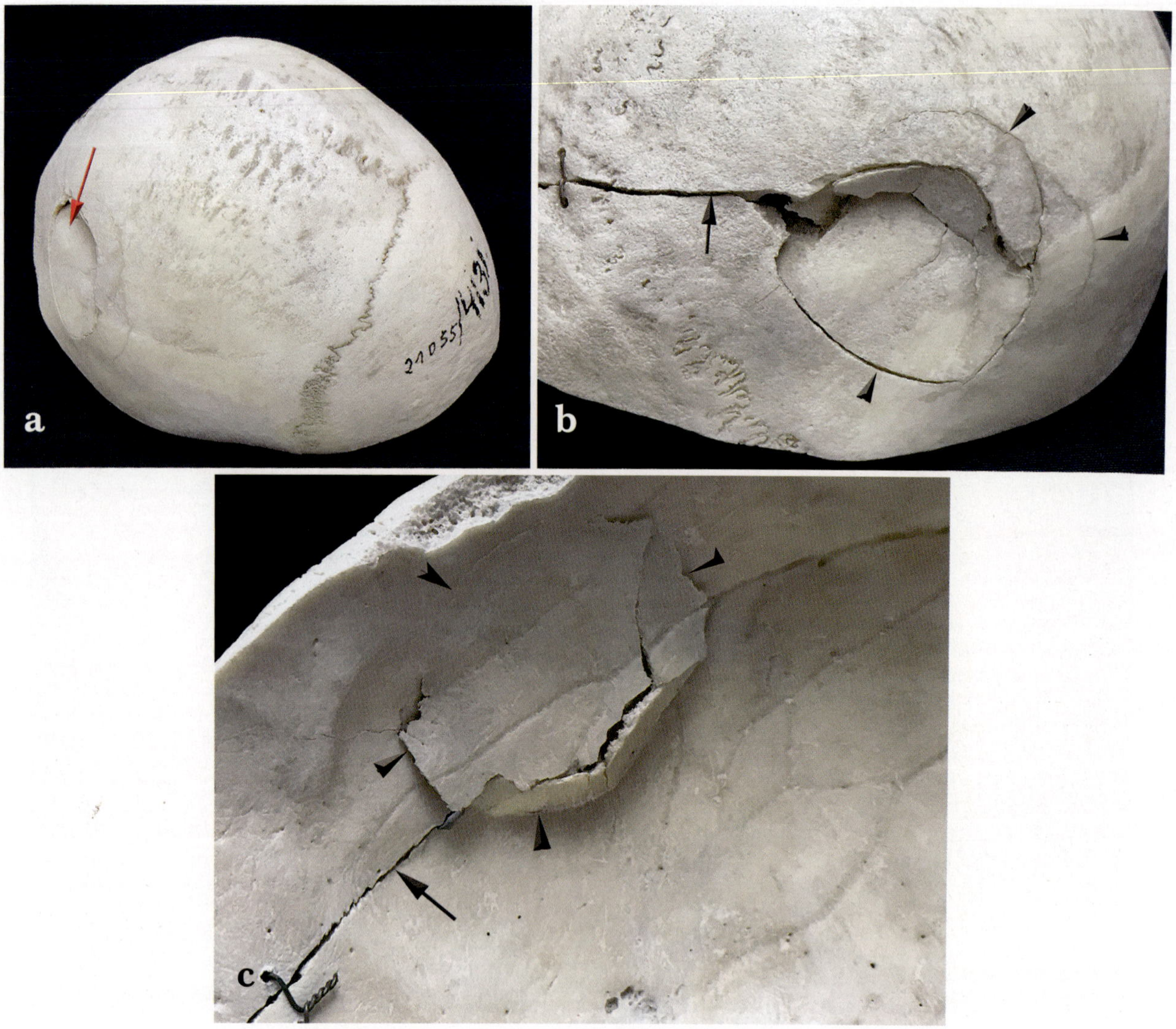


**Figure 375a-c.** (a) Anterior superior-inferior bone blunt force trauma (red arrow); (b) magnified ectocranial views of perimortem blunt force trauma to the posterior portion of the right parietal bone (black arrow and arrowheads); (c) endocranial view of the area of trauma showing that the point of impact resulted in the inward displacement of bone fragments; (b-c) show radiating fractures (black arrows) and concentric fractures (black arrowheads) in a 48-year-old male who died in 1894 (MN21.055/4131). See DiMaio and DiMai (2001); Kimmerle and Baraybar (2008); Watson-Jones (1944); Wedel and Galloway (2014) for more on skeletal trauma.

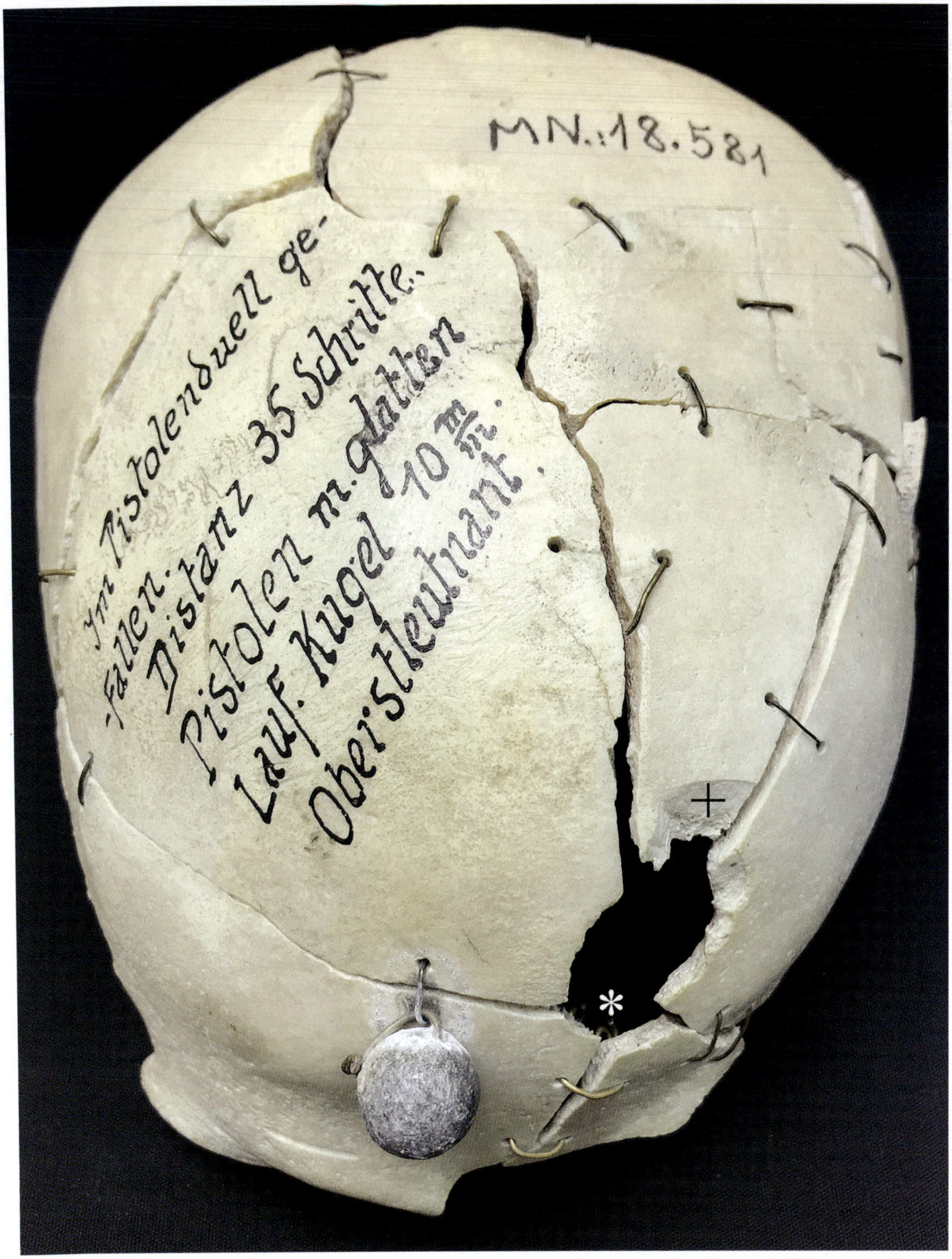


**Figure 376.** Fatal keyhole gunshot (musket ball) entry wound in the left frontal bone of a lieutenant colonel who died before 1918 (MN18.581). This gunshot wound was the result of a smooth bore (musket) pistol duel from a reported distance of 35 steps and reveals the characteristics of a keyhole injury with a circular point of entry (*). Keyhole gunshot entry wounds are so named because they resemble old keyholes and they exhibit features of both an entry and an exit wound. See DiMaio and DeMaio (2001); Smith et al. (2003); Spitz and Diaz (2020) for more on gunshot trauma.

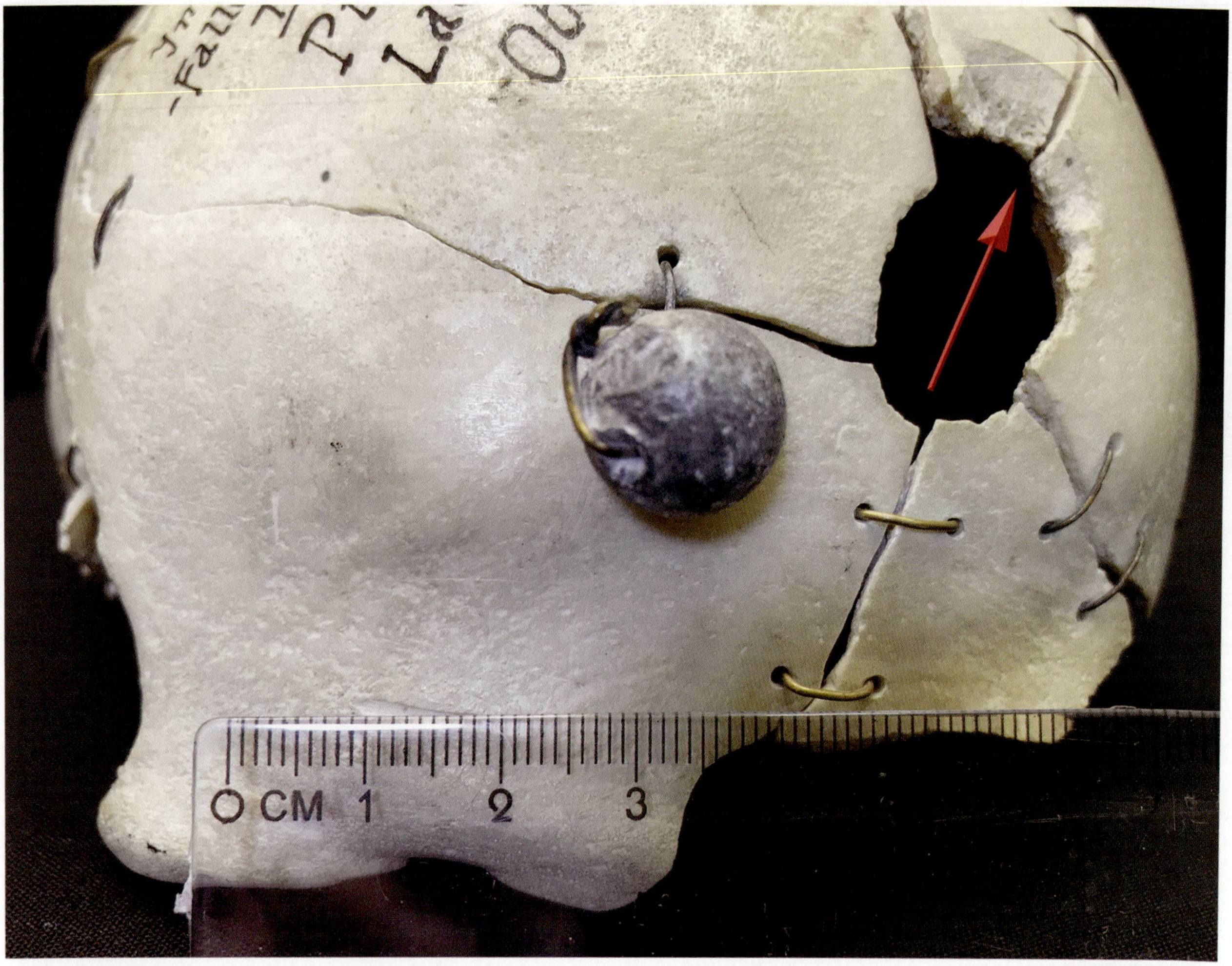


**Figure 377.** Magnified image of fatal keyhole gunshot (musket ball) entry wound in the left frontal bone that exhibits stellate fractures in a male who died before 1918 (MN18.581). This gunshot wound was the result of a smooth bore (musket) pistol duel from a distance of 35 steps. The injury in the frontal bone exhibits characteristics of a keyhole injury with a circular point of entry (*) and external beveling. The bullet traveled from the decedent's front to back (arrow) with part of the musket ball exiting the occipital bone (Figure 378). See DiMaio and DiMaio (2001); Kimmerle and Baraybar (2008); Smith et al. (2003); Spitz and Diaz (2020); Watson-Jones (1944); Wedel and Galloway (2014) for more on skeletal trauma.

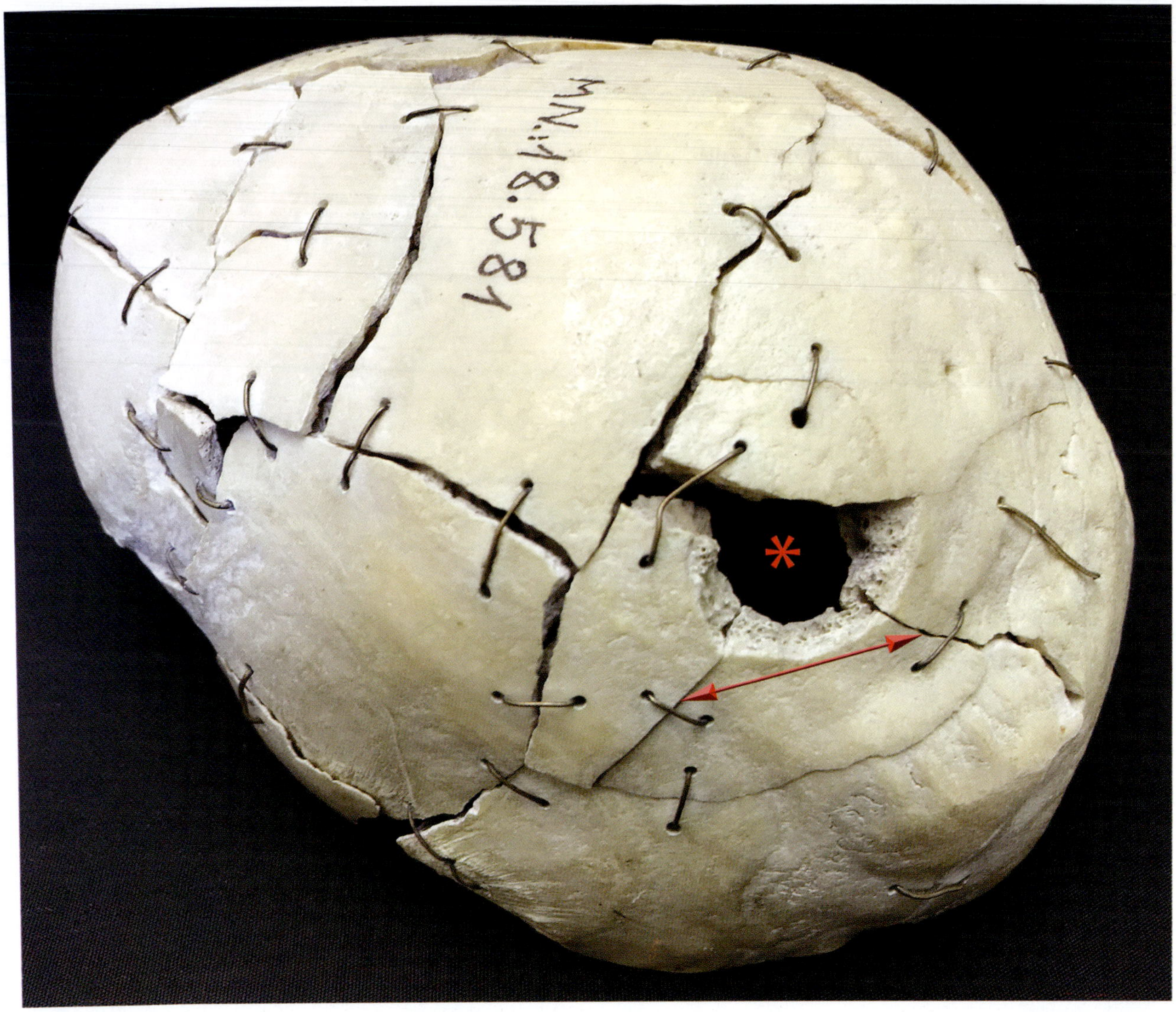


**Figure 378.** Exit wound (*) in the left parietal bone with external beveling/cratering from a musket ball (pistol duel at 35 paces) that entered the left frontal bone (MN18.581). Note the many radiating/radial and concentric fractures throughout the skull. External beveling forms when rectangular-shaped segments of bone fracture (concentric fractures) and separate between adjacent radiating fractures (double-ended arrow) along the margin of an external defect. Beveling such as this is actually the result of multiple segments of bone that have broken away around the gunshot exit wound. See DiMaio and DiMaio (2001); Kimmerle and Baraybar (2008); Smith et al. (2003); Spitz and Diaz (2020); Watson-Jones (1944); Wedel and Galloway (2014) for more on skeletal trauma.

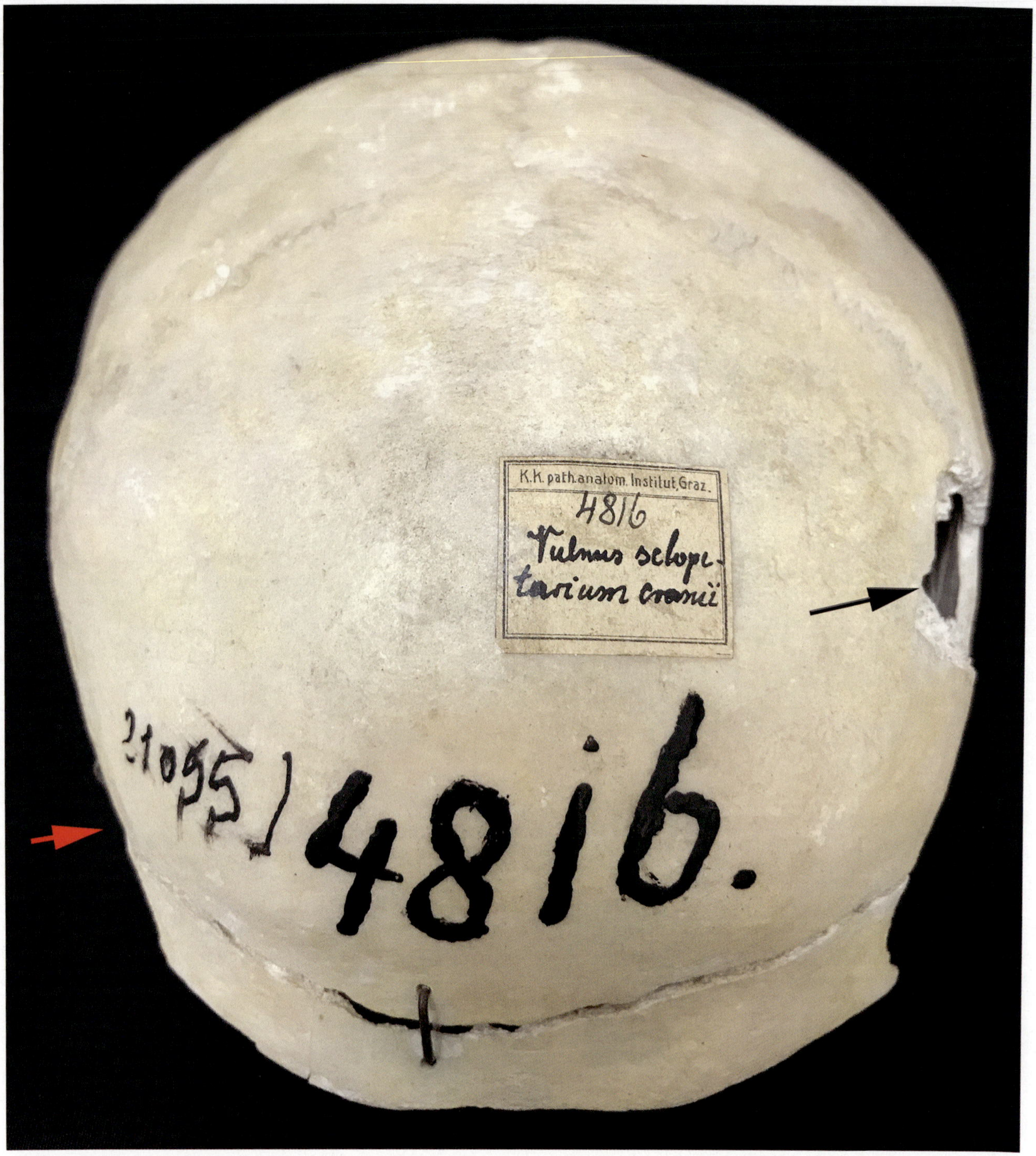


**Figure 379.** Entry (red arrow) and exit (black arrow) gunshot wounds in a 50-year-old male who died in 1902 (MN21.055/4816). Note the radiating fracture held together with wire that originated from the entry wound in the right temporal region and extends to the inferior margin of the exit wound in the left fronto-temporal region, reflecting Puppe's rule that a fracture will not cross an existing fracture (Amadasi et al. 2022; Madea and Staak 1988; Viel et al. 2008).

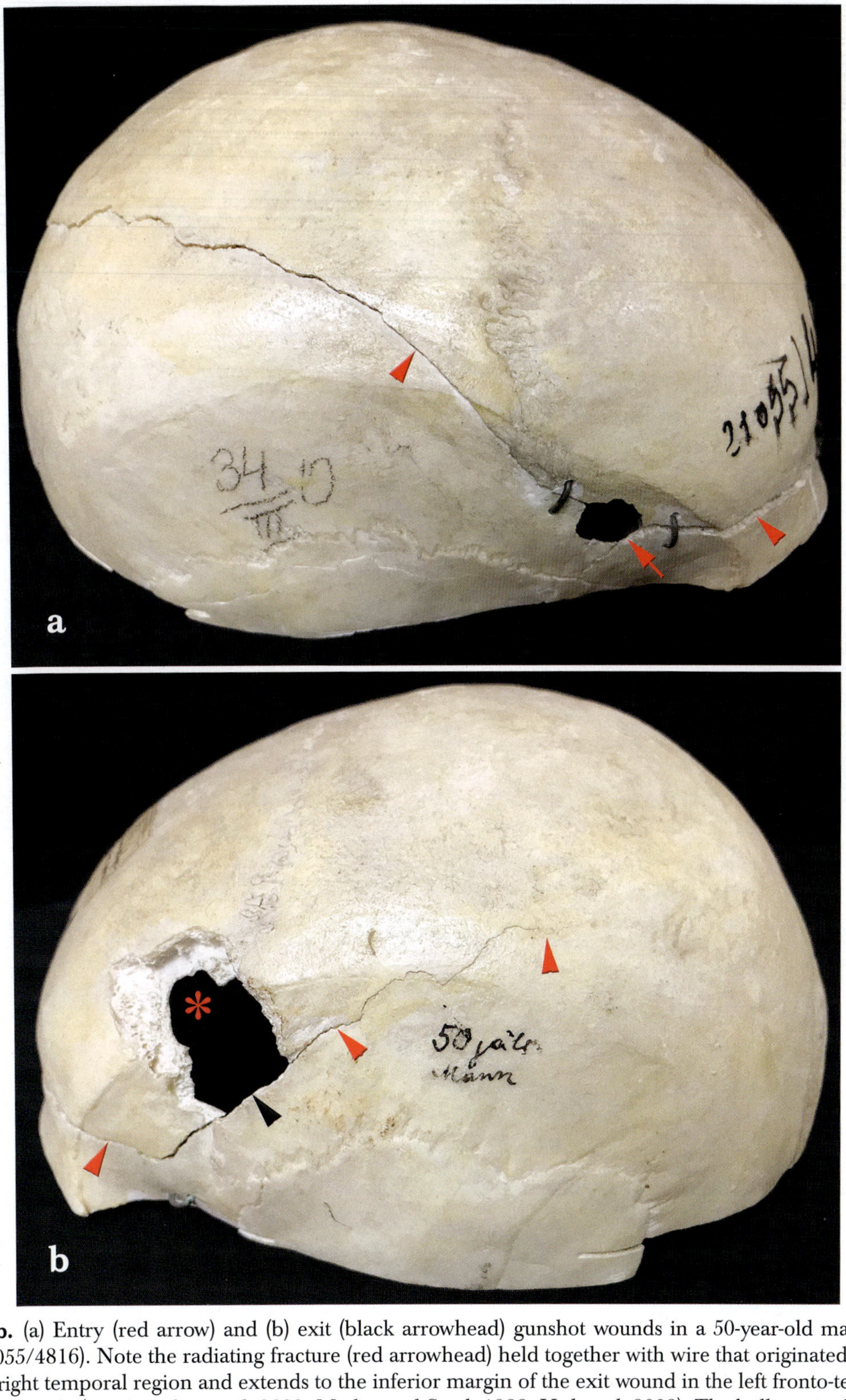


**Figure 380a-b.** (a) Entry (red arrow) and (b) exit (black arrowhead) gunshot wounds in a 50-year-old male who died in 1902 (MN21.055/4816). Note the radiating fracture (red arrowhead) held together with wire that originated from the entry wound in the right temporal region and extends to the inferior margin of the exit wound in the left fronto-temporal region, reflecting Puppe's rule (see Amadasi et al. 2022; Madea and Staak 1988; Viel et al. 2008). The bullet exited the left frontal bone leaving a round defect (*) and external beveling in the bone.

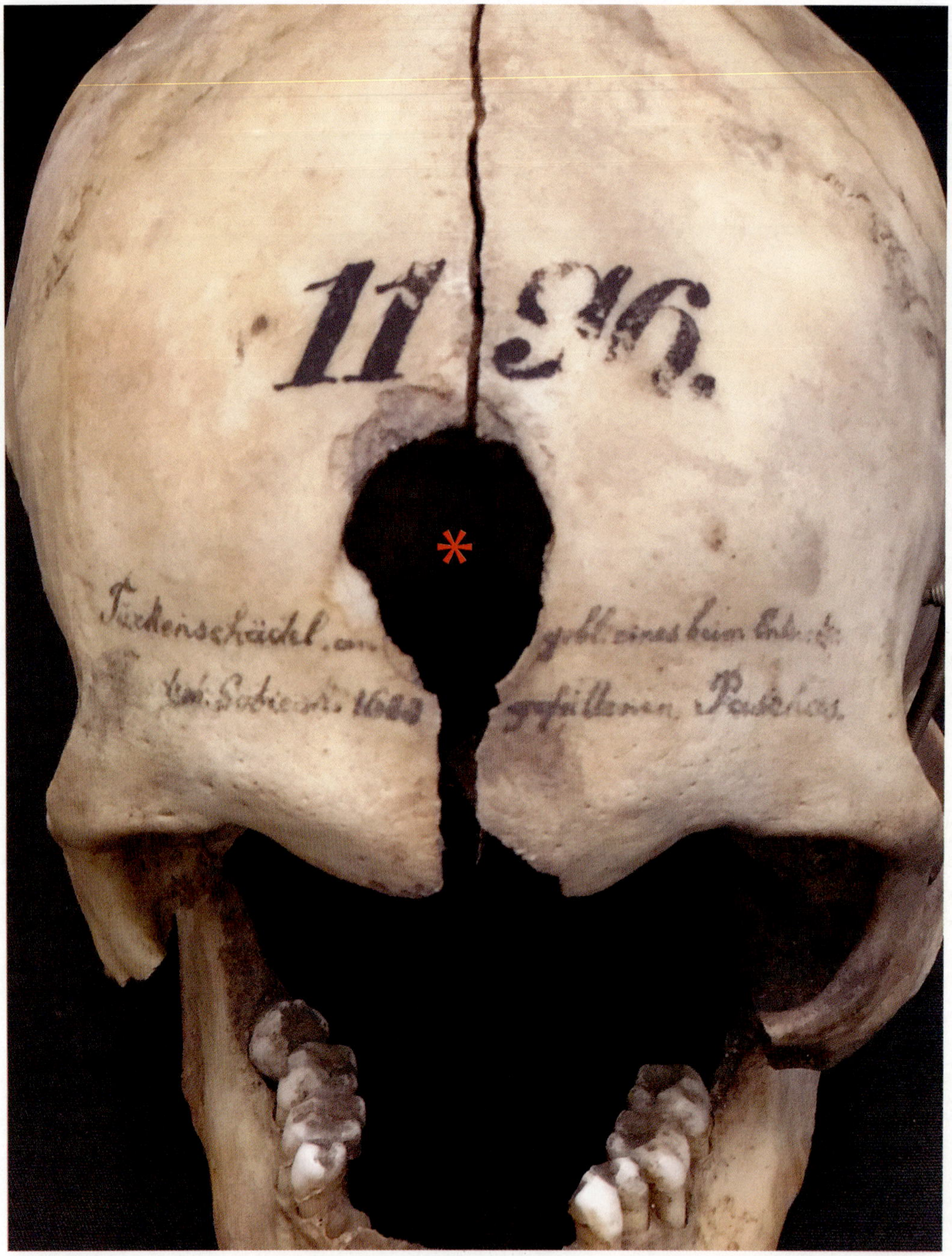


**Figure 381.** Gunshot entry wound (*) in the frontal bone of an adult male killed in the 1683 Battle of Vienna (MN1196). The musket ball entered the frontal bone and exited the occipital bone (Figure 382). Note the wide linear fracture extending from the superior margin of the gunshot wound.

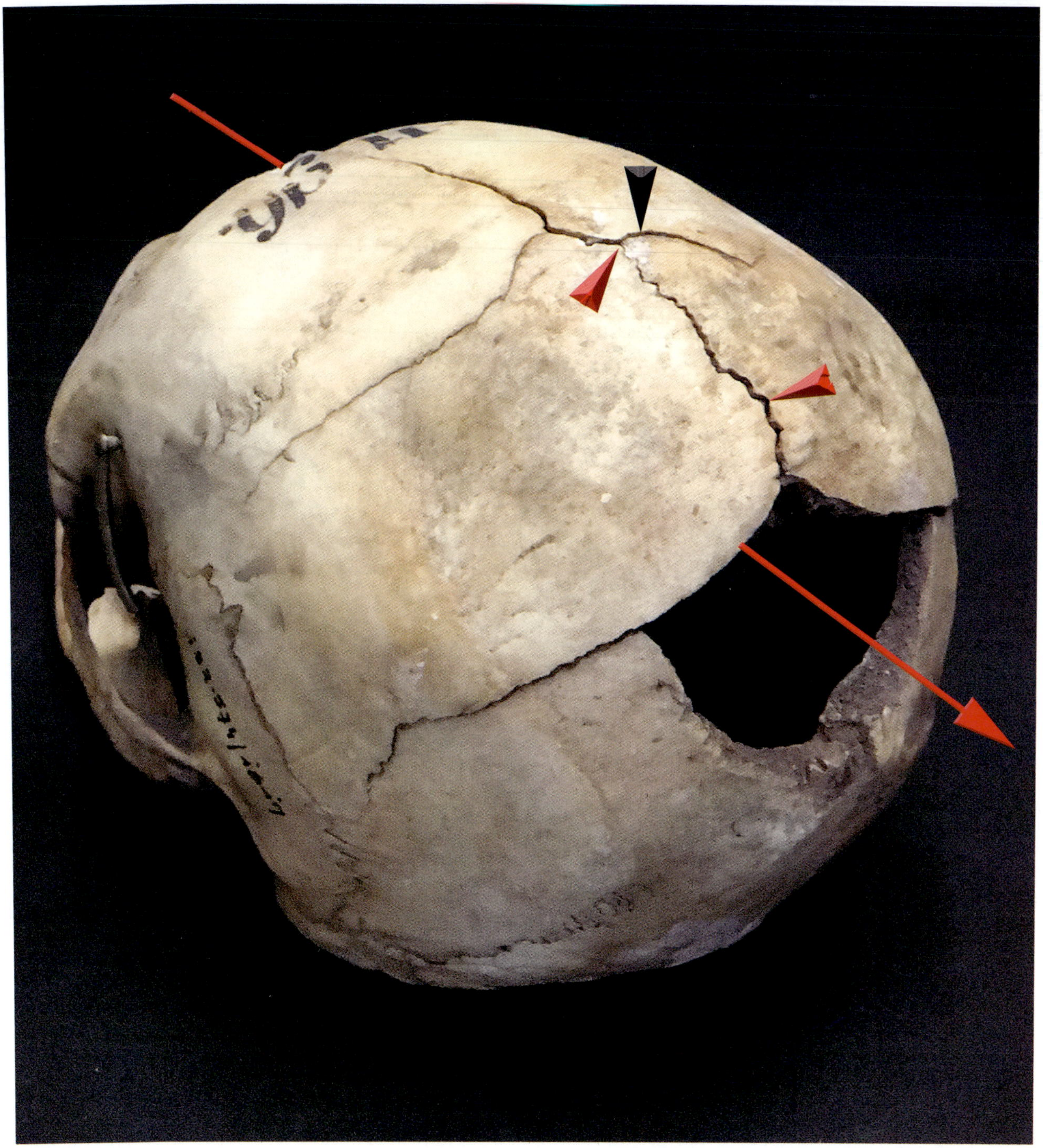

**Figure 382.** Gunshot entry wound through the frontal bone that exited through the parietal and occipital bones in an adult male who died in the 1683 Battle of Vienna (MN1196). Note the diastatic fracture (red arrowhead) along the posterior portion of the sagittal suture that likely occurred when the musket ball exited the occipital bone, opening the sagittal suture and posterior and terminating at the linear fracture (black arrowhead) formed by the entry wound. See DiMaio and DiMaio (2001); Kimmerle and Baraybar (2008); Smith et al. (2003); Spitz and Diaz (2020); Watson-Jones (1944); Wedel and Galloway (2014) for more on skeletal trauma.

**Figure 383.** Slicing injury to the right frontal and parietal bones with no evidence of healing in a probable adult who died in 1876 (MN21.055/2782). This slicing injury entered the decedent's skull traveling from his front to back, from above to below and from left to right. See DiMaio and DiMaio (2001); Vanezis (2021) for more on sharp force trauma.

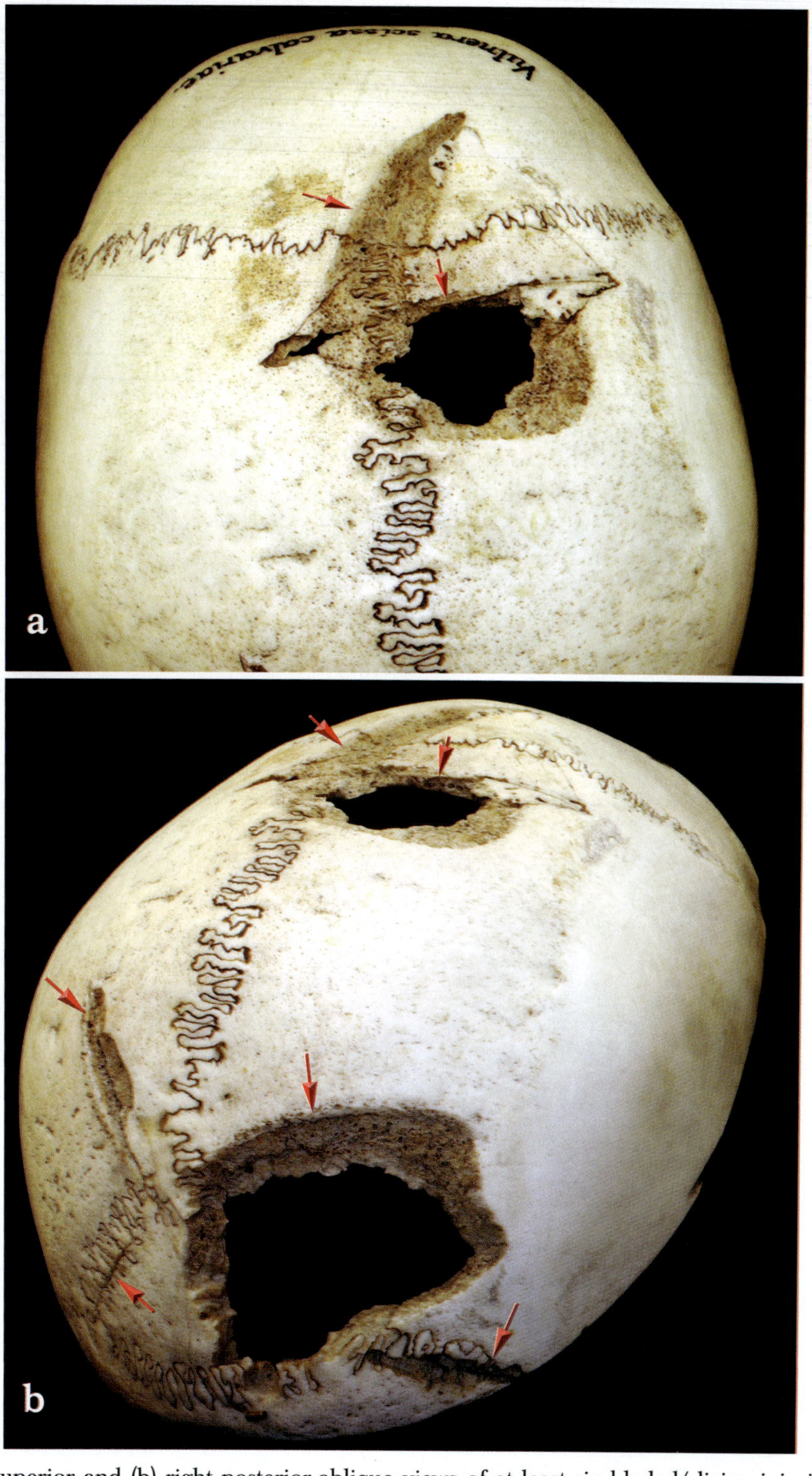


**Figure 384a-b.** (a) Superior and (b) right posterior-oblique views of at least six bladed/slicing injuries to the top of the skull (arrows), none of which exhibits evidence of healing in an unknown individual (MN5721). The morphology of these injuries suggest that the implement was a bladed (e.g., sword) rather than a chopping (e.g., axe) weapon. See DiMaio and DiMaio (2001); Vanezis (2021) for more on sharp force trauma.

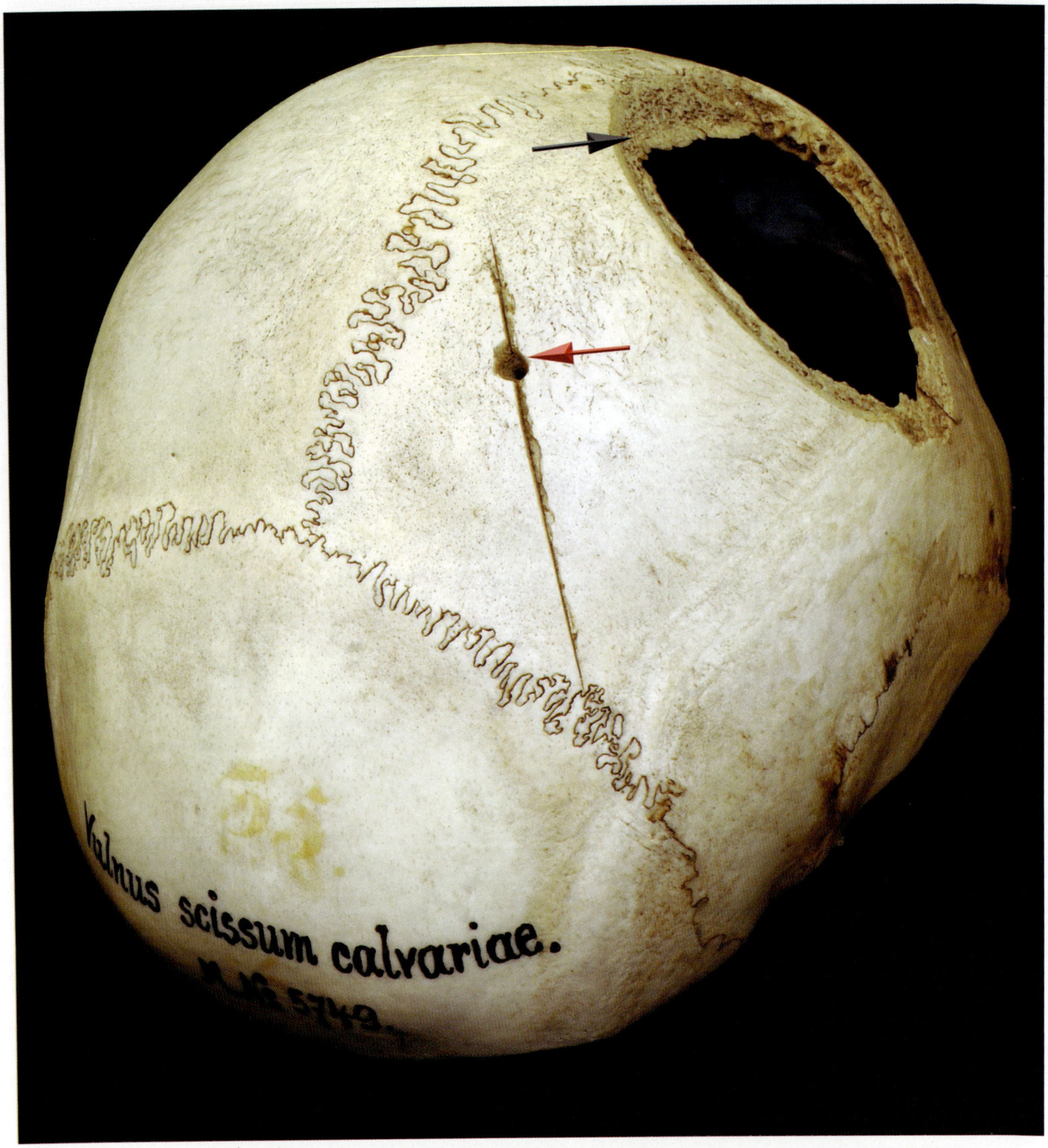


**Figure 385.** Two bladed, slicing injuries to the frontal and left parietal bones in an unknown adult (MN5749). One injury consists of a chopping blow with a conical penetrating defect (red arrow) affecting the anterior portion of the left parietal bone near the coronal suture and the other (black arrow) resulted in removal of a large portion of the left parietal bone near the occipital bone. See DiMaio and DiMaio (2001); Vanezis (2021) for more on sharp force trauma.

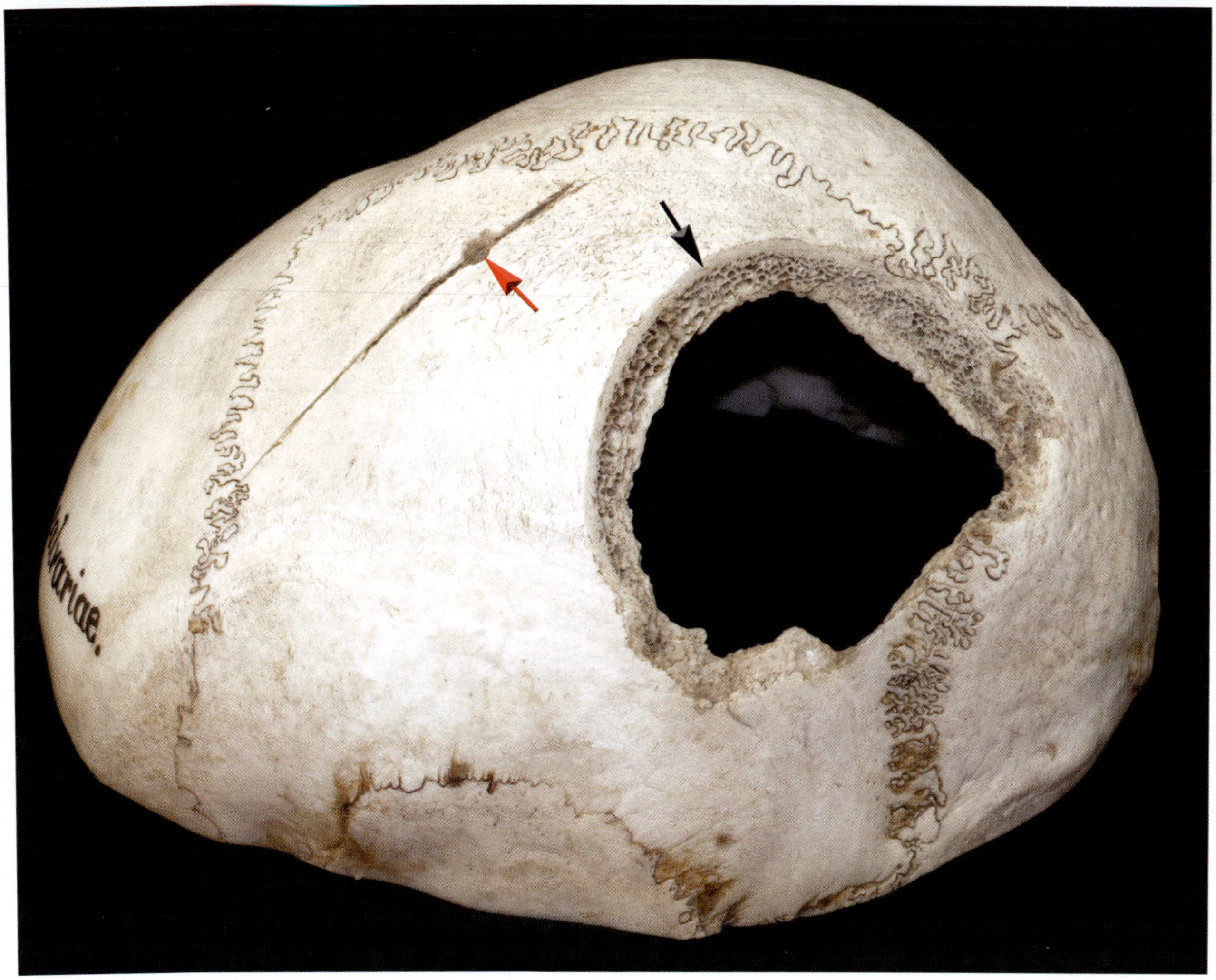

**Figure 386.** Two bladed, slicing injuries to the frontal and left parietal bones in an unknown adult (MN5749). One injury consists of a chopping blow with a penetrating conical component (red arrow) affecting the anterior portion of the left parietal bone near the coronal suture and the other (black arrow) resulted in removal of a large portion of the left parietal bone near the occipital bone. The directionality of the blade that caused the posterior defect was from the decedent's front to back, right to left, and from above to below. Note that the anterior portion of the large defect is smooth and clean suggesting entry by the blade and a jagged posterior margin along the lambdoidal suture suggesting its approximate exit. The absence of fractures connecting these two injuries makes it nearly impossible to determine the sequence of injuries. See DiMaio and DiMaio (2001); Vanezis (2021) for more on sharp force trauma.

**Figure 387.** Large perimortem sword or axe injury in the left parietal bone of an unknown adult (MN5722). This injury exhibits no evidence of healing suggesting that the individual probably died within 7-10 days of sustaining the injury. See Barbian and Sledzik (2008) for more on bone healing following cranial trauma, and DiMaio and DiMaio (2001); Vanezis (2021) for more on sharp force trauma.

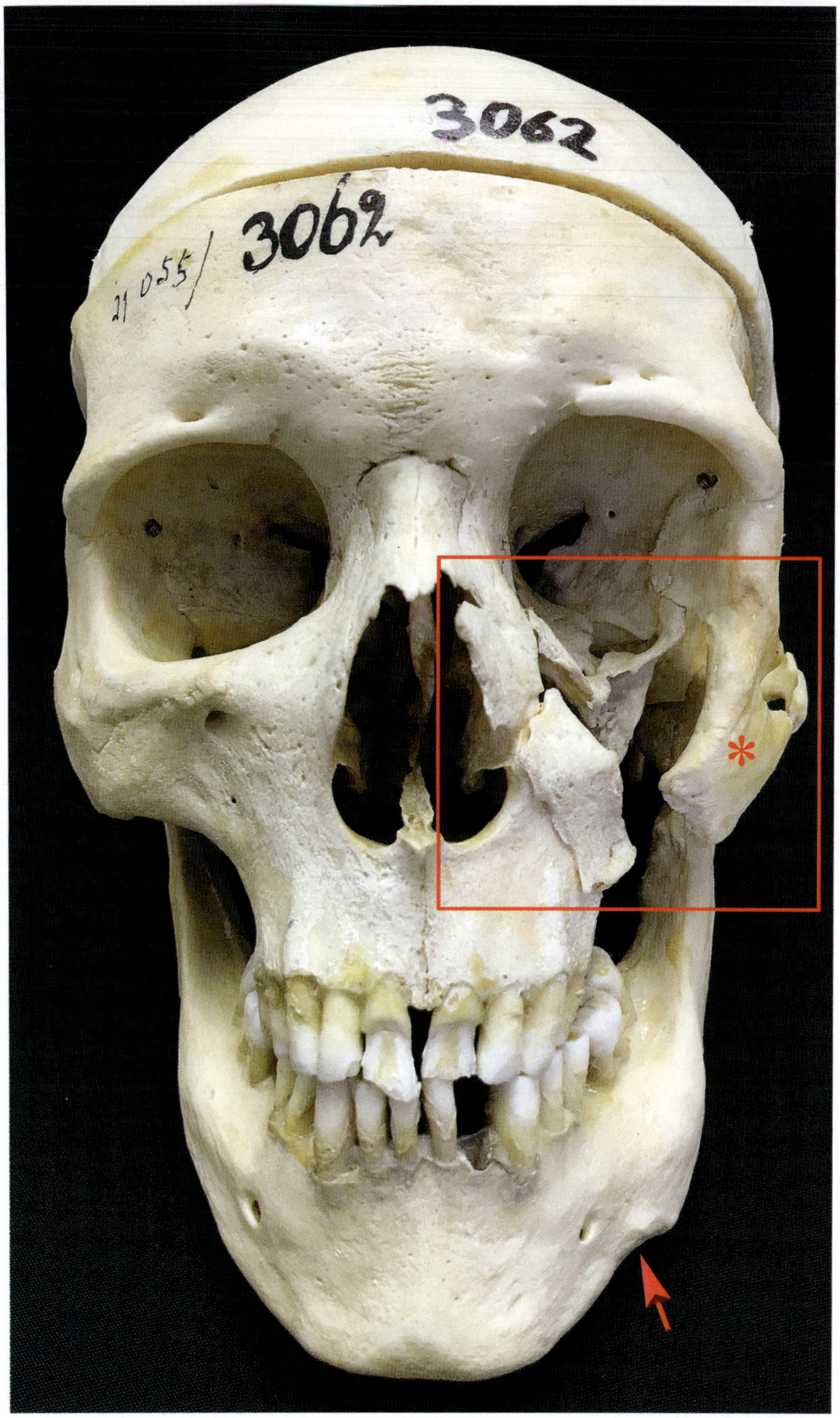


**Figure 388.** Healed blunt force trauma to the left facial region (square) and tripod fracture (*) of the left zygoma (square) and mandible (arrow) resulting in displacement of bones in a 38-year-old male who died in 1876 (MN3062). Note that the displaced bone fragments are fused/attached to other bones and their blunt broken edges show evidence of remodeling indicating he probably survived the event 7-10 days or more (Barbian and Sledzik 2008), if not months or years.

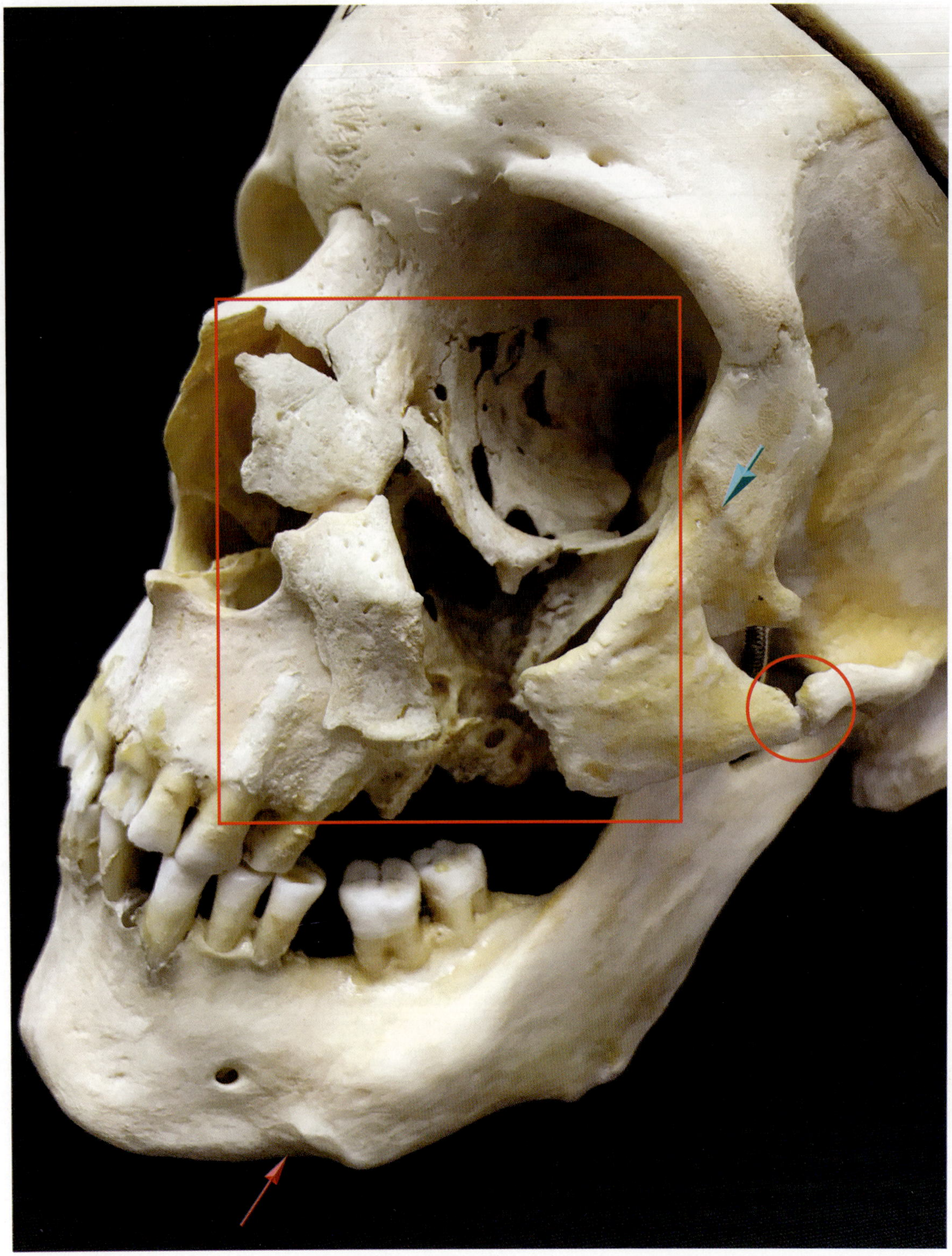

**Figure 389.** Healed blunt force trauma to the left facial region (square), zygomatic arch (circle), and mandible (arrow) resulting in tripod fracture (blue arrow) and displacement of bones in a 38-year-old male who died in 1876 (MN3062). The amount of bone healing indicates that he probably survived the event 7-10 days or more (Barbian and Sledzik 2008) and likely months or years.

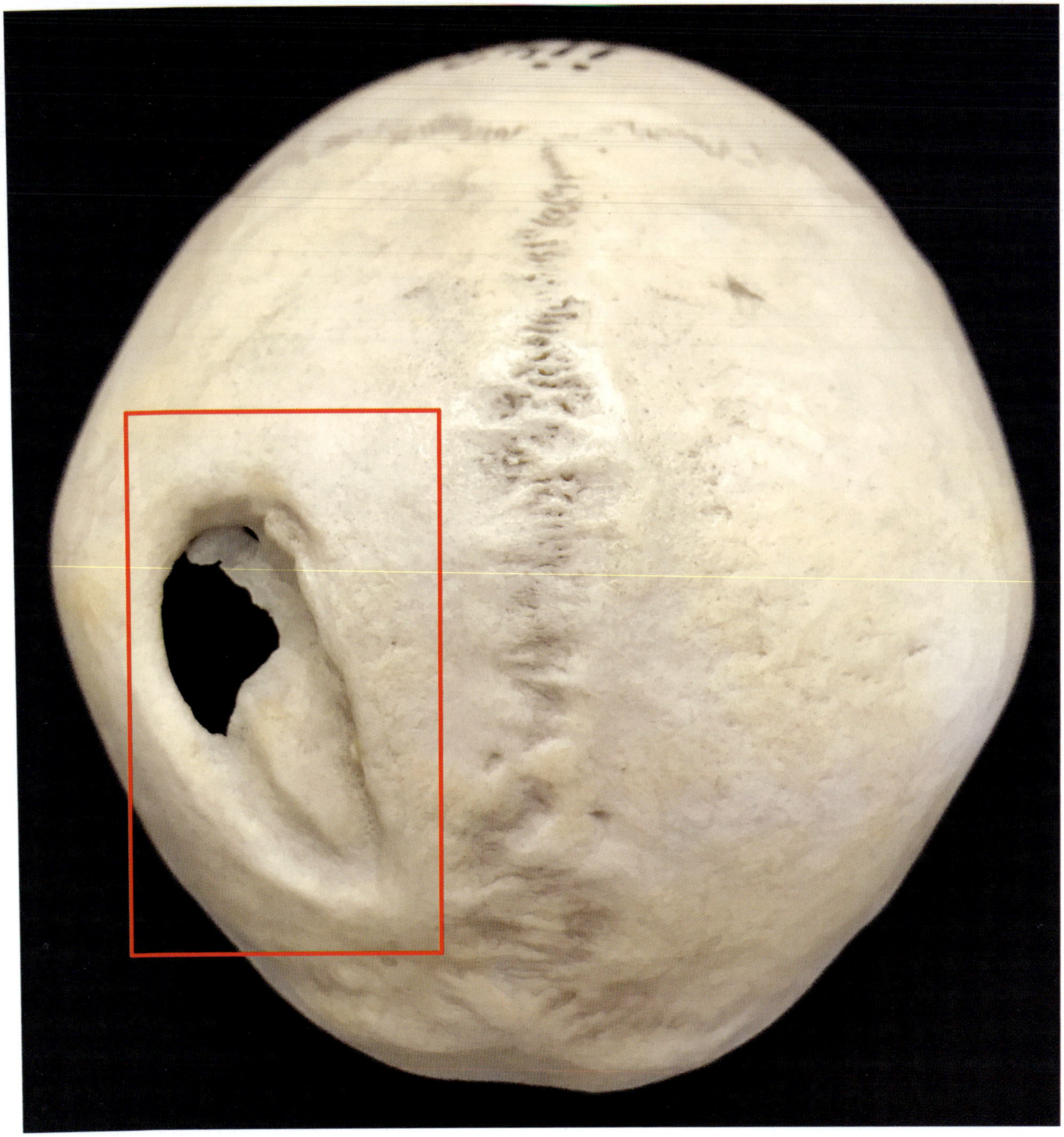

**Figure 390.** Healed blunt force trauma ("pond fracture") to the left parietal bone in an unknown adult who was kicked in the head by a horse and died in 1886. The angled kick resulted in an oval-shaped injury that perforated the inner table and resorbed the outer table and diploe (MN21.055/1142).

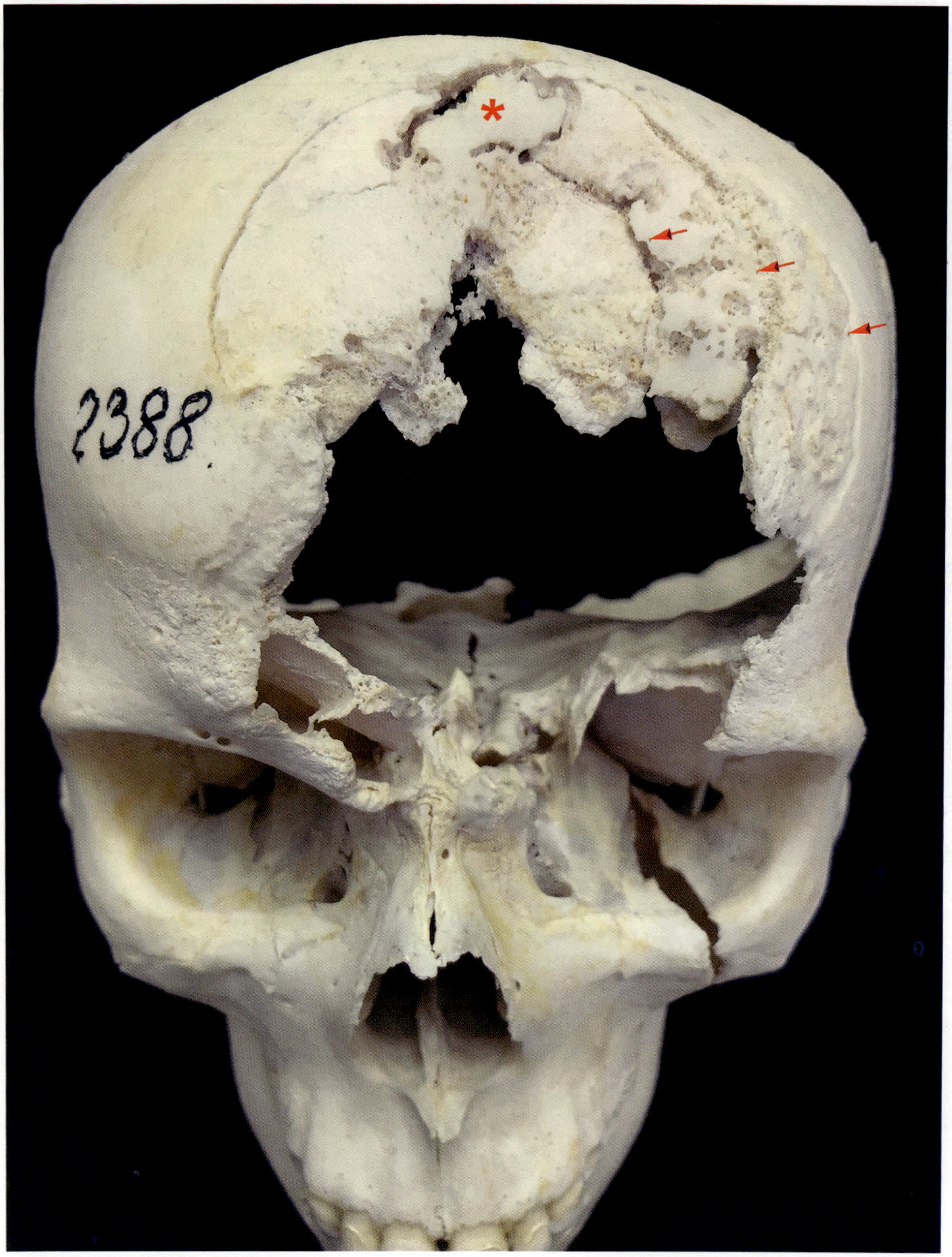


**Figure 391.** Comminuted blunt force trauma and subsequent infection (sequestrum *) of the frontal region of the cranium and left maxilla in a 29-year-old male who died of unknown causes in 1874 (MN2388). Note the concentric fractures and large broken area of the skull that is consistent with blunt force trauma, not bladed or a gunshot wound. Several comminuted fractures (arrows) are still visible despite the infection. See DiMaio and DiMaio (2001).

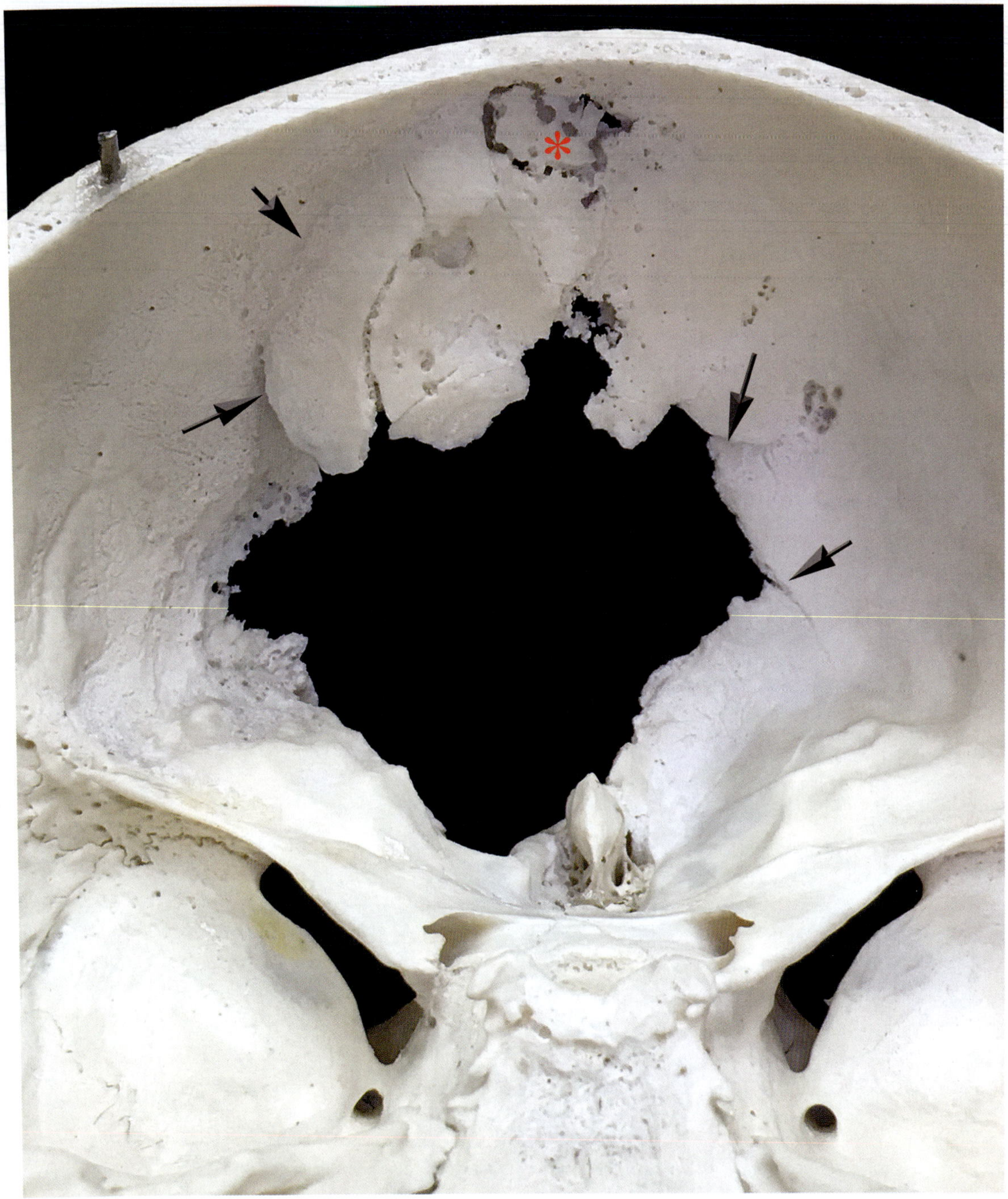

**Figure 392.** Endocranial view of the frontal region showing inwardly depressed bone fragments and fracture (arrows) indicating comminuted blunt force trauma in a 29-year-old male who died of unknown causes in 1874 (MN21.055/2388). Reactive bone, pitting and formation of a sequestrum (*) suggest infection and healing secondary to trauma of perhaps months or years.

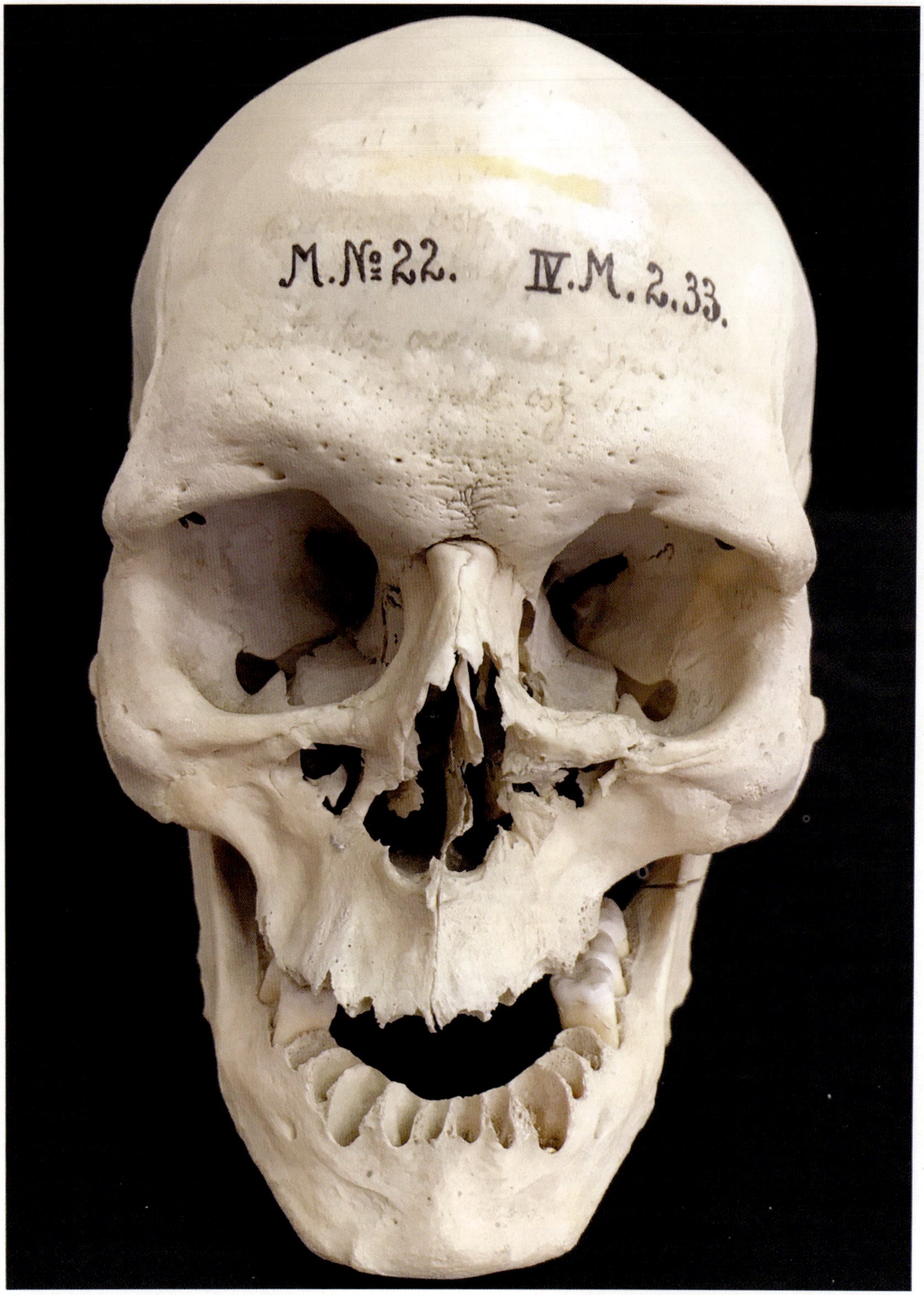


**Figure 393.** Healed blunt force trauma to the nasal region of unknown origin in an adult male who died in 1813 (MN22).

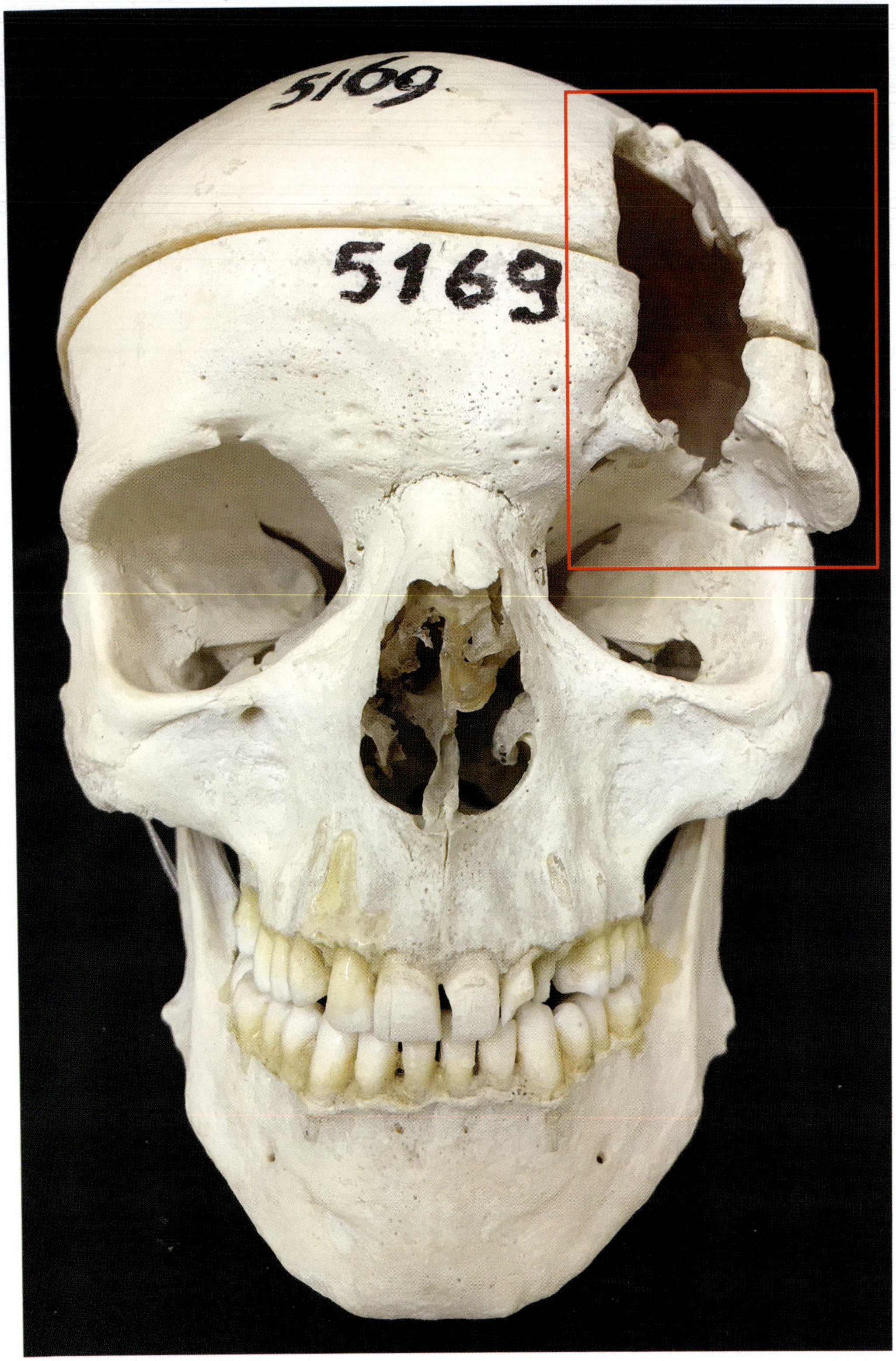


**Figure 394.** Healed trauma due to gunshot to the left frontal and orbit in a 35-year-old male who died in 1904 (MN21.055/5169). Note that portions of the left orbit and frontal bone are laterally displaced from the cranium.

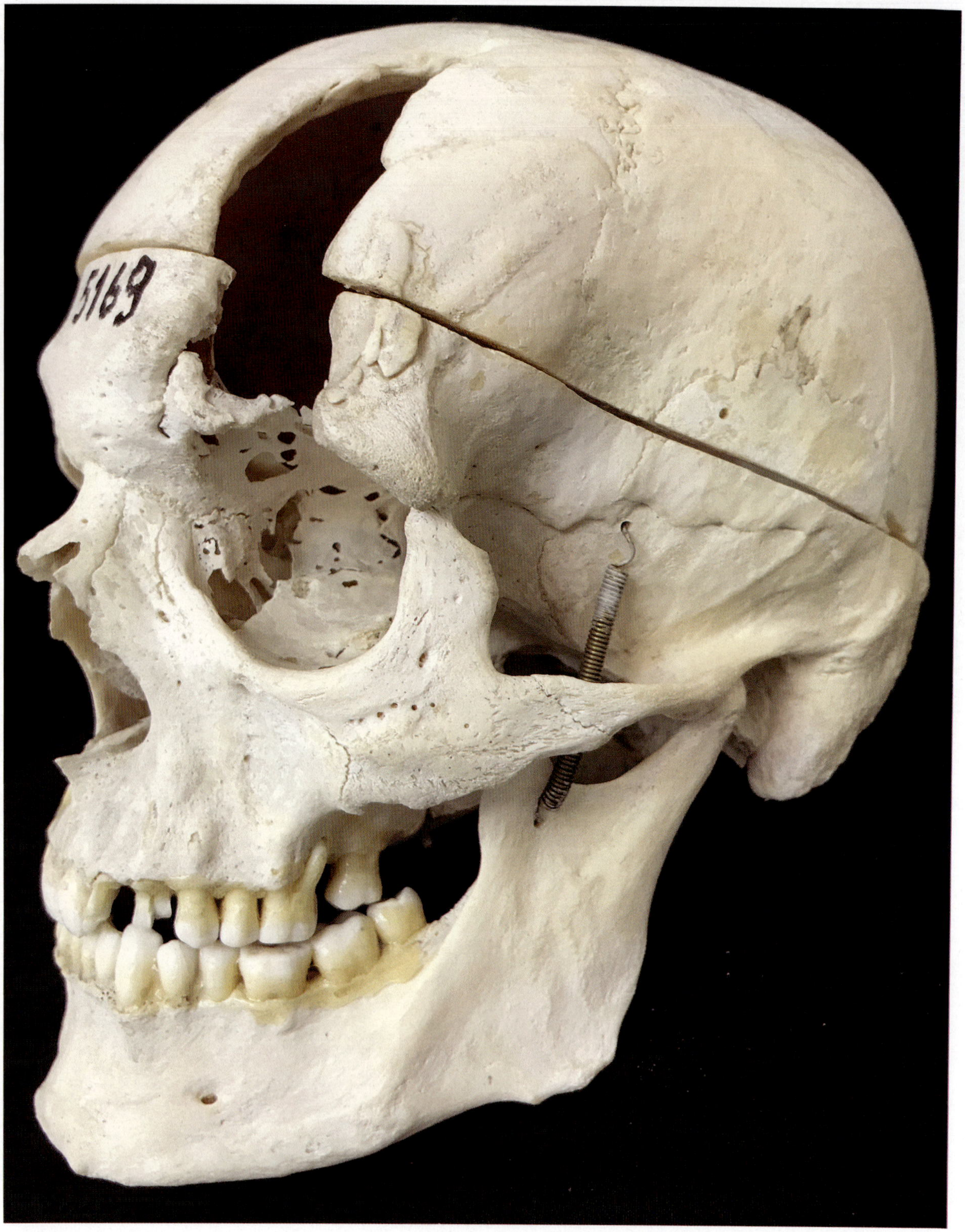


**Figure 395.** Left oblique view of healed trauma due to gunshot to the left frontal and orbit in a 35-year-old male who died in 1904 (MN21.055/5169). Note that the traumatic event resulted in lateral displacement of the left orbit and frontal bone.

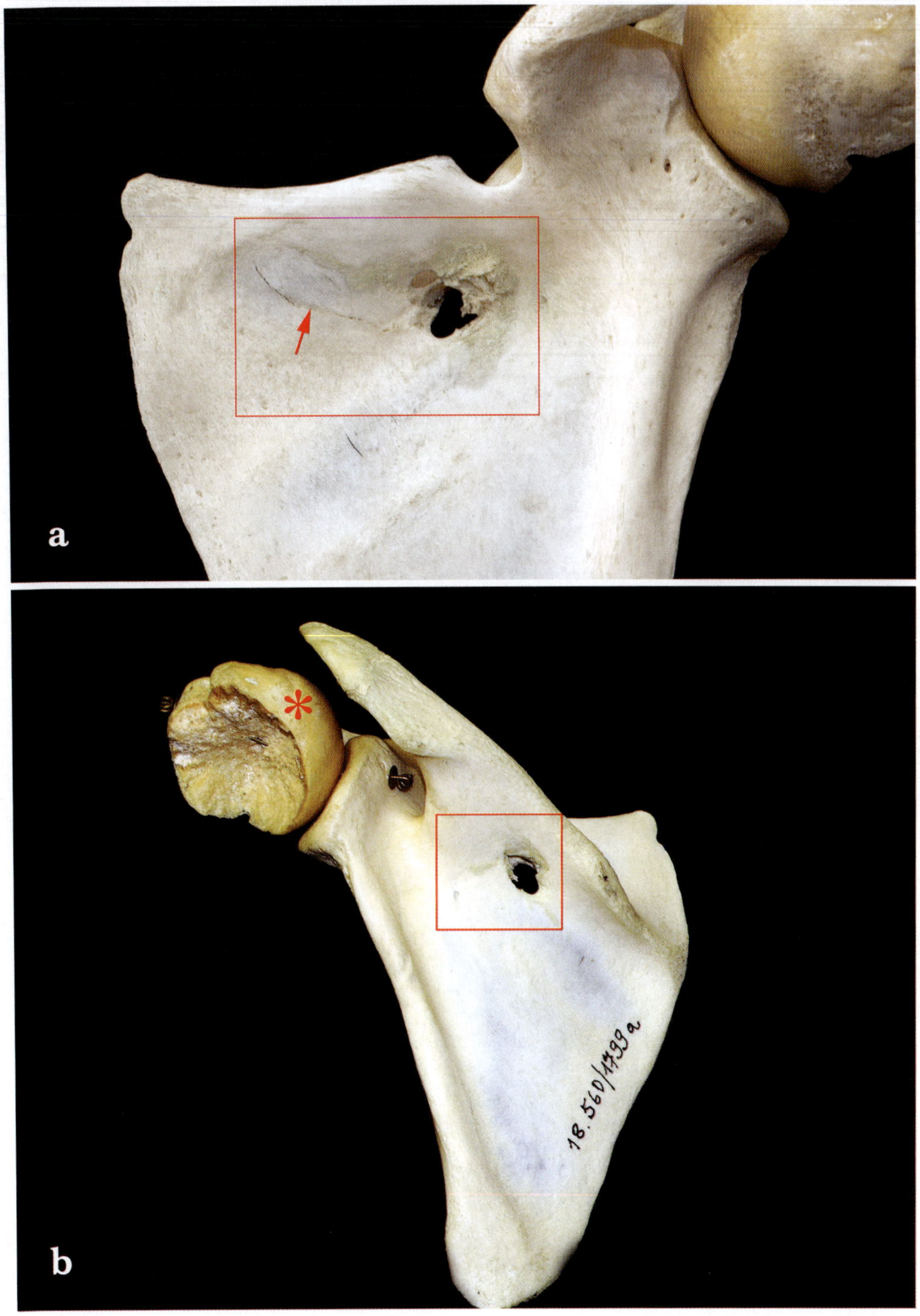


**Figure 396a-b.** (a) Anterior and (b) posterior views of healing fracture (arrow) of the left scapula caused by gunshot in a 20-year-old male who died in 1942 from a gunshot wound to the chest (MN21.060/5169). (a) Note healing around the injury and (b) that the proximal epiphysis (i.e., head) (*) of the humerus has not fully fused to the diaphysis, an indicator of subadult or young adult. See DiMaio and DiMaio (2001); Spitz and Diaz (2020) for more on gunshot trauma to the skeleton.

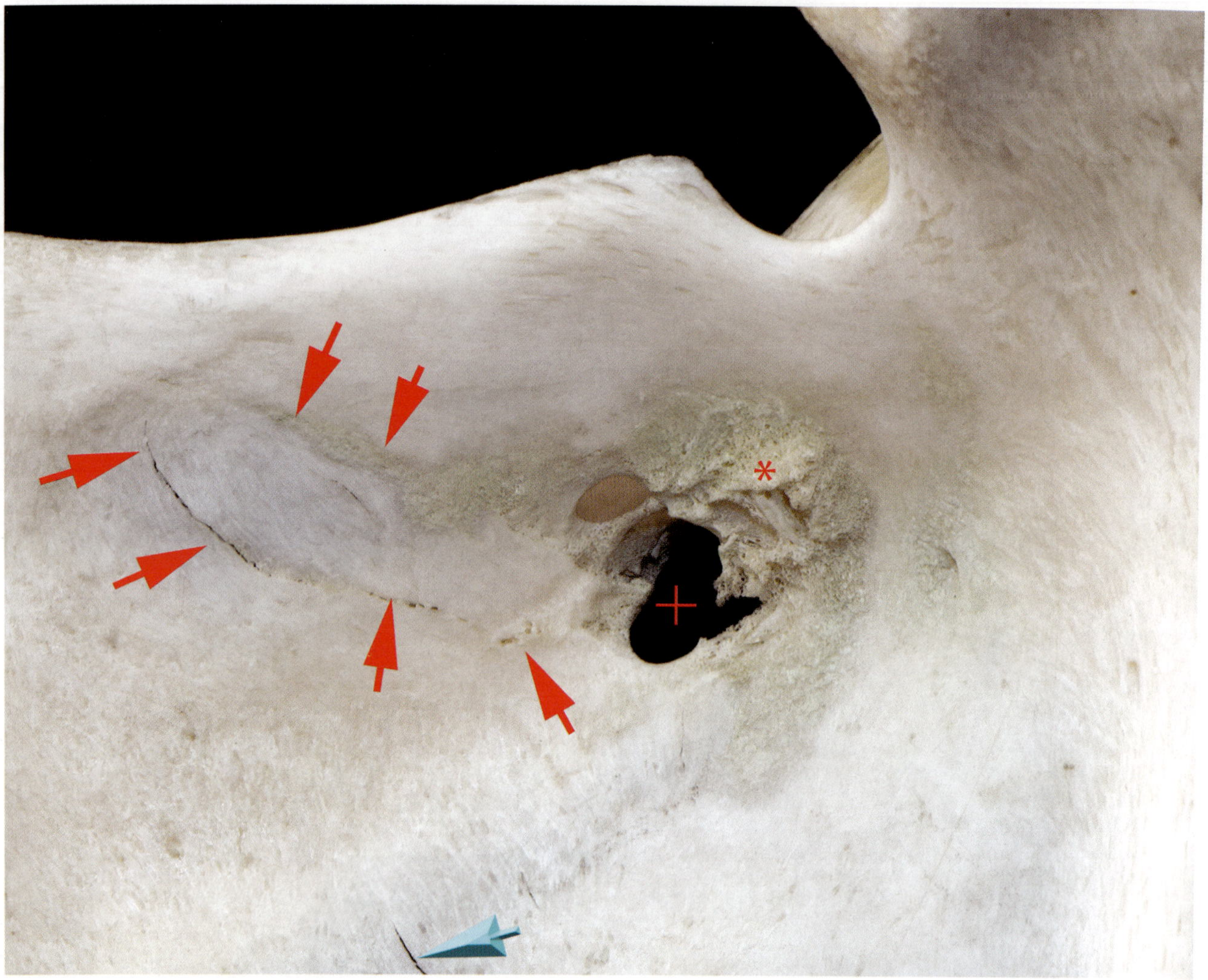

**Figure 397.** Closeup anterior view of healing fracture of the left scapula in a 20-year-old male who died of a gunshot wound to the chest in 1942 (MN18.560/1799a). Note the perforating defect (+), periosteal new bone (*), and radiating fractures (red arrows) with the presence of new bone indicating that this individual survived the injury for at least several weeks. Care should be exercised when identifying a perimortem fracture such as these, from breakage ("drying cracks") occurring postmortem, especially in the thin ilium and scapula, due to drying and shrinking of bone (blue arrow). The first author has noted the presence of postmortem cracks in defleshed bone within a few days or weeks. See DiMaio and DiMaio (2001); Spitz and Diaz (2020).

## Spine, Pelvis, Leg and Foot

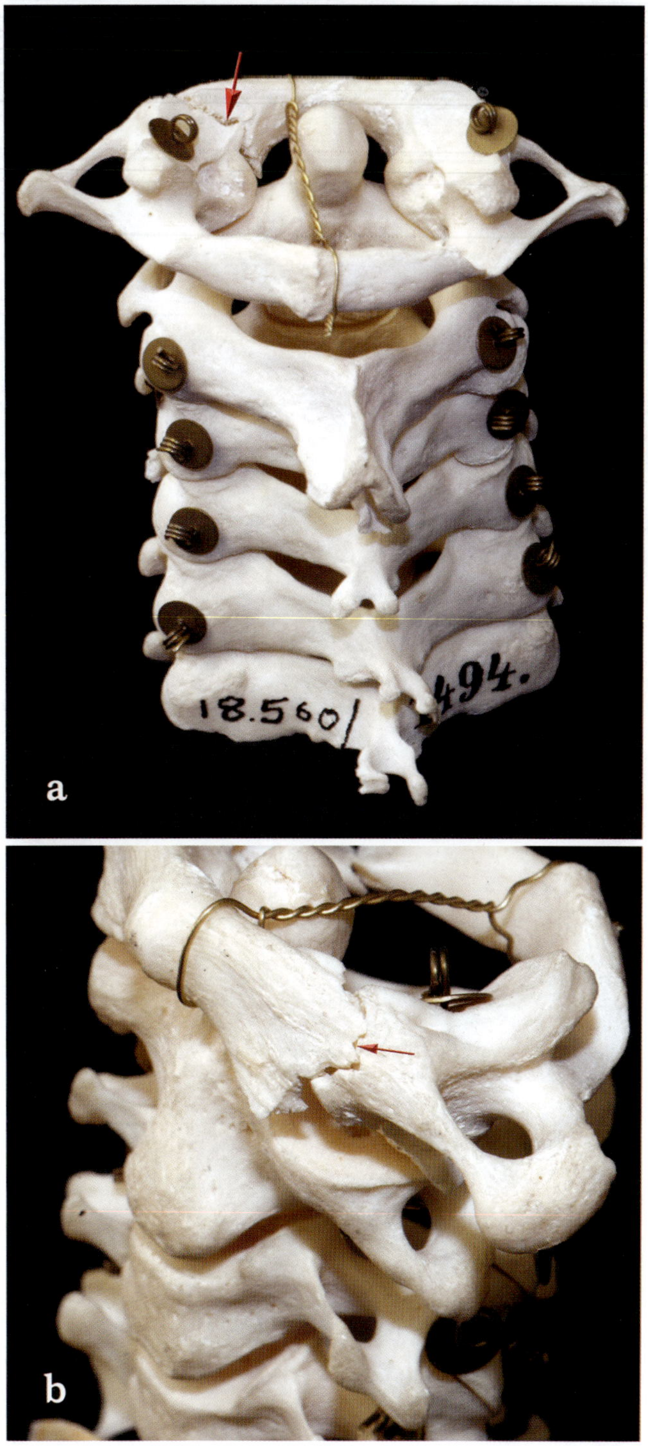


**Figure 398a-b.** (a) Posterior and (b) right oblique views of a perimortem Jefferson's fracture (arrow) through the right superior and inferior articular facets of the 1st cervical vertebra (C1, atlas) in a 17-year-old male who died in 1936 (MN18.560/1494). This type of fracture is usually associated with axial loading and possible head trauma, hence absence of healing suggesting that he died within days of the injury (P. Moroz, personal communication, November 17, 2022).

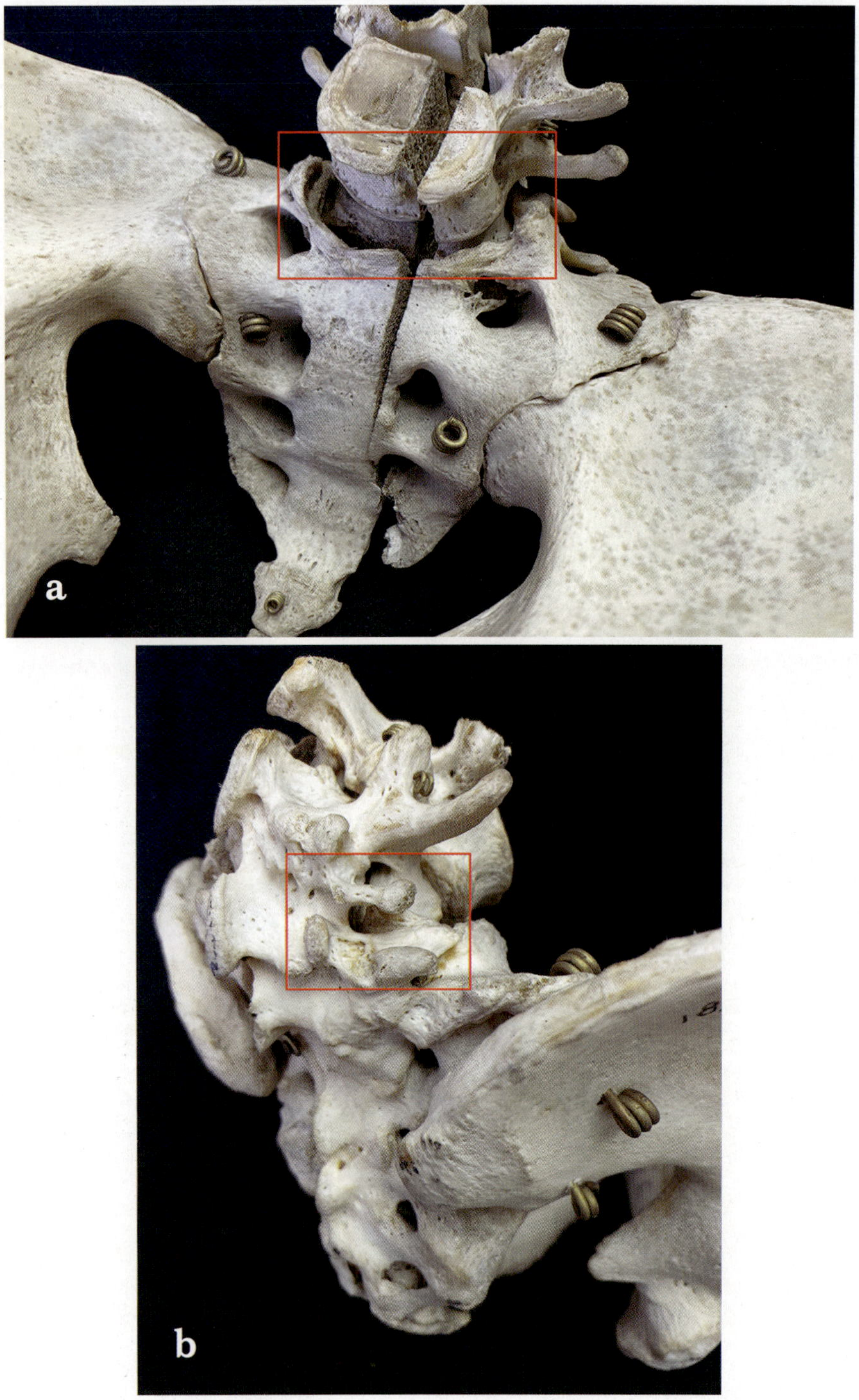


**Figure 399a-b.** (a) Left superior-inferior and (b) right lateral view of compression fracture of the fifth lumbar vertebra into the first sacral segment with remodeling ((b) rectangle) in an adult male who died in 1900 (MN18.560/297). "I wonder if this was a pathological fracture of the anterior vertebral body of L5, possibly from infection such as tuberculosis and causing kyphotic collapse? (P. Moroz, personal communication, November 17, 2022.) The vertical segment of bone was sawed/removed postmortem ((a) rectangle) to examine its anatomy. See Watson-Jones (1946) for more on lumbosacral fractures.

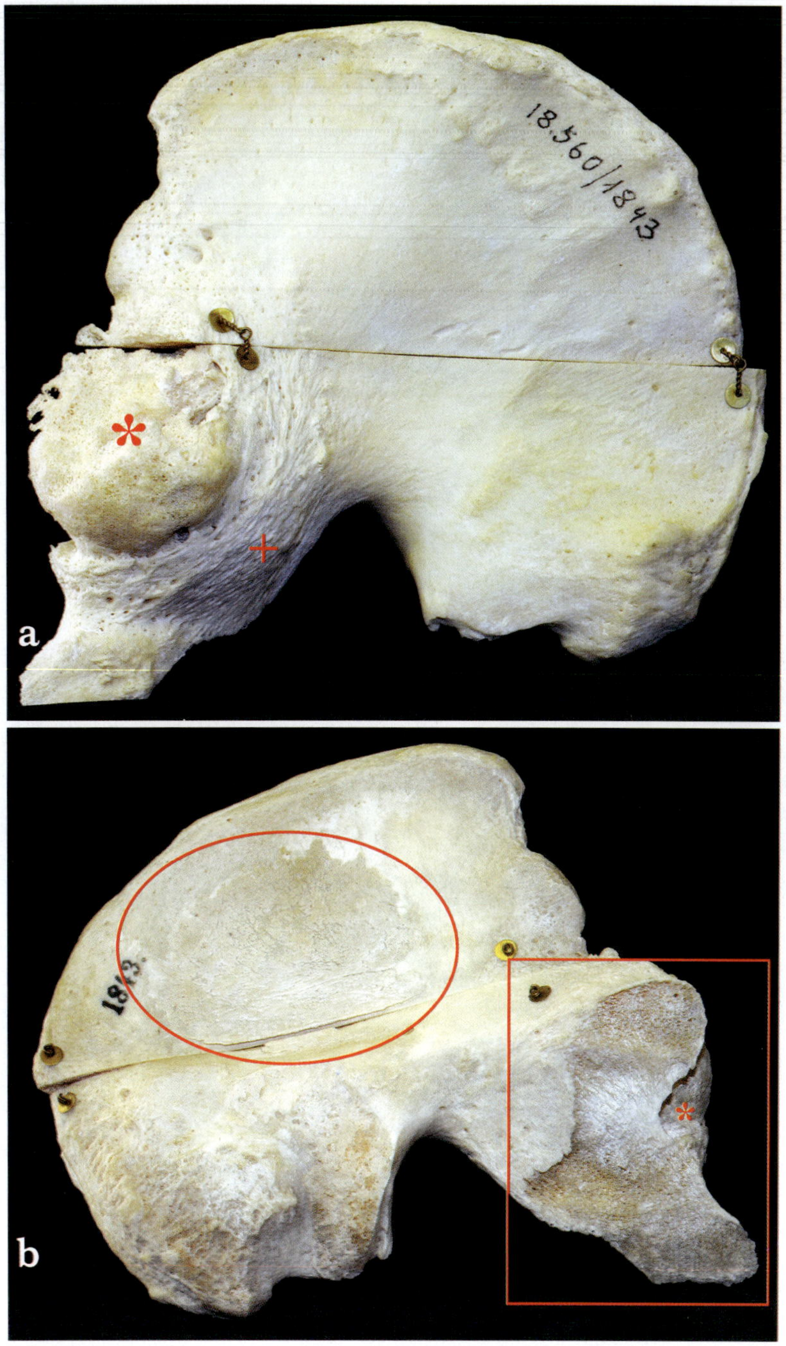


**Figure 400a-b.** (a) Lateral and (b) medial views of the left innominate of an adult male who sustained a fracture of the femoral neck (*) known as an intracapsular fracture and (b) pubic rami (rectangle) resulting in the formation of a thin layer of loosely attached bone (oval) filling part of the iliac fossa (MN18.560/1843). (a) Note the area of periosteal bone reaction (+) (scarring) encircling the acetabulum. Fracture of the femoral neck and pubic rami is a common occurrence in the elderly and can result from a fall from a standing height or from high energy trauma. See Aouad et al. (2022); Harinder et al. (2013); Watson-Jones (1944 and 1946) for more on hip fractures and Rana et al. (2009); Reeder (1993) for different types of periosteal responses in bone.

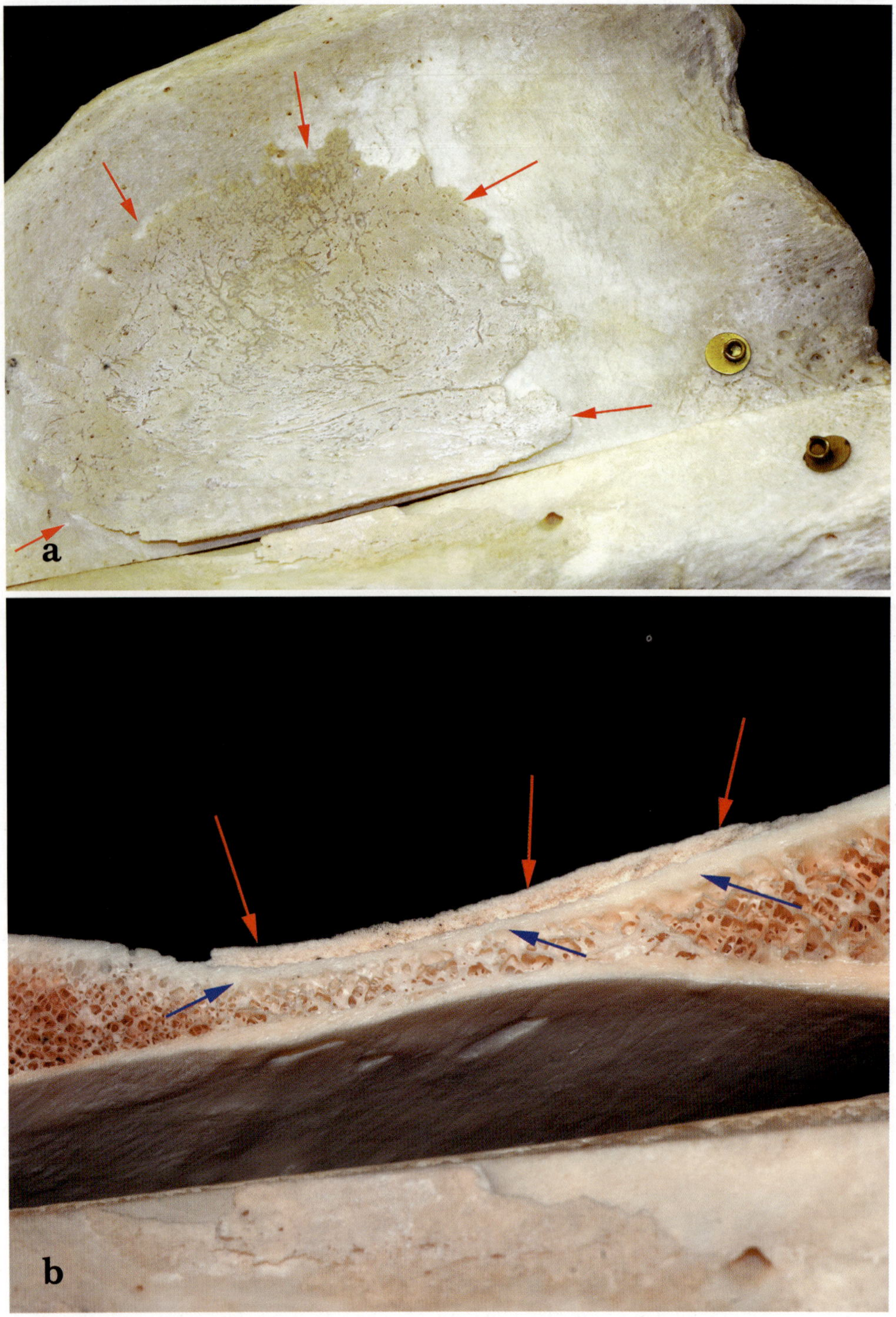


**Figure 401a-b.** (a) Medial and (b) magnified views of left innominate (pelvis) in an adult with a fractured left femoral neck and pubic rami bones (Figure 400) and a thin layer of loosely attached bone (arrows) with vascular markings on its surface filling part of the iliac fossa (MN18.560/1843). (a) This localized area of bone likely occurred in association with the fractured femur and pubic rami and may represent hemorrhage (ossified hematoma) or a vascular event. See Rana et al. (2009); Reeder (1993) for more on the formation and types of periosteal new bone. Regardless of its etiology, this is a rare dry bone finding in the iliac fossa.

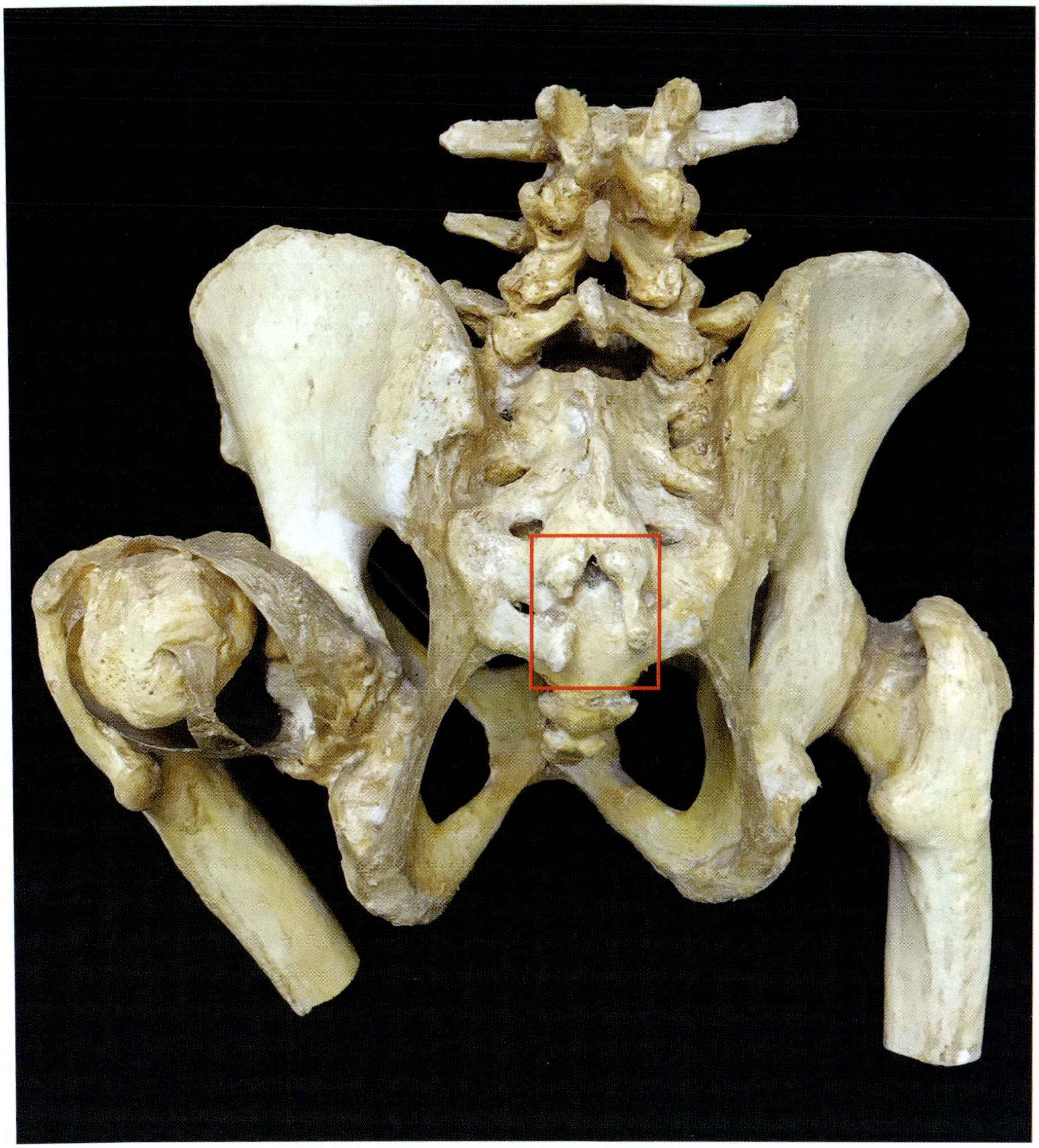

**Figure 402.** Posterior view of luxation (dislocation) of a posteriorly dislocated left hip, likely post-traumatic (P. Moroz, personal communication, November 17, 2022) in a 62-year-old male who died in 1876 (MN21.055/3262). Spina bifida (rectangle) is often used to describe/identify any incomplete closure of the sacral spinous processes and in this individual would be called spina bifida occulta, a very common occurrence. It is a small opening in the sacrum or lower lumbar spine posterior elements. It can be difficult to determine whether spina bifida was cystica or occulta in dry bone. See Watson-Jones (1946 and 1955); *Campbell's Operative Orthopaedics* (1956) for more on fractures, dislocations and other types of injuries of the hip.

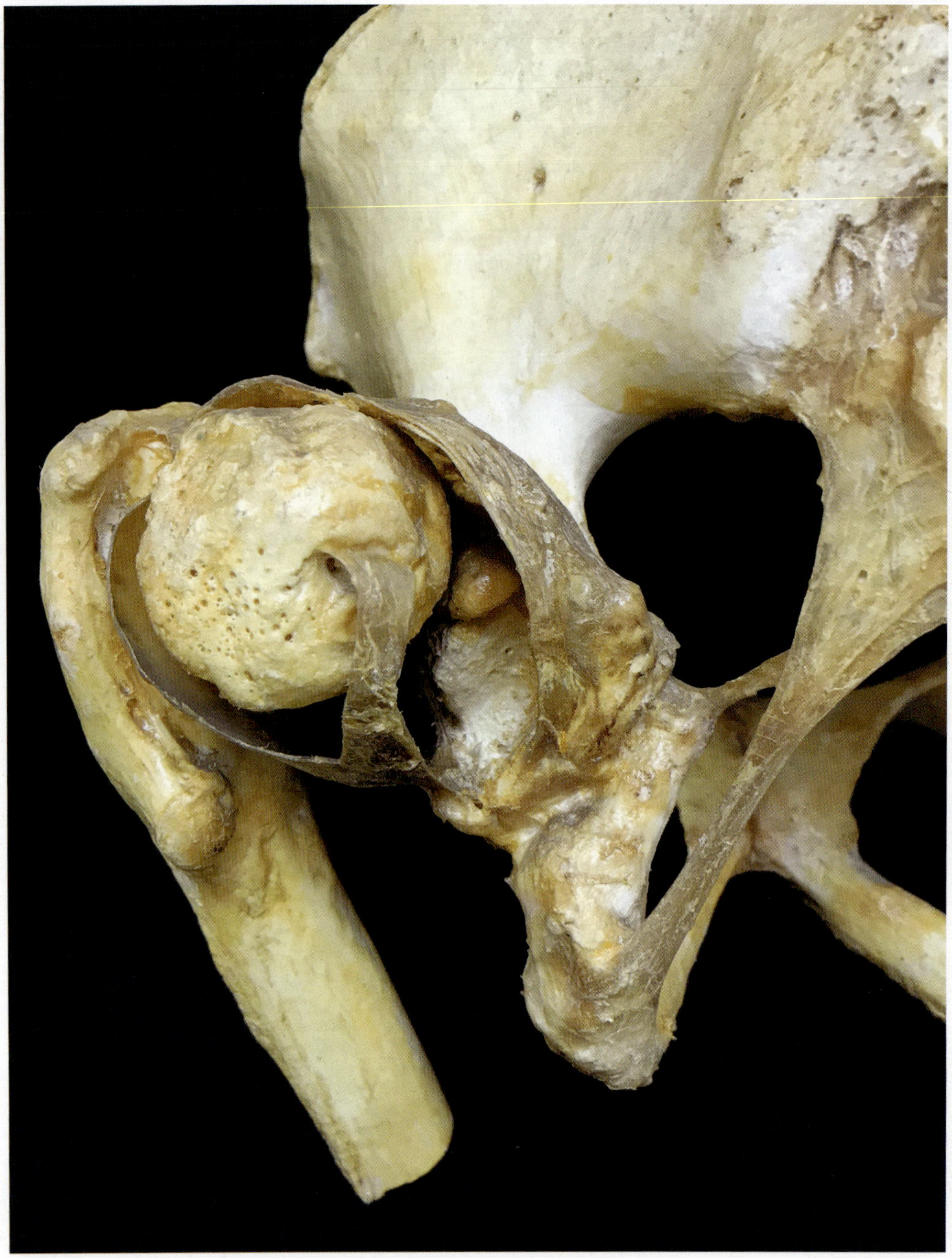

**Figure 403.** Magnified posterior view of complete dislocation of the left femur in a 62-year-old male who died in 1876 (MN21.055/3262). "Erratic posterior rim of acetabulum supports posterior dislocation since it breaks as the femoral head is forced posteriorly by the trauma. Fall from a height or car crash, maybe a horse and cart crash back then. High kinetic energy needed to dislocate a normal adult hip. Femoral head actually doesn't look as deformed as full-on avascular necrosis (AVN) would look like (still could be partial AVN), but femoral head fracture from dislocation does happen. Called Pipkin fracture classification" (P. Moroz, personal communication, November 17, 2022). See *Campbell's Operative Orthopaedics* (1956); Griffiths (1981); Watson-Jones (1946, 1955) for more on fractures, dislocations, and other types of injuries of the hip.

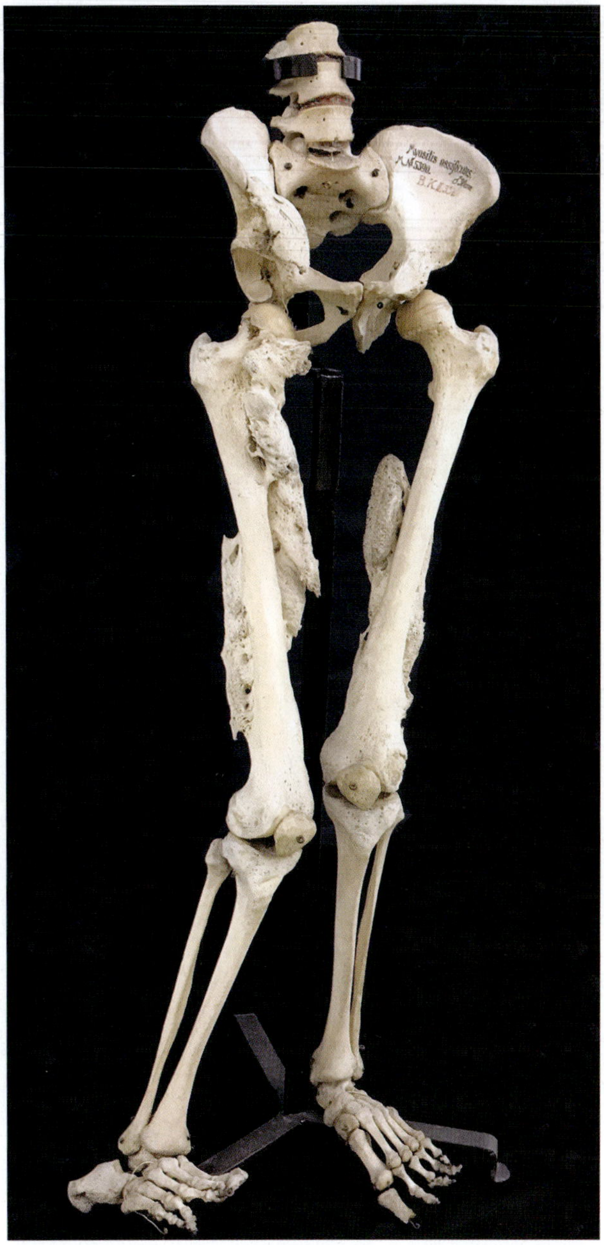

**Figure 404.** Myositis ossificans traumatica (MOT) of unknown etiology of the femurs and right hip in a 30-year-old male (MN5390). Myositis ossificans is an aberrant, self-limiting reparative process resulting in the formation of non-neoplastic benign heterotopic bone within skeletal muscles, especially large muscles (e.g., sartorius, adductor magnus, and biceps femoris), such as in this male. The disease has two forms, MO traumatica as a result of an acute or chronic traumatic event, burns, drug abuse, surgery, neuromuscular disorders, and, rarely, infections. The second form is MO progressiva which occurs early in life, resulting in diffuse heterotopic new bone and early death. One must use caution not to confuse MO with fibrodysplasia ossificans progressiva (FOP). See Geshchickter and Copeland (1936); Lacout et al. (2012); Lungu (2011); Meyers et al. (2019); Watson-Jones (1955; Volumes I and II) for more on these topics.

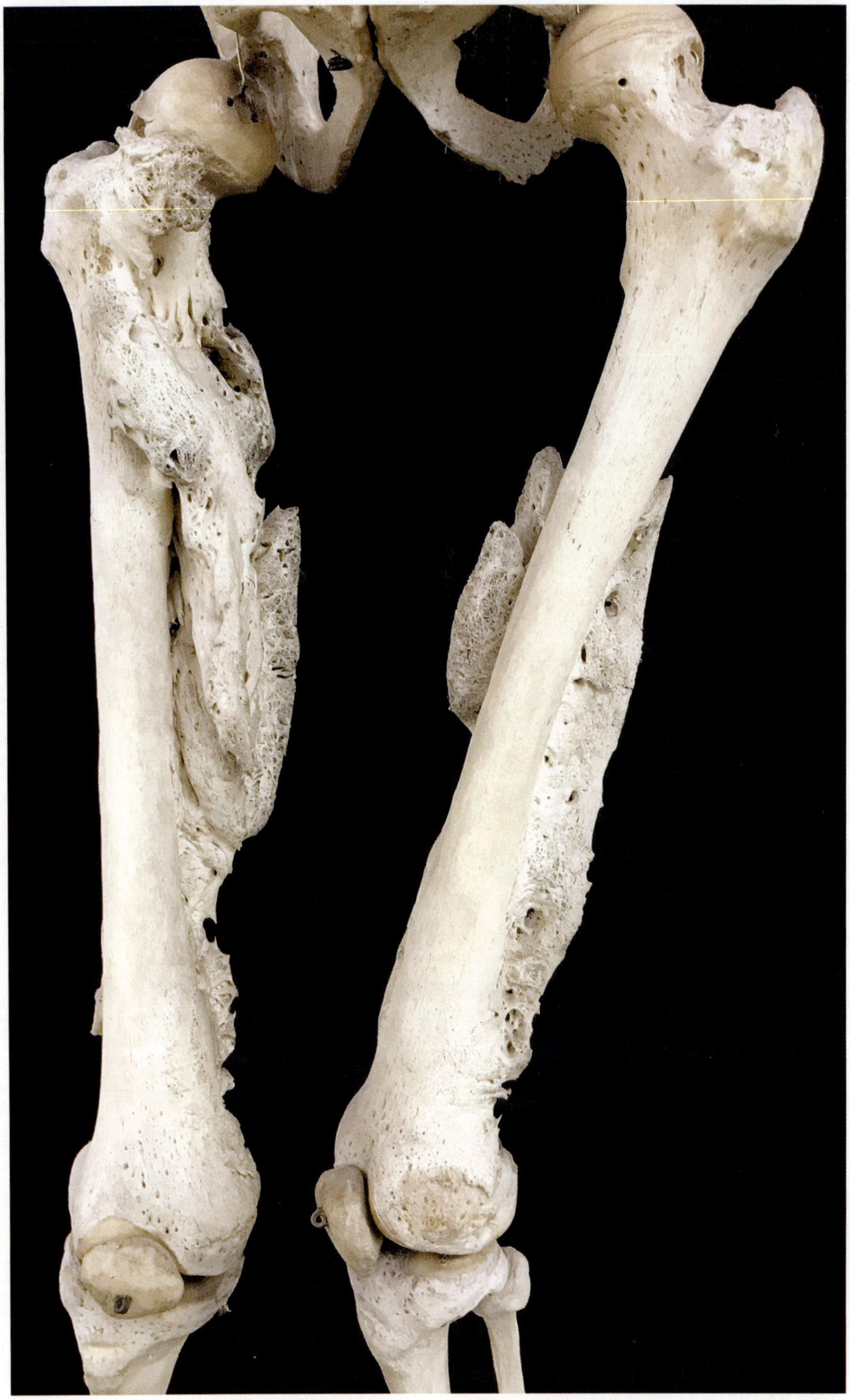

**Figure 405.** Magnified anterior view of myositis ossificans traumatica of the muscles along the femurs in a 30-year-old male (MN5390). See Geshchickter and Copeland (1936) for an excellent discussion of myositis ossificans and its laminated structure and Watson-Jones (1946 and 1955 Volumes I and II) for more on myositis ossificans of the legs and hip.

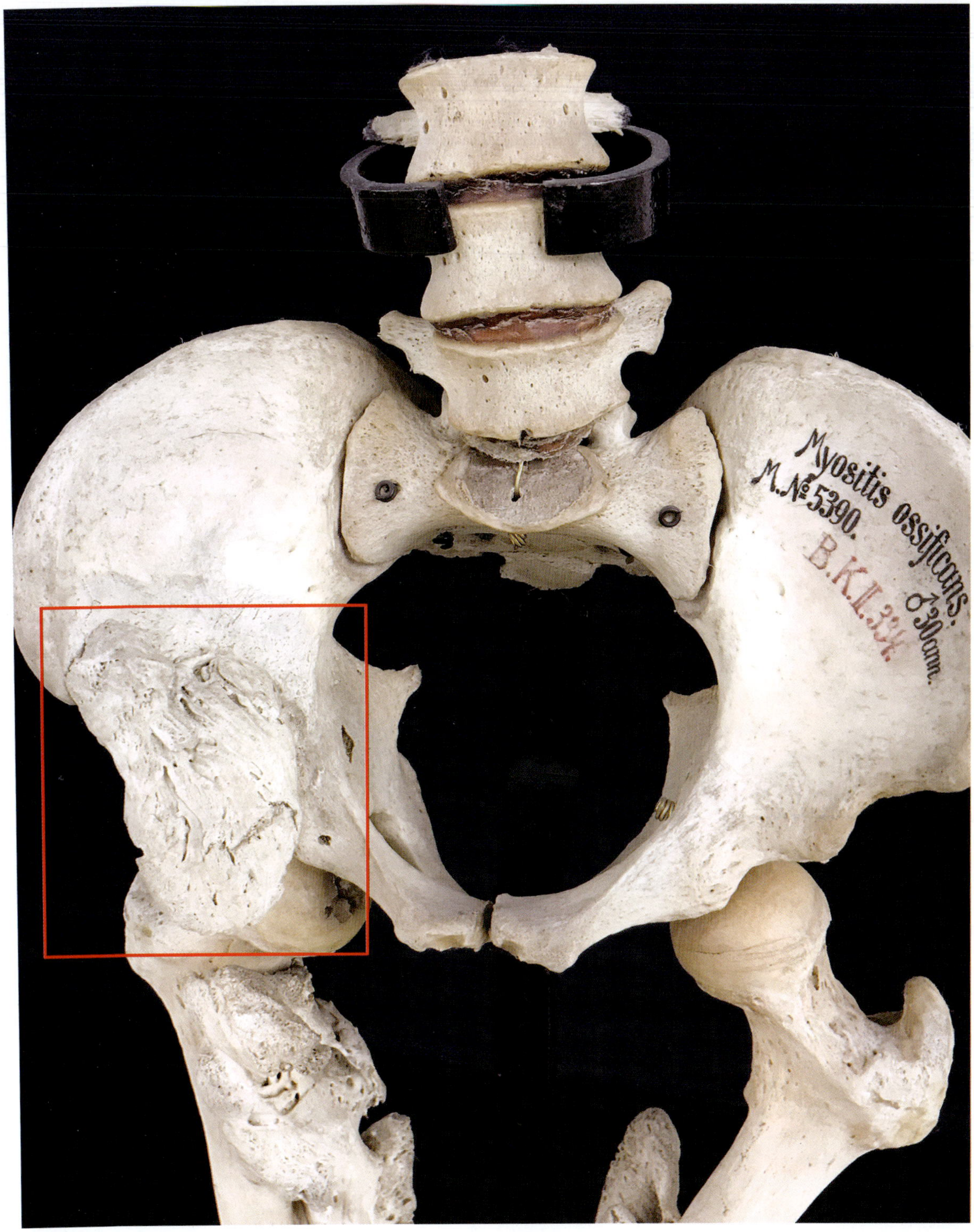


**Figure 406.** Myositis ossificans of the legs and right pelvis in a 30-year-old male (MN5390). Note heterotopic ossification (bone) along the femurs and anterior right ilium (box) for attachment of the hamstring muscles. See Watson-Jones (1946; Volume II) for more on myositis ossificans of the legs and pelvis and Watson-Jones (1955; Volumes I and II) for more on myositis ossificans in general.

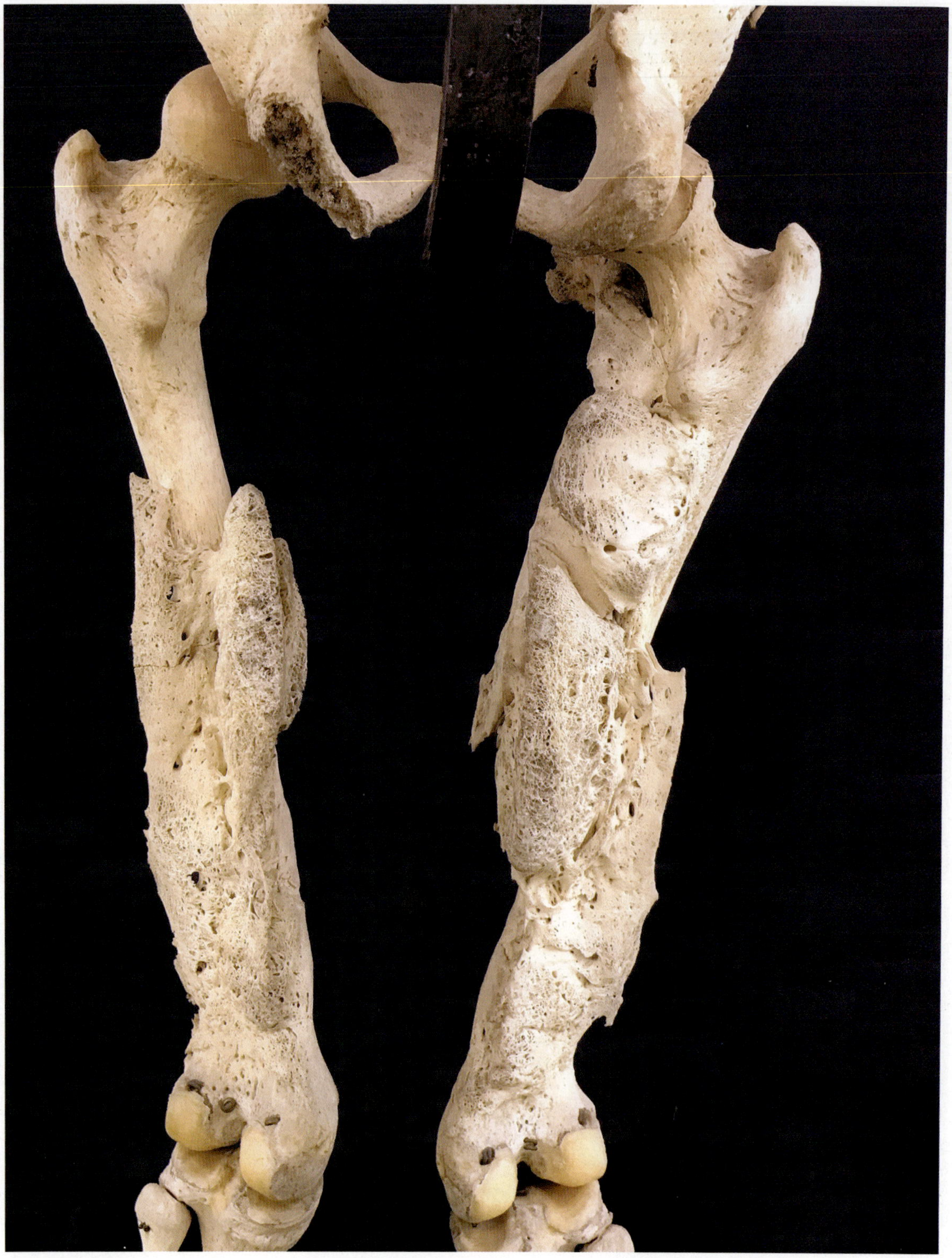

**Figure 407.** Posterior view of the femurs showing localized myositis ossificans traumatica (MOT muscles of the legs and right innominate) in a 30-year-old male (MN5390). See Geshchickter and Copeland (1936); Watson-Jones (1946, 1955) for more on MOT.

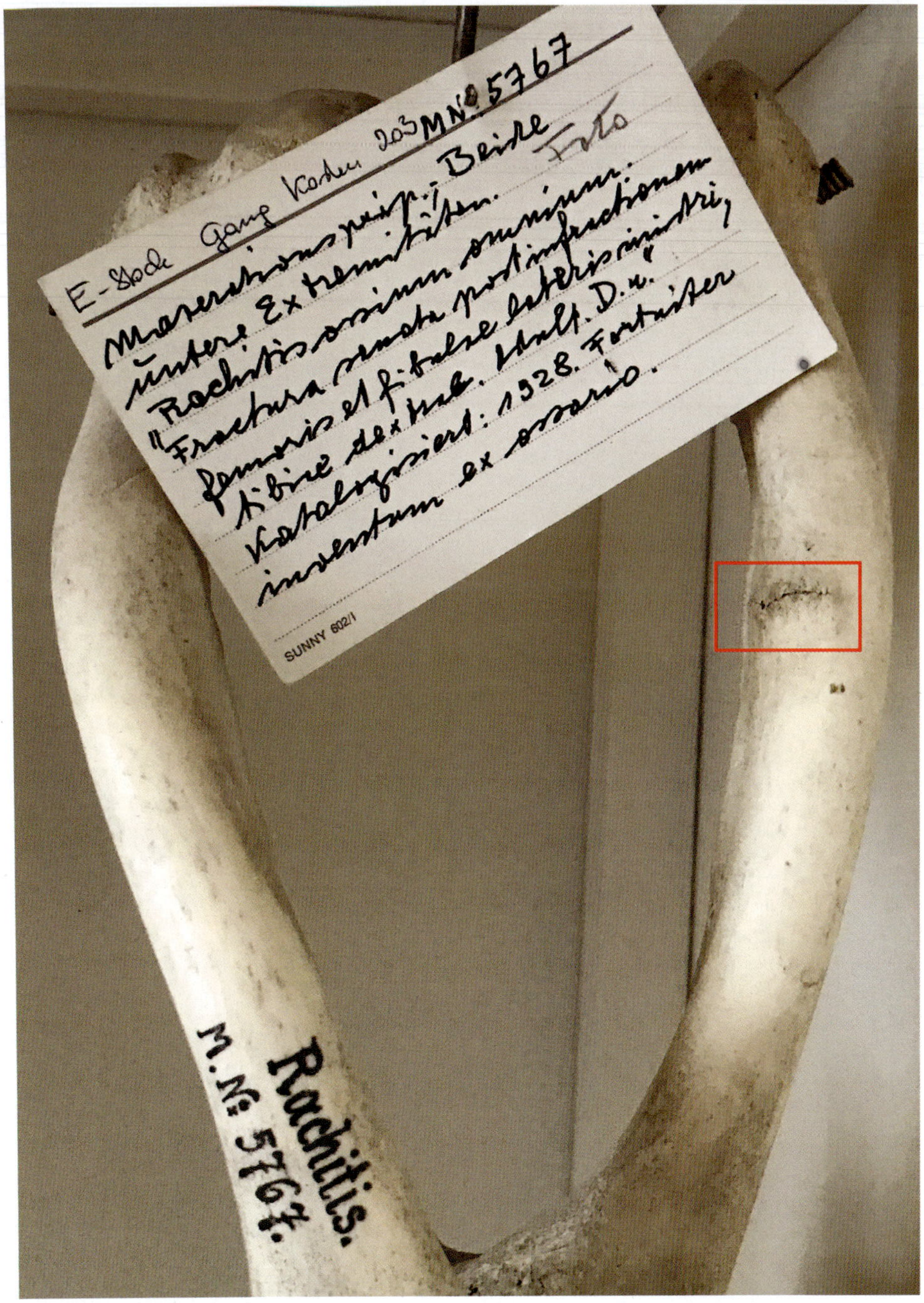


**Figure 408.** Rickets ("Rachitis" written on the femur), resulting in bowing of the femurs in an adult who died in 1828 (MN5767). Note the healing Looser's Zone fracture (rectangle), also known as insufficiency fracture, stress fracture, Milkman's fracture and pseudofracture that formed perpendicular to the length of the left femur due to failure of normal mineralization of bone and fracture through vessel foramina in the medial femur (Ahmad et al. 2010). Rickets usually refers to vitamin D deficiency in children and osteomalacia in adults. The museum card reads "Rachitis ossium omnium" which translates to "rachitis in all of the bones." See Watson-Jones (1955) for more on rickets.

**Figure 409a-b.** (a) Anterior and (b) posterior views of healed transverse fracture of the left patella with surgical wire to reunite the two broken pieces of bone in a 49-year-old female who died in 1931 (MN18.560/1193). "We still fix patella fractures like this. It's called a "tension-band" construct because the anterior patellar wire promoted compression of the fracture site, and therefore promotes healing, and so this is good. Inside the patella are two long Kirshner wires to which the circular wires are attached. They are hard to see." (P. Moroz, personal communication, November 17, 2022.) See Watson-Jones (1946); *Campbell's Operative Orthopaedics* (1956; Volume I) for more on fractures of the patella.

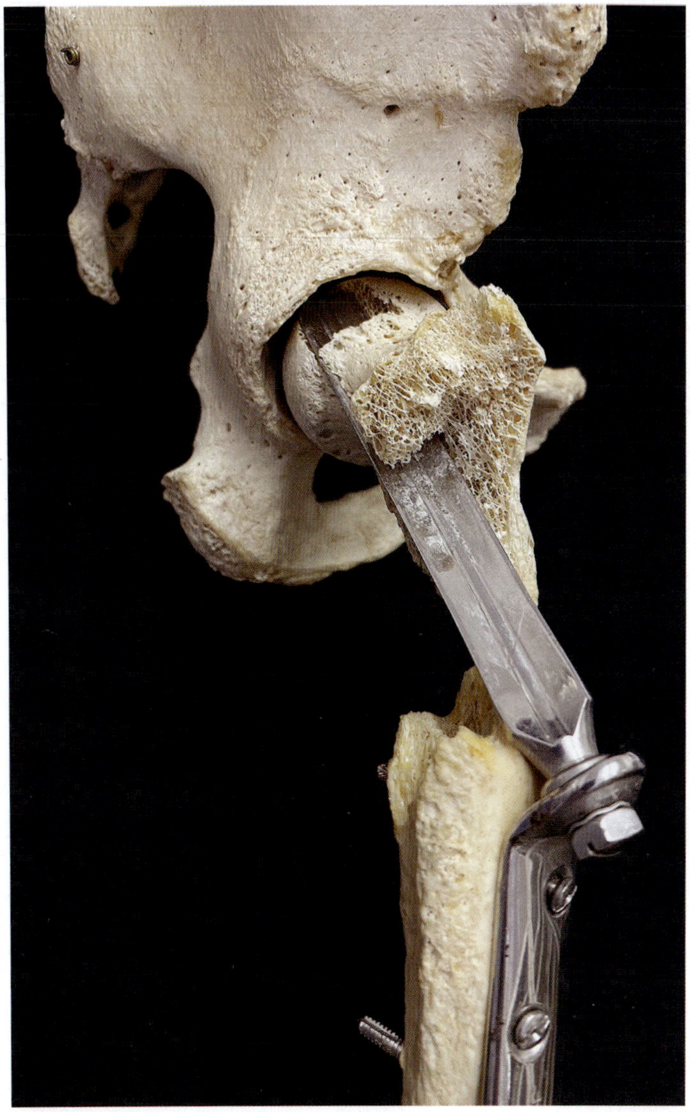

**Figure 410.** Right lateral view of an unknown adult right hip with intertrochanteric fracture of the femur that was repaired with a "tri-bladed" medical implant that would have been hammered into the proximal diaphysis of the femur and secured with screws. This individual presumably died in the 1950s or 1960s when the hospital that it came from opened (MN29.419). There is little or no evidence of healing and no additional information for this individual. "This was a common early hip fracture implant which was designed to allow the femoral head and neck to compress along the smooth edges of the tri-phlanged "nail." This compression would promote bony healing. The surgeon had to be careful not to put in too long a tri-phlange because if the bony head and neck compressed too much it would stick-out of the femoral head and into the acetabulum" (P. Moroz, personal communication, November 17, 2022). See Muckle (1977) for more on femoral and hip fractures and prostheses used prior to 1977, Watson-Jones (1944 and 1946) for more on hip fractures prior to the early 1940s and *Campbell's Operative Orthopaedics* (1956); Watson-Jones (1955) for surgical practices and implants up until the early 1950s.

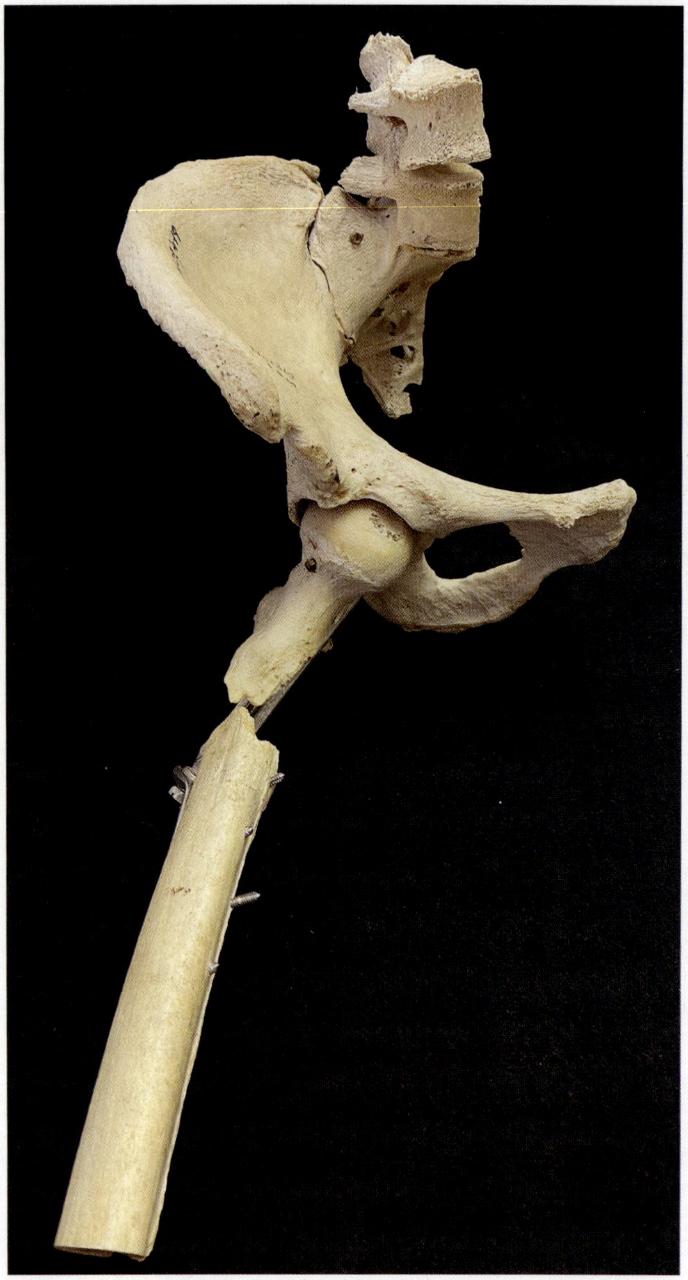

**Figure 411.** Anterior view of an adult right hip with intertrochanteric fracture of the femur repaired with a "tri-bladed" medical implant that would have been hammered into the diaphysis and secured with screws (MN29.419). There is no additional information available for this individual. See Rockwood and Green (1984) for more on fractures and healing. See Muckle (1977) for more on fractures of the femur and hip, Watson-Jones (1946) for fractures prior to the 1940s and *Campbell's Operative Orthopaedics* (1956); Watson-Jones (1955) for surgical practices and implants up until the early 1950s.

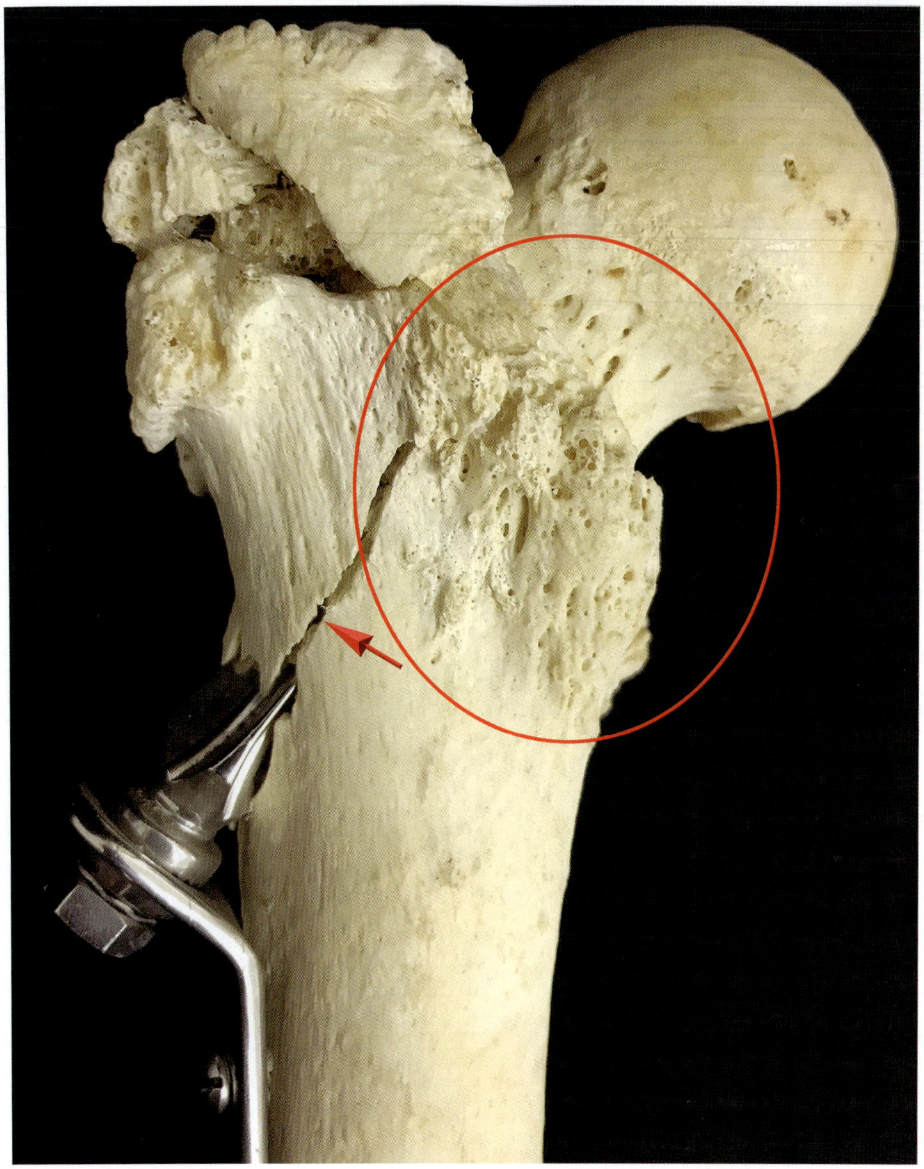

**Figure 412.** Anterior view of healing or healed comminuted fracture of the greater trochanter of the right femur with prosthesis to repair intertrochanteric fracture (oval), that resulted in secondary fracture (arrow) in an unknown adult who presumably died in the 1950s or 1960s (MN29.393). This is an old-style implant, possibly from the 1940's or 1950's. "The fracture extends from greater to lesser trochanters, hence inter-trochanteric hip fracture. In this type of fracture, they are more comminuted, and the tip of the greater trochanter is also broken off which is not good because the hip abductors start pulling that fragment away from other fragments over the few weeks after fracture. This may not be a screw going into the head, but instead, a blade plate that is hammered into the head by the surgeon and then screwed to the angled side plate to immobilize the hip fragments to allow it all to heal" (P. Moroz, personal communication, November 17, 2022). See Rockwood and Green (1984); Watson-Jones (1946, 1955) for more on fractures and rates of healing. See Muckle (1977); *Campbell's Operative Orthopaedics* (1956).

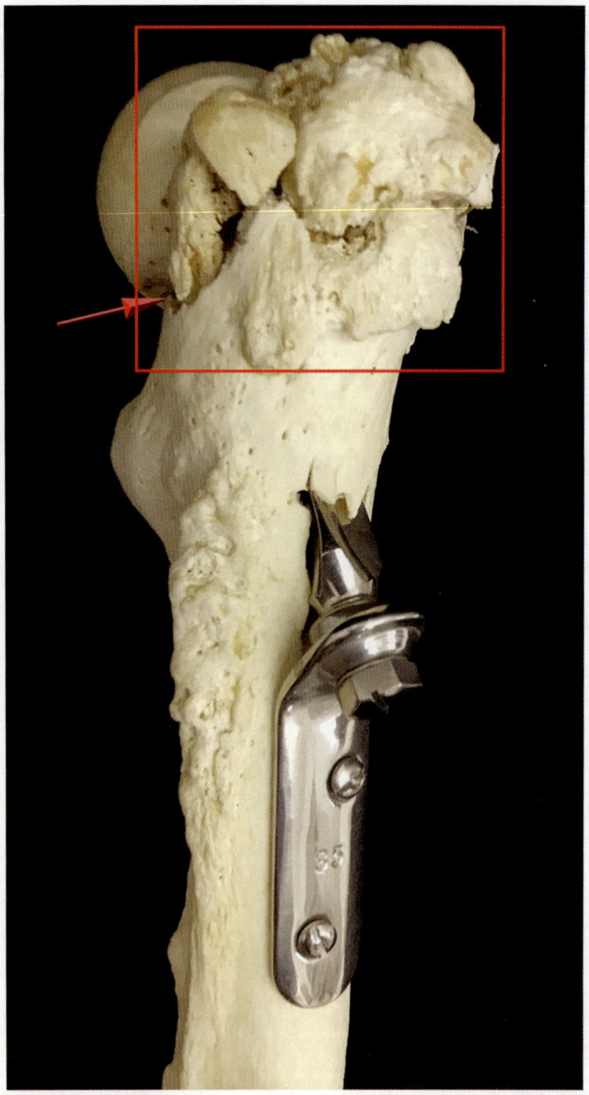

**Figure 413.** Lateral view of healing comminuted fracture of greater trochanter (square) and neck (arrow) of the right femur with prosthesis that resulted in secondary fracture, probably due to osteoporosis, in an unknown adult who presumably died in the 1950s or 1960s (MN29.393). The absence of healing/remodeling around the medical implant suggests that she did not survive the surgery. "The blade plate being smooth allows some controlled collapse of the femoral head and intertrochanteric fragments so they compress and this will promote healing of bone. Early screw implants, instead of a smooth blade plate, would grip the bone better, but would not allow for controlled compression and would actually promote non-unions by not allowing for collapse and compression. Surgeons later decided to do a sliding screw implant, which holds the femoral head well with the threaded screw and so won't penetrate the head, but will slide along a tube attached to the femoral side plate allowing for controlled collapse and boney compression which, again, promotes healing. Although the collapse causes some proximal femoral deformity, the bone heals and the head is basically in the socket" (P. Moroz, personal communication, November 17, 2022). See Rockwood and Green (1984); Muckle (1977); Watson-Jones (1946, 1955); *Campbell's Operative Orthopaedics* (1956) for more on fractures and rates of healing of the hip.

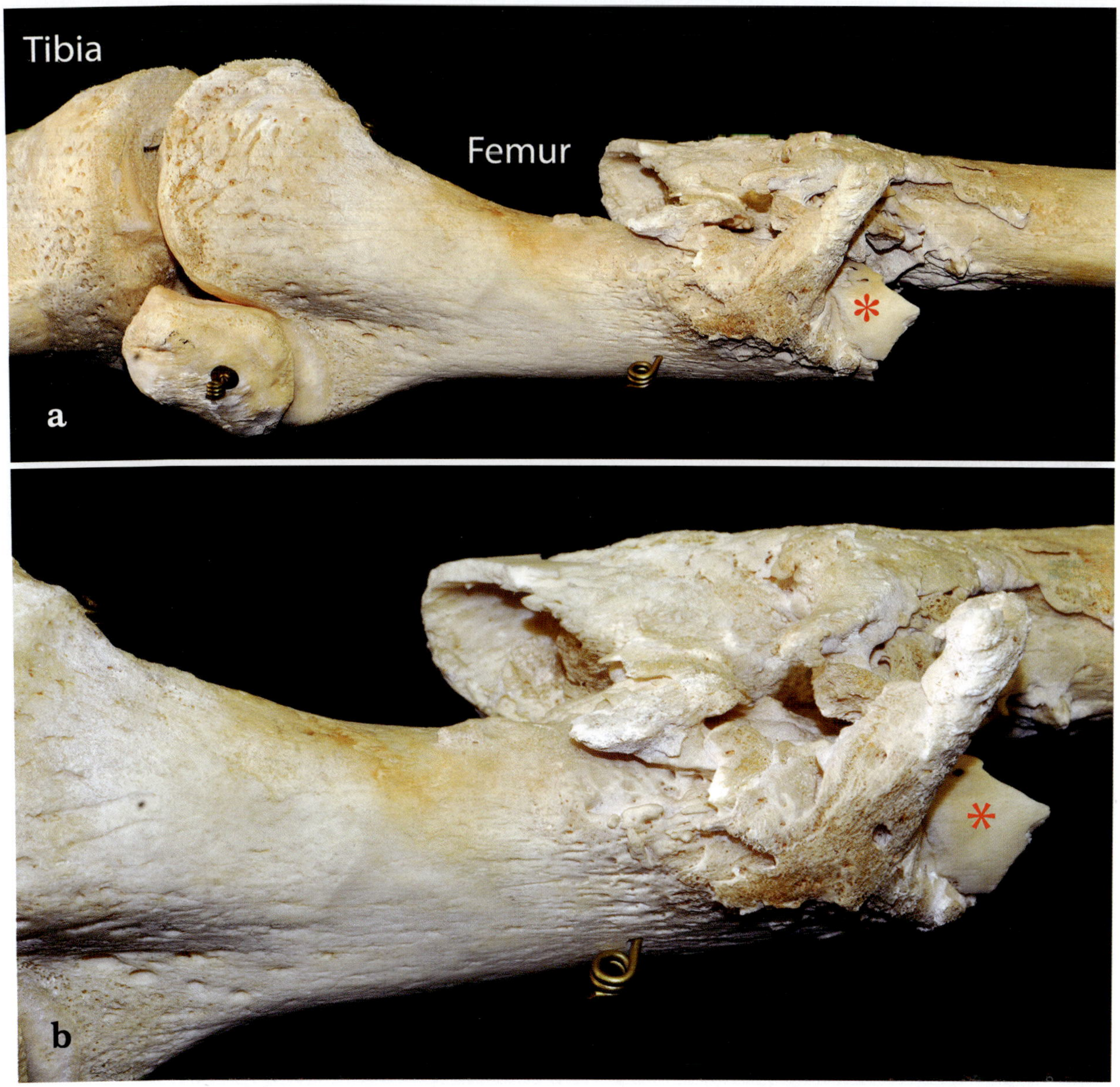


**Figure 414a-b.** (a) Medial and (b) closeup of a partially healed left femur "… as it would not be strong enough for a person to walk on without pain and a big risk of refracturing. Osteomyelitis in the days before antibiotics had a high death rate, 30-40% (P. Moroz, personal communication, November 17, 2022). Note the bone fragment/sequestrum (*) of the diaphysis in a 21-year-old male who died in 1939 (MN18.560/1705). See *Campbell's Operative Orthopaedics* (1956); Griffiths (1981); Rockwood and Green (1984); Watson-Jones (1946, 1955) for more on femoral fractures, infection and rates of healing.

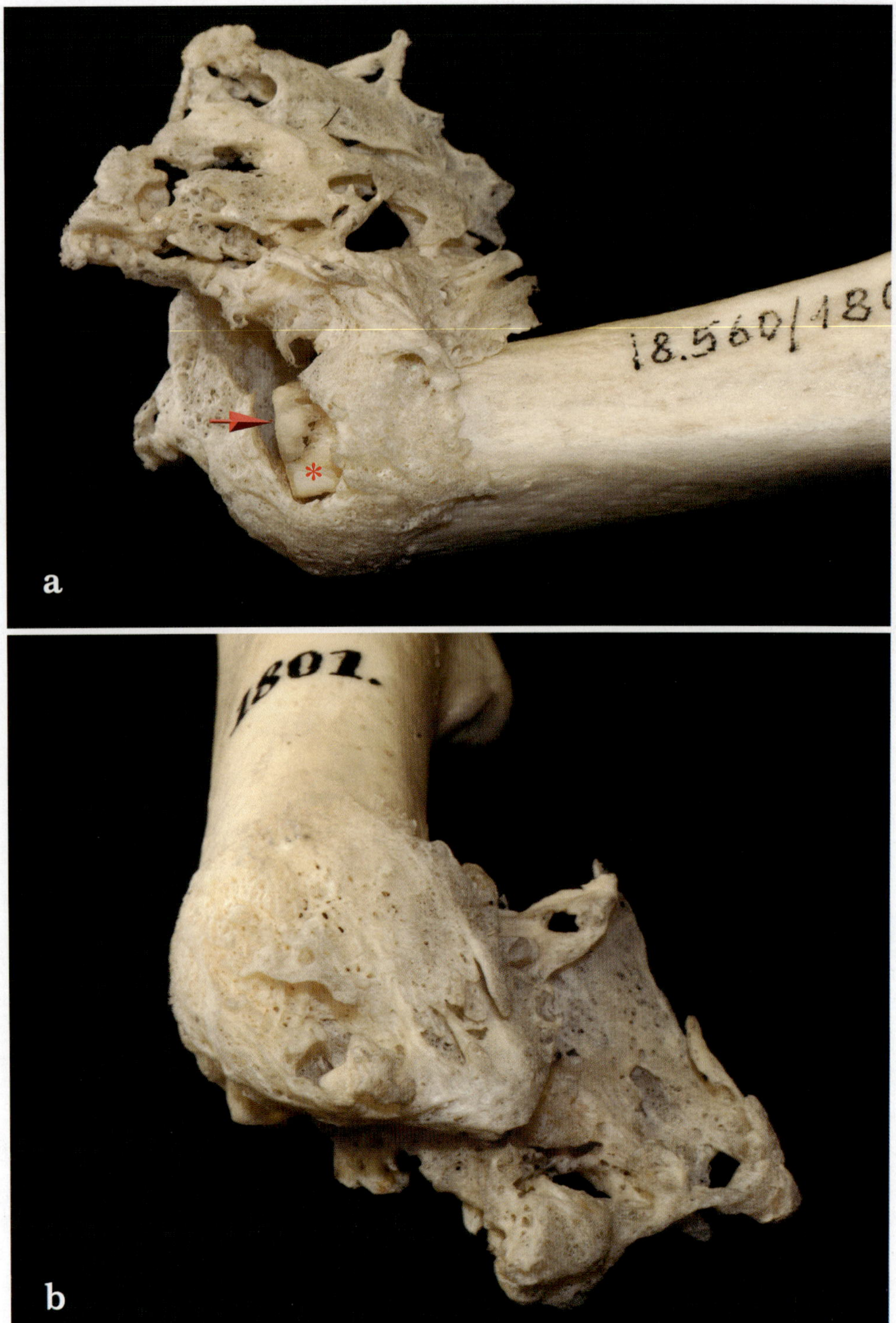


**Figure 415a-b.** Healed surgical amputation and heterotopic ossification (HO), or new bone formation, in the right distal femoral diaphysis/shaft of a 20-year-old male who died in 1942 (MN18.560/1801). (a) Note the site of amputation (arrow) of the distal femur and underlying diaphyseal bone (*) nearly encircled by heterotopic ossification (HO) that was almost certainly the result of surgery. See Rockwood and Green (1984); Watson-Jones (1946, 1955); *Campbell's Operative Orthopaedics* (1956) for more on femoral fractures and healing.

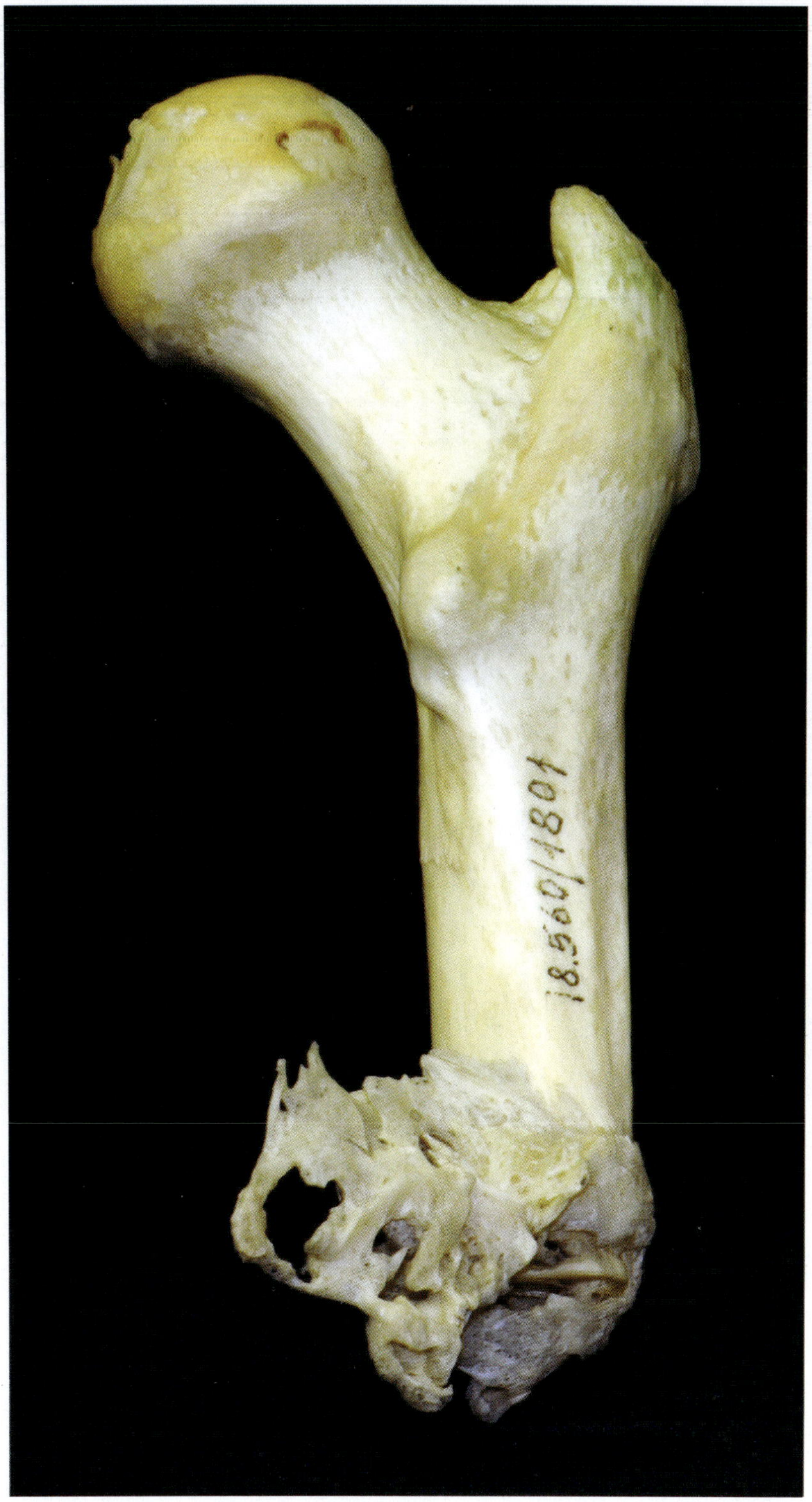


**Figure 416.** Healed surgical amputation and heterotopic bone formation in the distal end ("stump") of the right femoral diaphysis/shaft in a 20-year-old male who died in 1942 (MN18.560/1801).

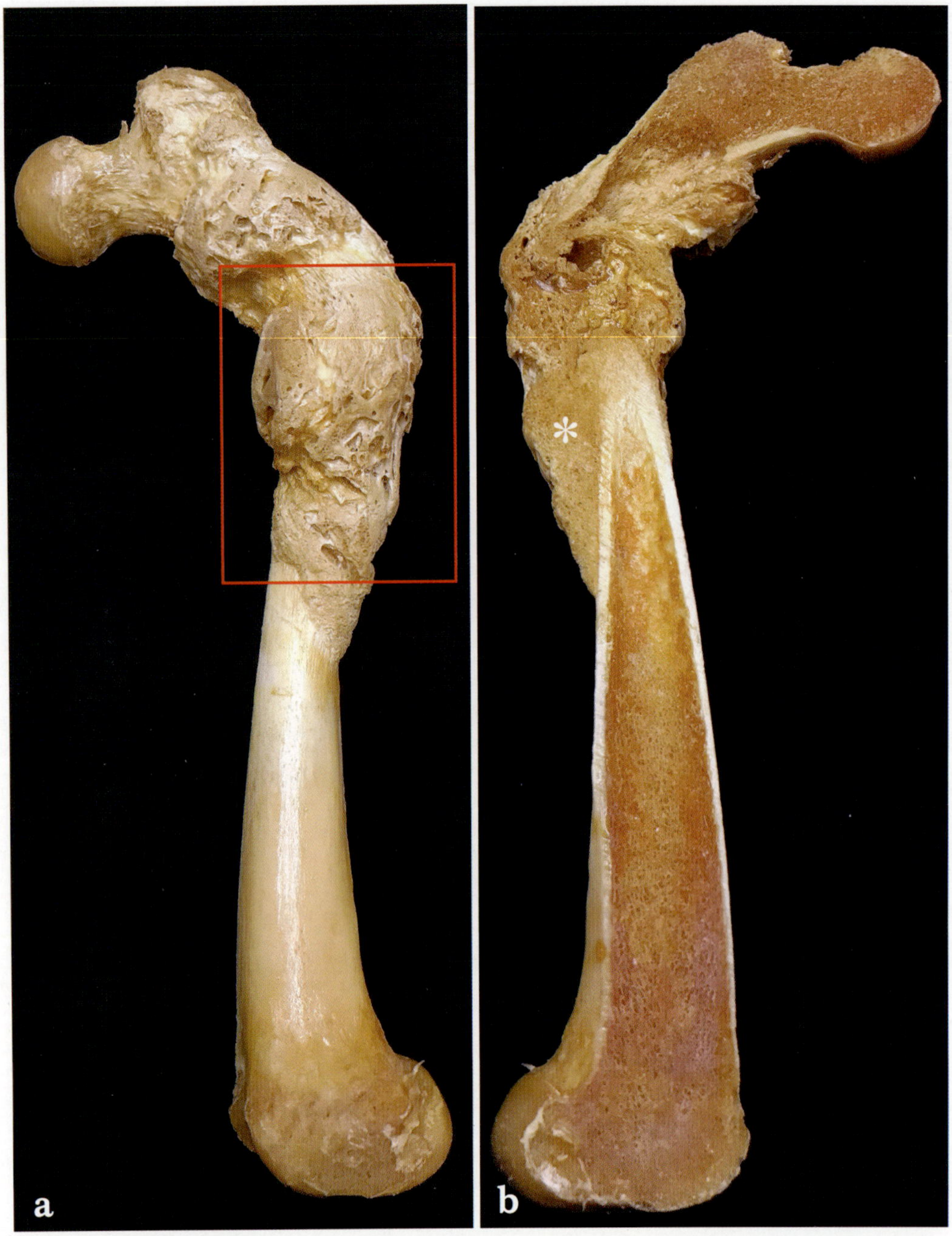


**Figure 417a-b.** (a) Anterior and (b) posterior views of healed fracture with angulation of the proximal one-third of a sawed left femur with a large callus (rectangle and *) in a 40-year-old female who died in 1930 (MN10.272). Unlike fibrous dysplasia that can result in fracture in this area of the femur, this spontaneous fracture occurred secondary to cancer of the thyroid. "(a) The distal femur appears to be laterally oriented whilst the proximal femur appears to be in a coronal, or anterior view. This means the pathological fracture not only healed angulated as seen, but also rotated internally by 90 degrees. Her foot would be pointing severely in. Surgeons will always try and maintain leg alignment while bones heal so that this does not happen" (P. Moroz, personal communication, November 17, 2022). See Rockwood and Green (1984); Watson-Jones (1946) for more on fractures and healing.

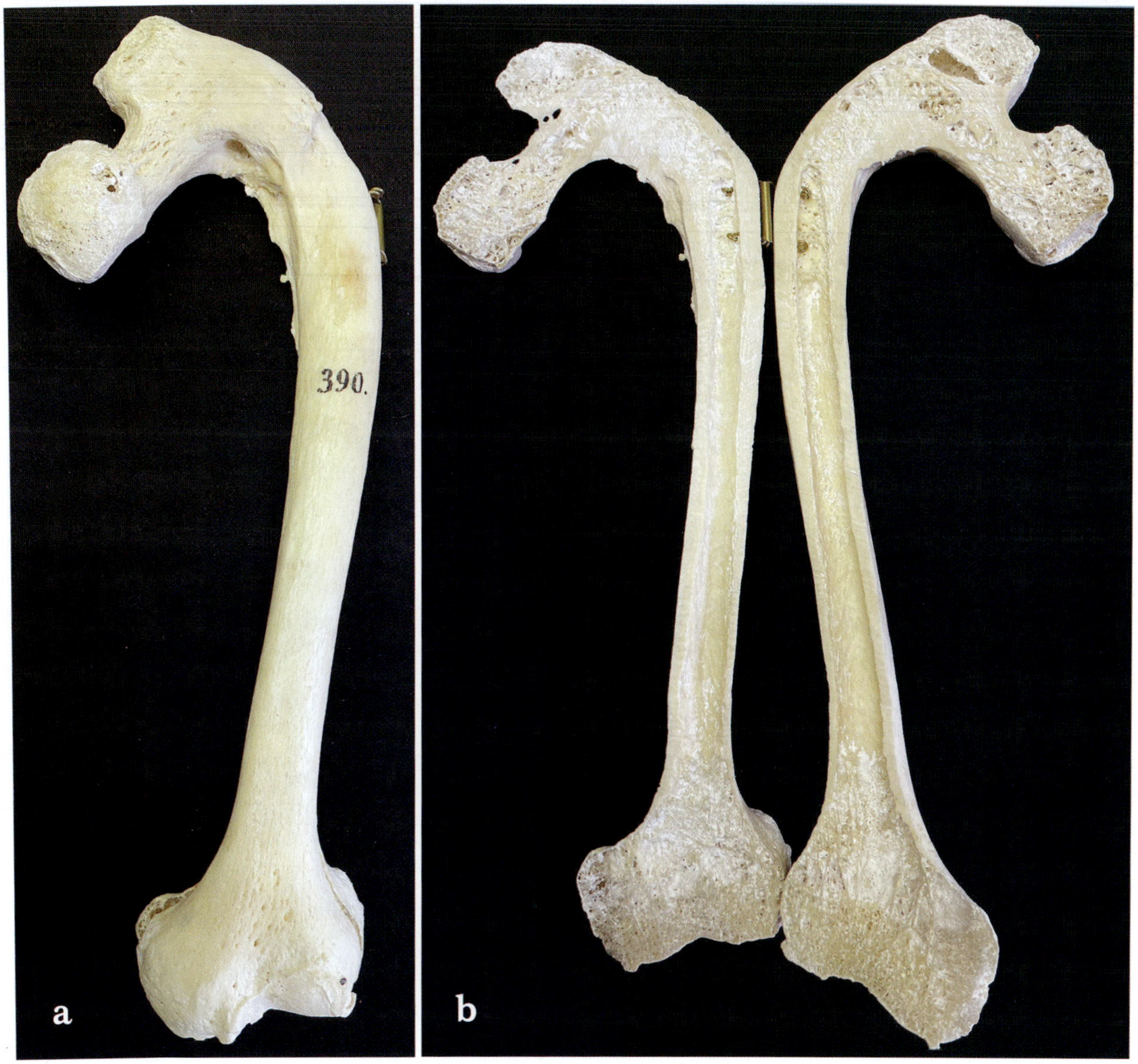


**Figure 418a-b.** (a) Anterior and (b) longitudinally sectioned left femur showing healed fracture and severe angulation of the proximal third of the bone in a 71-year-old male who died in 1900 (MN18.560/390). The fracture site has remodeled to the point that it can barely be discerned. This shepherd's crook deformity is likely the result of fibrous dysplasia, a rare genetic bone disorder resulting in insufficent bone that is prone to fracture (Al-Mouazzen 2013; NIH 2019). "Microfracture, then deformity, then another microfracture and further deformity, which is responding to gravitational stress of walking on weakened and ongoing deforming proximal femur. Today, surgeons would insert an intramedullary (IM) rod down the proximal femur, after correcting for (osteotomizing the shepherd's crook deformity). The IM rod would hold off the deforming force" (P. Moroz, personal communication, November 17, 2022). See Thomas and Kirsten (1985) for more on fibrous dysplasia and its distribution in the human skeleton. See Rockwood and Green (1984); Watson-Jones (1955) for more on fractures, fibrous dysplasia and healing.

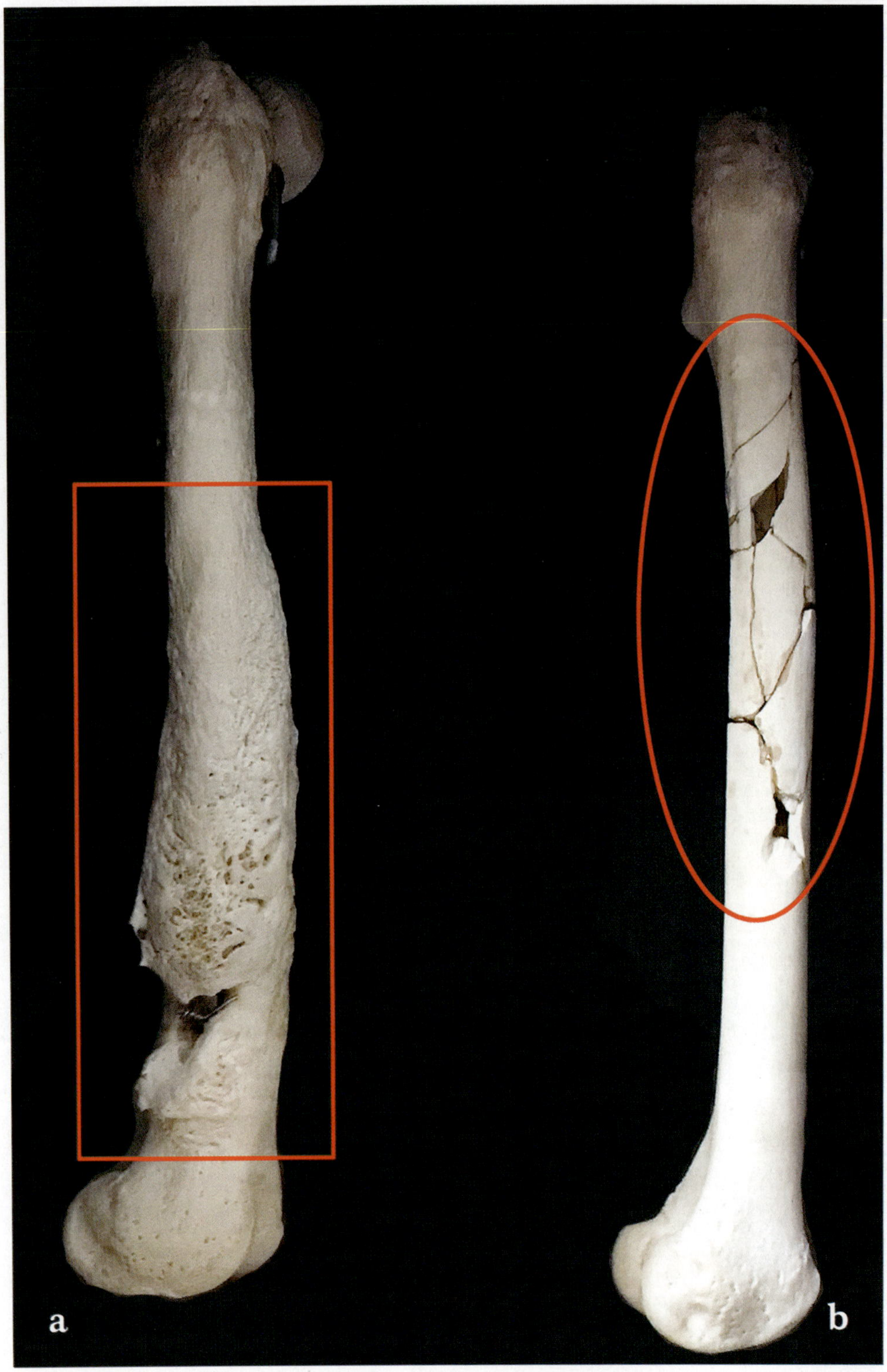


**Figure 419.** Battle injuries indicating (a) antemortem and (b) perimortem trauma to the diaphysis in two right femora. (a) A surgically repaired (the hardware is visible) femur with osteomyelitis that resulted in a large involucrum (rectangle) and (b) a femur that sustained massive perimortem comminuted fracture (oval) of the diaphysis with no evidence of healing. See Rockwood and Green (1984) for more on the biomechanics and classification of fractures and healing.

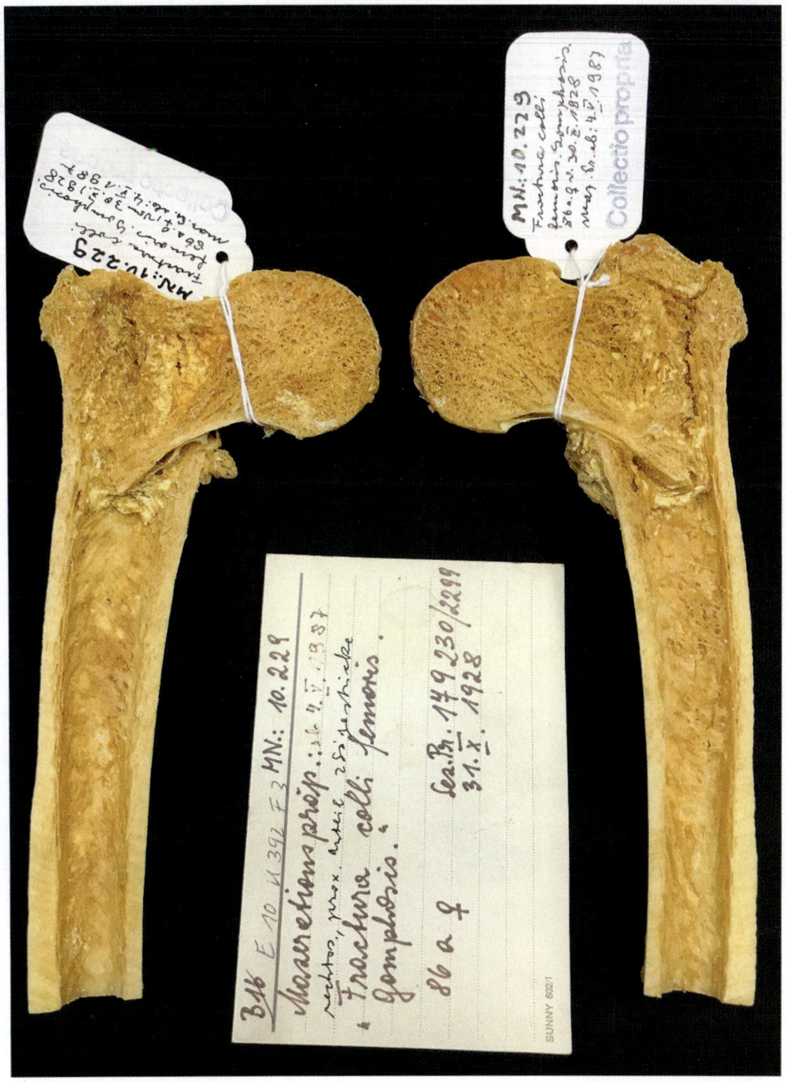


**Figure 420.** Longitudinally sawed left femur showing healed fracture in an 86-year-old female who died in 1938 (MN10.229). This intertrochanteric fracture is identified as a "gomphosis" (a thin fibrous joint) referring to it as a peg-and-socket joint found in the teeth and tooth sockets consisting of a thin fibrous joint that allows limited or no movement. "Gomphosis referenced here is the inferior spike of the proximal fragment being interdigitated into the intramedullary canal of the distal femoral fragment. The spike was in the "cup." This was an old "pre-implant" technique to engage certain types of proximal femoral fragments, which again allows for some bony immobility and compression and thus boney healing. An individual was lucky if the hip fracture was like this because they usually healed and were stable and got boney union. Or even fibrous union might be stable, as compared to a fracture where the inferior spike was outside the intramedullary canal. Or perhaps the trochanters were split in a big intertrochanteric or subtrochanteric fracture. Then it was grossly unstable and would continue to shorten, deform, was unable to be used for ambulation and was a "non-union" machine" (P. Moroz, personal communication, November 17, 2022). See Rockwood and Green (1984); Watson-Jones (1946) for more on fractures and healing.

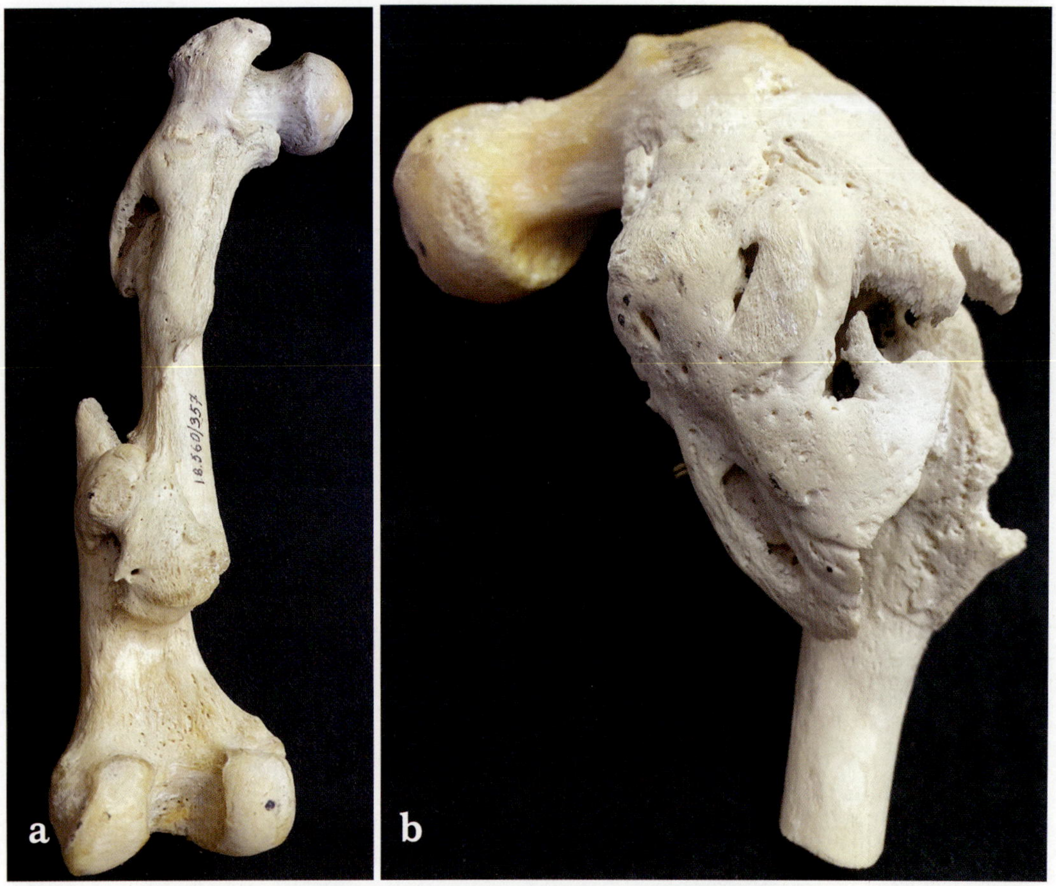


**Figure 421a-b.** (a) Posterior and (b) anterior views of healed fracture of the left femur in an adult who died before 1930 (MN18.560/357). (b) Note the large bony callus and heterotopic ossification (new bone formation) along its anterior surface. See Rockwood and Green (1984); Watson-Jones (1946) for more on fractures and healing.

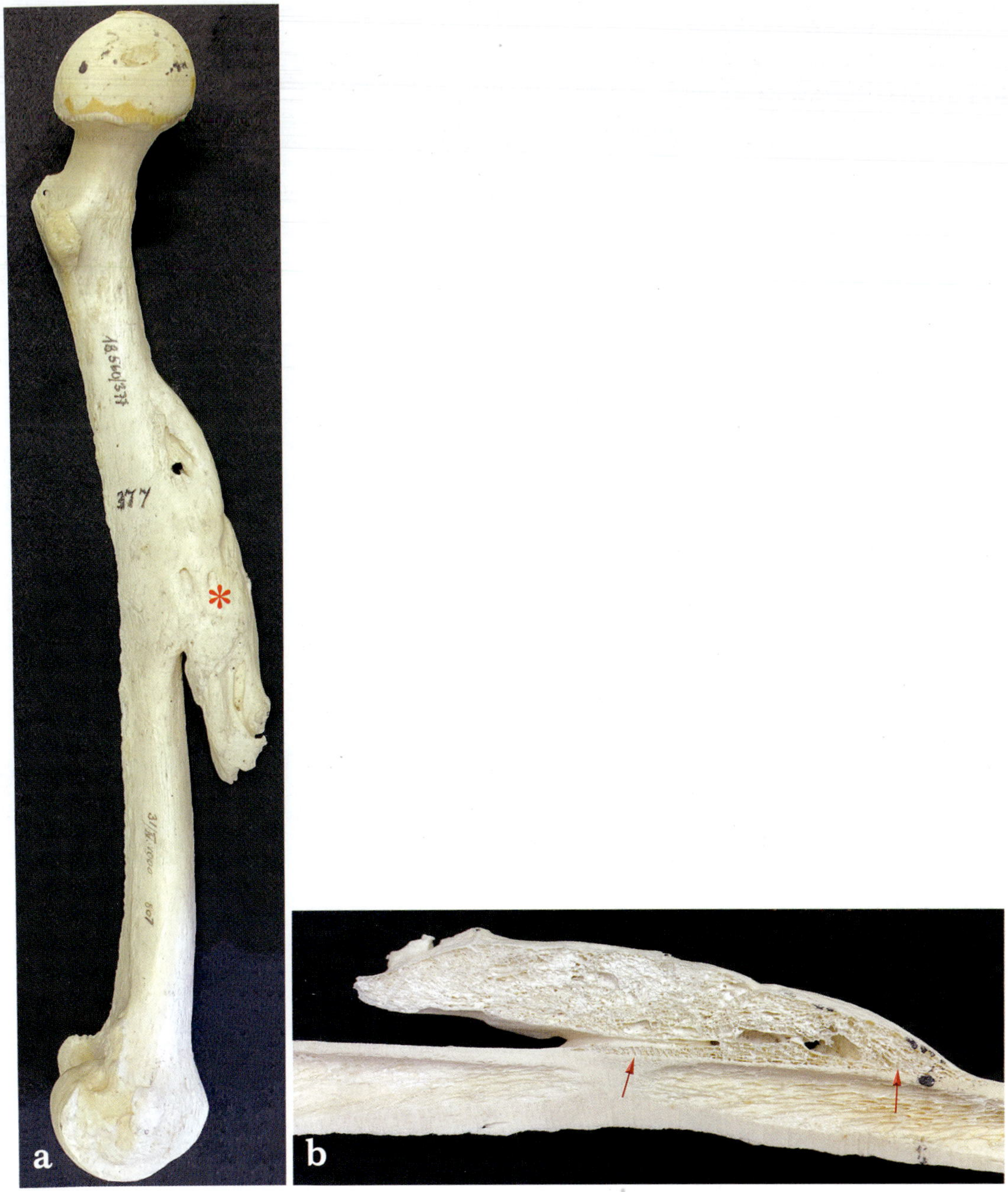


**Figure 422a-b.** (a-b) Myositis ossificans traumatica (MOT), also known as heterotopic ossification (HO) along the antero-medial surface of the left proximal femur in a 60-year-old female who died in 1900 (MN18.560/377). (b) Note that the laminated, mature heterotopic bone is highly trabecular and integrated into the cortex of the femur (b; arrows). HO forms in nonosseous soft tissue such as muscle, ligament and tendon as a result of a traumatic injury (Man et al. 2011; Meyers et al. 2019; Zychowicz 2013). See Rockwood and Green (1984); Geshchickter and Copeland (1936) for more on myositis ossificans traumatica and its laminated structure as seen in this individual and Watson-Jones (1946, 1955) for MOT in general.

**Figure 423.** Magnified view of myositis ossificans, also known as heterotopic ossification (HO) in the anterior surface of the proximal left femur in a 60-year-old female who died in 1900 (MN18.560/377). Note that the heterotopic ossification (bone) is highly trabecular and integrated into the cortex of the femur. See Geshchickter and Copeland (1936); Watson-Jones (1946, 1955) for more on MOT.

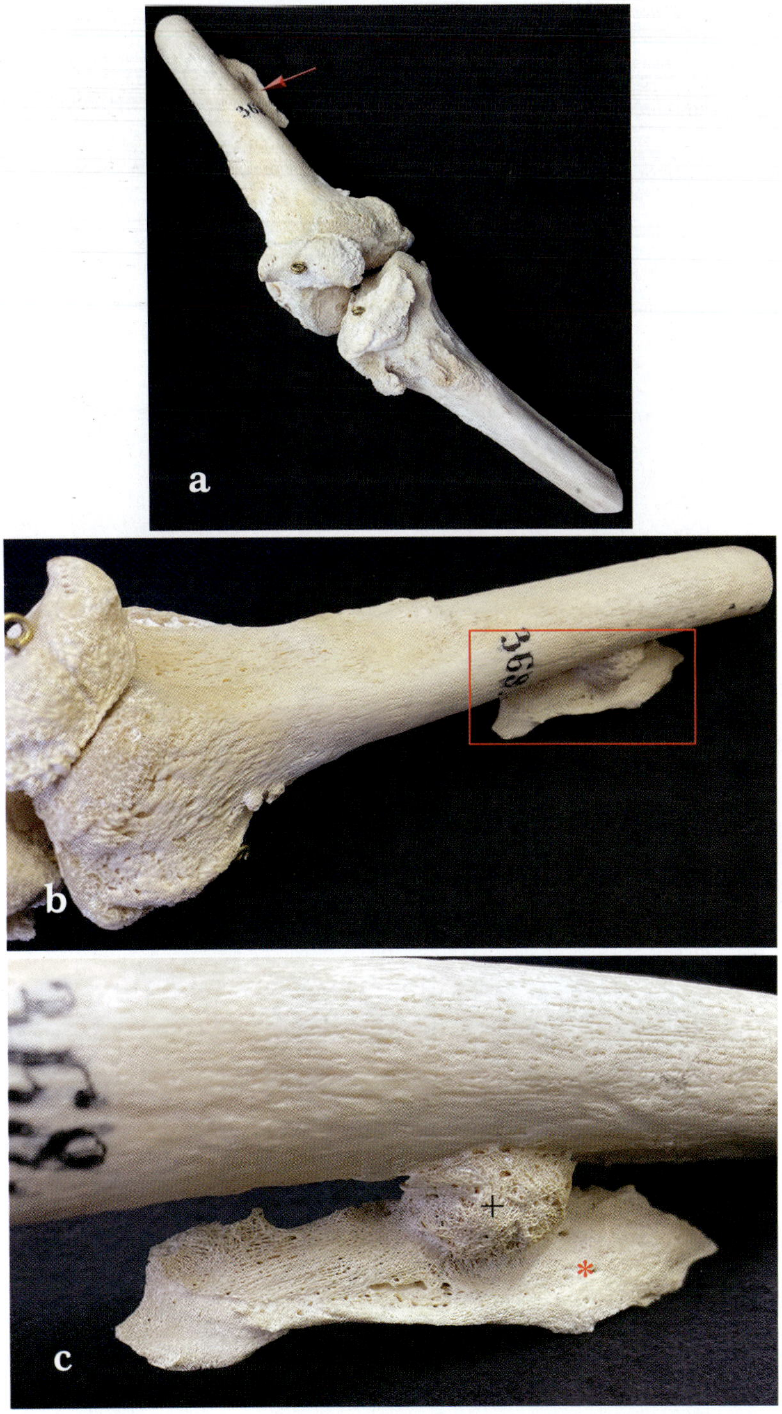


**Figure 424a-c.** (a-c) "Arthrophia tabica" (on museum card) of the left femur, patella and tibia with (a) destruction of the joint and (a-c) heterotopic ossification (HO) firmly attached to the postero-lateral surface of the femoral diaphysis in a 52-year-old male who died in 1917 (MN368). Note the sheet-like HO (a; arrow), (b; rectangle) and (c; *) attached to the diaphysis by a rounded area (c; +) of HO. HO forms in nonosseous soft tissue as a result of direct trauma, sickle cell disease, multiple sclerosis, tetanus and burns, among others (Meyers et al. 2019; Zychowicz 2013). See Rockwood and Green (1984) for more on heterotopic bone formation and healing.

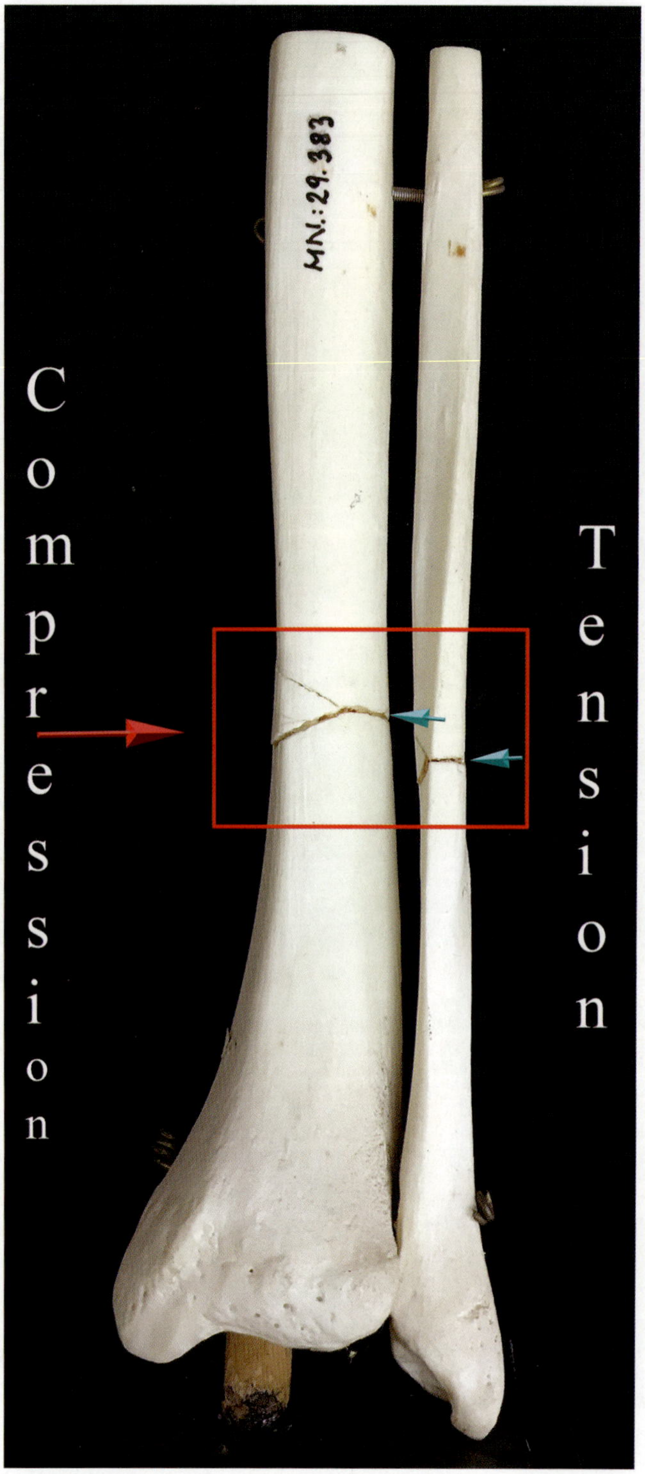


**Figure 425.** Posterior view of small butterfly fractures (rectangle) in a right tibia and fibula with no evidence of healing in an 18-year-old male who died in 1958 (MN29.383). The wedge-shaped pieces of bone indicate that the force was applied medial to lateral (red arrow) causing both bones to fail/fracture on the tension side resulting in a horizontal "Y". Although bone fails first on the tension side (blue arrows) and then the compression side, new research (Reber and Simmons 2015) revealed that butterfly fragments can form on the tension or compression side.

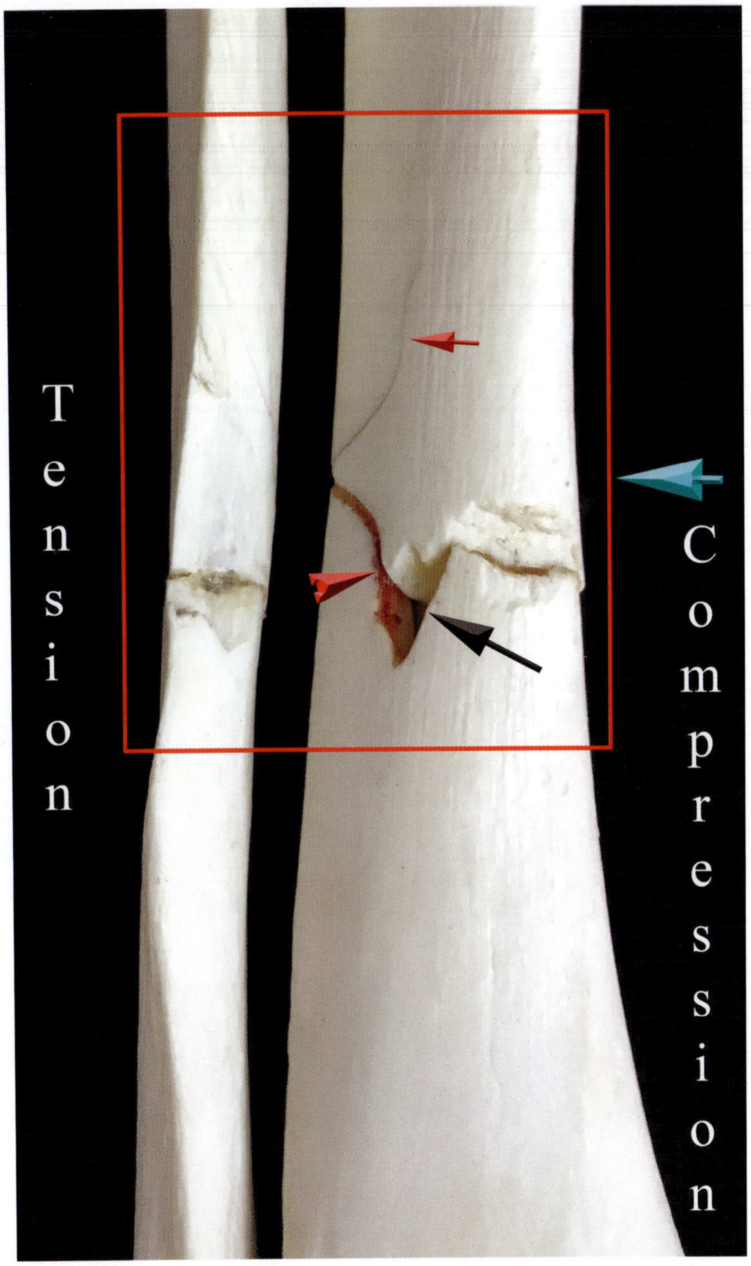


**Figure 426.** Anterior view of right tibia and fibula with butterfly fractures (rectangle) with no evidence of healing in an 18-year-old male who died in 1958 (MN29.383). Force was applied to the two bones from the compression side (impact side) (blue arrow) causing both bones to fail on the opposite tension side followed by the compression side. Butterfly fractures consist of three types of fractures that begin perpendicular to the long axis of the bone under tension, then form curved fractures (red arrow) due to shearing force, and then reverse direction (black arrow) resulting in jagged interlocking fractures due to compression (red arrowhead) (Mariyam et al. 2018) and one or more wedge-shaped fragments (see Toshika et al. 2022). Interpreting biomechanical forces associated with skeletal trauma is ever-evolving and can provide information on the direction of force causing the trauma. See Rockwood and Green (1984) for more on fracture types and classifications.

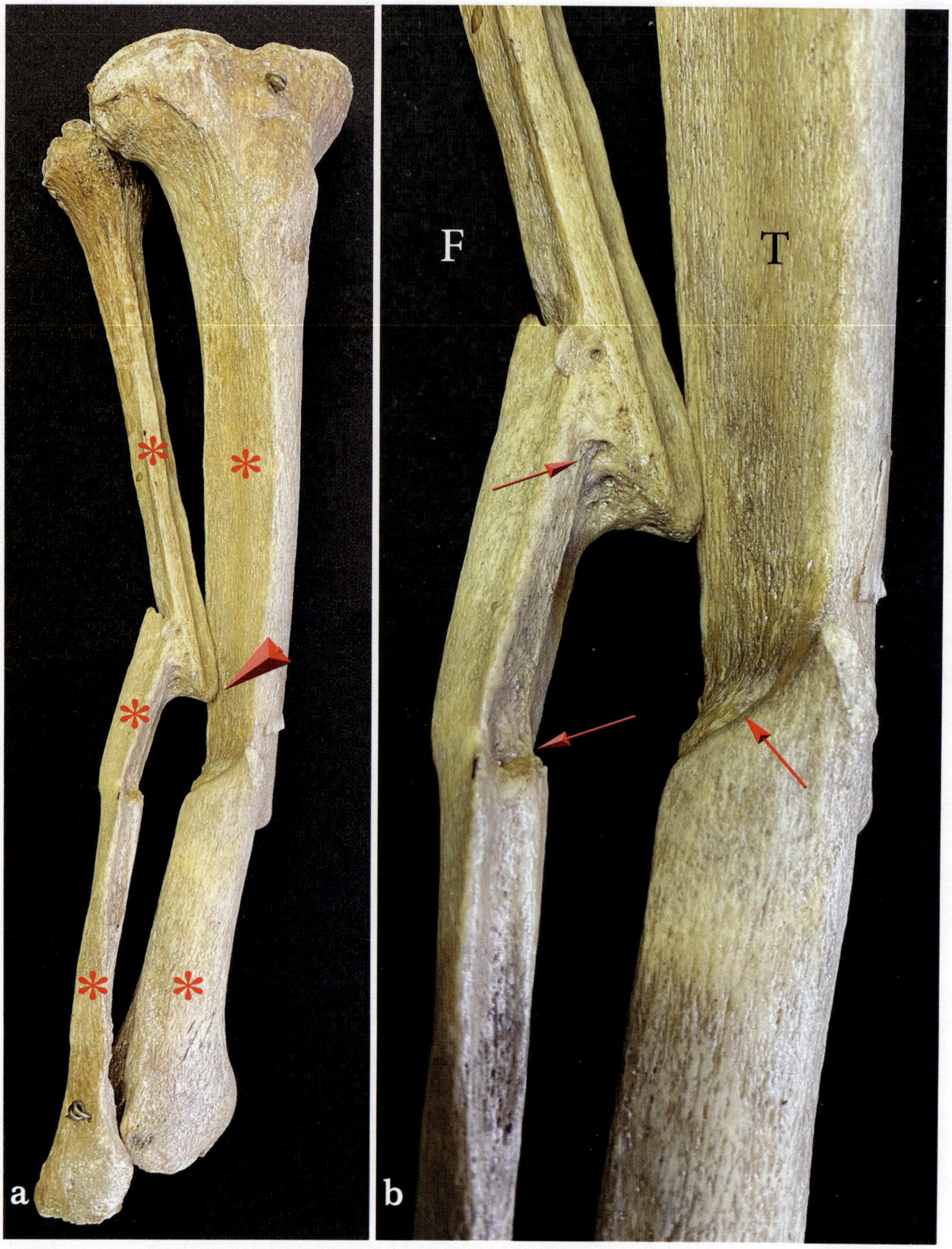


**Figure 427a-b.** (a) Anterior and magnified anterior views of healed fracture (arrows) of the right tibia (T) (in two pieces [*]) and fibula (F) (in three pieces [*]) in an adult who died before 1880 (MN18.560/71a). The superior fracture in the fibula moved medial until it came into contact with the tibia (arrowhead).

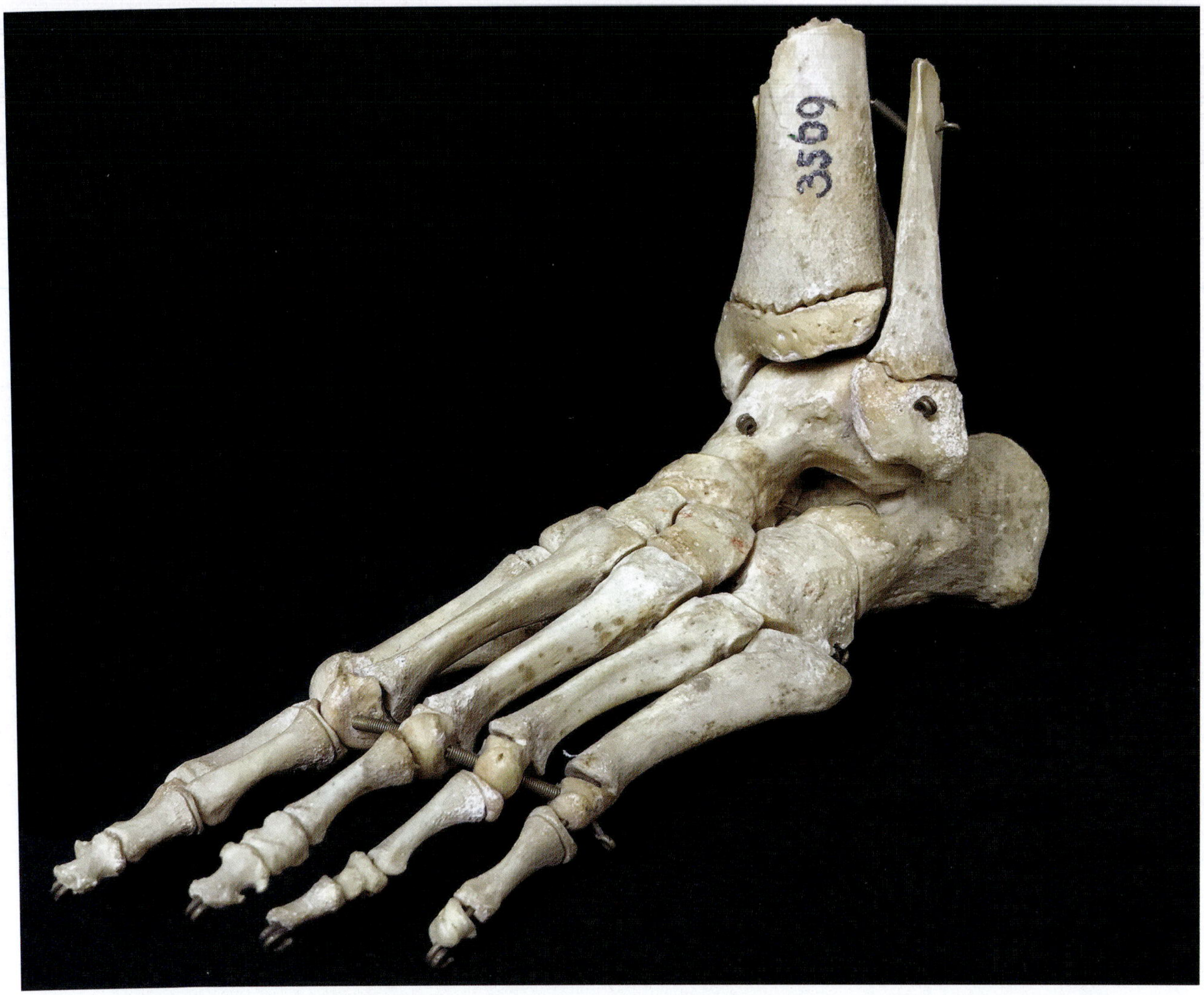


**Figure 428.** Oblique view of traumatic amputation of the lower leg and foot due to accident in an approximately 14 to 17-year-old male who died in 1880 (MN21.055/3569). There is no evidence of saw marks and the tibia and fibula exhibit jagged bending fractures reportedly resulting from traumatic amputation with some kind of an iron loop. The museum card reveals cruris (leg) ductile (breaks easily) indicating that the bone was easily broken or susceptible to fracture. Note that many epiphyses in the tibia, fibula, metatarsals, and phalanges are still unfused, indicating the individual was a subadult. The pattern of "butterfly" fractures indicates that stress was applied to the posterior surface of the lower leg (from behind to front). See Mariyam et al. (2018); and Reber and Simmons (2015) for more on the biomechanics of butterfly fractures. See Rockwood and Green (1984) for more on fractures and traumatic amputation.

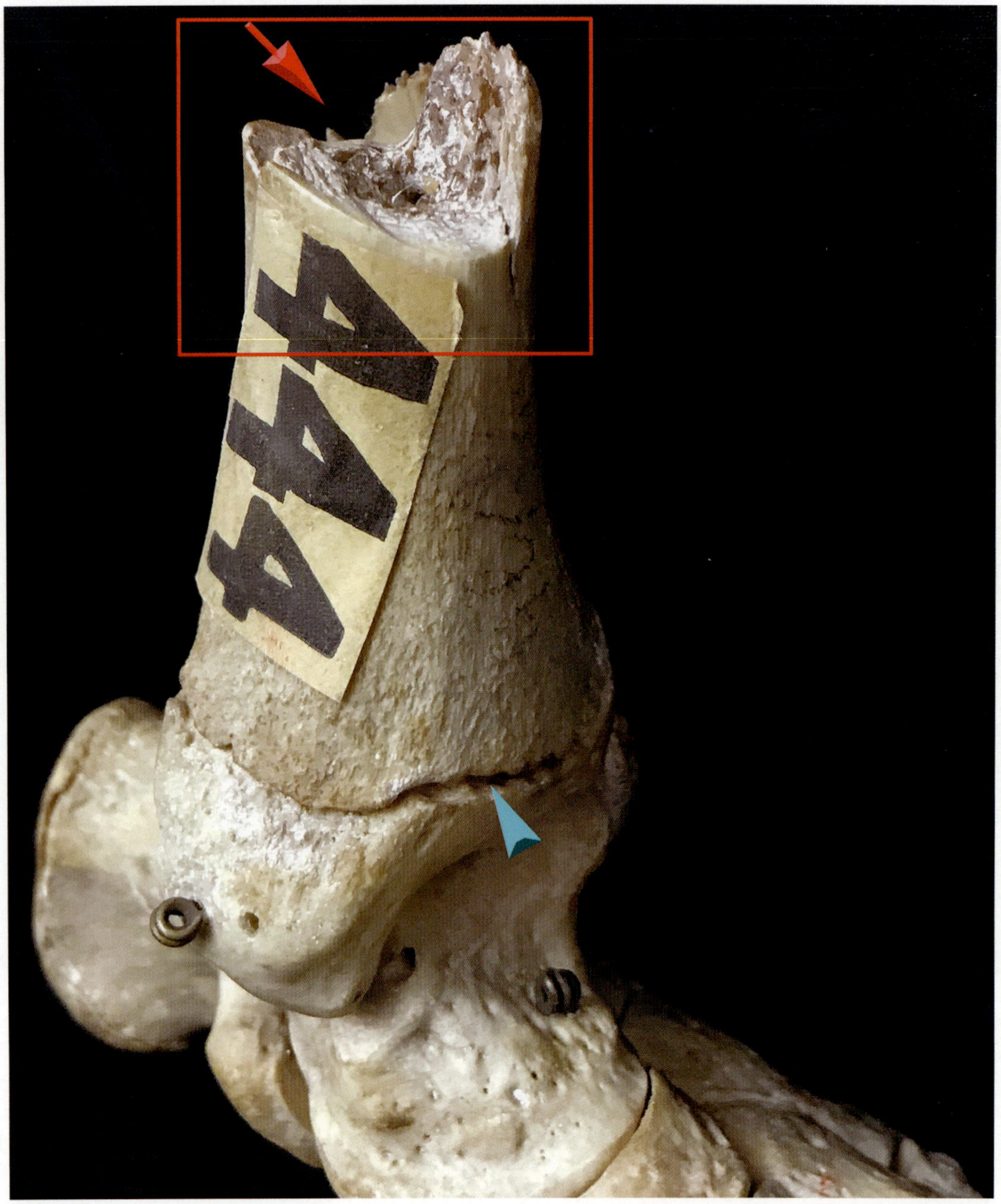


**Figure 429.** Medial oblique view of accidental traumatic amputation of the left lower leg in an approximately 14 to 17-year-old male who died in 1880 (MN21.055/3569). The museum card reveals cruris (leg) ductile (breaks easily) indicating that the bone either was easily broken or susceptible to fracture. There is no evidence of sawing or cutting, and the tibia and fibula exhibit jagged bending fractures (square) reportedly from traumatic amputation (Figure428) with some type of an iron loop (museum records). The pattern of "butterfly" fractures is consistent with stress to the posterior surface (red arrow) of the lower leg. Note the distal epiphysis or growth cap (arrowhead) indicating the individual was still growing (subadult). The jagged, interlocking bone in the tibia formed as a result of compression. See Mariyam et al. (2018); Reber and Simmons (2015) for more on butterfly fractures.

## Medical Implants

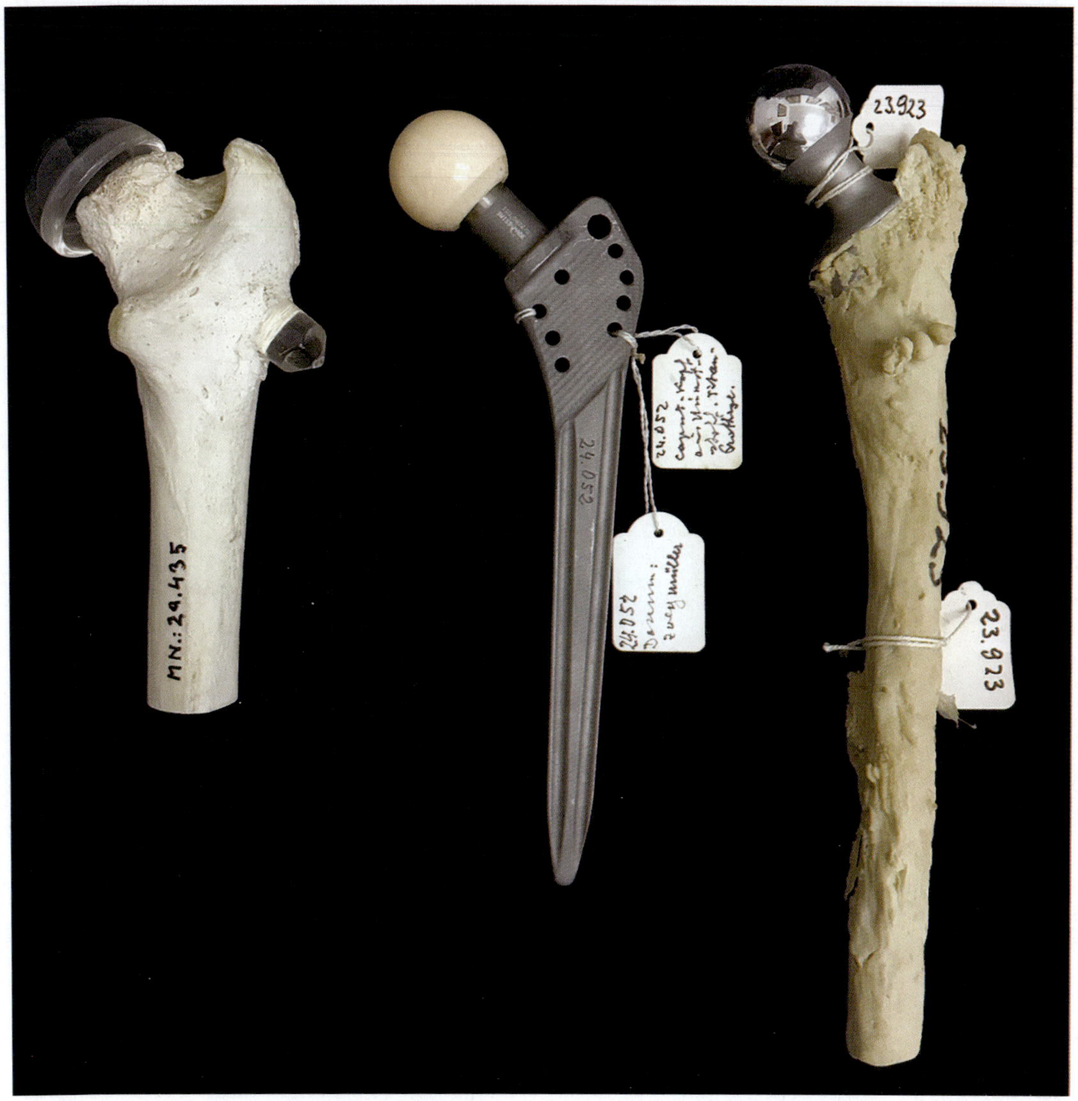


**Figure 430.** Femoral prostheses made of various materials including plastic on plastic, ceramic on metal, and titanium (photo from left to right). Ceramic implants, for the most part, were not used after the mid-1990s as they tended to fracture and crumble. "These are early generation proximal femoral prostheses. The left one a very early attempt to solve hip arthritis." (P. Moroz, personal communication, November 17, 2022). See Muckle (pp. 101-102; 1977) for more on femoral prostheses used prior to about 1977, especially the Austin Moore and Thompson type and *Campbell's Operative Orthopaedics* (1956) for implants and trauma up until the early 1950s. The femoral diaphysis in the photo right (23.923) appears to show evidence of infection.

# Chapter 6

# CASE REPORT OF MULTIPLE MYELOMA

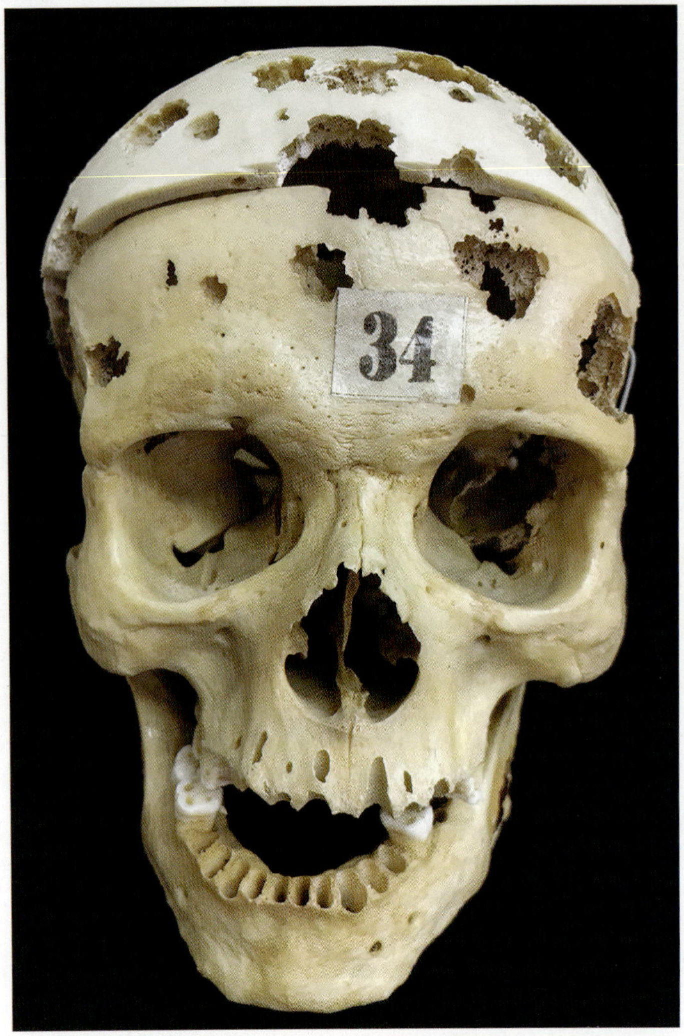


**Figure 431.** Frontal view of a 45-year-old male with multiple myeloma who was hospitalized at Wieden Hospital in Wien (Vienna) in December 1902, examined in February 1903 and died on June 21, 1903 (MN17.747/34). This man previously had typhus in 1873, was treated for multiple myeloma for two weeks at a hospital in 1902, and later checked into the Wieden Hospital. Jellinek (1904) reports that both of this man's femurs were broken while lifting him to put a bed pan under him. While the cause of multiple myeloma, a blood cancer, is unknown, genetics and environmental variables may be involved. See Jellinek (1904) for a detailed report with illustrations of this patient. Multiple myeloma results in many irregularly shaped lytic lesions, sometimes with reactive bone along the margin, throughout the skeleton in bone areas with marrow. See Bitelman et al. (2016); Reeder (1993); Rossi et al. (2021).

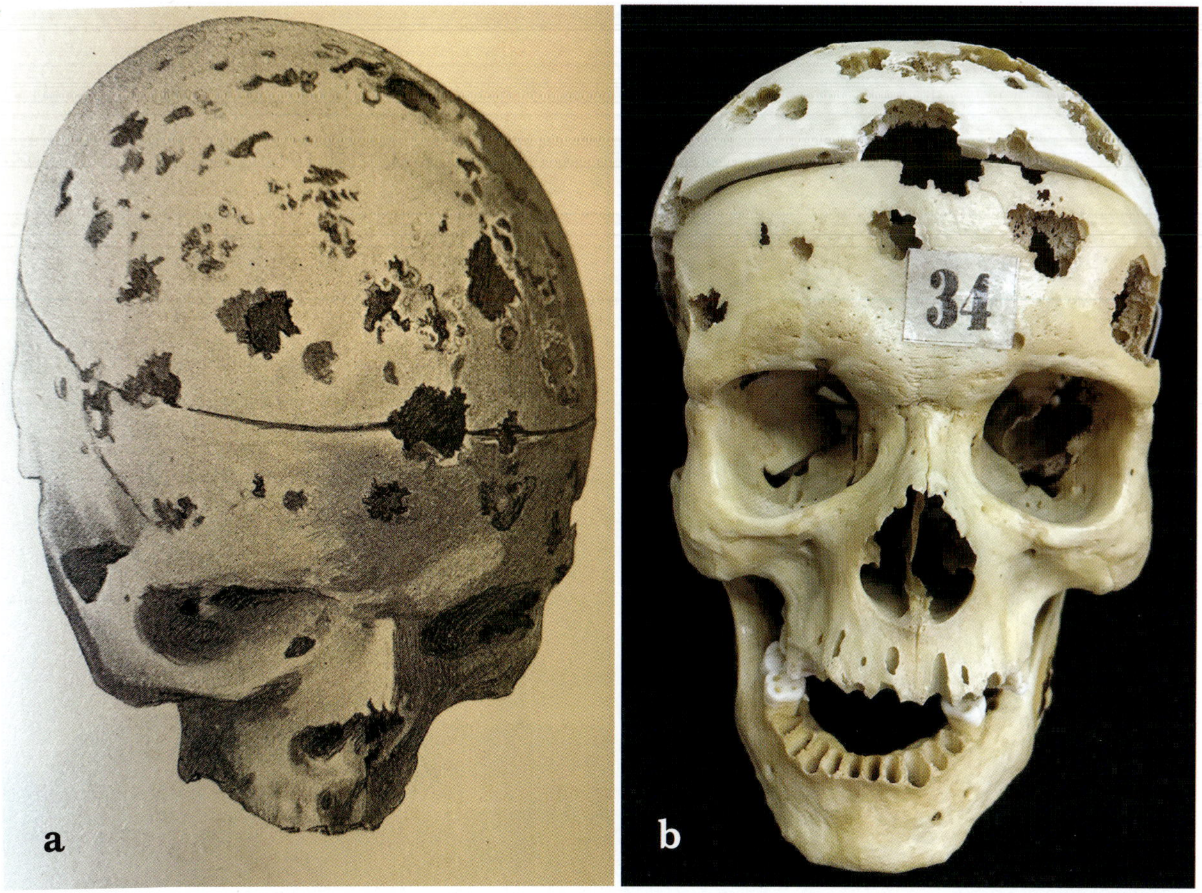


**Figure 432a-b.** (a) Medical illustration by Jellinek (1904) compared with (b) the actual skull showing the level of detail afforded by artistic rendering (MN17.747/34). Comparisons such as this give readers the opportunity to see how accurate and useful illustrations can be before photography became widely available and included in books, manuscripts and journals. Differentiating multiple myeloma from diffuse metastatic carcinoma can be extremely difficult as there often is no distinct pattern of lesions separating the two diseases (Kinney 1940; Limarzi 1951). See Bitelman et al. (2016); Reeder (1993); Rossi et al. (2021) for more on the features and distribution of myeloma lesions in the skeleton.

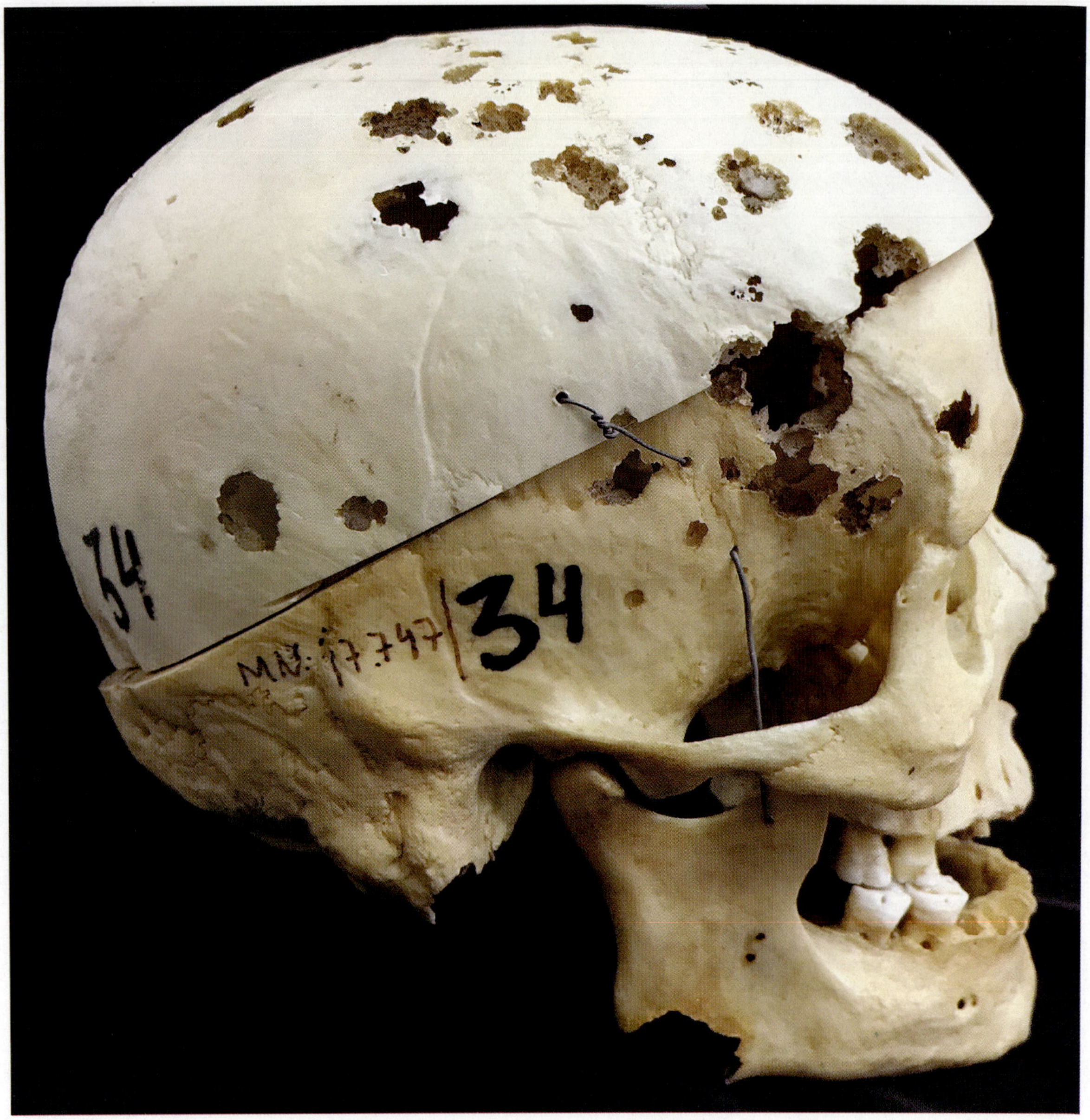


**Figure 433.** Right lateral view of multiple myeloma in a 45-year-old male who was examined in a hospital in 1903 and died later that year, showing lesions of varying size with scalloped margins in areas of diploe (MN17.747/34). Note both mandibular corpi exhibit large destructive lesions (see Figure 435). See Bitelman et al. (2016); Reeder (1993); Rossi et al. (2021) for more on the features and distribution of myeloma lesions in the skeleton.

**Figure 434.** Superior view showing lytic lesions of varying sizes and shapes in most of the vault in a 45-year-old male who was examined in a hospital in 1903 and died later that year (MN17.747/312). See Bitelman et al. (2016); Reeder (1993); Rossi et al. (2021) for more on the features and distribution of myeloma lesions in the skeleton.

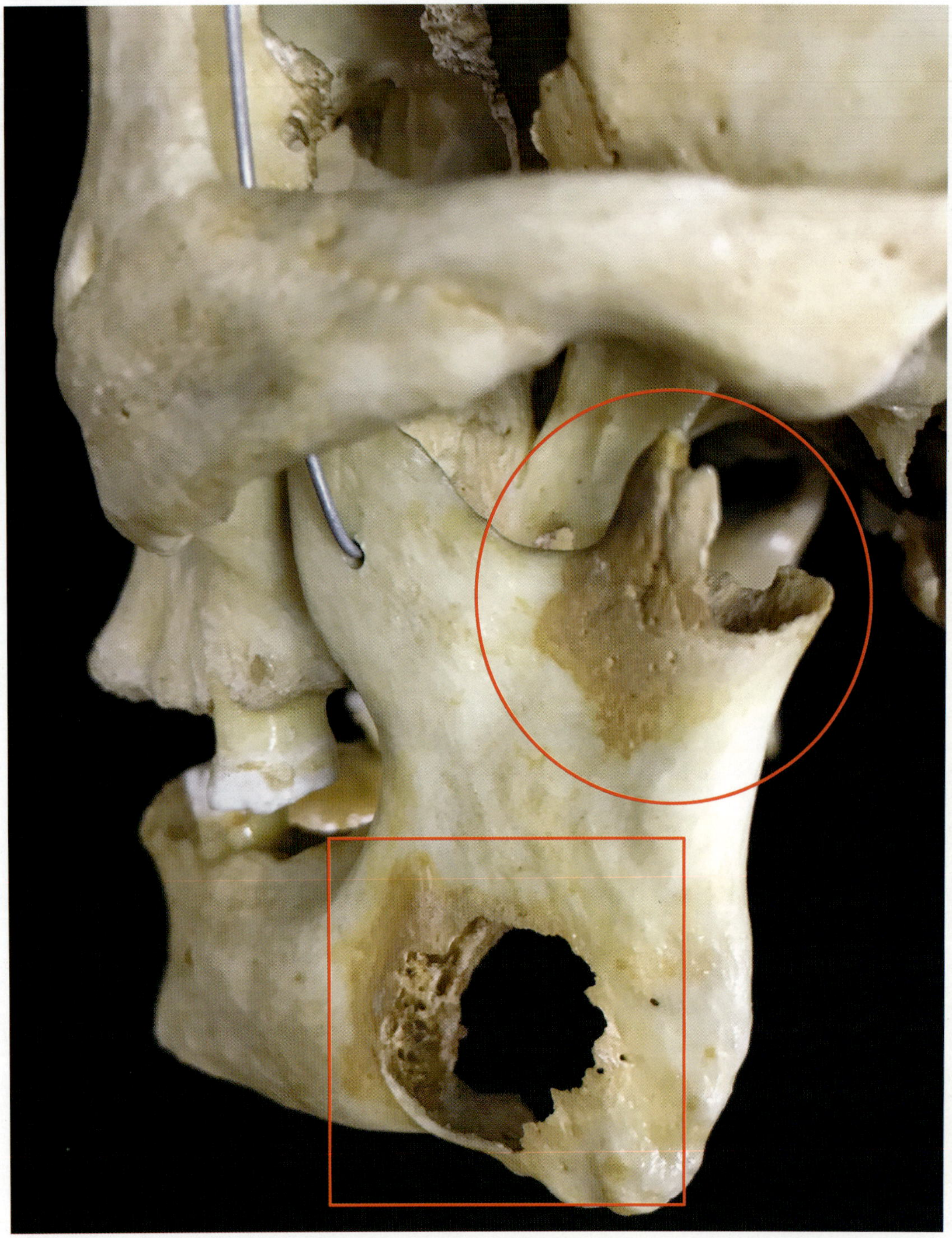

**Figure 435.** Large lytic lesion in the left posterior portion of the mandible (square) and condyle (circle) resulting in complete destruction of the latter in a 45-year-old male who was examined at a hospital in Wien in 1903 and died later that year (MN17.747/312). The dark areas surrounding the lesions indicate some remodeling or early destruction where the lesions would have progressed. See Bitelman et al. (2016); Reeder (1993); Rossi et al. (2021) for more on the features and distribution of myeloma lesions in the skeleton.

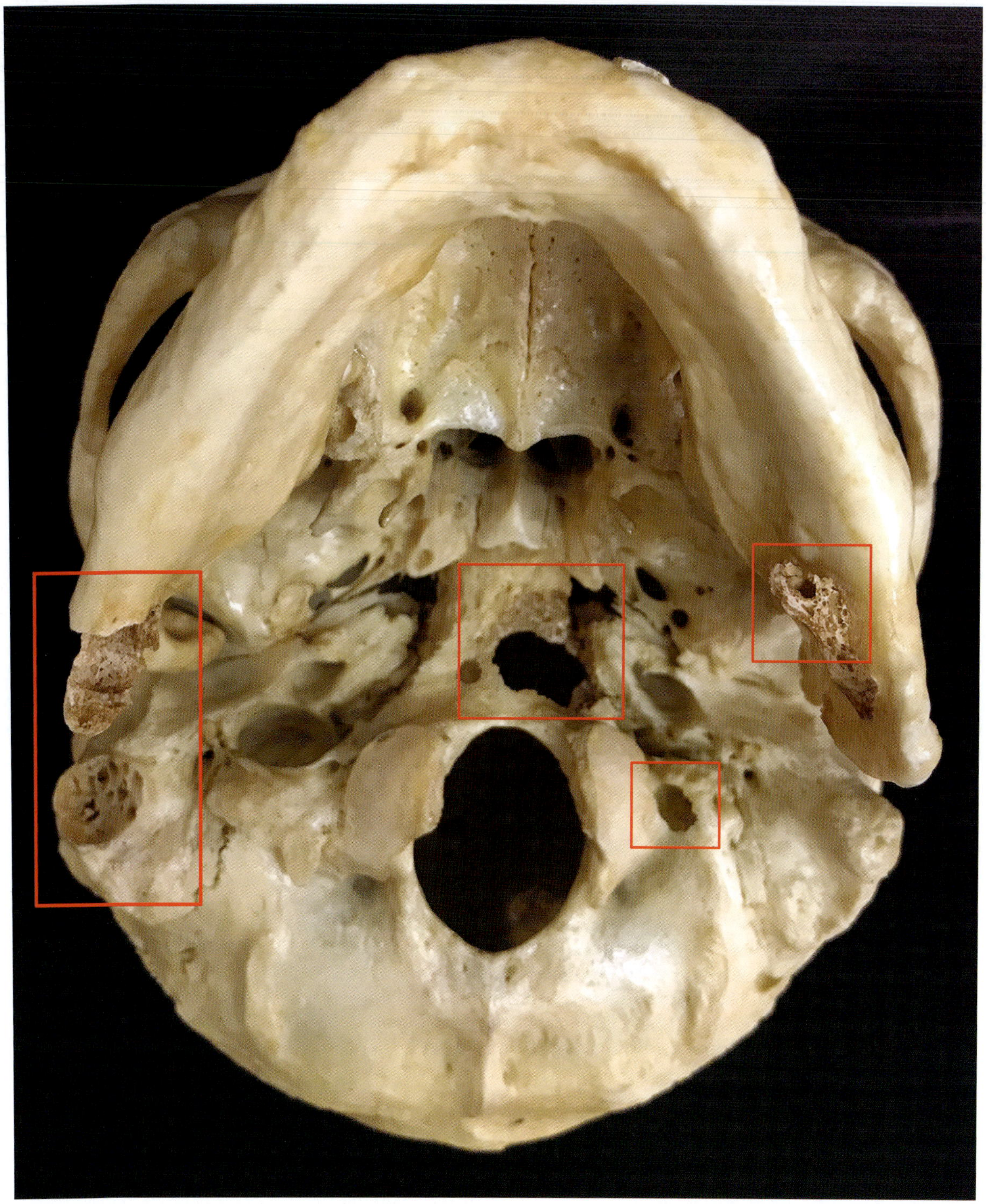

**Figure 436.** Inferior view of skull showing numerous lytic lesions (boxes) of the cranial base, including the right mastoid process and mandible in a 45-year-old male who was examined at a hospital in 1903 and died that same year (MN17.747/312). Myeloma usually does not appear as a lump or tumor, but the myeloma cells divide and expand within the bone marrow, damaging the bones and affecting the production of healthy blood cells (NHS Inform 2022). See Bitelman et al. (2016); Reeder (1993); Rossi et al. (2021) for more on the features and distribution of skeletal myeloma lesions.

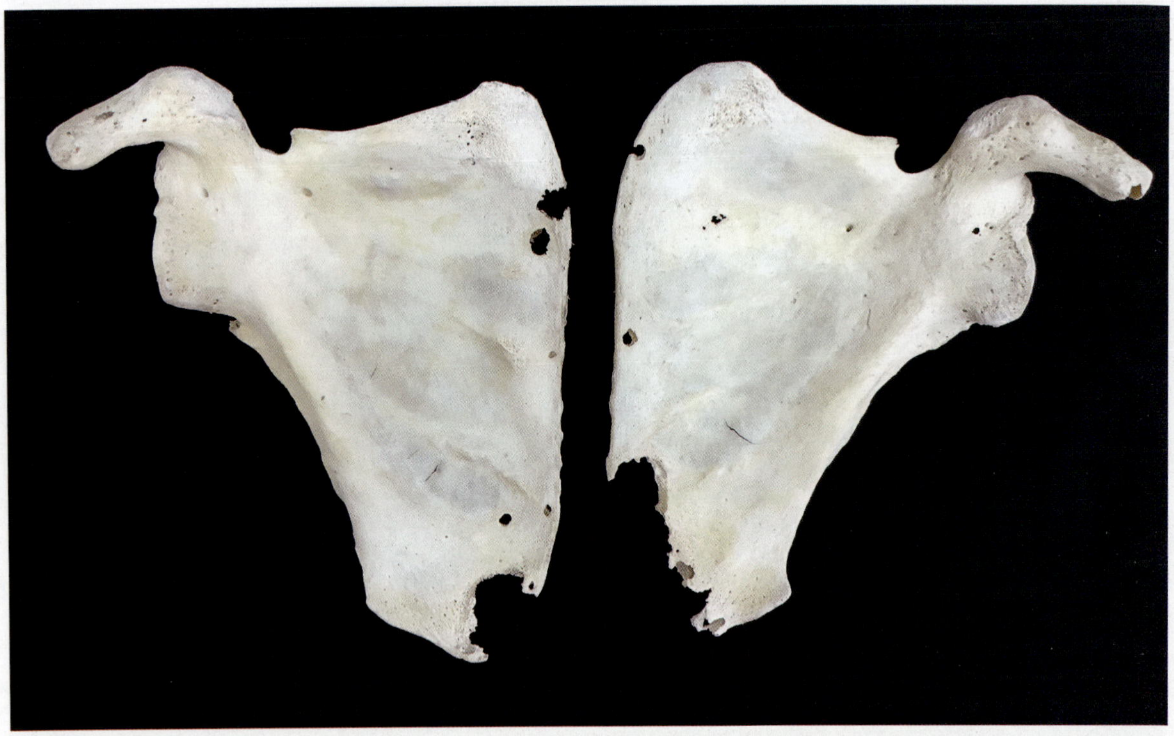

**Figure 437.** Right and left scapulae with numerous lytic lesions in the red marrow areas in a 45-year-old male who was examined and died at a hospital in 1903 (MN17.747/312). The lesions typically don't involve the thin portions of the scapula due to the absence of red marrow. Myeloma usually does not appear as a lump or tumor, but the myeloma cells divide and expand within the bone marrow, damaging the bones and affecting the production of healthy blood cells (NHS Inform 2022). See Bitelman et al. (2016); Reeder (1993); Rossi et al. (2021) for more on the features and distribution of myeloma lesions in the skeleton.

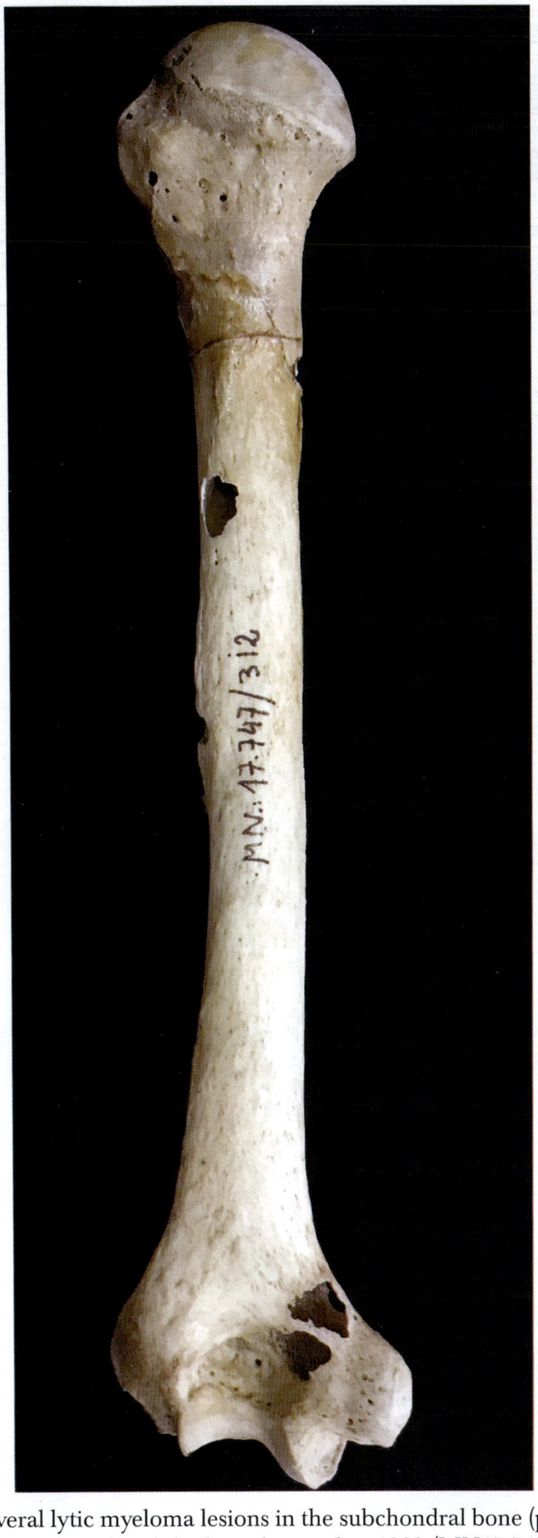


**Figure 438.** Left humerus with several lytic myeloma lesions in the subchondral bone (proximal joint), diaphysis and distal end in a 45-year-old male who was examined and died at a hospital in 1903 (MN17.747/312) (see Figure 441). See Yilmaz (2021) for more on myeloma in the humerus.

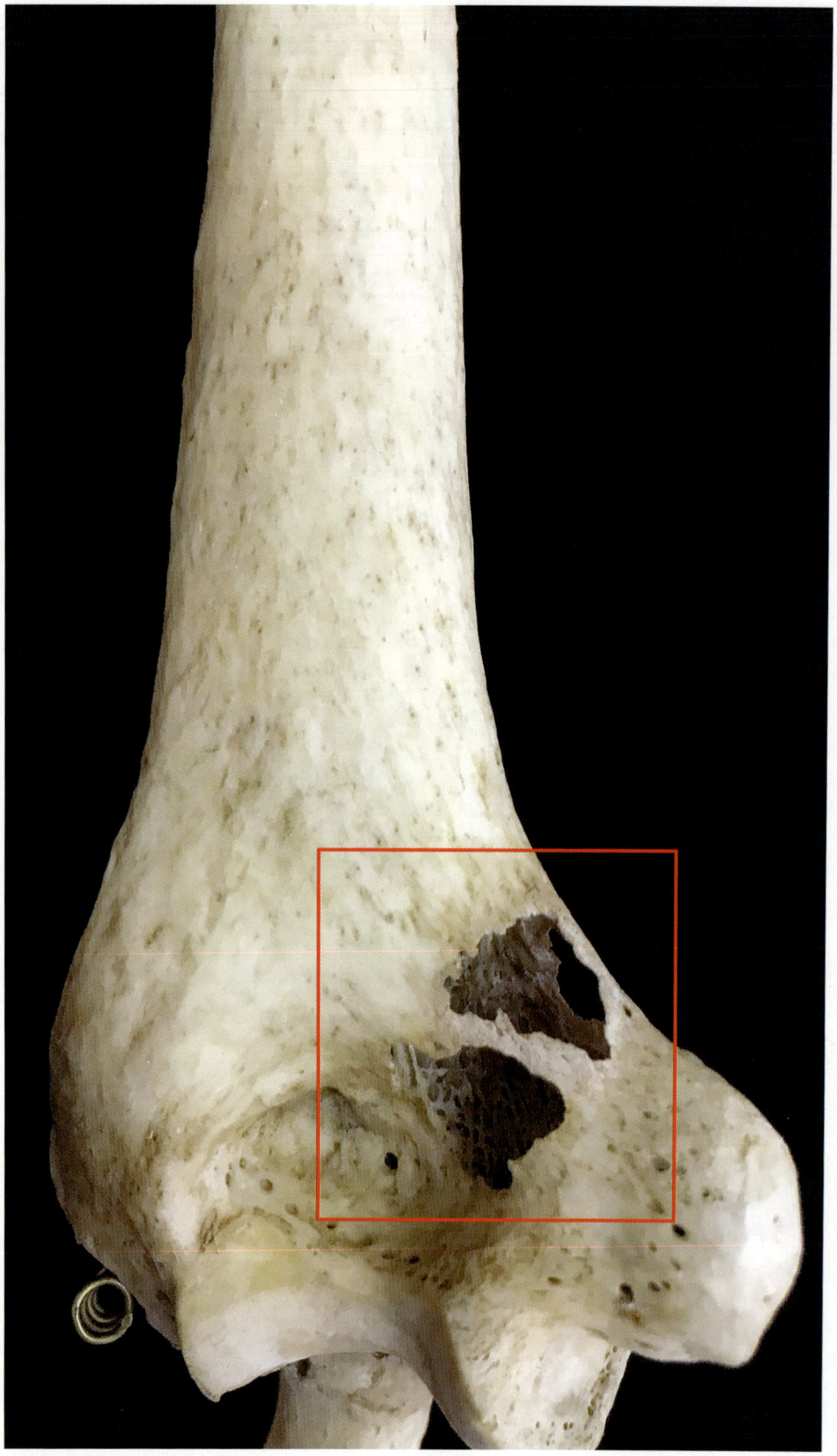

**Figure 439.** Magnified posterior view of distal left humerus showing large lytic lesions (square) with scalloped margins that perforated both the anterior and posterior cortex in a 45-year-old male who was examined and died in 1903 (MN17.747/312) (see Figure 441). See Yilmaz (2021) for more on myeloma in the humerus.

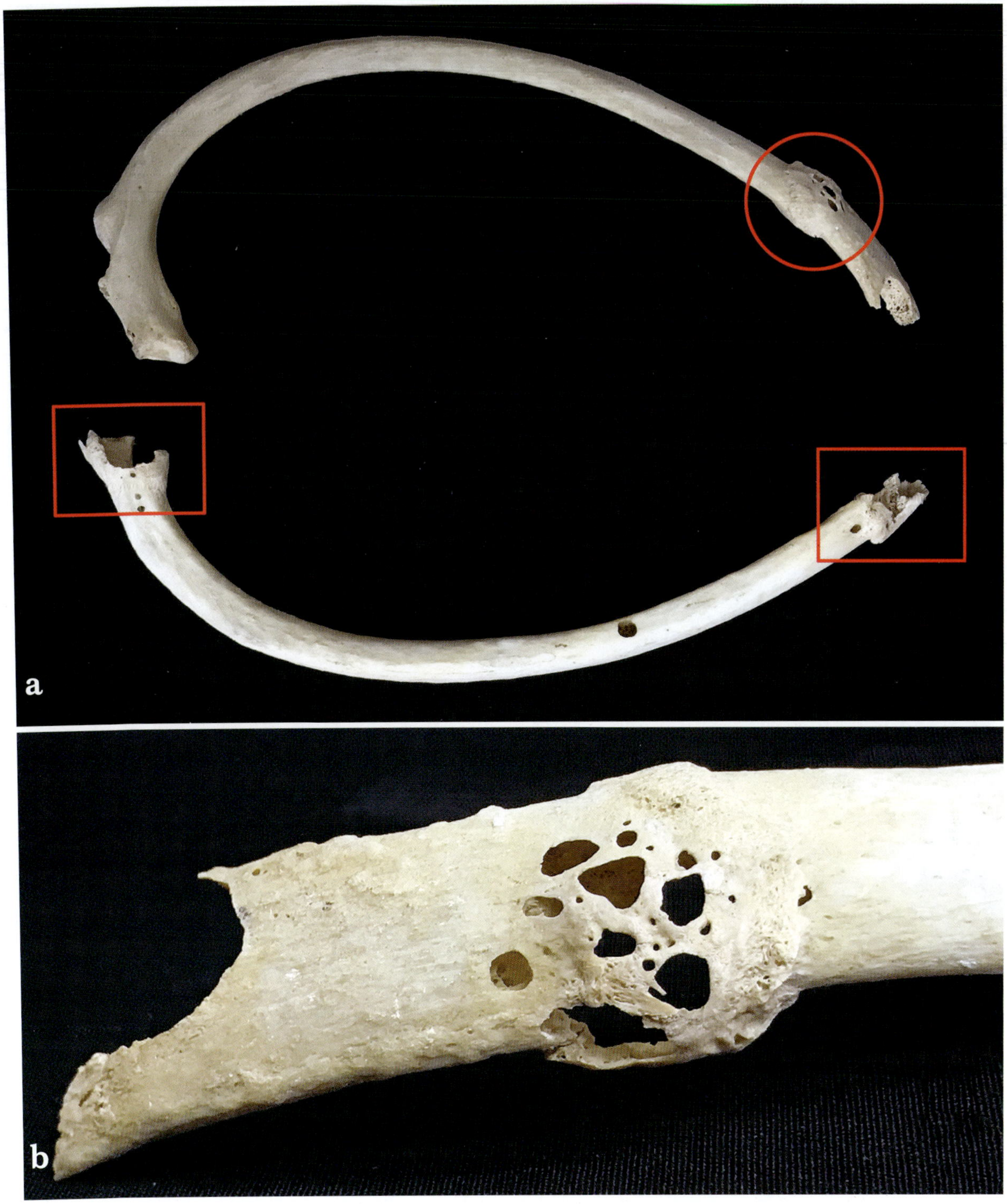


**Figure 440a-b.** (a) Left and right ribs showing multiple expansile rib lesions, some of which give the rib a "clubbed" appearance (rectangles) in a 45-year-old male examined in 1903 (MN17.747/312). (b) Magnified view of latticework-like lesion in a left rib (a; circle) that caused the cortex to enlarge and have a raised margin of reactive bone. The sternal end of one rib (b) exhibits a lytic lesion. Myeloma in ribs can result in extreme pain (NHS Inform 2022).

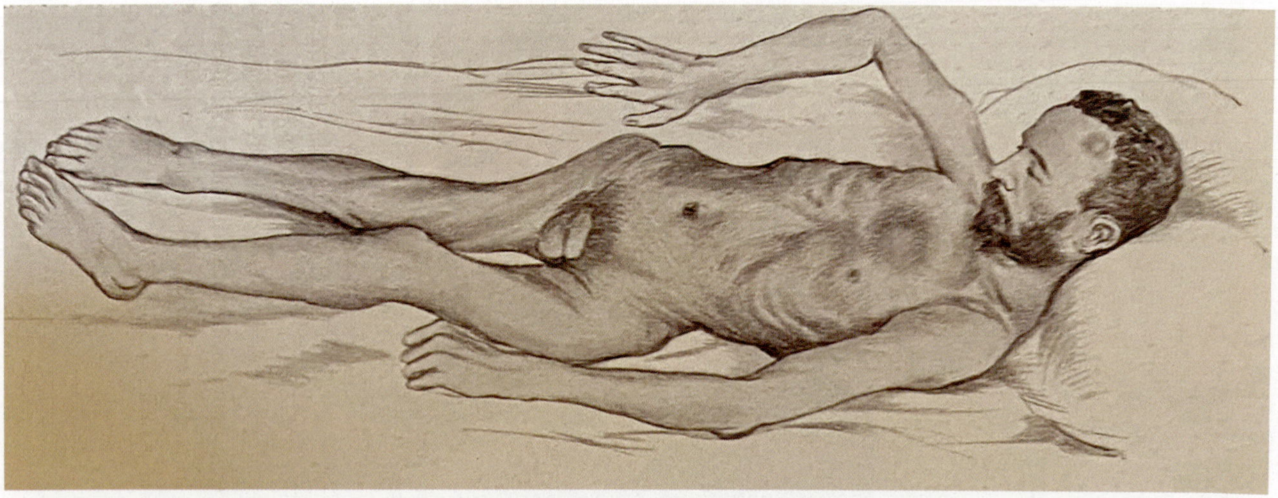

**Figure 441.** Illustration of the 45-year-old male with multiple myeloma while hospitalized in Wien in February 1903 (MN17.747/312) (Jellinek 1904). Note the wasted appearance of this man and lesion in his left frontal region near the scalp line.

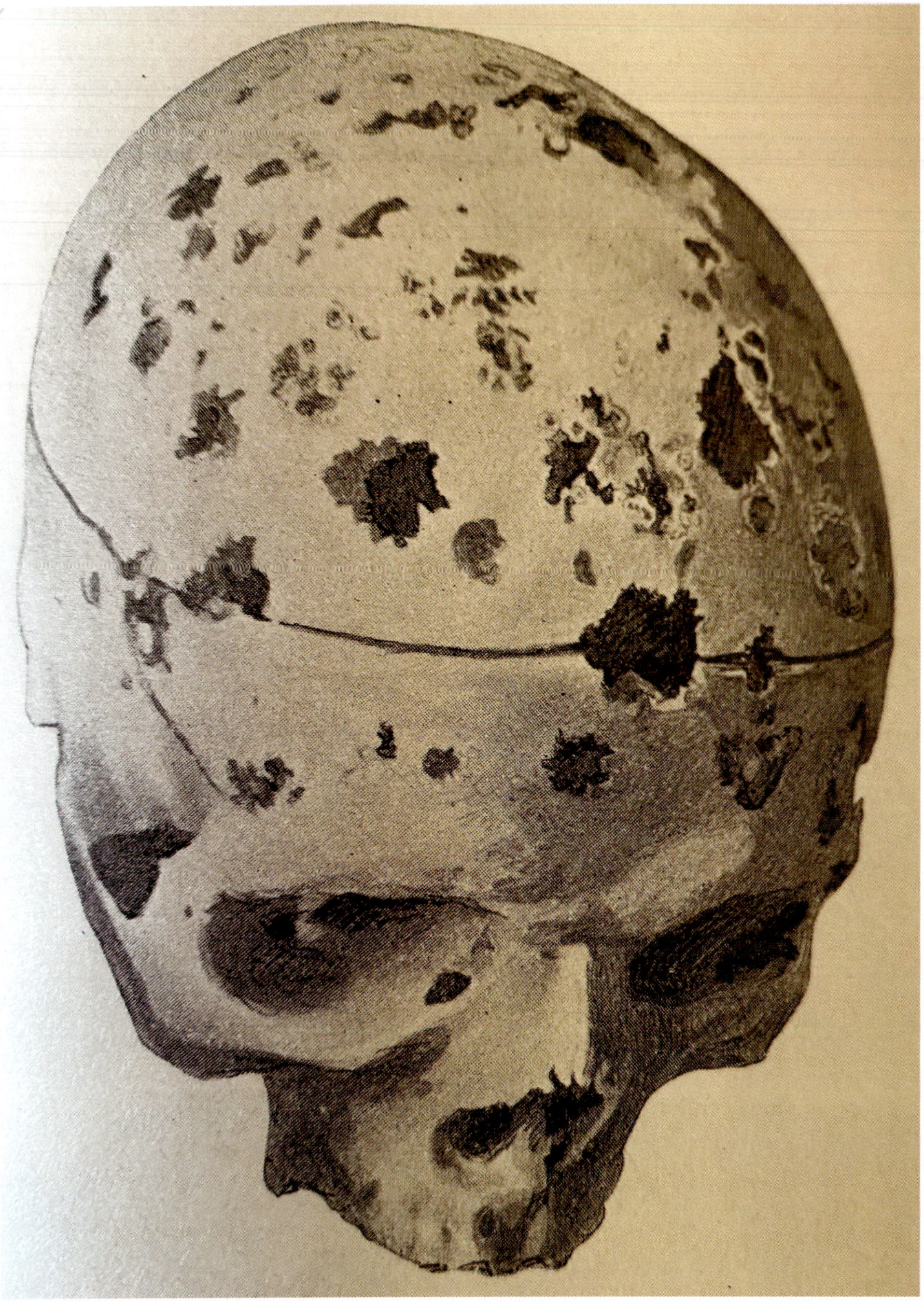

**Figure 442.** Illustration of numerous myeloma lesions in the skull of a 45-year-old male who was examined and died at a hospital in 1903 (MN17.747/312) (Jellinek 1904).

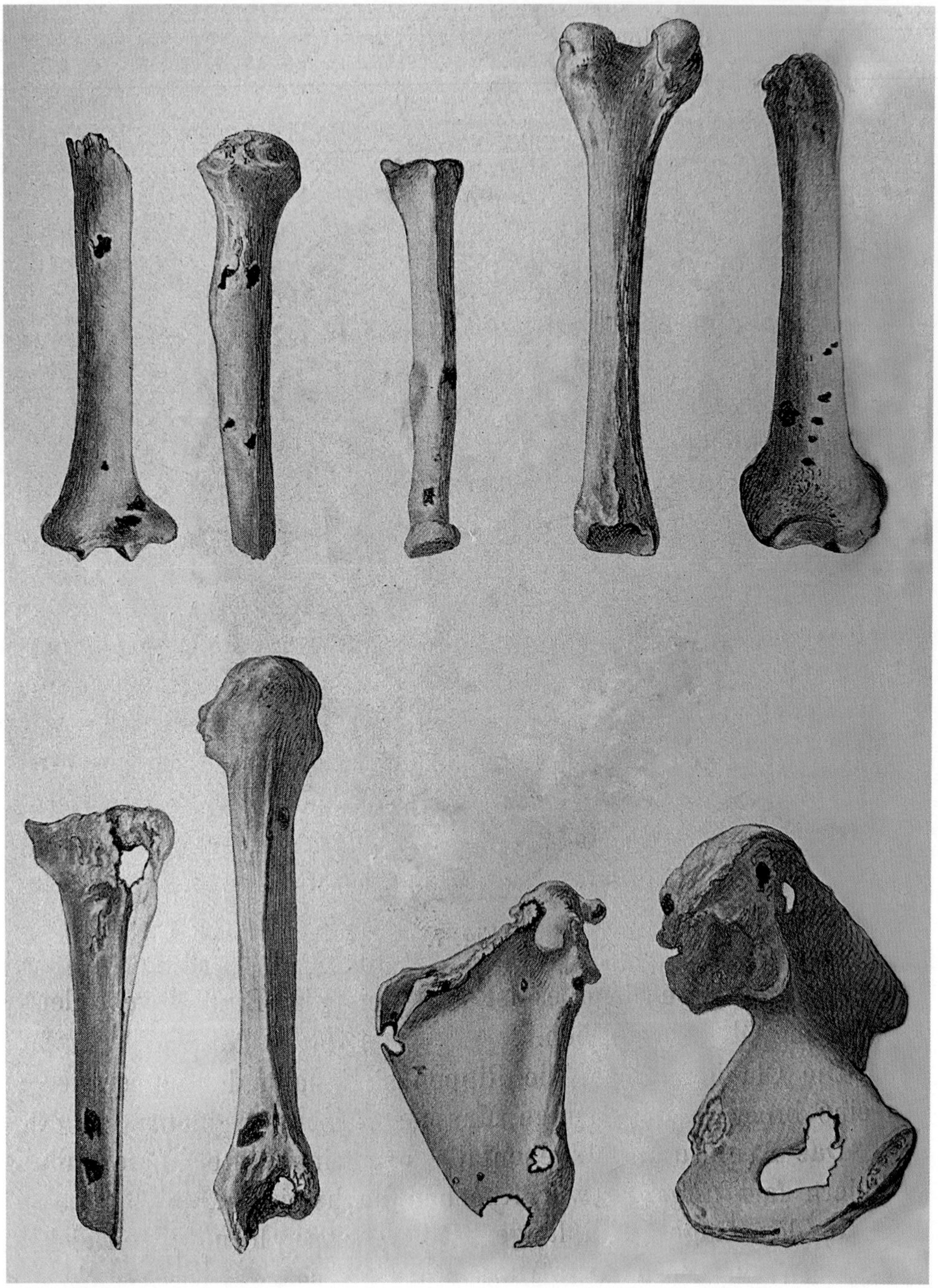

**Figure 443.** Illustrations of the myeloma lesions in some of the skeleton of a 45-year-old male who was examined and died at a hospital in 1903 (MN17.747/312) (Jellinek 1904).

# Chapter 7

# ANATOMICAL PREPARATION AND MEDICAL MOULAGE

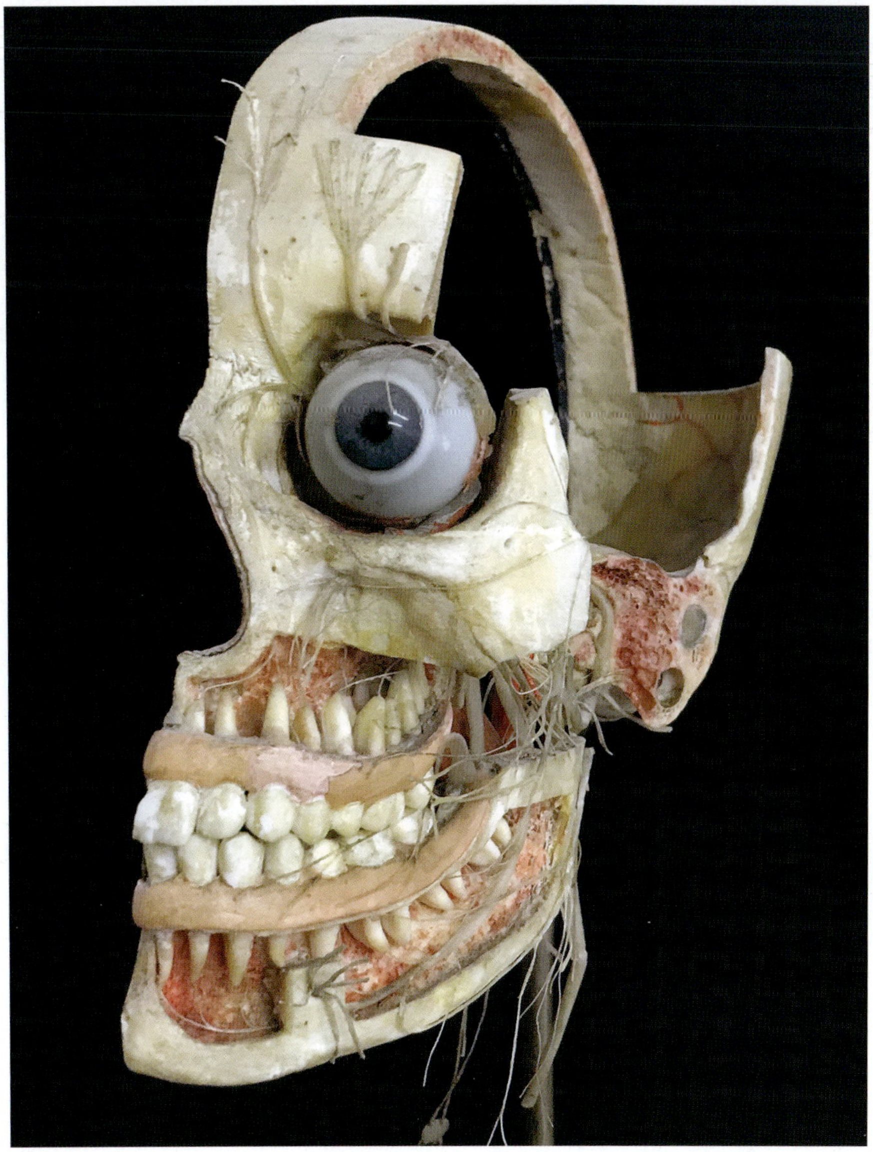

**Figure 444.** Left oblique view of anatomical preparation (not a moulage) of the left half of the skull revealing the alveolar, supraorbital, supratrochlear, and infraorbital nerves. These nerves and blood vessels are often associated with "accessory" frontal grooves.

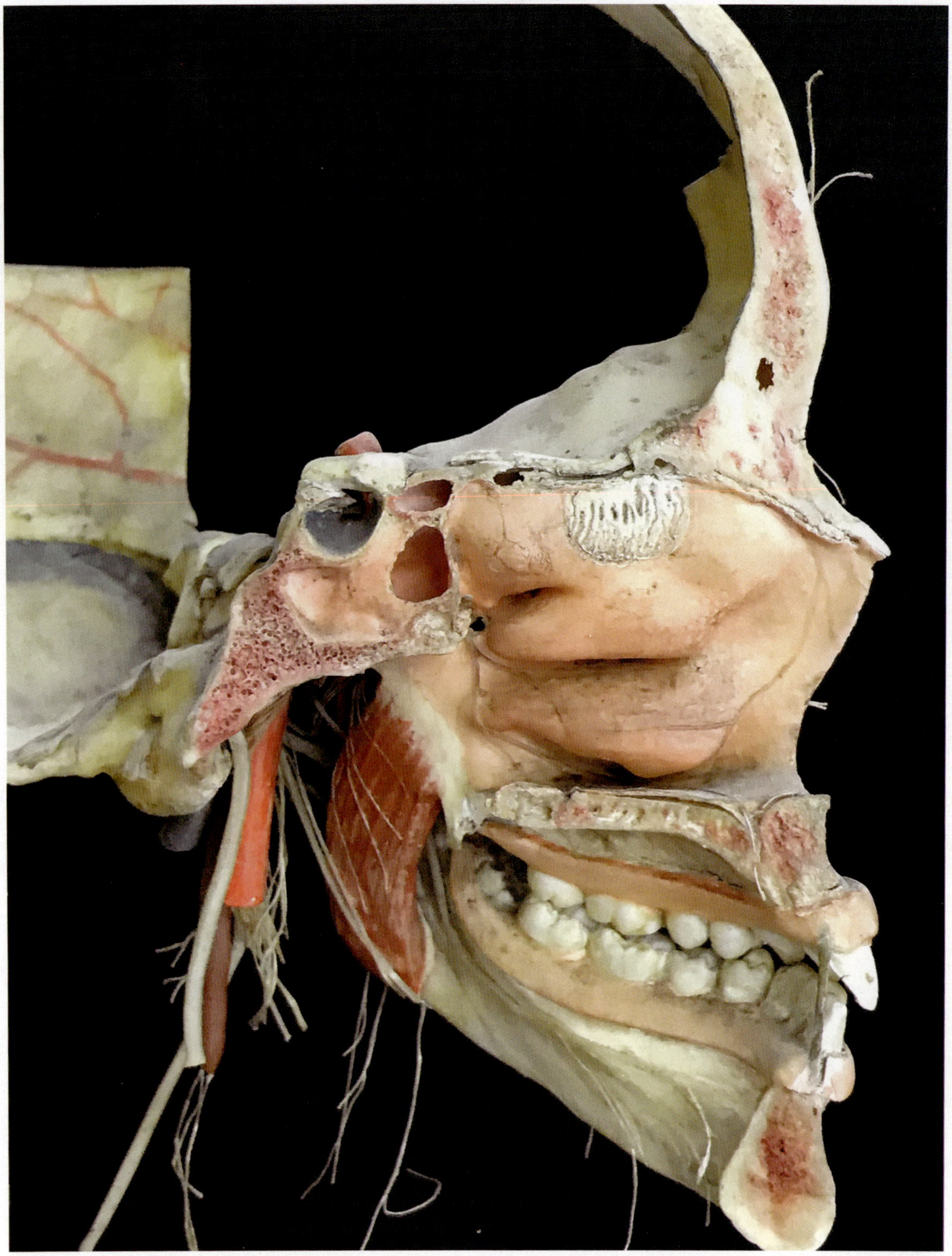

**Figure 445.** Right sagittal view of anatomical preparation of the left half of the skull revealing the alveolar, supraorbital, supratrochlear, and infraorbital nerves.

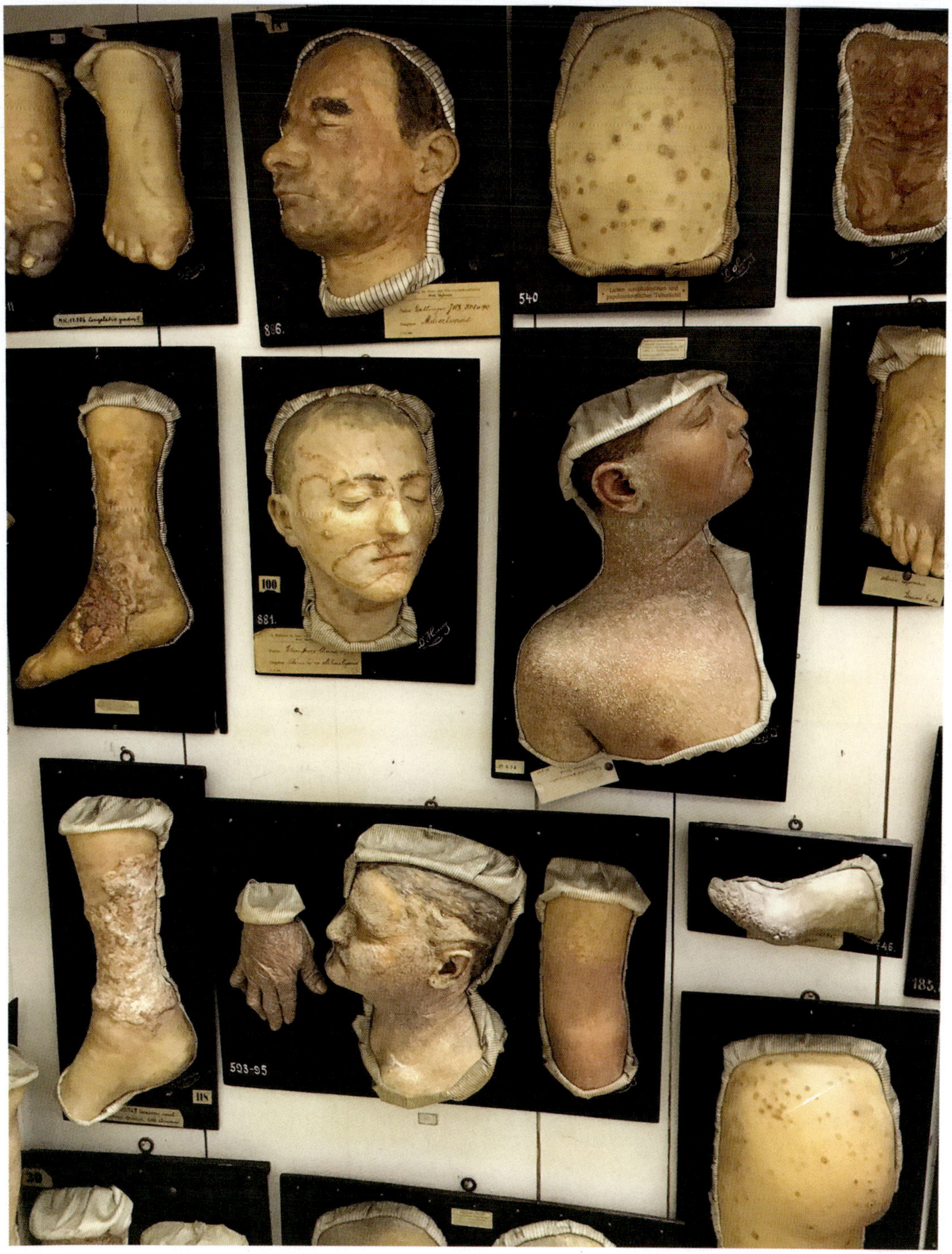

**Figure 446.** Wax medical moulages depicting symptoms of a variety of diseases. Moulages were primarily used to document and depict dermatological conditions in three dimensions, surpassing the detail of all other types of two-dimensional models, illustrations, and daguerreotype photography.

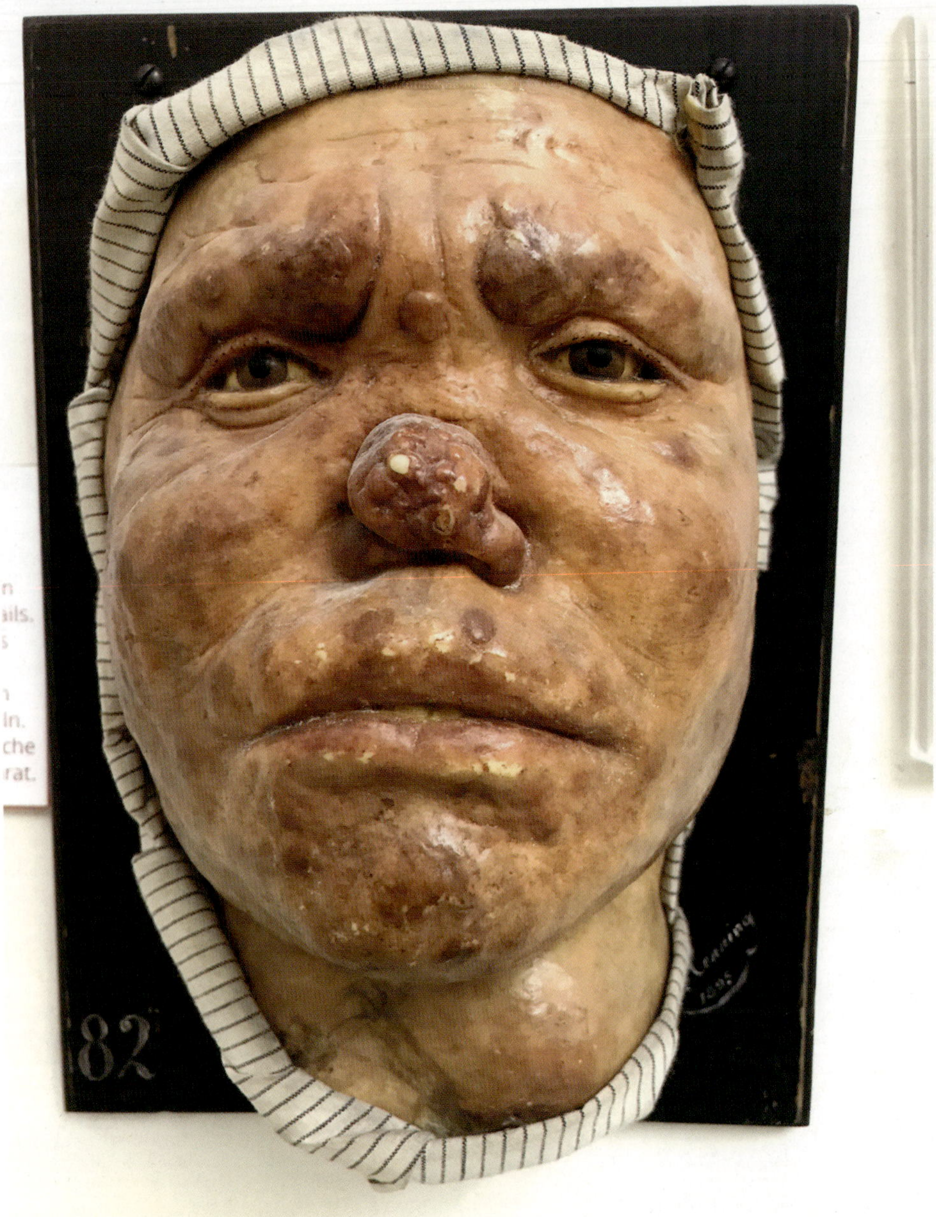


**Figure 447.** Tuberculosis ("Lepra tuberosa" on the card) consisting of rounded nodules on the skin in an adult female (MN82). This moulage was produced in 1895.

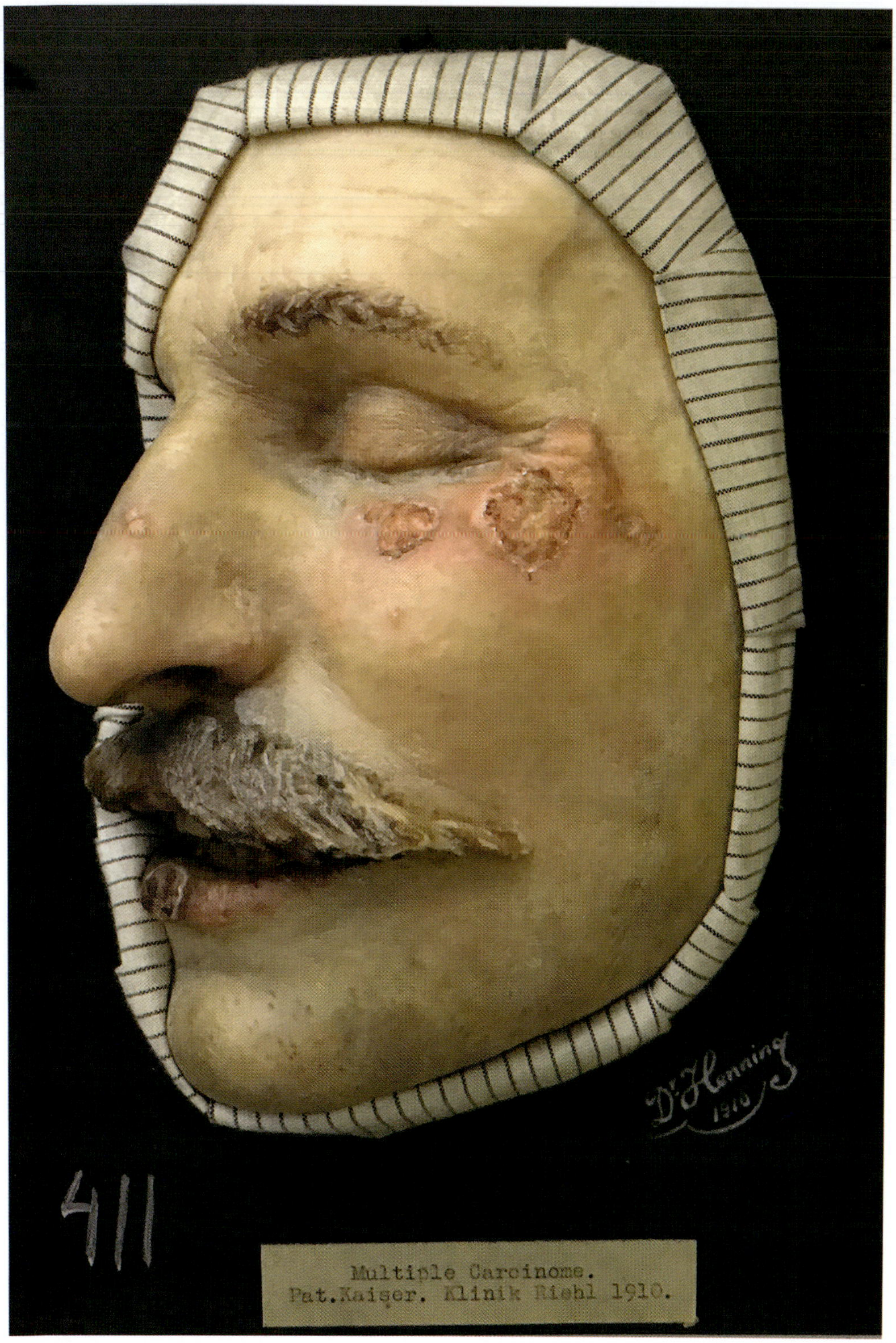


**Figure 448.** Multiple carcinoma (cancer) in an adult male prepared at the Klinik Riehl in 1910 (MN411).

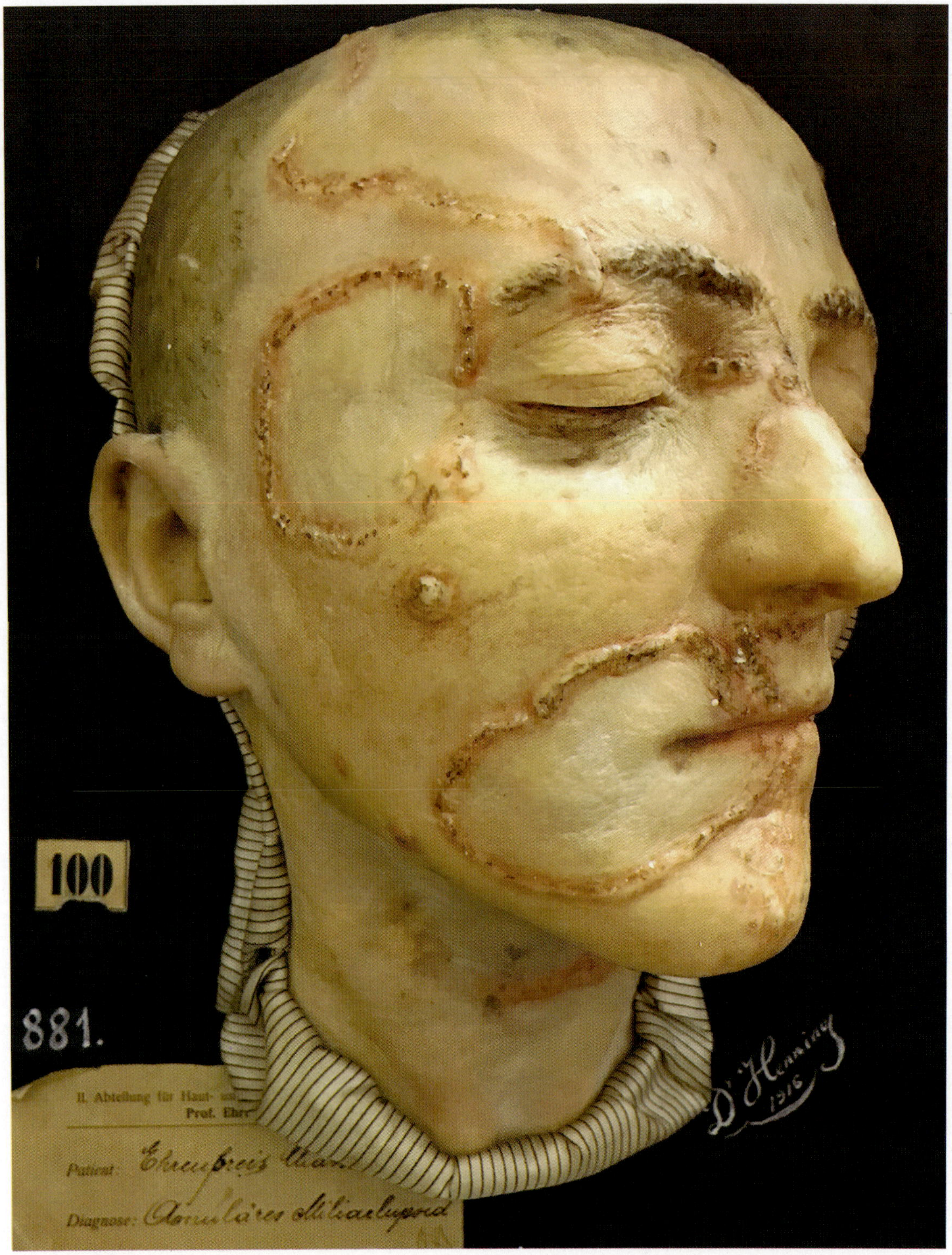


**Figure 449.** Lupus vulgaris (infectious) of the face (MN881). At the time this moulage was produced in 1916, it was not known whether the disease, "Annuläres Miliartuberkulosis" written on the bottom of the moulage for this individual, represented an annular type of Lupus vulgaris or an annular type of Lupus erythematosus (R. Panizzon and D.W. Johnson, personal communication, September 23, 2019).

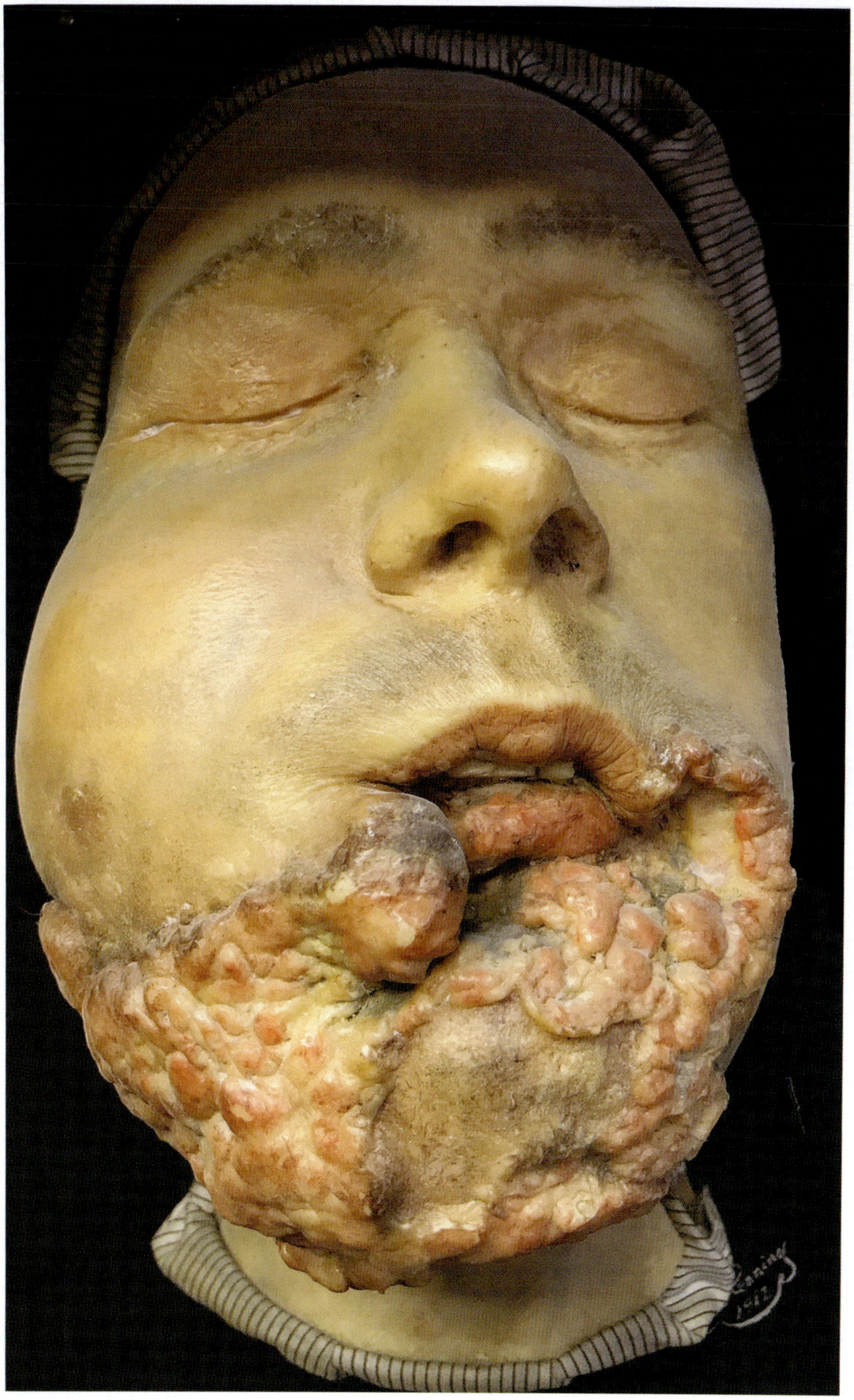

**Figure 450.** Wax moulage of carcinoma/cancer of the lower lip in an unknown individual prepared in 1912 (MN24.774).

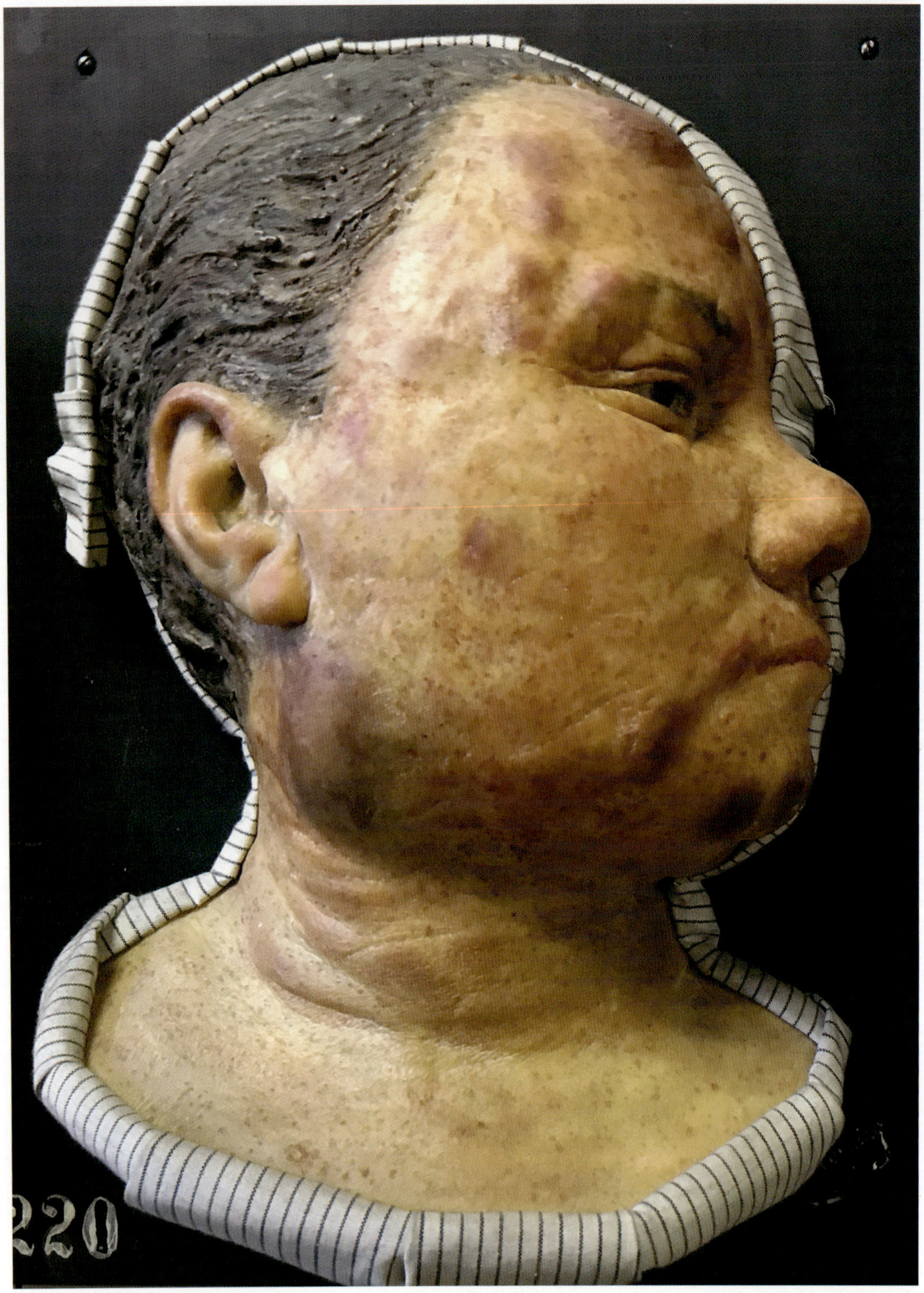


**Figure 451.** Mycosis fungoides, the most common type of blood cancer known as T-cell lymphoma in an adult female - this type of cancer usually affects individuals over 50 years of age (MN220). Cancerous cells within the scaly skin patches can form tumors and malignancies that can spread to other parts of the body.

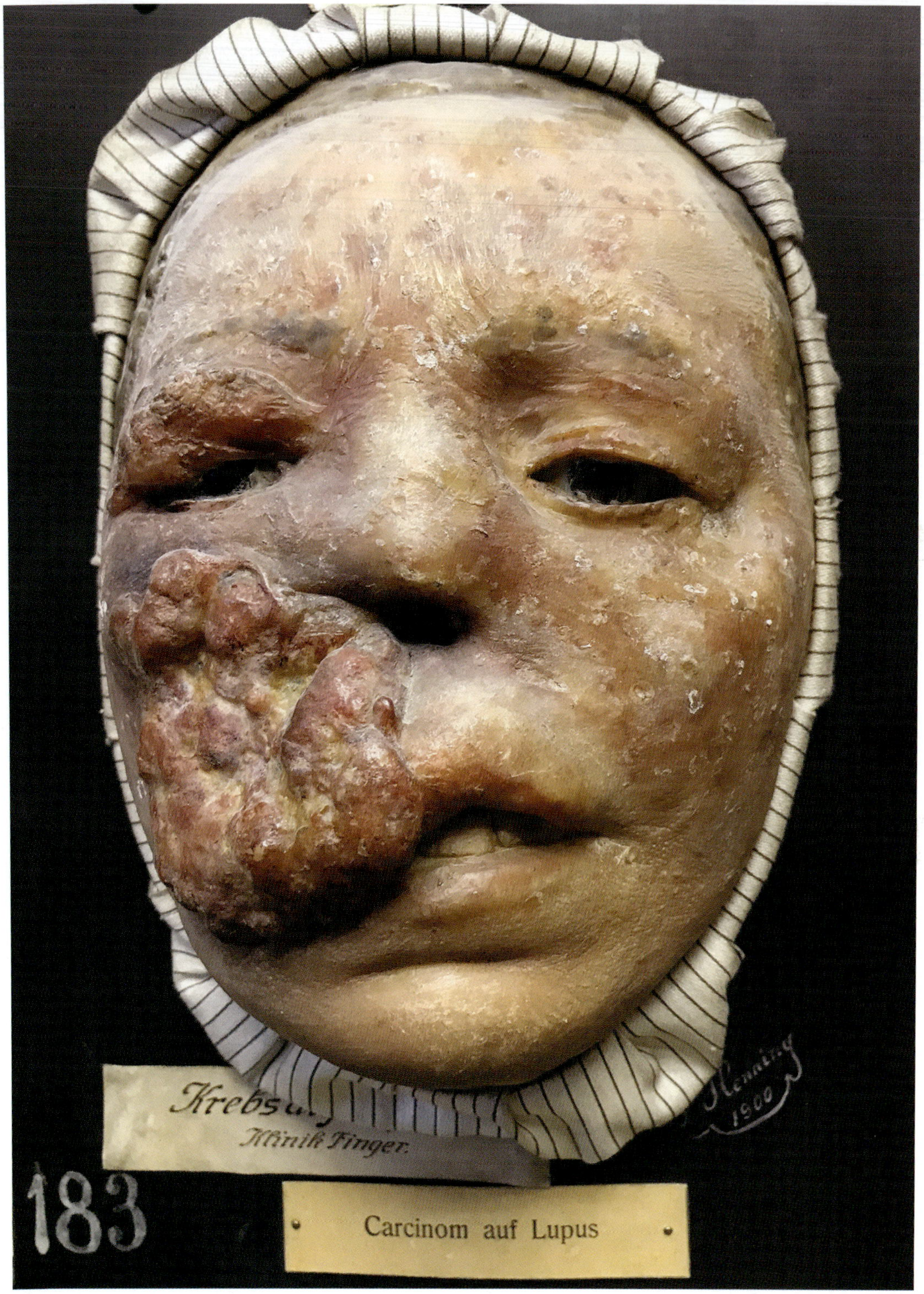


**Figure 452.** Moulage produced in 1900 of an adult female with carcinoma and lupus, the latter a chronic autoimmune disease (MN183). Lupus can increase the risk of developing cancer (Song et al. 2018).

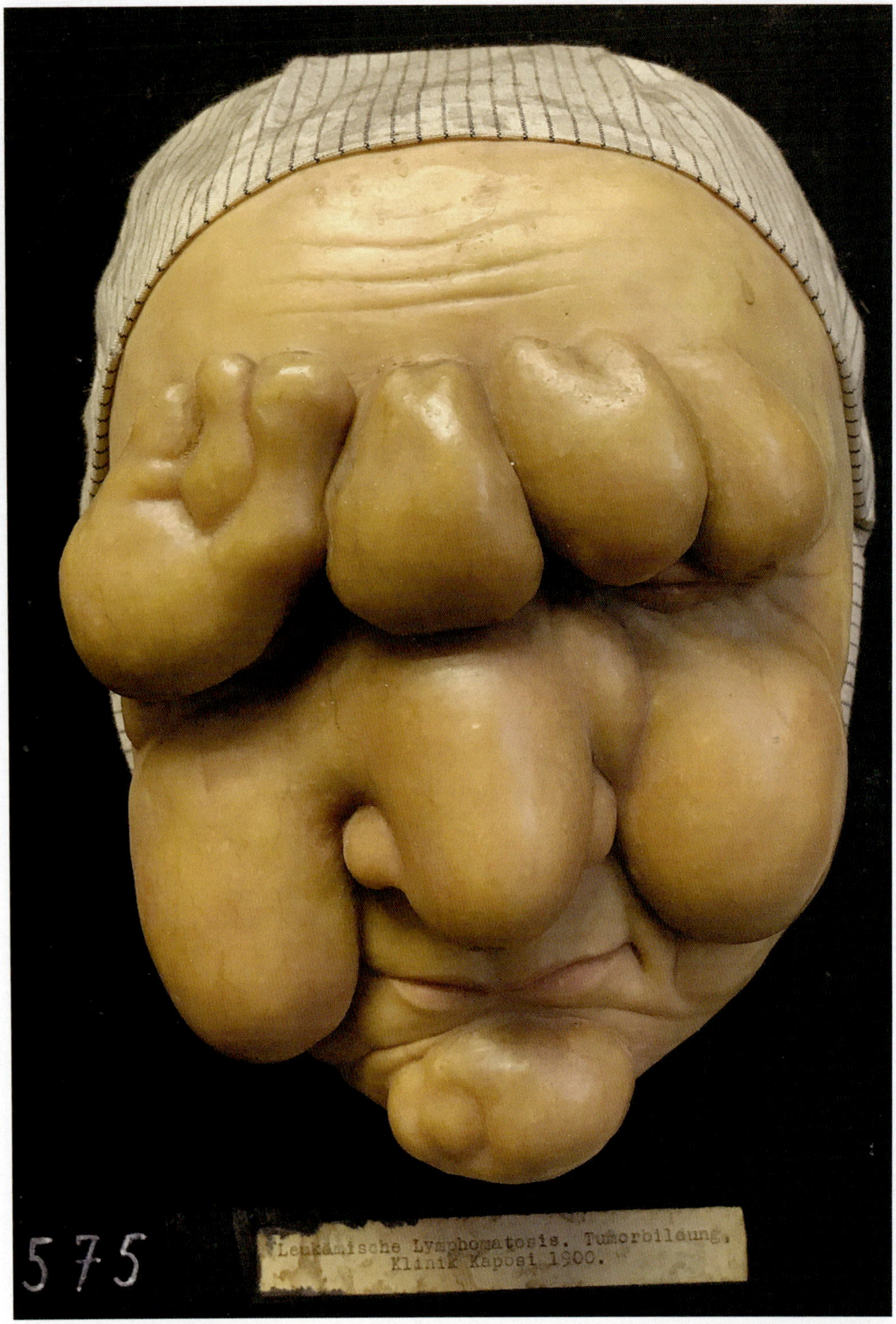


**Figure 453.** Lymphomatosis (in this individual, cutaneous T-cell lymphoma) in an adult female (MN575). The forehead, cheeks and chin in this individual exhibit large, pendulous cutaneous tumors due to cancer of the white (lymphocytes) blood cells. Moulages, wax models, used in medical and anatomical specialties reached their peak of three-dimensional realism in dermatology in the 1900s.

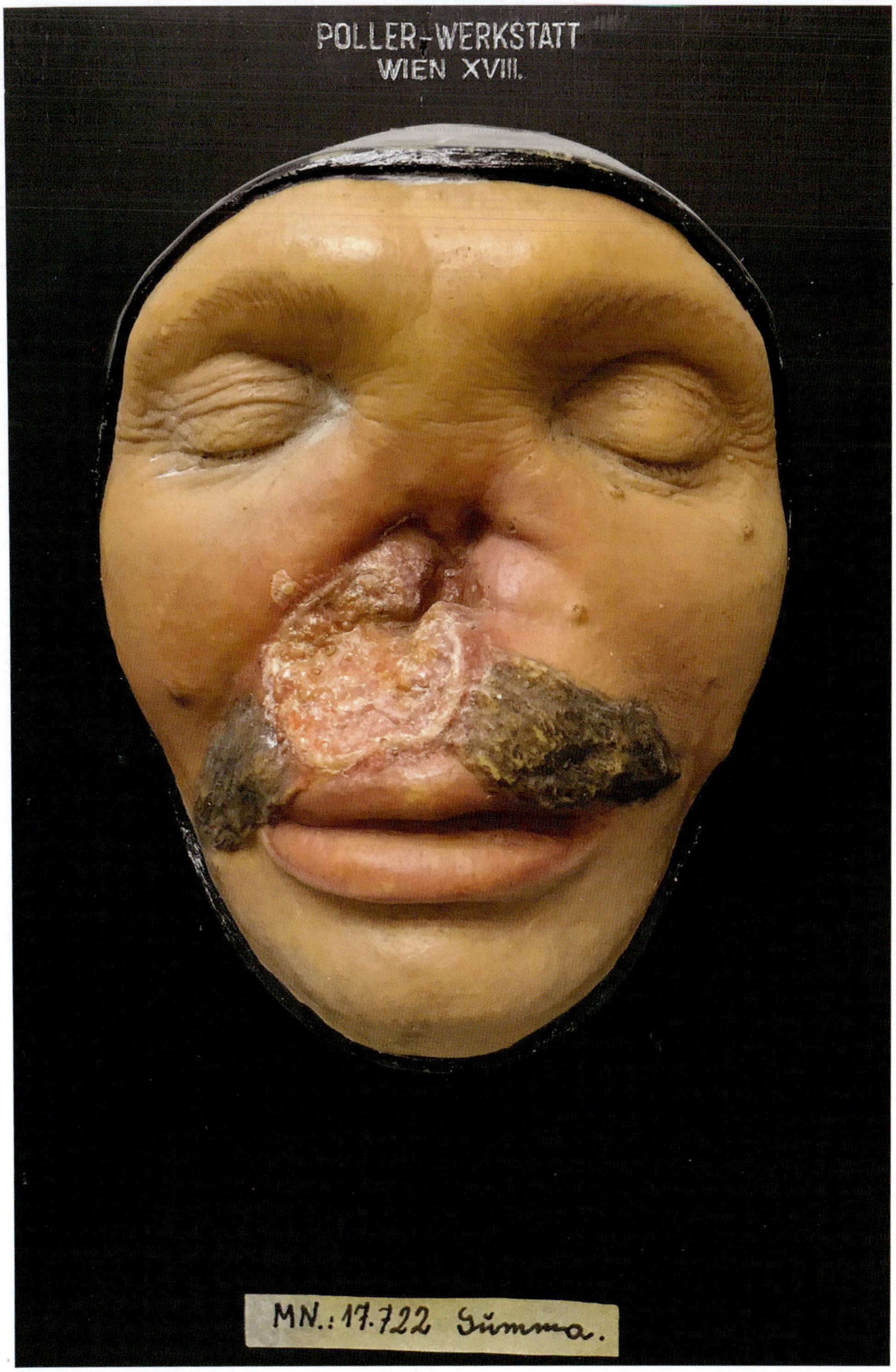


**Figure 454.** Adult male with syphilitic gumma of the nose (MN17.722). Gumma is an autoimmune response caused by spirochetal bacteria when the body fails to kill off the syphilis.

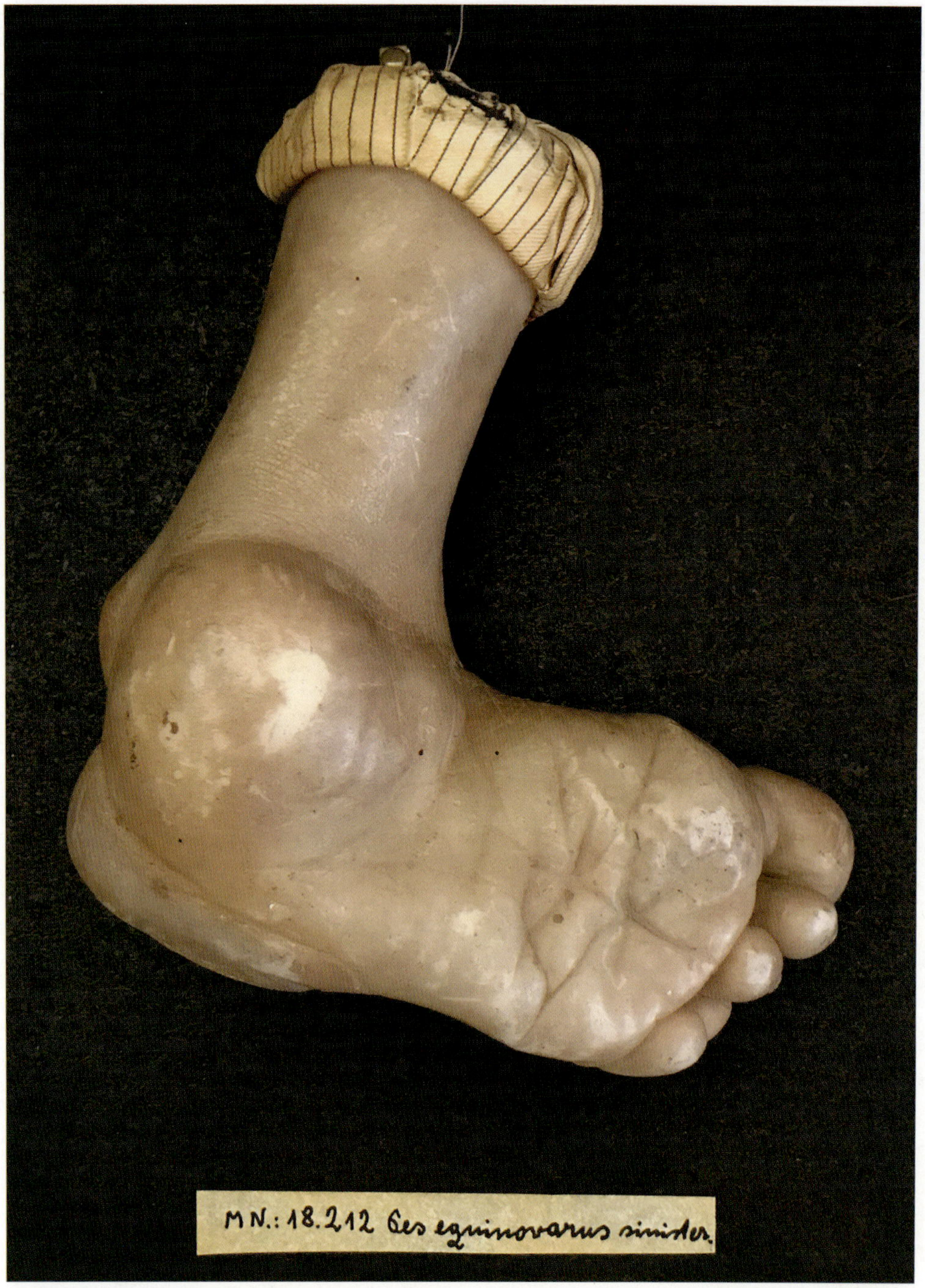


**Figure 455.** Clubfoot (pes equinovarus) of the left foot in an unknown individual (MN18.212). Clubfoot is a congenital condition (i.e., present at birth) of unknown etiology, possibly due to a combination of environmental and genetic factors where one or both feet are twisted/deformed.

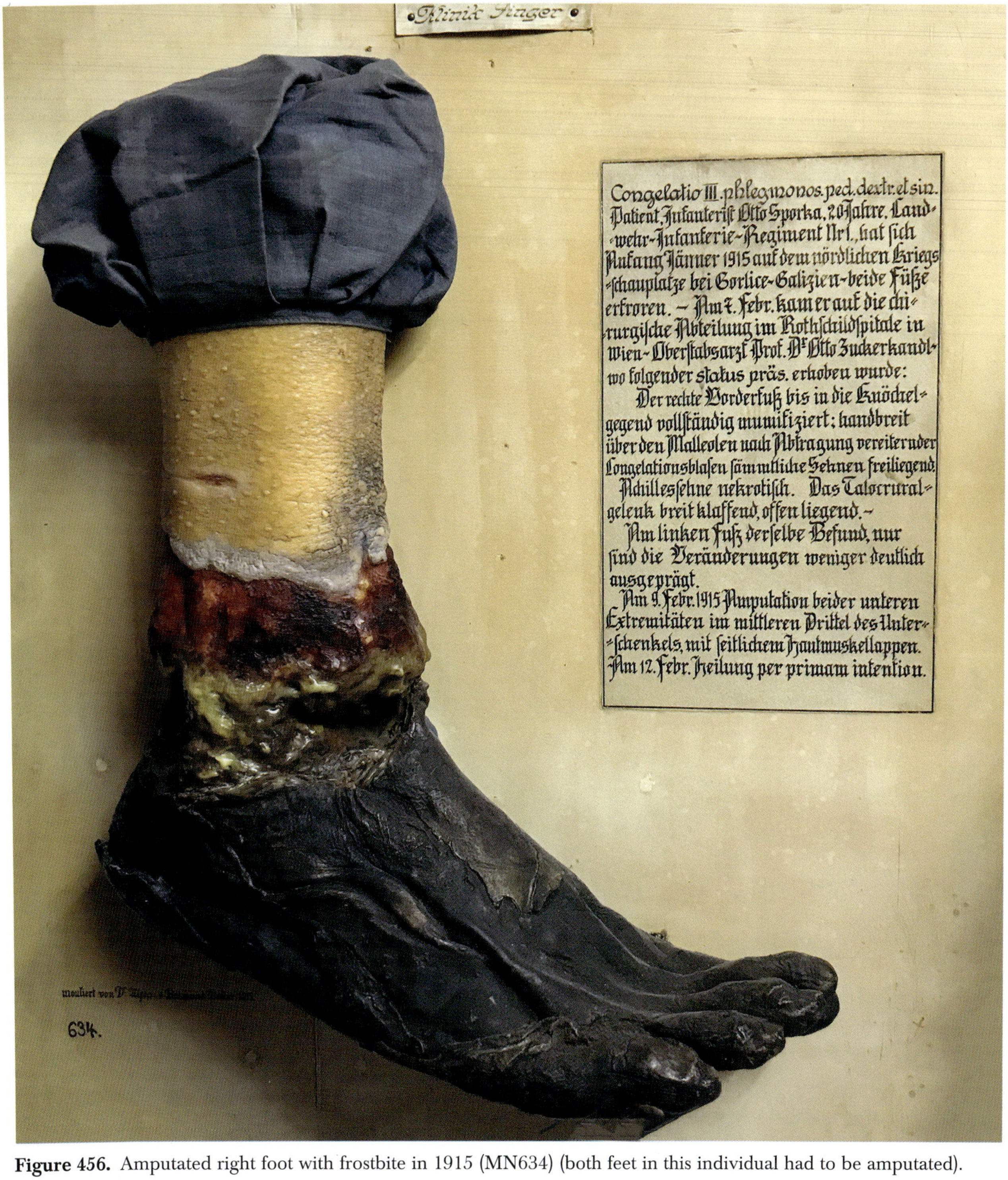


**Figure 456.** Amputated right foot with frostbite in 1915 (MN634) (both feet in this individual had to be amputated).

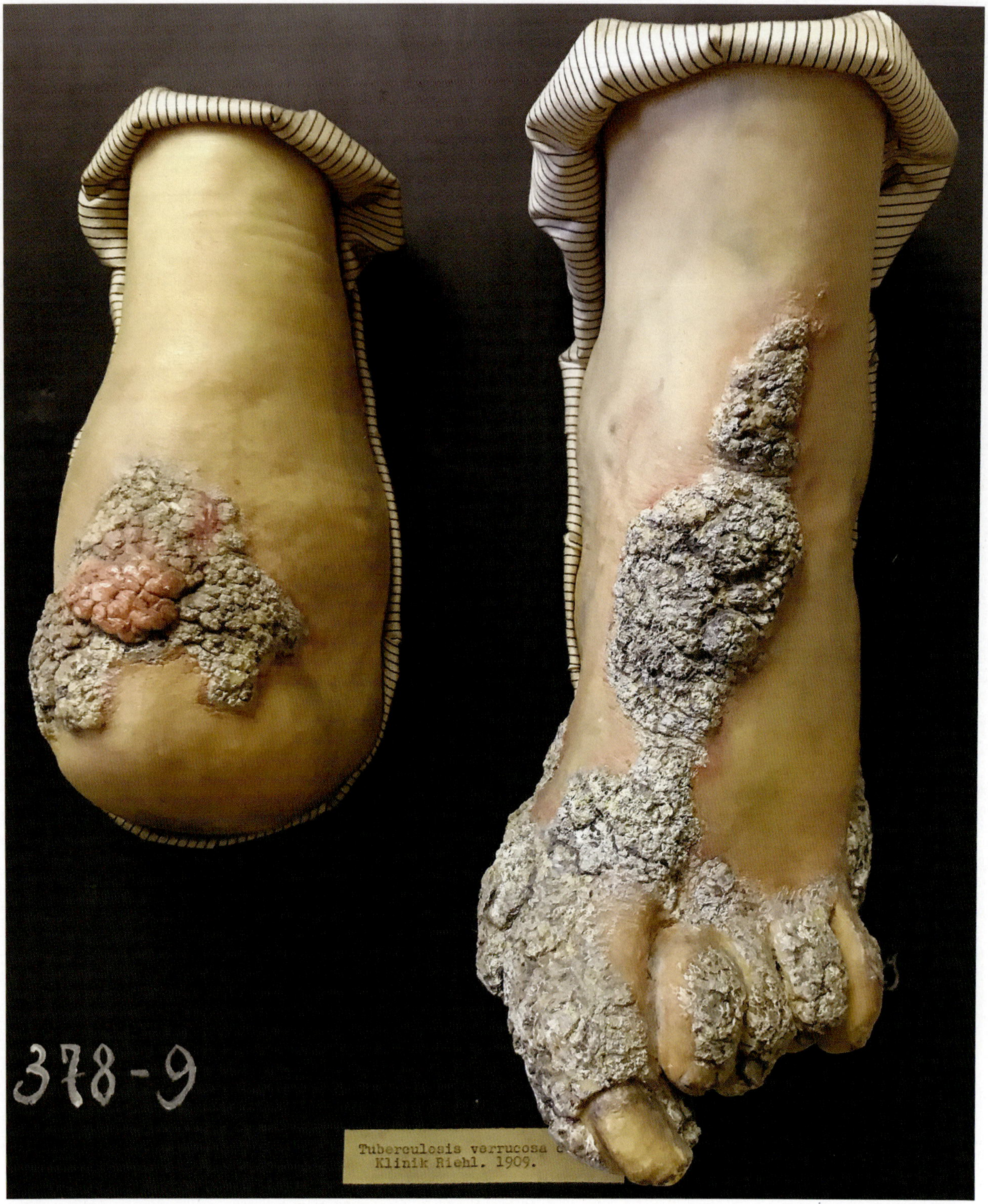


**Figure 457.** Tuberculosis verrucose, a form of cutaneous tuberculosis lichen (MN378-9).

**Figure 458.** Wax moulage of the stages of rare Syphilitic Rupia (malignant) secondary/tertiary syphilis, a bacterial skin infection resulting in crusted lesions (left) by Dr. Carl Henning in Vienna in 1904 (see Rajan et al. 2011). Left to right, this moulage shows the stages of healing of a large ulceration with a rupia crust (gumma). "Tertiary syphilis can develop skin ulcers or gumma. This may be one that went through the stages to healing, but doesn't mean that he or she was cured" (D.W. Johnson, personal communication, November 21, 2022). See Schlossberg et al. (1993) for more on Syphilitic Rupia.

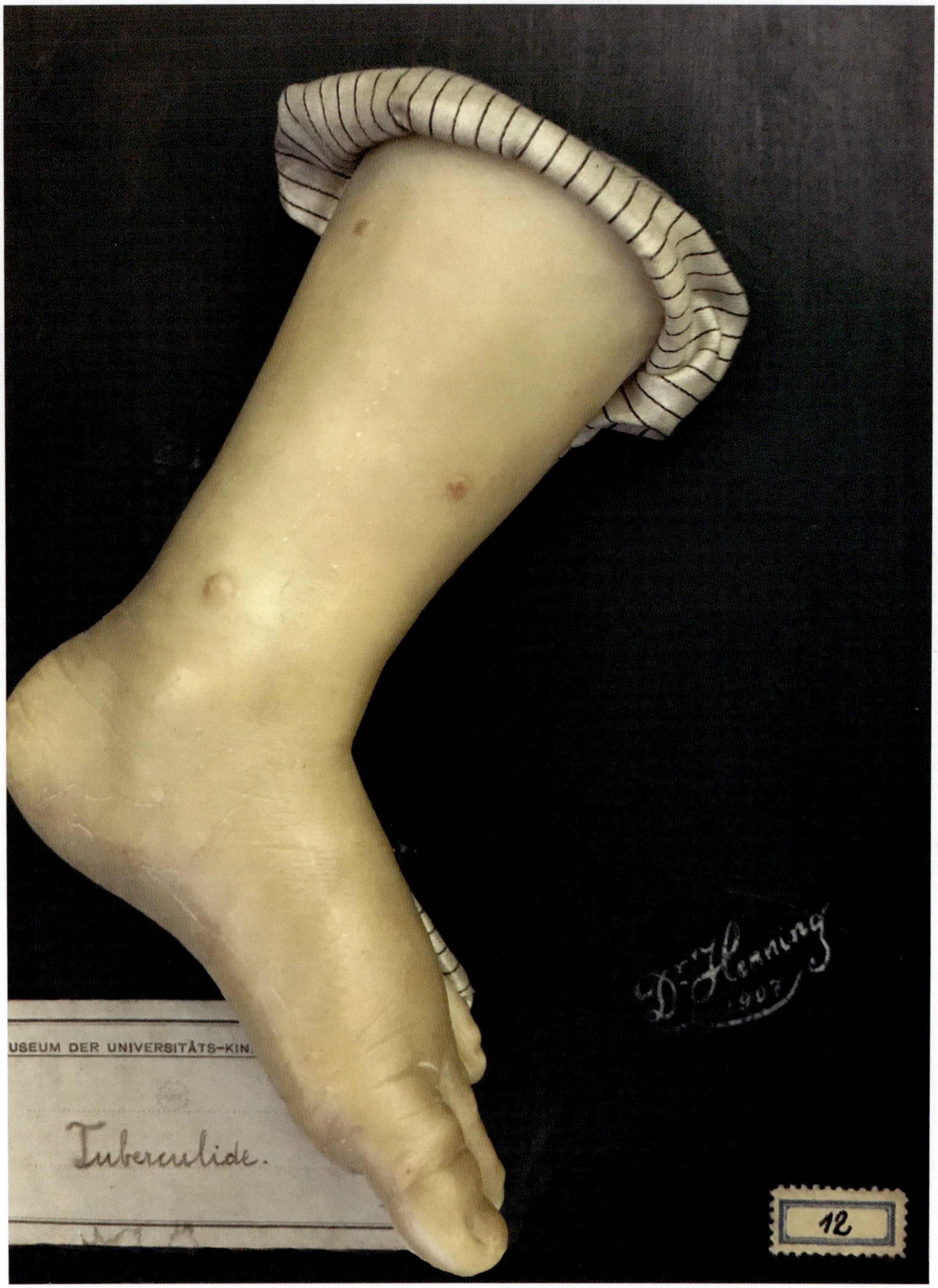


**Figure 459.** Wax moulage of Tuberculids ("Tuberculide"), in this individual of the left leg that are hypersensitivity reactions to mycobacterium tuberculosis, by Dr. Carl Henning in Vienna in 1907 (MN12).

## REFERENCES


Abegg C, Dutour O, Desideri J, Besse M. Cases of serpens endocrania symmetrica in young individuals from Neolithic Western Switzerland: Description and interpretation. *International Journal of Osteoarchaeology* 30(3):401-409, 2020.

Abdel-Aziz M. Congenital aural atresia. *Journal of Craniofacial Surgery* 24(4):e418-22, 2013.

Ahmad A-K, Peter G, Winston C, Rajiv N, Ponusamy S, et al. Charcot, the Man, the Joint and its Care. *Archives of Neurology and Neuroscience* 7(2):2020. ANN.MS.ID.00065

Ahmad F, Shah B, Nabi G, Sofi F. Looser's Zone. *Oman Medical Journal* 25(2):141, 2010. doi: 10.5001/omj.2010.38. PMID: 22125717; PMCID: PMC3215492.

Akanmode AM, Mahdy H. Macrosomia. [Updated 2022 Sep 6]. In: StatPearls [Internet]. Treasure Island (FL): StatPearls Publishing; 2022 Jan-. Available from: https://www.ncbi.nlm.nih.gov/books/NBK557577/

Alhazmi A, Vargas E, Palomo JM, Hans M, Latimer B, Simpson S. Timing and rate of spheno-occipital synchondrosis closure and its relationship to puberty. PLoS ONE 2017; 12(8):e0183305. https://doi.org/10.1371/journal

Allan E, Windson J, Stone C. Cleft lip and palate: Etiology, epidemiology, prevention and intervention strategies. *Anatomy and Physiology* (4):1–6, 2014.

Al-Mouazzen L, Rajakulendran K, Ahad N. Fibrous dysplasia, shepherd's crook deformity and an intracapsular femoral neck fracture. *Strategies Trauma Limb Reconstruction* 8(3):187-191, 2013. doi: 10.1007/s11751-013-0174-7. Epub 2013 Sep 14. PMID: 24036944; PMCID: PMC3800513.

Alwahab A, Kharsa A, Nugud A, Nugud S. Occipital Meningoencephalocele case report and review of current literature. *Chinese Neurosurgical Journal* 3:40-43, 2017. https://doi.org/10.1186/s41016-017-0104-5

Amadasi A, Franceschetti L, Magli F, Cappella A, Muccino EN, Bisogni K, Mazzarelli D, Cattaneo C. Two peculiar cases of cranial fractures running through craniotomy burr holes: may this be a kind of "exception" to the Puppe rule. *International Journal of Legal Medicine* **136**:1177-1180, 2022. https://doi.org/10.1007/s00414-022-02804-2

Amorosa JMH, Cleary-Goldman J, D`Alton ME. Physiologic Effects of Multiple Pregnancy on Mother and Fetus. In RA Polin, SH Abman, D Rowitch, WE Benitz, and WW Fox (Editors), *Fetal and Neonatal Physiology*, Fifth Edition, Volume 1, pp. 167-176. Elsevier Inc., 2017.

Aouad D, Sakhat G, Dahdouh R, Daher M, El Rassi G. Concomitant ipsilateral femoral neck and pelvic ring fractures in the setting of a low energy fall in an elderly patient: A case report. *International Journal of Surgery* Case Report 99:107661, 2022. doi: 10.1016/j.ijscr.2022.107661. Epub 2022 Sep 14. PMID: 36116309; PMCID: PMC9568747.

Armelagos GJ, Zuckerman MK, Harper KN. The science behind pre-Columbian evidence of syphilis in Europe: research by documentary. *Evolutionary Anthropology* 21(2):50-57, 2012. doi: 10.1002/evan.20340. PMID: 22499439; PMCID: PMC3413456.

Asenjo A, Uiberall E. Arterial angioma. *Journal of Neurosurgery* 3(3):259-264, 1946.

Assaf AT, Solaty M, Zrnc TA, Fuhrmann AW, Scheuer H, Heiland M, Friedrich RE. Prevalence of Stafne's Bone Cavity – Retrospective Analysis of 14,005 Panoramic Views. *In Vivo* 28(6):1159-1164, 2014.

Aufderheide AC. *The Cambridge Encyclopedia of Human Paleopathology.* Cambridge University Press; Reissue Edition, 2011.

Aufderheide AC, Rodriguez-Martin C. *The Cambridge Encyclopedia of Human Paleopathology.* Cambridge University Press, New York, 1998.

Bansal R, Jain A, Mittal S. Orofacial tuberculosis: Clinical manifestations, diagnosis and management. *Journal of Family Medicine and Primary Care* 4:335-41, 2015.

Barbian LT, Sledzik PS. Healing Following Cranial Trauma. *Journal of Forensic Sciences* 53(2):263-268, 2008. doi: 10.1111/j.1556-4029.2007.00651.x

Barnes E. *Developmental defects of the axial skeleton in paleopathology.* University of Oklahoma Press, Niwot, 1994.

Barnes E. *Atlas of Developmental Field Anomalies of the Human Skeleton: A Paleopathology Perspective.* Wiley-Blackwell, Hoboken, 2012.

Western AG, Bekvalac JJ. Hyperostosis frontalis interna in female historic skeletal populations: Age, sex hormones and the impact of industrialization. *American Journal of Physical Anthropology* 162(3):501-515, 2017. doi: 10.1002/ajpa.23133. Epub 2016 Nov 30. PMID: 27901271

Bellovits O. Genetical and epidemiological studies of polydactyly in Hungary. *Anthropologischer Anzeiger Jahrg* 61, H. 4, December, pp. 413-419, 2003. E. Schweizerbart'sche Verlagsbuchhandlung.

Berry RJA. A case of os parietale bipartitum in an Australian aboriginal skull. *Journal of Anatomy and Physiology* 44:73-82, 1910.

Bessell-Browne RJ, Thonell S. Bipartite parietal bone: A rare cause of plagiocephaly. *Journal of Medical Imaging and Radiation Oncology* 48(2):248-25, 2004.

Betsinger TK, Smith MO. A singular case of advanced caries sicca in a pre-Columbian skull from East Tennessee. *International Journal of Paleopathology* 24:245-251, 2019. https://doi.org/10.1016/j.ijpp.2019.01.003

Biehler-Gomez L, Tritella S, Martino F, Campobasso CP, Franchi A, Spairani R, Sardanelli F, Cattaneo C. The synergy between radiographic and macroscopic observation of skeletal lesions on dry bone. *International Journal of Legal Medicine* 133(5):1611-1628, 2019. doi: 10.1007/s00414-019-02122-0. Epub 2019 Jul 12. PMID: 31300917.

Bitelman VM, Lopes JA, Nogueira AB, Frassetto FP, Duarte-Neto AN. "Punched out" multiple myeloma lytic lesions in the skull. *Autopsy Case Report* 6(1):7-9, 2016. doi: 10.4322/acr.2016.020. PMID: 27284535; PMCID: PMC4880428.

Bloom RA, Libson E, Stoker DJ. The periosteal sunburst reaction to bone metastases: A literature review and report of 20 additional cases. *Skeletal Radiology* 16:629-634, 1987.

Bodian M. Oxycephaly. *Journal of the American Medical Association* 143(1):15–18, 1950. doi:10.1001/jama.1950.02910360017006

Bogdanović M, Radnić B, Savić S, Popović V, Durmić T. Copper-Beaten Skull Appearance as a Response of Chronically Increased Intracranial Pressure. *American Journal of Forensic Medicine and Pathology* 40(1):58-60, 2019.

Bolivar I, Whiteson K, Stadelmann B, Baratti-Mayer D, Gizard Y, Mombelli A, Pittet D, Schrenzel J; Geneva Study Group on Noma (GESNOMA). Bacterial diversity in oral samples of children in Niger with acute noma, acute necrotizing gingivitis, and healthy controls. PLoS *Neglected Tropical Diseases* 6(3):e1556, 2012. doi: 10.1371/journal.pntd.0001556. Epub 2012 Mar 6. PMID: 22413030; PMCID: PMC3295795.

Britannica, The Editors of Encyclopaedia. "Chang and Eng". *Encyclopedia Britannica*, 9 May. 2022, https://www.britannica.com/biography/Chang-and-Eng. Accessed November 21, 2022.

Brothwell DR and Sandison AT. *Diseases in Antiquity: A survey of the Diseases, Injuries and surgery of Early Populations.* Charles C. Thomas, Springfield, 1967.

Buchman J, Lieberman HS. Prevalence of syphilis of the bones and joints: Report of a statistical study at The Hospital for Joint Diseases, with a review of the literature. *Archives of Dermatology and Syphilology* 44(1):1-12, 1941.

Buikstra JE, Cook DC. Pre-Columbian Tuberculosis in West-Central Illinois: Prehistoric Disease in Biocultural Perspective, pp. 115-139. In: *Prehistoric Tuberculosis in the Americas* by JE Buikstra (Editor), Northwestern University Archeological Program, Scientific Papers, Number 5, Evanston, 1981.

Buzic I, Giuffra V. The paleopathological evidence on the origins of human tuberculosis: a review. *Journal of Preventive Medicine and Hygiene* 30;61(1 Suppl 1):E3-E8, 2020. doi: 10.15167/2421-4248/jpmh2020.61.1s1.1379. PMID: 32529097; PMCID: PMC7263064.

*Campbell's Operative Orthopaedics.* JS Speed and RA Knight (Editors), Volumes I and II, Third Edition. The C.V. Mosby Company, St. Louis, 1956.

Cavalli G, Praderio L. Charcot's Arthropathy of the Hip. *The Journal of Rheumatology* 40(10):1770, 2013. doi: 10.3899/jrheum.130266

Chalmers BP, Tibbo ME, Trousdale RT, Lewallen DG, Berry DJ, Abdel MP. Primary Total Hip Arthroplasty for Charcot Arthropathy is Associated With High Complications but Improved Clinical Outcomes. *Journal of Arthroplasty* 33(9):2912-2918, 2018. doi: 10.1016/j.arth.2018.04.002. Epub 2018 Apr 10. PMID: 29706256.

Chalmers J, Conacher WDH, Gardner DL, Scott PJ. Osteomalacia-A Common Disease in Elderly Women. *The Journal of Bone and Joint Surgery* (Br) 49(3):403-423, 1967.

Chae SY, Sim HB, Kim MJ, Jang YH, Lee SJ, Kim DW, Lee WJ. Button Osteoma: A Review of Ten Cases. *Annals of Dermatology* 27(4):394-397, 2015. doi: 10.5021/ad.2015.27.4.394. Epub 2015 Jul 29. PMID: 26273154; PMCID: PMC4530148.

Chan CYW, Naing KS, Chiu CK, Mohamad SM, Kwa MK. Pelvic obliquity in adolescent idiopathic scoliosis planned for posterior spinal fusion: A preoperative analysis of 311 lower limb axis films. *Journal of Orthopaedic Surgery* 27(2):1-10, 2019.

Cinalli G, Sainte-Rose C, Kollar EM, Zerah M, Brunelle F, Chumas P, Arnaud E, Marchac D, Pierre-Kahn A, Renier D. Hydrocephalus and craniosynostosis. *Journal of Neurosurgery* 88(2):209-214, 1998. doi: 10.3171/jns.1998.88.2.0209. PMID: 9452225.

Charcot JM. Sur quelques arthropathies qui parnessent dépendre a'une lésiondu cerveauou de la moëlle épinière. *Archives Physicle Normal Pathologie* 1:161-78, 1868.

Chaturvedi A, Klionsky NB, Nadarajah U, Chaturvedi A, Meyers SP. Malformed vertebrae: a clinical and imaging review. *Insights into Imaging* 9:343–355, 2018.

Christopher F. *A Textbook of Surgery by American Authors.* WB Saunders Company, Philadelphia, 1936.

Collmann H, Sorensen N, Krauss J. Hydrocephalus in craniosynostosis: a review. *Childs Nervous System* 21(10):902-912, 2005. doi: 10.1007/s00381-004-1116-y. Epub 2005 Apr 27. PMID: 15864600.

Cooper JM, Sanchez PJ. Congenital syphilis. *Seminars in Perinatology* 42(3):176-184, 2018.

Copp AJ, Adzick NS, Chitty LS, Fletcher JM, Holmbeck GN, Shaw GM. Spina bifida. *Nature Reviews Disease Primers* 30(1):15007, 2015. doi: 10.1038/nrdp.2015.7. PMID: 27189655; PMCID: PMC4898641.

Croft D, Sumayya, Almarzougi SJ, Morgan ML, Lee AG. Setting Sun Sign in: Schmidt-Erhurth U, Kohnen T (eds), *Encyclopedia of Opthalmology*, June, pp. 1-2, 2016.

Czajka CM, DiCaprio MR. What is the Proportion of Patients With Multiple Hereditary Exostoses Who Undergo Malignant Degeneration? *Clinical Orthopaedics and Related Research* 473(7):2355-2361, 2015. doi: 10.1007/s11999-015-4134-z. Epub 2015 Jan 13. PMID: 25582066; PMCID: PMC4457763.

D'Armiento M, Falleti J, Maruotti GM, Martinelli P. Diprosopus conjoined twins: radiologic, autoptic, and histologic study of a case. *Fetal Pediatric Pathology* 29(6):431-438, 2010. doi: 10.3109/15513815.2010.505628

Dass B, Puet T, Watanakunakorn C. Tuberculosis of the spine (Pott's disease) presenting as 'compression fractures'. *Spinal Cord* 40:604–608, 2002. https://doi.org/10.1038/sj.sc.3101365

De Lima JAP. Absence of the Auditory Canal and Other Abnormalities of the External Ear. *Journal of Anatomy and Physiology* 47(1):1-7, 1912.

De Paepe ME. Multiple Gestation: The Biology of Twinning. *Creasy and Resnik's Maternal-Fetal Medicine: Principles and Practice*, Eighth Edition, 5:68-80.e2, 2019. Elsevier, Cambridge.

Derry DE. Parietal perforation accompanied with flattening of the skull in an ancient Egyptian. *Journal of Anatomy and Physiology* 48(4):417-429, 1914.

Desai V, Priyadarshini SR, Sharma R. Copper Beaten Skull! Can It be a Usual Appearance? *International Journal of Clinical Pediatric Dentistry* 7(1):47-49, 2014. doi: 10.5005/jp-journals-10005-1233. Epub 2014 Apr 26. PMID: 25206238; PMCID: PMC4144051.

Desimpel J, Posadzy M and Vanhoenacker FM. (2017). The Many Faces of Osteomyelitis: A Pictorial Review. *Journal of the Belgian Society of Radiology* 101(1):1-10, 2017. doi: http://doi.org/10.5334/jbr-btr.1300

Dhillon MS, Nagi ON. Tuberculosis of the foot and ankle. *Clinical Orthopaedics and Related Research* 398:107-113, 2002. doi: 10.1097/00003086-200205000-00015. PMID: 11964638.

Dhirawani RB, Gupta R, Pathak S, Lalwani G. Frontoethmoidal encephalocele: Case report and review on management. *Annals of Maxillofacial Surgery* 4(2):195-197, 2014. doi: 10.4103/2231-0746.147140. PMID: 25593873; PMCID: PMC4293844.

Dias MS, SamsonT. Rizk EB, Governale LS, Richtsmeier JT. Identifying the Misshapen Head: Craniosynostosis and Related Disorders. *Pediatrics* 146(3):1-20, 2020.

DiMaio VJ, DiMaio D. *Forensic Pathology*, Second Edition. CRC Press, Boca Raton, 2001.

Drago L, Vecchi ED, Pasqualini M, Moneghini L, Bruno M. Rapid, progressive neuropathic arthropathy of the hip in a patient co-infected with human immunodeficiency virus, hepatitis C virus and tertiary syphilis: case report. *BMD Infectious Diseases* 11:159-162, 2011.

Dumitriu DI, Menten R, Clapuyt P. Pitfalls in the diagnosis of common benign bone tumours in children. *Insights Imaging* 5(6):645-655, 2014. doi: 10.1007/s13244-014-0356-y. Epub 2014 Sep 26. PMID: 25256565; PMCID: PMC4263798.

Durward A. A note on symmetrical thinning of the parietal bones. *Journal of Anatomy and Physiology* 63(3):356-362, 1929.

Edmonds LD, Layde PM. Conjoined twins in the United States, 1970–1977. *Teratology* 25:301-308, 1982.

Ekin EE, Yildiz HK, Altunrende ME. Vertebral Body Erosion in Geriatric Patient: Case Reports and Systematic Review of Literature. *Journal of Geriatric Medicine and Gerontology* 4:056, 2018. doi.org/10.23937/2469-5858/1510056. Accessed on November 20, 2022.

El-Mowafi DM. Contracted Pelvis: Obstetrics Simplified. *Geneva Foundation for Medical Education and Research*, Edited by Aldo Campana, September 27, 2017. https://www.gfmer.ch/Obstetrics_simplified/contracted_pelvis.htm

Eshed V, Latimer B, Greenwald CM, Jellema LM, Rothschild BM, Wish-Baratz S, Hershkovitz I. Button osteoma: Its etiology and pathophysiology. *American Journal of Biological Anthropology* 118(3):217-230, 2002.

Evans HL, Raymond AK, Ayala AG. Vascular tumors of bone: A study of 17 cases other than ordinary hemangioma, with an evaluation of the relationship of hemangioendothelioma of bone to epithelioid hemangioma, epithelioid hemangioendothelioma, and high-grade angiosarcoma. *Human Pathology* 34:680-689, 2003.

Figueiredo A, Ferreira R, Alegre C, Fonseca F. Charcot osteoarthropathy of the knee secondary to neurosyphilis: a rare condition managed by a challenging arthrodesis. *British Medical Journal Case Report* Aug 20;2018:bcr2018225337. doi: 10.1136/bcr-2018-225337. PMID: 30131415; PMCID: PMC6109733.

Fitzgerald DW, Sterling TR, Haas DW. Mycobacterium tuberculosis pp. 3129-3164. In: *Principles and Practice of Infectious Diseases*, Seventh Edition. Churchill Livingstone Elsevier, Philadelphia, 2010.

Foti PV, Attina G, Spadola S, Caltabiano R, Farina R, Palmucci S, Zarbo G, Zarbo R, D'Arrigo M, Milone P, Ettorre GC. MR imaging of ovarian masses: classification and differential diagnosis. *Insights Imaging* 7:21–41, 2016. https://doi.org/10.1007/s13244-015-0455-4

Foreman A. Why Footbinding Persisted in China for a Millennium. *Smithsonian Magazine*, February 2015.

Frantz CH, O'Rahilly R. Congenital skeletal limb deficiencies. *Journal of Bone and Joint Surgery* (American) 43A:1202-1204, 1961.

Freedman E. Leontiasis Ossea. *Radiology* 20(1):8-13, 1933.

Gaballah AH, Jensen CT, Palmquist S, Pickhardt PJ, Duran A, Broering G, Elsayes KM. Angiosarcoma: clinical and imaging features from head to toe. *British Journal of Radiology.* 2017 Jul;90(1075):20170039. doi: 10.1259/bjr.20170039. Epub 2017 May 4. PMID: 28471264; PMCID: PMC5594986.

Gaetano HM, Gaetano JS. Persistent open anterior fontanelle in a healthy 32-month-old boy. *Journal of American Osteopathic Association* 102(9):500-501, 2002.

Gaillard F, Worsley C. Hypertrophic osteoarthropathy. Reference article, *Radiopaedia.org.*, 2008 (accessed August 3, 2022). https://doi.org/10.53347/rID-147. doi: https://doi.org/10.53347/rID-1479

Gaebler-Spira D, Lipschutz R. Pediatric limb deficiencies. In: Alexander M, Matthews D (Editors). *Pediatric Rehabilitation*, Fourth Edition. New York, NY: Demos Medical; 2010.

Galhotra V, Ahluwalia P, Jodhka S, Sethi S, Gambhir RS. Effect of nutritional rickets on dental development in North Indian children: A prospective study. *Journal of Pediatric Dentistry* 3(3):88-91, 2015.

Galvis AE, Arrieta A. Congenital Syphilis: A U.S. Perspective. *Children* (Basel) Oct 29;7(11):203,2020. doi: 10.3390/children7110203. PMID: 33137962; PMCID: PMC7692780.

Garg RK, Somvanshi DS. Spinal tuberculosis: a review. *Journal of Spinal Cord Medicine* 34(5):440-454, 2011. doi: 10.1179/2045772311Y.0000000023. PMID: 22118251; PMCID: PMC3184481.

Gask GE. Discussion on fibrocystic disease in bone. Section on Orthopaedics and Section on Surgery. *Royal College of Surgeons, Joint Discussion* Number 4-1, pp. 973-983, 1934.

Geschickter CF, Copeland MM. *Tumors of Bone (Including the Jaws and Joints)*, Revised Edition. The American Journal of Cancer, New York City, 1936.

Giller DB, Giller BD, Giller GV, Scherbakova GV, Koroev VV, Kesaev OS, Imagozhev YG, Glotov AA. Video-assisted thoracoplasty in pulmonary tuberculosis: a retrospective study. *Journal of Thoracic Disease* 12(3):980-988, 2020. doi: 10.21037/jtd.2019.11.67. PMID: 32274167; PMCID: PMC7139062.

Gomez-Leon N, Pacheco-Barcia V, Ballesteros AI, Fraga J, Colomer R, Friera A. Skeletal muscle and solitary bone metastases from malignant melanoma: multimodality imaging and oncological outcome. *Melanoma Research* 28(6):562-570, 2018. doi: 10.1097/CMR.0000000000000466. PMID: 29975212; PMCID: PMC6221392.

Gouveri E, Papanas N. Charcot osteoarthropathy in diabetes: A brief review with an emphasis on clinical practice. *World Journal of Diabetes* May 15;2(5):59-65, 2011. doi: 10.4239/wjd.v2.i5.59. PMID: 21691556; PMCID: PMC3116009.

Griffiths H. *Basic Bone Radiology.* Appleton-Century-Crofts, Norwalk, 1981.

Ha KY, Deleon P, Deleon W. Invasive mucinous carcinoma of the breast. *Proceedings (Baylor University Medical Center)* 26(3):295-7, 2013. doi: 10.1080/08998280.2013.11928989. PMID: 23814397; PMCID: PMC3684304.

Hackett CJ. On the origin of the human treponematoses (Pinta, Yaws, Endemic Syphilis and Venereal Syphilis. *Bulletin of the World Health Organization* 29(1):7-41, 1963. PMID: 14043755; PMCID: PMC2554777.

Hackett CJ. Findings. In: Diagnostic Criteria of Syphilis, Yaws and Treponarid (Treponematoses) and of Some Other Diseases in Dry Bones. *Sitzungsberichte der Heidelberger Akademie der Wissenschaften Mathematisch-naturwissenschaftliche Klasse,* Volume 1976/4. Springer, Berlin, Heidelberg. https://doi.org/10.1007/978-3-662-06583-9_3

Hagglund G, Goldring M, Hermanson M, Rodby-Bousquet E. Pelvic obliquity and measurement of hip displacement in children with cerebral palsy. *Acta Orthopaedica* 89(6):652-655, 2018. doi: 10.1080/17453674.2018.1519104

Hameed A, Brady JJ, Dowling P, Clynes M, O'Gorman P. Bone disease in multiple myeloma: pathophysiology and management. *Cancer Growth Metastasis* 7:33-42, 2014. doi: 10.4137/CGM.S16817. PMID: 25187738; PMCID: PMC4133035.

Hanzlik E, Gigante J. Microcephaly. *Children (Basel)* 9;4(6):47, 2017. doi: 10.3390/children4060047. PMID: 28598357; PMCID: PMC5483622.

Harinder G, Salim P, Naveed A, Ajay S. Mutual exclusivity of neck of femur and pubic ramus fractures in the elderly. Is it a fact? *Current Orthopaedic Practice* 24(6):651-652, 2013. doi: 10.1097/BCO.0b013e31829fbff4

Hashmi JA, Al-Harbi KM, Ramzan K, Albalawi AM, Mehmood A, Samman MI, Basit SA. Novel splice-site mutation in the ASPM gene underlies autosomal recessive microcephaly. *Annals of Saudi Medicine* 36:391-396, 2016.

Hassin GB, Lukas C, Brown RO. Roentgenographic Bone Changes in a Case of Poliomyelitis. *Journal of the American Medical Association* LXV(17):1459-1460, 1915. doi: 10.1001/jama.1915.02580170047012

Hawes, MC, O'Brien JP. The transformation of spinal curvature into spinal deformity: pathological processes and implications for treatment. *Scoliosis* 1, 3:1-9, 2006. https://doi.org/10.1186/1748-7161-1-3

Hepburn D. Anomalies in the supra-inial portion of the occipital bone resulting from irregularities of its ossification, with consequent variations of the interparietal bone. *Journal of Anatomy and Physiology* 42(3):88-92, 1908.

Hershkovitz I, Greenwald C, Rothschild BM, Latimer B, Dutour O, Jellema LM, Wish-Baratz S. Hyperostosis frontalis interna: an anthropological perspective. *American Journal of Physical Anthropology* 109(3):303-325, 1999. doi: 10.1002/(SICI)1096-8644(199907)109:3<303::AID-AJPA3>3.0.CO;2-I. PMID: 10407462.

Hershkovitz I, Greenwald CM, Latimer B, Jellema LM, Wish-Baratz S, Eshed V, Dutour O, Rothschild BM. Serpens endocrania symmetrica (SES): a new term and a possible clue for identifying intrathoracic disease in skeletal populations. *American Journal of Physical Anthropology* 118(3):201-16, 2002.

Hooton EA. *The Indians of Pecos Pueblo: A Study of Their Skeletal Remains.* Yale University Press, New Haven, 1930.

Huang I, Leach JL, Fichtenbaum CJ, Narayan RK. Osteomyelitis of the Skull in Early-Acquired Syphilis: Evaluation by MR Imaging and CT. *American Journal of Neuroradiology* 28(2):307-308, 2007.

Hutchinson J. Clinical Lecture on Heredito-Syphilitic Struma: And on the Teeth as a Means of Diagnosis. *British Medical Journal* 1(20):515-517, 1861. doi: 10.1136/bmj.1.20.515. PMID: 20743880; PMCID: PMC2287664.

Ioannou S, Sassani S, Henneberg M, Henneberg RJ. Diagnosing Congenital Syphilis Using Hutchinson's Method: Differentiating Between Syphilitic, Mercurial, and Syphilitic-Mercurial Dental Defects. *American Journal of Physical Anthropology* 159:617–629, 2016.

Isaac S. "Phossy jaw" and the matchgirls: a nineteenth-century industrial disease. *Royal College of Surgeons of England*, 28 September 2018. https://www.rcseng.ac.uk

Ismunandar H, Himayani R, Al Farisi M. Rickets a Literature Review. *Medical Profession Journal of Lampung* 10(4): 644-653, 2021. https://doi.org/10.53089/medula.v10i4.163

Jellinek S. Zur klinischen Diagnose und pathologischen Anatomie des multiplen Myeloms. *Virchows Archiv für pathologische Anatomie und Physiologie und für klinische Medizin.* Band 177, Folge XVII, Band VII, pp, 96-132. Bruck and Verlag von Georg Reimer, Berlin, 1904.

JH Emerson Company. *The Evolution of "Iron Lungs,"* 1958.

Jolly MC, Sebire NJ, Harris JP, Regan L, Robinson S. Risk factors for macrosomia and its clinical consequences: a study of 350,311 pregnancies. *European Journal of Obstetrics Gynecology and Reproductive Biology* 10:111(1):9-14, 2003.

Kaufman MH. The embryology of conjoined twins. *Childs Nervous System* 20:508–25, 2004.

Keles-celik N, Kose O, Sekerci R, Aytac G, Turan A, Guler F. Accessory Ossicles of the Foot and Ankle: Disorders and a Review of the Literature. *Cureus* 9(11):e1881, 2017. doi: 10.7759/cureus.1881

Khamanarong K, Tuamsuk P, Woraputtaporn W, Namking M, Sawatpanich T, Toomsan Y, Iamsaard S. Incidence of metopism in adult Thai skulls. *International Journal of Morphology* 33(1):51-54, 2015.

Khanna PC, Thapa MM, Iyer RS, Prasad SS. Pictorial essay: The many faces of craniosynostosis. *Indian Journal of Radiology and Imaging* 21(1):49-56, 2011.

Kim HJ, Yang JH, Chang DG, Suk SI, Suh SW, Song KS, Park JB, Cho W. Adult Spinal Deformity: Current Concepts and Decision-Making Strategies for Management. *Asian Spine Journal* 14(6):886-897, 2020. doi: 10.31616/asj.2020.0568. Epub 2020 Dec 2. PMID: 33254357; PMCID: PMC7788366.

Kimmerle E, Baraybar. *Skeletal Trauma: Identification of Injuries Resulting From Human Rights Abuse and Armed Conflict.* CRC Press, Boca Raton, 2008.

Kinney LC. Multiple Myeloma. *Radiology* 35(6):467-475, 1940.

Kobylinska M, Malak R, Majewska K, Kędzia A, Samborski W. Assessment of anterior-posterior spinal curvatures in children suffering from hypopituitarism. *BMC Endocine Disorders* 19(1):137, 2019. doi: 10.1186/s12902-019-0462-4. PMID: 31829163; PMCID: PMC6907119. Accessed on November 20, 2022.

Koneru A, Hunasgi S, Manvikar V, Vanishree M. Nonsyphilitic occurrence of mulberry molars: A rare case report. *Journal of Oral Maxillofacial Pathology* Feb;23(Suppl 1):106-110, 2019. doi: 10.4103/jomfp.JOMFP_74_18. PMID: 30967737; PMCID: PMC6421924.

Koo SK, Jung SH, Moon JS, Son H. A Case of Nasal Swell Body (Septal Turbinate) Combined with Pneumatization of Perpendicular Plate of the Ethmoid Bone. *Journal of Rhinology* 23(1):70-73, 2016.

Kunwar A, Shrestha BM, Shrestha S, Paudyal P, Rawal S. Cyclopia with proboscis: A rare congenital anomaly. *Clinical Case reports* 9(7):e04466, 2021.

Krigsten WM, Phalen GS. Localized fibrocystic disease of the femur: A report of three patients treated by curettage and bone graft. *The American Journal of Surgery* 67(2):412-423, 1945.

Kryger M, Kosial W, Batko T. Bifid rib – usefulness of chest ultrasound. A case report. *Journal of Ultrasound* 13:446-450, 2013.

Lacout A, Jarraya M, Marcy PY, Thariat J, Carlier RY. Myositis ossificans imaging: keys to successful diagnosis. *Indian Journal of Radiological Imaging* 22(1):35–39, 2012. doi:10.4103/0971-3026.95402

LaFond RE, Lukehart SA. Biological basis for syphilis. *Clinical Microbiology Review* 19(1):29-49, 2006. doi:10.1128/CMR.19.1.29-49.2006

Layoun W, Testelin S, Devauchelle B. Cavernous hemangioma of the nasal bones. *Reviews Stomatologie Chirurgie Maxillo Facial* 104(4):235-238, 2003.

Lee YJ, Sadigh S, Mankad K, Kapse N, Rajeswaran G. The imaging of osteomyelitis. *Quantitative Imaging in Medicine and Surgery* 6(2):184-198, 2016. doi: 10.21037/qims.2016.04.01. PMID: 27190771; PMCID: PMC4858469.

Lei L, Wang R, Han D. Pneumatization of perpendicular plate of the ethmoid bone and nasal septal mucocele. *Acta Oto-Laryngologica* 124:2, 221-222, 2004. doi: 10.1080/00016480310015849

Le Gros Clark WE. On the Pacchionian bodies. *Journal of Anatomy and Physiology* 55(1):40-48, 1920.

Lehrer HZ, Maxfield WS, Nice CM. The Periosteal "Sunburst" Pattern in Metastatic Bone Tumors. *American Journal of Roentgenology* 108(1):154-161, 1970.

Lello GE, Makek M. Stafne's mandibular lingual cortical defect. *Journal of Maxillofacial Surgery* 13:172-176, 1985.

Lenchik L, Rogers LF, Delmas PD, Genant HK. Diagnosis of Osteoporotic Vertebral Fractures: Importance of Recognition and Description by Radiologists. *American Journal of Roentgenology* 183(4):949-958, 2004.

Levin EJ. Healing in congenital osseous syphilis. *American Journal of Roentgenology Radium Therapy and Nuclear Medicine* 110(3):591-597, 1970.

Lewis ME. Endocranial Lesions in Non-adult Skeletons: Understanding their Aetiology. *International Journal of Osteoarchaeology* 14:82-97, 2004.

Liang X, Liu T, Yuan C, Wang W, Liang P. The disappearance of femoral head and neck resulting from extensive bone defect caused by secondary syphilis: a case report and literature review. *BMC Musculoskeletal Disorders* 19:251-260, 2018. https://doi.org/10.1186/s12891-018-2152-1

Limarzi LR. Diagnostic and Therapeutic Aspects of Multiple Myeloma. *Medical Clinics of North America* 35(1):189-226, 1951.

Low CM, Stokken JK. Typical Orbital Pathologies: Hemangioma. *Journal of Neurological Surgery B Skull Base* 82(1):20-26, 2021. doi: 10.1055/s-0040-1722633. Epub 2021 Jan 19. PMID: 33777617; PMCID: PMC7987401.

Lungu SG. Myositis Ossificans – Two Case Presentations. *Medical Journal of Zambia* 38(2):25-31, 2011.

Luschka H. Die anomalen Articulationen des ersten Rippenpaares. Wiener Akad. Sitzungsberichte. *Math.-Natur.* Kl. 39:413-430, 1860.

Macé G, Castaigne V, Trabbia A, Guigue V, Cynober E, Cortey A, Lalande V, Carbonne B. Fetal anemia as a signal of congenital syphilis. *Journal of Maternal and Fetal Neonatal Medicine.* 2014 Sep;27(13):1375-1377. doi: 10.3109/14767058.2013.853288. Epub 2013 Nov 7. PMID: 24134545.

Madea B. *Handbook of Forensic Medicine.* John Wiley & Sons, West Sussex, 2014.

Madea B, Staak M. Determination of the Sequence of Gunshot Wounds of the Skull. *Journal of the Forensic Science Society* 28(5-6):321-328, 1988.

Maley A, Desai M, Parker S. Noma: A disease of poverty presenting at an urban hospital in the United States. *JAAD Case Report* 1(1):18-20, 2014. doi: 10.1016/j.jdcr.2014.10.001. PMID: 27075129; PMCID: PMC4802558.

Man SC, Schnell CN, Fufezan O, Mihut G. Myositis ossificans traumatica of the neck - a pediatric case. *Maedica (Bucur)* 6(2):128-131, 2011. PMID: 22205895; PMCID: PMC3239391

Manchester K. *The Archaeology of Disease.* University of Bradford, Bradford, England, 1983.

Mann RW. Enlarged Parietal Foramina and Craniosynostosis in an American Indian Child. *American Journal of Roentgenology* 154, MAR, 1990.

Mann RW. *Stafne's Defects of the Human Mandible.* Doctoral dissertation, Department of Anthropology, University of Hawaii, Honolulu, 2001.

Mann RW, Hunt DR. Non-metric Traits and Anatomical Variants that Can Mimic Trauma in the Human Skeleton. Forensic Science International, Special Issue (SI) on Trauma. *Forensic Science International* 301:202-224, 2019.

Mann RW, Hunt DR. *Photographic Regional Atlas of Bone Disease: A Guide to Pathologic and Normal Variation in the Human Skeleton*, Third Edition. Charles C. Thomas, Springfield, 2005.

Mann RW, Hunt DR, Lozanoff S. *Photographic Regional Atlas of Non-metric Traits and Anatomical Variants in the Human Skeleton.* Charles C. Thomas Publisher, Springfield, 2016.

Mann RW, Sledzik PS, Owsley DW, Droulette MR. Radiographic Examination of Chinese Foot Binding. *Journal of the American Podiatric Medical Association* 80(8):405-409, 1990.

Mann RW, Kobayashi M and Schiller AI. Biparietal Thinning: Accidental Death by a Fall from Standing Height. *Journal of Forensic Sciences* 62(5):1406-1409, 2017.

Mann RW, Tuamsuk P. Case Report: Differential Diagnosis of Severe Diffuse Skeletal Pathology in an Adult Thai Male, pp. 213-221. In: Mann RW and Hunt DR Hunt, *Photographic Regional Atlas of Bone Disease: A Guide to Pathologic and Normal Variation in the Human Skeleton*, Third Edition. Charles C. Thomas Publisher, Springfield, 2005.

Mansilla J, Amezcua-Guerra L, Vargas-Rodríguez A, Fernández-Tapia S, Leboreiro I, Resnick D, Martinez-Lavin M, Villanueva M, Gaytan E, Pineda C. Leontiasis ossea – A paleopathologic case report. *Journal of Clinical Rheumatology: Practical Reports on Rheumatic and Musculoskeletal Diseases* 13:269-272, 2007. doi: 10.1097/RHU.0b013e318156dd1a

Mantilla-Rivas E, Tu L, Goldrich A, Manrique M, Porras AR, Keating RF, Oh AK, Linguraru MG, Rogers GF. Occult Scaphocephaly: A Forme Fruste Phenotype of Sagittal Craniosynostosis. *Journal of Craniofacial Surgery* 31(5):1270-1273, 2020. doi: 10.1097/SCS.0000000000006440. PMID: 32282689; PMCID: PMC7343588.

Manske MC, Kennedy CD, Huang JI. Classifications in Brief: The Wassel Classification for Radial Polydactyly. (2017) *Clinical Orthopaedics and Related Research* 475:6:1740-1746. doi:10.1007/s11999-016-5068-9 - Pubmed

Mariyam II, Fenton TW, Deland T, Haut RC. Assessing Impact Direction in 3-point Bending of Human Femora: Incomplete Butterfly Fractures and Fracture Surfaces. *Forensic Sciences International* 63(1):38-46, 2018.

Markovic I, Bosnjakovic P, Milenkovic Z. Occipital Encephalocele: Cause, Incidence, Neuroimaging and Surgical Management. *Current Pediatric Reviews* 16(3):200-205,2020. doi: 10.2174/157339631566619101816 1535. PMID: 31656152; PMCID: PMC8193807.

Marx RE. Uncovering the cause of "phossy jaw" Circa 1858 to 1906: oral and maxillofacial surgery closed case files-case closed. *Journal of Oral Maxillofacial Surgery* 66(11):2356-2363, 2008. doi: 10.1016/j.joms.2007.11.006. PMID: 18940506.

Mavrogiannis LA, Wilkie AOM. Enlarged Parietal Foramina. 2004 Mar 30 [Updated 2019 Nov 27]. In: Adam MP, Mirzaa GM, Pagon RA, et al., editors. *GeneReviews®* [Internet]. Seattle (WA): University of Washington, Seattle; 1993-2022. Available from: https://www.ncbi.nlm.nih.gov/books/NBK1128/

McGahan JP, Nyberg DA, Mack LA. Sonography of Facial Features of Alobar and Semilobar Holoprosencephaly. *American Journal of Roentgenology* 154:143-148, 1990.

Mead LB, Millhouse PW, Krystal J, Vaccaro AR. C1 fractures: a review of diagnoses, management options, and outcomes. *Current Reviews in Musculoskeletal Medicine* 9(3):255-262, 2016. doi: 10.1007/s12178-016-9356-5. PMID: 27357228; PMCID: PMC4958388.

*Merriam-Webster.com.* 2011. https://www.merriam-webster.com (8 May 2011). Accessed June 26, 2022.

Meyers C, Lisiecki J, Miller S, Levin A, Fayad L, Ding C, Sono T, McCarthy E, Levi B, James AW. Heterotopic Ossification: A Comprehensive Review. *Journal for Bone and Mineral Research Plus* 3(4):1-19, 2019. e10172. doi: 10.1002/jbm4.10172

Migliorini F, Maffulli N, Trivellas A, Eschweiler J, Tingart M, Driessen A. Bone metastases: a comprehensive review of the literature. *Molecular Biology Report* 47(8):6337-6345, 2020. doi: 10.1007/s11033-020-05684-0. Epub 2020 Aug 4. PMID: 32749632.

Minisola S, Colangelo L, Pepe J, Diacinti D, Cipriani C, Rao SD. Osteomalacia and Vitamin D Status: A Clinical Update. *Journal of Bone and Mineral Research* 5(1):1-6, 2021. doi: 10.1002/jbm4.10447

Mirra JM. Angiosarcoma of bone: solitary and multifocal variants, pages 1382-1417. In *Bone Tumours: Clinical, radiologic and pathologic correlations.* Philadelphia, Lea and Febiger, 1989.

Mohamad AR, Koleri J, Hussain HMS, Al Soub H, Al Maslamani M. Recurrent skull vault actinomycosis: A case report and review of literature. *IDCases* 3:25:e01215, 2021. doi: 10.1016/j.idcr.2021.e01215. PMID: 34277352; PMCID: PMC8267539.

Mohammad D, Dhandapani S. Trigonocephaly: A simple modified technique. *Journal of Pediatric Neuroscience* 9(2):125-128, 2014.

Moller-Christensen V. *Bone Changes in Leprosy.* Munksgaard, Copenhagen, 1961.

Moller-Christensen V. Evidence of Leprosy in Earlier Peoples, pp. 295-306. In: DR Brothwell and AT Sandison (Editors), *Diseases in Antiquity.* Charles C. Thomas, Springfield, 1967.

Moller-Christensen V. *Leprosy Changes of the Skull.* Odense University Press, Odense, 1978.

Muckle DS. *Femoral Neck Fractures and Hip Joint Injuries.* John Wiley & Sons, New York, 1977.

Mumoli N. Lobster claw syndrome. *QJM: An International Journal of Medicine* 105(10):1017, 2012. https://doi.org/10.1093/qjmed/hcr176. Accessed November 20, 2022.

Mundlos S. Cleidocranial dysplasia: clinical and molecular genetics. *Journal of Medical Genetics* 36:177-182, 1999.

Munhoz L, Arita ES, Nishimura DA, Watanabe PCA. Maxillofacial manifestations of Proteus syndrome: a systematic review with a case report. *Oral Radiology* 37(1):2-12, 2021. doi: 10.1007/s11282-019-00416-y. Epub 2019 Nov 16. PMID: 31734933.

Munteanu O, Cîrstoiu MM, Filipoiu FM, Neamțu MN, Stavarache I, Georgescu TA, Bratu OG, Iorgulescu G, Bohîlțea RE. The etiopathogenic and morphological spectrum of anencephaly: a comprehensive review of literature. *Romanian Journal of Morphology and Embryology* 61(2):335-343, 2020. doi: 10.47162/RJME.61.2.03. PMID: 33544785; PMCID: PMC7864317.

National Library of Medicine, Medline Plus. *Noma.* https://medlineplus.gov/ency/article/001342.htm. Accessed July 27, 2022.

Nelson SJ, Ash MM. *Wheeler's Dental Anatomy, Physiology, and Occlusion,* Ninth Edition. Saunders Elsevier, St. Louis, 2010.

Neumann HW. The Paleopathology of the Archaic Modoc Rock Shelter Inhabitants. *Reports of Investigations,* No. 11. State of Illinois, Springfield, 1967.

*NHS Inform.* Scotland's National Information Health Service, 2022. https://www.nhsinform.scot/illnesses-and-conditions/cancer/cancer-types-in-adults/multiple-myeloma

NIH. Health Topics: Fibrous Dysplasia in-depth. Fibrous Dysplasia Overview. *National Institute of Arthritis and Musculoskeletal and Skin Diseases,* 2019. file:///C:/Users/Owner/Downloads/Fibrous%20Dysplasia%20en.pdf. Accessed July 9, 2022.

Nissanka-Jayasuriya EH, Odell EW, Phillips C. Dental Stigmata of Congenital Syphilis: A Historic Review With Present Day Relevance. *Head and Neck Pathology* 10(3):327-331, 2016. doi: 10.1007/s12105-016-0703-z. PMID: 26897633; PMCID: PMC4972761.

O'Connell JEA. Craniopagus twins: surgical anatomy and embryology and their implication. *Journal of Neurology, Neurosurgery, and Psychiatry* 39:1-22, 1976.

Ortner DJ. *Identification of Pathological Conditions in Human Skeletal Remains,* Second Edition. Academic Press, 2003.

Ortner DJ, Ponce P, Ogden A, Buckberry J. Multicentric osteosarcoma associated with DISH, in a 19th century burial from England. *International Journal of Osteoarchaeology* 22(2):245-252, 2012.

Ortner DJ. *Ortner's Identification of Pathological Conditions in Human Skeletal Remains,* Third Edition, JE Buikstra (Editor). Academic Press, 2019.

Ortner DJ and Putschar WGJ. *Identification of Pathological Conditions in Human Skeletal Remains.* Smithsonian Contributions to Anthropology, Number 28, Smithsonian Institution Press, 1981.

Osmanagaoglu MA, Aran T, Guven S, Kart C, Ozdemir O, Bozkaya H. Thoracopagus conjoined twins: a case report. *ISRN Obstetrics and Gynecology* 2011:238-360. doi: 10.5402/2011/238360. Epub 2010 Nov 28. PMID: 21660094; PMCID: PMC3099464.

Otis GA. *Medical and Surgical History of the War of the Rebellion,* Part II, Surgical History, Second Issue. Government Printing Office, Washington, 1877.

Palfi G, Maixner F, Maczel M, Molnar E, Posa A, Kristof LA, Marcsik A, Balazs J, Masson M, Paja L, Palko A, Szentgyorgyi R, Nerlich A, Zink A, Dutour O. Unusual spinal tuberculosis in an Avar Age skeleton (Csongrád-Felgyő, Ürmös-tanya, Hungary): A morphological and biomolecular study. *Tuberculosis (Edinb)* Jun;95 Suppl 1:S29-34, 2015. doi: 10.1016/j.tube.2015.02.033. Epub 2015 Feb 13. PMID: 25840822.

Palmar K, Berdine G. Thoracoplasty in tuberculosis. *The Southwest Respiratory and Critical Care Chronicles* 7(30):71-72, 2019.

Palmerini E, Leithner A, Windhager R, et al. Angiosarcoma of bone: a retrospective study of the European Musculoskeletal Oncology Society (EMSOS). *Scientific Reports* 10, 10853 (2020). https://doi.org/10.1038/s41598-020-66579-5

Park BH, Hwang E, Kim CH. Primary intraosseous hemangioma in the frontal bone. *Archives of Plastic Surgery* 40:283-285, 2013.

Parvizi J, Kim GK. *Benign Bone Tumors.* Benign Bone Tumors, Chapter 24, pp. 48-49. W.B. Saunders, 2010,

Paterson AM, Lovegrove FT. Symmetrical Perforations of the Parietal Bones: Including an Account of a Perforated and Distorted Cranium from the Liverpool Museum. *Journal of Anatomy and Physiology* 34(P2):228-37, 1900.

Persing JA, Jane JA, Shaffrey M. Virchow and the pathogenesis of craniosynostosis: a translation of his original work. *Plastic and Reconstructive Surgery* 83(4):738-742, 1989.

Petroulia V, Surial B, Verma RK, Hauser C, Hakim A. Calvarial osteomyelitis in secondary syphilis: evaluation by MRI and CT, including cinematic rendering. *Heliyon* 9;6(1):e03090, 2020. doi: 10.1016/j.heliyon.2019.e03090. PMID: 31938744; PMCID: PMC6953708.

Pigrau-Serrallach C, Rodrıguez-Pard D. Bone and joint tuberculosis. *European Spine Journal* 22 (Supplement 4): S556-S566, 2013. doi: 10.1007/s00586-012-2331-y

Pollock RA, Brown TW, Rubin DM. "Phossy jaw" and "Bis-phossy Jaw" of the 19th and the 21st Centuries: The Diuturnity of John Walker and the Friction Match. *Craniofacial Trauma Reconstruction* 8(3):262-270, 2015. doi: 10.1055/s-0035-1558452

Power D. *A System of Syphilis in Five Volumes* (2nd ed.). D Power and JK Murphy (Editors). Oxford Medical Publications, H. Frowde, London, 1914.

Rachlin A, Patel JC, Burns CC, Jorba J, Tallis G, O'Leary A, Wassilak SGF, Vertefeuille JF. Progress Toward Polio Eradication – Worldwide, January 2020–April 2022. MMWR **M**orb **M**ortal **W**eekly **R**eport/May (19):650-655.

Raikos A, Paraskevas GK, Yusuf F, Kordali P, Meditskou S, Al-Haj A, Brand-Saberi B. Etiopathogenesis of hyperostosis frontalis interna: a mystery still. *Annals of Anatomy* 193(5):453-458, 2011. doi: 10.1016/j.aanat.2011.05.004. Epub 2011 May 27. PMID: 21684729.

Rajakulasingam R, Botchu R, Vemuri VN, James SL, Subbarao K, Davies AM. Skull Imaging-Radiographs and CT revisited. *Neurology India* 68(4):732-740, 2020.

Rajan J, Prasad PV, Chockalingam K, Kaviarasan PK. Malignant syphilis with human immunodeficiency virus infection. *Indian Dermatology Online Journal* 2(1):19-22, 2011. doi: 10.4103/2229-5178.79864. PMID: 23130209; PMCID: PMC3481795.

Ramon Y, Samra H, Oberman M. Mandibular condylosis and apertognathia as presenting symptoms in progressive systemic sclerosis (scleroderma). Pattern of mandibular bony lesions and atrophy of masticatory muscles in PSS, presumably caused by affected muscular arteries. *Oral Surgery Oral Medicine Oral Pathology* 63(3):269-74, 1987.

Reeder MR. *Reeder and Felson's Gamuts in Bone, Joint and Spine Radiology.* Springer-Verlag, New York, 1993.

Rees AEJ, Vujanic GM, Williams WM. Epidemic of conjoined twins in Cardiff. *British Journal of Obstetrics and Gynaecology* 100(4):388-391, 1993.

Reis BL, Carvalho GTC, de Sousa AV, de Freitas WB, Brandao RACS. Primary hemangioma of the skull. *Arquivos de Neuro-Psiquiatria* 66(3-A):569-571, 2008.

Richardson ML. Chinese Foot Binding: Radiographic Findings and Case Report. *Radiology Case Report* 4(1):270, 2014. doi: 10.2484/rcr.v4i1.270. PMID: 27843528; PMCID: PMC5106533.

Rivera F, Mirazon L. New evidence suggesting a dissociated etiology for cribra orbitalia and porotic hyperostosis. *American Journal of Physical Anthropology* 164(1):76-96, 2017.

Rammos SK, Gardenghi B, Bortolotti C, Cloft HJ, Lanzino G. Aneurysms Associated with Brain Arteriovenous Malformations. *American Journal of Neuroradiology* 37(11):1996-1971, 2016. doi: https://doi.org/10.3174/ajnr.A4869

Rana RS, Wu JS, Eisenberg RL. Periosteal Reaction. *American Journal of Roentgenology* 193(4):W259-W272, 2009.

Roberts C, Manchester AK. *The Archaeology of Disease* (3rd ed.). The History Press, 2010.

Roberts T, Lawrence S, Beighton P. Cleidocranial dysplasia: a review of the dental, historical, and practical implications with an overview of the South African experience. *Oral Surgery, Oral Medicine, Oral Pathology and Oral Radiology* 115(1):46-55, 2013.

Rockwood CA Jr., Green DP. *Fractures in Adults*, Volume 1. JB Lippincott Company, Philadelphia, 1984.

Rodrigues G, Carnelio S. *Oral tuberculosis: A Review. In: Focus on Tuberculosis Research.* In: Smithe LT, editor. New York; Nova Science Publishers p. 172, 2005.

Rose AM, Cowen S, Jayasena CN, Verity DH, Rose GE. Presentation, Treatment, and Prognosis of Secondary Melanoma within the Orbit. *Frontiers in Oncology* 23(7):125, 2017. doi: 10.3389/fonc.2017.00125. PMID: 28690979; PMCID: PMC5481311.

Ross JA. Percivall Pott 1714-1788. *Paraplegia* 24:287-292, 1986.

Rossi UG, Ierardi AM, Cariati M. Multiple Myeloma: Lytic Bone Lesions of the Skull. *Acta Neurologica Taiwanica* 30(2):81-82, 2021. PMID: 34549394.

Rothschild BM, Hershkovitz I, Dutour O. Clues potentially distinguishing lytic lesions of multiple myeloma from those of metastatic carcinoma. *American Journal of Physical Anthropology* 105(2):241-250, 1998. doi: 10.1002/(SICI)1096-8644(199802)105:2<241::AID-AJPA10>3.0.CO;2-0. PMID: 9511917.

Rothschild BM, Zdilla MJ, Jellema LM, Lambert HW. Cribra orbitalia is a vascular phenomenon unrelated to marrow hyperplasia or anemia: Paradigm shift for cribra orbitalia. *Anatomical Record* (Hoboken)

304(8):1709-1716, 2021.

Ruhli FJ, Nicklisch N, Alt KW. A historical case of beaten-copper cranium. *Journal of Neurosurgery* 106:71-73, 2007.

Said, AS, Manji KP. Risk factors and outcomes of fetal macrosomia in a tertiary centre in Tanzania: a case-control study. *BMC Pregnancy Childbirth* 16:243-250, 2016. https://doi.org/10.1186/s12884-016-1044-3

Sakuma Y, Yamashita Y, Shiono O, Oridate N. Actinomycosis arising from the nasal cavity, with rare fatal progression. *British Medical Journal Case Report*, 2016 Feb 9;2016:bcr2015213747. doi: 10.1136/bcr-2015-213747. PMID: 26860115; PMCID: PMC5483556.

Salari N, Darvishi N, Heydari M, Bokaee S, Darvishi F, Mohammadi M. Global prevalence of cleft palate, cleft lip and cleft palate and lip: A comprehensive systematic review and meta-analysis. *Journal of Stomatology and Oral Maxillofacial Surgery* 123(2):110-120, 2022. doi: 10.1016/j.jormas.2021.05.008. Epub 2021 May 24. PMID: 34033944.

Salapura V, Zupan I, Seruga B, Gasljevic G, Kavcic P. Osteoblastic bone metastases from renal cell carcinoma. *Radiology Oncology* 48(3):243-246, 2014. doi: 10.2478/raon-2013-0034. PMID: 25177238; PMCID: PMC4110080.

Samal SK, Rathod S. Sirenomelia: The mermaid syndrome: Report of two cases. *Journal of Natural Science, Biology, and Medicine* 6(1):264-266, 2015.

Sanchez T, Stewart D, Walvick M, Swischuk L. Skull fracture vs. accessory sutures: how can we tell the difference? *Emergency Radiology* 17(5):413-418, 2010. doi: 10.1007/s10140-010-0877-8. Epub 2010 May 23. PMID: 20496093; PMCID: PMC2914264.

Sandre LB, Viandelli Mundim-Picolli, Fortes PF, Rodrigues LG, Bueno JM, Ferreira da Silva R. Prevalence of agenesis of frontal sinus in human skulls with metopism. *Journal of Forensic Odontostomatology* 2(35):20-27, 2017.

Saraf SK, Tuli SM. Tuberculosis of hip: A current concept review. *Indian Journal of Orthopaedics* 49(1):1-9, 2015. doi: 10.4103/0019-5413.143903. PMID: 25593352; PMCID: PMC4292318.

Schlossberg D, Morley J, Montero M, Krouse T. Rupia Syphilitica. *Archives of Dermatology* 129(4):514-515, 1993. doi:10.1001/archderm.1993.01680250130028

Schmale GA, Conrad EU III, Raskind WH. The natural history of hereditary multiple exostoses. *Journal of Bone and Joint Surgery* 76A:986-992, 1994.

Schmorl G, Junghanns H. *The Human Spine in Health and Disease*, Second American Edition. Grune & Stratton, New York, 1959.

Shapiro R. Anomalous Parietal Sutures and the Bipartite Parietal Bone. *American Journal of Radiology* 115(3):569-577, 1972.

She R, Szakacs J. Hyperostosis frontalis interna: case report and review of literature. *Annals of Clinical and Laboratory Science* 34(2):206-8, 2004. PMID: 15228235.

Sheikh MA, Ibrahim M, Tahir AN. Occipital Encephalocele and Review of Literature. *Pakistan Journal of Neurological Surgery* 18(1):62-66, 2014.

Shilpa K, Ranganath P, Sumathi S. Anencephaly: Incidence, Risk Factors and Biochemical Analysis of Mother. *International Journal of Current Research and Review* 10(11):20-26, 2018.

Siddiqui F, Vaqar S, Siddiqui AH. Lung Cancer. [Updated 2022 May 5]. In *StatPearls* [Internet]. Treasure Island (FL): StatPearls Publishing; 2022 Jan-. Available from: https://www.ncbi.nlm.nih.gov/books/NBK482357/

Singh G, Yap J. Fibrous dysplasia. Reference article (2008). *Radiopaedia.org.* (accessed November 9, 2022) https://doi.org/10.53347/rID-4915

Smith GE. The causation of the symmetrical thinning of the parietal bones in ancient Egyptians. *Journal of Anatomy and Physiology* 41(3):232-233, 1907.

Smith OC, Pope EJ, Symes SA. *Look Until You See: Identification of Trauma in Skeletal Material* pp. 138-159. In DW Steadman (Editor), Hard Evidence: Case Studies in Forensic Anthropology. Prentice Hall, Upper Saddle River, 2003.

Smith S, Boulgakow B. A case of cyclopia. *Journal of Anatomy and Physiology* 61(1):105-111, 1926.

Smith HS, Mohsin I. Painful boney metastases. *Korean Journal of Pain* 26(3):223-41, 2013. doi: 10.3344/kjp.2013.26.3.223. Epub 2013 Jul 1. PMID: 23861996; PMCID: PMC3710936.

Smith-Guzman NE, Toretsky JA, Tsaic J, Cooke RG. A probable primary malignant bone tumor in a pre-Columbian human humerus from Cerro Brujo, Bocas del Toro, Panama. *International Journal of Paleopathol-*

*ogy* 21:138-146, 2018.

Snapper I. *Bone Diseases in Medical Practice.* Grune & Stratton, New York, 1957.

Song L, Wang Y, Zhang J, Song N, Xu Xiaoyun, Yan L. The risks of cancer development in systemic luus erythematosus (SLE) patients: a systematic review and meta-analysis. *Arthritis Research & Therapy* 20:270, 2018. https://doi.org/10.1186/s13075-018-1760-3

Sorensen CJ, Norton BJ, Callaghan JP, Hwang CT, Van Dillen LR. Is lumbar lordosis related to low back pain development during prolonged standing? *Manual Therapy* 20(4):553-557, 2015. doi: 10.1016/j.math.2015.01.001. Epub 2015 Jan 14. PMID: 25637464; PMCID: PMC4469524.

Spitz WU, Diaz FJ. *Spitz and Fisher's Medicolegal Investigation of Death: Guidelines for the Application of Pathology to Crime Investigation,* Fifth Edition. Charles C. Thomas Publisher, Springfield, 2020.

Spranger JW, Brill PW, Poznanski A. *Bone Dysplasias: An Atlas of Genetic Disorders of Skeletal Development.* Oxford University Press, Oxford, 2002.

Srour ML, Marck K, Baratti-Mayer D. Noma: Overview of a Neglected Disease and Human Rights Violation. *American Journal of Tropical Medicine and Hygiene* 96(2):268-274, 2017. doi: 10.4269/ajtmh.16-0718. Epub 2017 Jan 16. PMID: 28093536; PMCID: PMC5303022.

Stafne E. Bone cavities situated near the angle of the mandible. *Journal of the American Dental Association* 29:1969-1972, 1942.

Sternbach G. Percivall Pott: Tuberculous spondylitis. *The Journal of Emergency Medicine* 14(1):79-83, 1996.

Steyn M, Buskes J. Skeletal manifestations of tuberculosis in modern human remains. *Clinical Anatomy, Special Issue on the Clinical Anatomy of Paleopathology* 29(7):854-861, 2016.

Steinbock RT. *Paleopathological Diagnosis and Interpretation: Bone Diseases in Ancient Human Populations.* Charles C. Thomas, Springfield, 1976.

Stibbe EP. Skull showing perforations of parietal bone, or enlarged parietal foramina. *Journal of Anatomy and Physiology* 63(2):277-278, 1929.

Stone JL, Goodrich JT. The craniopagus malformation: classification and implications for surgical separation. *Brain* 129:1084-270, 2006.

Storer SK, Skaggs DL. Developmental dysplasia of the hip. *American Family Physician* 74(8):1310-1316, 2006. PMID: 17087424.

Stuart-Macadam P. Porotic hyperostosis: new evidence to support the anemia theory. *American Journal of Physical Anthropology* 74(4):521-526, 1987.

Weir P, Suttner N J, Flynn P, McAuley D. Normal skull suture variant mimicking intentional injury. *British Medical Journal* 332:1020, 2006. doi: 10.1136/bmj.332.7548.1020

Teager SJ, Constantine S, Lottering N, Anderson PJ. Physiologic closure time of the metopic suture in South Australian infants from 3D CT scans. *Childs Nervous System* 35(2):329-335, 2019.

Teschler-Nicola M, Novotny F, Spannagl-Steiner M, Stadler P, Prohaska T, Irrgeher J, Zitek A, Daubl B, Haring E, Rumpelmayr K, Wild EM. The Early Mediaeval manorial estate of Gars/Thunau, Lower Austria: An enclave of endemic tuberculosis? *Tuberculosis* 95(1):S51-59, 2015.

Theodorou DJ, Theodorou SJ, Kakitsubata Y. Imaging of Paget Disease of Bone and Its Musculoskeletal Complications: Review. *American Journal of Roentgenology* 196:S64-S75, 2011.

Thomas C, Kirsten WH. *Sandritter's Color Atlas & Textbook of Macropathology.* Year Book Medical Publishers, Inc., Chicago, 1985.

Thomas MP, Stern LM, Morris LL. Etiologichetero geneity of neural tube defects. *New England Journal of Medicine* 294:365-369, 1976.

Tidy C. Benign Ovarian Tumours. *Patient, Oncology,* 2021. https://patient.info/doctor/benign-ovarian-tumours. Accessed November 11, 2022.

Toshika SE, Male CG, Heyns M. Interpreting impact direction: applying fractography to the analysis of butterfly fractures produced by blunt force trauma. *Australian Journal of Forensic Sciences* 54:1, 26-41, 2022. doi: 10.1080/00450618.2020.1781252

Tozoglu U, Caglayan F, Harorli A. Foramen tympanicum or foramen of Huschke: anatomical cone beam CT study. *Dentomaxillofacial Radiology* 41(4):294-729, 2012. doi: 10.1259/dmfr/62359484. PMID: 22517996; PMCID: PMC3728996.

Trevizan M, Filho N, Franzolin P, Braga SO, Consolaro A. Premaxilla: up to which age it remains separated from the maxilla by a suture, how often it occurs in children and adults, and possible clinical and therapeutic implications: Study of 1,138 human skulls. *Dental Press Journal of Orthodontics* 23(6), 16-29, 2018.

Tuite GF, Evanson J, Chong WK, Thompson DN, Harkness WF, et al. The beaten copper cranium: a correlation between intracranial pressure, cranial radiographs, and computed tomographic scans in children with craniosynostosis. *Neurosurgery* 39:691-699, 1996.

Tunnessen WW Jr. Persistent open anterior fontanelle. *JAMA* Nov 14:264(18):2450, 1990. doi: 10.1001/jama.264.18.2450. PMID: 2232007.

Turgut F, Turgut M, Başaloglu H, Başaloglu HK, Haberal A. Extremely rare type of conjoined twins: cephalothoracopagus deradelphus. *European Journal of Obstetrics and Gynecology and Reproductive Biology* 80(2):191-194, 1998. doi: 10.1016/s0301-2115(98)00131-6. PMID: 9846666.

Turk JL. Syphilitic caries of the skull--the changing face of medicine. *Journal of the Royal Society of Medicine* 88(3):146-148, 1995. PMID: 7752158; PMCID: PMC1295133.

Turner TB. The Race and Sex Distribution of the Lesions of Syphilis in 10,000 Cases. *Bulletin of the Johns Hopkins Hospital* 46(2):159-184, 1930.

Varadharajan E, Priya S, Prakash G, Mugundan A, Easwaramurthi P. Mucinous Carcinoma of the Breast with Neuroendocrine Differentiation. *Iranian Journal of Pathology* 10(3):231-236, 2015.

Valenzuela EN, Pietschmann P. Epidemiology and pathology of Paget's disease of bone – a review. *Wien Med Wochenschr* 167(1):2-8, 2017.

Vanezis P. *Pathology of Sharp Force Trauma.* CRC Press, Boca Raton, 2021.

Viel G, Gehl, A, Sperhake JP. Intersecting fractures of the skull and gunshot wounds. Case report and literature review. *Forensic Science, Medicine, and Pathology,* 5(1):22-27, 2008. https://doi.org/10.1007/S12024-008-9062-8

Viens NA, Watters TS, Vinson EN, Brigman BE. Neuropathic Arthropathy of the Hip as a Sequela of Undiagnosed Tertiary Syphilis. *Clinical Orthoppaedics and Related Research* 468:3126-3131, 2010. doi: 10.1007/s11999-010-1257-0

Viertel VG, Intrapiromkul J, Maluf F, Patel NV, Zheng W, Alluwaimi F, Walden MJ, Belzberg A, Youse M. Cervical Ribs: A Common Variant Overlooked in CT Imaging. *American Journal of Neuroradiology* 33:2191-2194, 2012.

Virchow R. Ueber die Natur der constitutionellsyphilitischen Affectionenem. *Virchows' Archives of Pathological and Anatomical Physiology* 15:217-336, 1858.

Vu HL, Panchal J, Parker EE, Levine NS, Francel P. The timing of physiologic closure of the metopic suture: a review of 159 patients using reconstructed 3D CT scans of the craniofacial region. *Journal of Craniofacial Surgery* 12(6):527-32, 2001.

Vyas T, Gupta P, Kumar S, Gupta R, Gupta T, Singh HP. Cleft of lip and palate: A review. *Journal of Family Medicine and Primary Care* 9(6):2621-2625, 2020. doi: 10.4103/jfmpc.jfmpc_472_20. PMID: 32984097; PMCID: PMC7491837.

Wahby B. Abnormal nasal bones. *Journal of Anatomy and Physiology* 38:49-51, 1903.

Waldron T. *Paleopathology.* Cambridge Manuals in Archaeology, Cambridge University Press, 2008.

Walker PL, Bathurst RR, Richman R, Gjerdrum T, Andrushko VA. The causes of porotic hyperostosis and cribra orbitalia: a reappraisal of the iron-deficiency-anemia hypothesis. *American Journal of Physical Anthropology* 139(2):109-125, 2009.

Walker A, El Demellawy D, Davila J. Rickets: Historical, Epidemiological, Pathophysiological, and Pathological Perspectives. *Academy of Forensic Pathology* 7(2):240-262, 2017. doi: 10.23907/2017.024. Epub 2017 Jun 1. PMID: 31239976; PMCID: PMC6474539.

Walker D, Powers N, Connell B, Redfern R. Evidence of skeletal treponematosis from the medieval burial ground of St. Mary Spital, London, and implications for the origins of the disease in Europe. *American Journal of Biological Anthropology* 156(1):90-101, 2015.

Watson-Jones R. *Fractures and Joint Injuries, Volume I,* Third Edition. The William and Wilkins Company, Baltimore, 1944.

Watson-Jones R. *Fractures and Joint Injuries, Volume II,* Third Edition. The William and Wilkins Company, Baltimore, 1946.

Watson-Jones R. *Fractures and Joint Injuries, Volumes I and II,* Fourth Edition. The Williams and Wilkins Company,

Baltimore, 1955.

Wedel VL, Galloway A (Editors). *Broken Bones: Anthropological Analysis of Blunt Force Trauma*, Second Edition. Charles C. Thomas, Springfield, 2014.

Wheeler BJ, Snoddy AME, Munns C, Simm P, Siafarikas A, Jefferies C. A Brief History of Nutritional Rickets. *Frontiers in Endocrinology* 10(795):1-4, 2019. https://doi.org/10.3389/fendo.2019.00795

Wiedijk JEF, Soerdjbalie-Maikoe V, Maat GJR, Maes A, van Rijn RR, de Boer HH. An accessory skull suture mimicking a skull fracture. *Forensic Science International* 260:e11-e13, 2016.

Winkens R, Guldemond F, Hoppener P, Kragten H, van Leeuwen Y. Pectus excavatum, not always as harmless as it seems. *British Medical Journal Case Reports*, 2009, bcr10.2009.2329. doi:10.1136/bcr.10.2009.2329

Winter RB, Pinto WC. Pelvic obliquity. Its causes and its treatment. *Spine* (Phila Pa 1976) 11(3):225-34, 1986.

Wu JK, Goodrich JT, Amadi CC, Miller T, Mulliken JT, Shanske AL. Interparietal bone (*Os Incae*) in craniosynostosis. *American Journal of Medical Genetics* 155(2):287-294, 2010.

Wyszynski DF (Editor). *Cleft Lip and Palate: From Origin to Treatment*. Oxford University Press, 2002.

Yamunadevi A, Pratibha R, Rajmohan M, Mahendraperumal S, Ganapathy N, Srivandhana R. First Molars in Permanent Dentition and their Malformations in Various Pathologies: A Review. *Journal of Pharmacy and Bioallied Sciences* 13(Suppl 1):S23-S30, 2021. doi: 10.4103/jpbs.JPBS_744_20. Epub 2021 Jun 5. PMID: 34447037; PMCID: PMC8375929.

Yang Y, Jian G, Wenbin M, Yongning L, Bing X, Zuyuan R, Changbao SU, Renzhi W. Primary Intraosseous Cavernous Hemangioma in the Skull. *Medicine*: March 2016 - Volume 95 - Issue 11 - p e3069. doi: 10.1097/MD.0000000000003069

Yilmaz E. Presentation of multiple myeloma occurring in the humerus after strain: a case report. *Clinical Rheumatology* 40(1):389-392, 2021. doi: 10.1007/s10067-020-05256-4. Epub 2020 Jul 1. PMID: 32613389.

Yochum TR, Rowe LJ. *Essentials of Skeletal Radiology*, Third Edition. Lippincott Williams & Wilkins, Baltimore, 2005.

Zychowicz ME. Pathophysiology of Heterotopic Ossification. *Orthopaedic Nursing* 32(3):173-177, 2013.

# INDEX

## C

**G**

**U**



**V**